RADAR ELECTRONIC COUNTER-COUNTERMEASURES

THE ARTECH RADAR LIBRARY
David K. Barton, Editor

Laser Radar Systems and Techniques
Christian G. Bachman
KRYBOK Associates, Inc.

MTI Radar
D. Curtis Schleher
Manager of Advanced Development,
Raytheon Company

Radar Anti-Jamming Techniques:
A Translation from the Russian
of Zashchita Ot Radiopomekh
M.V. Maksimov et al

Phased Array Antennas
Arthur A. Oliner
Head of the Department of Electronics and Electrophysics,
Polytechnic Institute of New York
George H. Knittel
Staff Member, MIT Lincoln Laboratory

Radar Detection and Tracking Systems
Shahen A. Hovanessian
Senior Scientist, Hughes Aircraft Company

Radar Signal Simulation
Richard L. Mitchell
Vice-President, MARK Resources

Radar System Analysis
David K. Barton
Consulting Scientist, Raytheon Company

Radar Technology
Eli Brookner
Consulting Scientist, Raytheon Company

RADARS – in seven volumes
I. Monopulse Radar
II. The Radar Equation
III. Pulse Compression
IV. Radar Resolution and Multipath Effects
V. Radar Clutter
VI. Frequency Agility and Diversity
VII. CW and Doppler Radar

David K. Barton,
Consulting Scientist, Raytheon Company

Significant Phased Array Papers
Robert C. Hansen
President, R.C. Hansen, Inc.

Synthetic Aperture Radar
John J. Kovaly
Consulting Engineer, Raytheon Company

RADAR ELECTRONIC COUNTER-COUNTERMEASURES

Stephen L. Johnston P.E.

ARTECH HOUSE

Cover graphic courtesy of General Electric Company,
Electronic Systems Division

Printed and bound in the United States of America.

Standard Book Number 0-89006-79-7

Library of Congress catalog card number: 79-18873

CONTENTS

FOREWORD

CHAPTER 1: Introduction to Radar Electronic Counter-Countermeasures 1

1.1 "Radar Electronic Counter-Countermeasures" 11
S.L. Johnston, *IEEE Transactions on Aerospace and Electronic Systems,* Vol. AES-14, No. 1, January 1978, pp. 109-117.

1.2 "The Role of ECCM in Friendly Radar Environment, EMC" 21
E. Freeman, *IEEE ELECTRO-76 Record,* Part 30, May 1976, pp. 30-4-1 through 30-4-5.

1.3 "Why ECCM? The Need for ECCM" 27
W.D. Meskill, *IEEE ELECTRO-76 Record,* Part 30, May 1976, pp. 30-1-1 through 30-1-4.

1.4 "Threats and Techniques" 31
R. Loomis, *Electronic Progress Magazine,* Vol. 17, No. 3, Fall 1975, pp. 17-25.

1.5 "Postulating an Airborne ECM Threat" 41
D.D. Anderson, *Microwave System News,* December/January 1976, pp. 41-47.

1.6 "Airborne Microwave ECM" 47
D.J. Steer, *Proceedings of the Military Microwaves Conference,* London, October 1978.

1.7 "Master Electronic Warfare Glossary" 55
S.L. Johnston, *Second International Countermeasures Handbook,* 1976, pp. 595-608.

1.8 "A Complete Bibliography: Soviet Literature on Radar and EW, Part 1" 69
S.L. Johnston, *Electronic Warfare Magazine,* Vol. 8, No. 8, September/October 1976, pp. 44-54.

CHAPTER 2: General ECCM 77

2.A ECCM Survey Papers

2.A.1 "Radar Interference and its Reduction" 81
D.B. Brick and J. Galejs, *The Sylvania Technologist,* Vol. 11, No. 3, July 1958, pp. 96-108.

2.A.2 "Tools and Techniques of Electronic Warfare" 95
J. Holahan, *Space/Aeronautics Magazine,* Vol. 33, April 1960, pp. 125-132.

2.A.3 "Electronic Dogfight: ECM vs. ECCM in Search Radar" 103
P.R. Dax, *Space/Aeronautics Magazine,* Vol. 33, April 1960, pp. 134-137.

2.A.4 "Counter-Countermeasures: Good Design Can Beat the ECCM Threat" 107
M. Simpson, *Space/Aeronautics Magazine,* Vol. 33, April 1960, pp. 138-142.

2.B ECCM Considerations

2.B.1 "Philosophy of ECCM Utilization" 113
S. L. Johnston, *Electronic Warfare Magazine,* Vol. 7, No. 9, May/June 1975, pp. 59-61.

2.B.2 "Radar ECCM: A European Approach" 115
M.F. Radford, *IEEE ELECTRO-76,* Rec. pt 30, pp. 30-6-1 to 5.

2.C ECCM Applications

2.C.1 "Guided Missile System ECM/ECCM" 121
S.L. Johnston, *Microwave Journal,* Vol. 21, No. 9, September 1978, pp. 20-24.

2.C.2 "Performance of Current Radar Systems in an EW Environment" 127
J.R. Prior and N. Woodward, *Proceedings of the Military Microwaves Conference,* London, 1978, pp. 39-50.

2.C.3 "ECCM from the Radar Designer's Viewpoint" 139
M.A. Johnson and D.C. Stoner, *IEEE ELECTRO-76 Record,* Part 30, May 1976, pp. 30-5-1 through 30-5-11.

CHAPTER 3: ECCM in the Transmitter 151

3.1 "Effect of ECCM Techniques on Transmitter Requirements" 157
W.L. Gardner, *Proceedings of the DDRE ECCM Symposium,* 1976.

3.2 "Analysis of ECCM Characteristics of Frequency Agile Radar Systems" 171
B. Bergkvist, *Proceedings of the Military Microwaves Conference,* London, October 1978, pp. 51-57.

3.3 "Combined X/K_A-band Tracking Radar" 179
L. Klaver, *Proceedings of the Military Microwaves Conference,* London, October 1978.

CHAPTER 4: ECCM in the Antenna 189

4.1 "Low Sidelobe Radar Antennas" 195
A.W. Rudge and P.R. Foster, *Proceedings of the Military Microwaves Conference,* London, October 1978.

4.2 "Effect of Fresnel Zone Blockage on Very Low Sidelobe Antennas" 205
R.L. Fante, *RADC Technical Report 77-361,* October 1977, DDC AD A051188.

4.3 "Tolerance Errors for Very Low Sidelobe Antennas" 219
R.L. Fante, *RADC Technical Report 77-360,* October 1977, DDC AD A049259.

4.4 "Antenna Sidelobe Reduction Using Spatial Filters" 225
R.J. Mailloux and A.C. Schell, *Proceedings of the DDRE Radar ECCM Symposium,* 1976.

4.5 "Conopulse Radar" 237
H. Sakamoto and P.Z. Peebles, Jr., *IEEE Transactions on Aerospace and Aeronautical Systems,* Vol. AES-14, No. 1, January 1978, pp. 199-208. Correction Vol. AES-14, No. 4, July 1978, p. 673c.

CHAPTER 5: ECCM in the Antenna: Sidelobe Blanker/Canceller 247

5.A Sidelobe Blankers

5.A.1 "A Sidelobe Blanking System Design and Demonstration" 251
P.O. Arancibia, *Microwave Journal,* Vol. 21, No. 3, March 1978, pp. 69-73.

5.B Sidelobe Cancellers

5.B.1 "Adaptive Arrays and Sidelobe Cancellers; A Perspective" 257
D.J. Chapman, *Microwave Journal,* Vol. 20, No. 8, August 1977, pp. 43-46.

5.B.2 "Adaptive Arrays" 261
S. Applebaum, *IEEE Transactions on Antennas and Propagation,* Vol. AP-24, No. 5, September 1976, pp. 585-598.

5.B.3 "Understanding and Optimizing Sidelobe Canceller Operation" 275
B.L. Lewis and J.P. Hansen, *NRL Report 7610,* October 25, 1973.

5.B.4 "Problems of Adaptive Sidelobe Suppression" 291
G. Reis and K. Krucker, *AGARD Conference Proceedings,* No. 197, 1976, pp. 23-1 – 23-12.

CHAPTER 6: ECCM in the Receiver (General) 303

6.1 "Naval Radar Anti-Jamming Techniques" 319
R.V. Alred, *Proceedings of the IEE,* Vol. 93, Part IIIa, No. 10, 1946, pp. 1593-1601.

6.2 "Designing Radar Receivers to Overcome Jamming" 329
J.C. Galenian, *Electronics Magazine,* Vol. 36, May 17, 1963, pp. 50-54.

6.3 "Desensitization in Transistorized PM/FM Receivers" 335
F. Hansen, *Proceedings of the 18th IEEE Vehicular Communications Conference,* 1967, pp. 78-86.

6.4 "Tackle Noise Variations at Video" 349
R.W. Cope, *Electronic Design Magazine,* Vol. 20, September 26, 1968, pp. 54-56.

6.5 "Digital MTI, a New Tool for the Radar User" 353
J.G. Ellis, *Marconi Review,* Vol. 36, No. 191, 4th Quarter, 1973, pp. 237-248.

6.6 "Conical Scan in Ground Clutter with Non-Coherent and Coherent MTI" 365
C.S. Sorkin and B.D. Steinberg, *Proceedings of the IRE East Coast Conference on Aeronautical Navigational Electronics,* 1958, pp. 35-45.

6.7 "Design Considerations for a Coherent (On Receive) MTI System for Conically Scanned Tracking Radar" 377
A.K. Subramanian, *Electro-Technology (India),* Vol. 18, No. 4, December 1974, pp. 95-98.

CHAPTER 7: ECCM in the Receiver Signal Processor 381

7.1 "Description and Performance Evaluation of the Moving Target Detector". 383
L. Cartledge and R.M. O'Donnell, *MIT Lincoln Laboratory Technical Report, ATC-69,* March 8, 1977, pp. 16-44, 210.

7.2 "The Application of SAW Technology to Radar ECCM". 413
P.H. Carr, *Proceedings of the DDRE Radar ECCM Symposium,* 1976.

7.3 "Anti-Clutter and ECCM Design Criteria for a Low Coverage Radar" 427
G. Petrocchi, S. Rampazzo, and G. Rodriqguez, *Proceedings of the International Conference on Radar,* Paris, December 1978, pp. 194-200.

7.4 "Performance of CFAR Devices in ECM Environments". 433
P.S. Tong and P.E. Steichen, *Proceedings of the DDRE Radar ECCM Symposium,* 1976.

7.5 "Noise Quality Optimizes Jammer Performance" 449
F.M. Turner, E.G. Ottoboni, and A.M. Imada, *Electronic Warfare Magazine,* Vol. 9, No. 6, November/December 1977, pp. 117-119.

CHAPTER 8: ECCM in the System, Operational, and Other ECCMs 453

8.1 "Improved Radar Designs Outwit Complex Threats" 457
A. Ivanov, *Microwaves Magazine,* Vol. 28, April 1976, pp. 54ff.

8.2 "OADARS (Optical Aids to Detection and Ranging Systems)" 463
A. V. Jelalian, D.A. Kawachi, and C.R. Miller, *IEEE ELECTRO-76 Record,* Part 30, pp. 30-3-1 - 30-3-5.

8.3 "Electronic Correlator for Passive Ranging" 469
K.H. Norsworthy, *Proceedings of the International Radar Countermeasures Conference,* London, September 1976, pp. 290-319.

CHAPTER 9: ECM/ECCM, ECCM Efficacy, Simulation, and Analysis 487

9.1 "Burnthrough Range, An Overrated Parameter" 493
S.L. Johnston, *Electronic Warfare Magazine,* Vol. 6, No. 6, November/December 1974, pp. 49-54.

9.2 "ECCM Improvement Factors (EIF)" 499
S. L. Johnston, *Electronic Warfare Magazine,* Vol. 6, No. 3, May/June 1974, pp. 41-45.

9.3 "Simulation to Obtain a Systems Measure of an Air Duel Environment" 503
A.A.B. Pritsker, R.C. Van Buskirk, and J.W. Wetherbee, *IRE Transactions,* Vol. EC-8, No. 1, March 1959, pp. 55-59.

9.4 "Simulator Checks Out Radio-Radar AJ Operation" 509
J. Lambert, *Space/Aeronautics Magazine,* Vol. 33, No. 4, April 1960, pp. 157-166.

9.5 "Radar/ECM Computer Modeling" 513
J. Cikalo and M. Greenbaum, *IEEE NAECON-75 Record,* pp. 431-437.

9.6 "Recent Innovations in Radar Simulation" 521
R.J. Hancock, P.D. Krolak, L.K. Wilson, and J.C. Cleary, *IEEE Region III SOUTHEASTCON-75 Record,* pp. 6E-4-1 – 6E-4-9.

9.7 "Applications of Computer Simulation Techniques to the Evaluation of Radar System Performance in ECM/ECCM Environments" 531
S.L. Gutsche, R.L. Bogusch, and F.W. Guigliano, *Proceedings of the DDRE Radar ECCM Symposium,* 1976.

FOREWORD

This reprint volume resulted from a suggestion of Mr. David K. Barton. It is the product of an intensive search, spanning nearly five years, of the international literature including both Engineering Index and Science Abstracts (Series B). About one-third of the reprint papers are by non-US authors from Chile, Denmark, Germany, India, Italy, Japan, Netherlands, Sweden, and several from the United Kingdom.

The papers in this volume cover a time span of over three decades, from 1946 to 1978. Three groups of papers from specialized conferences are included. Six papers comprised the Radar ECCM Session organized by this author and given at the IEEE ELECTRO-76 in Boston. That session received the best Technical Session Award at ELECTRO-76. Four papers were originally presented at the February 1976 DDRE Radar ECCM Symposium. Five papers were originally presented at the October 1978 Military Microwaves Conference, London, England.

This volume is a partial departure from the other ARTECH reprint volumes on radar technology in that it incorporates three types of papers: discussion, analysis, and hardware. Discussion papers are important in electronic warfare related matters since the philosophy of EW may be new to some radar engineers. History of EW as included in discussion papers is important in enabling the radar engineer to develop an appreciation for EW while benefitting from the experiences of the past. Finally, in some instances, suitable analytical papers on some radar ECCM topics have not been available prior to the publication of this volume. The papers reprinted here were selected from a host of candidate papers and are intended to represent a broad cross-section of Radar ECCM Technology. They were authored by workers in both radar and electronic warfare activities.

This author gratefully acknowledges contributions of many of his professional colleagues in both radar and electronic warfare. Special mention is also due to my wife for her patient assistance in the preparation of the manuscript. Assistance of the Georgia Institute of Technology Price Gilbert Memorial Library in obtaining copies of some of these papers is gratefully acknowledged.

BIBLIOGRAPHY

Each chapter in this book contains a bibliography and is organized under the topics listed at the end of the text for the chapter.

An extensive bibliography is essential for a technical book. It is even more important for this book since the literature on radar electronic counter-countermeasures is scattered through a host of publications. Radar ECCM includes a number of techniques which have been widely addressed for other reasons. These include (alphabetically) constant false alarm rate (CFAR), frequency agility, monopulse, moving target indicators (MTI), and pulse compression. Each of those has been separately and extensively indexed by Barton [1]. Additionally, each of those topics is the subject of an ARTECH reprint volume. Accordingly, the bibliographies in this volume will not contain the large number of items on the several topics covered in references [1-7].

Items in the bibliographies included here are either of more recent origin than [1-7] were in an obscure publication, are of special ECCM importance, were included in an extensive general topic in [1] (such as antennas, receivers, etc.), or are a patent or government technical report. Special bibliographies and special books are given at the beginning of a topic. Some entries will include comments by this author.

Abbreviations for publications cited here are generally as used in [1]. As an aid to the reader, abstracts in English for foreign language papers, some patents and dissertations are also included; e.g., SA 75-12345 denotes an abstract in Science Abstracts for 1975 with that number.

References

[1] Barton, D.K., *International Cumulative Index on Radar Systems,* IEEE Pub No J H 4675-5, 1978.

[2] Barton, D.K., *Radars, Volume I – Monopulse Radar,* Artech House Books, Dedham, MA, 1974.

[3] Barton, D.K., *Radars, Volume II – The Radar Equation,* Artech House Books, Dedham, MA,1974.

[4] Barton, D.K., *Radars, Volume III – Pulse Compression,* Artech House Books, Dedham, MA, 1974.

[5] Barton, D.K., *Radars, Volume VI – Frequency Agility and Diversity,* Artech House Books, Dedham, MA, 1977.

[6] Barton, D.K., *Radars, Volume VII – CW and Doppler Radar,* Artech House Books, Dedham, MA, 1978.

[7] Schleher, D.C., *MTI Radar,* Artech House Books, Dedham, MA, 1978.

Chapter 1

Introduction to Radar Electronic Counter-Countermeasures

Background

Radar, the use of electromagnetic energy for the detection and location of targets [1, 2] is a descendant of *radio* as evidenced by the origin of the term "radar"—Radio Detection and Ranging [3]. It is only natural therefore that many of the concepts of radar have their roots in radio. Indeed, this extends to the earliest days of use of electromagnetic waves for communication (radio) then known as *wireless.* It is thus instructive to recognize the influence of radio on radar.

Radar and radio may be susceptible to interfering signals or actions. This interference may be due to natural causes or it may be man-made. Some examples of natural interference are "static" created by lightning discharges or the aurora borealis, absorption, attenuation, and refraction of millimeter length waves by rainfall, etc. Man-made interference may either be intentional or unintentional. Examples of man-made interference include ignition electromagnetic noise from internal combustion engines, arcs from electrical machinery, dispersal of metallic strips to create false targets on a radar screen, and operation of two transmitters on the same frequency in the same place at the same time. In the latter case, either of the two transmitters could, under certain conditions, prevent, degrade or confuse reception of the radiation of the other transmitter by a receiver.

Intentional man-made interference used for military purposes was originally called jamming but is now known as electronic countermeasures (ECM) and will be formally defined later. Unintentional man-made radio interference is often called radio frequency interference (RFI).

It is interesting to note that military concern over effects of possible hostile electromagnetic interference date from the earliest days of *wireless.* Interference tests were conducted in conjunction with transmission of the first official message from a U.S. Naval vessel on November 2, 1899 [4]. The report of the Navy Radio Telegraph Board voiced their concern about the possible effects from the previously mentioned "two transmitter possibility." Marconi [4] was reported to share this concern, and he indicated in vague terms (due to his pending patent application) his solution for that problem (a tunable receiver!).

In that same time frame (circa 1900), history records the first actual employment of intentional man-made interference for non-military purposes. As Zahl recounted [5], Marconi had successfully used his *wireless* to report results of an American Cup yacht races off the Atlantic Coast in the Fall of 1899. In 1901 Marconi received a contract to report the International yacht races for the Associated Press. Another group, headed by Lee DeForest, also had a contract but theirs was from the Publishers Press Association for this race report. A third group, the American Wireless Telephone and Telegraph Co. with John Greenleaf Pickard as the principal engineer, was unable to obtain a sponsor but decided to "report" the race also. Indeed, as Zahl indicated, this was a fiasco for both Marconi and DeForest. The third group (Pickard) effectively jammed the other two groups by the simple procedure of continued transmittal of a series of dashes from their wireless. With a truly brilliant idea, Mr. Pickard had conceived a very clever means for "coding" his interfering dashes in such a way that his receiver could get his reports while denying reception by the other two competitors [6]. By pre-arrangement, the existence and meaning of this "code" was known only to Pickard's group. What's more, neither Marconi nor DeForest had the slightest suspicion that such a code was being employed against them. Hence the year 1901 saw what was probably the first use of ECM as an ECCM!

These episodes indicate the cat-and-mouse nature of intentional man-made interference, ways to employ it, and ways to combat it. This book is devoted to techniques for combatting effects of hostile interference on radar.

In this chapter, we will introduce the subject of Electronic Warfare and its three principal divisions: electronic countermeasures (ECM), electronic counter-countermeasures (ECCM) and electronic warfare support measures (ESM) with their interrelationships. ECCM and ECM are discussed. Several papers treat ECM in World War II, why ECCM is needed, and ECM threats and techniques. An extensive EW glossary and a bibliography of Soviet radar and EW books conclude this chapter.

The term *radar,* as used in this volume and as now defined by the IEEE [1, 2], actually includes both laser radar and those missile guidance systems which utilize electromagnetic radiation (other than for a command link only). While most of this book specifically refers to microwave radar, the material is also relevant to missile guidance systems and is extendable to laser radar, infrared (IR) and electro-optical (EO) missile guidance systems. ECCM aspects of missile guidance are discussed in two papers in this volume.

Synopsis of Reprint Papers in Chapter 1

1.1 "Radar Electronic Counter-countermeasures"— Johnston

This paper is the keystone of this volume. It defines electronic warfare and its principal subdivisions: electronic countermeasures, electronic warfare support measures and electronic counter-countermeasures. Through their interactions, it is shown that all elements can be used by both friendly and enemy forces. General effects of the various divisions of EW are given. Extensive tables list approximately 150 types of ECCM organized into ECCM in the transmitter, ECCM in the antenna, ECCM in the receiver and signal processor, ECCM in the system, and operational and other ECCMs. Each of these is the subject of a subsequent chapter in this volume. The last chapter in this volume, "ECCM Efficacy, Simulation and Analysis" is introduced by "ECCM Efficacy Assessment" in Paper 1.1, which was first presented at ELECTRO-76.

1.2 "The Role of ECCM in Friendly Radar Environment, EMC"— Freeman

As pointed out in Paper 1.1, Electromagnetic Compatibility (EMC), although not a division of electronic warfare, is very closely related to electronic counter-countermeasures. Freeman presents an excellent review of EMC with many of the EMC techniques used. It will be noted that many of those techniques are also used for ECCM against hostile interference and were so listed in Paper 1.1. A bibliography of the principal EMC references is included in this Paper 1.2, which was presented at ELECTRO-76.

1.3 "Why ECCM: The Need for ECCM"— Meskill

Under the heading "The Past is the Prologue," the importance of electronic countermeasures and counter-countermeasures is illustrated from several actions in the Battle of Britain, air combat over North Viet Nam, and the Yom Kippur War from which Meskill draws important lessons concerning ECM and ECCM. This paper was presented at ELECTRO-76.

1.4 "Threats and Techniques"— Loomis

This interesting paper was written from the ECM viewpoint rather than from the *ECCM* viewpoint. Consequently it will be necessary for the ECCM practitioner to switch gears mentally in reading this paper since the "threat" discussed by Loomis is the friendly environment, which includes radars and defensive missile systems. Loomis does not define the basic ECM techniques, but Steer (Paper 1.6 below) does. Loomis' paper contains radar scope photos of several specific ECMs and discusses the effects of ECM at various points inside a radar. From this paper, the radar designer can identify the necessity for many of the 150 ECMs listed by Johnston in Paper 1.1.

1.5 "Postulating an Airborne ECM Threat"— Anderson

Some threats are presented in rather sterile terms such as "power versus frequency" without further qualification. In this paper Anderson details a number of aspects of airborne jammers which enable the ECCM designer to better understand and interpret various ECM jamming threats.

As shown in Table VI of Johnston (Paper 1.1), jamming is one type of active ECM, the other being repeaters. The first is generally used for detection/range/velocity denial (Steer, Paper 1.6). The second type is for deception. Figure 1 in Dax (Paper 2.A.3) indicates that jammers may be further organized by bandwidth (narrow or wide), emission type, modulation characteristics, etc. Those factors may greatly affect jammer weight and size through modulator requirements; e.g., in a wide band pulsed noise jammer, as Dax points out.

The principal application of Anderson's paper is essentially to noise (denial) jammers, although as Steer indicated (Paper 1.6) some traveling wave tubes may be used for noise jamming or repeaters (pulsed). Another factor which enters into the characteristics of a jammer is its deployment; e.g., self screening, escort, or stand-off.

1.6 "Airborne Microwave ECM"—Steer

This paper is a valuable adjunct to the previous papers in this chapter by Loomis (Paper 1.4) and by Anderson (Paper 1.5). It primarily addresses aircraft-mounted denial jammers and deception repeaters (in the terminology of Tables VI and VII of Paper 1.1 on pages 16 and 17 of this volume) with a brief mention of chaff (passive ECM). These jammers are intended for use against both ground-based and airborne air defense radars. Jammer deployments considered are self screening, escort, and stand-off..Noise jammers (range/velocity denial) and repeater (deception) jammers (range and velocity gate pull-off, cover pulse and inverse gain) are treated. Applicability of these types against various radar types (e.g., surveillance) is shown. Jammer block diagrams and power output data on currently available (1978) jammer tubes are included. This paper served as the input ("threat") for the paper of Prior and Woodward (Paper 2.C.2). The F, I and J bands roughly correspond to S, X, and Ku-bands, respectively, in current IEEE standard radar letter band designations [8] as used by most radar engineers. This paper was presented at the Military Microwaves Conference in 1978.

1.7 "Master Electronic Warfare Glossary"—Johnston

Nearly every field of endeavor develops its own lexicon, and Electronic Warfare is no exception. This glossary which was compiled from a host of sources includes most of the more widely used EW terms. It will be noted that common *radar* terms are not included.

1.8 "A Complete Bibliography—Soviet Literature on Radar and EW"—Johnston

This title is a misnomer and is not the original title selected by the author, since the bibliography actually includes only books and pamphlets. Journal articles, conference papers, etc., are not included. Some of those are contained in Barton [7] and in the bibliography of this book. Translations of many of the Russian books are listed.

Reprinted papers in this chapter should give the reader a good foundation for the remainder of the book.

References for Chapter 1

[1] "IEEE Standard Radar Definitions," *IEEE Standard 686-1977,* Nov. 1977.

[2] "IEEE Standard Dictionary of Electrical and Electronic Terms," *IEEE Standard 100-1977,* New York, IEEE Press, 1977.

[3] "IEEE Standard Dictionary of Electrical and Electronic Terms," *IEEE Standard 100-1972,* New York, Wiley Interscience, 1972.

[4] Annual Report of the Secretary of the Navy, 1900, pp. 319-320 as reported in: Howeth, L.S., "History of Communications—Electronics in the United States Navy," U.S. Government Printing Office, Washington, 1963, pp. 32-34.

[5] Zahl, H.A., "The Seeds of Electronic Warfare," *Electronic Warfare Magazine,* vol. 3, no. 1, Winter 1971, p. 12.

[6] "Radioana" SRM 5-543, Mass. Institute of Technology, Cambridge, Mass., as reported in Howeth, *Op. Cit.,* pp. 38-39.

[7] Barton, D.K., *International Cumulative Index on Radar Systems,* IEEE Publication No. JH 4675-5, 1978.

[8] "IEEE Standard Letter Designations for Radar Bands," *IEEE Standard 521-1976,* Nov. 30, 1976.

Bibliography Topics for Chapter 1

ECM/EW History
ECM Techniques (General)
ECM: Chaff
EMC
ESM

BIBLIOGRAPHY FOR CHAPTER 1

The following special abbreviations are used in the bibliography to indicate sources of abstracts, translations, U.S. government reports, etc.:

DDC: Defense Documentation Center, Cameron Station, Alexandria, VA 22314 (for qualified users only; all others to NTIS below)

EI: Engineering Index prepared by United Engineering Center, 345 E. 47th St., NY, NY 10017.

IA: International Aerospace Abstracts, published by American Institute of Aeronautics and Astronautics, 750 3rd Ave., NY, NY 10017.

JPRS: Joint Publications Research Services; order from NTIS below.

NTIS: National Technocal Information Service, Springfield, VA 22151.

SA: Science Abstracts Series B, published by the Institute of Electrical Engineers, P.O. Box 8, Southgate House, Sterenage, Herts., England.

ECM/EW HISTORY

1. Books

Price, A., *Instruments of Darkness,* London, William Kimber, 1967, MacDonald and Jones, London, 1977.

2. Journal Articles

"Electronics Warfare: Report on Radar Countermeasures," Joint Board on Scientific Information, Office of the Publication Board, Office of Scientific Research and Development, War Dept., revised Oct 1945.

"Radar Countermeasures," *Engineer,* London, vol. 180, no. 4691, Dec 7, 1945, pp. 460-461.

"Electronic Warfare: A Report on Radar Countermeasures (extract)," *IRE Trans. Military Electronics,* vol. MIL-1, No. 1, March 1957, p. 31ff.

Blattner, D.J., "Electronic Countermeasures—the art of jamming," *Electronics World,* vol. 62, no. 6, Dec 1959, p. 47ff.

"Book Review: Instruments of Darkness," *Radar and Electronics Magazine,* Winter 1968, p. 23.

"A Short History of EW," *Navigator Magazine,* vol. XVI, no. 2, Winter 1968, pp. 1-4.

Price, A., "Afterthoughts on Instruments of Darkness," *Radar and Electronics Magazine,* Spring 1968, pp. 32-33.

Smith, H.F., "EW History," *Electronic Warfare Magazine,* vol. 1, no. 4, 1969, p. 18ff.

Simmons, R.F., "History and Principles of EW, pt. I" *Frequency Technology Magazine,* vol. 7, no. 11, Nov 1969, pp. 28-31; pt.II: vol. 7, no. 12, Dec 1969, pp. 28-31; pt. III: vol. 8, no. 1, Jan 1970, pp. 29-33.

"EW History in World War II," *Electronic Warfare Magazine,* vol. 4, Summer 1972, p. 20ff.

"EW History in World War II, pt. 2," *Electronic Warfare Magazine,* vol. 4, Conv. 1972, p. 20ff.

"The Silent War," *Electron. Electro-Optic and Infrared Countermeasures Magazine,* vol. 1, no. 1, 1975, p. 34ff.

Price, A., "Electronic Warfare in World War II," *Proc. Int. Countermeasures Conf., London, Sept '76,* EW Institute Ltd., Los Altos, CA. pp. 1-17.

ECM TECHNIQUES, GENERAL

1. Books and Bibliographies

Volzhin, A.N., and V.A. Yanovich, *"Radar Countermeasures,"* Moscow, Voyenizadt, 1960 (in Russian, trans. DDC AD 682311).

Boyd, J.A., D.B. Harris, D.D. King, and H.W. Welch, Jr., *Electronic Countermeasures*, Univ. of Mich. IST 71204, 1961, DDC AD 357877. (Also contains some ECCM).

Paliy, A.I., *Radio Warfare,* Moscow, Voyenizadt, 1961 (in Russian).

Schlesinger, R.J., *Principles of Electronic Warfare,* Prentice-Hall, 1961 (has chapters on ECM techniques and tactics, detection, electronic reconnaissance, radars, antennas, optimization, space era EW).

Vakin, S.A., and L.N. Shustov, *Principles of Jamming and Electronic Reconnaissance,* Moscow, Soviet Radio, 1968 (in Russian, trans. DDC AD 692642, -3).

Atrazhev, M.P., V.A. Il'yin, and N.P. Mar'yin, *Electronic Warfare,* Moscow, Voyenizdat, 1972 (in Russian, trans. DDC AD B 015529).

Paily, A.I., *Radio-Electronic Warfare,* Moscow Oborongiz, 1974, (in Russian).

Schoniger, K.R., and H.F. Eustace, eds., *The International Countermeasures Handbook,* first ed., 1975-6.

Eustace, H.F., ed., *The International Countermeasures Handbook,* second ed. 1976-7 (contains various information on ECM, ESM, EW bibliography: p. 579ff., etc.).

Johnston, S.L., "Ten Years of EW and Crow Caws," *Electron. Warfare Magazine,* vol. 8, no. 4, July/Aug 1976, p. 61ff. (cumulative index and cross index of first 10 years of Assn. of Old Crows Magazines).

Johnston, S.L., "Master Electronic Warfare Glossary," *Second Int. Countermeasures Handbook,* 1976, pp. 595-608 (reprinted in this volume).

Johnston, S.L., "A Complete Bibliography—Soviet Literature (books) on Radar and EW, Part I," *Electronic Warfare Magazine,* Sept/Oct 76, vol. 8, no. 8, p. 43ff (reprinted in this volume).

Technical Program (Proceedings) International Countermeasures Conference. London; EW Institute Ltd., Los Altos, CA, 1976.

Eustace, H.F., ed., *The International Countermeasures Handbook,* third ed., 1977-8, 1977.

Eustace, H.F., ed., *The International Countermeasures Handbook,* fourth ed., 1978-9 (contains various EW information, including an organized EW bibliography on over 40 topics, pp. 430-452).

VanBrunt, L.B., *Applied ECM, Vol I,* Dun Loring, VA, EW Engineering Inc., first ed., 1978 (covers radar ECM with descriptions of a large number of ECM techniques).

2. Journal Articles

"Radar Countermeasures Techniques," *Aero Digest,* Dec 15, 1945, p. 49ff.

Peters, R.G., "Radar Countermeasures," *Communications Magazine,* vol. 25, no. 12, Dec 1945, p. 50ff.

Stanford, L.H., "Fundamentals of Radar and Countermeasures," *J. Western Soc. of Engrs.,* vol. 50, no. 4, Dec 1945, pp. 167-76.

Fink, D.G., "Radar Countermeasures," *Electronics Magazine,* vol. 19, no. 1, Jan 1946, pp. 92-7.

McQuay, J., "Anti-Radar," *Broadcast Engr. J.,* vol. 13, no. 8, Aug 1946, p. 12.

Narasimhan, K., "Electronic Countermeasures as a Weapon of War," *India Air Force Quarterly,* Jan 1955.

"Electronic Countermeasures as a Weapon of War," *Mil. Rev.,* vol. 35, Oct 1955, pp. 83-7 (also discusses ECCMs).

Klass, P.J., "Exclusive Report on Countermeasures, pt. I: Avionics Fights its Own Silent War," *Aviation Week Magazine,* vol. 67, Nov 18, 1957, p. 50ff.

Klass, P.J., "Exclusive Report on Countermeasures, pt. II: Avionics War Aims at Deceipt, Confusion," *Aviation Week Magazine,* vol. 67, Nov 25, 1957, p. 102ff.

Blattner, D.J., "Electronic Countermeasures—The Art of Jamming," *Electron World,* vol. 62, no. 6, Dec 1959, p. 47ff.

Kovit, B., "Vast New Scale for Countermeasures in Space," *Space/Aeronautics,* vol. 33, no. 4, Apr 1960, pp. 147-154.

Hatcher, E.C., "Wide Bandwidth, High Power Handling Abilities Needed for ECM Antennas," *Space/Aeronautics,* vol. 33, no. 4, April 1960, pp. 171ff.

Magnuski, H., "Jamming of Communications Systems Using FM, AM and SSB Modulations," *IRE Trans. Mil. Electronics,* vol. 5, no. 1, Jan 1961, pp. 8-11.

McQuay, J., "How Good is Radar Jamming? Can We Block Enemy Radar?" *Radio-Electronics Magazine,* vol. 33, no. 11, Nov 1962, pp. 28-31.

Foa, Orienterar Om, "Electronic Warfare," *Forsvarets,* Forskingsanstalt, Stockholm 80, Sweden, 1967.

Wilson, R.O., "Marine Airborne EW-VMCJ," *Navigator Magazine,* vol. XVI, no. 2, Winter 1968, pp. 5-7.

"EW in NORAD," *Navigator Magazine,* vol. XVI, no. 2, Winter 1968, pp. 10-15.

Bull, J.G., "An Up to Date Evaluation of Pulse Jamming," *Navigator Magazine,* vol. XVI, no. 2, Winter 1968, pp. 16-20.

Mirman, IK. R., "Electronic Warfare," *Ordnance Magazine,* vol. 53, Nov-Dec 1968, pp. 297-301.

Miller, B., "Special Report: Electronic Warfare," *Aviation Week and Space Techno,* Aug 25, 1969, pp. 71-93; Sept 1, 1969, pp. 52-62; Sept 8, 1969, pp. 67-82; Sept 15, 1969, pp. 79-91.

"ECM Primer," *MicroWaves Magazine,* vol. 8, no. 11, Nov 1969, p. 44ff.

Espinosa, R.J., "Improved TWT's for Broadband Jamming," *MicroWaves Magazine,* Nov 1969, pp. 52-8.

Wolkstein, H.J., M. Freeling, and M.P. Puri, "The Loop TWT, Key to Range Deception," *MicroWaves Magazine,* vol. 8, no. 11, Nov 1969, pp. 73-9.

Haley, D.W., "Expendable Jammers," *Electronic Warfare Magazine,* vol. 2, April-May 1970, p. 4.

Aller, J.C., "Electronic Warfare Concept," *Naval War College Rev.*, vol. 22, May 1970, pp. 75-9.

Michael, A.L., "EW—A Tangible Resource," *Electronic Warfare Magazine*, vol. 2, conv. 70, pp. 5-6.

Scott, A.W., "Miniature TWT's That Fit Inside X-Band Waveguide," *Microwave J.*, vol. 13, no. 5, May 1971, pp. 53-67.

"Microwave Warfare—2: Spoofing Hostile Radars with Expendable Jammers," *MicroWaves Magazine*, vol. 10, no. 10, Oct 1971, p. 34ff.

Fitts, R.E., "Fundamentals of Electronic Warfare," *USAF Academy*, March 1, 1972.

Nelson, E.A., and D.D. Niebruegge, "Management Concepts for a Multifunction Aperture," *IEEE NAECON-73 Rec.*, p. 52. (abstract only)

Latour, C., "Electronic Warfare," *NATO's Fifteen Nations Magazine,"* vol. 19, Apr-May 1974, pp. 72-9.

Haas, G.N., "Digital Control of Active Jamming Systems," *IEEE NAECON Rec.*, 1974, pp. 161-7.

Segalen, J., "Electronic Warfare," *Recherche*, vol. 5, no. 46, June 1974, p. 528ff (in French).

McGee, J.J., Jr., "B-1 Radio Frequency Surveillance/ ECM Subsystem," *Def. Mgmt. J*, vol. 10, Sept 1974, pp. 34-5.

"Electronic Warfare," (several articles) *Aviation Week and Space Technol.*, Jan 27, 1975.

Belov, M., "New Factors in the Development of Modern Armies," *Soviet Military Rev.*, vol. 2, Feb 1975, pp. 10-13.

Laycock, D.E., "Improved TWT's for Electronic Countermeasures Application," *IEEE NAECON-75 Rec.*, pp. 411-5.

Feltz, J.F., "B-52G/H Protective Systems Test Range (PSTR)," *IEEE NAECON-75 Rec.*, pp. 416-23.

Chmiel, B.A., "AN/ALQ-117 Program Evaluation," *IEEE NAECON-75 Rec.*, pp. 438-41.

Matthews, A.R., "New Concepts for Electronic Warfare," *Naval Air Test Center Rpt. NATC-TM-75-1-CT*, July 1975, DDC AD A018868.

Dax, P.R., "Noise Jamming of Long Range Search Radars," *MicroWaves Magazine*, vol. 14, no. 9, Sept 1975, p. 54ff.

Meadows, L., "Electronic Countermeasures Technology—an Overview," *Electr. Prog. Magazine*, vol. 17, no. 3, Fall 1975, pp. 2-6.

Fusfield, R., "High Power Airborne Jammers," *Electr. Prog. Magazine*, vol. 17, no. 3, Fall 1975, pp. 7-16.

Loomis, R., "Threats and Techniques," *Electr. Prog. Magazine*, vol. 17, no. 3, Fall 1975, pp. 17-25 (reprinted in this volume).

Lowenschuss, O., "Signal Processing for ECM Systems," *Electr. Prog. Magazine*, vol. 17, no. 3, Fall 1975, pp. 26-31.

Black, A., "Multibeam Systems," *Electr. Prog. Magazine*, vol. 17, no. 3, Fall 1975, pp. 32-42.

Eustace, H.F., "Electronic Warfare, a Soviet View," *Microwave Syst. News Magazine*, vol. 5, no. 6, Dec/Jan 1976, p. 28ff.

Eustace, H.F., "ECM vs. Radar: A Lethal Scenario; Special Report, Radar Countermeasures," *Microwave Syst. News Magazine*, vol. 5, no. 6, Dec/Jan 1976, pp. 31ff.

Anderson, D.D., "Postulating an Airborne ECM Threat," *Microwave Syst. News Magazine*, vol. 5, no. 6, Dec/Jan 1976, pp. 41-7 (reprinted in this volume).

Bittencourt, R.P., "Electronic Warfare Technology," *Masters Thesis, U.S. Naval Postgraduate School*, June 1976, DDC AD A028210 (also discusses ESM and ECCM).

Fielding, J.G., "Signal Processing in Electronic Warfare," *IEE Conf. Pub. No. 144*, Sept 1976, pp. 112-120. (Considers signal processing in ECM, ECCM, and ESM.)

Walsh, B., "Design Considerations for ECM Systems in Attack and Fighter Aircraft," *Proc. Mil. Electron. Def. EXPO 76*, Oct 1976, Wiesbaden, Germany. pp. 46-59.

Johnson, G.R., "Jamming Passive Lobing Radars," *Electronic Warfare Magazine*, vol. 9, no. 2, Mar/Apr 1977, p. 75ff.

Johnson, G.R., "Jamming Low Power Spread Spectrum Radar," *Electronic Warfare and Def. Electronics Magazine*, vol. 9, no. 5, Sept/Oct 1977, p. 103ff.

Fawcette, J., "Commonopoly: Monopoly Updated," *Electronic Warfare and Defense Electronics*, vol. 9, no. 5, Sept/Oct, 1977, pp. 125ff. (Describes Adaptive Self-Protection Jammer: ASPJ.)

Turner, F.M., E.G. Ottobini, and A.M. Imada, "Noise Quality Optimizes Jammer Performance," *Electronic Warfare Magazine*, vol. 9, no. 6, Nov/Dec 1977, pp. 117ff (reprinted in this volume).

"What Jams What?" *International Countermeasures Handbook*, H.F. Eustace, ed., fourth ed., 1977, p. 411ff (has matrix of ECM types and radar types).

Stetson, R.M. and R.L. Brown, "Antenna Isolation Impact on ECM System Capability," *J. of Electron Def.*, vol. 1, no. 1, Jul/Aug, 1978, pp. 49ff.

Steer, D.J., "Airborne Microwave ECM," *Proc. Military Microwaves Conf.*, London, Oct 1978, pp. 29-38 (reprinted in this volume: description of ECM techniques and their applicability to various radars).

Andrews, R.S., "A Responsive High Power Jammer," *Proc. Mil. Microwaves Conf.,* London, Oct., 1978, pp. 57-61.

Anon., "Methodology for Specifying Jammer System Parameters," pp. 406ff in *The International Countermeasures Handbook,* H.F. Eustace, ed., 4th ed. 1978-9, 1978. (Does not consider the effects of ECCM in ECM calculations.)

VanBrunt, L.B., "Repeater Jammers," *Journal of Electron. Def.,* vol. 2, no. 1, Jan/Feb 1979, p. 47ff.

Van Brunt, L.B., "AGC Deception," *J. of Electron. Def.,* vol. 2, no. 2, Mar/Apr 1979, pp. 49-51.

ECM: CHAFF

1. Books

Stepanov, Y.G., *Radar Camouflage,* Moscow, Soviet Radio, 1968 (in Russian, partial translations: DDC AD 733507, JPRS 47734).

2. Journal Articles

Chu, L.J., "Analysis of Window (Chaff) and Related Matters," *Radio Research Lab Rpt.,* no. 4, Oct 22, 1942.

Bloch, F., M. Hammermesh, and M. Phillips, "Radar Reflections from Long Conductors," *J. Appl. Phys.,* vol. 17, no. 12, Dec 1946, pp. 1015-1020.

Van Vleck, J.H., F. Bloch, and M. Hammermesh, "Theory of Radar Reflections from Wires or Thin Metallic Strips," *J. Appl. Phys.,* vol. 18, Mar 1947, pp. 274-294.

Bloomhuff, J., and R. Grossel, "Testing Chaff," *Air Proving Ground Command Eglin AFB, Fla. Rpt. APGC TDR 62-28,* May 1962, DDC AD 277054.

Palermo, C.J., and L.H. Bauer, "Bistatic Scattering Cross Section of Chaff Dipoles with Application to Communications," *Proc. IRE,* vol. 53, no. 8, Aug 1965, pp. 1119-21c.

Kownacki, S., "Simulation of Radar Range and Doppler Effect by Means of a Stationary Target," *IEEE Trans.,* vol. AES-3, no. 1, Jan 1967, pp. 148-9c.

Kownacki, S., "Screening (Shielding) Effect of a Chaff Cloud," *IEEE Trans.,* vol. AES 3, no. 4, July 1967, pp. 731-4.

Shirman, Ya. D., "Optimum Detection of Radar Target in a Cloud of Passive Reflectors," *Radio Engrg. and Electron. Phys.,* vol. 15, no. 5, May 1970, pp. 799-806.

Dipare, A.L., "Chaff Primer," *MicroWaves Magazine,* vol. 9, no. 12, Dec 1970, pp. 46-7.

Popov, D.I., "Efficiency of the Coherent-Pulse Method of Signal Reception in the Presence of Passive Noise (Chaff)," *Telecom and Radio Engrg., pt. 2,* vol. 28, no. 11, 1973, pp. 92-4.

Richmond, J.H., L.M. Schwab, and R.G. Wickliff, "Tumble Average Radar Backscatter of Some Thin Wire Chaff Elements," *IEEE Trans.,* vol. AP-22, no. 1, Jan 1974, pp. 124-6c.

Puskar, R.J., "Radar Reflector Studies," *IEEE NAECON 74 Rec.,* pp. 177-83.

Wickliff, R.F. and R.J. Garbacz, "The Average Backscattering Cross Section of Clouds of Randomized Resonant Dipoles," *IEEE Trans.,* vol. AP-22, no. 3, May 1974, pp. 503-5c.

Tavis, M.T., "Total Radar Cross Section of Thin Wires at All Angles of Incidence," *J. Appl. Phys.,* vol. 46, no. 7, July 1975, pp. 3213-5.

Huber, J.C., Jr., "The AN/ALE-39 Countermeasures Set—Designed for Control by Radar Warning Receivers," *Proc. Int. Countermeasures Conf.,* London, Sept 1976, EW Institute Ltd., Los Altos, CA, pp. 321-8.

Mahaffey, M., "Electrical Fundaments of Countermeasures Chaff," *International Countermeasures Handbook,* H.F. Eustace, ed., second edition, pp. 512-517, 1976.

Coleman, E.R., "Chaff Meteorology," *International Countermeasures Handbook,* H.F. Eustace, ed., second edition, pp. 518-22, 1976.

Huber, J.C., Jr., "Smart Dispensing with the AN/ALE-39 and Derivative Systems," *Proc. Mil. Electron Def. EXPO-76,* Oct 1976, Wiesbaden, Germany, pp. 204-15.

Christmas, I.K., "Chaff Protection for Fast Patrol Boats," *Proc. Mil Electron Def. EXPO-76,* Oct 1976, Wiesbaden, Germany, pp. 216-24.

Wiseman, C.H., "Chaff Assists Anti-Ship Missile Defense," *Electronic Warfare Magazine,* vol. 9, no. 4, Jul/Aug 1977, pp. 73ff.

Silverman, G., "Chaff Dispersion and Polarization," *J. of Electron. Def.,* vol. 1, no. 3, Nov/Dec 1978, pp. 39ff.

ELECTROMAGNETIC COMPATIBILITY (EMC)

2. Journal Articles

Busch, C.W., "Suppression of Mutual Interference Between Two Radar Equipments Working in the Same Band," *Elektronische Rundschau,* vol. 9, no. 2, Feb 1955, pp. 50-2 (in German, trans. abstract SA 55-3941).

Morelli, M., and J. Molnar "Radar to Radar Interference," *Proc. 2nd Conf. on Radio Interference Reduction,* Armour Research Foundation, Chicago, Mar 6-7, 1956.

Campbell, R.D., "Radar Interference to Microwave Communication Services," *Elec. Eng.,* vol. 77, no. 10, Oct 1958, pp. 916-921.

Chase, K.H. and S.O. Newman, "The Radar Interference Problem," *Sperry Eng. Rev.*, vol. 11, no. 3, Oct 1958, pp. 21-27.

Newman, S.O. and K.H. Chase, "The Radar Interference Problem," *Proc. NEC,* vol. 14, 1958, pp. 811-820.

Campbell, R.D., "Radar Interference to Microwave Communication Services," *AIEE Trans.*, vol. 77, pt. 1, no. 39, Nov 1958, pp. 717-722.

Ryan, A.H., "Control of Microwave Interference," *IRE Trans., vol. RFI-1 No. 1,* May 1959, pp. 1-10.

Wall, R.A., "Radar's Running Rabbits," *Electronics,* vol. 32, no. 27, July 3, 1959, p. 58.

Chase, K.H. and J.L. Pierzga, "Reducing Mutual Radar Interference," *Electronics,* vol. 32, no. 28, July 10, 1959, pp. 39-43.

Robbins, R.L., "Radar Interference With Microwave Radio," *Bell Lab Rec.*, vol. 38, no. 4, Apr 1960, pp. 142-145.

Johnson, R.C., "Radar Search Antennas and RFI," *IEEE Trans.*, vol. EMC-6, no. 2, July 1964, pp. 1-8.

Hudson, C.L. and W.R. Limberg, "Loss of System Effectiveness Due to Electromagnetic Interference," *Proc 10th IEEE Tri-Service Conf. on Electron-Compatibility,* Nov 1964, pp. 28-57.

Wibbe, J.L., "Radar EMC in the Space Age," *IEEE Trans.*, vol. EMC-7, no. 1, March 1965, pp. 31-34.

Katz, L., "PPI Interference Prediction," *IEEE Trans.*, vol. EMC-7, no. 2, June 1965, pp. 119-24.

Marek, F.L. and M.D. Aasen, "Oscillator Stability and Electromagnetic Compatibility," *IEEE Int. Conv. Rec.*, vol. 14, pt. 2, 1966, pp. 114-124.

Cifventes, E.G., L.G. Dominguez, L.F.J. Irribarre, "Electromagnetics Compatibility in Military Aerials and Electronic Systems," *Electronia Fis. Aplic.*, vol. 9, no. 36, Oct-Dec 1966, pp. 469-476 (in Spanish, trans. abstract SA 67-19608).

Golovnya, Ya. P., "Analysis of Sources of Pulse Radio Interference," *Telecomm and Radio Engrg Pt. 2,* vol. 22, no. 11, Nov 1967, pp. 140-2.

Turesin, M., "Electromagnetic Compatibility Guide for Design Engineers," *IEEE Trans.*, vol. EMC-9, no. 3, Dec 1967, pp. 139-145.

Gajsek, J., "Sources of Radio Interference and Criteria for its Suppression," *Elektrotech Vestnik* (Yugoslavia) vol. 35, no. 4-5, Apr-May 1968, pp. 85-7 (in Slovenian, trans. abstract SA 68-29712).

Gajsek, J., "Measures Against Radio Interference in Yugoslavia," *Elektrotech Vestnik,* vol. 35, no. 4-5, Apr-May 1968, pp. 88-90 (in Slovenian, trans. abstract SA 68-29713).

Sachs, H.M., and M.J. Feil, "An Evaluation of EMC Performance of Pulse-Doppler Radar Systems," *Proc. IEEE Electromag. Compatability,* Symp Proc, vol. ELMCS, 1970, pp. 30-39.

Blythe, J.H., "Separation of Radars on Common Frequencies by Pulse-Repetition Frequency Discrimination," *GEC J of Sci and Technol,* vol. 37, no. 4, 1970, pp. 157-62.

Newhouse, P.D., "Simplify EMC Design," *MicroWaves Magazine*, vol. 9, no. 5, May 1970, pp. 59-62.

Murray, J.P., "Electromagnetic Compatability," Ch 29 in *Radar Handbook* (M.I. Skolnik, ed) McGraw-Hill, 1970.

Vinogradov, Ye. M., V.I. Vinokurov, and I.P. Kharchenko, "Statistical Estimate of Electromagnetic Compatability," *Izv VUZ Radioelektron,* vol. 15, no. 4, Apr 1972, pp. 475-481. (in Russian, abstract SA 74-11595).

Fishbein, W., R. Olesch, and O. Rittenbach, "Radar Interference Reduction Techniques," *AGARD Conf Proc no 159 on Electromagnetic Noise Interference and Compatability,* 21-25 Oct 1974, pp. 22-1 thru 23.

Ral'nikov, V.I. and I.P. Kharchenko, "A Mathematical Model for Electromagnetic Compatibility of Radar Installations," *Izv VUZ Radioelektron,* vol. 17, no. 12, Dec 1974, pp. 72-6 (in Russian trans. abstract SA 75-16709).

Moore, R.L., "Passive ECM Applied to False Target Elimination," *Int. Radar Symp. Rec.*, Apr 1975, pp. 458-462. (Used radar absorbing material on building near radar to eliminate false target).

Freeman, E., "The Role of ECCM in Friendly Radar, EMC," *IEEE ELECTRO-76 Conv Rec, pt 30,* pp. 30-4-1 thru 5, May 1976 (reprinted in this volume).

Fisbein, W.E., R.E. Olesch and O.N. Rittenback, "Methods and Apparatus for Eliminating Interference in Radar Systems," *U.S. Patent No. 4,010,468,* Mar 1977.

"Reduction of Interference Between Pulse Radar Stations by Pulse Repetition Frequency Discrimination (p r f d)," *CCIR Study Group Doc,* Spec Preparatory Mtg (WARC-79) Doc P/26-E, United Kingdom, 29 May 1978.

ELECTRONIC WARFARE SUPPORT MEASURES (ESM)

1. Books

Vakin, S.A., and L.N. Shustov, *Principles of Jamming and Electronic Reconnaissance,* Moscow, Soviet Radio, (summary: *Electron Warfare Magazine,* vol. 7, no. 3, May/June 1975, p. 65ff) 1968, (in Russian translation DDC AD 692642, 3).

2. Journal Articles

Stansfield, R.G., "Statistical Theory of D.F. Fixing," *Proc. IEE*, vol. 94, pt III A, 1947, pp. 762-70.

Sparagna, J.J., "Monopulse Direction Finding Techniques, Part I," *CROW CAWS Magazine*, vol. 2, no. 4, June 1968, p. 6ff.

Edelcreck, G., "Effects of Antenna and Frequency Scan on Detection Probability in Reconnaissance Receivers," *CROW CAWS Magazine*, vol. 2, no. 4, June 1968, p. 18ff.

Kaplan, A., "Detection and Analysis of Frequency Hopping Radar Signals," *Electronic Warfare Magazine*, vol. 1, no. 4, Conv 1969, p. 13ff.

Margerum, D.L., "Pinpointing Location of Hostile Radars," *MicroWaves Magazine*, Nov 1969, pp. 60-64.

Bullock, L.G., G.R., Oeh, and J.J. Sparagna, "An Analysis of Wideband Microwave Direction-finding Techniques," *IEEE Trans.*, vol. AES-7, no. 1, Jan 1970, pp. 188-203.

Margerum, D.L., "Passive ECM Part I (Direction Finding)," *Electron Warfare Magazine*, vol. 2, no. 2, Apr/May 1970, p. 8ff.

Perry, J.L., "Log Amplitude Direction Finding," *Electron Warfare Magazine*, vol. 2, no. 2, Apr/May 1970, p. 11ff.

Margerum, D.L., "Passive ECM Part II (Direction Finding)," *Electron Warfare Magazine*, vol. 2, no. 3, Summer 1970, p. 5ff.

Kivi, A., and H. Shnitkin, "Broadband Direction Finding—Side Looking Radar," *Electron Warfare Magazine*, vol. 2, no. 4, Conv 1970, p. 32ff.

Casperson, J.C., "Panoramic Receiver," *Electron Warfare Magazine*, vol. 2, no. 4, Conv 1970, p. 35ff.

Oleari, G., "A Building Block Reconnaissance Receiver," *Electron Warfare Magazine*, vol. 3, no. 1, Winter 1970, p. 8ff.

Bullock, L.G., et al, "Passive ECM Emitter Location Techniques," *Microwave J.*, vol. 14, no. 5, May 1971, p. 45.

Chrzanowsi, E.J., "Detection Capability of an ECM Receiver vs. Detection Capability of a Radar: ECM Application Note," *Microwave J.*, vol. 14, no. 5, May 1971, pp. 58-60.

Rygiel, J.C., "Spectrum Analyzer Capabilities for the Applications of the 70's: ECM Application Note," *Microwave J.*, vol. 14, no. 5, May 1971, pp. 61-65.

Wiley, R.G., "Return of IFR (Instantaneous Frequency Indicating Receiver)," *Electron Warfare Magazine*, vol. 3, no. 4, Conv 1971, p. 7ff.

Roubadeaux, D.L., "Panoramic Receiver Operation," *Electron Warfare Magazine*, vol. 3, no. 4, Conv 1971, p. 21ff.

Byrom, C., "Multi-Purpose Electromagnetic Receiving System," *Electron Warfare Magazine*, vol. 4, no. 1, Winter 1972, p. 7ff.

Morin, H.J., and F.J. Enge, "Netting of ESM Platforms," *Electron Warfare Magazine*, vol. 4, no. 1, Winter 1972, p. 10ff.

McElhowe, R.D., and R.D. Tortora, "Small Sample Method for Rapid CEP From Direction Finding," *Electron Warfare Magazine*, vol. 4, no. 2, Spring 1972, p. 6ff.

Bullock, L.G., G.R. Oeh, and J.J. Sparagna, " Monopulse Direction Finding, Part I," *Electron Warfare Magazine*, vol. 4, no. 1, Winter 1972, p. 34ff. Part II: vol. 4, no. 2, Spring 1972, pp. 12ff.

Brown, G.G., "Direction Finding Using Short Baseline Time of Arrival Techniques," *Electron Warfare Magazine*, vol. 4, no. 3, Summer 1972, p. 14ff.

Keenly, R.R., "New Systems Concepts for Increased ELINT Intercept," *Electron Warfare Magazine*, vol. 5, no. 1, Feb 1973, p. 63ff.

"Multi-mode ELINT Receiver Developed," *Electron Warfare Magazine*, vol. 5, no. 2, Apr 1973, p. 59ff.

Eby, J.C., Jr., "A Digital ELINT Parameter Analyzer," *IEEE NAECON-73 Rec.*, pp. 53-9.

Harper, T., "Are Rotary D F Systems Outdated?," *Electron Warfare Magazine*, vol. 6, no. 2, Mar/Apr 1974, p. 11ff.

Shilladay, R.W., "Video Detectors for Broadband D F," *Electron Warfare Magazine*, vol. 6, no. 3, May/June 1974, p. 51ff.

Mabie, C., and A. Sock, "Automatic Scan Period Determination Algorithms for Simultaneous Measurement on up to Ten Sources with Minimal ESM System Performance Degradation," *IEEE NAECON-74 Rec.*, pp. 168ff.

Krueger, C.H., Jr., and J.B.Y. Tsui, "Applications of Surface Wave Acoustic Filter to Channelized Receivers," *IEEE NAECON Rec.*, 1975, pp. 424-430.

Ernandes, F.A., "New Surviellance System Offers Automatic Intercept and Response," *Electron. Warfare Magazine*, vol. 7, no. 5, Sept/Oct 1975, p. 75ff.

O'Donnell, R.V., "Read-Through-While-Jamming Perspectives and Prospects," *Electron. Warfare Magazine*, vol. 7, no. 5, Sept/Oct 1975, p. 111ff.

Oliver, T., "Multi-Purpose Receiver System Features Automatic Operation," *Electron. Warfare Magazine*, vol. 7, no. 5, Sept/Oct 1975, p. 115ff.

Klaassen, C.F., Jr., "ECM Waveform Analysis System," *Electron. Warfare Magazine,* vol.7, no. 5, Sept/Oct 1975, p. 133ff.

Robbins, L.W., "Dynamic Range Considerations in Electronic Countermeasures Receivers," *Microwave System News,* Dec/Jan 1976, pp. 53-61.

"The Channelized Receiving System," *Microwave System News,* Dec/Jan 1976, pp. 63-70.

"ALR-59 Passive Detection System," *Electron. Electro-Optic Infrared Countermeasures,* vol. 2, no. 4, Apr 1976, pp. 23-5.

Fielding, J.G., "Signal Processing in Electronic Warfare," *IEE Conf. Pub. No. 144,* Sept 1976, pp. 112-120. (Considers signal processing in ECM, ECCM, and ESM.)

Mullen, P.N., J. Bodey, and J. Bleakley, "R-5000 Microwave Receiving System," *Proc Mil Electron. Def EXPO-76,* Oct 1976, Wiesbaden, Germany, pp. 385-94.

Rappolt, F.A., and N.L. Stone, "Receivers for Signal Acquisition," *Proc Mil Electron Def EXPO-76,* Oct 1976, Wiesbaden, Germany, pp. 403-20.

Ernst, B.F., "Automatic VHF-UHF Triangulation System for Reconnaissance Applications," *Proc Mil Electron Def EXPO-76,* Oct 1976, Wiesbaden, Germany, pp. 421-34.

Rappolt, F.A. and N. Stone, "Receivers for Signal Acquisition," *Microwave J.,* vol. 20, no. 1, Jan 1977, p. 29-33.

Mountcastle, P.H., and W.A. O'Berry, "Antenna Scanning Apparatus," *U.S. Patent No. 4,010,472,* Mar 1, 1977, (Relfector antenna with two feeds; one for active tracking radar and one for passive ESM receiver).

Lundy, D., "SAW Filters Shrink Channelized Receivers," *Electronic Warfare and Def. Electronics,* vol. 9, no. 5, Sept/Oct 1977, pp. 169ff.

Rowe, P.L., "An Advanced ESM System for Ship Defense," *Proc. Mil. Microwaves Conf.,* London, Oct 1978, pp. 3-8.

East, P.W. and K.W. White, "Complexity/Performance Trade-offs in IFM Receiver Design," *Proc. Mil. Microwaves Conf.,* London, Oct 1978, pp.16-25.

Hartman, R., "ESM Receiver Overview," *Proc. Mil. Microwaves Conf.,* London, Oct 1978, pp. 26-28.

Backus, P.H., "Electronic Warfare Support Measures," *Jnl of Electron Def.,* vol. 2, no. 1, Jan/Feb 1979, p. 21ff.

Radar Electronic Counter-countermeasures
S.L. Johnston
IEEE Transactions on Aerospace and Electronic Systems, Vol. AES-14, No. 1, January 1978
Pages 109-117

1.1

Introduction

Intense development of radar and its extensive employment in World War II by both sides inevitably led to creation of electronic means for combating radar [1]-[4]. Those antiradar means were originally called radar countermeasures (RCM) but are generally known today as electronic countermeasures (ECM). The two principal World War II ECMs were jamming and chaff.

Means to permit radar to operate successfully in an ECM environment were originally known as antijamming (AJ) but are now known as electronic counter-countermeasures (ECCM). ECM and ECCM are the principal divisions of electronic warfare (EW). Electronic warfare has developed into a very complex field; in fact, an organization which is devoted exclusively to EW, the Association of Old Crows (AOC), was formed in 1964. The AOC was named after nature's own imitator, the crow.

One U.S. book [5], a recently declassified book on ECM [6], and a large number of journal articles on EW have been published in the U.S. Several Soviet books on EW [7]-[14] are also in existence. Most Soviet radar books have a chapter on EW [15]. In U.S. literature, principal EW interest appears to be focused on ECM, possibly due to World War II. During that war, the Allies had a very large ECM program [3], [4]. The U.S. ECCM effort was considerably smaller, however, because of the military situation. Allied pioneer use of microwave radar (itself an ECCM) was contrasted with Axis use of decimeter radar. Heavy use of ECM by the Allies caused Germany to engage in an extensive crash ECCM program [4]. Current ECCM techniques were largely born in World War II.

World War II ECM as reported by Price [1] and others [2]-[5] were recently updated by Meskill [16] to show the importance of ECCM to radar designers of today. This paper intends to provide a survey of open literature radar ECCM techniques.

ECCM and Electronic Warfare

ECCM is one division of the field called electronic warfare. The other divisions are ECM and electronic warfare support measures (ESM). ECCM is also related to another field—electromagnetic compatibility (EMC). ECCM involves countering hostile interference while EMC involves unintentional friendly interference. The EW tree is shown in Fig. 1 [17]. Official definitions [18] of EW, ECM, ESM, and EMC are listed below.

Electronic warfare (EW) is military action involving the use of electromagnetic energy to determine, exploit, reduce, or prevent hostile use of the electromagnetic spectrum and action which retains friendly use of the electromagnetic spectrum.

Electronic countermeasures (ECM) is that division of electronic warfare involving actions taken to prevent or reduce an enemy's effective use of the electromagnetic spectrum.

Electronic counter-countermeasures (ECCM) is that

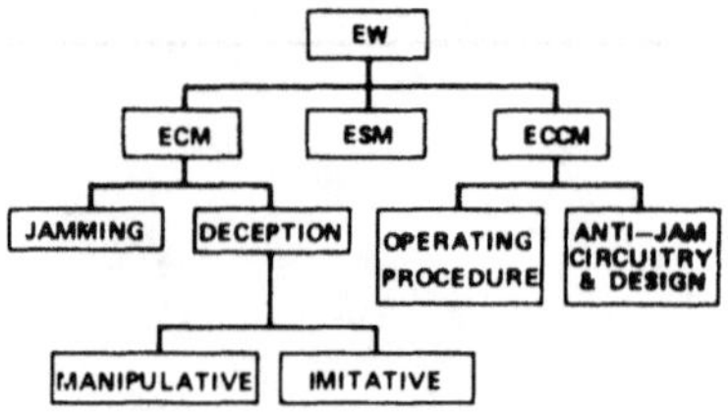

Fig. 1. The EW Tree.

Fig. 2. The Interactions of Electronic Warfare.

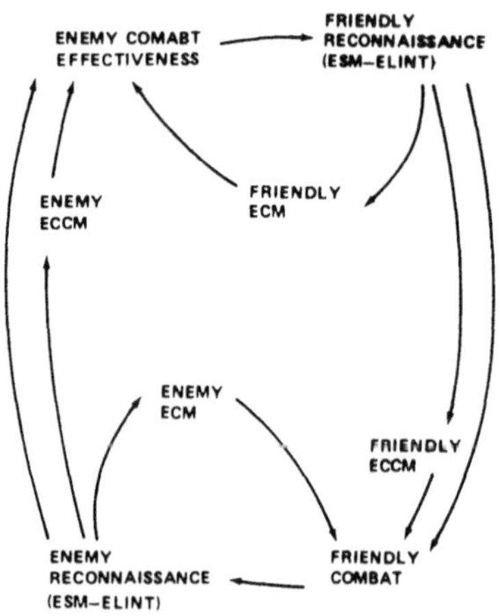

division of electronic warfare involving actions taken to insure friendly effective use of the electromagnetic spectrum despite the enemy's use of electronic warfare.

Electronic warfare support measures (ESM) is that division of electronic warfare involving actions taken to search for, intercept, locate, record, and analyze radiated electromagnetic energy for the purpose of exploiting such radiations in the support of military operations. Thus, electronic warfare support measures provide a source of electronic warfare information required to conduct electronic countermeasures, electronic counter-countermeasures, threat detection, warning, avoidance, target acquisition, and homing.

Electromagnetic compatibility (EMC) is the ability of communications electronics equipments, subsystems, and systems to operate in their intended operational environments without suffering or causing unacceptable degradation because of unintentional electromagnetic radiation or response.

It is important that these definitions be noted carefully because their application can be confusing. It should be noted that some devices or techniques can be either an ECM or an ECCM or both. ECCM may be either friendly or hostile. A particular device or technique may be an ECCM while also providing EMC [20].

The interactions of electronic warfare are shown in Fig. 2 [21]. Electronic intelligence (ELINT) is the intelligence information product of activities engaged in the collection and processing, for subsequent intelligence purposes of foreign, noncommunications electromagnetic radiations emanating from other than nuclear detonations and radioactive sources [19]. The merry-go-round of Fig. 2 is an electronic representation of Newton's law of action and reaction. It clearly shows that it is not meaningful to consider ECM or ECCM alone.

Previous ECCM Work

ECCM has been addressed extensively in open literature. The author has identified over 200 references in journal articles, conference papers, and books. These references are in many languages, such as German [22], Russian [23]-[25], etc. The one U.S. book on EW [5] in existence has limited treatment of ECCM. Skolnik [26] devoted seventeen pages to ECCM. Brick and Galejs [27] presented an excellent description of approximately twenty ECCM techniques eighteen years ago. Many of their techniques were described originally in the MIT Radiation Laboratory series, but had not been prominently identified as ECCM (or AJ) techniques. Many ECCM techniques, e.g., monopulse, have other benefits and have been integrated into radar technology for other reasons. The IEEE International Cumulative Index on Radar Systems [14], [28] listed approximately sixty references on ECCM, per se, not including monopulse, MTI, CFAR, et al. In 1973, John C. May [29] gave a lecture on basic ECM and ECCM considerations in the Boston Section IEEE Series on Modern Radar Techniques. Most of the Soviet books on radar have a chapter on EW, e.g., [23]. A previous paper of the author [30] which discusses the effects of ECCM included a list of approximately 150 ECCM types. Philosophy of ECCM utilization was discussed in a previous paper by the author [31]. Operational ECCMs were briefly addressed by Lampos [32]. Despite the availability of extensive information on ECCM in the open literature, a comprehensive treatment of ECCM has not been made to date. Adequate treatment requires a full book. This paper, however, will address an overview of ECCM. Individual ECCM techniques will not be discussed but some reference to the literature will be cited. Many of the techniques are well known to radar engineers for other reasons and are described in Skolnik [33].

The radar engineer should consider several aspects of ECCM. These include not only ECCM techniques themselves, but also general effects of the elements of EW, identification of specific ECCMs for use against specific ECMs, and assessment of the efficacy of ECCMs. These topics also will be covered briefly in this paper.

Effects of Elements of EW

Before considering ECCM techniques, it will be beneficial to consider the general effects of the elements of EW. This will help to put things in a better perspective.

ECM

The various types of ECM will generally have these effects:

denial of detection
operator confusion or deception
delay in detection or tracking initiation
tracking of an invalid target
overloading of computer (excessive number of targets)
denial of measurement of target position, range rate
target tracking loss
errors in values of target position, range rate.

ECCM techniques must, therefore, counteract those effects.

ESM

There are two groups of ESM effects: those in support of ECM and those in support of ECCM.

A. ESM effects in support of ECM. These include provision of the following data:

radiation characteristics of radars: frequency, modulation characteristics, antenna pattern, etc.
angular position location of radars
identity, type, and purpose of radars
information for ECM/ARM (antiradiation missile) employment
weapon targeting information.

It should be noted that ESM does not provide location of a bistatic radar receiver. ESM is subject to possible confusion by dummy transmitters.

B. ESM effects in support of ECCM. These include provision of the following data:

warning of possible hostile activity (alerting)
angular locations of hostile radars
information on radiation characteristics of jammers
information for ECCM selection
data for home on jam/track on jam (HOJ/TOJ) employment
weapon targeting information
information for weapon avoidance.

ESM is subject to confusion from spoofers and standoff jammers (SOJ).

ECCM

ECCM effects are essentially the countering of the effects of ECM above. These ECCM effects include:

prevention of receiver saturation
constant false alarm rate (CFAR)
enhancement of S/J ratio
directional interference discrimination
rejection of invalid targets
maintenance of track.

It should be noted that a principal objective of ECCM is to permit the radar to accomplish its intended function while minimizing the amount of information transmitted to the enemy [34]. It is important to identify the specific benefits of an individual ECCM technique. Since ECMs often are used in combination, e.g., jamming and chaff in World War II [3], [4], the radar designer may have to employ a combination of ECCMs [30], [31].

ECCM Techniques

A radar engineer is usually trained in electronic circuitry. It is only natural, therefore, to think of ECCM techniques in terms of electronic circuitry. There are two categories of ECCM, electronic and operational. The former includes things inside the radar, while the latter includes how the radar is used. These two categories must be used in combination. As Lampos [32] indicated, only the latter group can be readily changed in a fielded radar.

Electronic ECCM Types

A previous paper [30] alphabetically listed approximately 150 ECCM types. These ECCM types can be divided into four subgroups: those in the transmitter, those in the antenna, those in the receiver-signal processor, and those in the "system." This division is in accord with the usual breakdown of a radar. Some ECCM techniques, e.g., monopluse, have effects in more than one part of a radar; other ECCM techniques, e.g., radar frequency, could be considered as part of several different electronic ECCM subgroups. Groupings used herein are those selected by the author for various reasons. No particular significance should be attached.

A. *ECCM in the Transmitter*

This subgroup may be further subdivided into those concerning the radar frequency, power, waveform, and pulse repetition (actually part of the waveform). ECCMs in the transmitter are listed in Table I.

B. *ECCM in the Antenna*

This subgroup also may be further subdivided into those concerning main-lobe pattern characteristics, sidelobe characteristics, and angular measurement techniques. ECCMs in the antenna are listed in Table II.

Sidelobe cancellation in Table I has evolved from a very simple technique into a very sophisticated technology with the application of adaptive processing. Two special issues of the *IEEE AP-S Transactions* [36], [37] have been devoted to this important development. Gabriel [37] presented an excellent overview of this in the latest issue.

C. *ECCMs in the Receiver-Signal Processor*

This subgroup is the most numerous of all the subgroups; in fact, it is larger than all the other subgroups combined. This popularity arises from several factors.

1. It is generally easier and cheaper to make changes to a fielded radar in the receiver than in the antenna or the transmitter. A receiver ECCM may be a handful of transistors while raising radar peak power output may be very costly.

TABLE I

ECCM in the Transmitter

Frequency
- Fast Manual Frequency Shift.
- Fine Frequency.
- Frequency Agility.
- Frequency Diversity.
- Frequency Shift (Variable Frequency).
- Higher Frequency.
- Multi-frequency Radar (Multi-band Radar).
- New Frequency Bands.
- Pulse-to-Pulse Frequency Shift.

Power
- Increased Power Output.
- Power Management (Power Output Control).

PRF
- Interpulse Coding, Pulse Position Modulation.
- Jitter PRF (PRF Change).
- Linear Intrapulse FM (CHIRP).
- Modulated PRF.
- Modulation on Pulse.
- Nonlinear Intrapulse FM.
- Pulse Burst Mode.
- Pulse Compression and Stretching (CHIRP).
- Pulse Frequency Modulation.
- PRF Agility.
- Short Pulse Radar.
- Staggered PRF.
- Variable PRF.

Waveform
- Coded Waveform.
- CW
- FM
- Noise/Spread Spectrum (Pseudo Noise).
- Pulse Doppler.

Other
- Diplexing.

TABLE II

ECCM in the Antenna

Angular Measurement Techniques
- Adaptive Angle Scan Modulation.
- Compensated Conical Scan.
- Hybrid Monopulse.
- Lobe on Receiver Only (LORO).
- Monopulse.
- Monopulse MLC.
- Multisimul Antenna.
- Phased Array Radar.
- Scan Rate Amplitude Modulation.
- Variable Scan Rate.

Main-Lobe Pattern Characteristics
- Bistatic Radar.
- Circular Polarization.
- Increased Angular Resolution.
- Mainlobe Cancellation (MLC).
- Polarization Diversity/Selector.
- Polarization Main Lobe Cancellation.

Sidelobe Characteristics
- Sidelobe Blanker (SLB).
- Sidelobe Canceller (SLC).
- Sidelobe Reduction/Suppression.
- Sidelobe Suppression by Absorbing Materials.

TABLE II (continued)

Others
- Dummy Antenna/Dummy Load.
- Single Lobe Blanking (SLB).

2. Receiver ECCMs are passive, i.e., their presence is not disclosed to an enemy.

3. Recent advances in signal processing, especially use of digital techniques, have permitted great improvements in radar receivers.

4. Receiver ECCMs are generally more versatile.

5. Receiver ECCMs are more amenable to operator adjustment for maximum effectiveness than those in the antenna or transmitter.

6. There are many varieties of some ECCMs, e.g., MTI, CFAR, and Dicke-Fix.

Receiver ECCMs also may be further subdivided into predetection, detection, postdetection, PRF considerations, and displays. These are given in Table III. As general considerations, the receiver linearity and dynamic range should be maximized [38], [39]. Receiver desensitization should be avoided [40].

The importance of digital signal processing, e.g., fast Fourier transform (FFT) techniques [41], is noteworthy. Cartledge and O'Donnell [42] further treated advanced signal processing technique which are meaningful in an ECCM role. They used automatic adaptive thresholding and fully coherent processing as a digital replacement for the Dicke Fix.

D. *ECCM in the System*

ECCM in the system includes both radar data manipulation and analysis and the use of target data from other sources. These are listed in Table IV. The first subgroup was made possible by recent advances in digital technology. Laser radars were covered both in the U.S. [43] and U.S.S.R. [44]. The second subgroup is outside the radar, but data from those sources may greatly add to the usefulness of a radar in an EW environment.

Operational and Other ECCMs

Most of these are not in the direct cognizance of the radar engineer. He should be familiar with them and be able to make provisions for their use. However, this group may be further subdivided into operator, operating methods, radar deployment, ESM, and "other." These are listed in Table V. The most important operational ECCM is the operator himself. Proper training is mandatory. There is a limit to automatic circuitry. There is no substitute for a well-trained operator.

ECM-ECCM Matrix

There are approximately fifteen types of ECM. (Some are variants of basic types, e.g., spot and barrage noise.) It is instructive to consider possible ECCM types for each ECM type. An ECM-ECCM matrix is shown in Table VI.

TABLE III

ECCM in the Receiver-Signal Processor

A. *General Considerations*
- Desensitization.
- Dynamic Range.
- Linearity.

B. *Predetection*
- Balanced Mixer/Image Suppressor.
- Broad Band Receiver.
- Coherent Long Pulse Discrimination.
- Compressive IF Amplifier.
- CW Jamming Canceller.
- Dicke Fix:
  - Clark Dicke-Fix (Cascaded Dicke-Fix).
  - Coherent MTI Dicke-Fix.
  - Craft Receiver.
  - Dicke Log Fix.
  - IF Canceller MTI Dicke-Fix
  - IF Dicke-Fix (CFAR) (Zero Crossings, Dicke-Fix (CFAR).
  - Instantaneous Frequency Dicke-Fix.
  - Noncoherent MTI Dicke-Fix.
  - Video Dicke-Fix (CFAR).
- Frequency Preselection (Narrow Band Widths).
- Gain Control Techniques:
  - Anticlutter Gain Control.
  - Automatic Back Bias.
  - Automatic Gain Control (AGC).
  - Automatic Noise Leveling.
  - Dual Gated AGC.
  - Fast AGC.
  - Gated AGC.
  - Instantaneous AGC (IAGC).
  - Jam Attenuator Control.
  - Manual Gain Control.
  - Pulse Gain Control.
  - Sensitivity-Time Control (STC).
- Guard Band Blanker.
- IF Diversity, IF Limiter.
- Integration (Coherent):
  - Coherent IF Integration.
  - Coherent IF Integration (Moving Target).
  - Coherent IF Integration (Stationary Target).
  - FM Delay Line Integration.
- Jamming Cancellation Receiver.
- Kirbar Fix.
- Lamb Suppressor.
- Lin-Log IF (Log Receiver).
- Local Oscillator Off.
- Matched Filtering.
- Moving Target Indicators (MTI):
  - Area MTI (Velocity Filter).
  - Cascaded Feedback Canceller MTI.
  - Clutter Gated MTI.
  - Coherent MTI.
  - MTI-CFAR.
  - Noncoherent MTI.
  - Pseudo Coherent MTI.
  - Pulse Doppler.
  - Re-entrant Data Processor.
  - Single-Delay Line Canceller MTI.
  - Three-Pulse Canceller.
  - Two-Pulse Canceller.
- Narrowband Limiter.
- Passive Receiver for IF Strobe.
- Shielding.
- Swept Local Oscillator Receiver (SLOR).

TABLE III (continued)

C. *Detection*
- Aural Detection.
- Correlation Detection.
- Detector Back Bias.
- Detector Balanced Bias.
- Double Threshold.
- Jammer Tracked by Azimuth Crossing.
- Least Voltage Coincidence.
- Video Correlator.
- Zero Crossing Counter.

D. *Post Detection*
- Audio Limiter.
- Autocorrelation Signal Processing.
- Automatic Cancellation of Extended Targets (ACET).
- Automatic Video Noise Leveling.
- Conical Scan Guard Band.
- Constant False Alarm Rate (CFAR):
  - Automatic Threshold Variation.
  - Cross-Gated CFAR.
  - Dispersion Fix CFAR.
  - IF Dicke-Fix CFAR.
  - Instantaneous Frequency CFAR.
  - MTI CFAR.
  - Unipolar Video CFAR.
  - Video Dicke-Fix CFAR.
  - Wideband CFAR.
  - Zero Crossings CFAR.
- Cross Correlation Signal Processing.
- Fast Time Constant.
- High Video Pass.
- Integration (Noncoherent);
  - AM Video Delay Line Integration.
  - Delay Integration.
  - Noncoherent (Video) Integration.
  - Pulse Integration.
  - Video Delay-Line Integration.
- Log FTC.
- Log Video.
- Logical ECCM Processing.
- Post Canceller Log FTC.
- PRF Considerations:
  - High PRF Tracking
  - PRF Discrimination.
  - Pulse Edge Tracking.
  - Pulse Interference Eliminator (PIE).
  - Pulse Interference Suppression and Blanking (PISAB).
  - Pulse Length Discrimination.
  - Pulse Shape Discrimination.
  - Pulse Width Discrimination (PWD).
  - Random Pulse Discrimination.
- Range Gating of Video.
- Video Correlator.
- Video Discrimination.
- Wide Band Limit.
- Widepulse Blanking (WPB).

E. *Displays*
- Azimuth vs. Amplitude.
- Baseline Break.
- Electronic Implementation of Baseline Break.
- Evaluation vs. Integrated Log (EVIL).
- Jam Strobe (Jamming Amplitude vs. Azimuth) (JAVA).

TABLE IV
ECCM in the System

Radar Data Manipulation/Analysis
- Acceleration Limit.
- Azimuth Blanking.
- Azimuth Gain Reduction.
- Beam-to-Beam Correlation.
- Doppler-Range Rate Comparison.
- Jammer Finder.
- (Manually) Aided Tracking.
- Manual Rate Aided Tracking.
- Passive Angle Tracking (PAT).
- Passive Detection and Tracking.
- Prediction Computer.
- Pulse-to-Pulse Correlation.
- Range/Angle Rate Memory.
- Range Rate Memory.
- Velocity Tracker.

Target Data from Other Sources
- Acoustic Listening.
- D/F on VHF Communications.
- Dual Radars.
- Laser Radar.
- Passive Tracking/HOJ/TOJ.
- Use of Data from Remote Battery.
- Use of Range from Acquisition Radar with Optical Tracker Instead of Tracking Radar.
- Visual/IR Tracking.

TABLE V
Operational and Other ECCMs

Operator
- Training.

Operation Methods
- Appropriate Assignment of Normal Operating Frequencies to Various Radars.
- Intermittent Operation of Radars.
- Minimization of Operating Time.
- Reservation of Certain Frequencies for Emergencies.
- Use of Combined ECCM's to Meet Two Combined ECM's.
- Use of Dummy Transmitters to Draw ECM to Other Frequencies.

Radar Deployment
- Optimum Siting.
- Overlapping Deployment.
- Radar Mobility.
- Remote Location of Antenna to Protect Operators.

ESM
- Monitor A/C Communications.

Other
- Attack Chaff Drop A/C.
- Close Liaison Between Development Engineers and the Field.
- Multiple Weapon Types.
- Rapid Exploitation of Capture of Hostile EW and Radars.
- Strong QRC ECCM Research Program.
- Tight Security on Our Radar Development.

TABLE VI
ECM-ECCM Matrix: ECM Type and Possible ECCM Techniques

A. ACTIVE DENIAL

CW Jamming
Frequency diversity, frequency agility, fast time constant, gain controls, pulse width discrimination, sidelobe cancellation, TOJ/HOJ.

Long Pulse Jamming
All of the above plus integration.

Spot Noise Jamming
Frequency diversity, frequency agility, integration, Dicke-Fix CFAR.

Barrage Noise Jamming
Dicke-Fix CFAR, log FTC, automatic video noise level, local oscillator off, higher ERP, longer time on target.

Medium and Fast-Sweep Jamming
Dicke-Fix CFAR, variable IF bandwidth, integration, pulse compression, frequency agility, guard band blanking.

Impulse Jamming
Dicke-Fix CFAR

Short Pulse Jamming
Integration, pulse compression.

Sidelobe Repeaters
PRF jitter and integration, pulse compression, sidelobe blanking, bistatic radar.

B. PASSIVE DENIAL

Chaff
MTI, Doppler radar, improved resolution (narrow beam width, short pulse, pulse compression), log FTC, gain controls, low scan rate on search.

Radar Absorbing Materials
Higher ERP, longer time on target.

C. ACTIVE DECEPTION

Repeaters
Complex radar waveform, spread spectrum technique, frequency agility, PRF jitter and integration, sidelobe blanking.

Range and Velocity Gate Stealers
PRF jitter, leading edge tracking, acceleration limiting, Doppler/range rate checking.

Scan Modulation Jamming
Monopulse, lobe on receiver only, variable conical scan frequency.

Angle Gate Stealers (TWS Radar)
Lobe on receive only.

D. PASSIVE DECEPTION

Decoys
Target/decoy trajectory characteristics, multistic radars, bistatic operations.

Chaff (Tracking Radar)
Higher resolution in range and angle speedgate loop, MTI, leading/trailing edge tracking.

This table does not list all of the 150 ECCM types of tables I through V. It is an attempt at simplification of a very complex topic. Many qualifying considerations have been omitted in the interest of simplicity. It was prepared from a compilation of several references [45]-[50]. A limited U.S.S.R. ECM-ECCM matrix [23] was also noted. Some ECCMs can only be used in certain types of radar, e.g., CFAR is applicable to search radars. A radar engineer should determine the effect of various ECMs on his radar; then he should seek ECCMs which can counteract those effects.

Johnson and Stoner [39] presented vivid visual demonstration of the effectiveness of several ECCMs against ECMs listed in Table VII. The spectacular photographs in their paper clearly show that some rather difficult ECMs *can* be defeated. Of course this is only round one in the continuing battle between ECM and radar.

Radford [51] showed that the European approach to ECCM differs from the U.S. concept. Even so, there is an excellent appreciation for ECCM in the European radar community.

TABLE VII

Demonstration of Selected ECM-ECCM Combinations

ECM	*ECCM*
Swept CW	Coherent Sidelobe Canceller (CSLC)
False Target Repeater	Sidelobe Blanker
High Duty Cycle Pulse Jammer	Sidelobe Blanker and CSLC
Chaff	MTI and Adaptive Threshold (CFAR)
Barrage Noise Jamming	Sidelobe Suppression Clutter Suppression (MTI) and Adaptive Threshold (CFAR)
Mutual Interference (EMC)	Pulse-to-Pulse Correlation

ECCM Efficacy Assessment

The radar engineer must consider the performance of ECCMs whether they be electronic or operational. This has not received adequate attention in the literature. There are several aspects of ECCM efficacy assessment.

Quantitative Measure of ECCM Effects

There is a need for a quantitative measurement of ECCM effects [52]. Performance of an automobile may be given in miles per hour, miles per gallon of fuel, acceleration, etc. Several measures of ECCM effects have appeared in the literature. These include the improvement factor for MTI, processing gain, cancellation ratio (for sidelobe cancellers), subjamming visibility, etc. The author proposed the parameter ECCM improvement factor (EIF) [30], [52] for this purpose. This term has recently been added to the IEEE standard definitions on radar [53]. A measure such as the EIF will permit ready representation of the effects of some ECCMs in the performance analysis of radars. Most of these parameters have limitations [30].

Measurement of Effectiveness of ECCMs

Quantitative representation of the effect of an ECCM is inadequate as a measure of effectiveness, or MOE in the language of operations researchers. A modern radar is merely a sensor in its usual application today. Some EW practitioners use burn-through range as the MOE of a jammer or radar. The fallacy of this criterion was pointed out [54] and better criteria were cited. A Russian paper [55] presented the parameter interference resistance, but this was shown [56] to be a variant of the ECCM improvement factor and inadequate as an overall MOE. Mitchell [34] and Schlesinger [5] indicate the need for an MOE in the ECM environment. Among the better MOEs suggested [54] were change in the number of aircraft destroyed or surviving, change in first engagement range, etc. One of the first studies on the effectiveness of ECM in World War II used change in the number of aircraft lost. Another Russian book [24] has a very interesting chapter on this topic.

Adequate Simulation Models

The many ECCM techniques briefly mentioned in this paper indicate there is a need for simulation models to adequately include these ECCM techniques. This problem has been addressed [52], [54]. Unfortunately, an adequate simulation model is much more complex than a radar slide rule or an ECM slide rule because it must involve the whole weapon system. This necessity for complexity is regrettable but unavoidable. It is a price of modern weapons technology.

Adequate Data for Simulation Models

Adequate data are needed for simulation models [51], [52], and unfortunately such data do not exist frequently. It should be noted also that adequate data on basic radar do not exist frequently either. In fact, currently there is no standard list of characteristics for radars. That problem is presently being addressed by the IEEE Aerospace and Electronic Systems Society (S-AES) Radar Systems Panel. It is hoped that a standard list of radar characteristics will be promulgated soon.

Conclusion

A very large number of ECCMs has been conceived. These have been briefly listed in the accompanying tables. Many of them are already well known to many radar engineers, although they may not have been thought of as ECCMs. It is hoped that this brief overview of ECCMs will be of benefit to radar engineers.

Acknowledgment

The author gratefully acknowledges contributions to this paper by many colleagues. D.K. Barton, past chairman of the Radar Systems Panel suggested this paper; O.B. Mitchell, of Emerson Electronic, and Lee Davies and Franklin Turner, both of Standard Research Institute, furnished much information and assistance.

References

[1] Alfred Price, *Instruments of Darkness.* London: William Kimber, 1967, MacDonald and Jones, London, 1977

[2] Harry F. Smith, "EW history," *Electronic Warfare Magazine,* vol. 1, no. 4, 1969, pp. 18 ff.

[3] "Electronic warfare: A report on radar countermeasures" (extract), *IRE Trans. Military Electronics,* vol. MIL-1, p. 31, March 1957.

[4] "Electronics warfare: Report on radar countermeasures," Joint Board on Scientific Information, Office of the Publication Board, revised Oct. 1945.

[5] Robert J. Schlesinger, *Electronic Warfare.* Englewood Cliffs: Prentice-Hall, 1961.

[6] J.A. Boyd, D.B. Harris, D.D. King, and H.W. Welch, Jr., *Electronic Countermeasures.* Ann Arbor: University of Michigan, IST 71204, R 1961.

[7] S.A. Vakin and L.N. Shustov, *Principles of Jamming and Electronic Reconnaissance.* DDC AD 692642 and 692643. Moscow: Soviet Radio, 1968 (in Russian; trans AD692 642, -3).

[8] ———, "Principles of jamming and electronic reconnaissance," *Electronic Warfare Magazine,* vol. 4, Spring 1972, pp. 38-39.

[9] M.P. Atrazher, *Bor' Ba S Radio Elektronnymi Sredstvami (Combat by Electronic Means),* Moscow: Soviet Radio, 1972.

[10] A.N. Volzhin and V.A. Yanovich, *Radar Countermeasures.* Moscow: Voyenizdat, 1960 (in Russian; trans. AD

[11] Y.G. Stepanov, *Radar Camouflage,* Moscow: Soviet Radio, 1968 (in Russian; trans AD 733507 partial; JPRS: 47734 partial).

[12] A.I. Paliy, *Radiovoina (Radio-Electronic Warfare),* Moscow: Oborongiz, 1974.

[13] ———, *Radiovoina (Radio Warfare).* Moscow: Voyenizdat, 1963.

[14] D.K. Barton, "International cumulative index on radar systems updated," *IEEE Trans. Aerosp. Electron. Syst.,* vol. AES-13, May 1977, pp. 336 ff.

[15] S.L. Johnston, "A complete bibliography: Soviet Literature on Radar and EW. Part I," *Electronic Warfare Magazine,* vol. 8, Sept.-Oct. 1976, pp. 43 ff.

[16] W.D. Meskill, "Why ECCM: The need for ECCM," *ELECTRO-76 Convention Record,* pt. 30, pp. 30-1-1 to 4, Boston, Mass., May 11-14, 1976.

[17] "Electronic warfare definitions," *Electronic Warfare Magazine,* vol. 2, no. 4, 1970, p. 29.

[18] Department of Defense, Joint Chiefs of Staff, "Dictionary of military and associated terms," JCS Pub-1, Sept. 1974.

[19] "Dictionary of U.S. Army terms," (AD), Department of Army Regulation 310-25, Sept. 15, 1975 w/change 1, May 15, 1977.

[20] E.R. Freeman, "The role of ECCM in friendly radar environment EMC," ELECTRO Record, 1976, pt 30, pp 30-4-1 to 5.

[21] Major Richard E. Fitts, "Fundamentals of electronic warfare," U.S. Air Force Academy, March 1, 1972 edition.

[22] F. Farner, Stroesignale in Radargeraeten und Methoden zu iher Unterdrueckung, "Interference signals in radars and methods for their suppression," *Lousanne Bulletin d-Association sursse Des Electriciens,* vol. 61, pp 725-732, Aug. 8, 1970.

[23] A.I. Leonov and K.I. Fomichev, *Monopulse Radar.* DDC AD 742696. Moscow Soviet Radio, 1962 (in Russian, trans JPRS 15381).

[24] V.N. Tipugin and V.A. Veytsel, *Radio Control.* JPRS 15281.

[25] M.M. Maksimov. *Pomekhoustoĭchivost' Mnogokanal' Nykh Komanduyka Radioliniĭ Upravlenia (Interference Immunity of Multichannel Radio Control and Guidance.* Moscow: Soviet Radio, 1970.

[26] Merrill I. Skolnik, *Introduction to Radar Systems.* New York: McGraw-Hill, 1962, pp. 554-567.

[27] Donald B. Brick and Janis Galejs, "Radar .nterference and its reduction," *The Sylvania Technologist,* vol. 11, July 1958, pp. 96-108.

[28] D.K. Barton, "International cumulative index on radar systems," *IEEE Trans. Aerosp. Electron. Syst.,* vol. AES-11, May 1975, p. 428.

[29] John C. May, "Basic ECM and ECCM considerations," General Electric Co., presented in the IEEE Boston Section Series on Modern Radar Techniques, Nov. 13, 1973.

[30] Stephen L. Johnston, "ECCM improvement factors (EIF)," *Electronic Warfare Magazine,* vol. 6, May-June 1974, pp. 41-46.

[31] ———, "Philosophy of ECCM utilization," *Electronic Warfare Magazine,* vol. 7, May-June 1975, pp. 59-61.

[32] N.J. Lampos, "Electronic warfare defense planning tactics," *Electronic Warfare Magazine,* vol.6, Nov.-Dec. 1974, pp. 44-47.

[33] Merrill I. Skolnik, ed., *Radar Handbook.* New York: McGraw-Hill, 1970.

[34] O.B. Mitchell, "ECCM: Effective deployment and employment," *Electronic Warfare Magazine,* vol. 6, Nov.-Dec. 1974, pp. 36-38.

[35] *IEEE Trans. Antennas and Propagation* (special issue on active and adaptive antennas), vol. AP-12, Mar. 1964.

[36] *IEEE Trans. Antennas and Propagation* (special issue on adaptive antennas), vol. AP-24, Sept. 1976.

[37] William F. Gabriel, "Preface to special issue on adaptive antennas," *IEEE Trans. Antennas and Propagation,* vol. AP-24, Sept. 1976, pp. 573-574.

[38] J.C. Galenian, "Designing radar receivers to overcome jamming," *Electronics,* vol. 36, May 17, 1963.

[39] M.A. Johnson and D.C, Stoner, "ECCM from the radar designer's viewpoint," ELECTRO Record, 1976, pt. 30, pp. 30-5-1 to 11.

[40] Ray Moore, "Designing communications receivers for good strong-signal performance," *Ham Radio Magazine,* Feb. 1973, pp. 6-15.

[41] H.M. Halpern and R.P. Perry, "Digital matched filters using fast fourier transforms," *EASCON Rec.* 1971, pp. 222-230.

[42] L. Cartledge and R.M. O'Donnell, "Description and performance evaluation of the moving target detector," Project Report ATC-69, Lincoln Laboratory MIT, Mar. 8 1977.

[43] Albert V. Jelelian, "Optical aids to detection and ranging systems (OADARS)," ELECTRO Record 1976, pp. 30-3-1 to 5.

[44] V.I. Petrovskiy and O.A. Pozhidayev, *Laser Radars.* Moscow: Publ. House, of Ministry of Defense, 1969 (in Russian; trans. JPRS 50123).

[45] Lee E. Davies, "Handbook of analysis of electromagnetic systems," Stanford Research Institute (unpublished), 1974

[46] Richard W. Hanford, "Radar testing for a war environment," *IRE Natl. Conv. Rec.,* vol.7, pt. 5, pp. 208-217.

[47] ———, "Radar testing for a countermeasures environment," Proc. Natl. Aeronautics Electronics Conf., May 1959.

[48] Murray Simpson, "Counter-countermeasures: Good design can beat the ECM threat," *Space/Aeronautics,* vol. 33, Apr. 1960, pp. 138-142.

[49] Charles L. Bachtel, "The EMC/ECCM syndrome," *Signal Magazine,* Oct. 1973, pp. 25-29.

[50] Peter R. Dax, "Electronic dogfight: ECM vs ECCM in search radar," *Space/Aeronautics,* Apr. 1960, pp. 134-137.

[51] M.F. Radford, "Radar ECCM: A European approach," *IEEE Trans. Aerosp. Electron. Syst.,* vol. AES-14, no. 1, Jan. 1978. ELECTRO Record, 1976, pt. 30, pp. 30-6-1 to 5.

[52] Stephen L. Johnston, "Military radar: Weapon system analysis," presented Sept. 18, 1970 as lecture in the Short Course, Principles of Radar, Georgia Institute of Technology. Published as Lecture Notes.

[53] "IEEE standard definitions: Radar and navigation aid terms," *IEEE Standard Radar Definitions:* IEEE Std 686-1977, Nov. 9, 1977.

[54] Stephen L. Johnston, "Burnthrough range: An overrated parameter," *Electronic Warfare Magazine*, vol. 6, Nov.-Dec. 1974, pp. 49-53.
[55] P.I. Baklashov, "Radar jamming and protection of radar stations," Oct. 14, 1968, JPRS 46664.
[56] Stephen L. Johnston, "Radar ECCM improvement factor, interference resistance and burnthrough range," (submitted to *Electronic Warfare Magazine*).

Stephen L. Johnston (S'42–A'44–M'51–SM'53) was born in Oklahoma City, Ok., on October 13, 1923. He received the B.E.E. degree with honors and the M.S.E.E. degrees in 1948 and 1949, respectively, from the Georgia Institute of Technology, Atlanta, Ga. During World War II he served in the U.S. Army Signal Corps where he attended the Southern Signal Corps Radar School. Subsequently, he provided radio communications for U.S. radio broadcast correspondents with General MacArthur.

Mr. Johnston was a research assistant at the Engineering Experiment Station at Georgia Tech from 1947 to 1950. He entered Federal Civil Service in 1950 at the Joint Long Range Proving Ground, Patrick AFB, Fla. He has been employed by the U.S. Army Missile Command at Redstone Arsenal, Ala., since 1951 where he is now Chief of the Plans/ Technology Team in the Advanced Systems Concepts Office, U.S. Army Missile Research and Development Command. His areas of interest include electronic warfare, weapon system simulation, penetration aids for ballistic missiles, radar analysis, target signature characteristics, ballistic missile trajectory and guidance analysis, counter-mortar radars, magnetic bubbles, and submillimeter wave radar. Mr. Johnston is a member of the Radar Systems Panel of the IEEE Aerospace and Electronic Systems Society (AES), and Chairman of the Radar ECCM Committee of that panel. He created the term ECCM improvement factor which was recently incorporated into the "IEEE Standard Definitions on Radar and Navigational Aid Terms." His session on Radar ECCM received the best technical session award at ELECTRO-76.

Mr. Johnston is an Associate Fellow of AIAA, a member of the Association of Old Crows and president of his local AOC chapter, and a registered professionsl engineer. He is a member of Phi Kappa Phi.

The Role of ECCM in Friendly Radar Environment ECM

E. Freeman

IEEE ELECTRO-76 Record, Part 30, May 1976
Pages 30-4-1— 30-4-5

1.2

GENERAL

Electromagnetic Compatibility (EMC) is an "umbrella" term used to describe those studies, techniques and relationships which contribute to the state wherein electrical and electronic systems can operate in the same environment without interference from each other. It includes the subclasses known as Electromagnetic Interference/Radio Frequency Interference (EMI/RFI), Hazards of EM Radiation to Ordnance (HERO) and Radiation Hazards (RADHAZ), and spectrum Management (Frequency Allocation and Assignment).

The elements of the situation are encompassed in the well-known simple representation of the source-path-sink type shown in figure 1. The sources can be categorized as natural or man-made. Of the latter variety the sources can be further categorized in many ways, all of which are significant from both a technical and management viewpoint. Among these are intentional - unintentional which is most pertinent to the panel, broad-band vs. narrow-band, pulsed vs. CW, operationally required vs. spurious, high-level vs. low-level, etc. The sources of natural interference are basically terrestrial triboelectricity, and extra-terrestrial cosmic noise. Examples of many sources of all varieties are found in Table 1.

The paths by which the interfering energy is transferred are shown in figure 2. These paths are also the basis for categorization. The major breakdown is radiated vs. conducted interference. In the category of radiated energy are the antenna, coupled path (most analogous to the ECCM situation) as well as all combinations of antenna, wire and case coupling. Conducted energy includes paths due to earth ground links and the "common mode" problem.

The receptors of electromagnetic energy may be categorized as natural and man-made. The human body is a natural receptor and the subject of considerable study in the RADHAZ area. The man-made receptors may be classified as communications-electronics receivers, amplifiers, industrial and consumer devices and ordnance.

System Susceptibility

In order to rate the susceptibility of receivers to in-band interference a convenient index can be developed based on their sensitivity and bandwidth, e.g.,[1]

$$RS = \sqrt{\frac{Bk^2}{4RFKT}}$$

where, B = bandwidth (Hz)

k = proportionality factor

R = resistive component of the equivalent input impedance in ohms

F = equivalent noise figure of the receptor

$KT = 4(10)^{-21}$ watts/Hz of bandwidth

RS_c is an indication of the interference-to-noise ratio (INR) that would occur due to a coherent interference signal. Using this method, typical audio or sensor amplifiers would score an RS of 95 dB, HF AM receivers about 195 dB and radar receivers 230-240 dB, which is in the extremely susceptible region. This does not take the spurious response intermodulation or cross-modulation characteristics of the receivers into account. There are EMC figures-of-merit being developed that do take all the significant aspects of EMI into consideration.

The antenna-to-antenna interference problem exists both on-site and in widely separated cases, sometimes loosely described as "near-field" and "far-field" situations. This latter area of EMC is the one that overlaps the ECCM field. The possible effects of such interference are summarized in Table 2. In simple radar systems the symptoms of such an interference problem could be the familiar "running rabbits" type spiral or a PPI scope. The effects include masking of actual targets, delays in identifying targets and human fatigue. With the advent of automatic detection systems, constant false alarm rate processors, and a variety of interference reduction techniques, almost all of the apparent interference is eliminated but more insidious effects may be encountered in the form of desensitization not obvious to the observer. High level interference signals may cause an effective loss in detection range with no evidence of their presence on any display. This is especially true of PD radars due to their AGC circuits.

EMC "Fixes"

For the purposes of this discussion, the remedies used to alleviate EMC problems fall into two classes, stationary and dynamic. The stationary, or fixed, category includes frequency allocation, grounding, bonding, shielding and most filtering. These areas are not so straightforward as is supposed by many designers. Most of the EMI problems encountered are due to non-application or mis-application of these safeguards. Many good design guides are available that cover the subject. The dynamic fixes are those that are applied based on the situation. These include frequency selection, blanking, cancelling, video processing, and all those features included in the ECCM domain. Some of the features especially designed for interference elimination are:

Harmonic Suppressions Filters There are many types of high power effective and absorptive type filters designed for suppressing harmonic and out-of-band signals in high power transmitters.

Bandpass Filters Preselectors for radar receivers are an example of tunable bandpass filters. These usually must have a somewhat wider bandwidth than the receiver. Fixed bandpass filters are also used to exclude out-of-band interference.

Sector Blanking This is a rather drastic fix in which the receiver or transmitter is turned off when the antenna covers a specific sector. Another example is where an electronically scanned phased away is programmed to avoid a certain solid angle. Of course, no signals can be received from the blanked sectors.

Pulse Blankers These devices gate receivers off at the time an interference pulse is expected. The trigger is supplied via wired paths, or in certain specialized cases, by special receivers. There are blankers that operate at RF, IF and video. While they eliminate certain interference effects they also reduce the probability of target detection.

PRF Discriminators A pulse-repetition-frequency discriminator utilizes a delay line and a coincidence circuit to eliminate all interference signals that do not have an interpulse period equal to that of the victim radar. The discriminator, usually operating at video, receives a target pulse; one pulse repetition period later, a second target pulse arrives at the input to the coincidence circuit just as the first leaves the delay line. In this scheme all except the first pulse are processed. Interference signals having PRFs equal to multiples of the receiver pulse rate are also passed through the discriminator. Several techniques for taking advantage of energy in a series of pulses are used. Integrators sum the recirculated signal from the delay line with the new input signal. Correlators multiply the two signals. In some cases the signal enhancement is used to allow relatively high detection-threshold settings that exceed the amplitude of the nonsynchronous signals. The high detection threshold may be used in lieu of or in conjunction with coincidence circuits as an interference-reduction device.

Pulse-width Discriminators Pulse-width discriminators differentiate the video pulse to produce a positive and negative spike at the leading trailing edges, respectively. The two spikes, separated by the pulse width t, are sent through two parallel channels, one inverting the signal and the other delaying it by the pulse width. Each signal in each channel contains a coincident spike of the same polarity if the incoming pulse had a width equal to the delay. The paired spikes are passed by a coincidence circuit to the video processor.

Tunability Although the ability to tune over wide portions of the RF spectrum is not usually thought of as an interference-reduction device, this capability is frequently the most effective means of achieving compatibility in congested environments. In the past, many radar systems have been designed to operate as fixed·tuned devices, and in some cases this practice continues. Although tunability imposes some additional cost penalties, such costs are usually well justified during the service life of the equipment. The Department of Defense Joint Frequency Allocation Panel has adopted tunability as a key evaluation factor in approving new system applications.

The operational or dynamic aspects of EMC control include the use of frequency shifting to avoid interference. In this aspect, radars are very difficult for frequency managers to handle. Since the receiver and transmitter are usually interconnected and independent of other systems they are mostly free to wander within their tunable range wherever they wish. In addition, due to the low duty cycle aspects of many radar types it is difficult to "make a case" for not assigning more than one to the same frequency. After all, 400 pulse per second is not a very high level of interference. The question of how many to assign, if such assignment were contemplated, is then important. Another aspect is the degree of operational flexibility sacrificed by such a scheme and its effect on ECCM planning.

Another interesting phenomenon is the relative insensitivity of certain important interference effects to specific frequency. These include RADHAZ and high power effects. Among these high power effects are the detection of pulse envelopes in circuit non-linear junctions such as solder joints. The problems affect many types of systems including other than those with RF sections.

Sidelobe Reduction Much interference is coupled into a radar receiver via relatively high antenna sidelobe gain levels. Antenna design techniques are available to reduce the scope of this problem. Johnson[2] has achieved a 10 db reduction in median sidelobe level at C and S bands by installing blinders on either side of the feed horn. Design guidance is also available for other techniques such as corrugated E-plane feed horns and tunnel antennas.[3]

Sidelobe Blanking This technique utilizes a low-gain omnidirectional antenna operating in conjunction with the main antenna. Signals from both antennas are fed to separate receivers and their outputs compared. Any comparisons showing greater power levels in the low-gain channel are cancelled, thus eliminating all returns from those sectors of the main antenna having gain levels less than the omnidirectional antenna. A gain control in the omnidirectional-antenna channel allows adjustment of the degree of sidelobe cancellation attained.

Pulse Synchronization If all radar systems operating in a congested area trigger simultaneously, their high-level signals will appear at neighboring receivers while TR tubes are still conducting, thus preventing damage. Pulse-synchronizing systems are available for this purpose and may take one of several forms. Over short distances triggering signals from an arbitrarily designated "master" radar may be distributed via coaxial cable to local systems. Over longer distances an audio signal (sinusoidal) may be transmitted over voice circuits to provide synchronization. In either case, the slave transmitters must operate at the same PRF as the master or at some multiple or submultiple of the PRF. When multiples are involved, the lower-PRF systems must be blanked during the extra pulse intervals to avoid displaying interference. If the lower-PRF systems are gated, as in tracking systems, the probability of extra pulses appearing in the gate may be low enough to preclude the need for additional blanking. If expected pulse levels are extremely high, crystal burnout may occur in the victim receiver even though the receiver is gated off at video or IF.

Pulse Shaping The RF energy existing in the spectral sidebands of a pulse signal constitutes a potential source of degrading interference and an undesirable use of the available spectrum. The narrow pulse widths and steep leading edges required for high resolution cause wider and wider spectra. Minimizing spectrum occupancy wherever possible is a necessary fundamental step toward spectrum conservation and compatibility. One of the problems in utilizing pulse shaping is the necessary use of output stages, that allow for this, e.g., Klystions and TWT's. The use of magnetrons, though less costly in dollars is sometimes extravagant in terms of Hertz.

EMC Analysis

EMC Analysis at the level of "far-field", antenna-coupled interference is almost identical to standard weapon systems analysis for EW purposes. The most significant differences are that the EMC interfering signals are "dumb", that the culprits in some cases, are as anxious to avoid problems as the victim, and that the problem involves very large numbers of systems. This has lead to many versions of automated models designed to "cull" through an environment and isolate specific problems. This is done by numerous interations of the one emitter to one susceptor model. The following steps of the analysis are based on good engineering judgement and data availability and should be identical to that performed for ECCM.

One of the most difficult problems to solve is the "operational configuration" question. This has to do with the switch settings of the systems of interest. In the cases I have been concerned with, this has been, and, as far as I know, still is largely undefined. This, in turn, leads to very "iffy" analysis. Decision-makers do not like to have answers to questions such as; Do we have an interference problem?, to be couched in terms such as -- Yes, if switch A is up and B is down -- No, if switch C is in position 2. One way to get around the problem was to couch the answer in probabilistic terms. This, unfortunately, can lead to disastrous consequences, since answers such as; You can operate satisfactorily at the .95 level of confidence based on normal operating conditions, might neglect to state that you cannot operate at all when you really need the system which is 2% of the time during which you will experience 90% degradation of performance.

SUMMARY & CONCLUSIONS

In conclusion the fields of EMC and ECCM have two common meeting grounds. The first is that area of ECCM design where EMC practices, standards, and conventions must be met or lived-with as in the case of all electronic systems. The second is the special area of antenna-coupled interference where, except for minor differences, the analysis and many of the fixes are identical. The use of ECCM features to combat EMI is an old story but not well documented. The mutual effects of EMI and ECCM circuits has been studied but not in great detail nor is there a great deal of available documentation.

Problems faced by both communities include the lack of sufficient data and resources. These problems could be considerably alleviated by bridging the gap betwuen the communities as I hope this session will help to achieve. It is also quite obvious that the consideration of ECCM features is of overriding significance in EMC analysis but, in many cases, is not adequately covered.

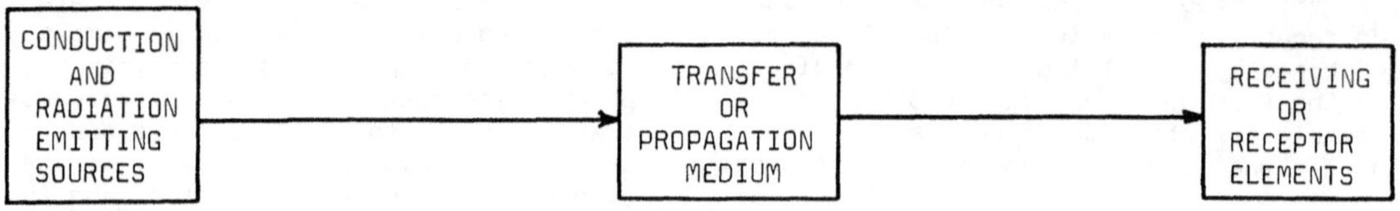

Sources	Medium	Receptors
Communication Transmitters	Space Separation	Communication Receivers
Radar & Telemetry Transmitters	Shielding Materials	Radar or Telemetry Receivers
Navigation Transmitters	Absorptive Materials	Navigation Receivers
Receiver Local Oscillators	Filters	Digital Computers
Motors, Switches, Power Lines, Fluorescent Lights Engine Ignition, Diathermy, Dielectric Heaters, Arc Welders	Cabling and Ground Inter-Couplings	Sensitive Indicators & Relays
Natural Sources: Atmospherics, Galactic Noise, Lightning	Power-Lines	Ammunition and Ordnance
		Human Beings (Biological Hazards)

Three Basic Elements of an Emitting-Susceptibility Situation

Figure 1

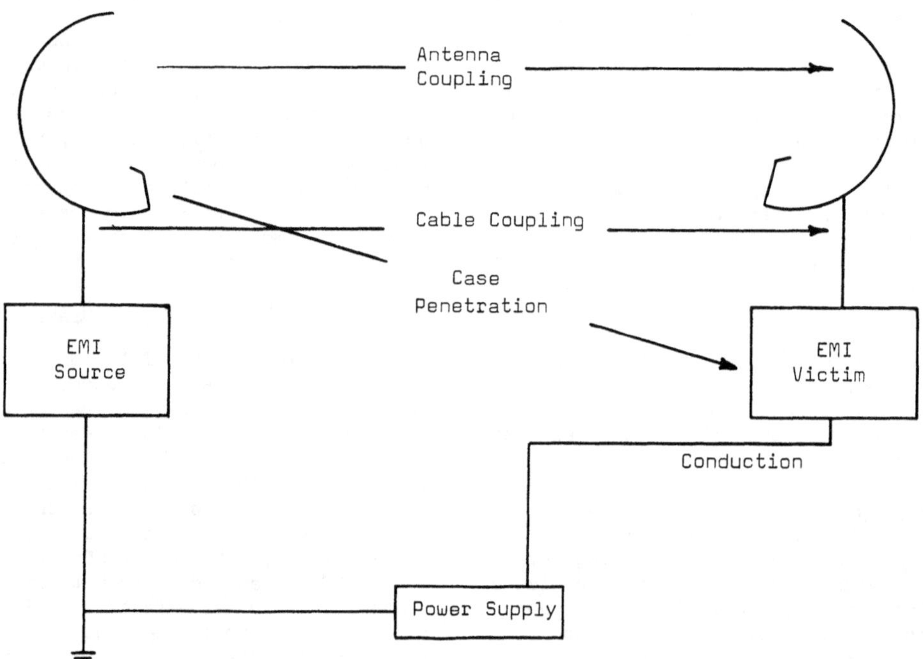

$$I/N = G_t \cdot G_r \cdot \int P_t(f) G_r(f) df / L_p L_r L_r$$

One-to-One EMI Situation

Figure 2

TABLE I

Example of Sources of Electromagnetic Interference

1. NATURAL SOURCES

1.1 Terrestrial Sources:
- a. Atmospherics (thunderstorms around the world)
- b. Lightning discharges (Local storms)
- c. Precipitation Static
- d. Whistlers

1.2 Extra-Terrestrial Sources:
- a. Cosmic Noise
- b. Radio Stars
- c. Sun:
  - Disturbed
  - Quiet

2. MAN-MADE SOURCES

2.1 Electric-Power:
- a. Conversion (Step Up/Down)
  - Faulty/Dirty Insulators
  - Faulty Transformers
- b. Distribution
  - Faulty/Dirty Insulators
  - Faulty Transformers
  - Faulty Wiring
  - Pick-up and Re-radiation
  - Poor Grounding
- c. Generators
- d. Transmission Lines
  - Faulty/Dirty Insulators
  - Pick-up and Re-radiation

2.2 Communications-Electronics (C-E):
- a. Broadcast
  - MF Amplitude Modulation
  - VHF/FM
  - VHF/UHF TV
- b. Communications (non relay)
  - Aeronautical Mobile
  - Amateurs (Hams)
  - Citizens Radio
  - Facsimile
  - HF Telegraphy
  - HF Telephony
  - Land Mobile
  - Maritime Mobile
  - Radio-Control Devices
  - Telemetry
  - Telephone Circuits
  - Wireless Microphone
- c. Navigation (non radar)
  - Aircraft Beacons
  - Instrument Landing Systems
  - Loran
  - Marker Beacons
  - Omega
  - VOR/TACAN/VORTAC
- d. Radar
  - Air Search
  - Air Surface Detection
  - Air Traffic Control
  - Harbor
  - Mapping
  - Police Speed Monitor
  - Surface Search
  - Tracking/Fire Control
  - Weather
- e. Relay Communications
  - Ionospheric Scatter
  - Microwave Relay Links
  - Satellite Relay
  - Tropospheric Scatter

BIBLIOGRAPHY

1. White, D.R.J., "Electromagnetic Interference and Compatibility", Vol. 3, Don White Consultants, Inc., 1973.
2. Johnson, R.C. "Reducing Wide-Angle Sidelobes in Radar Antennas", Microwave Journal, August, 1966.
3. Freeman, E.R., & Smith, K.E., "Techniques for the Suppression of Sidelobe and Backlobe Energy from High-Gain Radar Antennas", NWL CR-14, April 1969, DDC # AD 859 938L.
4. Skolnick, M.E., "Radar Handbook" 2nd Edition, McGraw-Hill, 1970.
5. "Electromagnetic Compatibility Design Guide for Avionics and Related Ground Support Equipment", NAVAIR AD1115, Naval Air Systems Command 2nd Edition, February, 1976.
6. AFSC Design Handbook DH 1-4, "Electromagnetic Compatibility", Air Force Systems Command, Aeroneutical Systems Division, Wright-Patterson AFB, Ohio.
7. Heisler, K.G., and Howitt, H.J., "RADC Interference Notebook", RADC-TR-66-1.
8. Johnston, S.L., "Radar ECCM Techniques," Electro 76, May, 1976.

Why ECCM? The Need for ECCM
W.D. Meskill
IEEE ELECTRO-76 Record, Part 30, May 1976
Pages 30-1-1—30-1-4

1.3

SUMMARY

It is the earnest desire and most profound wish that neither this country nor any other country will ever have to use electronic countermeasures or electronic counter-countermeasures again. On the other hand it would be the height of folly not to be prepared to do so. Therefore, the United States must take those steps necessary to equip, exercise, and test its baseline EW capability so that if it were ever called upon to do so, could achieve that guaranteed success or at the worst, parity. Admiral Thomas Moorer, former US Chairman of the Joint Chiefs of Staff, recognized the importance of the matter when he said, "If there was a World War III the winner would be the side that can best control and manage the electromagnetic spectrum." I could not agree more.

WHY ECCM?

When considering the military capabilities of a country, in this era and at this time, there is little need to convince an international audience of electrical engineers that there is a vital requirement for Electronic Countermeasures and Electronic Counter-Countermeasures. This concept is accepted without challenge. The documented evidence of grim successes or failures of war ventures in our times is replete with case studies of electronic advances which help sway the tide of battle. A close examination of World War II, the Southeast Asia conflict, and the more recent 1973 Yom Kippur War provide ample proof of the effectiveness of Electronic Warfare applied in earnest. Nevertheless, it is germane to review the topic. Not so much to prove the value of countermeasures but rather to examine the known facts about military electronic warfare so that an orderly plan can evolve which will cope with the application of EW against this or any other country.

THE PAST IS PROLOGUE

The importance of electronic countermeasures and counter-countermeasures was conclusively demonstrated in World War II. The following chart indicates the moves and countermoves that were made in the Battle of Britain.[1]

THE BATTLE OF BRITAIN

THRUST	PARRY
In earliest use of electronic guidance, Luftwaffe bombers with loop antennas lead mass raids on English cities on beams from Nazi Lorenz short wave stations.	British meaconing receiver and transmitter spaced 5-10 miles apart, repeat Lorenz beams lead to German bomber crews, eliminating or falsifying their bearings. Confused crews land on British soil.
German bombers navigate to London between dotted and dashed beams from two Headache intercommunicating transmitters on French coast.	British Aspirin receiving - transmitting system retransmits Headache beams to new location.
German Ruffian propaganda station narrows its beam before each German air raid down to three degrees which bombers follow to London. Bomb release points are determined by intersection of a second three degree beam.	Long fooled by Ruffian, British finally use Bromide masking beacon concept to make pilot Ruffian beam useless for navigation, cause many bombs to be dropped in channel.
Luftwaffe bomber squadron outfitted with all available navigation aids drops incendaries on London as markers for following bombers with less navigation equipment.	British light Starfish waste fires outside London causing up to 90 percent of bombs to be dropped on these false targets. In the Battle of Britain radar control helps 700 RAF fighters defeat some 2,000 German bombers.

It should be noted that the British developed an electronic counter-countermeasure against each new German tactic and were able to do so quickly and effectively. Of this battle, Sir Winston Churchill was later to write, "This was a secret war, whose battles were lost or won unknown to the public, ... unless British science had proven superior to German, ... we might well have been defeated and destroyed."[2]

NORTH VIETNAM

A quarter of a century later, in the skies over North Vietnam, the United States was to find itself engaged in a similar war ... United States Air Force and Naval aircrews were required to penetrate and bomb areas where the most sophisticated and well-trained air defense complex ever encountered lay in waiting. A seven year period of electronic thrust and parry ensued. The chart below indicates some of the moves and counter moves.

AIR COMBAT OVER NORTH VIETNAM

THRUST	PARRY
Soviet SA-2 missiles, combined with radar controlled anti-aircraft artillery and occasional MIG fighters, challenge US aircraft over North Vietnam. First SAM fired in anger downs one F-4, damages others.	Three days later, USAF strike force supported by Marine EF-10Bs destroy SAM site. EB-66 becomes first aircraft with jammers and radar homing and warning (RHAW) systems. Remodeled F-100s and F-101s (later F-105s and F-4s) begin WILD WEASEL missions to seek out and destroy enemy radars and missile sites.
North Vietnam during 1965-1967 steadily increases number and types of radars and increases proficiency at GCI sites. Introduction of integrated air defense system with various radar controlled AAA systems and mobile SAM systems takes toll on US fighter bombers. NVN SAMs force EB-66 to fly further and further away from F-4 and F-105 fighter bombers.	USAF employs EB-66 to jam NVN radars with telling effect. Heavy NVN SAM build-up results in introduction of self-protection ECM pods on fighter bombers. Loss rate declines dramatically.
April 1968, air war over NVN ceases. Air Defense system and SAM logistic support builds. ECCM developments in SA-2 system occur. SA-7 (Strella IR seeker) introduced as new weapon.	USAF air attack strategy maximizing use of all ECM assets refined. Electro-optic, laser, and infrared countermeasures used for first time to pinpoint and destroy hidden and difficult targets in South Vietnam and Ho Chi Minh trails.
History's heaviest barrages to date of anti-aircraft artillery coupled with 1,000 SAMs are fired at US aircraft attacking Hanoi and Haiphong.	729 B-52 sorties plus numerous F-111s, and USAF/USN tactical aircraft strike military targets in and around Hanoi and Haiphong; civilian casualties held to a minimum while B-52 bombardment forces rapid resumption of peace negotiations; as a result of effective ECM, loss rates for B-52s are less than anticipated.

The air war over North Vietnam in late 1972 was to be most aptly described as a savage air battle. "It was the first full scale ECM war between massive offense and massive defense. Both sides improvised tactics and equipment as the tide of battle surged. Hanoi tracked strobes from B-52 emissions, triangulated sources, computed flight paths and then fired salvos as great as 100 SAMs along the bomber's predicted path. The North Vietnamese also relayed tracking data from remote radars to SAM sites whose radars were being jammed. New frequencies were used.

The attackers were also changing their tactics. They abandoned their traditional stream for simultaneous approaches around the compass that foiled flight path prediction. Jamming techniques were more skillfully applied. Fighter-bombers attacked the SAM sites between B-52 raids. By the fifth day, the North Vietnam defenses were beginning to sag and no B-52s were lost for two operational days. By the eleventh day, the northern defenses were shattered and B-52s roamed the skies with impunity. Hanoi could no longer track B-52s with its radar, get MIG interceptors airborne, or launch any SAM attack. The damage to the north was devastating. On December 30th Hanoi decided to resume serious peace negotiations."[3]

In simply understood terms, the lessons the Air Force and Navy learned were these. The Air Force found the proper application of ECM drove loss rates down. When the North Vietnamese countered with radar or tactical fixes, the loss rates climbed again, to be driven down again by renewed ECM techniques, illustrating the dynamic character of electronic warfare. The Navy concluded from the war that it would have lost five strike aircraft to SAMs for every one actually lost between 1965 and 1972 had they not had ECM protection.[4]

THE YOM KIPPUR WAR

In October 1973, it became apparent that the lessons learned in the Southeast Asian conflict were a two way street. In a surprise attack, a 1,000 tank Arab force moved across the Syrian Front under the protection of mobile SA-3 and hand carried SA-7 surface-to-air missile systems augmented by quad-mounted radar-directed ZSU 23-4 anti-aircraft guns firing up to 1,000 rounds per minute. The Arab forces took a heavy toll in the first hours of the Golan Heights battle as the Israeli strike aircraft sought to halt the Arab armor.

The SA-6 surfaced as the major technological surprise of the October war and presented a formidable challenge to the countermeasures designers. It employed a microwave radar applying doppler techniques that was especially effective against low flying aircraft. Until the Israelis changed air strike tactics, brought in self-protection pods, and used helicopter-borne jammers, the issue was undecided. Gradually the combination of superior EW tactics and effective countermeasures against the Arab systems overcame the initial surprise. For their part, the Israelis recalling the sinking of their destroyer, Elat, in 1967 by a Russian developed Styx missile, developed a rapid blooming chaff which effectively screened them from the weapons system's radar. As a result, they scored several successes against Syrian-operated missile-carrying ships.[5]

CLASSIC DUELS LESSONS LEARNED

Clearly it can be seen that in the intervening years between 1941 and 1973, electronics countermeasures technology has advanced by giant steps. Nevertheless those historic air combat operations over the English channel revealed the same basic lessons as did those conflicts employing more complex systems at a much later time in history. It would be well to consider some revealing commonalities of the three historic duels and study them for future use.

-- No countermeasure is effective forever. It is time sensitive and time perishable.

-- Countermeasures are most effective when used in a "surprise" mode. The wise battle planner should act accordingly.

-- Men are the ultimate difference in an EW engagement. A good system with an unskilled operator will always lose to a marginal system with an alert, knowledgeable technician at its controls.

-- Countermeasures, in themselves, are useful in military operations. Coupled with imaginative tactics, countermeasures are decisive factors for victory.

THERE IS NO MONOPOLY ON ECM/ECCM INNOVATIVENESS

The classic duels have ceased, at least for the present. What goes on is the advance of countermeasure technology. The relatively simple thrusts and parries of World War II have been replaced by complex electronics developments which seem to be expanding in new areas at almost an exponential pace. Consider, for example, electro-optical countermeasures. Growing European military emphasis on the application of advanced electro-optical techniques in solving the challenges of night time surveillance, accurate weapons delivery, missile launch warning, and infrared/optical countermeasures was evident at the 1975 Paris air show.[6] There is a wide-spread conviction that the USSR and its allies have a capability in the E-O arena about equal to those of the NATO countries. Even a casual reading of the USSR text book "Principles of Jamming and Electronic Reconnaissance" by S.A. Vakin and L.N. Shustov reveals an incisive knowledge of all facets of Electronic Warfare including electro-optical and infrared systems. It is reasonable to assume that comparable E-O equipment in the US and NATO inventories is also being fielded by Warsaw Pact forces and there follows the need for effective countermeasures.[7] The technological advances in E-O will undoubtedly follow the same historic course and trials that radar countermeasures underwent. And the obvious corollary is that there is every reason to expect and believe that in the future no one nation will have exclusive technological expertise in advanced countermeasure techniques.

WHAT WILL TIP THE BALANCE?

From the preceding commentary and more importantly, from the overwhelming evidence of the historical wartime EW duels that have been studied, two inescapable conclusions emerge. First, EW will be employed with telling effect by both sides in any future conflict. Secondly, neither side in this conflict will enjoy a technological monopoly on the systems that are used or the tactics and techniques that are applied. What then will tip the balance in this complex struggle?

It seems to me that there is a certain fundamental approach and logic which can be brought to bear on this question which will almost certainly guarantee success or, at the very worst, achieve parity in this deadly aspect of survival.

(1) A baseline EW capability geared toward the known technological threat that the enemy possesses must be established, maintained, constantly improved upon as the state-of-the-art demands. This baseline EW capability must be employed and trained through periodic exercises and tests. The proficiency and ability of the system will be in direct proportion to the realism of the training.

(2) An information gathering capability must be maintained to stay aware of potential adversary's capabilities and limitations. Moreover, if a conflict does ensue, this intelligence collection capability must be keyed to recognize the "surprises" the enemy springs and estimate the impact they will have. Early recognition and timely notification is imperative for successful countermeasures.

(3) There must be a finely tuned capability established to dynamically respond to these "surprises". A combined military/civilian team to immediately design and deploy offsetting countermeasures equipment, tactics, or both to overcome surprises is vital.

Putting this into perspective, it can be said, the ability of a nation to control the three "R's" (Recognition, Reaction, Resolution) will tip the ECM/ECCM balance. Recognize "surprises" and correctly estimate their effect. React immediately by alerting corrective forces to respond. Resolve the situation by a military/scientific group effort geared toward a dynamic solution-motivated mission.

REFERENCES

1. Grumman Aerospace Horizon Magazine (Supplement), Fall 1974, Vol 101, No. 2.

2. Churchill, Winston S., Their Finest Hour, Boston: Haughton Mifflin Col, 1949), pp 51-54.

3. Editorial - Aviation Week and Space Technology, p 7, Feb 12, 1973.

4. U.S. Strives to Cope with Growing Threats. Aviation Week and Space Technology, p 42, Jan 27, 1975.

5. L. Meadows. Electronic Countermeasures Technology - An Overview. Electronic Progress, Volume XVII, No. 3, pp 2-3, Fall 1975.

6. Barry Miller, Advanced Electro-Optics Demand Grows. Aviation Week and Space Technology, p 48, June 16, 1975.

7. Harvey Rinn and Joseph Savino. Electro-Optical EW Seen Key to New Developments. Electronic Warfare, Vol 7, No. 6, Nov/Dec 1975.

Threats and Techniques
R. Loomis
Electronics Progress Magazine, Vol. 17, No. 3, Fall 1975
Pages 17-25

1.4

INTRODUCTION

In the design and development of an electromagnetic countermeasures system, two factors are of fundamental importance, the threat and the ECM technique. This article addresses the methods used to define the threat and to develop countermeasures techniques to nullify that threat. The threat consists of the hostile electronic equipments comprising the environment in which the vehicle to be protected must operate. The specific actions used to degrade the effectiveness of the hostile electronic equipment comprise ECM techniques.

THREAT MODELS

The threat model is used as a means of systematically organizing all of the available data that defines the hostile system. It is composed of three separate entities: signal models, system models, and scenarios. The signal model deals with the characteristics of the emission from a single radiating element. The system model is a summation of all radiating characteristics and includes other pertinent factors, such as missile range, warhead lethality, and reaction time. The scenarios define the types and locations of hostile and friendly systems. The hostile scenarios are typically made up of mixed system types acting under coordination.

Figure 1 illustrates the signals which might be used by a threat system. The threat model for this system would include nine signal models as part of an overall system model. Figure 2 illustrates a scenario. Although both of these models are products of the author's imagination, they contain key elements which proved to be major factors in the 1973 Middle East war.

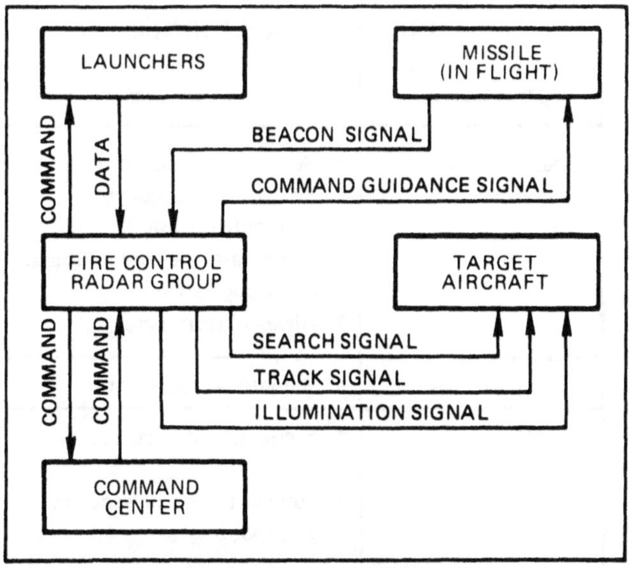

FIGURE 1 *Hypothetical Threat System Configuration*

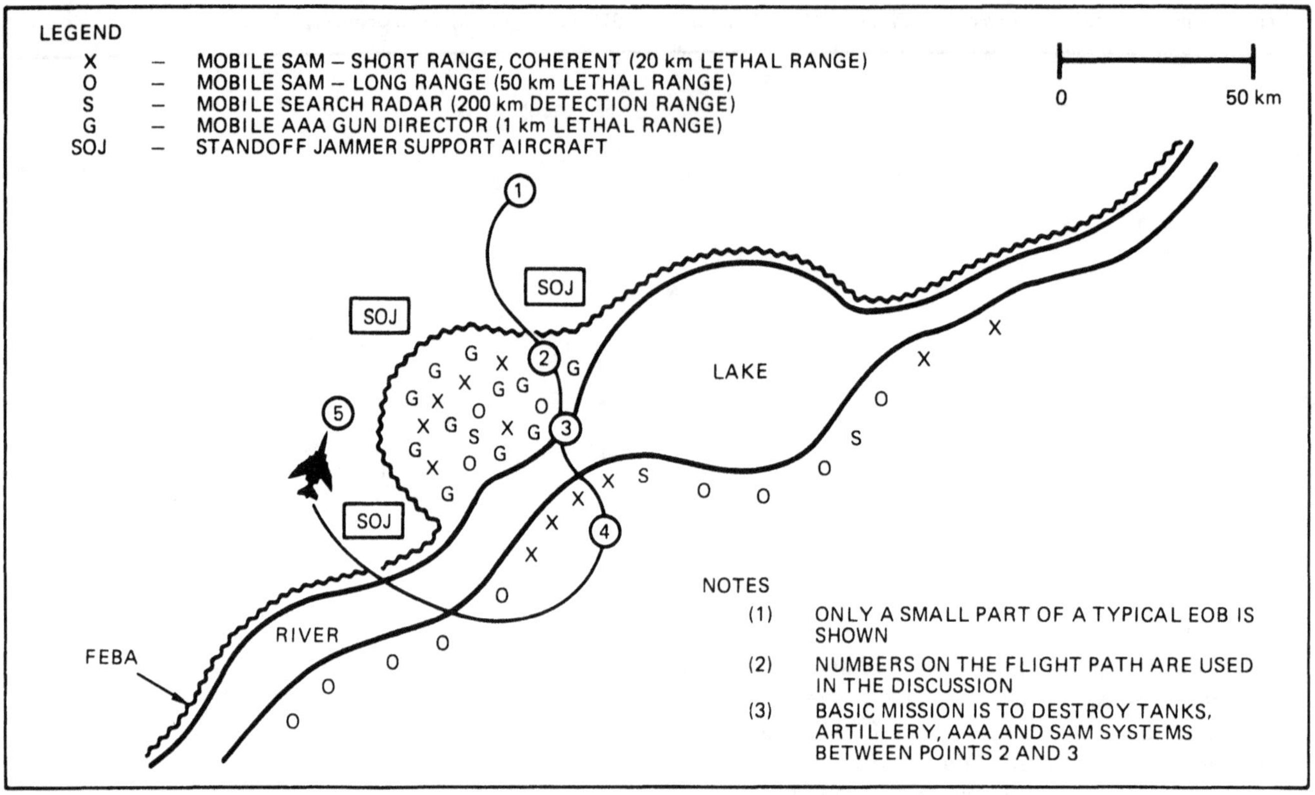

FIGURE 2 *Possible Scenario*

TABLE 1
THREAT DATA SOURCES

Data Source	Comments
Open Foreign Literature	• Several years out of date • Screened for security (willingly released) • Inaccurate – either deliberately or through error • Always incomplete
Covert Sources	• Sometimes excellent
ELINT	• Frequently difficult to interpret • Subject to hostile security and spoofing
Captured Equipment	• Time required to evaluate • They know you have it • Usually obsolete
Technology Extrapolation	• Requires knowledge of hostile technology, design practice and stated military requirements

SIGNAL MODELS

The source for most signal model work is electronic intelligence (ELINT) data which is provided by a network of sensory devices which receive and analyze RF emissions. Other data acquisition sources are listed in Table 1. These data must be extended using technical judgment based on our understanding of hostile technology, design procedures, and objectives. Extrapolation based on U.S. technology, design procedures, and tactics is dangerous because hostile designers work from different technological, economic and political bases.

Table 2 lists some of the signal model characteristics of a hypothetical emitter that combines the search and track functions in one radar and therefore represents two threat signals.

SYSTEM MODELS

ECM equipments interfere with one or more of the individual signals used by a weapon system in order to degrade the performance of the entire system. This not only requires data on the signals selected for ECM action but a considerable amount of additional information. The sources for this type of data are similar to those used to obtain the signal data. Referring to the simplified model of a surface to air missile system,

TABLE 2
PARTIAL LISTING OF TYPICAL SIGNAL MODEL CHARACTERISTICS

Parameter	Value	Comments
Carrier Frequency	9.0-9.5 GHz	1. Seven fixed frequencies selectable available 2. Manual frequency change, radar shuts down for one second for change 3. Operation possible from 8.7 to 9.6 GHz without major change 4. No multiple frequency transmission observed
Effective Radiated Power	90 dBW	
Antenna Gain	35 dB	1. First sidelobe down 25 dB 2. Far sidelobes (average) down 40 dB
Search Scan	Azimuth 360 deg	1. Active conical scan at all times 2. Elevation sector selectable 3. All search under manual control
Pulse Repetition Rate	1000 pps 2000 pps	1. 10 percent jitter observed for both search and track
Pulsewidth	Search 1.0 μs Track 0.5 μs	1. No intrapulse coding

shown in Figure 1, several additional factors are needed to define the appropriate ECM for use against it.

Detection range is needed to define the point at which ECM response should be initiated. If the response is initiated too early, it provides early warning to the hostile radar and permits the radar operators to optimize their counter action. Initiating response too late permits the radar to acquire and establish track. This makes the ECM task more difficult because it is generally easier to maintain track on an aircraft in the presence of jamming than to acquire track—especially if the radar system incorporates an optical back-up system. Maximum and minimum lethal range values for the weapon are also key factors in defining ECM control logic, and in establishing the required ERP, and mission tactics.

ECCM capability is the most uncertain and frequently the most important threat system characteristic.

The type, resolution, and characteristics of the threat system display are often major factors in selecting appropriate ECM parameters. For example, the difference between a fully automatic range detection circuit and an operator using an expanded "A" oscilloscope involves a change (on the order of 20 dB) in required effective radiated power and a different type of modulation waveform.

Tactics in using the weapon system are frequently critical. These include such factors as mobility, deployment relative to other hostile systems, and radiation doctrine.

SCENARIOS

Figure 2 illustrates a small portion of a scenario similar, in some respects, to those which occurred during part of the 1973 Middle East war. Scenario characteristics have a very strong influence on ECM system design requirements and therefore, the definition of the scenarios to be used in the ECM system design process is a critical task. Space precludes detailed discussion of scenarios, but a few comments are necessary as background to the ECM techniques discussion:

- Actual combat almost never involves just one weapon system against one aircraft (the one-on-one situation).
- The ECM system must deal simultaneously with a rapidly changing combination of threat systems.
- Actual combat involves very rapid changes in both the RF environment and the lethal situation (i.e., which systems are dangerous now).

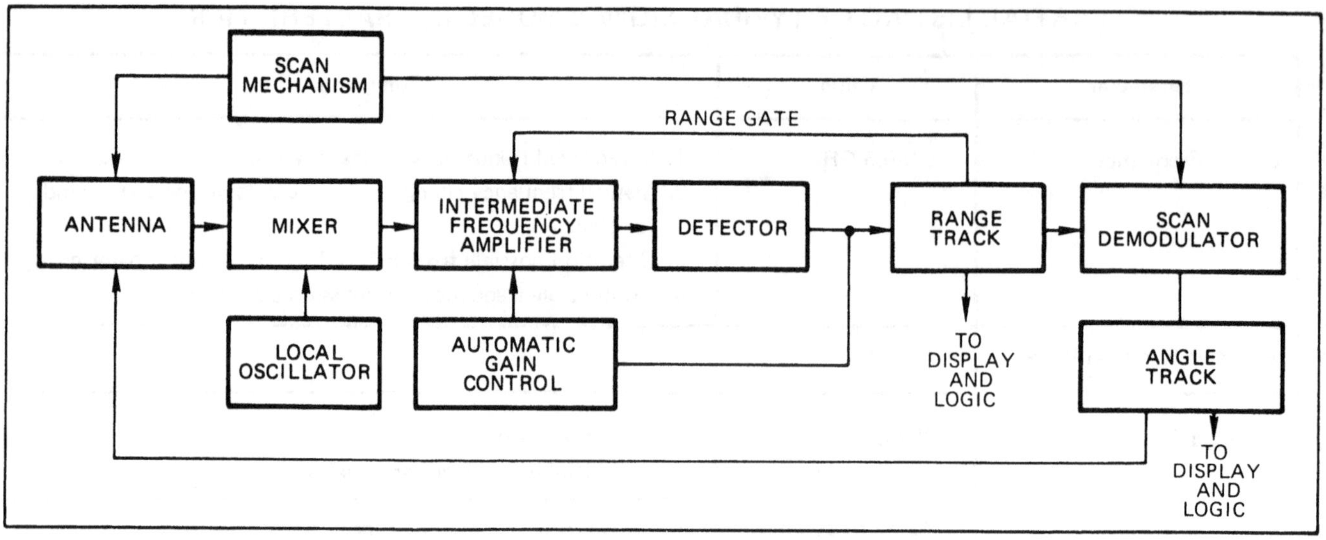

FIGURE 3 *Radar Receiver Model Simplified Block Diagram*

ECM TECHNIQUES

The term ECM technique is usually applied to the use of modulation imposed on an RF carrier. However, RF modulation is only one countermeasure technique available for the defense of the aircraft. Other techniques include choice of flight path and altitude, maneuvers, chaff, radiation absorbent material, coordinated/cooperative attack, and fire suppression. The latter are included in the following discussion only to the extent that they influence the modulation parameters of the ECM techniques.

RADAR SYSTEMS

The functional block diagram illustrated in Figure 3 represents a conical scan tracking radar of the type frequently used for antiaircraft artillery (AAA) fire control, missile guidance, and Airborne Intercept (AI). This radar receiver is noncoherent and of the general type to be expected for the "track signal" of Figure 1. A very similar block diagram, with the "range track" replaced by a "velocity track" unit could serve as the missile seeker in Figure 1.

BASIC ECM TECHNIQUES

ECM techniques are built up using one or more of a few basic effects in combination or sequence. The following discussion covers the more important of these effects and a few of the means used to achieve them.

Antenna

The first point of ECM attack is the radar antenna. The ECM designer can attempt to shift the Poynting vector or, more accurately, the way in which the antenna will operate on the Poynting vector in developing angle data. The more common methods of accomplishing this form of degradation are:

1) low altitude flight to take advantage of radar limitations due to clutter and multipath,
2) chaff which provides the radar with a multiple target problem, a clutter problem and an attenuation problem,
3) active expendables which provide an effect very similar to chaff and can be repeaters, transponders, or both,
4) cross-polarization—in which the ECM set radiates energy at a polarization nearly orthogonal to the radar polarization and exploits the pattern differences of some antenna types,
5) cooperative countermeasures in which two or more ECM platforms provide the radar with an enhanced multiple target problem,
6) dual coherent source ECM—which acts to change the nature of the phase front at the radar antenna,
7) decoys which are electronically augmented to provide a radar return very similar to normal aircraft skin return,
8) clutter enhancement in which the ECM set illuminates the ground so as to enhance the radar's clutter rejection problem, and
9) stand-off jamming which acts to "blind" the radar.

A number of other techniques, such as inverse gain, velocity gate pulloff and range gate pulloff are modulations which act on the radar's processing circuitry.

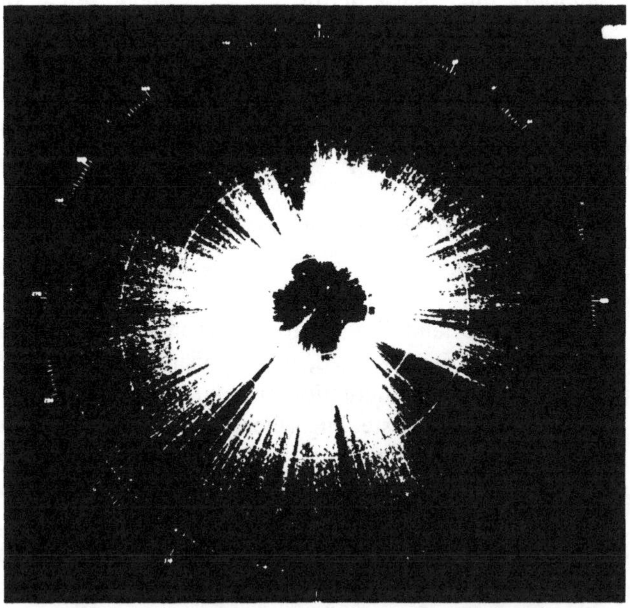

FIGURE 4 *Effective High Power Noise Jamming*

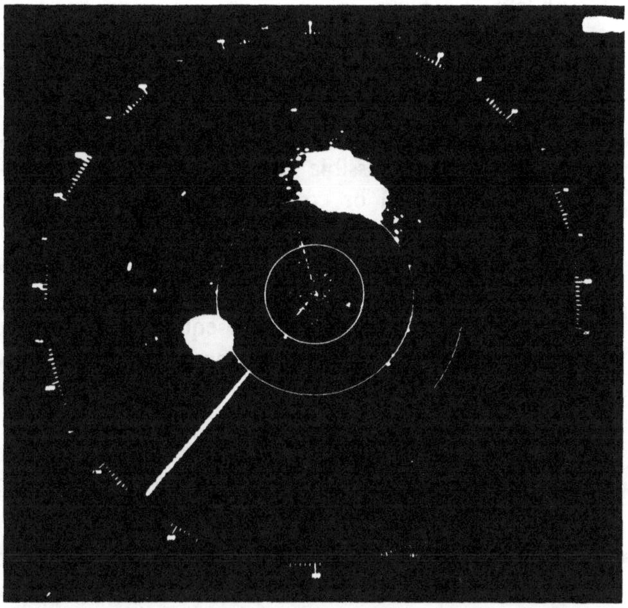

FIGURE 5 *Noise Jamming Chaff, Decoys and Clutter Enhancement*

Figure 4 illustrates the effect of high power noise jamming in the case where the radar is attempting to retain full operating capability (i.e., the operator, using a plan position indicator (PPI) oscilloscope, has not turned down the gain). The display indicated is a dual display in which the first two range rings are driven by a moving target indicator (MTI) receiver and the rest of the display by a conventional pulse receiver. This figure shows a total win for the jammer. The MTI is totally saturated and, therefore, useless. The conventional pulse circuitry shows an infinite number of targets, with no ability to select the right one.*

The "holes" in the jamming pattern are caused by radar sidelobe nulls combined with terrain masking effects and have a significant effect on stand-off jamming requirements.

A radar operator faced with the condition shown in Figure 4 would reduce the gain of his receiver and would accept reduced detection capability to obtain a strobe on the jammer as shown in Figure 5. As in Figure 4, the first two range rings are driven by MTI circuitry for the PPI display while the others are driven by conventional pulse receiver circuitry. The "targets" in the first two range rings are actually temporary artifacts where the MTI drops out of saturation. They are not real targets. The strobe provides jammer azimuth accurately—but no range data. The discrete targets could be real (large) targets, chaff or active expendables. The large "blurs" could be chaff drops, enhanced clutter, groups of active expendables or deceptive jamming aircraft.

Figure 6 is similar to Figure 5, but serves to illustrate the effects of multiple jammers, some on penetrating aircraft. Here again, the radar operator has been forced to reduce his receiver gain in order to get any data. In this case, however, there are several strobes, one or more of which may be an attacker. All are using high duty cycle jamming waveforms.

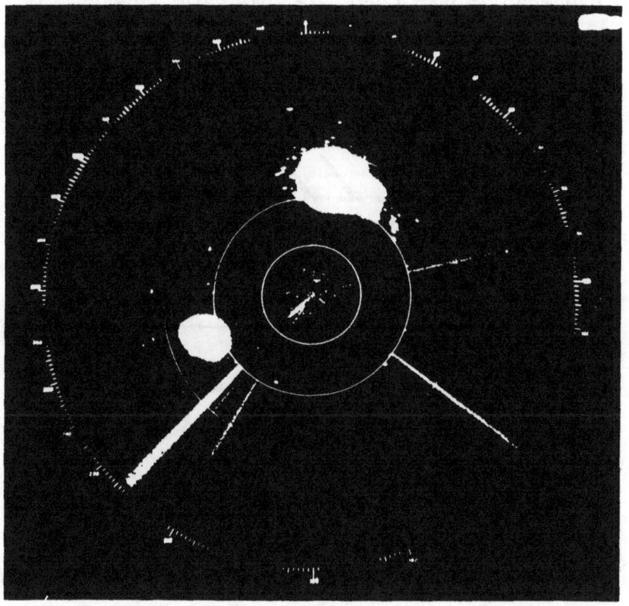

FIGURE 6 *Multiple Jammer Effects*

* There was a B-52 radar target present when this photograph was taken.

Mixer

The radar mixer and preamplifier act to translate the incoming RF signal to a convenient intermediate frequency for amplification and processing. This translation is a nonlinear process which is subject to ECM attack. Some of the possible attacks are: a) saturation—using either pulsed or continuous signals (with adequate power, it is sometimes possible to burn out either the mixer or preamplifier); b) dual frequency jamming which generates spurious IF frequencies; or c) offset frequency jamming to distort subsequent processing of clutter or chaff. ECCM in this area can make effective attack on this unit very difficult.

IF and AGC

The basic ECM attack against the IF amplifier circuitry involves some form of periodic saturation. The AGC circuit is subject to similar, but more subtle forms of attack. The AGC circuit acts to average the signal (plus noise) over a selected interval and adjust gain settings to provide a preselected output signal level. Some of the more common attacks on the AGC circuit are: 1) range gate stealing—discussed later, 2) waveform distortion—where the jamming signal is significantly different from the expected signal (e.g., a very short pulse when the AGC expects CW), 3) band edge signals designed to capture the AGC without providing data, and 4) amplitude modulation designed to continuously modulate the AGC. These ECM attacks act through (and on) these circuits, but their effects are more obvious as part of the reaction of the range, velocity, or angle tracking circuits. Figure 7 illustrates the effect of one type of ECM attack on the AGC circuit as it affects the signal used for angle tracking in a conical scan radar. There are two basic effects: 1) the jamming puts holes in the expected signal when the pulses appear, and 2) the amplitude of the data carrying sinusoid is reduced which produces a reduction in the angle track loop gain, and results in poor angular accuracy.

Range Track

The basic attack elements against conventional range tracking are:

- Noise jamming where an essentially continuous signal is generated, usually using FM to induce amplitude fluctuation in radar IF.
- Range gate stealing where a repeater introduces progressively greater delay in an amplified signal to pull the radar range gate away from the skin return.
- False targets which can be generated by onboard equipment, chaff, active expendables, decoys or clutter enhancement.

The basic effect of noise jamming signals on a PPI display (Figures 4, 5, and 6) is generated by an AM signal which is almost always present at the input to the range track (and range display) circuit. Figure 8 illustrates the power density spectrum of such a jammer. Note that power density spectra are long term averages and can be achieved in many ways. And although the bulk of the power is concentrated at the selected frequency, spurs and a wide spectrum, low level emission is also present.

One of the major ECM problems in this area is the "A scope" display, which provides a radar operator with signal amplitude versus time. This display is shown in Figure 9 with two real targets clearly visible.

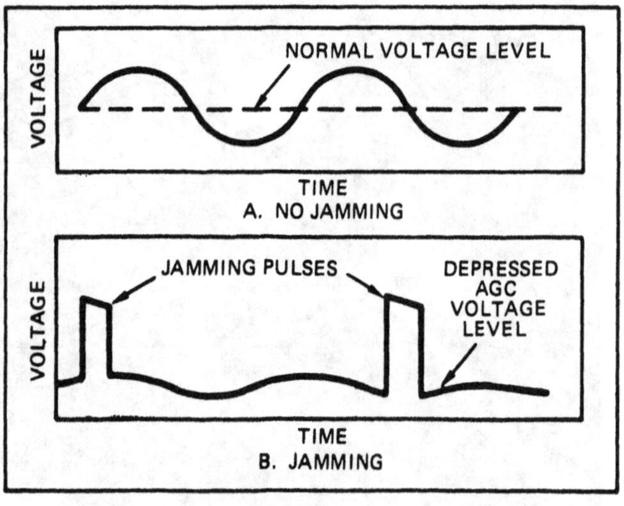

FIGURE 7 *AGC Suppression Jamming Effects*

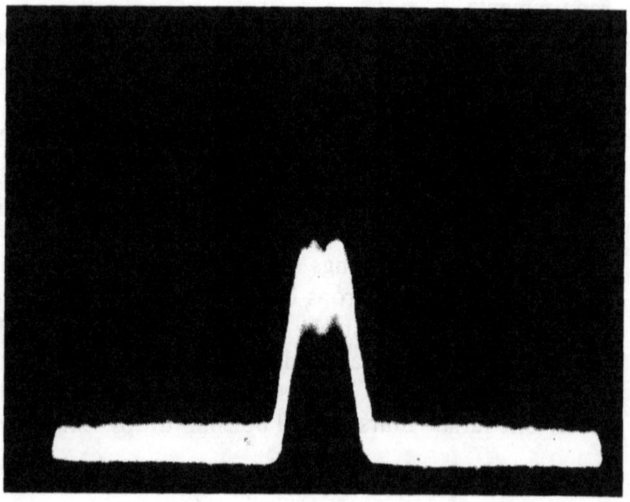

FIGURE 8 *Typical Noise Jammer Spectrum*

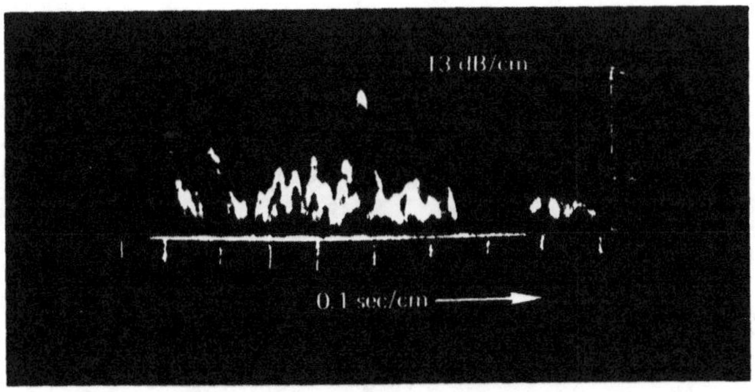

FIGURE 9 *"A Scope" Display without Jamming*

In a real situation, these targets would move in a manner permitting the operator to easily and confidently identify them as targets. The real return generally tends to add to the noise and move in a regular manner creating an appearance which has been described as the "mouse under the rug." This effect can be defeated by appropriate ECM action, as shown in another "A scope" display, Figure 10. Figure 11 illustrates the "A scope" displays for two other forms of attack on the range acquisition and range track function.

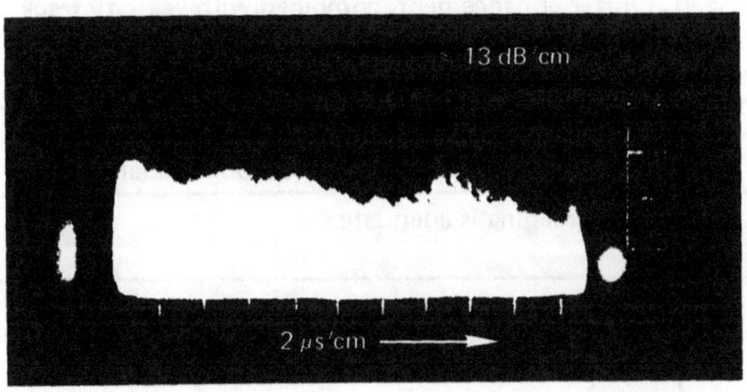

FIGURE 10 *"A Scope" Display with Effective Range Denial Jamming*

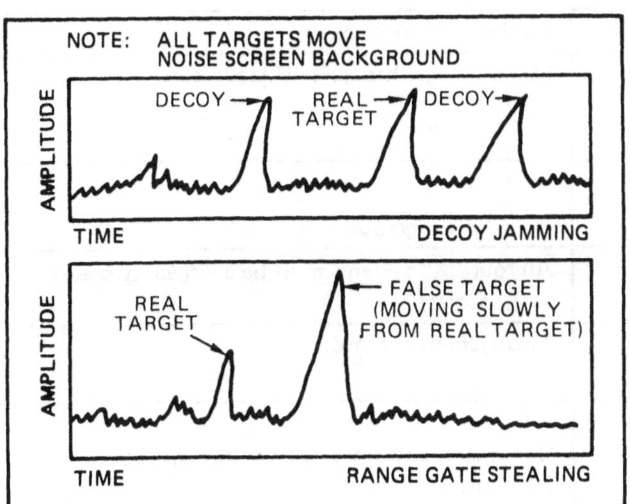

FIGURE 11 *Decoy and Range Gate Stealing Appearance on "A Scope" Display*

Synchronous Demodulator/Scan Filter

The radar model in Figure 3 is a conical scan unit and all of the attacks described as affecting the antenna can also produce effects in the synchronous demodulator. In addition, the following attacks are pertinent:

1) Inverse Gain in which the ECM set detects the AM caused by normal small radar tracking errors (active con-scan) and applies an AM modulation to an amplified signal which is 180 degrees out of phase.
2) Inverse AM in which the ECM set AM modulates the amplified radar signal at the radar con-scan frequency.
3) Swept square wave in which the repetition frequency of a rectangular AM modulation is slowly changed to move across the region of uncertainty in the victim lobing frequency.

Angle Track

The radar angle track is at the end of the processing chain and, consequently, is affected by successful degradation of any of the preceding functions. The specific actions which can be directed against angle track involve such actions as maneuvers and the timing of other attacks.

ECM USAGE

Considering the scenario shown in Figure 2, the penetrating force would be using a variety of ECM techniques simultaneously and in sequence. It is assumed that the stand-off jammer (SOJ) aircraft have the capability to engage all radars which can affect the aircraft mission and continue to do so throughout the mission.

At Point 1, the penetrators would not use their ECM equipment and would take maximum advantage of terrain masking and low altitude flight in order to conceal the timing and direction of the attack. The SOJ tactics and turn-on of on board ECM would be coordinated to enhance this effect.

At Point 2, the mission would be attacking tanks, artillery, AAA and the short range SAM system. Warning systems on board the aircraft would alert the pilots when the radars were acquiring the aircraft and this would initiate the ECM response. Note that several systems beyond those under attack must be engaged.

At Point 3, the aircraft would have reached a point of minimum coverage from the SOJ, because the SOJ would be operating into the backlobes of many of the defense radars, and the penetrators would change their ECM techniques.

Point 4 is the vicinity of strikes against two more missile systems and another change would occur.

At Point 5, the mission would be beyond lethal range and ECM action would be terminated. Table 3 summarizes possible ECM actions and purposes at each point and Table 4 summarizes assumed system parameters.

The general tenor of the preceding discussion could have misled the reader to the erroneous conclusion that ECM equipment can provide the pilots with a very high level of protection under any circumstances and, given adequate data, can provide complete protection. It is

TABLE 3
POSSIBLE ECM ACTIONS AND PURPOSES

Location	Radar Type	ECM Action	Comments
Point 1	All	• SOJ Radiation – high density noise pulses	SOJ radiation initiated when penetrators can first be detected
		• SSJ Radiation – none, low altitude flight	Three penetrating aircraft
Point 2	AAA	• AM modulation at con-scan frequency on noise carrier by all penetrators	Joint attack on Poynting vector, range track, AGC and angle track
	Short Range SAM	• AM modulation swept across con-scan frequency combined with doppler noise (repeater)	Clutter enhancement, combined with velocity track and angle track attack – all ECM directed against missile
		• Noise blinking at slow rates, not synchronized	Range track attack, clutter enhancement and Poynting vector attack against ground radar
	Long Range SAM	• Low altitude flight, no radiation	SOJ screening is adequate
	Search Radar	• SOJ radiation only	Penetrators do not radiate against search radars
Point 3	AAA	Same as Point 2	Directed against nearest "G"; previous one destroyed by strike
	Short Range SAM	Same as Point 2	Previous "X" destroyed
	Long Range SAM	Inverse scan on noise carrier; chaff, active expendables	SOJ coverage through backlobes is marginal
Point 4	AAA	None	No AAA dangerous
	Short Range SAM	Same as Point 2	All four "X" systems must be engaged; two are strike targets
	Long Range SAM	Same as Point 3	Two systems engaged
Point 5	All	Turn off all radiation	Note that many systems must be engaged during withdrawal

TABLE 4
ASSUMED SYSTEM CHARACTERISTICS

Search Radar	• Fixed RF frequency • Fixed scan over 360 deg • MTI • ECCM – sidelobe blanking
Mobile Short Range SAM	• Coherent ground tracking • CW illumination for homing seeker • Con-scan on receive only at different lobing frequencies for ground radar and missile seeker • ECCM – Home-on-Jam (missile) and Track-on-Jam (ground radar)
AAA	• Two alternate frequencies available with one second required to change • Noncoherent • Con-scan angle tracking • MTI • Fixed PRF • ECCM – manual operators
Mobile Long Range Sam	• Dual frequency, simultaneously pulsed • Active track-while-scan • Noncoherent • ECCM – jittered PRF, manual operators

true that improvements in ECM technology and an increased emphasis on ECM have already provided greatly improved performance against existing threats and an ability to react rapidly to changes in those threats. Further improvements are rapidly being introduced. However, analogous improvements are occurring in the threats, which necessitate the vigorous developments of new generations of ECM techniques and tactics. When implemented as a carefully integrated part of the overall military force structure, these new ECM techniques will play an increasingly important role in future conflicts.

Postulating an Airborne ECM Threat
D.D. Anderson
Microwave System News, December/January 1976
Pages 41-47

1.5

What are the projected capabilities of the enemy's airborne jammers during the next five years? A partial answer to that question can be extracted from this article, which sets forth unclassified guidelines for specifying such elements as tube type, antenna configuration, cooling capacity, and pod weight.

Donald D. Anderson
Foreign Missiles Evaluation Group
U.S. Army EW Laboratory (ECOM)
White Sands Missile Range, NM 88002

Whether based on foreign or U.S. technology, most airborne ECM threat information is not detailed enough to allow any flexibility in its use. Typically, the threat is given in watts versus frequency bands; and the user does not know, for example, if the numbers include antenna gain or system losses. Without such information, the user can only utilize the threat as it is.

With the information provided in this article, the reader will be able to postulate his own ECM threat based on U.S. technology. He will be able to calculate the power of airborne jammers at any frequency from 0.2 to 12 GHz and make a state-of-the-art estimate or project future threats (1980). He will also know tube power, antenna gain, system losses, prime power requirements, and total system weight.

Selecting the Tube. Many factors are to be considered (Table I) when choosing a tube for an airborne ECM system; but for postulating a threat, power-bandwidth versus frequency curves present the best data.

Figures 1, 2, and 3 present curves that show the 1975 state-of-the-art in average power for microwave tubes (0.2 to 12 GHz). These curves also include the author's estimates of the power levels that can be expected by 1980, which are based

Table I
Factors Affecting Tube Selection

Power Output	Weight
Frequency	Size
Tuning Range	Overall Power Consumption
Tuning Rate	Stability
Instantaneous Bandwidth	Linearity
Gain	Operating Temp. Range
Noise Figure	Altitude Requirements
Dynamic Range	Lifetime
Efficiency	Reliability
Power-Bandwidth Product	Shock
Voltage Gain-Bandwidth Product	Vibration
Cooling	Exotic Requirements

POSTULATING AN AIRBORNE THREAT

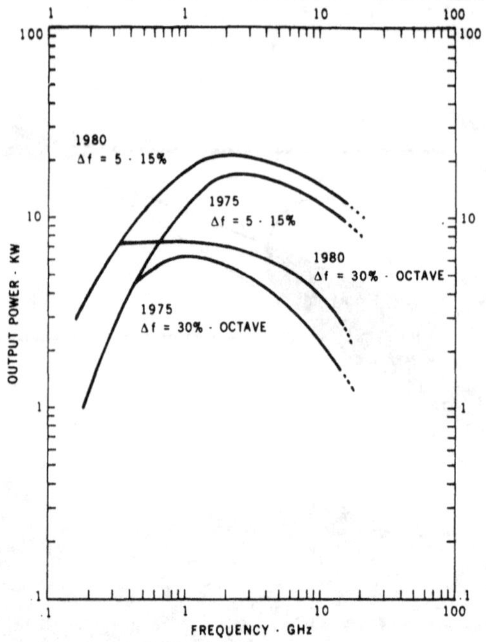

Figure 1. Traveling-wave tube.

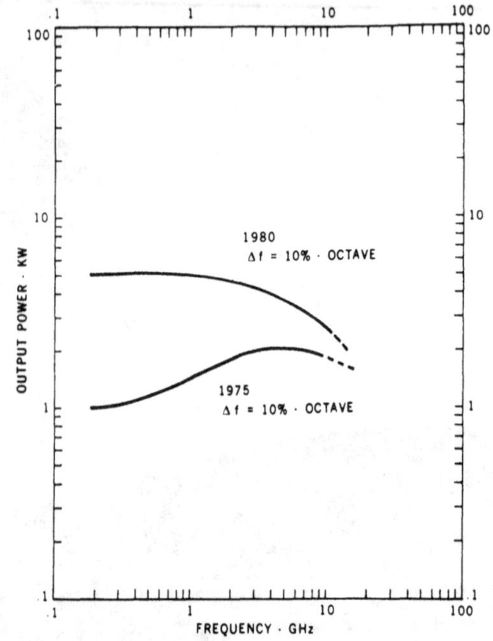

Figure 3. Cross-field amplifier (CFA).

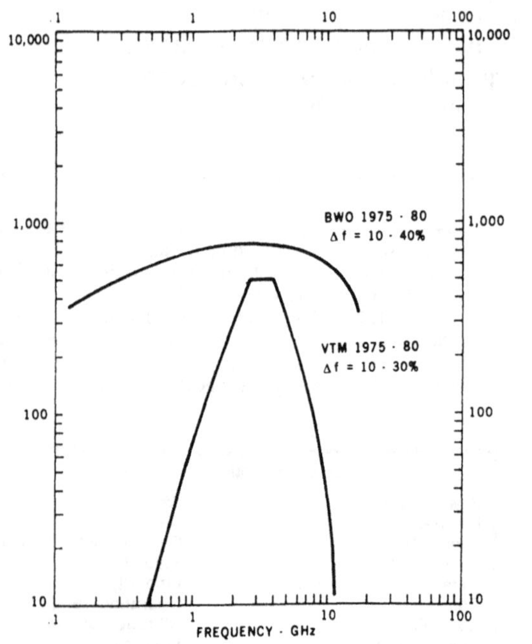

Figure 2. Backward-wave oscillator (BWO) and voltage-tuned magnetron (VTM).

on the assumption that over the next 5 years funding for developing higher power tubes will increase. If funding remains at the 1975 level, there will be little change over the 1975 curves.

The CW power curves and multiplying factor given below are used for estimating peak power values for traveling-wave tubes (TWT) and cross-field amplifiers (CFA):

TWT - 10 times average power (Figure 1) at 5% duty cycle
- 50 times average power (Figure 1) at 2% duty cycle

CFA - 5 times average power (Figure 3) at 10% duty cycle
- 10 times average power (Figure 3) at 2% duty cycle

Based on these figures, TWTs are the best choice for high power in the 1975–1985 time frame.

Table II is a partial list of CW TWTs that have saturated outputs ≥1 kW. The list includes manufacturer, tube number, frequency power, and tube weight.

Prime Power Requirements. Because of the high tube power being considered, the aircraft may require auxiliary power generators. Two types—ram air turbine and gas turbine—are considered the most practical for the 1975–1985 time frame. The weights of these generators are as follows:

Ram Air Turbine Generators—One lb/kW for airspeed greater than Mach 0.5 (limited to 90 kW). Higher power is possible with higher lb/kW.

Gas Turbine Generators (including fuel)—Five lb/kW (limited to 400 kW for large aircraft use).

An aircraft using a gas turbine generator will have to carry extra fuel to offset not only the weight of the generator but also the weight of extra fuel required for the generator.

Higher power can also be achieved by using more than one generator or by operating at less than a 100% duty cycle.

To estimate prime power requirements, use a value 10 times the power of the final tube. For example, a 1-kW TWT system will require 10 kW of prime power. A 10-kW ram air turbine generator will weigh 10 pounds. A 10-kW gas turbine generator, operating for 2 hours, will weigh 100 pounds.

POSTULATING AN AIRBORNE THREAT

Table II
One Kilowatt or Higher TWTs

Manufacturer*	Model No.	Frequency (GHz)	Bandwidth (%)	Saturated Output Power (kW)	Weight (lb)
MIC	MA 2050	.2– .4	67	10	100
MIC	MA 2015	.2– .4	67	1	50
MIC	MA 2003	.3– .9	100	1	26
MIC	MA 2014	.4– 1.0	86	1	26
MIC	MA 2024	.5– 1.0	67	2	27
MIC	MA 2017	.5– 1.0	67	5	45
MIC	MA 2030	.6– 1.2	67	1	23
MIC	MA 2019	.7– 1.4	67	1.8	24
WAJ	WJ 305	.7– 1.4	67	2.5	48
WAJ	WJ 340	1.0– 2.0	67	1	18
MIC	MA 2055	1.0– 2.0	86	1.7	33
HUP	551H	2.0– 4.0	67	1	22
MIC	MA 2038	2.5– 3.5	33	1	25
WAJ	AJ 493	2.6– 5.2	67	1	25
WAJ	WJ 468	4.0– 8.0	67	2	55
WAJ	WJ 3605	4.5– 6.0	29	2	51
HUP	656H	4.5– 6.5	36	1	20
MWE	M 5600	5.9– 6.4	8	8	150
HUP	614H	5.9– 6.4	8	8	95
VAR	VTC 6363A1	5.9– 6.4	8	3	135
HUP	745H	7.9– 8.4	6	1.2	42
HUP	792H	7.9– 8.4	6	5	66
MWE	M 5731	7.9– 8.4	6	3	70
MWE	M 4444	7.9– 8.4	6	12.5	90
MWE	M 5654	8.0–10.0	22	1	70
VAS	VTX 6380A1	8.4– 9.8	15	1	115
HUP	722H	8.5– 9.5	11	1	45
HUP	710H	8.5– 9.6**	8	9	60

*MIC–Microwave Associates, WAJ–Watkins-Johnson, VAR–Varian, HUP–Hughes Aircraft, MWE–Teledyne-MEC, VAS–Varian/TWT Division.

**U.S. microwave manufacturers do not list any kilowatt TWTs with bandwidths greater than 5% between 9.6 and 15 GHz. However, Hughes makes a 1-kW TWT at 15.0–17.5 GHz that weighs 65 lbs. Use this weight for TWTs operating between 9.6 and 12 GHz.

Table III
Cooling Capability

Method	Capacity (watts/cu. in.)
Self-cooling without heat sink	3
Self-cooling with heat sink	5
Force gas convection	5 to 9
Force liquid convection	10 to 40
Expendable	50 to more

Cooling Requirements. There is no simple way to present the state-of-the-art in cooling techniques or to forecast future trends. A cooling system depends on the operational environment and physical constraints. What would work at 10,000 feet may not be adequate at 15,000 feet. What can cool a 2-kW CFA tube may not cool a 2-kW TWT. Depending on the cooling method used, the cooling capability versus size can vary from 3 watts per cubic inch to values over 50 watts (Table III).

To "ball park" the weight for a pod-mounted cooling unit, assume that the power to be dissipated is 80% of the prime power. Use this dissipated power to calibrate the weight of the cooling unit. The weight will depend on expected airspeed: 3 lb/kW dissipated for airspeeds between Mach 0.6 and 1.0 or 12 lb/kW dissipated for airspeeds between Mach 1 and 2. These weights are not expected to change over the next 5 years.

Pod Weight Consideration. If the jammer is to be mounted in a pod, the pod weight must be considered. A pod's weight will depend on size and operational environment. Pods weigh from 100 to 500 pounds. For estimating pod weight, use the following numbers:

100 lb for tube power less than 1 kW
300 lb for tube power between 1–4 kW
500 lb for tube power over 4 kW

Calculating Total System Weight. Add the weights of the output tube, prime power, cooling unit, and pod (if used). Double this value to obtain total system weight.

System Losses Consideration. There will be some losses between the output tube, antenna, and radome. For estimating the effective radiated power (ERP), these losses must be included. Assume a system loss of 1 dB if antenna is less than 2 feet from output tube. Assume 3 dB for distance over 2 feet.

Selecting the Antenna. The antenna may be the power-limiting component in an ECM system. Power-handling capability of an antenna depends on a number of interrelating design factors, including size, material, and altitude. Some examples of airborne antennas are given in Tables IV and V.

To increase the average power-handling capability, the antenna can be pressurized. Table VI shows how many times the power can be increased by using different gasses at different pressures.

Therefore, an antenna rated at 100 watts could be made to operate a 2.2 kW.

POSTULATING AN AIRBORNE THREAT

Also, antennas may be used in numerous ways to obtain the required ERP (Figures 4, 5, and 6).

Table IV
Examples of Antennas

Type	Frequency (GHz)	Input Power (max) Weight, Size
Spiral	.24– .50	2 kW, 16 lb size: 36 in. diameter: 1.5 in.
Scimitar	.35– .95	1 kW, 9 lb size: 11.5 x 14 in.
Slotted Cylinder	2.50– 3.50	500 w, 1.5 lb size: 7.3 x 5.5 in.
Flush Mount	2.50– 3.90	1 kW, 4.3 lb size: 3 x 12.8 x 13.3 in.
Flush Mount	4.20– 5.80	1 kW, 2.2 lb size: 3 x 10.2 x 12.4 in.
Blade	4.45– 6.55	1 kW, 9 lb Size: 10.4 in. diameter: 11 in.
Scimitar	7.20–10.50	500 w, 15 oz size: .86 x 5.5 in.
Slotted Waveguide	8.45– 9.65	1 kW, 1 lb size: 37 in. diameter: 5.5 in.

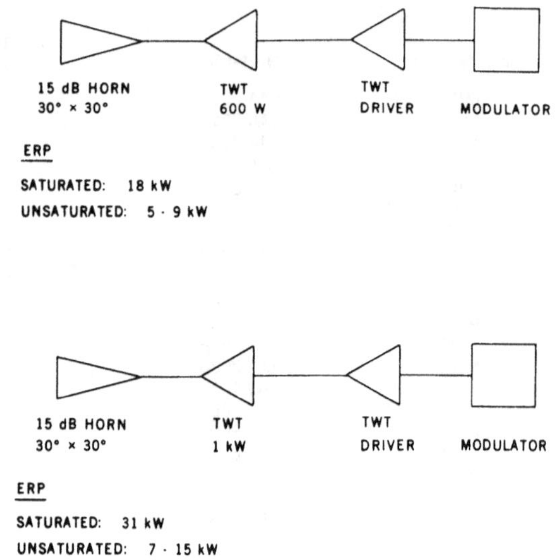

Figure 4. Possible jammer arrangements for saturated ERPs of 18 and 31 kW.

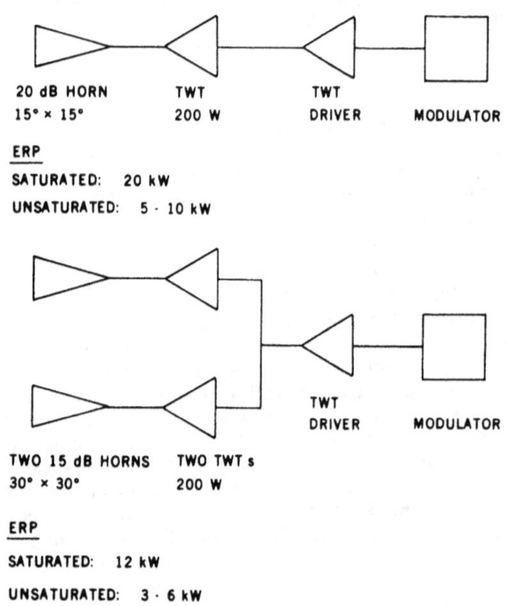

Figure 5. Possible jammer arrangements for saturated ERPs of 20 and 12 kW.

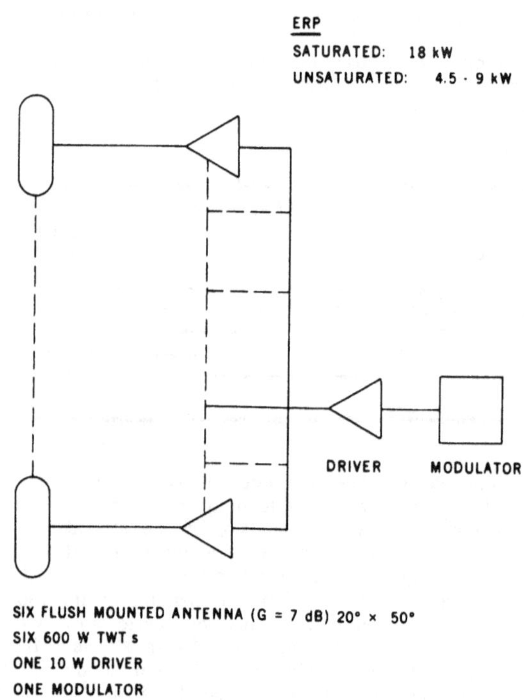

Figure 6. Another jammer arrangement for saturated ERP of 18 kW.

POSTULATING AN AIRBORNE THREAT

Table V
Standard Gain Horn

Frequency	10 dB Gain			15 dB Gain			20 dB Gain		
Dim: (inches)	A	B	C	A	B	C	A	B	C
2.60– 3.95	7.5	3.6	4.9	15.3	6.0	8.0	29.8	10.9	14.9
3.95– 5.85	4.4	2.3	3.0	9.4	3.7	5.3	15.1	6.6	9.1
5.85– 8.20	3.2	1.6	2.3	6.5	2.6	3.7	12.2	4.7	5.6
7.05–10.00	2.6	1.3	1.8	5.2	2.1	3.0	10.7	3.8	5.3
8.20–12.40	2.0	1.3	1.7	5.5	2.1	2.8	10.1	3.8	5.0
Elev x Az	50° x 50°			32° x 32°			15° x 15°		

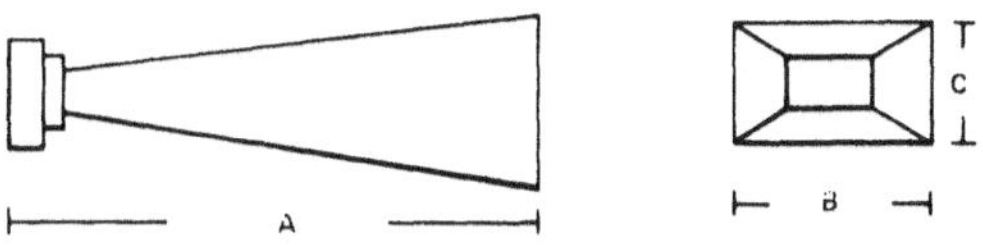

Table VI
Pressurized Antenna

Pressure	Dry Air	Sulphur Hexafluoride
1 atm abs	1 time	4 times
2 atm abs	2 times	16 times
3 atm abs	4 times	20 times
4 atm abs	6 times	22 times

Conclusions. With the above information, the reader can postulate the ERP, estimated prime power and cooling requirements, and calculate the system weight of an airborne jammer at any frequency from 0.2 to 12 GHz. Postulations can be made for 1975 or for 1980. By putting limits on tube power, system weight, prime power, cooling, or antenna gain, the reader can make trade-offs that will affect the ERP.

An example of how to use the above information is as follows: Assume that the ERP and weight of a CW, pod-mounted jammer is to be calculated. The jammer is to operate between 2 to 4 GHz, at speeds between Mach 1 and 2, and will have an antenna distance of one foot.

The highest power tube in the 2- to 4-GHz band is a 4-kW TWT (Figure 1). From Table II, a 2- to 4-GHz TWT will weigh around 25 pounds.

The prime power is 10 times the tube power or 40 kW. A 40-kW ram air turbine will weigh 40 pounds.

Since the jammer is operated at airspeed above Mach 1, the cooling unit will weigh 12 lb/kW dissipated. The power dissipated is 80% of the prime power (40 kW), or 32 kW. Therefore, the cooling unit will weigh 12 × 32, or 384 pounds.

Pod weight will be 300 pounds.

The total subsystem weight is: tube (25 lbs) + power (40 lbs) + cooling (384 lbs) + pod (300 lbs) = total (749 lbs).

Total system weight is 2 × 749, or approximately 1,500 lbs.

The ERP will depend on the antenna used and system losses. If a 15-dB horn is used and system losses are 1 dB (system gain of 25.1), the saturated ERP will be 4 kW × 25.1 or approximately 100 kW. ■

Airborne Microwave ECM
D.J. Steer
Proceedings of the Military Microwaves Conference, London, October 1978

1.6

ABSTRACT

The paper examines techniques currently used in airborne ECM. It covers the philosophy of jamming and gives details of the various types of jammers used against specific radar systems. Jammer transmitters are discussed and a brief survey of the power output of travelling wave tubes (TWT) is included. This leads to an assessment of the jammer powers likely to confront Western radar designers. The general design features of a responsive noise jammer are presented and it is suggested that this type of jammer is the basis of selfscreening (SSJ) escort (EJ) and standoff jamming (SOJ). The paper ends with a brief mention of chaff.

1 INTRODUCTION

The words "Electronic Warfare" are currently commonplace and the subject now pervades almost all weapon system designs. Although the first evidence of electronic warfare being practised goes back to World War I the great rise in the importance of electronic warfare has been comparatively recent. The events of the 1973 Yom Kippur war and the electronic sophistication of the weapons used therein demonstrated clearly that electronic warfare is an essential feature of all modern weapon systems.

2 ELECTRONIC COUNTERMEASURES

Electronic warfare is defined as "the struggle to control the use of the electromagnetic spectrum" but is often used loosely to describe any electronic equipment or techniques contributing to or against a weapon system. This paper will address the subject of airborne microwave ECM, i.e., electronic countermeasures carried by aircraft platforms to counter the electronics of opposing enemy systems, e.g., radars. The principal elements of airborne ECM are active measures, i.e., jammers, and passive, principally "Chaff".

2.1 Airborne Jamming—Philosophy

Why is there a need for such jamming? Simply expressed, if the array of weapon defences opposing an air operation employs no electronic technique there is little point in ECM. But if these defences have radars which are vital to their operation (e.g., early warning radars for detection, radars for directing guns or SAMs etc) and especially if the defence effectiveness or attrition against air operation in the absence of any ECM is high, then ECM becomes an essential factor for aircraft survivability.

3 THE OPPOSING DEFENCES

The types of jammer used and their design will depend on the opposing electronic threat. A typical defensive system will consist mainly of:

On the Ground

early warning radars	for the detection at the furthest possible range of approaching enemy aircraft.
G C I/SAM directing radars	for target tracking and allocation of aircraft targets to fighter or SAM or guns.
Acquisition radars	part of SAM or AAA systems for acquiring targets.
Fire Control	target tracking radars for SAM and AAA guns. (Anti-Aircraft Artillery).

In the Air

Fighter aircraft with AI radars and possibly Airborne Warning and Control Systems (AWACS) or Airborne Early Warning (AEW) aircraft.

These types of radar are a network linked by a communications and control system. The total system can be degraded by jamming at any stage. The resistance to jamming of the total system will depend on the factors such as redundancy or multiplicity of the systems as well as the actual jamming/signal ratios achievable. Current Western philosophy, motivated primarily by the need to achieve a definite effectiveness at minimum cost, is to jam the teeth that can bite you—the so-called terminal threats, i.e., the tracking radars of Fire Control systems. There is much to be said for jamming the surveillance radars as well, particularly if there are few of these radars and they are not very diverse in radio frequency when they become attractive targets. But if the surveillance network is made up of a very large number of radars well spread in radio frequency, then such jamming is less attractive and more costly.

The aim of all airborne jamming of defensive systems is to protect attacking aircraft either by denying information on their bearing and/or range (noise jamming) or by deceiving the radar defences that they are elsewhere (deceptive jamming). Before proceeding to the details of equipment design we will examine the general types of radar jammer.

4 GENERAL TYPES OF JAMMING

A typical pulse radar emits pulses and the operator or automatic detection circuits seek to identify received pulses which are replicas of those transmitted in the presence of (receiver) noise. Thus the waveforms most likely to give confusion or target obliteration are pulses of the same form as those transmitted, or noise. In general there are two types of jammers: pulse jammers and noise jammers.

4.1 Noise Jammers

Noise jammers are usually noise modulated CW sources—that is to say, the jamming is more or less continuous (though there are special intermediate cases). Almost any CW power source could be used and indeed, the backward wave oscillator and the voltage tuned magnetron have been used as noise jammers. However, for applications where the noise jammer is required to operate over a 10% band or over an octave of radio frequency, the travelling wave tube amplifier is usually employed. The aim is to obtain a high level of noise in the victim receiver to exceed the signal power in an aircraft echo. Since the jammer power travels over one path between jammer aircraft and radar, whereas the radio signal travels over two such paths out and back, the radar signal power varies inversely with the fourth power of the range, whereas the jammer power at the victim varies inversely with the square of the range. At long ranges it is easy for the jammer power to swamp the aircraft signal. As range is reduced, jammer and signal powers become equal at a range roughly equivalent to the so-called burn-through range, and at ranges shorter than this the target also will be of greater strength than the noise jamming and should become visible.

Clearly it is important for the jammer to radiate its maximum CW power. If amplitude modulation were applied to a TWT working in its saturated mode, there would be a drop in output power as the tube was, in effect, modulated down. This is why noise is almost always applied as frequency modulation since the mean output power then remains constant. Amplitude modulation is produced in the receiver by conversion of the frequency transients at the receiver input. There is some loss in this process too.

4.2 Range/Velocity Denial

So the primary features of noise jammers are almost continual denial of range (or velocity in the case of a doppler radar) down to the burnthrough range. The direction of the jammer will not be denied by a noise jammer, which will give rise to a jamming spoke on a PPI.

It is possible to apply further amplitude modulation to the jamming transmitter to provide angle distortion against conical scan tracking radars. If the conical scan frequency is known, the jammer output is modulated at this frequency. This will produce angular errors in simple conical scan radars, though there will be a net reduction in jammer power caused by these modulations.

In general noise jammers have two possible modes of rf operation; barrage and responsive. In the barrage case the frequency deviation of the fm noise modulation is sufficiently large to encompass the radio frequencies of all intended victims. This of course leads to dilution of jammer power lowering the power density in watts per MHz, and it is a mode of operating only suitable where the jammer is presented with a large number of victims spread in frequency or where a radar is truly pulse-to-pulse frequency agile and the jammer is forced to spread its power across the frequency spread of the rf agility.

In the responsive noise case the jammer makes best use of the available power by ensuring that most of it enters the victim radar rf bandwidth. To do this the jammer must use a receiver to detect the radar pulses and measure the rf. A low powered oscillator is rapidly tuned to the correct frequency and amplified by the TWTA. The bandwidth of the fm noise modulation is made greater than the largest rf bandwidth anticipated for the victim radar. The speed at which these actions take place is important. If the radar is a steady rf tracking radar, then perhaps many milliseconds can be taken to find the victim frequency and start the jamming. To jam scanning beam radars (e.g., surveillance radars), the jammer must identify the radar rf and adjust its jamming to it in much less than one beam dwell time preferably during one radar PRI which calls for high speed rf measurement. Noise jammer powers are discussed later.

4.3 Repeater Jammers

These form the other main jamming class and are really deception jammers. Consider a receive aerial connected to an amplifier where output is fed to a TWTA and thence to a transmitting aerial. Any pulse received from a radar will be received, amplified and transmitted back to the radar. If however amplitude or temporal modulations are applied we have a powerful deception technique. Low frequency amplitude modulation applied to the signal will modulate the output signal which can easily be many times the radar skin echo power (the TWTA is now working in a pulsed condition and the peak power can easily be 1 KW). Such a signal will capture the radar agc circuits and depress the true skin echo. Against simple conical scan radars, the angle modulation if correctly chosen will cause large errors in angle tracking.

4.4 RGPO

If the received pulse can be delayed either in a delay line or by replicating it in SAW oscillator, then the repeater can be used in a transponder mode to give pulses which can be delayed with respect to the radar skin echo. The large amplitude jammer pulse will capture the agc level and depress the skin echo. By suitably increasing the delay time it will slowly move off to a false range, dragging with it the automatic range track loop of a tracking radar. Having moved the range say one mile away, the false repeater target is switched off dumping the range strobe at a range where there is no target and causing loss of lock and a re-cycle of the system through search to acquisition and lock-on. This technique is particularly applicable to automatic tracking system, e.g., in a missile, but not so effective when human operators are concerned.

4.5 VGPO

In the same way that a TWTA designed to amplify pulses can give deception, so can a TWTA designed to amplify incoming CW signals. Without any deception, modulation signals will be amplified and re-transmitted faithfully in terms of frequency spectrum. However there is a so-called serrodyne technique whereby a variable voltage is applied to the helix of the TWT slow wave structure. This interacts with the travelling wave to give a frequency shift to the signal, ie direct frequency variation. This gives the capability to apply pseudo Doppler shift to the transmitter signal. By gradually increasing the frequency shift the velocity tracking gate of a Doppler radar can be captured, moved off in frequency, and dumped (like RGPO). This is VGPO.

4.6 Cover Pulse Jamming

This is the so-called "smart noise" jammer and is a means of obtaining responsive noise jamming for a short span of range. A repeater type jammer is used in the transponder mode, i.e., the received signal is used to set on the frequency of a lower powered oscillator. Several radar pulses are received and the PRI of the radar is established. The time of arrival of the subsequent radar pulses is predicted and the oscillator is gated on for a period of perhaps 5 microseconds ahead of the next predicted pulse time to 5 microseconds later. The oscillator is noise modulated (fm) and this signal is amplified and transmitted. The system works well against radars with steady prf and enables a low powered repeater to respond to a number of threats by time sharing. Complete range denial is not achieved.

4.7 Inverse Gain Jamming

This is a method used principally against scanning radars. The gain of a repeater is varied inversely with the radar signal strength received being a minimum when the radar beam is pointed straight at the jammer and a maximum when the jammer is in the radar side lobes. Thus the return from the skin echo is distorted in its azimuthal appearance.

Table I summarises the effectiveness of all the above jamming techniques against various radars.

TABLE 1

Jamming Technique	RADAR TYPE			
	Surveillance	Acquisition	Tracker	Doppler Tracker
NOISE (RVD)	†	†	†	†
RGPO			†	
VGPO				†
AD			†	
Cover Pulse		†	†	
I G	†	†		

5 JAMMER TRANSMITTERS

Section 4.1 suggests that modern jammers could use a travelling wave tube amplifier as transmitter power tube. These tubes produce power of the correct order required, and cover a wide frequency range varying from 10% of the mean operating frequency to a full octave. Wide bandwidth is advantageous when the opposing threat radars are diverse in frequency and to cover new defensive radars which may arise in the lifetime of the jammer.

5.1 Power Output

As stated before Travelling Wave Tubes are used in jammers because of their broad bandwidth compared with other power sources. There are two main focusing techniques, the periodic permanent magnet (PPM) being is generally preferred for airborne use because of its lower weight. Where weight and primary power are secondary considerations a solenoid focused TWT may be more suitable. The principal slow wave structures in use are the helix and the coupled cavity. For use over octave bandwidths, say 2-4, 4-8 or 8-16 GHz, a PPM focused helix tube is best. However if the application only calls for 10%-30% bandwidth, a coupled-cavity tube would give an increase of 2 or 3 times in power output.

TWTs are designed for either CW or pulse use. A CW tube is used for noise jammer or for deception (e.g., VGPO) of a CW or very high prf pulse doppler radar. The so-called dual mode tubes are basically CW tubes with a pulsed-up capability of 3 dB to 6 dB higher than the CW rating during pulsed conditions. Pulse TWTs are used in deception repeaters.

A survey of TWTs currently in manufacturers' catalogues gave the results in Figure 1. These tubes are all of octave or more bandwidth, mainly PPM but some solenoid focused.

As shown, CW power levels range from about 2 kilowatts at 3 GHz to 100 Watts at 12 GHz. Since tube manufacturers are continually seeking to increase RF power and efficiency by various technical advances such as depressed collector operation, more efficient cooling systems and reduction of rf losses, it would seem prudent to assume that radars entering service in 5 years time will be confronted with CW power levels of:

1000 watts at F Band
500 watts at I Band
200 watts at J Band

Pulse tubes currently available have peak powers of more than 1 kilowatt over the range 2-16 GHz.

6 TYPICAL RESPONSIVE NOISE JAMMER

Figure 2 shows the block diagram of a responsive noise jammer. In principle it uses a receiver to obtain signals from radars illuminating the aircraft. The received data is examined for priority threats and, depending on their occurrence, decisions are made on the transmission of jamming and the modulation to be used. The tunable oscillator is rapidly adjusted to the threat frequency and its output is applied to the TWT amplifier. In general when the transmitter is on, the receiver is blanked. The diagram shows separate transmit and receive aerials though it is possible to use a common aerial with a T/R switch. If the receiver is of the superheterodyne type it will be possible to receive signals via the receive aerial at frequencies well removed from the frequency being jammed, but, to obtain further data on the status of the radar being jammed using simpler receivers, the transmitter must be switched off to give the so-called lookthrough period. The lookthrough time is kept short because the victim radar will be unjammed, but in this time the receiver must obtain at least one or two pulses from the radar. A general aim is to keep the lookthrough time as low as 1%, i.e., a ratio of 1 to 100 for receiver ON/OFF time ratio, with the receiver ON time being about 1 pulse recurrence interval (PRI). Since radar PRI varies widely with type, e.g., tracking radars have short PRI of about 500 microseconds or less whereas long range surveillance radars may have PRI of about millisecond or more, it is useful to vary the look-through period according to the threat received.

Scanning radars call for a fast reaction from the jammer. The radars have beam dwell times on the order of 10-20 milliseconds, and jamming should be started with 1 or 2 PRI of the radar being received.

The noise bandwidth of the jammer (the rf excursion of the carrier by the fm noise modulation) will exceed the effective receiver bandwidth of the radar—10 MHz being a typical figure for the latter.

The system described has one tunable oscillator so far dealing with one threat. Additional tunable oscillators could be used in parallel to drive the TWT either by time sharing or power sharing. Either method causes a reduction in the power density (watts/MHz) received by each individual radar. The aerial gains are usually low (around 3 dB) since ideally a broad all-round aerial pattern is desired. The typical system described could be of octave bandwidth. For a narrower bandwidth system a different type of TWT could be used giving higher output power.

7 RUSSIAN ECM

The USSR must be well aware of the general nature of EW techniques from the many Western technical publications. On their part publications on monopulse radar [1] and jamming principles [2] show a detailed understanding of the jamming methods previously described. We conclude that they appreciate the importance of electronic countermeasures and we should expect these techniques to be used by them.

8 THE ECM THREAT

Using the typical basic noise jammer of section 6 it is convenient to consider airborne jammers in a modular context. Thus the system as described could be configured for an attack (bomber) aircraft such as the A-10 or Jaguar aircraft optimised for use against terminal threats, i.e., anti-aircraft defences. This would be a self-screening or self-protection jammer (SSJ).

Aircraft of similar type and performance could be equipped specifically for electronic countermeasure support to the attacking aircraft. These aircraft such as the F111, or the Russian Brewer E would be escort jammer (EJ) not carrying weapon loads. They could be equipped with 2-4 units of the responsive noise jammer module. The units could be arranged to reinforce in power over a particular band or to give increased frequency coverage. They would again be mainly targeted against terminal threats with some devoted to surveillance/acquisition elements.

Very large aircraft such as the HERCULES or the Russian CUB 'C' can be used as stand-off jammers (SOJ). These aircraft could be positioned well beyond the opposing air defences (say 100 kilometres) and could direct their jamming at the surveillance radars with the aim of confusion and deception. The SOJ aircraft could carry 10 or more of the modular jammers. These jammers can be used responsively as described or could be used to provide noise modulated power over a given bandwidth in a continuous manner.

9 CHAFF

This passive electronic countermeasure possesses the advantages of being relatively cheap, consuming no electrical power, and being broadband. It is generally used in a corridor mode, i.e., by continuously laying trials from lead aircraft to hide the approach of following aircraft. It can also be used by attacking aircraft from defence against terminal threat radars. In this case cartridges of chaff are fired which distract the radar causing errors in range and angle. We should anticipate the continued use of chaff as a countermeasure (in conjunction with jamming) and corridor densities with echoing areas of $10^{-6} M^2/M^3$ should be expected.

REFERENCES

[1] A I Leonov, K I Fomichev — Monopulse Radar.

[2] S A Vakin, L N Shustov — Principles of Jamming and Electronic Reconnaissance.

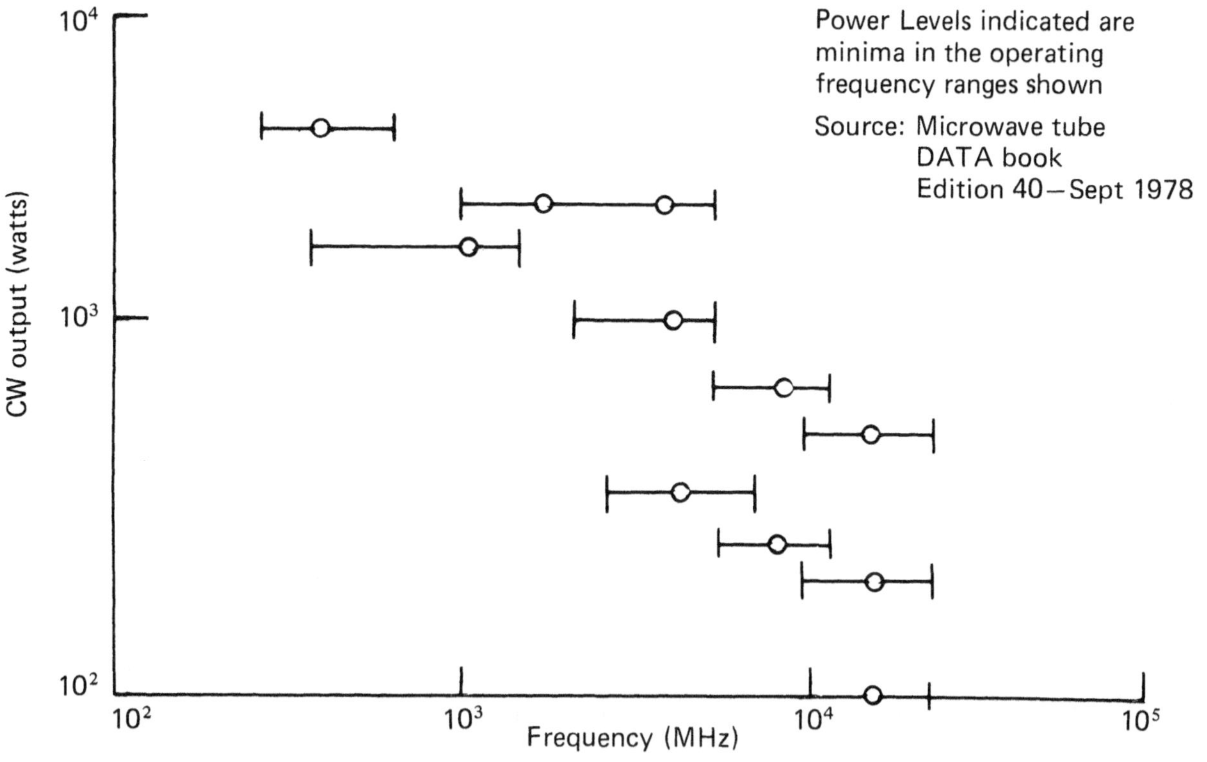

Figure 1 Broadband CW TWTS Currently Available

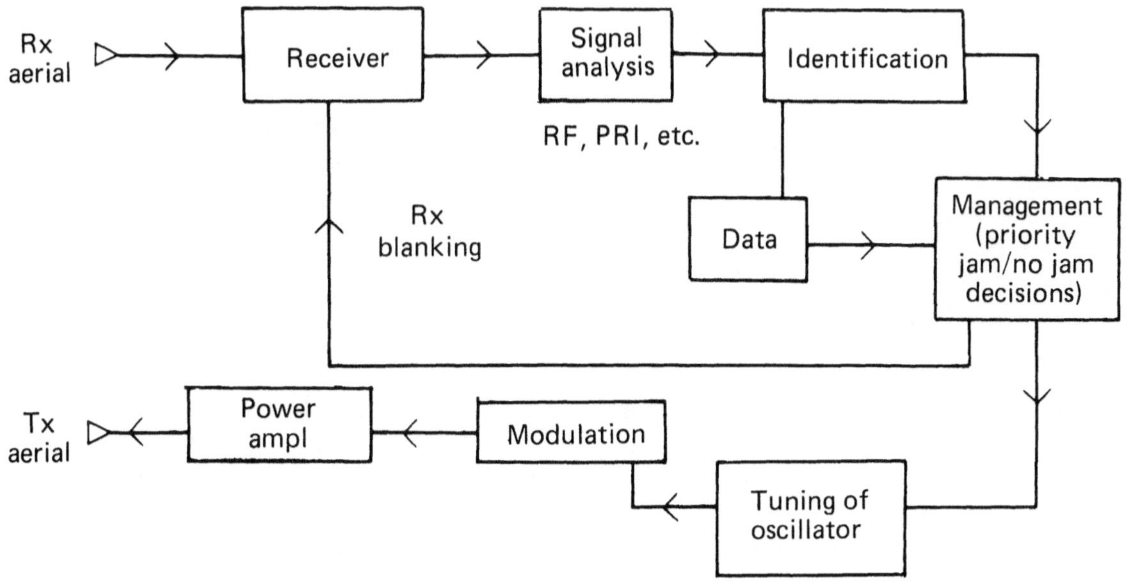

Figure 2 Simplified Block Diagram of Responsive Noise Jammer

Master Electronic Warfare Glossary
S.L. Johnston
Second International Countermeasures Handbook, June 1976
Pages 595-608

1.7

Master Electronic Warfare Glossary

Stephen L. Johnston
Huntsville, Alabama

Most fields of endeavor have developed special terms; some adopted from other fields, some adapted or redefined from other fields, and many new terms. Electronic Warfare is no exception to this. Many "EW" terms originated in radio and radar. At present, there are a number of documents which contain definitions of EW terms. Some of these are exclusively devoted to "EW." There are differences in definitions in some of these documents.

This Glossary was limited to either pure EW terms or those terms which have strong EW implications. Many general terms will be found in the *IEEE Standard Dictionary* (ref. 1). Most radar terms will be found in *IEEE Standard 172* (ref. 2) which in now being incorporated into the forthcoming revisions of ref. 1. Where a term in this Glossary appears in ref. 1 or ref. 2, these definitions were generally used here; since the IEEE definitions are internationally recognized and are usually more precise than those of the services. In some cases services and IEEE definitions were combined here.

The author gratefully acknowledges the assistance of Richard E. Fitts (ref. 3) and several others in obtaining material for this Glossary. Correspondence on any of these definitions will be welcomed.

References

1. IEEE Standard Dictionary of Electrical and Electronics Terms, *IEEE STD 100-1972*, New York, Wiley Interscience, 1972.
2. IEEE Standard Definitions: Radar and Navigational Aid Terms, *IEEE No. 172*, 1975.
3. Richard E. Fitts, private communication.

A

Accidental Jamming—Accidental jamming is interference due to transmission by friendly equipment.

ACET—Automatic Cancellation of Extended Targets.

Acoustic Jamming—The deliberate radiation or reradiation of mechanical or electro-acoustic signals with the objectives of obliterating or obscuring signals which the enemy is attempting to receive and of deterring enemy weapons systems.

Active Electronic Countermeasures—The impairment of enemy electronic detection, control or communications devices/systems through deliberate jamming or deception.

Active Jamming—Intentional radiation or reradiation of electromagnetic waves with the object of impairing the use of a specific portion of the electromagnetic wave spectrum.

AGC—See Automatic Gain Control.

Aided Tracking—A system of tracking a signal in azimuth, elevation, or range, in which a constant rate of motion of the tracking mechanism is maintained by mechanical means so that an equivalent constant rate of motion of the target can be followed.

AJ—See Anti-Jamming (see ECCM).

AM-CW Jamming—Jamming in which a carrier wave is modulated at a constant recurring rate. The recurrence rate of amplitude modulation is variable, and reflects a noticeable change in the radar scope. Because of the spiraling or chaining effect produced, AM-CW jamming is easily identified.

Angel—1. Radar returns which are not desired in a particular situation and which are caused by atmospheric inhomogeneities, refractive index discontinuities, insects, birds, or unknown phenomena. Note: Originally when some physical target could not be identified through direct visual observation, all echoes from such unknown causes were designated as "angels." 2. A type of confusion reflector often having the reflecting material suspended from parachutes or balloons to give delayed descent.

Angle Jamming—ECM technique, when azimuth and elevation information from a scanning fire control radar present in the modulation components on the returning echo pulse is jammed by transmitting a jamming pulse similar to the radar pulse but with modulation information out of phase with the returning target angle modulation information.

Angle Noise (in radar)—The noise-like variation in the apparent angle of arrival of a signal received from a target, caused by changes in phase and amplitude of target scattering sources and including angular components of both *Clint* and *Scintillation Error*.

Antenna Cross Talk—A measure of undesired power transfer through space from one antenna to another. Ratio of power received by one antenna to power transmitted by the other, usually expressed in decibels.

Anti-Clutter Circuits (in radar)—Circuits which attenuate undesired reflections to permit detection of targets otherwise obscured by such reflections.

repeater jammers.

Bistatic Radar—A radar using antennas at different locations for transmission and reception.

Blanketing—Action of a powerful radio signal or interference rendering a receiving set unable to receive desired signals.

Blanking—1. The cutting off of the electronic beam in a cathode-ray tube during the time when the picture is not being formed. The beam is blanked while the spot is returning to the starting position (normally right to left). 2. Process of making a channel or device noneffective for a desired interval. See Gating.

Blinking—1. ECM technique employed by two aircraft separated by a short distance and within the same azimuth resolution so as to appear as one target to a tracking radar. The two aircraft alternately spot jam, causing the radar system to oscillate from one plane to another, making an accurate solution of fire control problem impossible. 2. A method of providing information by modifying the signal at its source so that the signal presentation on the display at the receiver alternately appears and disappears; for example, in loran, blinking is used to indicate that a station is malfunctioning.

Blocking—The process of obscuring guidance signals by active jamming.

Broad Pulse Jamming—Transmission of broad pulses for control system jamming when little is known about the command pulse group. For example, a broad pulse might cover a whole group of command pulses, thus jamming that command.

Burnthrough Range—The distance at which a specific radar can discern targets through the external interference being received. (The interference is normally associated with jamming.)

Buzzer—An unclassified brevity code word signifying electronic jamming type of active ECM.

C

Camouflage Factor—1. The ratio of the RMS noise power to the peak signal power required to provide effective jamming. 2. The ratio of the RMS noise power to the peak pulse power of the radar for effective jamming. 3. The ratio of RMS noise power to peak signal power where the RMS noise power is determined on the basis of the bandwidth of the receiver up to the second detector.

Capture—Where the jammer takes control over the guidance signal by active jamming.

Capture Effect—The tendency of a receiver to suppress the weaker of two signals within its passband.

CFAR—See constant false alarm rate.

Chaff—An airborne cloud of lightweight reflecting objects typically consisting of strips of aluminum foil or metallic coated fibers which produce clutter echoes in a region of space.

Chirp—A pulse compression technique characterized by linear frequency modulation on pulse (LFMOP).

Chirp Mode—The chirp mode enables the operator to transmit the outgoing pulse in steps of increasing or decreasing frequency. This feature is similar to frequency agility, except that each pulse is broken into smaller sections and these smaller sections are transmitted at separate frequencies. The radar system compiles these smaller sections back into the original pulse before being sent to the receiver, thus providing a coding that is difficult for jammer to match.

Chirp Radar—Radar in which a swept-frequency signal is transmitted, received from a target, and then compressed in time to give a final narrow pulse called the chirp signal. It has high immunity to jamming and an inherent rejection of random noise signals.

Circularly Polarized Jamming—The techniques of radiating jamming energy in both planes of polarization simultaneously. With this method, there is a loss of 3 dB of effective power in either plane, but the enemy cannot cross-polarize his antenna to escape jamming.

Clipped Noise Modulation—A clipping action is performed to increase the bandwidth of the jamming signal. Results in more energy in the sidebands, correspondingly less energy in the carrier, and an increase in the ratio of average power to peak power.

Clutter Elimination—The clutter eliminator circuit discriminates against any target echo that exceeds three times the transmitted pulsewidth, and will not display them on the indicator. It is normally employed on the lower beams of a high frequency radar. This will allow targets above a preset signal strength to be presented, while the clutter (land) will be eliminated.

Clutter Gating—This is a technique which provides switching between MTI and normal videos. It results in the normal video being displayed in regions with no clutter and the MTI video being switched in only for the clutter areas. Clutter gating is achieved automatically by the PW discrimination or the use of storage tubes. It also can be achieved by a manually operated range azimuth gate. The clutter gate vastly increases the effectiveness of noncoherent MTI against chaff.

Coast—A radar memory feature that causes the range or angle systems to continue to move in the same direction and at the same speed that an original target was moving; used to prevent lock-on to a stronger target near the target being tracked.

Coherent MTI (in radar MTI)—A system in which the target echo is selected on the basis of its Doppler frequency when compared to a local reference frequency maintained by a coherent oscillator.

Coherent Repeater Jammer—A jammer that uses the phase information of the received radar signal in creating false targets.

Coherent Video—Bipolar video obtained from a synchronous (coherent) detector. See Coherent MTI.

Coherent Video Mode—In the CV mode, the radar transmits and receives twice. The receiver samples the returning echoes and only those that coincide in time will be presented to the indicator.

COMINT—Communications Intelligence.

COMSEC—See Communications Security.

Communications Countermeasures—Electronic countermeasures used specifically against communications.

Communications Deception—Use of devices, operations, and techniques with the intent of confusing or misleading the user of a communications link or a navigation system.

Communications Jamming—The part of electronic jamming used against a medium that employs electromagnetic radiation to convey information from one person or headquarters to another.

Communications Security (DoD)—The protection resulting from all measures designed to deny unauthorized persons information of value which might be derived from the possession and study of telecommunications, or to mislead unauthorized persons in their interpretation of the results of such possession and study. COMSEC includes: 1. cryptosecurity; 2. transmission security; 3. emission security; and 4. physical security of communications security material and information.

1. *Cryptosecurity*—The component of communications security which results from the provision of technically sound cryptosystems and their proper use.
2. *Transmission security*—The component of communications security which results from all measures designed to protect transmissions from interception and exploitation by means other than cryptoanalysis.
3. *Emission security*—The component of communications security which results from all measures taken to deny unauthorized persons information of value which might be derived from intercept and analysis of compromising emanations from crypto-equipment and telecommunications systems.
4. *Physical security*—The component of communications security which results from all physical measures necessary to safeguard classified equipment, material, and documents from access thereto or observation thereof by being within a friendly power.

Confusion Reflector—Reflector of electromagnetic radiation used to create echoes for confusion against radar, guided missiles, and proximity fuses. See Chaff.

Constant False Alarm Rate Receiver—A radar receiver with automatic detection circuits designed to produce a constant number of erroneous target detections independent of noise level at the receiver input. CFAR techniques are intended primarily to prevent receiver saturation and overload, and to present clean video information to the display and a constant noise level to an automatic detector. A device that accomplishes the above objectives may respond to the signal-to-noise ratio; for example, rather than the absolute signal level above a fixed threshold. CFAR does not usually permit the detection of a target if the target is weaker than the jamming, but it does attempt to remove the confusing effects of the jamming.

Correlation Detection (Modulation Systems)—Detection based on the averaged product of the received signal and a locally generated function possessing some known characteristics of the transmitted wave. *Notes:* 1. The averaged product can be formed, for example, by multiplying and integrating or by the use of a matched filter whose impulse response, when reversed in time, is the locally generated function. 2. Strictly, the above definition applies to detection based on cross correlation. The term correlation detection may also apply to detection involving autocorrelation, in which case the locally generated function is merely a delayed form of the received signal.

Correlation Direction Finder—Satellite station separated from a radar to receive jamming signals. By correlating the signals received from several such stations, range and azimuth of many jammers may be obtained.

Countermeasures—That form of military science which by the employment of devices and/or techniques has as its objective the impairment of the operational effectiveness of enemy activity. See also Electronic Countermeasures.

"Criss-Cross"—The name given to that pattern produced on an ½'A" scope by CW modulated with a medium frequency signal. Criss-Cross appears as diagonal lines in two directions above the baseline.

Cross-Eye Jamming—A technique of jamming whereby two beams of energy are transmitted in the direction of the target; in a manner so that the beams are crossed between the target and the transmitters. With such a technique, it is difficult for the target to determine the points from which the transmissions are originating.

Cross-Gated CFAR—This CFAR technique is employed to achieve the fast switching required for an optimum combination of normal and MTI modes. In this case, the MTI video signals are used to "gate" on the normal video when the MTI indicates a target in clutter. CFAR action is achieved by the wideband as in the zero-crossing and DICKE FIX CFARs.

Cross Modulation—A type of intermodulation due to modulation of the carrier of the desired signal by an undesired signal wave.

Crystal-Video Receiver—A broadband receiver consisting of a crystal detector, followed by a high gain video amplifier. Selectivity is determined only by the antenna as there is no tuned circuits.

CW Jamming—The transmission of constant-amplitude, constant-frequency, unmodulated jamming signals to change the signal-to-noise ratio of a radar receiver.

D

DBB—See Detector Back Bias, Detector Balanced Bias.

Dead Time—An interval following response to one signal or event during which a system is unable to respond to another.

Deceiver—An ECM equipment which attempts to deceive or mislead a radar by emitting a pulse-like signal similar to the radar signal.

DECM—Deceptive ECM (see Electronic Deception).

Deception—Those measures designed to mislead the enemy by manipulation, distortion, or falsification of evidence to induce him to react in a manner prejudicial to his interests (see Electronic Deception, Imitative Deception, or Manipulative Deception).

Deception Controlled Devices—The techniques of deliberately radiating or reradiating electromagnetic transmissions for the purpose of generating false control data within an enemy-controlled device.

Deception Device—A device that works to confuse unfriendly signals so as to make them either unusable or misleading.

Decoy—A device or devices used to divert or mislead enemy defensive systems so as to increase the probability of penetration and weapon delivery.

Delayed Opening Chaff—Chaff which blooms at a specific elapsed time after it is dispensed.

Detector Back Bias (DBB)—This technique performs much the same function as IAGC, except that it operates at the diode detector instead of the IF amplifier.

Detector Balanced Bias—Controlling circuit used in radar systems for anti-clutter purposes.

DICKE FIX—The "DICKE FIX" is a technique that is specifically designed to protect the receiver from fast sweep jamming of the carcinotron type. The basic configuration consists of a broadband limiting IF amplifier, followed by an IF amplifier of optimum bandwidth. The limit level is preset at approximately the peak amplitude of receiver noise, the bandwidth may vary from 10 to 20 MHz, depending on the jamming environment. This device provides excellent discrimination against fast sweep jamming (10–500 MHz), usually something on the order of 20 to 40 dB, without appreciable loss in sensitivity. However, strong CW jamming will seriously degrade the performance of the DICKE FIX because the CW signal captures the radar signal in the limiter.

DICKE FIX CFAR—Constant False Alarm Rate.

DICKE FIX MTI CFAR—This is an MTI CFAR technique, similar to DICKE FIX CFAR. Limiting and narrowbanding follow wideband amplification, phase detection, and cancellation so as not to impair the MTI performance.

Diplex—Two transmitters operating alternately on approximately the same RF and using a common antenna. The normal procedure is to pulse each transmitter at one-half of the desired PRF, 180° out-of-phase. The advantage is that higher peak power per transmitter is possible because each transmitter is operating at one-half the normal duty cycle.

Diplex Operation—Simultaneous transmission or reception of two signals using a specified common feature, such as a single antenna or a single carrier.

Direct Amplified Noise—Direct Amplified Noise (DINA) without a carrier frequency is used for the purpose of increasing (saturating) the radar receiver's noise level; this is something called video stealing. The biggest danger from this type of jamming is that the radar operator may not realize he is being jammed when the automatic gain control (AGC) or the automatic noise leveling (ANL) are employed. This is a barrage type of jamming with a bandwidth normally in excess of 50 MHz. DINA appears as a black wedge when the radar is operated with AGC or ANL. It appears as bright, high intensity wedges when the manual gain is employed.

Down-Link Jamming (DLJ)—Some command guidance missiles carry a beacon (down-link) which is used by the parent radar to track the missile. If this beacon reply can be hidden from the parent tracking radar, the missile guidance solution can be defeated. Hence, down-link (beacon) jamming is intended to screen the missile beacon signal from the parent radar's view.

Dummy Antenna—A device that has the necessary impedance characteristics of an antenna and the necessary power-handling capabilities, but that does not radiate or receive radio waves. *Note:* In receiver practice, that portion of the impedance not included in the signal generator is often called dummy antenna.

Dummy Load (Radio Transmission)—A dissipative but essentially nonradiating substitute device having impedance characteristics simulating those of the substituted device.

Duplex—In radar, a condition of operation when two identical and interchangeable equipments are provided—one in an active state, and the other immediately available for operation.

Duplication Jamming—See Imitative Jamming.

Dynamic Range—1. The difference, in decibels, between the overload level and the minimum acceptable signal level in a system or transducer. *Note:* The minimum acceptable signal level of a system or transducer is ordinarily fixed by one or more of the following: noise level, low-level distortion, interference, or resolution level. 2. Ratio of the specified maximum signal level capability of a system or component to its noise or resolution level, usually expressed in decibels. 3. Figure of merit in electronic counter-countermeasures receiver system.

E

ECCM—See Electronic Counter-Countermeasures.

ECCM Improvement Factor—The power ratio of the ECM signal level required to product a given output signal from a receiver using an ECCM technique to the ECM signal level producing the same output from the same receiver without the ECCM technique.

Echo Intensifier—A device, located at the target, that is used to increase abnormally the amplitude of the reflected energy.

ECM—See Electronic Countermeasures.

Effective Confusion Area—The amount of chaff whose radar cross-sectional area equals the radar cross-sectional area of the particular aircraft at a particular frequency.

Electromagnetic Compatibility—The capability of electronic equipment and systems to operate in the intended environment at designated levels of efficiency without degradation due to unintentional interference.

Electromagnetic Cover and Deception—The suppression, control, alteration, or simulation of electromagnetic radiations associated with friendly systems, equipment, devices or weapons components to deny any enemy a source of knowledge as to the location of combat elements, or mislead him as to their capabilities and intentions. Electromagnetic cover and deception tactics include emission control, signal suppression, profile alteration, and electronic deception.

Electromagnetic Interference—Impairment of the reception of a wanted electromagnetic signal caused by an electromagnetic disturbance.

Electromagnetic Intrusion—The intentional insertion of electromagnetic energy into transmission paths in any manner with the objective of deceiving operators or of causing confusion. See also Electronic Warfare.

Electromagnetic Reconnaissance—See Electronic Reconnaissance.

Electromagnetic Test Environment (AF)—A range complex of radars at Eglin Air Force Base, Florida, operating in different frequency bands and modes to provide a very flexible facility for evaluating aircraft antenna patterns, reflectivity measurements, infrared, reconnaissance, airborne interceptors, and electromagnetic warfare devices and techniques.

Electronic Confusion Area—Amount of space that a target appears to occupy in a radar resolution cell, as it appears to that radar beam.

Electronic Counter-Countermeasures (ECCM)—That division of electronic warfare involving actions taken to ensure friendly effective use of the electromagnetic spectrum despite the enemy's use of electronic warfare.

Electronic Countermeasures (ECM)—That division of EW involving actions taken to prevent or reduce an enemy's effective use of the electromagnetic spectrum.

Electronic Deception (DoD)—The deliberate radiation, reradiation, alteration, absorption, or reflection of electromagnetic radiations in a manner intended to mislead an enemy in the interpretation of or use of information received by his electronic systems. There are two categories of electronic deception: 1. *Manipulative deception*—The alteration or simulation of friendly electromagnetic radiations to accomplish deception. 2. *Imitative deception*—The introduction of radiations into enemy channels which imitate his own emissions.

Electronic Defense Evaluation—A mutual evaluation of radar(s) and aircraft with the aircraft trying to penetrate the radars area of coverage in an electronic countermeasures environment.

Electronic Intelligence (ELINT)—The intelligence information product of activities engaged in the collection and processing, for subsequent intelligence purposes, of foreign, noncommunications, electromagnetic radiations emanating from other than nuclear detonations or radioactive sources.

Electronic Intelligence Parameter Limits List (EPL)—A compilation of identified signals which have been assigned ELINT notations.

Electronic Interference—Any electrical or electromagnetic disturbance that causes undesirable response in electronic equipment. Electrical interference refers specifically to interference caused by the operation of electrical apparatus that is not designed to radiate electromagnetic energy.

Electronic Jammers—1. *Expendable*—A transmitter designed for special use such as dropped behind enemy lines. 2. *Repeater*—A receiver-transmitter device which, when triggered by enemy radar impulses, returns synchronized false signals to the enemy equipment. The returned impulses are spaced and timed to produce false echoes or bearing errors in the enemy equipment. See Expendable and Repeater Jammers.

Electronic Jamming—The deliberate radiation, reradiation, or reflection of electromagnetic energy with the object of impairing the use of electronic devices, equipment, or systems being used by the enemy.

Electronic Order of Battle—A listing of all the electronic radiating equipment of a military force giving location, type, function and other pertinent data.

Electronic Reconnaissance—The detection, identification, evaluation, and location of foreign, electromagnetic radiations emanating from other than nuclear detonations or radioactive sources.

Electronic Security (ELSEC)—The protection resulting from all measures designed to deny to unauthorized persons information of value which might be derived from their interception and study of friendly noncommunications electromagnetic radiations; for example, radar.

Electronic Tuning—1. Altering the frequency of a reflex klystron oscillator by changing the repeller voltage. 2. Frequency changing in a transmitter or receiver by changing a control voltage, rather than circuit components.

Electronic Warfare (DoD)—Military action involving the use of electromagnetic energy to determine, exploit, reduce, or prevent hostile use of the electromagnetic spectrum and action which retains friendly use of the electromagnetic spectrum. There are three divisions within electronic warfare: electronic warfare support measures, electronic countermeasures and electronic counter-countermeasures.

Electronic Warfare Intelligence (ELINT)—Electronic warfare intelligence is the product resulting from the collection, evaluation, analysis, integration, and interpretation of all available information concerning foreign nations or areas of operations which are significant to EW.

Electronic Warfare Support Measures (ESM)—That division of electronic warfare involving actions taken to search for, intercept, locate, and identify immediately radiated electromagnetic energy for the purpose of immediate threat recognition. Thus, electronic warfare support measures provide a source of information required for immediate action involving electronic countermeasures, electronic counter-countermeasures, avoidance, targeting, and other tactical employment of forces.

Electro-Optic Counter-Countermeasures (EOCCM)—Actions taken to ensure the effective friendly use of the electro-optic spectrum despite the enemy's use of countermeasures in that spectrum.

Elevation Versus Integrated Log (EVIL)—The EVIL equipment is functionally very similar to the TCU except that EVIL is used in conjunction with the FPS-6 and FPS-26 height finder radars. This addition provides a capability for obtaining an elevation angle report on target aircraft that are jamming with sufficient intensity to deny range or height information to the active radar. This report is transmitted to the SAGE DC, where it is converted directly to a height for a particular target. With the addition of EVIL, a total passive-only air defense capability is available if the amount of jamming still exceeds that required to effectively deny active data to the air defense system.

ELINT—See Electronic Intelligence.

Emanation Security—The protection resulting from all measures designed to deny unauthorized persons information of value which might be derived from unintentional emissions from other than telecommunications systems.

EMCON—See Emission Control.

Emission Control (EMCON)—1. The management of electromagnetic radiations to counter an enemy's capability to detect, identify, or locate friendly emitters for exploitation by hostile action. 2. Controlling the radiation of an active system such as a radar so that it emits RF energy only when absolutely necessary to perform its mission.

Emission Control Orders—Orders, referred to as EMCON orders, used to authorize, control, or prohibit the use of electronic emission equipment.

Emission Security—The protection resulting from all measures designed to deny unauthorized persons information of value which might be derived from intercept and analysis of compromising emanations from other than cryptoequipment and telecommunications system.

EMP—See Electromagnetic Pulse.

EOB—See Electronic Order of Battle.

ER—See Electronic Reconnaissance.

ESM—See Electronic Warfare Support Measures.

Essential Elements of Information (EEI)—The critical items of information regarding the enemy and his environment needed by the commander by a particular time to relate with other available information and intelligence in order to assist him in reaching a logical decision.

EW—See Electronic Warfare, Early Warning.

EWO—See Electronic Warfare Officer.

Extraneous Emission—Any emission of a transmitter or transponder, other than the output fundamental, plus only those sidebands intentionally employed for the transmission of intelligence.

Extraneous Response—Any undesired response of a receiver, recorder, or other susceptible device, due to the desired signals, undesired signals, or any combination or interaction among them.

F

FAGC (Fast AGC)—See Fast Automatic Gain Control.

False Alarm—An erroneous radar target detection decision caused by noise or other interfering signals exceeding the detection threshold.

False-Alarm Number—The average number of possible detection decisions during the false alarm time.

False Alarm Probability—The probability that noise or other interfering signals will erroneously cause a target detection decision.

False-Alarm Time—The average time between false alarms—i.e., the average time between crossings of the target decision threshold by signals not representing targets. In the early work of Marcum, it is the time in which the probability of one or more false alarms is one-half, but the usage is deprecated.

False Echo Device—Device that produces an echo different in character or time from that normally observed.

Fast Automatic Gain Control (FAGC)—This is an AGC scheme characterized by a response time that is long with respect to a PW and short with respect to the target. An ultrafast FAGC will reduce the CW capture effect on the DICKE FIX.

Fast-Time Constant Circuit (in radar)—1. A circuit with short time-constant used to emphasize signals of short duration to produce discrimination against extended clutter, long-pulse jamming, or noise. 2. A resistance-capacitance differentiating network with a time constant about equal to the transmitted pulse width and employed in the video portion of the receiver to provide discrimination against jamming with low modulating frequencies. FTC is effective against CW, long jamming pulses, swept jamming, and clutter. It is also fairly effective against chaff corridors. Based on time, the radar receiver will allow only pulses equal to its own pulsewidth to pass and be presented on the indicators as targets. Because of this, only the leading edges of long pulses will be displayed.

Fence—1. Line or network of early warning radars. 2. Concentric steel fence erected around a ground radar transmitting antenna to serve as an artificial horizon and suppress ground clutter that would otherwise drown out weak signals returning at a low angle from the target.

Ferret—An aircraft, ship, or vehicle especially equipped for the detection, location, recording, and analyzing, of electromagnetic radiations.

FM-By-Noise Modulation—A method of frequency modulating with the use of random noise. Considered a highly effective jamming method against AM and fixed-tuned FM receivers. Not very effective against continuously tunable FM receivers; careful tuning can defeat a great portion of the jamming signal. For this reason, FM-By-Noise is not considered optimum as a type of modulation for the jamming of FM receivers.

FM-CW Jamming—This is a highly efficient jammer that utilizes two variants in producing a jamming signal. First, the frequency modulation (FM) rate is variable; and second, the jamming signal is swept in frequency through and usually beyond the radar's bandpass (also variable). FM-CW jamming appears on a radar screen as shifting noise spokes.

FM Jamming—Technique consisting of a constant amplitude RF signal that is varied in frequency about a center frequency to produce a signal over a band of frequencies.

Frequency Agility—This term refers to a radar's ability to change frequency within its operating band. Pulse-to-pulse frequency shift, or changing the transmitter frequency radically during every interpulse is the ultimate in frequency agility.

Frequency Azimuth Intensity—Type of radar display in which frequency, azimuth, and strobe intensity are correlated.

Frequency Diversity—Method of transmission and/or reception using a number of frequencies simultaneously to minimize the effects of selective fading, deliberate jamming, or interference.

Frequency Time Intensity—Type of radar display in which the frequency, time, and strobe intensity are correlated.

"FRUIT"—In a radar beacon system, there is a type of interference called "FRUIT," caused by beacon replies to interrogation asynchronous with the observer's interrogator. The largest amount of this interference is received through the sidelobes of the interrogating antenna, but it can become dense enough to cause false target indications.

FTC—See Fast Time Constant.

G

Gain (transmission gain)—The increase in signal power in transmission from one point to another under stated conditions. *Note:* Power gain is usually expressed in decibels.

Gain (Manual)—The receiver gain control allows the operator to vary the receiver sensitivity. It is not designed as an AJ feature; however, when properly employed it may greatly reduce the effects of jamming. The radar detection capability is also reduced by an equal amount.

Gain Control—A device for adjusting the gain of a system or transducer. The gain control is designed to prevent the gain from reaching a value that will cause saturation or instability of the system. See Gain Margin.

Gain Margin (control system, feedback) (loop transfer function for a stable feedback system)—The reciprocal of the gain at the frequency at which the phase angle reaches minus 180 degrees. *Note:* Gain margin, sometimes expressed in decibels, is a convenient way of estimating relative stability by Nyquist, Bode, or Nichols diagrams. For systems with similar gain and phase characteristics in a conditionally stable feedback system, gain margin is understood to refer to highest frequency at which the phase angle is minus 180 degrees.

George Box—An amplitude sensitive device, employed in an IF amplifier. This unit rejects small jamming signals not of sufficient amplitude to operate its ciruits. However, jamming signals of sufficient amplitude are not affected.

Ghosts—1. In passive detection, the intersection points of lines of position which do not represent actual targets but are only crossover points of multiple plotted lines of position from two or more detection stations. 2. Unwanted signal appearing on the screen of a radar indicator, caused by echoes which experience multiple reflections before reaching the receiver.

Glint (in radar)—1. The random component of target location error caused by variations in the phase front of the target signal (as contrasted with *Scintillation Error*). Glint may affect angle, range of Doppler measurements, and may have peak values corresponding to locations beyond the true target extent in the measured coordinate. 2. Electronic countermeasures that use the scintillating, or flashing effect of shuttered or rotating reflectors to degrade tracking or seeking functions of an enemy weapons system.

Ground Photo—An unclassified general call sign for ground radar stations engaged in active ECM.

Gull—A floating radar reflector used to simulate surface targets for deceptive purposes.

H

High Pass Filter—A filter having a single transmission band extending from some cutoff frequency, not zero, up to infinite frequency.

High Video Pass (HVP)—HVP is an improved or super FTC, and is effective against CW, swept, pulse jamming and clutter/chaff. Based on time, the radar receiver will pass much shorter pulses than FTC, and will then present these on the indicator. It is good for distinguishing between targets that are too close together to be resolved as individual targets.

HOJ—See Home on Jam.

Home-on-Jam (HOJ)—A method of passive guidance designed to use the jamming signal emitted by the target to track the target in angle.

I

IAGC (Instantaneous AGC)—See Instantaneous Automatic Gain Control.

IFC—Instantaneous Frequency Correlator.

IFRU (Interference Rejection Unit)—A tunable filter or wave-trap which can be adjusted to reject any frequency within the IF passband and allow the remainder of the passband curve to remain intact. It is adjusted to reject an interference signal and as such is a form of anti-jamming.

Imitative Deception—The introduction of radiations into enemy channels which imitates his own emissions.

Imitative jamming—The jamming technique of transmitting a signal identical to the original guidance signal.

Infrared Counter-Countermeasures—Actions taken to effectively employ our own infrared radiation equipments and systems in spite of the enemy's actions to counter their use.

Infrared Countermeasures—1. Countermeasures used specifically against enemy threats operating the infrared spectrum. 2. Actions taken to prevent or reduce the effectiveness of enemy equipment and tactics employing infrared radiation are termed infrared countermeasures.

Instantaneous Automatic Gain Control (IAGC)—1. That portion of a system that automatically adjusts the gain of an amplifier for each pulse so as to obtain a substantially constant output pulse peak amplitude with varying input pulse peak amplitudes, the adjustment being sufficiently fast to operate during the time a pulse is passing through the amplifier. 2. (radar) A quick-acting automatic gain control that responds to variations of mean clutter level, or jamming over different range or angular regions, avoiding receiver saturation (also known as sack-bias). IAGC automatically adjusts the gain of the radar receiver so that strong signals do not block adjoining weak signals. IAGC is not quick enough to block short pulse jamming, but is capable of reducing the effects of long pulse jamming or extended clutter.

Instantaneous Frequency—1. The time rate of change of the angle of an angle-modulated wave. *Note:* if the angle is measured in radians, the frequency in hertz is the time rate of change of the angle divided by 2π. 2. Zero-crossing constant false alarm rate device takes the form of the intermediate frequency DICKE FIX constant false alarm rate, except the phase detector is replaced with a frequency discriminator.

Instantaneous Frequency CFAR (IFCFAR)—This CFAR technique is similar to the IF DICKE FIX CFAR, except for having a phase detector instead of a frequency discriminator. The primary use of this technique, in conjunction with rapid random frequency changes, is to make chaff signals appear noise-like so they can be handled in a conventional manner.

Intercept Receiver—Special calibrated receiver which can be tuned over a wide-frequency range to detect and measure radio signals transmitted by the enemy. Also called Search Receiver.

Interference (Electronic)—An electrical or electromagnetic disturbance that causes undesirable responses on electronic equipment. *Electrical interference* refers specifically to interference caused by the operation of electrical apparatus that is not designed to radiate electromagnetic energy.

Interference Spectrum—The frequency distribution of the jamming interference in the propagation medium external to the receiver.

Intermediate Frequency DICKE FIX CFAR—This CFAR technique also uses a broadband limiting IF amplifier that precedes a coherent detection and narrowband video filter. The limit level is preset well below the receiver noise level. Consequently any increase in noise power at the input will not affect the output noise level. A usable signal-to-noise ratio is established by the ratio of the filter bandwidths, usually on the order of 10 to 1.

Intermediate Frequency Jamming—Form of CW jamming that is accomplished by transmitting two CW signals separated by a frequency equal to the center frequency of the radar receiver IF amplifier.

Intrusion—The intentional insertion of electromagnetic energy into transmission paths in any manner with the objective of deceiving operations or of causing confusion.

Inverse-Gain Repeater Jammer—A form of repeater in which the jammer creates false targets by varying the output power inversely with the strength of the received radar signal.

IRCM—See Infrared Countermeasures.

J

Jaff—Expression for the combination of electronic and chaff jamming.

Jam Attenuator Control—The Jam Attenuator Control is used to prevent receiver saturation from any strong signals, including electronic jamming, chaff, or clutter; it also permits the determination of the bearing and elevation of jammers that would cause wide sectors of the scope to be obscured. The Jam Attenuator Control reduces all signals equally, jamming as well as targets. This control should be used in the jammed sector only and not for an entire antenna revolution. Targets in the jammed sector would only be seen if they were stronger than the jamming. However, the jammed sector may be reduced enough in size to allow the operator to determine either bearing or elevation.

Jammer—Transmitter used to jam radio or radar transmission.

Jammer Finder—Radar which attempts to obtain the range of the target by training a highly directional pencil beam on a jamming source. Also called burnthrough.

Jammers Tracked by Azimuth Crossing—Semiautomatic strobe processing and tracking device. Permits automatic detection and tracking on azimuth only data gathered from the jamming signals emanating from an airborne vehicle.

Jamming—See Acoustic Jamming, Barrage Jamming, Electronic Jamming or Spot Jamming.

Jamming (in radar)—1. A form of ECM in which noise or noise-like signals are transmitted at frequencies in the receiver bandwidths of a radar to obscure the radar signal. 2. The deliberate radiation, reradiation, or reflection of electromagnetic energy with the object of impairing the use of electronic devices, equipment, or systems by an enemy.

Jamming to Signal Ratio—The inverse of the signal-to-jamming ratio.

J/S—Jamming-to-Signal Ratio.

Jam Resistant Data Link—An electronic data transmission system which uses various techniques to reduce the effect of enemy electronic countermeasures (jamming) on its operational use by friendly elements.

Jam Strobe—Jam Strobe is also called JAVA (Jamming Amplitude Versus Azimuth). This circuit generates a marker on the PPI to indicate signal strength as a function of bearing. It does this by sampling the jamming intensity once each repetition period. In addition to showing the direction of the jammer, it also indicates the severity of main and sidelobe jamming.

Jittered Pulse Recurrence Frequency—Term used to describe the random variation of the pulse repetition period. Provides a discrimination capability against repeater type jammers.

L

Large Scale ECM Mission—Conduct of active airborne ECM by six or more aircraft working as a unit.

LIFMOP—Linearly Frequency Modulated Pulse. See Chirp, Pulse Compression.

Limiter—A device which prevents its output signal from exceeding a predetermined value.

Lin-Log Amplifier—An automatic gain control amplifier that operates in a linear manner for low amplitude input signals but responds in a logarithmic manner to high amplitude input signals.

Lobe-on-Receive-Only (LORO)—A passive scanning technique in which a non-scanning beam is used to illuminate the target and a nonradiating, receiving antenna is scanned. Normally used with conical, lobe-switching or unidirectional sector scans.

Local Oscillator OFF—This is a simple expedient of shutting off the local oscillator during barrage jamming. The barrage jammer will not be seen unless there is a beating signal such as a target; therefore, targets will appear. Since targets will not appear in directions where no jamming arises, either an automatic azimuth switch or an additional receiver-display is required.

Logarithmic Fast Time Constant Log (LOG-FTCL)—This device consists of a logarithmic IF amplifier followed by an FTC circuit. The LOG-FTC combination is very effective in removing variations (in the video output noise level) caused by spot noise, wideband noise, and slow sweep noise modulated AM jamming.

Logarithmic Receiver—A receiver whose response approximates the logarithm of the strength of the incoming signal. A special type of receiver having a large dynamic range of automatic gain control which gives considerable protection against receiver saturation by strong jamming on interference signals. Useful against weather, clutter, chaff, and spot jamming.

Logarithmic Video (LOG-Video)—Log-Video reduces strong signals to prevent receiver saturation at a logarithmic or varying rate.

Look-Through—1. When jamming a technique whereby the jamming emission is interrupted irregularly for extremely short periods to allow monitoring of the victim signal during jamming operations. 2. When being jammed, the technique of observing or monitoring a desired signal during interruptions in the jamming signals. 3. A visual display which continuously monitors the target signal in time and frequency. It derives its primary importance from the desirability for rapid acquisition, setting-on, and frequency of a communication signal while jamming.

LORO—See Lobe on Receive Only.

M

Masking—1. In air defense, the use of additional transmitters to hide a particular electromagnetic radiation as to location of source and/or purpose of the radiation. 2. The condition where a part of the missile blocks a portion of the warhead emission, fuze sensing area, or electromagnetic radiation area.

Manipulative Communications Deception (MCD)—The use of friendly communications to falsify information which a foreign nation can obtain from communications analysis.

Manipulative Deception—The alteration or simulation of friendly electromagnetic radiations to accomplish deception.

Manipulative Radar Deception—Accomplished by deliberately radiating or reradiating electromagnetic signals for the purpose of misleading the enemy.

Manual Rate-Aided Tracking—Radar circuit which tracks individual targets by computing the velocity from position fixes inserted manually into the circuitry.

Matched-Filter Receiver—A radar receiver whose signal processing circuits are designed to discriminate against signals with characteristics different from those of the transmitted RF pulse.

Meaconing—A system of transmitting actual or simulated radio navigational signals for the purpose of confusing navigation.

Mechanical Jamming—See Chaff.

MIJI—See Meaconing, Intrusion, Jamming, and Interference.

Misleading—The process of decoying with active jamming.

MLC—Main Lobe Cancellation.

Modulated CW Jamming—A CW carrier waveform that has been modulated with some other signal such as noise, low, medium, or high frequencies, and is transmitted for the purpose of deception. May be AM, FM, PM, pulse, etc.

Modulated PRF (MPRF)—The deliberate modulation of the interpulse spacing in a pulse train; this includes PFM in which the modulation wave is used to frequency modulate a series of pulses. See PRF Jitter and PRF Stagger.

Modulation on Pulse (MOP)—The modulation, in frequency or phase, of an RF carrier wave during the pulse, either in a discrete or continuous pattern. Normaly employed for pulse compression.

Monopinch—Antijam application of the monopulse technique where the error signal is used to provide discrimination against jamming signals.

MTI CFAR—This is a technique that provides a CFAR capability in an MTI receiver. The cancellation of ground clutter is not impaired during ECM action.

Multiband Radar—This type of radar uses simultaneous operation on more than one frequency band through a common antenna. This technique allows for many sophisticated forms of video processing and requires that the jammer must jam all channels at the same time.

Multiple Frequencies—Simultaneous transmission on more than one frequency.

Multitarget Generator Jamming—This generator takes the radar's PRF, scan rate, and antenna lobe pattern, and computes when to transmit false targets at the radar's frequency. The targets appear as true targets, but normally only about 20 percent of the targets are programmed to give a logical course and speed. The numerous targets saturate the tracking operator's capability. Multitarget generation jamming as identified by the sudden appearance of multiple targets and/or many targets either stationary or on illogical courses and speeds or maneuvers such as 90° turns at high speeds but with no displacement. Also, targets may appear in back and sidelobe positions.

Music—In air intercept, a term meaning electronic jamming.

N

Narrowband Linear (NBL)—NBL is not really an antijamming feature except that it connects the manual gain control into the system. It is used to vary the receiver gain level at a linear rate.

Narrowband (NB-1)—NB-1 narrows the receiver frequency making it more selective. It limits the target signals and the jamming signal to a set level of amplitude and reduces the level of a jamming signal if the jammer is not tuned to the radar's exact frequency.

Narrowband (NB-2)—NB-2 is the same as NB-1 except that the receiver sensitivity is set 20 dB higher, which increases the resolution of the radar. NB-2 will not exclude all of the jamming signals from display because of the higher sensitivity of the radar.

Navigation Countermeasures—The detection and evaluation of enemy electronic aids to navigation, and the use of jamming and deception to interfere with enemy use of such aids.

Noise—1. Any unwanted disturbance within a dynamic electrical or mechanical system, such as undesired electromagnetic radiation in any transmission channel or device. 2. Uncontrolled random disturbances which arise in a guided missile system as a result of various physical phenomena.

Noise Jamming—Noise jamming is direct (straight), AM, or FM noise on a carrier frequency that has a highly variable bandwidth for the purpose of increasing (saturating) the radar receiver's noise level. Noise jamming is extremely effective against most radars. It may be swept through a frequency spectrum at various rates or spot speeds. Noise jamming is identified by video saturation at all ranges in a sector of the scope; the size of the sector is dependent upon the power and range of the jammer. The techniques of taking random noise and radiating it for the purpose of jamming. May be amplified and radiated directly, or a carrier may be modulated with the noise.

Noise Modulated Jamming—Random electronic noise that appears at the receiver as background noise and tends to mask the desired radar echo or radio signal. See Noise Jamming.

Noise Source—A device employed for the generation of noise. Tubes used for this purpose include photo-multiplier and gaseous discharge types.

Noncommunications Jamming—Electronic jamming used against electronic devices other than those that are used as a means of communications. Examples are: navigational aids, radar, guided missiles, pilotless aircraft, proximity fuzes, and similar devices.

O

Off-Line Jamming (Off-Target)—Jamming from a station away from the vicinity of the target. Necessarily must be active jamming.

On-Line Jamming—Use of the jamming device in line with the target and the radar set. May be either active or passive jamming.

"Out-Phasing" Antenna System—A second antenna or group or antennae used with the basic antenna system in such a manner as to pick up jaming signals; then, with certain phasing adjustments the jamming signal is essentially cancelled as far as the receiver is concerned.

Optical Countermeasures—Applications of electronic countermeasures in the visible light portion of the electromagnetic spectrum. Actions taken to prevent or reduce an enemy's effective use of the visible spectrum. Also called visual countermeasures.

P

Passive Angle Tracking (PAT)—A target may be tracked "passively" if that target emits an electromagnetic radiation; i.e., jamming radio, radar signal of sufficient duration that a DF bearing may be obtained. The emission from the target is DFd in azimuth and/or elevation. No range information will be available unless cross DF techniques are used by two or more passively tracking sites.

Passive Detection and Tracking—By combining azimuth data on jamming strobes from several stations, intersections are obtained which indicate the position of the jammers. The number of ghosts can be reduced by increasing the number of friendly station intersections and obtaining elevation angles of strobes when available.

Passive Electronic Countermeasures—Electronic countermeasures based on the reflection, absorption or modification of the enemy's electromagnetic energy. This distinction between active and passive countermeasures is not currently used, but it is based on the presence or absence of an electronic transmitter.

Passive Jamming—The techniques of deception by employing confusion reflectors to return spurious and confusing signals to the transmitting radar set.

PECM—Passive ECM. See Passive Electronic Countermeasures.

Penetration Aids—Techniques and/or devices employed by aerospace systems to increase the probability of weapon system penetration of an enemy defense. Examples are: low altitude profiles, trajectory adjustments, reduced radar cross sections of attack vehicles, improved vehicle hardness to effects of defense engagements, terrain avoidance radar, bomber defense missiles, decoys, chaff, electronic countermeasures, etc. Penetration aids are used by an offensive system to penetrate enemy defenses more effectively. Also called PENAIDS.

Phantom Signals—Signals appearing on the screen of a cathode-ray tube indicator, the cause of which cannot readily be determined and which may be caused by circuit fault, interference, propagation anomalies, jamming, etc.

Phantom Target—1. An echo box, or other reflection device, that produces a particular blip on the radar indicator. 2. A condition, maladjustment, or phenomenon (such as a temperature inversion) that produces a blip on the radar indicator resembling blips of targets for which the system is being operated.

PIE—See Pulse Interference Elimination.

PLD—Pulse Length Discrimination. (See PWD).

Polarization Diversity—This technique involves the variation of polarization (such as horizontal, vertical, circular, or elliptical for radar use) either simultaneously or singly.

Power Out Control—With Power Out Control the transmitter peak power is variable to give higher average power, longer detection ranges, and increased burnthrough range.

PRF Agility—See Frequency Agility.

PRF Change (Jittered PRF)—The PRF is rapidly varied at a random rate so that false targets appear to jitter or appear fuzzy on the scope. An alternative to jittered PRF is to change the PRF momentarily. This causes the false targets to change their position on the scope.

PRF Stagger—The technique of switching PRF or PRI to different values on a pulse-to-pulse basis such that the various intervals follow a regular pattern. This is useful in compensating for blind speeds in pulsed MTI radars. Interpulse intervals which differ but follow a regular pattern.

Pulse Compression—The coding and processing of a signal pulse of long time duration, to one of short time duration and high range resolution, while maintaining the benefits of high pulse energy. This technique uses matched filter techniques for discriminating against signals that do not correspond to the transmitted code. Its implementation involves stretching the transmitted pulse and compressing the received pulse. It permits an increase in average transmitted power (without increase in peak power) with no loss in range resolution. Pulse compression is a special form of pulse coding and correlation.

Pseudonoise—A modulation technique resulting in low signal selectibility and low vulnerability to ECM.

Pulse Discriminator—Device which responds only to a pulse having a particular characteristic, such as duration, or period.

Pulse Interference Eliminator—Device which removes pulsed signals which are not precisely on the radar operating frequency.

Pulse Interference Suppression and Blanking (PISAB)—This PRF device is an automatic interference blanker. It will blank all video signals that are not synchronous with the radar PRF. It does not require any trigger and operates on both normal and MTI modes. It is effective against random pulse signals.

Pulse Jitter—Random variation of interpulse interval.

Pulse Modulated Jamming—Use of jamming pulses of various widths and repetition rates.

Pulse Width Discriminator (PWD)—This device measures the pulse length of video signals and passes only those whose time duration falls into some predetermined design tolerances. PWD offers good discrimination against long pulse jamming and ECM with low frequency noise modulation. It affords little or no discrimination against short pulses and HF noise modulations. It will give a small gain against barrage jamming, similar to that of a matched filter in the video.

PWD—See Pulse Width Discriminator.

Q

QRC—See Quick Reaction Capability.

Quick Reaction Capability (QRC)—To satisfy urgent operational requirements in support of the service mission. Applies only to HQ-designated electronic warfare, reconnaissance, and intelligence programs for which special management procedures are used.

R

"**Rabbits**"—Interference from another radar on or near the frequency of the receiving radar. Shows on the indicators as interference at the rate of the relative PRF or the interfering radar.

Radar Absorbent Material (RAM)—Radar absorbent material used as a radar camouflage device to reduce the echo area of an object.

Radar Countermeasures—See Electronic Countermeasures.

Radar Deception—See Electronic Deception.

Radar Decoy—Reflecting object used in radar deception, having the same reflective characteristics as a target.

Radar Homing—Homing on the source of a radar beam.

Radar Homing and Warning (RHAW)—Typically, a radar homing and warning system consists of an airborne, wideband crystal video receiver designed to intercept, identify, and display the direction to pulse-type emitters. See Radar Warning Receiver.

Radar Intelligence—1. Intelligence concerning radar or intelligence derived from the use of radar equipment. In this sense, the term has been used with several specific meanings. These are: a. That aspect of electronic intelligence that deals with radar. b. Intelligence concerning the radar aspects of a radar mission, especially a radar-bombing mission. c. Radar target intelligence derived from information procured by means of radar, particularly with regard to bomb damage assessment and bomb scoring. 2. Organization or activity that deals with such intelligence.

Radar Reconnaissance—Reconnaissance by means of radar to obtain information on enemy activity and to determine the nature of terrain.

Radar Silence—An imposed discipline prohibiting the transmission by radar of electromagnetic signals on some or all frequencies.

Radar Warning Receiver (RWR)—A wideband crystal video receiver (usually airborne) designed to intercept, identify, and display the type of and direction to radar emitters. Not a homing sysem. See also Radar Homing and Warning.

Radiation Intelligence (RINT)—Intelligence derived from the collection and analysis of noninformation bearing elements extracted from the electromagnetic energy unintentionally emanated by foreign devices, equipments, and systems, excluding those generated by the detonation of atomic/nuclear weapons.

Radio Countermeasures—See Electronic Countermeasures.

Radio Deception—The employment of radio to deceive the enemy. Radio deception includes sending false dispatches, using deceptive headings, employing enemy call signs, etc. See also Electronic Deception.

Radio Frequency Interference (RFI)—See Electromagnetic Interference.

Radio Frequency Interference—Unintentional interfering signals present in electronic equipment. Although these signals are designated as radio frequency there is no implication that they are transmitted by electromagnetic radiation through space and enter the equipment through an antenna. Radio frequency interference may enter the equipment through the case or may be conducted along power lines or other wires that enter the equipment.

Radio Recognition—The determination by radio means of the friendly or enemy character, or the individuality, of another.

Radio Silence—A period during which all or certain radio equipment capable of radiation is kept inoperative. (In combined or United States joint or intraservice communications, the frequency bands and/or types of equipment affected will be specified.)

Railing—1. Pertains to radar pulse jamming at high recurrence rates (50 to 150 kHz). It results in an image on a radar indicator resembling fence railing. 2. The name given to that pattern produced on an "A"-scope by CW modulated with a high frequency signal. Railings appear as a series of vertical lines resembling target echoes along the baseline.

Rainbow—A technique which applies pulse-to-pulse frequency changing to identifying and discriminating against decoys and chaff.

RAM—See Radar Absorbing Material.

Random Noise—Electromagnetic energy having no particular modulation or pattern. Random noise may be generated by either natural atmospheric phenomena or by electromagnetic radiating devices.

Random Noise (fluctuation noise)—Noise that comprises transient disturbances occurring at random. *Note:* The part of the noise that is unpredictable except in a statistical sense. The term is most frequently applied to the limiting case where the number of transient disturbances per unit time is large, so that the spectral characteristics are the same as those of thermal noise. Thermal noise and shot noise are special cases of random noise.

Random Pulse Jamming—The technique whereby a pulse transmission system is pulsed irregularly by random noise signals.

Range Gate Capture—Electronic countermeasures technique, using a spoofer radar transmitter to produce a false target echo that can make a fire control radar move off the real target and follow the false one.

Read-through—The continuous recovery of the target modulation in an audio channel. A means for rapid evaluation of the effectiveness of a jamming effort, but also of potential value in secure communication systems.

Reconnaissance—A mission undertaken to obtain, by visual observation or other detection methods, information about the activities and resources of an enemy or potential enemy; or to secure data concerning the meteorological, hydrographic, or geographic characteristics of a particular area.

Recovery Time—The time for a part of a receiver to recover to a zero-signal condition after receiving a jamming pulse of saturation intensity. If recovery is accomplished before the next controlling pulse arrives, the jamming is ineffective.

Repeater Jammer—Equipment to confuse or deceive the enemy by causing his equipment to present false information. This is done by a system that intercepts and reradiates a signal on the frequency of the enemy equipment, the reradiated signal being modified to present erroneous data on azimuth, range, number of targets, etc.

RFI—See Radio Frequency Interference.

RHAW—See Radar Homing and Warning.

RINT—See Radiation Intelligence.

ROB—Radar Order of Battle.

Rope—An element of chaff consisting of a long roll of metallic foil or wire which is designed for broad, low-frequency response. See Chaff.

Rope-Chaff—Chaff which contains one or more rope elements. See Chaff.

Rotating Polarization—The result of a rotating feed. This should not be confused with circular polarization where the electric field rotates about the axis of propagation at the radar frequency.

RPD—Random Pulse Discrimination.

RWR—See Radar Warning Receiver.

S

Saturating Signal (AF)—In radar, a signal of an amplitude greater than the dynamic range of the receiving system.

Sawtooth Modulated Jamming—Electronic countermeasure technique when a high-level jamming signal is transmitted, thus causing large AGC voltages to be developed at the radar receiver that, in turn, causes target pip and receiver noise to completely disappear.

Self-Screening Range—Range at which a target can be detected by a radar in the midst of its jamming mask, with a certain specified probability.

Sensitivity Time Control (STC)—Programmed variation of the gain (sensitivity) of a radar receiver as a function of time within each pulse repetition interval or observation time in order to prevent overloading of the receiver by strong echoes from targets or clutter at close ranges. STC reduces the gain of the radar receiver for detection of close-in targets. It is particularly effective in removing close-in clutter and strong nearby signals. When adjusting STC, particular attention must be paid to close-in ranges so that small reflecting surfaces are not eliminated.

Sidelobe Blanking (SLB)—A device which employs an auxiliary wide-angle antenna and receiver to sense whether a received pulse originates in the sidelobe region of the main antenna and to gate it from the output signal if it does. This technique uses an omnidirectional antenna and compares relative signal strength between the omni and the radar antenna. The omnichannel (plus receiver) has slightly more gain than the sidelobes of the normal channel, but less gain than the main beam. Therefore, any signal that is greater in the omnichannel must have been received from a sidelobe, and therefore is blanked. This technique is effective in removing spoofer signals with a duty cycle up to approximately 50 percent.

Sidelobe Canceller (SLC)—A device which employs one or more auxiliary antennas and receivers to allow linear subtraction of interfering signals from the desired output if they are sensed to originate in the sidelobe of the main antenna. This technique employs the same antenna and receiver configuration as the SLB, except that a gain matching and canceling process takes place. Extraneous signals entering the sidelobe of the main antenna canceled while the targets remain. This type of system exhibits cancellation on the order of 20 dB against a single noise jammer. With multiple jammers at various azimuths, the performance of this device rapidly deteriorates.

Sidelobe Jamming—Jamming through a sidelobe of the receiving antenna in an attempt to obliterate the desired signal received through the main lobe of the receiving antenna at fixed points.

Sidelobe Suppression—The suppression of that portion of the beam from a radar antenna other than the main lobe.

Signal Intelligence (SIGINT)—A generic term which includes both communications intelligence and electronic intelligence.

SIGINT—See Signal Intelligence.

Signal Security (DoD)—A generic term which includes both communications security and electronic security.

Signal-to-Jamming Ratio—The ratio of the signal power to the jamming (ECM) power at a particular point in a system. This ratio is often expressed in decibels.

Sine-Wave Modulated Jamming—Jamming signal produced by modulating a CW signal with one or more sine waves.

Single Beam Blanking (SBB)—The SBB feature is used by phased array radars as an alternative method of BBC. It is effective to some degree against many multiple target generators and swept frequency jammers. Because of overlapping beamwidths, a target signal will appear in more than one beam as the beams are scanned past a true target. When jamming signals are transmitted along one beam, that beam is blanked by the radar receiver.

S/J—Signal-to-Jamming Ratio.

Small Scale ECM Mission—Conduct of active airborne ECM by a single aircraft or by 2 to 5 aircraft working as a unit.

SOJ—See Stand-Off Jamming.

SORO—Scan on Receive Only. See Lobe on Receive Only.

Spoking (RADAR)—Periodic flashes of the rotating time base on a radial display. Sometimes caused by mutual interference.

Spoof—Electronic warfare measures that cause an enemy to be misled and commit errors in tactics.

Spoofer—In air intercept, means a contact employing electronic or tactical deception measures.

"Spoofing"—A type of deception by using an electronic device to transmit a "target" echo. The spoofing transmitter must operate at the same frequency and PRF as the radar to be deceived. The radar main pulse triggers the spoofing transmitter which, after a delay, transmits a false echo.

Spot Jamming—The jamming of a specific channel or frequency.

SSJ—Self-Screening Jammer.

Standoff Jamming (SOJ)—An ECM support aircraft orbits in the vicinity of the intended target. As the fighter-bomber pilot starts his strike penetration, the ECM aircraft directs jamming against all significant radars in the area. This technique provides broad frequency band ECM without affecting performance of the strike aircraft.

Staggered PRF—Staggered PRF allows an increase in MTI blind speeds such that no zeros exist in the velocity response at lower velocities. In a two-period mode, the usual "blind speed" or occurrence of a zero in the velocity response is multiplied by a factor which is a function of the ratio of the two repetition periods.

STC—See Sensitivity Time Control.

Stream—1. Dispensing of chaff (solid/random interval/bursts). 2. An unclassified brevity code word signifying chaff dispensing type of active ECM.

Subjamming Visibility—Relates to how well a particular radar antijam technique can see through jamming signals.

Susceptibility—Capability of or capacity for being acted upon. A device is susceptible to any force to which it responds. A system is susceptible to noise jamming if it is of such a nature to admit noise jamming.

Sweep Jammer—Electronic jammer which sweeps a narrow band of electronic energy over a broad bandwidth.

Sweep-Lock-on-Jammer—A transmitter in which a narrow band jamming signal can be tuned over a broad frequency band and the signal locked on a particular frequency.

Sweepthrough—Jamming transmitter that sweeps through a radio-frequency band and jams each frequency breifly, producing a sound like that of an aircraft engine.

Swept Local Oscillator Receiver (SLOR)—SLOR is a modification to the 0A3751/FST-2. The 0A-3751, together with minor modifications to the FST-2, basically provides a capability for receiving and processing jamming energy, differentiating main lobe antenna response from sidelobe responses and for determining and transmitting an accurate estimate of the jammer's azimuth to the BUIC NCC and/or SAGE direction center for further processing.

Synchronized Pulse Jamming—The technique of attempting to insert jamming pulses into a receiver each time the receiver gate opens.

Synchronous Pulsed Jamming—This jammer matches exactly the PRF of the victim radar; it then transmits multiples of the PRF and is most effective if the jammer also matches the PW of the radar. Synchronous pulsed jamming is easily recognized since the spacing between successive target lines is equal and each target line is the same in depth, from the center outward. The width of the jammed sector is dependent upon the range of the jammer from the radar.

T

Tactical Electronic Warfare—That application of electronic warfare to tactical air operations. Tactical electronic warfare encompasses the three major subdivisions of electronic warfare: electronic warfare support measures, electronic countermeasures, and electronic counter-countermeasures.

Tangential Sensitivity on Look-Through—The strength of the target signal, measured at the receiver terminals, which would give a signal pulse twice the apparent height of the noise.

Target Susceptibility—The degree to which a target is capable of being affected by outside forces. A missile is susceptible to energy transmissions to which it reacts.

Target Vulnerability—The resistance to destruction which a target possesses with respect to a weapon which intercepts it.

Time Discrimination Circuits—Employed in video amplifiers for anti-jamming. Pulses may be discriminated according to their widths, their location with respect to the expected repetition rate, the rate of change in this position, and the coincidence of one or more pulses.

TOJ—See Track on Jam.

Track on Jam—A method of passive target tracking using the jamming signal emitted by the target.

"Tramlines"—The name given to that pattern produced on an "A"-scope by CW modulated with a low frequency signal. Tramlines appear as a number of horizontal lines above the baseline.

Two Signal Jamming (also called Straddle Jamming)—A method of jamming whereby two signals are transmitted on two RF frequencies only slightly separated. Effective against certain types of radar where receiver bandwidth is narrow enough to defeat noise jamming.

U

Unintentional Radiation Exploitation (URE)—Exploitation for operational purposes of noninformation bearing elements of electromagnetic energy unintentionally emanated by targets of interest.

Unipolar Video CFAR—This CFAR technique takes advantage of the baseline-break phenomenon in a slightly different form. The device consists of a wideband IF amplifier, followed by an envelope detector and a video limiter. The proper detection bandwidth is established by a low-pass video filter. This technique generally is not as acceptable as the DICKE FIX CFAR.

Unmodulated CW Jamming—Unmodulated CW jamming is the transmission of a high-power carrier frequency that causes an overload effect to occur in the radar receiver. CW jamming is used against bandpass radars that have a limited tuning capability. Unmodulated CW jamming can be identified by a blackening of the scope background (no video present) in a wedge-shaped sector, or by a solid brightening of a wedge or sector, normally exceeding one bandwidth.

URE—See Unintentional Radiation Exploitation.

V

Video Correlator—Radar circuit that enhances automatic target detection capability, provides data for digital target plotting, and gives improved immunity to noise, interference, and jamming.

Video DICKE FIX CFAR—This CFAR technique also provides for a linear mixer to precede a video limiter. The technique is used in conjunction with MTI receivers, and its operation is essentially the same as the preceding fix.

Video Discrimination—Radar circuit used to reduce the frequency band of the video amplifier stage in which is used.

Video Integrator—1. An electric counter-countermeasures device used to reduce the response to non-synchronous signals such as noise and useful against random pulse signals and noise. 2. A device which uses the redundancy of repetitive signals to improve the output signal-to-noise ratio by summing the successive video signals.

Vulnerability—The characteristics of a system which cause it to suffer a definite degradation (incapability to perform the designed mission) as a result of having been subjected to a certain level of effects in an unnatural (manmade) hostile environment. EW vulnerability is the measure of the systems inability to perform in the face of hostile ECM. EW vulnerability can only be measured in its intended operational environment (actual or simulated) and must take into account such factors as: susceptibility of the system, interceptibility by hostile ESM activities, and the nature and extent of the hostile EW threat.

W

Warning Receiver (Electronic Warfare)—A receiver with the primary function of warning the user that his unit is being illuminated by an electromagnetic signal of interest.

Wideband Constant False Alarm Rate Receiver (WB-1)—WB-1 is used against individual or combinations of rapidly swept EM-CW, noise or CW. This mode has nonlinear limiting and gives poor resolution of overlapping targets.

Wideband Limit (WB-2)—WB-2 is used against narrow pulse (Spike) or swept type jamming. This mode uses wideband limiters preceding narrowband filters to prevent filter ringing. It gives better resolution of overlapping targets.

Wide Pulse Blanking (WPB)—The WPB circuit eliminates clutter and noise pulses from the video display in much the same manner as the clutter eliminator circuit.

Window—1. Strips of frequency-cut metal foil, wire or bars which may be dropped from aircraft or missiles, or expelled from shells or rockets as a radar countermeasure. A confusion reflector. 2. A passive radar deception or confusion device; usually consisting of some metallic structure, designed in size and shape to effectively reflect impinging signals, so as to simulate a true target.

Z

Zero-Crossing Counting—This CFAR technique uses a wideband limiting IF amplifier followed by a zero-crossing counter that indicates the target when the rate of crossings falls below a predetermined value. The limiting action of the wideband amplifier and the random nature of the noise permit the use of a fixed threshold detection level that is independent of the jamming signal amplitude.

A Complete Bibliography—Soviet Literature on Radar and EW, Part I
S.L. Johnston
Electronic Warfare Magazine, Vol. 8, No. 8, September/October 1978
Pages 44-54

1.8

Here is a compendium of approximately 300 Russian language books on radar and electronic warfare. Part I is based on subject and key words. Part II is alphabetic by author.

This bibliography represents a search made of the Library of Congress for holdings of Russian language books on radar, EW and related topics. Searches were also made in DDC (Defense Documentation Center) and JPRS (Joint Publications Research Service) for English language translations of the Russian language books. Occasionally the DDC and JPRS searches identified books which may not be in the Library of Congress. Review of the bibliographies in the English language translations also contained additional Russian language books.

From time to time digests of Russian language books on radar and EW have appeared in *Electronic Warfare* Magazine. These articles, while informative, covered only a few of the available books. The IEEE International Cumulative Index on Radar Systems (Ref. 1) cited a number of journal articles in various languages on ECM/ECCM. Several Russian language articles were included. As indicated in Reference 2, Reference 1 did not adequately cover foreign journals/books.

This list is thus a composite of lists from several sources. It includes pamphlets, patents and monographs as well as "books." Following the example of Reference 1, many of the books listed here deal with special topics related to radar; e.g. antennas, rather than a complete radar system.

This list should not be considered as complete in the special topics. The specific special topic books listed should be of special interest since they were referenced in Russian language radar books. The level of books listed here varies from juvenile level thru highly theoretical treatments of special radar topics. Some books are military training texts (training manuals). Some are historical. Still others address policy considerations of radar employment. Those books which are of doubtful engineering interest are marked with an asterisk (*) in the left-hand column. Books with an asterisk were included for completeness since they may be of interest to some readers of this article. A few Russian language translations of English language books were also included.

Stephen L. Johnston
Professional Engineer
Huntsville, Alabama

This list of Russian language books on radar/EW is given in two forms. Part I is a special list and Part II is alphabetic by author. The special list is a combination subject/title key word list. Since a number of the books do not appear to be in the Library of Congress, often only the information in the title was available for preparation of Part I. Several instances of multiple translations of the same Russian language title were found. Frequently, noticeable variations in the title were observed. Some of these are given in parenthesis in this list. Where possible, books were compared by referring to the transliterated title, publisher and date. In Part I, each book has been entered only once. Users of this list should refer to several different headings (subjects/title key words) to find books on a desired topic.

Subject classification of a book based on only one word is recognized as being of dubious validity; but a proper complete cataloging of these books would be an extensive undertaking. Many descriptive words of similar meaning were noted in these titles. These include: basic, fundamentals, principles of, theory of, etc. Many of such books are probably very similar. Books of the same descriptive word are grouped together here in Part I. This, of course, is not fully meaningful, especially since both the author and the translator probably made arbitrary choice of the descriptive words.

No attempt has been made here to identify those radar books that have chapters on EW. Such an effort is planned for a later date when more translations are identified.

Identification call numbers for English language translations are given for about 60 of the approximately 300 Russian language books listed, as well as the Library of Congress call numbers of most books. Some of the translations give only contents or a partial translation.

Assistance of Mr. Steve Dresner of the Library of Congress, and title translation and suggestions of Mr. David K. Barton (Raytheon) are gratefully acknowledged.

References

1. "International Cumulative Index on Radar Systems," IEEE Trans. Aerospace and Electronic Systems, Vol AES-11 No. 3, pp 428ff, May 1975.
2. S. L. Johnston, "Radar Literature Sources," IEEE Trans. Aerospace and Electronic Systems, Vol AES-12, No. 1, pp 71–73, January 1976.

Part I—Subject Listing

Footnote *Denotes short monographs of books of doubtful engineering interest

Library of Congress No. and Accession No. of Translation	Author, Title and Publisher
	AIRCRAFT RADAR
JPRS 46626 (partial)	Yermolaev, G. I. et al., *Fundamentals of Radar, and Radar Equipment for Aircraft,* Moscow: Mashinostroyeniye, 1967. *(Radio Altimeters and Intercept Radar)*
	AIR/ANTI-AIR DEFENSE
UF 625.Dy*	Druzhinin, V. V., *The Radio Technical Troops of the National Air Defense Science,* 1968
UG 632.K7*	Krysenko, G. D., *Modern Systems for Antiaircraft Defense,* Moscow: Voyenizdat, 1966. Methods & Means For Conducting Military Activities
UG 610.L4*	Leonov, A. I., *Anti-Missile Defense Radar,* Moscow: Voyenizdat, 1967, 135 p.
	Peresada, V. P., *Anti-Aircraft Guided Missiles,* Moscow: Voyenizdat, 1961.
	AMPLIFIERS
TK 6590.A5L8 JPRS 52891 (partial)	Lukoshkin, A. P., *Radar Amplifiers for Wideband Signals,* Moscow: Soviet Radio, 1964, 254 p.
	Mamonkin, I. G.: *Pulse Amplifiers,* State Power Engineering Publishing House, 1958
	Rizkin, A. A.: *Fundamentals of Amplifier Circuit Theory,* Soviet Radio, 1958
AD 672 334	Shutenko, M. S., *Low Noise Ultra High Frequency Amplifiers,* Moscow, 1966.
	Yevtyanov, S. I.: *Transient Processes in Receiver-Amplifier Circuits,* State Publishing House of Literature on Communications and Radio Problems, 1948
	ANALOG-DIGITAL CONVERSION
TK 7887.5.F5* JPRS 32957	Filippov, V. G., *Analog-to-Digital Conversion,* Moscow: Voyenizdat, 1965, 143 p.
	ANTENNAS
TK6565. A656	Aizenberg, G. Z., *VHF Antennas,* Moscow, Communications, 1957.
TK 6590.A6B4*	Beketov, V. I., *VHF Antennas,* Moscow, Voyenizdat, 1957, 119 p. (paperback).
TK6565.A63391962 AD 650519	Belotserkovskiy, G. B., *Antennas,* Moscow: Oborongiz, 1962, 492 p.
TK 6590.A6G3*	Galimov, G. K., *Designing Optimal Dual-Reflector Antennas with Scanning Patterns,* Moscow: College of Mines, 1969, 121 p.
TK 6590.A6K3	Karus, A. P., *Duplex Antennas,* 1957.
TK6590.A6N4*	Nelepets, V. S., *Meter Wave Antennas,* Moscow: Voyenizdat, 1957, 70 p.
JPRS 47643 (partial)	Rezikov, G. B., *Antennas for Aircraft and Spacecraft,* Soviet Radio, Moscow
JPRS 44020 (contents)	Zhuk, M. S. and Molochkov Y. B., *Design of Antenna Feeder Devices,* Moscow, Izd-Vo "Svyaz", 1967
	(No Author) *Antenna Switches,* Soviet Radio, 1950, (May be translation of U.S. book)
	ASYMPTOTIC METHODS
	Vakman, D. A.: *Asymptotic Methods in Linear Radar,* Soviet Radio, 1962
	CW RADAR
TK 6580.S86	Supryaga, N. P., *CW Radar,* Moscow: Voyenizdat, 1963, 121 p.
TK 6575.V5	Vinitskiy, A. S., *Fundamentals of CW Radar,* Moscow: Soviet Radio, 1961, 494 p.
	CALCULATIONS IN RADAR
TK 6578.G6	Golev, K. V., *Calculation of Radar Range,* Moscow: Soviet Radio, 1962, 204 p.
	Vystoskiy, B. F. and Voynich, B. A., *Fundamentals of Engineering Calculation of Radar Systems,* Moscow: Av. Inst. 1966.
TK 7885. R3	Rezorenov, N. G., *Electronic Calculating Machines and Tracking Systems,* 1961
	CAMOUFLAGE OF RADAR
UG632.S83 AD 733507 part JPRS 47734 part	Stepanov, Y. G., *Radar Camouflage,* Moscow: Soviet Radio, 1968.
	CATHODE RAY TUBES
	(No Author) *Cathode-ray Tubes and Indicators,* Parts I and II, Soviet Radio, 1950, (May be translation of U. S. book)
	CENTIMETER WAVES
	(No Author) *Bases of Technology of Centimeter Waves in Radar,* Soviet Radio, 1951 (May be translation of U. S. book)
	COMMUNICATIONS THEORY
	Kharkevich, A. A.: *Outlines of General Communication Theory,* Gostekhizdat, 1955
	COMPLEX SIGNALS
TK 6575.V28 Trans: TK 6575.V283, 1968b	Vakman, D. E., *Complex Signals and Uncertainty Principle in Radar,* Moscow: Soviet Radio, 1965; New York, Springer, 1968, 303 p.
TK 6580.L43	Leonenko, I. M., *Methods of Forming and Processing Complex Radar Signals,* Central Scientific Research Institute for Patent Information & Technical Economic Studies
	COMPUTERS
	Dobrogurskiy, S. O., Kazakov, V. A., Titov, V. K.: *Computers,* Oborongiz, 1959
	Fel'dbaum, A. A.: *Computers in Automatic Systems,* Publishing House of Literature on Physics and Mathematics, 1959
	Kitov, A. I. and Krinitskiy, N. A.: *Electronic Digital Computers and Programming,* Publishing House of Literature on Physics and Mathematics, 1959
	Zhdanov, G. M.: *Computers Part II Integrating-Differentiating Devices,* Moscow Power Engineering Institute, 1954

CONTROL, AFC

	Kaplanov, M. R. and Levin, V. A.: *Automatic Frequency Control*, 1953 (Gosenergoizdat)

CONTROL, AGC

	Tartovskiy, G. P.: *Dynamic Systems of Automatic Gain Control*, State Power Engineering Publishing House, 1957

CONTROL, AUTOMATIC

	Gitis, E. I.: *Electrical Radio Automatics*, State Power Engineering Publishing House, 1959
	Pugachev, V. S.: *Theory of Random Fluctuations and its Applications to Automatic Control Problems*, State Publishing House of Technical and Theoretical Literature, 1957
	Solodovnikov, V. V.: *Introduction to Statistical Dynamics of Automatic Control Systems*, State Publishing House of Technical and Theoretical Literature, 1952
	Solodovnikov, V. V.: *Statistical Dynamics of Linear Automatic Control Systems*, Fizmatgiz, 1960
	(No Author): *Fundamentals of Automatic Control*, Mashgiz, 1959

CONTROL, RADIO

TK 6553.B29	Barsukov, F. I., *Radio Control*, 1962
UG 632.G8	Gutkin, L. S., *Principles of Radio Control of Pilotless Objects*, Moscow: Soviet Radio, 1959, 383 p.
UG 632.R32	Gutkin, L. S., *Radio Control of Guided Missiles and Space Vehicles*, Moscow: Soviet Radio, 1968, 679.
TK 6570.C6G87	Gutkin, L. S. (ed.), *Radio Control*, Moscow: Soviet Radio, 1970, 324 p.
UG 632.M3 NASA-TT-F367	Maksimov, M. B. and Gorgonov, G. I., *Radio Control of Rockets*, Moscow: Soviet Radio, 1964.
TK 7800.N65, 1971,3	Plonshii, A. F., *Radio Control*, 1971

COUNTERMEASURE, RADAR

UG 485.A77	Atrazher, M. P., *Combat By Electronic Means*, 1972.
TK 6580.V6 AD 682311	Volzhin, A. N., and Yanovich, V. A., *Radar Countermeasures*, Moscow: Voyenizdat, 1960, 134 p.
UG 485.R39	Razingar, A., *Electronic Countermeasures*, 1971

DETECTION

TK 6553.K593	(Kliuev, NF) Klyuyev, N. P., *Detection of Pulse Signals by Means of Digital Integration*, Moscow: Soviet Radio, 1963.
TK 6580.M55*	Mishchenko, Yu. A., *Zones of Detection*, Moscow: Voyenizdat, 1963, 94 p.
TK 6580.S5 AD 651073	Shirman, Ya, D. and Golikov, V. N., *Basic Theory of Detection of Radar Signals and Measurement of Their Parameters*, Moscow: Soviet Radio, 1963, 277 p.
TK 6580.K36*	Katsenbogen, M. S., *Detector Characteristics*, Moscow: Soviet Radio, 1965.

DICTIONARY

TK 6634.S4	Bargin, B. G. and Buchinskiy, A. S., *Seven-Language Dictionary of Television, Radar and Antennas*, Moscow: Fizmatgiz, 1961, 244 p.

DIGITAL METHODS

JPRS 47627 TK 6580.S2 JPRS 47055 (partial) JPRS 47397 (contents) also 45884 (contents)	Samsonenko, S. V., *Digital Methods of Optimal Processing of Radar Signals*, Moscow: Voyenizdat, 1968, 319 p.
TK 6580.K79	Kuz'min, S. Z., *Digital Processing of Radar Information*, Moscow: Soviet Radio, 1967, 400 p.

DIRECTION FINDING/DISPLAYS

TK 6590.A6G6*	Gorin, B. S., and Spivak, P. U., *Direction Displays*, Moscow: Voyenizdat, 1960, 108 p.
TL 507.M64, Vol 105*	Pestriahov, V. B., (ed), *Study of the Accuracy and Noise Stability of Phased Radio Direction Finders* (Collection of Articles), 1959.

DOPPLER SYSTEMS

	Sytina, N. V.: *Self-Contained Doppler Radio Navigational Instruments*, Soviet Radio, 1957
JPRS 46230 contents JPRS 47102 partial TK 6580.T88 AD 698161	Tuzov, G. I., *Filtering Selection and Processing of Information in Doppler Systems*, Moscow: Soviet Radio, 1967, 255 p.

ECONOMY AND RADAR

H39V82. N 25, 1954 JPRS 47591 (partial)*	Trofinov, K. N., *Radar and its Use in the National Economy—1954*, Znanie. Moscow.

ELECTROMAGNETIC WAVES

	Vaynshteyn, L. A.: *Electromagnetic Waves*, Soviet Radio, 1957

ELECTRONICS

	Kaptsov, N. A.; *Electronics*, Publ. House for Technical-Theoretical Literature, 1954

ELECTROVACUUM DEVICES

	Vlasov, V. F.; *Electrovacuum Devices*, Communications Publications House, 1949

EMITTING SYSTEM

	Zelkin, E. G.: *Design of an Emitting System by the Directional Pattern*, Gostekhizdat, 1963

FLUCTUATION

	Bunimovich, V. I.: *Fluctuating Processes in Radio Receivers*, Soviet Radio, 1951
TK 6563.G87, 1972	Gutkin, L. S.: *Theory of Optimum Methods of Radio Reception in the Presence of Fluctuating Noise*, Soviet Radio, 2nd Edition, 1972
TK 6553.K272	Kanevskiy, Z. M. and Finkel'shteyn, M. I., *Fluctuating Interference and Detection of Pulse Radio Signals*, Moscow: Gosenergoizdat, 1963.
	Stranovich, R. L., *Selected Problems in the Theory of Fluctuations in Radar*, Moscow: Soviet Radio, 1961.

GENERATORS, FREQUENCY

TK 6565.0789	Bychkov, S. I., *Stabilization of Frequency of Generators*, Moscow: Soviet Radio, 1962.
JPRS 44146 (contents)	Zakharov, A. M.: *Decimeter Wave Generators, Moscow, Izd-Vo- "svyaz", 1967.*

GEODOSEY IN RADAR

TA 601.P7 TA 601.P68	Provorov, K. L., *Radio Geodesy:* Lecture Course, Novosibirsk Inst. Geodijii, 19xx; Moscow: Nedra, 1965, 278 p.

GUIDANCE, RADAR

UG 632. T5 JPRS 15381	Tipugin, V. N. and Veytsel', V. A., *Radio Guidance*, Moscow: Soviet Radio, 1962, 749 p.

HARMONIC ANALYSIS

	Zayezdnyy, A. M.: *Harmonic Synthesis in Radioengineering and Electrical Engineering,* Leningrad Electrical Engineering Communications Institute, 1957.

HIGH FREQUENCY COMPONENTS

TK 6585.V55 AD 764333	Vlasov, V. I., and Berman, Ya. I., *Designing High-Frequency Components for Radar Systems.* Leningrad: Sudostroyeniye, 1961, 356 p.
VM 480.V53 AD 764333	Vlasov, V. I., and Berman, Ya. I., *Designing High-Frequency Devices for Radar Stations,* Leningrad: Sudostroyeniye, 1972, 368 p.

HISTORY, RADAR

UG 610.A6*	Anisin, I. I., *The Orange Snow Storm*, 1962
TK 6575.B7*	Brenev, I. V., *The Paths of Radar's Development:* A Shorthand Record of Lectures, Leningrad, 1948.
TK 6576.C45*	Chestnov, F. I., *The Invisible Guide*, 1957
TK 6576.C47*	Chestnov, F. I., *The Far Seeing Echo*, 1968
UG 610.D7*	Drigo, G. A., *The Enemy Has Been Forestalled and Vanquished,* Personal Experience of an Excellent Worker, 1962
TK 6575.D9*	Dzhigit, I. S., *Radar: A Shorthand Record of Public Lectures,* Moscow, 1948
UG 610.K47	Khoroshilov, P. E., *It Began Thusly—* History of the Creation and Testing of Apparatus for the Radiolocation of Aerial Objects in the USSR in the Early 1930's, 1970.
TK 6575.L6*	Lobanov, M. M., *From the History of Radar*, Moscow: Voyenizdat, 1969, 212 p.
TK 6575.075*	Oschchepkov, P. K., *Life and Dreams:* The Memoirs of an Engineer Inventor, Designer and Scholar, 1967.
TK 6576.T73*	Trofimov, K. N., *A Matter of Great Importance,* Moscow: Soviet Radio, 1959, 255 p.

INDICATORS

	Rakov, V. I., *Indicating Devices of Radar Stations,* SUDPROMGIZ, Leningrad, 1962
AD 821242	Slutskiy, V. Z. et al., *Principles of Radio Engineering and Radar,* Vol. 4, *Indicators, Rectifiers and Semiconductors,* Moscow: Voyenizdat, 1967, 425 p. (see Vol. 1, Kalashnikov).

INFORMATION THEORY

JPRS 47692 (contents) TK 6580.K6 AD 688562 JPRS 46772 (better contents)	Kogan, I. M., *Information Theory and Problems of Short-Range Radar,* Moscow: Soviet Radio, 1968, 144 p.

INTERFERENCE, RADAR

TK 6580.F33	Fal'kovich, S. E., *Reception of Radar Signals in a Background of Fluctuating Interference,* Moscow: Soviet Radio, 1961, 310 p.
TK 6563.G87	Gutkin, L. S., *Theory of Optimum Methods of Radio Reception in the Presence of Fluctuating Interference,* Moscow: Gosenergoizdat, 1961.
TK 6570.C6M35	Maksimon, M. B., *Interference Immunity of Multichannel Radio Control and Guidance,* Moscow: Soviet Radio, 1970, 341 p.
TK 6580.M58	Mityashev, B. N., *Determining Time Delay of Pulsed Signals in Interference,* Moscow: Soviet Radio, 1962, 198 p.
TK 6575.L46	Leonenko, I. M., *Methods of Preventing Radar Interference,* 1966
UG 610. T7	Trofimov, K. N., *Radar Interference,* Moscow: DOCAAF, 1962, 74 p.

INTEGRATION/INTEGRALS

	Khurgin, Ya K., and Yakovlev, V. P.: *Methods of the Theory of Integral Functions in Radar Physics, Communications Theory and Optics,* Fizmatgiz, 1962
	Krayzmer, L. P.: *A Device For Accumulating Discrete Information,* State Power Engineering Publishing House, 1961
JPRS 38037	Lezin, Yu. C., *Optimal Filters and Integrators for Pulsed Signals,* Moscow: Soviet Radio, 1963.
TK 6580.T75	Tsivlin, I. P., *Electronic Range Finder with Double Integration,* Moscow: Soviet Radio, 1964, 183 p.

JAMMING, RADAR

TK 6575.563 JPRS 46664	Baklashov, P. I., *Radar Jamming and Protection of Radar Stations against Interference,* Chap. XI in Pedak, 1968.
UG 610.V3 AD 692642 Vol 1 AD 692643 Vol 2	Vakin, S. A. and Shustov, L. N., *Principles of Jamming and Electronic Reconnaissance,* Moscow: Soviet Radio, 1968, 444 p.

LASER RADARS

JPRS 50123	Petrovskiy, V. I. and Pozhidayev, O. A.: *Laser Radars,* Publishing House of the Ministry of Defense, 1969, 104 p.

LINES, TRANSMISSION

TK 6587.M3	Markov, I. P., *Transmission Lines.* Military Publishing House, Moscow, 1958.
	(No Author) *Lines of Transmission of Centimeter Waves,* Parts I and II, Soviet Radio, 1951 (May be translation of U.S. Book).

MAGNETRONS

K7872.V3B9	Bychkov, S. I., *Magnetron Transmitters,* Moscow: Voyenizdat, 1955.
	Fisk, Khagsturi, Gatman, *Magnetrons,* Soviet Radio, 1948.

	(No Author) *Magnetrons in the Centimeter Range,* Part I, Soviet Radio, 1950, (May be translation of U.S. Book).

MARINE RADARS

VG 76.L3 AD 685415	Latinsky, S. M., *The Deviations of Marine Radars,* Leningrad: Sudostroyeniye, 1966.

MEASUREMENT, RADAR

TK 6580.B35	Basharinov, A. E., *The Use of Wideband Sounding Signals in The Measurement of Range and Speed:* Synopsis of Lectures, 1965.
TK 6575.K65*	Korosielev, A. A., *Automatic Coordinate Measurement,* Moscow: Voyenizdat, 1961, 100 p.
TK 6580.L9	Lykov, I. A., *Coordinates Determined by Radar,* 1957.
VK 560.S27	Saibel', A. G., *Basic Theoretical Accuracy of Radio Engineering Methods of Position Determination,* Moscow: Oborongiz, 1958, 52 p.

MEMORY DEVICES

	Prokudin, V. A.: *Memory Devices For Digital Computers,* Soviet Radio, 1961.

METEOROLOGY, RADAR/METEORS

QB 743.F5*	Fialko, E. I., *Radar Methods of Observing Meteors,* Moscow: Soviet Radio, 1961, 111 p.
TK 6575.K64 AD 765363	Korol', O. G. and Chernyak, R. D., *Basic Radar and Meteorological Equipment,* Leningrad: Gidrometeoizdat, 1971, 333 p.
AD 402589	Nelepets, V. S. and V. S. Stepanenko: *Radar Methods of Meteorological Observations,* (Gidrometerologicheskoye) Tzdatel' Stvo, Leningrad, 1961, 175 p.
QC 973.5.57 JPRS	Stepanenko, V. D., *Radar in Meteorology* (2nd ed.), Gidrometeorizdat, 1973, 343 p.

MODULATORS

TK 6578.G37	Garber, I. S., *Magnetic Pulse Modulators,* Moscow: Soviet Radio, 1964, 160 p. (paperback).

MONOPULSE RADAR

TK 6592.M6K88	Kuzin, E. V., *Monopulse Radar,* Moscow: Voyenizdat, 1969, 100 p. (paperback).
TK 6592.M6L4 AD 742696	Leonov, A. I. and Fomichev, K. I., *Monopulse Radar,* Moscow: Soviet Radio, 1970, 392 p.
TK 6592.M6S94 AD 698160 (FSTC) AD 765545 (FTD)	Sviridov, E. F., *Comparative Efficiency of Monopulse Direction-Finding Systems,* Leningrad: Sudostroyeniye, 1964, 115 p.

MOVING TARGET INDICATORS

TK 6575.B35* AD 651619 (FTD) AD 706486 (FSTC)	Bakulev, P. A., *Radar Methods of Selecting Moving Targets,* Moscow: Oborongiz, 1958, 98 p.
TK 6580.B3 AD 644144	Bakulev, P. A., *MTI Radar,* (Radar Location of Moving Targets) Moscow: Soviet Radio, 1964, 335 p.
UG 610.V5 AD 826182	Vishin, G. M., *Selection of Moving Targets,* Moscow: Voyenizdat, 1966, 274 p.
TK 6580.V56	Vishin, G. M., *Multiple-Frequency Radar,* Moscow: Voyenizdat, 1973, 92 p.

NAVIGATION RADAR

VK 397.A4	Aizinov, M. M., et al., *Radio Engineering and Radio Navigation Equipment,* Leningrad: Morekoy Transport, 1962, 474 p.
VK 560.N7	Gerasimov, N. V. et al., *New Techniques of Ship Navigation,* Moscow: Transport, 1973, 264 p.
TK 6575.R27	Mirimanov, R. E. (ed.), *Radar and Radio Navigation,* Moscow: Vsesoyuzmyi Inst., 1972, 397 p.

NOISE

TK 6580.B36*	Basharinov, A. E., *Separation of Radar Information from Received Signals in Noise.* Moscow: Energet. Inst., 1966, 61 p.
McGraw-Hill (1959)	Kotel' nikov, V. A., *The Theory of Optimum Noise Immunity,* Moscow: Gosenergoizdat, 1956.
TK 6565.V3 Trans: TK 6565.V3B	Vaynshteyn, L. A. and Zubakov, V. D., *Extraction of Signals from Noise,* Moscow: Soviet Radio, 1960; Prentice-Hall, 1962, 381 p.

NON-LINEAR METHODS

	Itskhoki, Ya S.: *Nonlinear Radio Engineering,* Soviet Radio, 1955.
	Kharkevich, A. A.: *Nonlinear and Parametric Phenomena in Radio Engineering,* State Technical Press, 1956.

NUMERICAL METHODS

TK 6580.L54	Likarev, V. A., *Numerical Methods and Design of Radar,* 1973.

OPERATOR, RADAR

TK 6575.F6* AD 758253	Fomin, L. A., *A Manual for Radar Operators,* Moscow: DOSAAF, 1970, 235 p.
TK 6580.L38	Latinsky, S. M., *Theory and Practice of Radar System Operation,* Moscow: Soviet Radio, 1970, 432 p.
TK 6580.S55*	Smotritskii, E. A., *The Radar Operator,* 1957.

OSCILLATIONS

	Gorelik, G. S.: Oscillations and Waves, Fizmatgiz, 1959.
	(No Author) *Generation of Electric Oscillations of Special Form,* Parts I and II, Soviet Radio, 1951, (May be translation of U.S. books).

OSCILLOGRAPHS

TK 7872.C27M3	Markov, I. P., *Electronic Oscillographs,* Military Publishing House, Moscow, 1959.

OVER THE HORIZON RADAR

TK 6580.M54 JPRS 56925	Mishchenko, Yu. A., *Over-the-Horizon Radar,* Moscow: Voyenizdat, 1972, 96 p.

PHASE MODULATION/RELATIONS

	Aseyev, B. P.: *Phase Relations in Radioelectronics, Svyaz'izdat Press, 1954.*
TK 5101.P387 JPRS 47692 (contents)	*Pestryakov, V. B., Phase-Modulated Radio Systems,* Moscow: Soviet Radio, 1968, 466 p.
TK 5102.5. V28	Vakman, D. E., *Regular Method of Synthesizing Phase-Coded Signals,* Moscow: Soviet Radio, 1967, 96 p.

PLASMA DEVICES

JPRS 36,717 (partial)	Bernashewskiy, G. A., et al. *Plasma and Electronic Super-High Frequency Amplifiers and Oscillators.* Sovietskoye Radio, Moscow, 1965.

POLARIZATION OF SIGNALS

TK 6580.K3 JPRS 46624 (partial)	Kanareikin, D. M., *Polarization of Radar Signals,* Moscow: Soviet Radio, 1966, 440 p.

PROBABILITY THEORY

	Kunin-Barkovskiy, I. V.: *Probability Theory and Mathematical Statistics in Engineering,* State Publishing House of Technical and Theoretical Literature, 1955.

PROCESSING, SIGNAL

TK 6575.S48 AD 756599	Sloka, V. K., *Problems of Processing Radar Signals,* Moscow: Soviet Radio, 1970, 256 p.

PROPAGATION

	Arenberg, A. G., *Propagation of Decimeter and Centimeter Waves,* Moscow: Soviet Radio, 1957.
	Feynberg, Ye. L., *Propagation of Radio Waves Along the Earth's Surface,* Moscow: USSR Acad. Sci., 1961.
Microfilm Slavic 962 TK	Vvedenskiy, B. A. and Arenberg, A. G., *Problems in the Propagation of UHF Waves,* Moscow: Soviet Radio, 1948.

PULSED DEVICES

TK 7835.B38	Belotserkovskiy, G. B., *Basic Pulse Techniques and Radar,* Leningrad: Sudostroyeniye, 1965, 458 p.
TK 7835.G8*	Gusev, V. V., *Pulse Generators,* Moscow: Voyenizdat, 1958, 100 p.
TK 7835.185	Itskhoki, Ya. S., *Pulse Techniques,* Moscow: Soviet Radio, 1949, 295 p.
	Itskhoki, Ya. S: *Pulsed Devices,* Soviet Radio, 1959.
	Meyerovich, L. A. and Zelichenko, L. G.: *Pulse Engineering,* Soviet Radio, 1953.
	Morugin, L. A.: *Pulse Devices With Delayed Feedback,* Soviet Radio, 1961.
	Petrovich, N. T. and Kozyrev, V. A.: *Generation and Conversion of Electrical Pulses,* Soviet Radio, 1954.
	Tsypkin, Ya. Z., *Theory of Pulse Systems,* Fizmatgiz, 1958.
	Vasil' yev, V. P. et al., *Designing Radar Pulsing Equipment,* Gosenergoizdat, 1963.

RADAR (GENERAL)

TK 6575.C48, 1955	Chestnov, F. I., *Radar* 2nd Ed., 1955.
TK 6575.C48, 1958	Chestnov, F. I., *Radar,* 1958.
TK 6575.M35*	Matlin, I. I., *Radar*: A Textbook For Soldiers and Sergeants, 1960.
481T (Microfilm Slavic)	Shamshur, V. I., *Radar,* 1949.
TK 6575.S6	Solodyazhnikov, N. N., *Radar,* Moscow: Gosenergoizdat, 1956, 471 p. Textbooks for Technical Schools
TK 6575.V6*	Vodopyanov, F. A., *Radar,* Moscow: Gosplanizdat, 1946, 159 p.
TK 6575.T7*	Trofimov, K. N., *Radar,* Moscow: Voyenizdat, 1957, 102 p.

RADAR APPLICATIONS

TK 6576.K85*	Kurhandskii, S. D., *Radar and Its Military Applications,* 1959.
TK 6550.L5 AD 613964	Listov, K. M. and Trofimov, K. N., *Radio and Radar Engineering and Its Applications,* Moscow: Voyenizdat, 1960, 423 p.
UG 610.S55	Sluchevskiy, B. F., *Radar and Its Applications,* Moscow: Voyenizdat, 1962, 258 p.
TK 6576. T7, 1956*	Traskin, K. A., *Radar Engineering and Applications,* Moscow: Voyenizdat, 1956, 139 p.

BASIC RADAR

TK 6575.K7 AD 299208 JPRS 16690 (partial) NEW FTD TRANS	Kromin, G. S. and Katkov, E. A., *Basic Radar Techniques,* Moscow: Voyenizdat, 1956. FTD-ID(RS)T-0102-76
TK 6563.L4463	Levichev, V. G. et al., *Basic Radio and Radar: Vol. 3, Transmitting and Receiving Apparatus,* Moscow: Voyenizdat, 1962, 494 p. (See Vol. 1, Kalashnikov)
TK 6575.M5	Migulin, V. V., *Lectures on Basic Radar,* Moscow Univ. Press, 1958, 120 p.
TK 6575.N4	Nelepets, V. S. and Belotserkovskiy, G. B., *Basic Radar,* Moscow: Oborongiz, 1954, 303 p. Textbook for Radar Equipment Designers
TK 6575.N43	Nemets, A. A. and Fedotov, V. I., *Basic Radar and Television,* Moscow: Vyssh. Shkola, 1971, 350 p.
TK 6575.S63 JPRS 47864 JPRS 46664 (partial)	Pedak, A. M. and Druzhinin, V. V., *Handbook of Basic Radar Engineering,* Moscow: Voyenizdat, 1967, 766 p. See Behloshov.

RADAR EQUIPMENT

TJ 7M63vyP163*	Basharinov, A. E., (ed.), *Radar Equipment,* Moscow: Energet. Inst., 1973, 158 p. (Collection of Articles)
TK 6575.B42	Belotserkovskiy, G. B., *Radar Equipment,* Moscow: Oborongiz, 1961, 431 p.
TK 6575.G74*	Grigor'iants, V. G., *Introductory Course in Radar Equipment,* Moscow Univ., 1962, 177 p.
TK 6553.K74	Krivitskiy, B. Kh., *Automatic Systems in Radio and Radar Equipment,* Moscow: Gosenergoizdat, 1962.
TK 6585.V9 AD 678823	Vystoskiy, B. F. and Kharybin, A. I., *Radar Equipment,* Moscow: Oborongiz, 1960.

RADAR FUNDAMENTALS/ FOUNDATIONS

	(No Author) *Concise Foundations of Radar,* Soviet Radio, 1951, (May be translation of U.S. Book)
TK 6575.B64	Bogomolov, A. F., *Radar Fundamentals,* Moscow: Soviet Radio, 1954, 302 p.
TK 6575.B68*	Breytbart, A. Ya., *Brief Fundamentals of Radar,* Moscow: Soviet Radio, 1951, 134 p.
TK 6550.075	Kalashnikov, A. M., (ed.), *Fundamentals of Radio Technology and Radar,* 4 Vols, 1961.
TK 6550.0752	Kalashnikov, A. M., (ed.), *Fundamentals of Radio and Radar,* Second Edition, 1962–6.
TK 6550.0753	Kalashnikov, A. M., (ed.), *Fundamentals of Radio and Radar,* Third Edition, 1965.
TK 6575.S32 AD 298779 N64-24634	Saibel', A. G.', *Radar Fundamentals,* Moscow: Soviet Radio, 1961, 384 p.

TK 6575.S46	Sivers, A. P.: *Fundamentals of Radar,* 1956.
TK 6575.S465 AD 727869	Sivers, A. P., et al., *Fundamentals of Radar,* Leningrad: Sudostroyeniye, 1959, 350 p.
TK 6575.08	(No Author), *Fundamentals of Radar Technology,* translated from English.

RADAR HANDBOOK

AD 716660	Tsylov, V. Ya, et al., *Handbook of Radar Engineering Fundamentals,* 1967.
TK 6550.S67	Zimin, G. V., *Handbook-Workbook of Basic Radio and Radar,* Moscow: Voyenizdat, 1967, 527 p.

RADAR, JUVENILE

TK 1975* Microfilm Slavic	Kubarkin, L. V., *The Invisible Scouts: Radar* (Part of Polytechnical Library for Young People), 1959.

MODERN RADAR

TK 7800.N65, 1972	Smirnov, G. A. and Panov, V. I., *Modern Radar,* Moscow: Znaniye, 1972.

PASSIVE RADAR

	Maryshkin, Ye. A., *Radiometry,* Moscow: Voyenizdat, 1961.
TK 6580.N5 AD 750855	Nikolaev, A. G. and Pertsov, S. V., *Radiometry: Passive Radiolocation,* Moscow: Soviet Radio, 1964, 334 p.

RADAR AND PICTURES

TK 6576.R38	Razmakhnin, M. K., *Radar Without Formulas but With Pictures,* Moscow: Soviet Radio, 1971, 127 p.

RADAR PRINCIPLES

TK 6575.B38, 1949*	Bazhanov, S. A., *What is Radar?,* Moscow: Voyenizdat, 1949, 182 p.
TK 6550.0754	Kalashnikov, A. M. and Stepuk, Ya. V., *Principles of Radio Engineering and Radar,* 4th ed., Vol. 1: *Oscillatory Systems,* Moscow: Voyenizdat, 1972. See Vol. 2, Kalashnikov; Vol. 3, Levichev; Vol. 4, Slutskiy.
	Kalashnikov, A. M. and Slutskiy, V. Z., *Principles of Radio Engineering and Radar,* Vol. 2: *Electronic Vacuum Tube Devices and Pulse Techniques,* Moscow: Voyenizdat, 1972. See Vol. 1 above.
TK 6575.P44*	Perlya, Ya. Z., *How Radar Works,* Moscow: Oborongiz, 1955, 118 p.
	(No Author) *Parts and Elements of Radar Stations,* Parts I and II, Soviet Radio, 1952, (May be translation of U.S. Book)
	(No Author) *Principles of Radar,* Parts I and II, Soviet Radio, 1949, (May be translation of U.S. Books)

PRACTICAL RADAR

TK 6576.A7	Artem' ev, I. A., *Practical Radar,* Moscow: Detskoye lit-ry, 1960, 286 p.

RADAR PROBLEMS

TK 6580.V36 AD 771650	Vasin, V. V. and Stepanov, B. M., *Book of Problems in Radar,* Moscow: Soviet Radio, 1969, 178 p.
TK 6580.Z3	Aksel' rod, S. M., *Problems in Radio and Radar Engineering,* Moscow: Gosenergoizdat, 1962, 414 p.
	(No Author) *Problems of Radar Technology,* Collection of Translations for 1952–1955, Foreign Literature Publishing House.

RADAR RECEIVERS

	Chistyakov, N. I., et al., *Radar Receivers,* Moscow: Communications, 1959.
	Fradkin, S. L., *Principles of Theory and Calculation of Radar Receivers,* Moscow: Mashinostroyeniye, 1969, 199 p.
	Sivers, A. P.; *Radar Receivers,* Soviet Radio, 1952.
JPRS 56143	Sivers, A. P., *Radar Receivers: Design and Planning,* Moscow: Soviet Radio, 1959.
	(No Author) *Receivers for Radar Stations,* Parts I and II, Soviet Radio, 1949, (May be translation of U.S. Book)

RADAR RELIABILITY

TK 6580.B7	Braudo, S. I., *Preserving Reliability of Radar Equipment,* Moscow: Soviet Radio 1965, 470 p.

RADAR, STORV

TK 3855* Microfilm Slavic	Plotnikov, L. T., *Repair of the "STVOR" Radar,* 1963.

RADAR, SHORT RANGE

TK 6580.K58	Kogan, I. M., *Short-Range Radar,* Moscow: Soviet Radio, 1973, 272 p.

RADAR, SIDE LOOKING

AD 787070	Reutov, A. P. and Mikhailov, B. A., *Side-looking Radar,* Moscow: Soviet Radio, 1970, 360 p.

RADAR SURVEY

UG 610.S7*	Stepanov, B. M., *Radar Survey,* Moscow: Voyenizdat, 1959, 64 p.

RADAR SYSTEMS

TK 6575.K63 AD 771077	Kontorov, D. S. and Golubev-Novozhilov, Yu. S., *Introduction to Radar System Engineering,* Moscow: Soviet Radio, 1971, 367 p.
TK 6575.R28	Vasin, V. V., et al., *Radar Systems: Theory and Design Principles,* Moscow: Soviet Radio, 1970, 680 p.

RADAR TECHNICAL CHARACTERISTICS

TK 6580.G7	Grigor'iants, V. G., *Technical Characteristics of Radar Systems,* Moscow: Voyenizdat, 1963, 238 p.

RADAR TECHNOLOGY

TK 6261 Microfilm Slavic	Belousov, V. M., *Radar Technology:* Survey of Foreign Patents, 1963.
	(No Author) *Radar Technology,* Parts I and II, Soviet Radio, 1949, (May be translation of U.S. Book)
	(No Author) *Technology of Radar,* Parts I and II, Voyenizdat, 1949 (May be translation of U.S. Book)
	(No Author) *Bases of Radar Technology,* Parts I and II, Oborongiz, 1951, (May be translation of U.S. Book)

RADAR THEORY

TK 6575.D77 AD 673149	Dulevich, V. E. (ed.), *Theoretical Principles of Radar*, Moscow: Soviet Radio, 1964, 731 p.
	Karpeng'ye, M., *Modern Theory of Radar*, Moscow: Soviet Radio, 1965.
TK 6573.M6	Moscow Aviation Inst., *Theory and Technique of Radar*, Moscow: Mashinostroyeniye, 1966.
JPRS 56143 JPRS 47690, 1966 (part)	Saibel', A. G., *Radar Theory and Techniques*, Moscow: Mashinostroyeniye, 1971.
	Saybel, A. G., *Radar Theory and Techniques*, Mashinostroyeniye Publishing House, Moscow, 1966.
TK 6575 S455	Shirman, Ya. D., *Basic Theory of Radar*, Moscow: Soviet Radio, 1970, 560 p.

RADAR TEXTBOOK

TK 6575.M64*	Motorichev, I. A. and Smirnov, V. N., *Illustrated Textbook of Radar*, Moscow: Voyenizdat, 1957, 151 p.

RADAR AND TRANSPONDERS

TK 6580.S54*	Smirnov, G. D. and Gorbachev, V. P., *Radar Systems with Active Transponders*, Moscow: Voyenizdat, 1962, 113 p.

RADAR VISIBILITY

VK 560.P4	Peresada, V. P., *Radar Visibility of Marine Objects*, Leningrad: Sudostroyeniye, 1961, 158 p.

RADIO ENGINEERING CIRCUITS

	Gonorovskii, I. S., *Radio Engineering Circuits and Signals*, Soviet Radio, 1964.

RADIO ENGINEERING FUNDAMENTALS

	Aseyev, B. P., *Fundamrntals of Radio Engineering*, Svyaz'izdat, 1947.
	Gonorovskiy, I. S., *Fundamentals of Radio Engineering*, State Publishing House of Literature on Communications and Radio Problems, 1957 Svyaz'izdat.
	Kotel'Nikov, V. A. and Nikolayev, A. M., *Fundamentals of Radio Engineering*, State Publishing House of Literature on Communications and Radio Problems, 1950.
	Krylov, N. N., *Theoretical Fundamentals of Radio Engineering*, Morskoy Transport, (Marine Transportation Press), 1953.

RADIO ENGINEERING HANDBOOK

	Spirenin (ed.), *Handbook on Radio Engineering*, Gosenergoizdat, 1950.

RADIO, JUVENILE

TK 6550.R28*	Berg, A. I., (ed.), *Radio for Radio Enthusiasts*, Collection of reports on selection and design, published in RADIO 1968–1970, 1974.
TK 6550.R33115*	Kabanov, N. I. (ed.), *Material for the Symposium of Young Scientists and Specialists*, Novosibirsk, 1968, 176 p.

RADIO PROBLEMS

TK 6553.G68	Gorinshtein, A. M., *The Numerical Solution of Problems in Radio & Communications Engineering*, 1972.

RADIO RECEIVERS

	Siforov, V. I., *Radio Receivers*, Voyenizdat, 1954.
	Izyumov, N. M., *Radio Reception*, Voyenizdat, 1954.

RADIO SYSTEMS ENGINEERING

TJ7.M63vyp117	Penin, P. I. (ed.), *Radio Systems Engineering*, Moscow: Voyenizdat, 1972, 170 p. Collection of Articles.
TK 6575.R29*	Kazarinov, Yu. M., et al., *Radio Systems Engineering*, Moscow: Soviet Radio, 1968, 496 p. Textbook for Tech schools

RADIO TRANSMITTERS

	Drobov, S. A., *Radio Transmitters*, Voyenizdat, 1951.
JPRS 44146 (contents)	Yevsey, L'Vovich Okun': *Radio Transmitting Equipment*, Leningrad Izd-Vo "Sudostroyeniye", 1967.

RADIO TRANSLATIONS

TK 7800.R42	(No Author) *Radio and Electronics Abroad:* Collection of translations and synopses from foreign periodicals, bimonthly-began 1952.

RANDOM PROCESSES

K6553.L45	Levin, B. R., *Theory of Random Processes and Its Application in Radio Technology*, Moscow: Soviet Radio, 1960. 2nd ed. 1957.

RECONNAISSANCE

UG 610.K3	Kalashnikov, E. V., *Reconnaissance and Target Indication*, 1967.

RECORDING

	Ioffe, A. F., *Applications of Magnetic Recording*, State Power Engineering Publishing House, 1959.

REFLECTION, GROUND

JPRS 47664 (partial) JPRS 47108 (partial)	Zubkovich, S. G., *Statistical Characteristics of Radio Signals Reflected from the Ground*, Moscow: Soviet Radio, 1968, 224 p.

■

Chapter 2
General ECCM

Summary

Three groups of "general" ECCM papers are presented. The first group addresses surveys of multiple ECCM applications; e.g., both ECCM in the transmitter and ECCM in the receiver. The second group illustrates special ECCM considerations, while the third group concerns ECCM in systems. The remaining chapters in this book will follow the general order of ECCM treatment of Paper 1.1.

Synopsis of Reprint Papers in Chapter 2

A. ECCM Survey Papers

2.A.1 "Radar Interference and Its Reduction"—Brick and Galejs

This is a classic paper, now twenty years old. It addresses both ECCM in the receiver and ECCM in the transmitter with extensive references. Many of the ECCM techniques discussed in this paper originated at the MIT Radiation Laboratory during World War II and were reported in that laboratory's post-war series of radar books. Brick and Galejs organized ECCMs according to their *basic* function: spatial selectivity, frequency selectivity, amplitude selectivity, time selectivity and signal selectivity. This should be contrasted with Johnston's approach, (Paper 1.1) used in this reprint volume, which focuses on principal point in the radar where the ECCM is applied. Both approaches have merit. Numerous ECCMs for each basic ECCM method are listed.

2.A.2 "Tools and Techniques of Electronic Warfare"-- Holahan

Both ECM and ECCM are addressed. Several deception ECM techniques are described. These should be contrasted with ECM techniques treated in Chapter 1. The paper contains a small ECM/ECCM matrix (cf Table VI, Paper 1.1). Of the several ECCMs discussed, Holahan makes a case for frequency agility as the most effective.

2.A.3 "Electronic Dogfight: ECM vs ECCM in Search Radar"—Dax

As Tables VI and VII of Paper 1.1 indicate (see pages 16 and 17 of this volume), ECM can take many forms. Figure 1 in this paper shows basic-jamming (more properly active denial ECM) methods. Applicability of these ECMs to search radar is treated by this paper with various ECCMs. ECCMs discussed include prevention of receiver saturation, frequency agility, instantaneous gain control, moving target indicators (against chaff) and passive direction finding. It is pointed out that the ECM designer has many problems such as achievement of long pulse delays in repeater jammers.

2.A.4 "Counter-countermeasures: Good Design Can Beat the ECM Threat"—Simpson

This paper is similar to the two previous papers. Scope displays of multiple false target deception ECM are illustrated. Effects of receiver saturation under jamming conditions are shown. Rapid tunability of the radar (frequency agility) is stated to be one of the strongest ECCM design tools. Utilization of cosine tapered primary distribution of antenna illumination is shown to lower sidelobes by 7 dB at the expense of lower main beam gain and broadened main beam width. Improvement of signal detection is indicated to result from linear integration of several radar sweeps.

B. ECCM Considerations

2.B.1 "Philosophy of ECCM Utilization"—Johnston

Several interesting points are made on various aspects of ECCM: ECCMs may be either electronic or *non* electronic. A complete radar could be considered to be an ECCM. ECCM should be an integral part of the design of a radar.

2.B.2 "Radar ECCM: A European Approach"—Radford

This important paper by a British author indicates several deficiencies of phased array (electronic scanning) radars in an ECM environment as compared to mechanically rotating search radars: narrower frequency bandwidth, higher sidelobes, higher losses, and higher costs. Radford states that both types of radars have constraints but of different forms. This paper was first presented at ELECTRO-76.

C. ECCM Applications

2.C.1 "Guided Missile System ECM/ECCM"– Johnston

A generic guided missile system is presented. Typical ECM methods with representative ECCM techniques are discussed for target detection radars, tracking radars, missile guidance systems and missile command transmission. Extensive use is made of over 20 of the reprint papers in this volume.

2.C.2 "Performance of Current Radar Systems in an EW Environment"– Prior and Woodward

This paper discusses the likely performance of current radar systems in the threat environment posed to NATO and indicates the trend for the future. Radars considered include those used for long range early warning (ADGE), Airborne Early Warning (AEW), Airborne Intercept (AI) and surface-to-air missile system (SAM) acquisition and tracking. Typical present-day characteristics for each type radar are defined, their vulnerability to the EW threat is discussed and the benefits to be obtained from ECCMs are pointed out. Theoretical and operational factors are considered. An indication is given of the future trends in the design of such systems where relevant to their operation in an ECM environment, and the role of the more modern ECCMs arising from new technologies is described. This paper is based on the ECM "threat" as detailed by Steer (Paper 1.6). It was presented at the Military Microwaves Conference, October 1978.

2.C.3 "ECCM From the Radar Designers's Viewpoint" – Johnson and Stoner

This paper details the application of a number of ECCM techniques in a current radar. Most of the techniques described here employ analog, rather than digital, mechanization. Radar scope photo performance results show that ECCM can be effective. This paper was first presented at ELECTRO-76.

Bibliography Topics for Chapter 2

General ECCM

BIBLIOGRAPHY FOR CHAPTER 2

ECCM TECHNIQUES, GENERAL

1. Books

Schlesinger, R.J., *Principles of Electronic Warfare,* Prentice-Hall, 1961.

Skolnik, M.I., *Introduction to Radar Systems,* Mc-Graw-Hill, 1962 (ECM and ECCM: section 12.10, pp. 559-67).

Nemirovskiy, M.S., *Noise Immunity of Radio Communication,* Moscow: Energiya, 1966, in Russian.

Vasin, V.V., O.V. Vlasov, et al, *Radar Equipment (Theory and Principles of Construction),* Moscow, Sovetskoye Radio, 1970, in Russian. (Chapter 18: Interference immunity of radar systems).

Zyuko, A.G., *Noise Immunity and Efficiency of Communications Systems,* Moscow: Svyaz, 1972, in Russian.

Maksimov, M.V., M.P. Bobnev, L.N. Shustov, B. Kh. Krivitskiy, G.I. Gorgonov, V.A. Il'in, B.M. Stepanov, *Zashcita ot Radiopomekh, (Radar Anti-Jamming Techniques),* Moscow, 1976, Sovetskoye Radio. To be published in English by Artech House, Dedham, MA, Fall 1979. (EMC, extensive ECCM, 366 p., 214 refs., several of these reprinted in this volume.)

2. Journal Articles

Koller, L.R., "Identification of Naval Spoofs," *Radio Research Lab Rpt 411-129,* DDC ATI 222 MI Reel 293, Feb 29, 1944.

"Antijamming is an Art," *Radar Magazine,* vol. 2, May 1944, p. 17.

"A-J Medicine Show," *Radar Magazine,* vol. 5, Sept 1944, pp. 40-42.

Sard, R.D., "Survey of German Radar from the Countermeasures Point of View," *Radio Res Lab Rpt 411-95,* Sept 4, 1944 DDC ATI 15357 (Gives characteristics of approximately 100 German World War II radars—land, sea and air— and describes ECCM features of each radar).

"RCM: Tailored to Fit," *Radar Magazine,* vol. 8, Feb 1945, pp. 38-41 (has photograph of German ECCM retrofit on Wurzburg radar.)

Lusted, L.B., "Installation of Antijamming Devices in the German Ground Radar Equipment FUSE 62," *American British Laboratory Div 15 Rpt 1045-TM5,* Mar 21, 1945, DDC ATI 14104.

O'Brien, R.S., and R.A. Soderman, "Intelligence Information in RCM Effectiveness in the E.T.O. and Appendixes I-V," *American British Lab Harvard Univ Radio Res Lab Cambridge, Rpt R-1045-MR-15,* June 1945, DDC ATI 62978 (lists and describes German World War II radar ECCM techniques).

Mitchell, J.K., "History of US Navy Radar Anti-Jamming," 1 Feb 1946, DDC AD493649.

Alred, R.V., "Naval Radar Anti-Jamming," *J IEE pt III A*, vol. 93, no. 10, 1946, pp. 1593-601 (reprinted in this volume).

"Defeat Enemy Interference," *Air Force Magazine*, vol. 33, no. 6, June 1950, pp. 43-5.

Kober, C.L., "Interference and Interference Suppression of Reflections in Electro-Magnetic Fields," *OST. INGEN-ARCH*, vol. 5, no. 1, 1951, pp. 1-11 (in German, trans. abstracts SA 50-4286 and 51-4385).

"Jamming and Anti-Jamming," *The Arty Schl, Ft. Bliss, Texas RPT M-36503, R*, 1951, no. 5080.

Kromin, G.S., and K. Katkov, "Radar Interference and Protection Against It," pp. 430-67 in *Fundamentals of Radar Engineering*, Moscow, 1956, Voyenizdat, (in Russian, partial trans. PRS 16690, DDC AD 299208).

Brick, D.B., and J. Galejs, "Radar Interference and its Reduction," *Sylvania Technologist*, vol. 11, no. 3, Jul 1958, pp. 96-108 (reprinted in this volume).

Hanford, R.W., "Radar Testing for a War Environment," *IRE Nat Conv Rec.*, vol. 7, pt. 5, 1959, pp. 208-217 (Table I contains list of many ECCM's for various ECMs).

Hanford, R.W., "Radar Countermeasures Summarized for Reference," *Military Electronics Magazine*, vol. 6, June 1959, p. 29ff.

Holahan, J., "Tools and Techniques of Electronic Warfare," *Space/Aeronautics Magazine*, vol. 33, Apr 1960, pp. 126-32 (reprinted in this volume).

Dax, P., "Electronic Dogfight: ECM vs ECCM in Search Radar," *Space/Aeronautics Magazine*, vol. 33, Apr 1960, pp. 134-7 (reprinted in this volume).

Simpson, M., "Counter-Countermeasures: Good Design can Beat the ECM Threat," *Space/Aeronautics Magazine*, vol. 33, Apr 1960, pp. 138-42 (reprinted in this volume).

Davies, L.E., "Some General Considerations Regarding Radar ECCM Requirements," *Weapons Systems Laboratory Stanford Res Inst Rpt WSL Tech Note 30*, Mar 1961.

Archer, D.H., "Secondary Surveillance Radar," *Electronic Engr.*, vol. 33, nrs. 401, 402, Jul 1961, pp. 414-20 and Aug 1961, pp. 490-4.

Hedemark, R., "Methods for Protection of Radar Receivers Against Interference," *Elektrotek Tidss-Krift*, vol. 76, no. 1, Jan 1963, pp. 3-10 (in Norweigian, trans. abstract SA 63-5400).

Bacon, F.W., Jr., "An Introduction to Radar Sensors," *Naval Engrs J.*, vol. 77, no. 3, June 1965, pp. 437-47.

Kawecki, H., "Radar Jamming and its Reduction," *Problemy Elektron. Telekomun.*, Metody Wspolczenej Radiol, no. 11, 1966, pp. 168-85 (in Polish, trans. abstract SA 69-6610) (contains 30 references).

Babanov, YuN., et al, "Development of Methods for Protecting Radio Communications Systems from Impulse Noise," *Telecomm and Radio Engr., Part II*, vol. 22, no. 11, Nov 1967, p. 82-8 (contains 92 references).

Baklashov, P.I., "Radar Jamming and Protection of Radar Stations," Moscow, 14 Oct 1968 (in Russian; trans: JPRS 46664).

Farner, F., "Interference Signals in Radar Equipment and Methods for Their Suppression," *Bull. Assoc. Suisse Elect* (Switzerland), vol. 61, no. 16, Aug 8, 1970, pp. 725-32 (in German, trans. abstract SA 70-38820).

Mishra, S., "Radar—Past, Present, Future," *Inst Engrs (India) I & T. E.*, vol. 51, May 1971, pp. 123-4.

Carpentier, M.H., "Influences of the New Radar Techniques Upon Operational Equipment," *Radar and Electronics Magazine*, Spring/Summer 1971, pp. 6-12.

Sundarababu, A., *Fundamentals of Radar*, Asia Publishing House, London, 1972 (interference elimination, p. 163: Fast Time Constant i.e., high pass filter, IAGC,STC; radar countermeasures, pp. 484-5).

Bachtel, C.L., "The ECM/ECCM Syndrome," *Signal Magazine*, vol. 28, Oct 1973, pp. 25-9.

Riengi, T., "ECCM – A Poor Country Cousin," *Electronic Warfare Magazine*, vol. 6, no. 6, Nov/Dec, 1974, p. 35.

Hyatt, B.B., "Total Scenario," *Electronic Warfare Magazine*, vol. 7, no. 1, Jan/Feb 1975, p. 122.

Baier, W., "Considerations Concerning Interference-Rejecting Communications Systems," *Siems Forsch and Entwicklungsber*, vol. 4, no. 2, 1975, pp. 61-7 (in German, trans. SA 75-25398).

Singarayar, S., and K. Natarajan, "Swamping Radar Interference," *J. Inst. Engr.* (india), vol. 55, no. ET 2-3, Apr 1975, pp. 80-1.

DiNardo, U., "Adaptivity Criteria of a Modern Air Traffic Control Radar," *Riv Tech Selenia*, vol. 2, no. 4, 1975, pp. 17-32 .

Johnston, S.L., "Philosophy of ECCM Utilization," *Electronic Warfare Magazine*, May/June 1975, vol. 7, no. 3, pp. 59ff (reprinted in this volume).

Ivanov, A., "Improved Radar Designs Outwit Complex Threats," *MicroWaves Magazine*, Apr 1976, pp. 54ff (reprinted in this volume).

Meskill, W.D., "Why ECCM: The Need for ECCM," *IEEE ELECTRO-76 Conv Rec May 1976, pt 30*, pp. 30-1-1-thru 4 (reprinted in this volume).

Fielding, J.G., "Signal Processing in Electronic Warfare," *IEE Conf. Pub. No. 144,* Sept 1976, pp. 112-120. (Considers signal processing in ECM, ECCM, and ESM).

Stoner, D.C., "Adaptive MTI and Signal Processing Techniques for Long Range Chaff and Weather," *IEE Conf Pub.,* no 155, Radar 77, London, Oct 1977, pp. 195-8.

Johnston, S.L., "Radar Electronic Counter-Countermeasures," *IEEE Trans.,* vol. AES-14, no. 1, Jan 1978, pp. 109-117 (reprinted in this volume).

Radford, M.F., "Radar ECCM: A European Approach," *IEEE Trans.,* vol. AES-14, no. 1, Jan 1978, pp. 194-8 (reprinted in this volume).

Johnson, M.A., and D.C. Stoner, "ECCM from the Radar Designers Viewpoint," *Microwave J.,* vol. 21, no. 3, Mar 1978, pp. 59-63 (reprinted in this volume).

Johnston, S.L., "Guided Missile System ECM/ECCM," *Microwave J.,* vol. 21, no. 9, Sept 1978, pp. 20ff (reprinted in this volume).

Prior, J.R., and N. Woodward, "Performance of Current Radar Systems in an EW Environment," *Proc. Mil Microwaves Conf.,* London, Oct 1978, pp. 39-50 (reprinted in this volume).

Goulder, M.E., "ECCM," guest editorial, *J. of Electron. Def.,* vol. 1, no. 3, Nov/Dec 1978, pp. 12ff.

Backus, P.H., "The Nature and Management of ECCM," *J. of Electron. Def.,* vol. 1, no. 3, Nov/Dec 1978, pp. 25ff. (Uses several tables and figures from papers that are reprinted in this volume).

Radar Interference and Its Reduction
D.B. Brick and J. Galejs
The Sylvania Technologist, Vol. 11, No. 3, July 1958
Pages 96-108

2.A.1

ABSTRACT

The effects of interference on radar operation are discussed with primary emphasis on r-f radiated interference. A summary of the various types of interference, that is, active, passive, radiated and conducted, is given. Various criteria used in discriminating between interfering and desired signals are discussed briefly along with some of the better known interference reduction techniques. Several of the more novel techniques designed to reduce various types of r-f radiated interference against conventional pulse radars are then described in some detail. These include side-lobe blanking, frequency discrimination, MTI, pulse length selection, interpulse period selection, optimum filtering, delay-line video integration, and some detection enhancement methods. Finally, some unconventional modes of radar operation with interference suppression capabilities are discussed. These include triangular FM systems, binary-coded phase-modulated radar transmissions, pulse compression, and noise correlation systems.

I. Introduction

Electromagnetic interference can be a serious and costly problem. It has therefore evoked considerable attention from civil and military authorities who are concerned with the utilization of an electromagnetic spectrum that is overcrowded, even in the absence of interference. Of particular importance to national defense and airways and marine navigation are the effects of interference on radar operation. For example, expensive missiles continue to be deflected due to the effects of spurious interfering signals. During certain periods, airline navigation systems become hopelessly fouled. Even though these periods are temporary, their effects are quite serious and their frequency of occurrence is increasing rather than decreasing.

The subject of radar interference is much too extensive to be covered in one paper. Hence, it is the intent of this paper to cover only one aspect of this subject, r-f radiated interference, and to treat only some of its more important aspects in a rather heuristic manner. First, some of the causes and effects of interference on radar operation are discussed, with emphasis on pulsed, scanning radars. Then, some techniques designed to reduce interference in conventional pulsed radars are described, and finally, some of the more novel, unconventional modes of radar operation and their interference-rejection capabilities are discussed.

II. Interference and its Effects on Radar Operation

Radar interference can be described as either active or passive, depending upon the manner in which the interfering signals are generated.

Active interference is caused by (1) independent electromagnetic generators such as other radars, radios, countermeasures equipment or other man-made devices intended to generate radio-frequency signals external to the equipment itself; (2) electrical machinery, X-ray or diathermy equipment, or other man-made generators not intended to generate radio frequencies external to the equipment itself, and (3) natural disturbances such as electrical storms, aurora, sun activity, and other extraterrestrial activity. The signals generated may be of the CW, FM, AM, pulsed, railing, impulsive, or noise or noiselike types. They can arrive at the radar through radiation or conduction paths. Insofar as the radar is concerned, they may appear in r-f, i-f, or video frequency regions.

Passive interference is that which is not caused by independent electromagnetic generators. Reflections from nearby obstacles, surrounding geography, and weather (precipitation and clouds) are prime examples of r-f passive interference. These examples are sometimes all grouped under the heading of clutter. I-F and video passive interference are usually caused by paths of stray leakage in the i-f and video sections of the receiver and reflections or scattering from nearby obstacles. This is sometimes referred to as mutual coupling between the particular obstacle and the radar. The primary path at these frequencies is usually conduction and induction rather than radiation.

Proper application of good receiver design techniques such as balancing and shielding will reduce the effects of i-f and video interference to a tolerable level. Reduction of r-f interference is not as well-known an art. The emphasis in this paper, as stated previously, is on r-f radiated interference.

Figure 1 shows the effects of various types of interference on radar operation as displayed on a Plan Position Indicator (PPI). Figure 1(a) may be used as a reference for an idealized "clean" PPI (that is, no interference) with a single target at 230° in azimuth, at a range of 75 nautical miles. Figure 1(b) shows unsynchronized pulse interference, or railing, that is, a pulse train that is not synchronized with the radar-generated pulse train. Both the desired target (at 72

Radar Interference

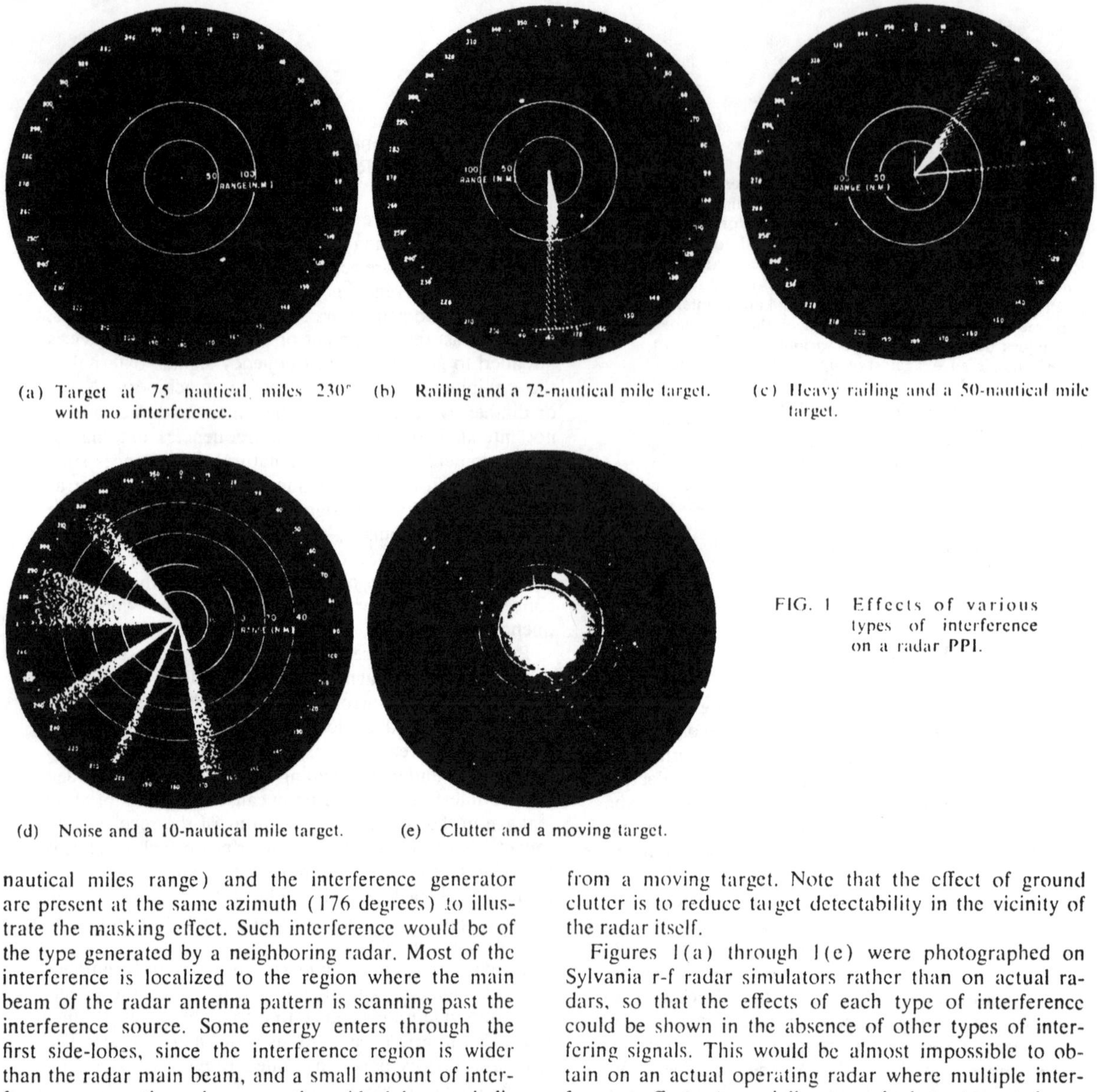

(a) Target at 75 nautical miles 230° with no interference.

(b) Railing and a 72-nautical mile target.

(c) Heavy railing and a 50-nautical mile target.

(d) Noise and a 10-nautical mile target.

(e) Clutter and a moving target.

FIG. 1 Effects of various types of interference on a radar PPI.

nautical miles range) and the interference generator are present at the same azimuth (176 degrees) to illustrate the masking effect. Such interference would be of the type generated by a neighboring radar. Most of the interference is localized to the region where the main beam of the radar antenna pattern is scanning past the interference source. Some energy enters through the first side-lobes, since the interference region is wider than the radar main beam, and a small amount of interference enters through some other side lobes as indicated by the faint lines at 130 and 220 degrees. In Fig. 1(c), where the strength of the interference is increased and the 50 nautical mile range target and interference generator are again at the same azimuth, the side-lobe effect becomes marked. Even stronger interference is shown in Fig. 1(d). Here the interference is of the noisy type. The masked target is at 280 degrees, 10 nautical miles, and the interference source is also at 280 degrees. It can be seen from this figure that side-lobe interference can be an important factor in masking or degrading the detectability of targets. In Fig. 1(e) the effect of ground or geographical clutter on a long-range radar is shown in a multiple exposure photograph of a PPI display. The diagonal line of dots running from the upper-left to lower-right corners is the return from a moving target. Note that the effect of ground clutter is to reduce target detectability in the vicinity of the radar itself.

Figures 1(a) through 1(e) were photographed on Sylvania r-f radar simulators rather than on actual radars, so that the effects of each type of interference could be shown in the absence of other types of interfering signals. This would be almost impossible to obtain on an actual operating radar where multiple interference effects (especially ground clutter) are almost always present simultaneously.

III. R-F Interference Reduction Methods

Methods for reducing r-f radiated interference usually make use of one or several characteristics of the desired radar signals or undesired interfering signals in order to distinguish them. Some of the techniques used are:

1. *Spatial Selectivity*. The direction of arrival or path by which the signal enters the radar is utilized to select desired from undesired signals. This method is useful if the interference is arriving by a leakage path, on the radar side lobes, or from some unknown direction that can be discriminated against.

Brick and Galejs

2. *Frequency Selectivity*. Possible differences in frequency between the desired and interfering signals are utilized for discrimination. This method can be employed to discriminate against wide band, off-frequency, and impulsive signals, and with recourse to doppler shifts introduced by radial velocity differences, it can be used for velocity discrimination between the target and the source of interference, particularly if the interference is of the passive type (clutter).

3. *Amplitude Selectivity*. Amplitude characteristics are employed to discriminate, for example, against high energy or impulsive interference.

4. *Time Selectivity*. The time of arrival of the signal or time separation between successive returns is utilized for distinguishing between interference and target returns. This can be used against signals that are not synchronized with the radar signal, such as railing or random pulse trains in a pulsed radar.

5. *Signal Selectivity*. The signal and interference are separated by means of signal statistics, shape, spectrum (closely related to frequency selectivity), or polarization. In some cases the peculiarities of the interference are utilized, while in other cases the radar signal is made as nearly unique as possible to reduce the probability of and degree of similarity between interference and signal. This latter method provides the basis for the unconventional modes of radar operation to be discussed in Section V. Time Selectivity and Signal Selectivity are closely related.

Table I lists several techniques for each method of interference reduction. The techniques in italics appear in more than one category. Asterisked techniques will be examined in more detail in Sections IV and V; those not asterisked will be briefly discussed at this point.

Shielding and side-lobe reduction are two obvious methods of applying spatial selectivity. Shielding can reduce interference entering through undesired paths in all portions of the receiving system with the exception of the antenna proper. It should be supplemented by proper design, to reduce the possibility of unbalanced currents, and adequate filtering. Side-lobe reduction is the straightforward, although frequently the brute-force, approach to reducing side-lobe interference. Its implementation is, in many cases, unfeasible.

Proper tuning of the various components of the radar is of prime importance in reducing the effects of off-frequency interference. Also, if the interference occupies only a narrow-frequency band, the capability to tune the radar easily would allow the operator to tune away from the undesired signal. Proper design of the receiver filters and balanced mixing can provide significant improvement in many cases. The latter is especially suggested when the interference is not monochromatic.

Limiting[1] and logarithmic amplifier*[2] techniques, in themselves, are helpful in reducing the transient and

* The logarithmic amplifier has a gain characteristic which is a logarithmic function of the input for large signals and linear for small signals.

TABLE I

R-F INTERFERENCE REDUCTION

	Method	Techniques
	Spatial Selectivity (Utilizes direction or entrance path to select desired from undesired signal)	Shielding, Side-Lobe Reduction, *Side-Lobe Blanking**
	Frequency Selectivity (Discriminates against wide-band off-frequency and impulsive signals; also uses velocity discrimination)	Proper Tuning, Tunability, Filtering, Balanced Mixing, Frequency Discrimination,* *Guard Band Techniques*,* *Lamb Suppressor*,* Moving Target Indicator,* Doppler Radars*
	Amplitude Selectivity (Uses amplitude characteristics to discriminate against, e.g., high energy or impulsive interference)	Limiting, *Guard Band Techniques*,* *Lamb Suppressor*,* *Side-Lobe Blanking*,* Log Amplifier, *Instantaneous Automatic Gain Control*, *Detector Balanced Bias*.
Closely Related	Time Selectivity (Uses time of arrival of signal or time separation between successive returns to distinguish between interference and target return)	Pulse Interference Separator and Blanker,* *Delay-Line Integration*,* *Lamb Suppressor*,* Sensitivity Time Control
Closely Related	Signal Selectivity (Employs signal statistics, shape, spectrum, or polarization)	*Instantaneous Automatic Gain Control*, Fast Time Constant, *Detector Balanced Bias*, Circular Polarization, Optimum Filtering,* *Guard Band Techniques*,* Pulse Shape Discrimination,* Detection Enhancement (Coherent Detection, "Mouse Under the Rug"), Unconventional Modes of Operation,* *Delay-Line Integration**

other auxiliary effects such as blooming on the display due to strong signals of the continuous or impulsive type; however, they do not assist in canceling out these signals nor in improving detectability when they are present. These methods are often very valuable as adjuncts to other techniques such as MTI or the Lamb Suppressor. Instantaneous Automatic Gain Control[2] (IAGC) is a gain-control circuit in which the time constant is of the same order of magnitude as the pulse-width in a pulsed radar. Detector Balanced Bias[3] (DBB) is a gain control whose characteristics are similar to IAGC except that the control is applied as a bias to the second detector.[4] Both have the effect of reducing the response to strong signals and hence reducing the deleterious effects due to these signals in a manner similar to the limiting and logarithmic amplification techniques. In addition, IAGC or DBB may improve detectability in the presence of strong interference if a loss of sensitivity is incurred because the strong interfering signal biased the amplifier or detector, respectively, into a saturation region.* IAGC and DBB also appear in the listing of signal selective devices in Table I. This occurs because of their short time constants, by virtue of which they act as differentiators (high-pass filters). Hence, they discriminate against long constant-amplitude returns, such as clutter, by providing responses only at positions of rapid changes in amplitude.

Sensitivity Time Control[5] (STC) is the only device in the time selectivity category that will not be described in detail. It discriminates against near-by strong returns such as clutter by increasing the receiver gain from the time-of-arrival of near-by returns to a time-of-arrival corresponding to some fixed minimum range where full receiver gain is desired. Except for the reduction of saturation effects, no improvement in detectability is obtained.

The two techniques in the category of signal selectivity that will not be discussed in detail in the following sections are Fast Time Constant[6] (FTC) and Circular Polarization[7] (CP). The FTC circuit is a differentiator or high-pass filter with a time constant of the order of a pulse width placed in the video portion of the radar. Its action is identical to that of IAGC and DBB in that it discriminates against long constant-amplitude returns; however, no graduated gain control other than on-off is achieved.

Circular Polarization involves transmitting a circularly-polarized wave and receiving the same polarization (where the direction of polarization is defined with respect to the direction of propagation). Clouds and rain consist of almost spherically symmetric particles and hence the returns from these will be polarized primarily in the opposite sense from the incident wave. Consequently, these will not be received. Aircraft are not symmetric and hence will return some signal in the original polarization (again defined with respect to the direction of propagation). Thus, a large portion of the weather return may be canceled and aircraft returns are received with some reduction in strength. It must be emphasized that this technique does not always work according to the simple theory just described since in certain situations weather returns will contain large amounts of the original polarization. Circular Polarization is normally used on a trial and error basis.

* Best detectability is obtained when the receiver has a linear gain characteristic.

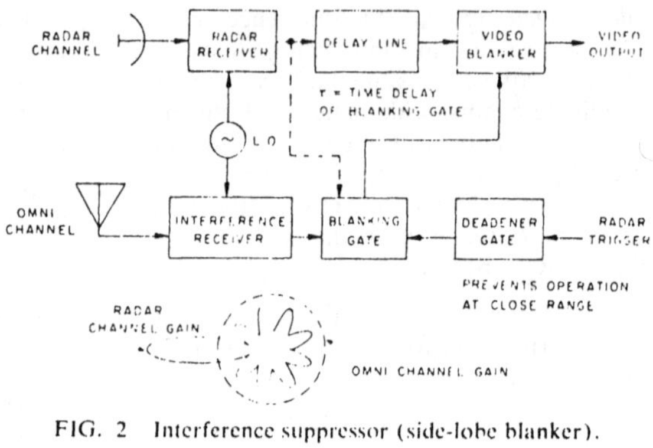

FIG. 2 Interference suppressor (side-lobe blanker).

IV. Some Interference Reduction Techniques For Pulse Radars

1. *The Side-Lobe Blanker*

As shown in Fig. 1, interference arriving through the radar side lobes may result in serious degradation of radar performance. A device designed to combat low duty cycle (pulsed), side-lobe interference is shown in Fig. 2.[8]

The technique consists of adding to the normal radar receiver an identical auxiliary receiver, called the interference receiver, which is fed by the same or a well-synchronized local oscillator. The gain and AGC of the interference receiver are adjusted to bear a constant relationship to that of the normal receiver. The interference receiver is fed by an antenna whose gain characteristic is omnidirectional in the azimuth plane and, when applied to a scanning radar, closely approximates the radar antenna pattern in the vertical direction. The relationship of the two channel azimuthal gain functions is shown in the lower part of Fig. 2 (in highly idealized form). The gain of the interference channel is adjusted to be larger than that of any side lobe of the normal channel but much smaller than the main-lobe gain.

In the blanking-gate generator, the outputs of both receivers are compared in amplitude. Should the signal in the auxiliary channel be larger than the signal in the radar channel, a blanking gate is generated which prevents the signal in the radar receiver from passing on to the display. This corresponds to the case of a side-lobe signal since, as can be seen from the gain patterns in Fig. 2, a side-lobe signal will be larger at the interference receiver output than at the normal output. If the signal

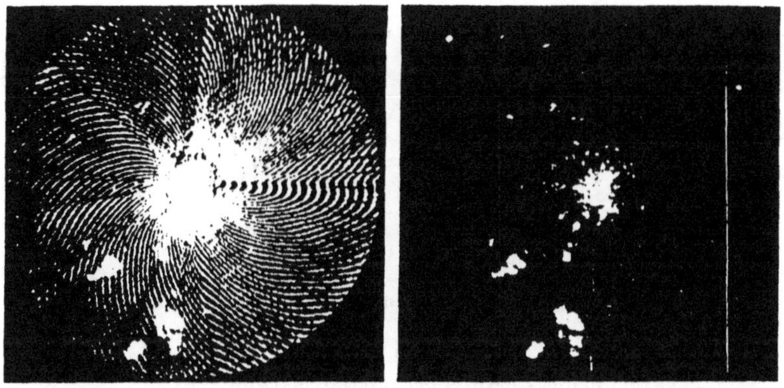

(a) PPI, 100-mile sweep; normal operation. (b) PPI, 100-mile sweep; suppressor in operation.

FIG. 3 Photographs illustrating performance of interference suppressor.

in the normal channel is larger, this indicates a main-lobe signal and no blanking gate is generated.

The delay line is introduced to compensate for the inherent delay of the blanking gate. The radar trigger prevents operation during periods immediately following the transmission of a radar pulse, since strong returns at short ranges and the transmitted pulse could cause triggering of the blanking circuit.

The circuit, of course, suffers a slight loss in sensitivity due to periods of blanking. This loss increases with the duty cycle of the interference.

Figure 3 illustrates the degree of improvement afforded by this system. Figure 3(a) shows severe railing (unsynchronized pulse trains) occupying the whole PPI, that is, entering all side lobes. After the side-lobe blanker was inserted, Fig. 3(b), the scope was clear of railing except for the direction of the interfering source at 180 degrees.

2. *Frequency Discrimination*[9]

The desired signals and off-frequency interference can be distinguished on the basis of frequency response. If any auxiliary receiver with a frequency response like that shown in Fig. 4(a), where the null is at the radar i-f frequency, were added in parallel with the radar receiver as in the previous device, and its output were compared with that of the normal radar output, off-frequency interference would provide a larger output in the auxiliary channel than in the normal channel. When this happens a blanking pulse is generated. If the normal output is greater than the auxiliary output, no blanking is introduced. This technique is limited to interfering signals of both short time duration, so that desired pulses are not blanked, and fairly narrow spectrum, so that frequency discrimination can be obtained.

This device can be combined with side-lobe blanking if (1) the auxiliary receiver frequency characteristic is made to have a response at the i-f frequency in addition to a larger response at the side frequencies as shown in Fig. 4(b)*, and (2) this receiver is used as the interference receiver with the omnidirectional antenna and other circuitry of Fig. 2. The auxiliary channel gain at the center frequency must be adjusted, with respect to that of the radar receiver, so that the gain is larger than any radar side lobe and much smaller than the main lobe.

Different combinations and subsequent processing of frequency discrimination circuitry can be devised. They all may be categorized under the heading of guard-band techniques.

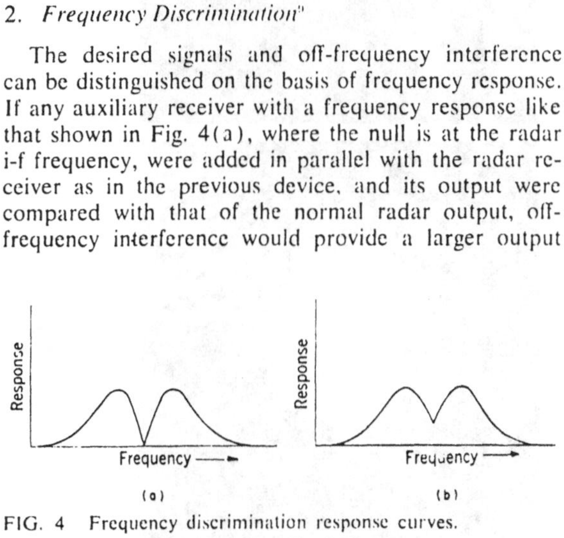

FIG. 4 Frequency discrimination response curves.
(a) Unmodified frequency discriminator.
(b) Modified frequency discriminator.

3. *The Lamb Noise-Silencing Circuit*

Lamb,[10] in 1936 and Schukin,[11] in 1946, proposed the circuit shown at the top of Fig. 5 for reducing the effects of strong off-frequency impulsive and pulse-like interference. This device, which is closely related to the side-lobe blanker shown in Fig. 2, consists of the addition of a wide-band amplifier followed by a limiter in front of the normal amplifier. In order to describe its operation, the spectral representations in Fig. 5 are required. This circuit is designed to operate against off-frequency interference that is intense enough for the spectral sidebands of the signal to interfere with normal radar reception.

Such an interfering spectrum is shown by the dashed curve, Fig. 5(a). The envelope of the normal radar pulse spectrum is shown as a solid line. The two vertical dashed lines delineate the normal i-f passband. It can be seen here that the first sideband of the interference signal causes difficulty. The frequency responses of the wide-band amplifier and narrow-band amplifier are illustrated in Fig. 5(b). The wide-band amplifier is designed to accept most of the interfering spectrum. After limiting, if both signals are not present simultaneously and if the limit level is above the amplitude of the desired pulse, the amplitude of the interfering pulse is reduced relative to the radar signal and therefore its sidebands are reduced to tolerable levels. In

* This can be accomplished by adding a normal radar i-f amplifier in parallel with the off-frequency amplifier.

Radar Interference

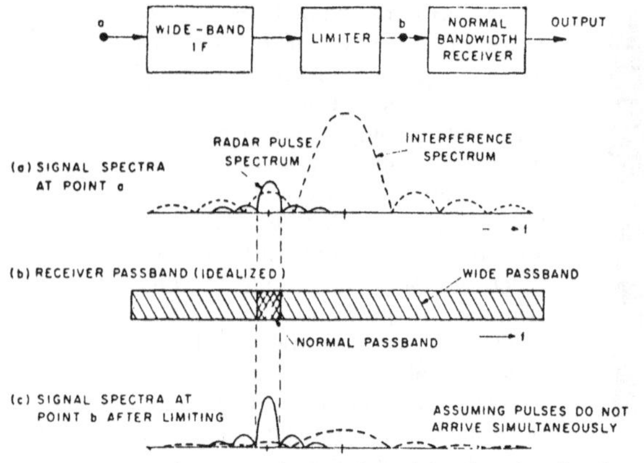

FIG. 5 Block diagram and operation of Lamb noise-silencing circuit for reducing off-frequency pulsed interference.

Fig. 5(c) the relationship after limiting is shown in somewhat exaggerated form.

4. *The Moving Target Indicator Circuit*[12,13]

The normal method of discriminating against clutter and slow moving weather returns is with a Moving Target Indicator (MTI) circuit. Such a circuit discriminates on the basis of radial velocity, or doppler shift differences. Radar returns from obstacles with sufficient radial velocity with respect to the radar will be accepted; others will be attenuated. MTI exists in many different forms. What is perhaps the simplest and oldest of these is described in the following paragraphs. The circuitry of such a simple delay-line MTI circuit is shown in Fig. 6.

After limiting (or lin-log amplification) to remove amplitude variations and coherent detection* the signal is passed into the delay-line unit. Here the signal is delayed by one repetition period, $1/f_R$, and is subsequently subtracted from the undelayed signal. This circuit cancels signals with doppler shifts at zero frequency and multiples of f_R, the PRF.

The frequency response of the delay-line circuit is shown in Fig. 7. It has a $\sin^2(\pi f_d/f_R)$ shape, where f_d is the doppler frequency $(2V_r f_{rf}/c)$. V_r is the radial velocity, c is the velocity of light and f_{rf} is the r-f frequency. The spectrum of an infinite pulse train is also shown. Notice that the spectral lines occur at the doppler

* In a coherent detector the i-f signal is mixed with the radar-generated i-f frequency to produce coherent or bipolar video whose amplitude depends upon the phase difference between the two mixed signals. When there is a frequency difference the signal is amplitude modulated with the difference frequency.

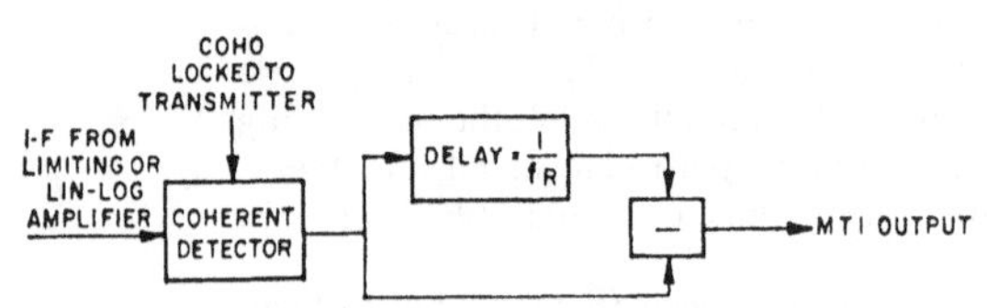

FIG. 6 Block diagram of simple delay-line MTI circuit.

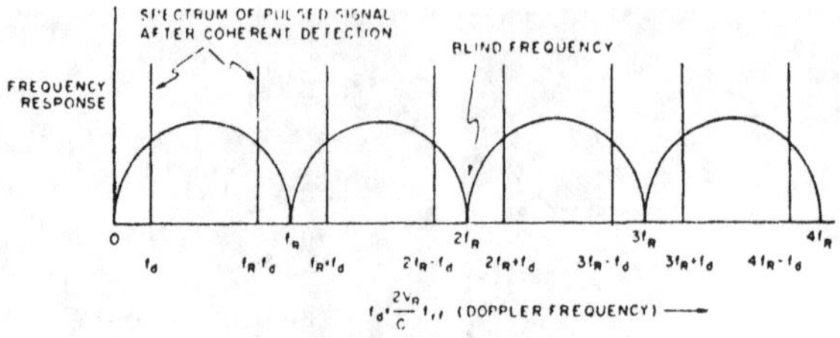

FIG. 7 Simple MTI response curve.

frequency and at multiples of f_R plus and minus the doppler frequency.

The simple delay-line MTI circuit does not work in the exact manner desired because targets with velocities corresponding to doppler shifts at multiples of the PRF are canceled. These are called blind velocities. Furthermore, the frequency cut-off around zero is gradual; hence some clutter may leak through and some slower moving targets may be attenuated. The improvement offered by the simple delay line MTI circuit is illustrated in Fig. 8. Figure 8(a) shows a badly cluttered PPI. In Fig. 8(b) the photograph is over-exposed which tends to mask the effectiveness of the MTI, although it can be seen that the bad patches of clutter have been removed.

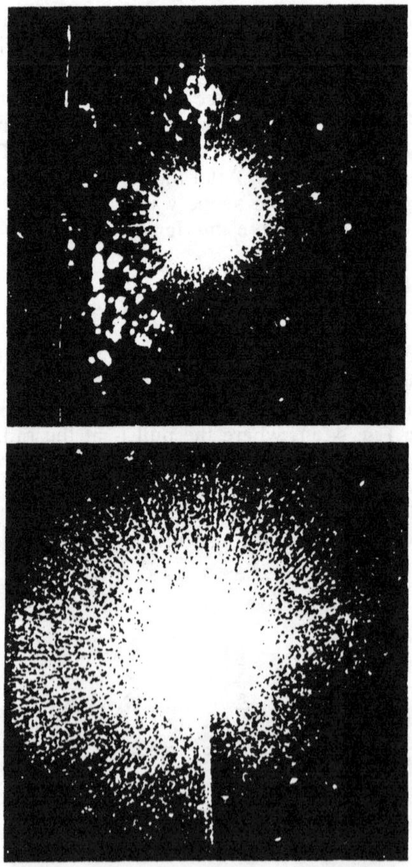

(a) Normal scope, 35-mile sweep.

(b) Scope with MTI, 35-mile sweep.

FIG. 8 Comparison of normal and MTI PPI presentations.

Brick and Galejs

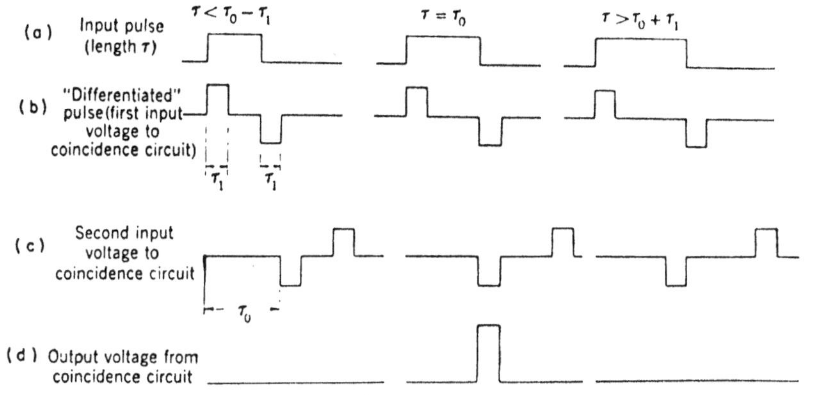

← FIG. 9 Waveforms illustrating pulse-length section.

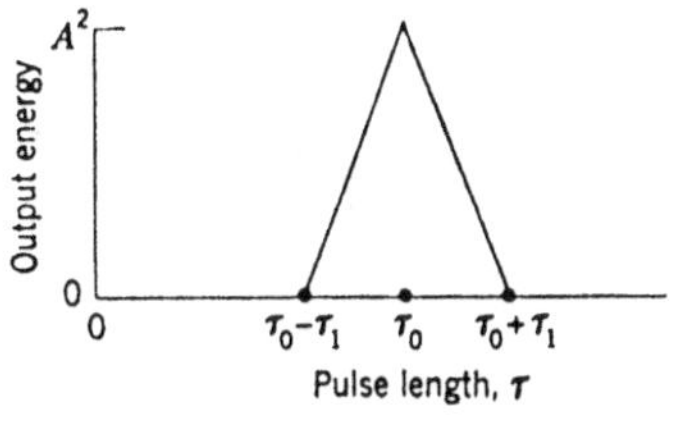

FIG. 10 Response for pulse-selection circuit for input pulses of length τ.

5. *Pulse Length Selection*[14]

A well-known device, which makes use of part of the information contained in the pulse shape itself, is the pulse length selector. Within certain tolerances pulses of undesired lengths are discarded by this circuit.

One method of accomplishing this is shown in Fig. 9. Pulses of length τ_0 are desired. Pulses of length τ, where τ is greater than $\tau_0 + \tau_1$ or less than $\tau_0 - \tau_1$, are discarded. The tolerance τ_1 is inserted to account for perturbations introduced on the desired pulse lengths by propagation and target-scattering characteristics, receiver-introduced variations, and variations in the transmitted pulse shape.

In operation, the received signal (a) is delayed by a time τ_1 and subtracted from the original pulse (b). This operation is called delay-line differentiation since the resulting pulse appears to be similar to the original signal differentiated. This "differentiated" output is then delayed by the desired pulse length τ_0, (c), and is used in a coincidence or gating circuit where the other input is the undelayed "differentiated" signal. In (d) it may be seen that a response is obtained only for pulses of length τ, where $\tau_0 + \tau_1 < \tau < \tau_0 - \tau_1$.

The response of this circuit is illustrated in Fig. 10. This device provides significant improvement in the presence of low duty-cycle pulsed interference of pulse lengths different from τ_0.

6. *Interpulse Period Selection*

Another method for distinguishing desired from undesired pulse trains consists in the use of the interpulse period. This technique is employed in the Pulse Interference Separator and Blanker. The operation of this device against a series of pulses that is not synchronized with the radar transmission (that is, railing) is shown in Fig. 11, where the solid pulses are the desired pulses and dashed pulses are the interference. The incoming signal (a) is delayed by an interpulse period, *T*, and used to gate the incoming pulse train. In other words, the receiver is gated by signals received one pulse period earlier. This device is very effective against low duty cycle, unsynchronized interference.

7. *Optimum Filtering*[15,16]

In the presence of noise or noiselike interference it is important to utilize, to the fullest extent, the information available in the desired signal. An important contribution to the proper utilization of the signal is provided if the i-f passband shape is optimized. Van Vleck and Middleton,[15] and D. O. North[16] discovered independently that the proper shape of the "optimum" or "matched" filter for maximum utilization of the input signal in the presence of noise of constant spectral intensity should be such that the frequency response of the filter is the complex conjugate of the Fourier transform of the desired signal (except for a possible linear phase shift). Such a filter maximizes the output signal-to-noise ratio for a given input signal-to-noise ratio. In the time domain, the impulse response of the optimum filter is the image (or time negative) of the desired signal obtained by reflecting the waveform with respect to the instant when the signal-to-noise ratio is maximized.

The relationship for a hypothetical signal is shown in Fig. 12. Note that if the signal pulses are symmetric in time, the optimum filter impulse response is identical to the signal (with the possible exception of a time delay in the time domain or linear phase shift in the frequency domain).

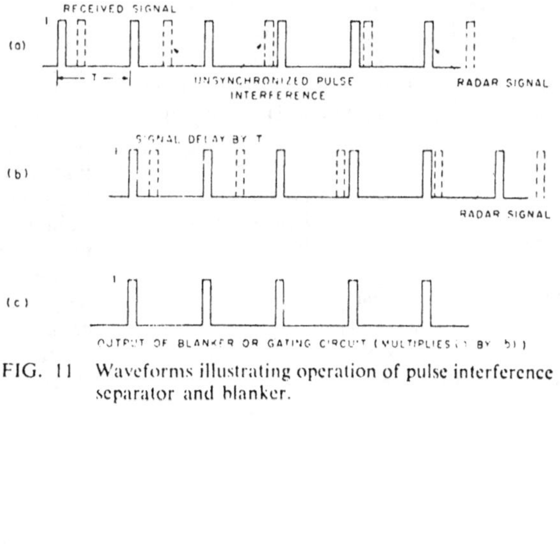

FIG. 11 Waveforms illustrating operation of pulse interference separator and blanker.

Radar Interference

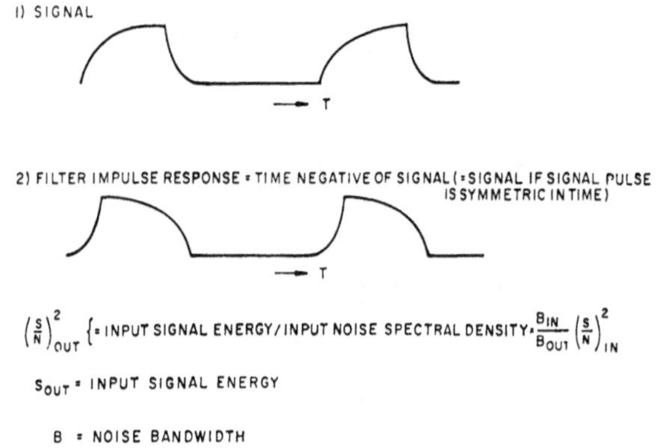

FIG. 12 Optimum filter for maximizing signal-to-noise ratio.

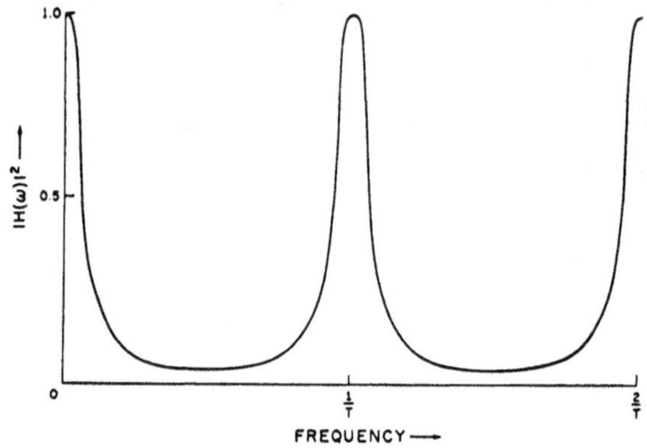

FIG. 14 Normalized frequency response of the recirculating delay line ($K = 0.7$).

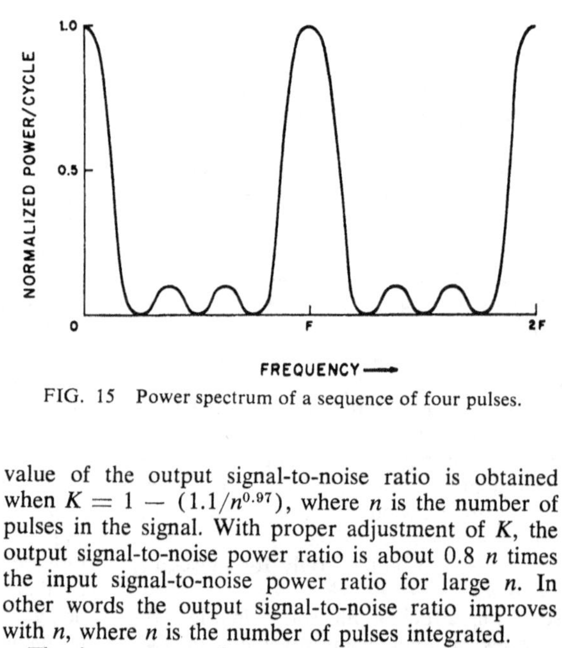

FIG. 15 Power spectrum of a sequence of four pulses.

When the filter is matched, the maximum output signal amplitude is equal to the *energy* in the input signal, and the output signal-to-noise ratio is equal to the input signal energy divided by the input noise spectral power density, assuming that the input noise spectrum is uniform. The filter has the unusual property of transforming all of the energy in the original signal to usable signal amplitude at the output. The resulting signal-to-noise power ratio is also equal to the input signal-to-noise power ratio multiplied by the ratio of input noise bandwidth (that is, bandwidth prior to the filter) to the noise bandwidth of the filter itself.

The optimum filter criterion has been general knowledge for some years and attempts have been made to approach it in practice. Owing to the difficulty involved in implementing the theory, however, exact applications have not been attempted until recently.

8. *Video Integration*[17]

Because of the difficulty of instrumenting an optimum filter, other methods of improving the signal-to-noise ratio have often been utilized. A particularly successful technique is video integration, which has been best accomplished by means of a delay-line integrator such as the one shown in Fig. 13. Essentially, the device delays each pulse by integer multiples of the interpulse period, $T = (PRF)^{-1}$, and adds them to the previous pulses. Hence, as time progresses from the start of the pulse train, the output of the device builds up. It stacks pulses on top of one another. The feedback factor, K, must be less than unity to prevent the device from breaking out in self-oscillations triggered by noise. The maximum

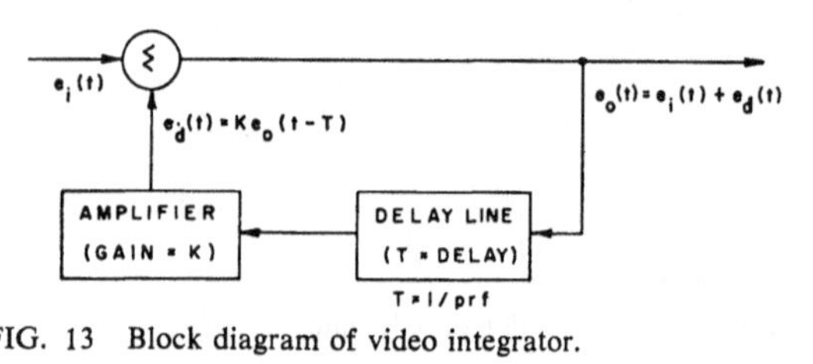

FIG. 13 Block diagram of video integrator.

value of the output signal-to-noise ratio is obtained when $K = 1 - (1.1/n^{0.97})$, where n is the number of pulses in the signal. With proper adjustment of K, the output signal-to-noise power ratio is about 0.8 n times the input signal-to-noise power ratio for large n. In other words the output signal-to-noise ratio improves with n, where n is the number of pulses integrated.

The improvement is comparable to what could be obtained with an optimum filter. The reason for this will become evident from observation of Figs. 14 and 15. In Fig. 14 the frequency response of a recirculating delay line, $K = 0.7$, is shown. This value of K is about optimum for $n = 4$. In Fig. 15 the power spectrum of four infinitely narrow pulses, that is, $n = 4$, is shown. The optimum-filter criterion for this case states that the power spectrum of the signal should be identical to the filter frequency response $|H(\omega)|^2$. The optimum delay line is close to this. As the number of pulses in the pulse train increases, the small lobes in Fig. 15 increase in number and decrease in size and the main lobes become narrow as $1/n$. The width of the spectral lines of the frequency response of the delay line (Fig. 14) decreases in a slightly slower manner with increasing n; hence as n increases, the approach to the optimum filter criterion is closer.

Brick and Galejs

9. *Detection Enhancement*

In addition to the aforementioned techniques designed specifically to discriminate between desired signals and interference, proper use of the available information by the second detector and the radar display is an important factor in improving operation in an interference environment. This is especially true in the presence of noise.

Most radar receivers do not use the best method of detection, that is, coherent detection,[18] for interference rejection purposes. The advantages of a coherent detector are obvious from an examination of Figs. 16 and 17. In both of these figures the abscissa is the signal-to-noise ratio at the input to the detector, indicated by small s; the ordinate is proportional to the ratio of output-to-input signal-to-noise ratios. The curves are shown for three detectors: the square-law detector, the linear detector and the coherent detector. The asymptotic values of the curves for large input signal-to-noise ratios are also indicated. These curves were computed by Smith.[18] In Fig. 16 the output signal-to-noise ratio shown, S_0, is defined as the ratio of the output signal power to the rms fluctuation power when no signal is present. This particular definition of signal-to-noise ratio indicates the shift in output when the signal is applied. The advantage of the coherent detector is evident in Fig. 16 for input signal-to-noise ratios of only less than one. Since this is the range where noise interference creates the largest problem, the advantage of the coherent or synchronous detector is obvious. This advantage is even more evident when one examines the actual output signal-to-noise or output signal-to-fluctuation ratio defined when both signal and noise are present. This is the factor S_s shown in Fig. 17. Here, the advantage of the coherent detector remains for the whole range of input signal-to-noise ratios. Thus, it is seen that a significant advantage is offered through the use of coherent detection. However, an efficient utilization of coherent detection requires an accurate a priori knowledge of the phase and frequency of the input signal.

The type of display itself is also important in radar systems using human operators for observation of displays. The "Mouse Under the Rug" technique mentioned by Lawson and Uhlenbeck[19] and discussed by Javid[20] provides a method of improving operator detectability of a signal in noise-like interference. In essence, this technique utilizes instead of a PPI, which is a range azimuth scanning system with signal indications provided by Z-axis intensity modulation, an A-scope, which is a range versus signal amplitude type display. During World War II, radar operators found it more advantageous to observe the characteristic break in the A-scope baseline of the display occurring in the presence of the signal than to observe the peaks of the signals. The former would provide a much more sensitive detector for pulsed signals buried in noise-like interference. This occurs because the peaks of the signal and noise are jagged and uneven, but the baseline shift, due to the shift in the average value of the probability distribution of the output when signal is present in the noise, is smooth and much sharper than the peak. The nomenclature, "Mouse Under the Rug," is quite apropos for this characteristic. Although sensitivity increases of as much as 10 db have been observed, the technique without further instrumentation is inconvenient to use in a scanning radar. Javid[20] has described a method of providing this instrumentation. His primary purpose was to reduce ambiguities due to higher order time around returns in high PRF radars.

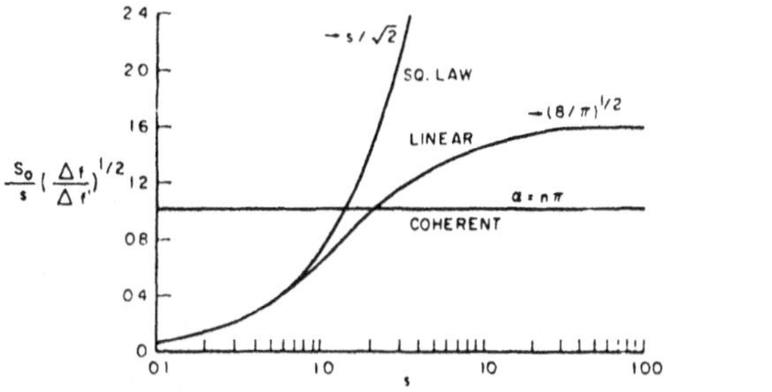

FIG. 16 Signal/noise ratio S_0 at output of detector ($\Delta f' << \Delta f$).

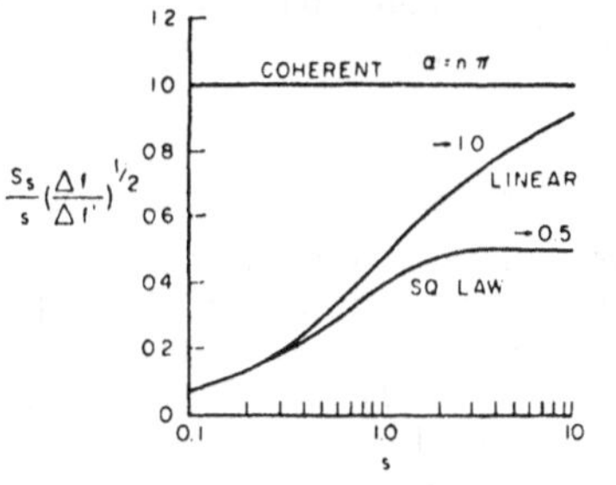

FIG. 17 Signal/noise ratio S_s at output of detector ($\Delta f' << \Delta f$).

V. Unorthodox Radar Systems for Improved Interference Suppression

Radar systems capable of suppressing several types of interference use complex waveforms characterized by large bandwidth and time duration products. These systems provide good range and velocity resolution which are proportional to the bandwidth and time duration of the signal, respectively. Typical wideband signals are

Radar Interference

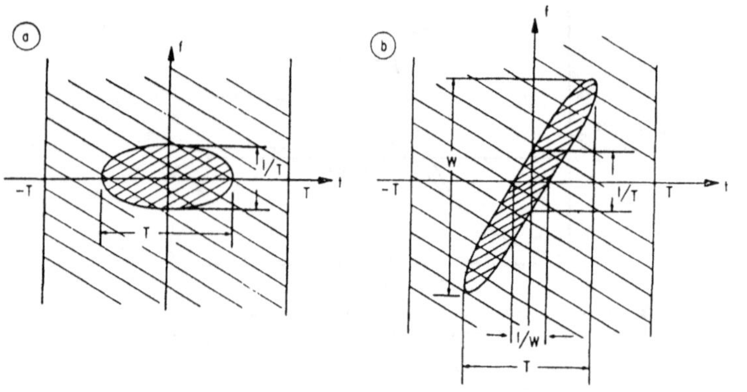

FIG. 18 Frequency-time uncertainty plot for a single pulsed sinusoid and a linear sweep FM pulsed carrier.

FIG. 19 Frequency-time uncertainty plot for a binary coded pulsed waveform.

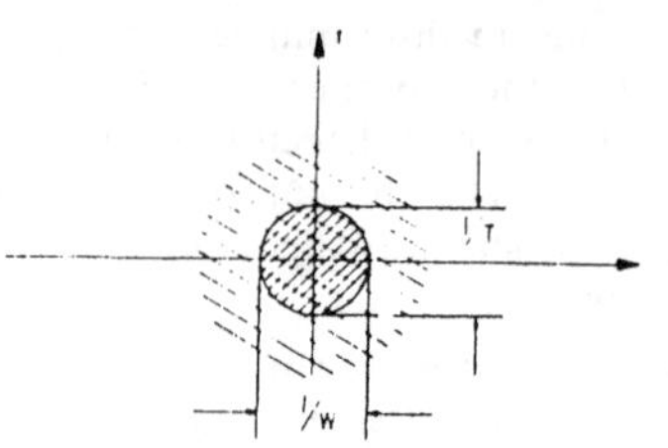

generated by frequency modulation, pulse coding and noise modulation. The interference suppression in such radar systems is based on waveform, spectral or statistical discrimination.

1. *Waveform Considerations*

The capabilities of complex radar waveforms and their limitations can be estimated from the ambiguity plots of Woodward.[21,22] Woodward has introduced the cross-correlation function between a radar signal and the same signal shifted in time and frequency, as a measure of ambiguities in range and velocity that occur in the use of these radar waveforms. A target with range delay t_0 and doppler shift f_0 cannot be resolved from a target with parameters $(t_0 + t)$ and $(f_0 + f)$ if the cross-correlation function

$$X(t,f) = \int_{-\infty}^{\infty} u(\tau)\, u(t + \tau) \exp(-2\pi i f \tau) d\tau$$

is close to unity. The three-dimensional plot of the correlation functions can be reduced to two dimensions by shading parts of the f-t plane. The cross-hatched areas in the frequency-time plane of Figs. 18 and 19 represent regions in which the cross-correlation function is close to unity and where signals with different delays and frequencies cannot be distinguished in the presence of small amounts of noise. The lined area represents regions in which the cross-correlation function is still sizable and where different signals are indistinguishable in the presence of larger amounts of noise. Thus, the shape of the cross-hatched area represents the possible range and velocity ambiguities associated with that particular radar waveform. Using an analogy with the pulse radars, the ambiguity contour indicates the combinations of range and velocity gates which will be simultaneously activated after the reception of the radar waveform.

Figure 18(a) presents the uncertainties for a single pulsed sinusoid. If time accuracy is important, the pulse length, T, is made small and the resulting loss of accuracy in frequency is accepted. A waveform with linearly increasing or decreasing frequency was suggested to provide a means of accurately determining time and frequency. Fig. 18(b) illustrates the characteristics of this waveform. It may be seen that a number of combinations of t and f are equally probable. The accuracy of the short pulse, $1/w$, where w is the total frequency excursion, can be obtained only if the frequency, that is, the radial velocity of the target, is accurately known. If the frequency is not accurately known, the range accuracy decreases to that of the long pulse, T. Both the range and velocity can be determined accurately if the pulse is frequency-modulated with a triangular wave rather than a ramp. This, in effect, yields a second measurement with a frequency-time representation similar to that in Fig. 18(b), but with the ellipse lying in the second and fourth quadrants. Now, range and frequency can both be accurately determined, but only if the two measurements are known to apply to the same target. In the case of multiple targets the ambiguity plot contains more than two intersecting lines, and it is difficult to determine which intersections represent true targets and which represent ghosts.

Figure 19 presents the uncertainties for a coded pulse or noise modulated nonperiodic waveform. The ambiguity plot is a single dot for sufficiently long durations and wide bandwidths of the signal. This waveform gives the same accuracy of determining target speed and range in one measurement as the FM/CW waveform yields in two measurements without introducing multiple target ambiguities.

Modified ambiguity contours can also be used to determine the degree of interference rejection afforded by a particular radar waveform. The cross-correlation function between the radar waveform and the interference waveform plotted in the f-t plane should be negligible relative to the primary correlation peak of the radar waveform, in order to achieve an effective interference rejection.

2. *Triangular FM Systems*

Triangular FM/CW systems[23,24] transmit continuously. These systems provide the lowest possible peak powers for constant average powers in the transmission.

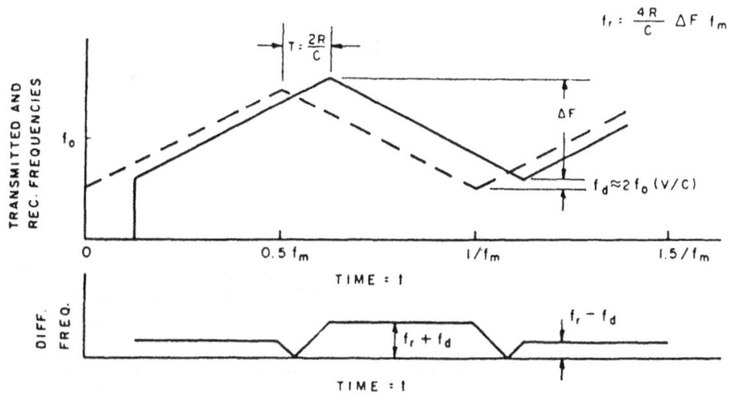

FIG. 20 Frequency waveforms of FM/CW radars.

The differential frequency between the transmitted and the received signals is measured in such systems during the up and down sweeps of frequency. As may be seen from Fig. 20 the sum of the two measurements is proportional to the range, while the difference is proportional to the velocity. The range and velocity accuracy depends on the accuracy of frequency measurement. Long modulation repetition periods permit accurate range and velocity determination if narrow-band filters are used. The theoretical limit to the measurement accuracy is given by target motion during the observation time. The differential frequencies $(f_r + f_d)$ and $(f_r - f_d)$ will be frequency modulated because (1) the doppler shift f_d is proportional to the instantaneous carrier frequency which changes with time, and (2) the range shift f_r is proportional to the instantaneous target range, which changes with time for moving targets. The ultimate resolution is reached if the frequency sweep of the differential frequency signals during the observation time is of the order of the filter bandwidth. However, system drift and instability or waveform nonlinearities may preclude the realization of such a resolution. Also, FM/CW radar systems have a rather limited multiple target handling capability. In the presence of multiple targets several values of $(f_r + f_d)$ are measured during the downsweep of the carrier frequency and several values of $(f_r - f_d)$ are measured during the upsweep of the carrier frequency. A number of combinations of f_r and f_d are equally likely unless a priori information is available to reject certain target ranges or velocities.

The FM/CW radars are sensitive to signals in a narrow frequency band swept along the transmitted waveform. Only a small fraction of the total energy of wide band noise-like or pulsed interference signals is accepted by the narrow band filters. Single frequency CW or nonsynchronous swept-frequency interference signals are similarly rejected.

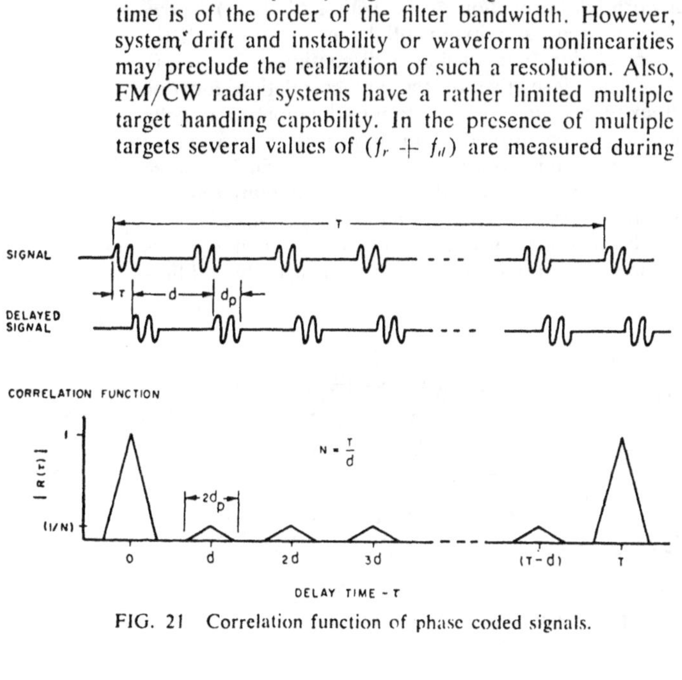

FIG. 21 Correlation function of phase coded signals.

3. *Binary-Coded Phase-Modulated Waveform*

A radar waveform consisting of pulses whose phase changes from pulse to pulse according to the binary code selected has been described by Siebert[22] as an approximation to the ideal radar waveform which should provide a simultaneous and accurate indication of range and velocity. Shift register codes described by Huffman[23] would constitute a possible coding waveform. The individual elements of such codes assume the value of $+1$ or -1. The code is repetitive and has a period of $N = 2^n - 1$ elements, where n is an integer. The binary codes have autocorrelation functions

$$\overline{x_i x_{i+k}} = \begin{cases} 1 \text{ for } k = 0, N, 2N, \ldots \\ 1/N \text{ for } k \neq 0, N, 2N, \ldots \end{cases} \quad (1)$$

An example for $N = 7$ is the sequence 1, 1, 1, -1, -1, 1, -1. Such binary codes can be conveniently generated in free-running binary feedback networks or shift registers.

The object of coding a radar transmission is to "tag" each transmitted pulse element with $+1$ or -1 information corresponding to the binary code selected; to receive the tagged pulse train, to distinguish between properly and improperly coded sequences; and to selectively integrate only the properly coded sequences. Because of the autocorrelation property shown above, interference like spurious individual pulses, uncoded sequences of pulses, and coded sequences with the wrong time delay will produce signal outputs that are N times less than the output produced by the properly coded sequences with the required time delay.

The application of the above codes is limited by the spurious responses that they generate. These codes are optimum for an infinite repetition sequence of pulses in the absence of doppler shifts. Such an idealized range response is proportional to the autocorrelation function of the code depicted in Fig. 21. In a practical system only a finite number of code elements is available that can be received successively. The gradually increasing number of received pulse elements precludes an averaging operation indicated in Eq. (1) and causes higher spurious correlation peaks even for zero doppler shift.

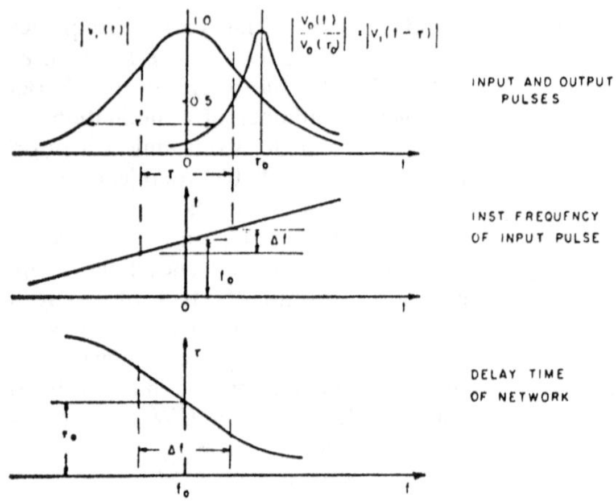

FIG. 22 Pulse compression.

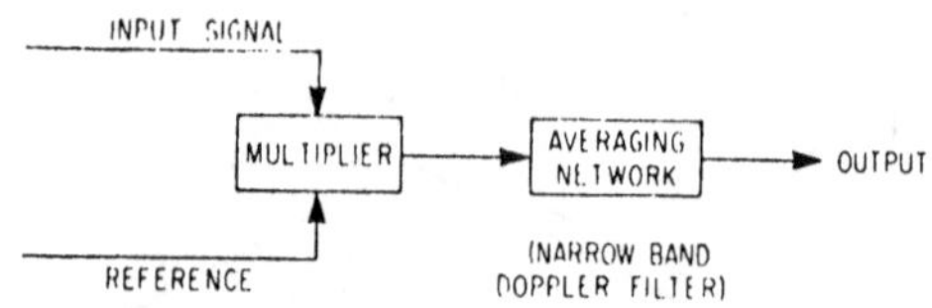

FIG. 23 Block diagram of basic correlator.

A limitation of the linearly swept FM waveform is evident from Fig. 18(b). It is not possible to distinguish between targets with delay times and doppler shifts that result in overlapping *f-t* plane contours. This may affect the ability of the system to discriminate between closely separated targets of approximately equal velocities.

4. *Pulse Compression*

Pulse compression designates a procedure in which wide frequency-modulated r-f pulses are compressed into narrow pulses of increased amplitude.[26] This procedure was evolved in Germany during the last war. The patent of Hüttmann on pulse compression dates back to 1940.[27] The theory outlined by Krönert is based on the theoretical investigations of this problem made by Professor W. Cauer. An identical procedure is described in a U.S. patent by Darlington.[28]

A wide linear frequency-modulated pulse can be compressed by filters whose delay time varies with frequency but whose attenuation is independent of frequency. One operates on the skirt of the filter where delay decreases with increasing frequency if the frequency of the r-f carrier increases linearly with time as shown in Fig 22. The earlier input oscillations of lower frequency will be delayed by the filter, while the oscillations of higher frequency entering the filter later will be advanced relative to the lower frequency oscillations. The pulse at the filter output is compressed and has an increased amplitude if the delay characteristic of the filter is properly matched to the pulse shape and its frequency modulation. For a given amount of energy the pulse compression of μ $(0 < \mu < 1)$ corresponds to an increase in instantaneous power by a factor of $1/\mu$ and an increase in pulse amplitude by the factor $1/\sqrt{\mu}$. The numerical examples given by Krönert[26] for Gaussian-shaped pulses indicate the feasibility of achieving compression ratios of $\mu = 0.125$ to 0.025.

The interference rejection in linear FM pulse-compression radars is based on waveform discrimination similar to that in FM/CW radars. Also, the increased average power for constant peak power and constant range resolution affords higher interference rejection than is feasible in single-frequency pulsed radars.

5. *Noise Correlation*

Continuous wide-band noise signals are transmitted in noise-correlation radar systems.[29] The received signal is correlated with a delayed duplicate of the transmitted signal.[30] The target is at a range corresponding to the delay time that results in maximum cross-correlation. Only the target returns with the proper delay time provide correlation. Uncorrelated noise, pulse trains or spurious individual pulses cause only a negligible correlator output. Therefore, the system is capable of rejecting such interference. The correlator may consist of a mixer followed by a narrow band i-f amplifier or doppler filter as indicated in Fig. 23. The range response is proportional to the autocorrelation function which is related to the power spectrum of noise by the Fourier transform. Typical autocorrelation functions and noise spectra are shown in Fig. 24 for equal half-power bandwidths of the noise. It may be seen that a flat noise power spectrum results in an oscillatory range response. A noise spectrum of $(\sin x/x)^2$ results in a triangular autocorrelation function. Noise with a $(1 + x^2)^{-1}$ power spectrum gives an exponential range response. If the autocorrelation function has a discontinuity at $\tau = 0$, the second derivative of the autocorrelation function will exhibit a spike or an approximation to a delta function which provides an improved indication of $\tau = 0$ or an improved range resolution.[29] Such a range response can be achieved by twice differentiating the correlator reference signal as indicated in Fig. 25.

Each range gate of a noise-correlation radar requires that the identical noise waveform be delayed by a differ-

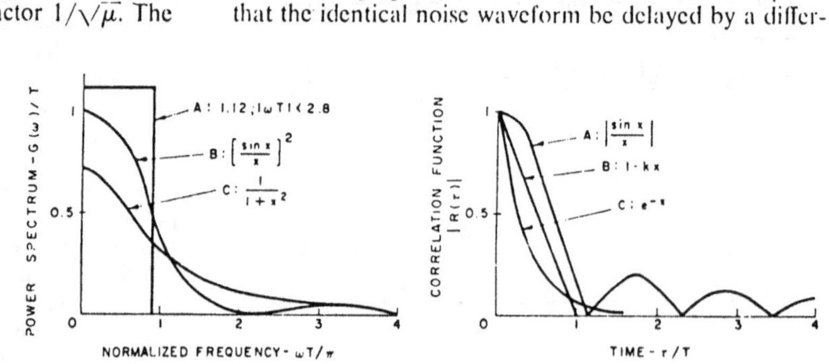

FIG. 24 Power spectra and autocorrelation function.

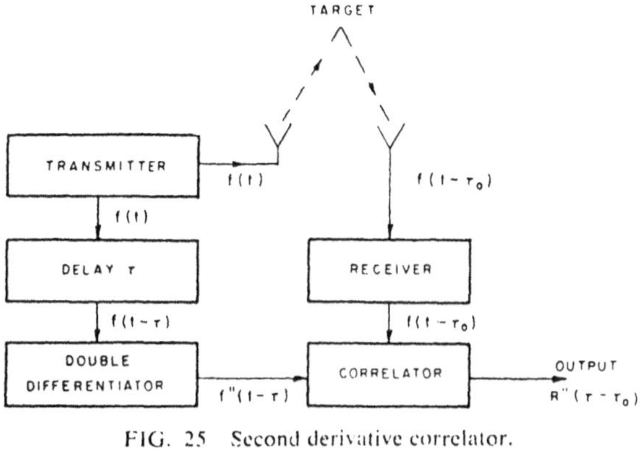

FIG. 25 Second derivative correlator.

ent amount. The number of delay lines and also the number of correlation mixers is equal to the number of range gates, which increases the complexity of noise radars that examine a large number of range cells simultaneously. Furthermore, the delayed noise waveform must duplicate the original noise waveform in all frequency components in order to obtain maximum correlation. This imposes severe linearity and frequency response requirements on the delay lines and the associated amplifiers.

6. *Interference Suppression*

The radar techniques described earlier discriminate against narrow band continuous or wide band low-duty-ratio interference. Only a small fraction of the interfering signal energy is accepted by the radar receiver. A significant amount of interference can be achieved only with wide band large-duty-ratio or continuous noise-like signals. If the interfering signal does not match the fine structure of the radar signal, the interfering signal is still rejected unless it has a considerable power advantage. To duplicate the fine structure of the radar signal, intentional interference requires complex detection and waveform analysis equipment. Even if such equipment is available, the interfering transmitter will have difficulty in keeping up with the radar which may change its signal structure periodically.

ACKNOWLEDGMENTS

The authors wish to acknowledge the assistance provided them by the references listed. In particular, appreciation is expressed to the McGraw-Hill Book Company, Inc. for permitting reproduction of parts of the following figures from the Massachusetts Institute of Technology Radiation Laboratory Series: 16.1 from Volume I and 12.15, 12.16, 12.17, 12.18 and 12.19 from Volume XXIV; in this paper these figures are numbered 8, 2, 3, 4, 9 and 10 respectively. The aid of the authors' associates at Sylvania, whose work on interference reduction contributed greatly to the contents of this paper, is also gratefully recognized.

REFERENCES

1. J. L. Lawson and G. E. Uhlenbeck, "Threshold Signals", *Rad. Lab. Series*, Vol. 24, McGraw-Hill Book Company Inc., New York, 1950, p. 341.
2. *Ibid.*, p. 306.
3. *Ibid.*, pp. 302, 308, 345.
4. *Ibid.*, p. 308.
5. *Ibid.*, p. 311.
6. *Ibid.*, p. 313; also L. N. Ridenour, "Radar System Engineering", *Rad. Lab. Series*, Vol. I, McGraw-Hill Book Company Inc., New York, 1947, p. 459.
7. *Final Engineering Report on Effects of the Type of Polarization on Echo Characteristics*, Antenna Lab., Ohio State University Research Foundation, 15 June 1953 to 30 September 1954.
8. Lawson and Uhlenbeck, *op. cit.*, p. 360.
9. *Ibid.*, p. 362.
10. J. F. Lamb, "A Noise Silencing I.F. Circuit for Superheterodyne Receivers", *QST*, February, 1936.
11. A. N. Schukin, *Bull. Acad. Sci. USSR*, Ser. Phys. 10, No. 1, pp. 49-56, 1946.
12. Lawson and Uhlenbeck, *op. cit.*, pp. 324 ff.
13. Ridenour, *op. cit.*, pp. 626 ff.
14. Lawson and Uhlenbeck, *op. cit.*, p. 363.
15. J. H. Van Vleck and D. Middleton, "A Theoretical Comparison of Visual, Aural and Meter Reception of Pulsed Signals in the Presence of Noise", *J. Appl. Phys.*, Vol. 17, No. 11, pp. 940-971, November, 1946.
16. D. O. North, "An Analysis of the Factors which Determine Signal-Noise Discrimination in Pulsed Carrier Systems", RCA Laboratories Report PTR-6-C, June, 1943.
17. J. Freedman and J. Margolin, "Signal-to-Noise Improvement Through Integration in a Delay-Line Filter System", M.I.T. Lincoln Laboratory, T.R. No. 22, 13 May 1953.
18. R. A. Smith, "The Relative Advantages of Coherent and Incoherent Detectors: A Study of Their Output Noise Spectra Under Various Conditions", Monograph No. 6, Radio Section, *Proc. IEE*, 1951.
19. Lawson and Uhlenbeck, *op. cit.*, p. 248.
20. M. Javid, "Performance Characteristics of Baseline Scanning Ambiguity Filters for High PRF Radars", Unpublished Ph.D. Thesis, McGill University, August, 1955.
21. P. M. Woodward, *Probability and Information Theory, with Applications to Radar*, McGraw-Hill Book Company Inc., New York, 1955, pp. 118-125.
22. W. McSiebert, "A Radar Detection Philosophy", *I.R.E. Trans.*, Vol. IT-2, No. 3, pp. 204-221, September, 1956.
23. D. G. Luck, *Frequency Modulated Radar*, McGraw-Hill Book Company Inc., New York, 1949.
24. L. N. Ridenour, "Radar System Engineering", *Rad. Lab. Series*, Vol. I, McGraw-Hill Book Company Inc., New York, 1947, p. 143.
25. D. A. Huffman, "The Synthesis of Linear Sequential Coding Networks", *Proc. Third London Symposium on Information Theory*, pp. 77-95, September, 1955.
26. R. Krönert, "Pulse Compression", *Nachrichtentechnik*, Vol. 7, pp. 148-152, 162, April 1957. For an abstract of this article see *Proc. IRE*, Vol. 46, Abstract No. 72, p. 518, February, 1958.
27. E. Hüttmann, "A Procedure for Distance Measurements", German Patent 768068, issued 22 March 1940.
28. S. Darlington, "Pulse Transmission", U.S. Patent 2678997, issued May 12, 1954; Application Filed 31 December 1949.
29. R. Bourret, "A Proposed Technique for the Improvement of Range Determination with Noise Radar", *Proc. of IRE*, Vol. 45, p. 1744; December 1957.
30. For an analysis of correlation detectors see R. M. Fano, "Signal-to-Noise Ratio in Correlation Detectors", M.I.T., Research Laboratory of Electronics, TR. 186, February 19, 1951.

Tools and Techniques of Electronic Warfare
J. Holahan
Space/Aeronautics Magazine, Vol. 33, April 1960
Pages 125-132

2.A.2

2

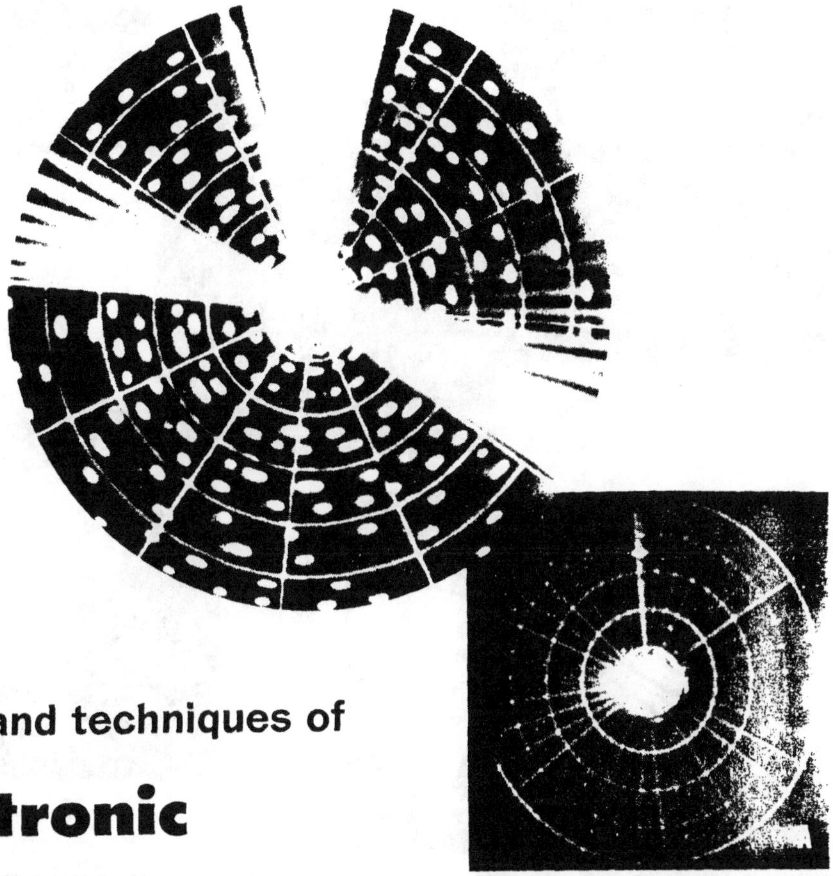

WHICH is the target? Drawing by Sperry Gyroscope's Countermeasures Group (top left) shows how a PPI would look with a combination of noise and false-target jamming. The noise is coming from three locations. The reason for the broad seg-

Tools and techniques of electronic warfare

- Sophisticated ECM designs are still in the lab stage
- RF-power-generating system is the key to good ECM
- Advanced radars are ECM tools in themselves

by **James Holahan,** Electronics Editor

OUR ELECTRONIC countermeasures (ECM) today, whether in operation or in development, are aimed mainly at radar and communications. There is also a medium-sized and growing program in infra-red ECM, and the first steps have been taken on ECM against radio navigation aids.

Radar, because it is the military's principal means of data gathering, accounted for the greatest portion by far of our ECM effort and dollars. And the major area in radar ECM has been airborne systems for manned bombers.

The trend in ECM design is away from brute-force noise jamming and toward "educated" and integrated systems. The final goal is automatic, self-programed designs.

Today, though, a SPACE/AERONAUTICS industry survey shows, the sophisticated ECM techniques with a

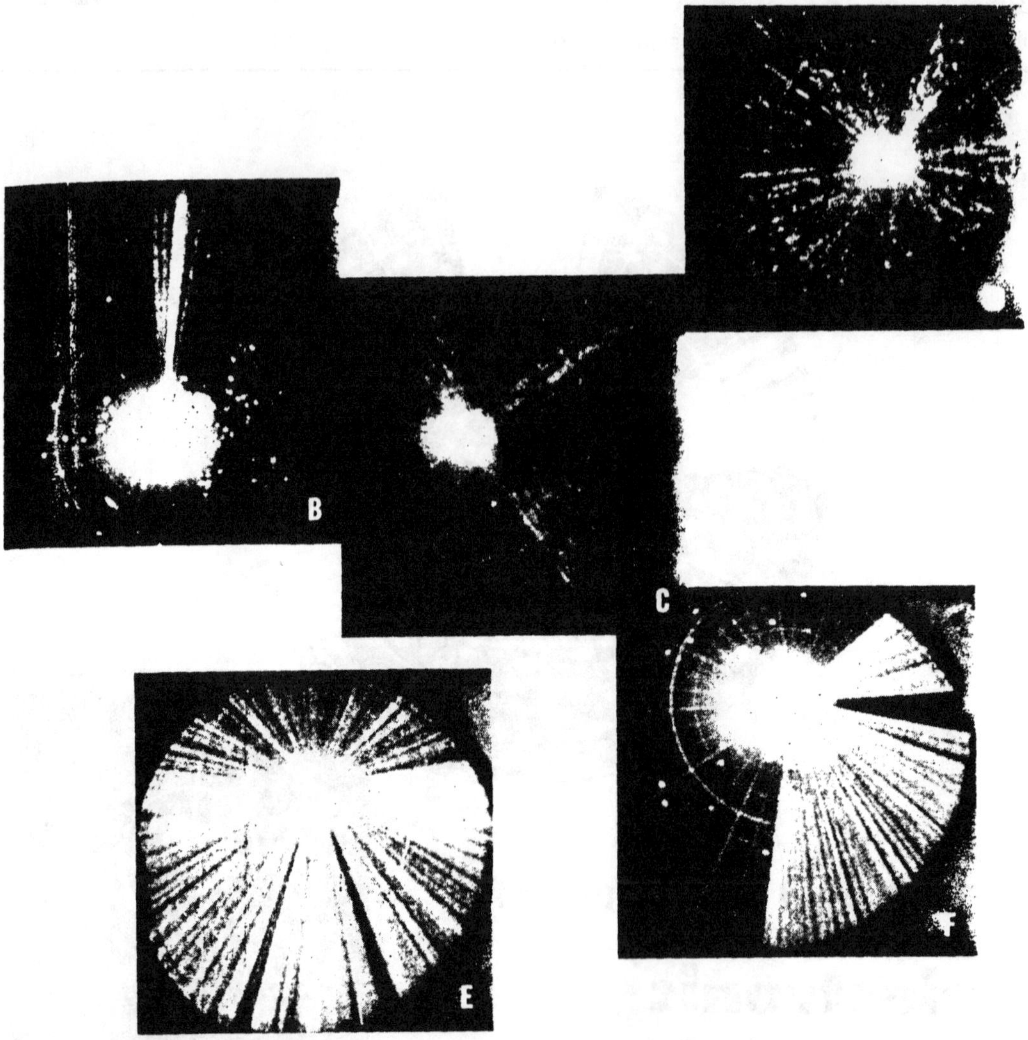

ments is that the jamming is entering the radar's side lobes as well as main lobes. Bottom left and above: Typical noise jamming against a single search radar is shown by photos taken by ECM engineers at GE Light Military Electronics' Advanced Research Center. Both A and B show strobes caused by the jammer. These are not very effective. (The target is plainly visible in A.) The higher degree of jamming in C and D could be due to increased jammer power or radar sensitivity or to poor radar antenna characteristics. Complete jamming is shown in E The effect is the same in F, except that the jamming is limited to certain sectors.

few exceptions are still in the lab stages. For our operation systems, we are relying almost exclusively on brute-force methods—noise jamming and chaff dispensers.

The only approach so far to an integrated ECM system is that used on the Convair B-58. Developed over five years ago and now being produced by Sylvania, this system is made up of three subsystems: radar sensor, jammers, and chaff dispenser. Its operation can be either automatic or manual.

Radar ECM may be very arbitrarily divided into two types by their results: confusion and deception. In both cases, obviously, the aim is to blank out, delay, or down-grade the enemy's radar-derived data. The difference between the two cases arises because the enemy normally knows when he is being confused but not when he is being deceived. (In deception, he is fed false data that he thinks are correct.)

Deception techniques include repeater jammers, transponders, inverse gain, and gate stealers.

Confusion techniques include active, wide- and narrow-band noise jammers, multiple-target generators, decoys, and chaff. Most known active confusion jammers rely on penetrating the major and minor lobes of the radar's radiation.

Barrage jamming is the common name given to wide-band noise generators. Its advantage is that the jammer needs to know little or nothing about the radars it is attempting to jam. Barrage jamming also is effective against all known types of modulation and signal processing, provided it has a healthy power advantage over the radar signals. On the other hand, barrage jamming wastes power, interferes with friendly electronic measures, and is detected easily by passive direction finders.

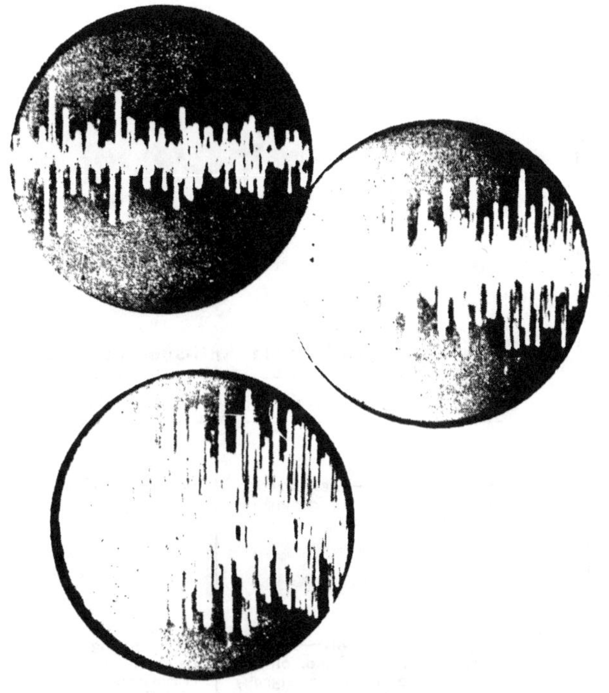

FIGURE 1: Photos of Soviet jamming of a Voice of America program broadcast from Tangier, Morocco, and received in Switzerland shows signal transmitted in the 21-mc band (left), signal just visible in noise shortly after jamming had started (right), and jamming at full intensity (bottom).

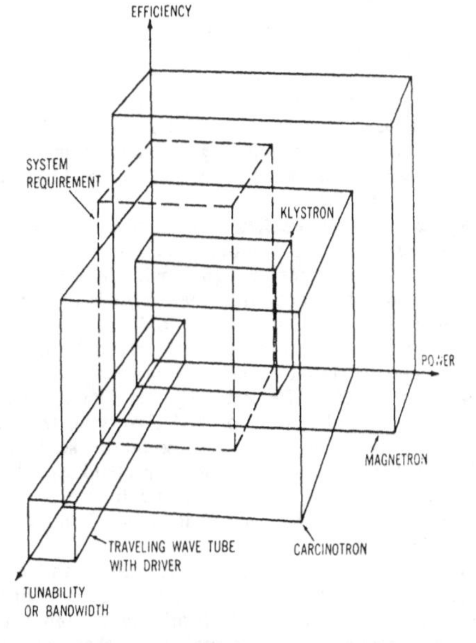

FIGURE 2: Choice of RF power source is the major problem confronting the jammer designer. Here microwave sources are compared for tunability, power, and efficiency with an arbitrary system requirement. None fully meets all the requirements, so the designer must make the best compromise. Other considerations are modulability, weight and size of the tube and of associated equipment (power supplies, etc.), and reliability in severe environment.

The main reason for the wide use of noise modulation in ECM is that any intelligence transmission is a probability function of the signal-to-noise ratio. Early receivers could be jammed even when the jamming-to-radar-signal ratio (J/S) was below 1 db at the receiver input. With newer receivers, this ratio in many cases has to be many decibels in favor of the jammer.

Known schemes for letting the receiver recover the signal from noise trade time for detectability. Given enough time, the receivers theoretically can see through any amount of noise jamming. Practically, it is limited to very short integrating periods, because of factors like target velocity, scan rate, hits per pulse, and the design problems of long-time integrators.

Jamming on one or several specific frequencies is called "spot jamming." To conserve energy, the jamming signal is tailored to the bandpass of the victim. Many types of modulation are used: AM, FM, noise, stepped tones, CW, etc.

The oldest confusion device, chaff, still ranks high with the military. New ways of coating, packaging, and dispensing it are being worked on. Forward-shot chaff can be used to deceive range and velocity tracking gates. Most of USAF's chaff testing is done at Malabar, Fla., on a range operated for RADC by Radiation, Inc. This company is also doing extensive work on missile and aircraft reflectivity measurements.

The multiple-target generator, another important confusion device, sends back many returns during the interpulse interval of a pulse radar for each received pulse. The radar display would thus show a string of targets for each true return. By penetrating the radar's sidelobes, these can be spread out all over the scope, confusing the operator and saturating any data processing system connected to the radar.

ECM engineers at Rome (N. Y.) Air Development Center have made detailed studies of the displays that most befuddle radar operators. One of the most confusing displays, they say, is made up of a series of concentric, jagged patterns—it gives the operator no contrasting background to work against, as would a completely blacked or white-out scope.

The synchronous rings are generated by synchronizing the jamming signal to some multiple of the radar's PRF. The jagged effect is produced by getting slightly out of synch. This technique, claims RADC, gives an effective jam power many times greater than the power that is actually used.

"Look-through" techniques, which let you receive the enemy's radar while you're jamming his receiver, have been developed by both BuWeps and WADD for following frequency changes during jamming. Truly simultaneous looking and jamming raises many technical problems, which mainly have to do with isolation. The methods in use are believed to rely on time-sharing.

Deception devices, as a rule, are more complex than confusion jammers, but also much more efficient in their use of radiated power. The main problem is building these devices in packages as small and reliable as the confusion jammers.

Typical deception schemes are:

- *Repeater Jammers*—These receive radar signals as they are transmitted, amplify them, remodulate them with false information, and almost instantaneously retransmit them on precisely the same frequency at which they were received. Against radars designed to recognize their own wave shape, a true repeater would have to duplicate that shape.
- *Transponder Jammers*—These may be tuned, partially tuned, or untuned. They repeat the impinging radar signal with some delay. Conceivably they could send back either a much larger signal return (indicating a much larger radar cross-section or closer range) or noise jamming.
- *Range Gate Stealers*—These are used when the target senses that it has been acquired by a tracking radar. The stealers return signals with false delay times, thus "telling" the radar that the target is at a range at which it really isn't. The delay could be lengthened or shortened so much that the range servo limits are exceeded and the radar reverts to the search mode.
- *Velocity Gate Stealers*—Against Doppler tracking radars of the pulse, CW, or FM-CW type, frequencies indicating false Doppler shifts could be sent back to falsify the attacker's speed or to pretend that he isn't moving at all. Other methods could take advantage of the small number of targets that known Doppler tracking radars can handle.
- *Angle Track Breaking* — Elevation and azimuth track in sequential-lobing (conical-scan) antenna systems can be broken by transmitting a signal at the nutating frequency. This signal will either confuse the radar antenna servo or prevent it from following.
- *Radar Absorbing Materials* (RAM)—RADC and Naval Research Lab scientists have been searching for years for a material that will shrink the radar cross-section of aircraft and missiles and make them partially or fully invisible. The problem is finding a material that will absorb impinging radar signals and/or scatter them from the vehicle. Such a material must be (1) thin and light, or it will seriously compromise the flight characteristics of the vehicle; (2) rugged enough for high speed environments; (3) five or six octaves wide in bandwidth in the microwave region.

Even partial invisibility helps

Several classified materials look promising, though they appear to be feasible only for subsonic vehicles. However, a hard-to-see "slow" vehicle may be much better as a weapon than a fast one that sticks out like a sore thumb on the radar screen.

Even partial invisibility is helpful, says the Air Research & Development Command. It points out that, if an RAM-coated missile could attenuate the power of impinging radar signals by just 12 db (so that 94 per cent of the signal power is absorbed), this would have the same effect as cutting the range of the observing radar in half. A vehicle with a reduced radar cross-

Table I: Possible Anti-Jam Techniques*

Basis of Technique	Technique
Spatial selectivity—uses direction or entrance path to distinguish signal from jamming	side lobe reduction, side lobe blanking
Frequency selectivity—discriminates against wide-band, off-frequency, and impulsive signals; also uses velocity discrimination	proper tuning, tunability, filtering, balanced mixing, frequency discriminators, Lamb suppressor, MTI, Doppler shift detection
Amplitude selectivity—uses amplitude characteristics of signal to discriminate against such interference as high energy or impulsive signals or noise	limiting, Lamb suppressor, side lobe blanking, log amplifier, IAGC, detector balanced bias
Time selectivity—uses time of arrival of signal or time separation between successive returns to distinguish between jamming and target returns	pulse interference separator and blanker, delay line integration Lamb suppressor, STC
Signal selectivity—uses signal statistics, shape, spectrum, or polarization	IAGC, FTC, detector balanced bias, circular polarization, optimum filtering, pulse shape discrimination, coherent detection, delay line integration, some unconventional modes of operation like complex waveforms, triangular FM systems, binary-coded PM waveforms, pulse compression noise correlation

* Adapted from D. B. Brick & J. Galejs, "Radar Interference and Its Reduction," Sylvania Technologist, July '58.

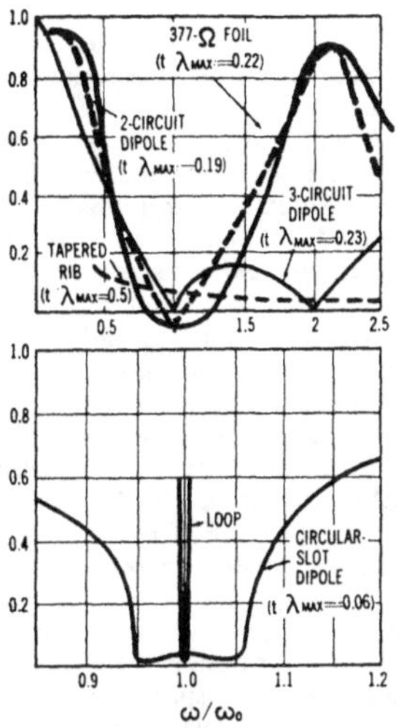

REFLECTION coefficient as a function of frequency for several absorber types for which the ratio of thickness to maximum wavelength at which $|r|$ equals 1 is given. The wavelength is 11 cm at $\omega/\omega_0 = 1$ and 5.5 cm at $\omega/\omega_0 = 2$. Absorbers of these types are being studied at the University of Gottingen, Germany, for RADC.

section also would need less jamming power to screen itself, as the jamming education shows.

Most of the absorbers under consideration are based on transmission line theory. A promising new class is based on the electric dipole moment of the molecules of certain liquids. The problem in this case is finding the right binder material.

- *Radar Camouflage*—RAM, corner reflectors, Luneberg lenses, and transponders can be used to mask ground targets from aerial radars or to set up dummy replicas of landmarks or targets. Buildings can be contoured to give stronger or weaker returns—certain features can be softened with RAM, others can be accentuated with reflectors.
- *Inverse Gain Control*—In this technique, signals are sent back whose strengths are inversely proportional to those of the impinging radar signals.

RF power generation is the key

The ECM engineer must have a thorough technical knowledge of the equipment he is trying to counter—not only of the sensor (i.e., the radar) but also of the entire system in back of it (automatic programing, data processing, etc.). One pitfall, warns Don Beilman, supervisor of ECM engineering at the Advanced Electronics Center of GE's Light Military Electronics Dept., is to design your ECM device around a weakness of the defense that can be eliminated by relatively simple design fixes. What you should do, says Beilman, is to aim at an inherent weakness of the defense that can't be corrected except by major modification or a completely new design approach.

The keystone of ECM design is the RF-power-generating system. Whether it is airborne or ground-based, this system more often than not will be power-limited. The greatest need therefore is for high average power generating devices to cover the usable UHF and microwave spectrum.

True, not all jammers must radiate high powers—actually, a few watts goes a long way under certain conditions, because of the $1/R^2$ *vs* $1R^4$ advantage of the jammer (see *"Good Design Can Beat the ECM Threat," p. 138*). However, the trend in radars is toward much higher powers and to techniques that increase sub-jamming visibility. The direct way of coping with these trends is higher ECM transmitter powers.

To make the most efficient use of the available energy, the ECM designer often tries to concentrate the power he does have in space, frequency, or time. Usually discrimination in any of these categories requires a knowledge of the defense radar's characteristics or intelligent signal analysis circuits.

In ECM as in radar, two types of power-generating systems are used: the power oscillator and the modulated oscillator-power amplifier (MOPA). The former uses magnetrons or carcinotrons and, of late, barratrons (a type of magnetron operated in an unstable mode).

MOPA modulation at low levels

To date, the fixed-tuned (narrow- and wide-band) jammers use the magnetron. Its advantages are simplicity, low cost, availability, low weight, and high efficiency (over 70 per cent). Where rapid frequency tuning is desired, the carcinotron is used. It has a high power output (100-800 kw average), is highly efficient (over 50 per cent), and tunes rapidly (to 100 mc/usec). It is quite expensive, however.

The main drawback of power oscillators for ECM is that they have to be modulated at high levels. This means high voltages and large power supplies. It also restricts the designer in the number and type of modulation schemes he can use.

In the MOPA, a low-power oscillator like an O-type backward-wave oscillator (BWO) or a voltage-tuned magnetron (VTM) drives a string of high-power traveling-wave tubes (TWTSs) in cascade. The BWO and VTM are voltage-tuned and can be rapidly scanned over an octave in frequency. Modulation is done at low levels, so there's virtually no limit to the tricks the designer can apply. (*Figure 2* compares characteristics of the magnetron, carcinotron, TWT, and kylstron. The last tube is of limited usefulness in ECM, because of its relatively small electronically tuned range.)

The trend in ECM transmitters has been to the MOPA. It was clearly shown by the three most advanced systems yet planned by USAF: those for the Convair B-58, Boeing B-52H, and NAA B-70. (Only the first of these is still in being; the other two have been canceled.)

The "electronic shield" being produced by Sylvania for the B-58 apparently respresents the first real effort to develop an integrated, advanced airborne ECM system. Though, neither USAF nor Sylvania has said so publicly, the "special microwave tubes" that the company announced it developed for the system are TWT power amplifiers and 0-type BWOs, according to industry sources.

The canceled ALQ-27, which was developed by Sperry for the B-52H, shaped up as one of the major markets for TWT makers. A family of rugged TWT amplifiers and distributed amplifiers (for below UHF) was designed specially for this system.

The ECM system proposed by Westinghouse for the B-70, observers comment was due to take up where the ALQ-27 left off. It's a safe bet, therefore, to assume that the system was built around the TWT.

The cancellation of the ALQ-27 and B-70 ECM systems may mean at least a temporary setback for sophisticated airborne ECM. WADD will manage a replacement system for the B-52H that apparently will use available hardware and few advanced techniques. Many

Table II: ECM vs ECCM

Countermeasure	AJ Counter-countermeasure
Low-PRF random pulse jam	pulse delay and comparison with pulse on next sweep and blanking
Wide-band noise Very-high-PRF random pulse jam	increased look time and integrating; antenna pattern with high sidelobe discrimination
CW jam	methods for avoiding saturation (IAGC and the like, followed by FTC and third detector)
Pulse repeater jam	PRF jitter, then as for random pulse jam
Short pulse jam	limiting at output of wide-band front end
Slow sweep-through jam	as for random narrow pulse jam
Fast sweep-through jam	as for wide-band noise
Decoys	discrimination on basis of reflectivity, mass, and size
Chaff	Doppler shift, acceleration gating
Range gate stealing	Doppler shift, narrow range gate, acceleration gating
Angular track breaking	simultaneous lobing techniques

Iron Curtain. Today they are estimated to have several thousand high-powered jamming transmitters (with outputs as high as ½ Mw) connected by a highly efficient intercept and control network (*Fig. 1*). Some of these transmitters send their signals over sky waves, others over ground waves. The Red jammers have amazing frequency agility—if VOA jumps to a new frequency, they are right on top of it in seconds.

VOA has studied several possibilities for piercing this electronic curtain, such as higher power, frequency jumping, "cuddle-up" (getting next to an important Russian broadcast station in frequency, so that, in jamming VOA, the Reds would also jam their own station), and propagation diversity. VOA admits it has no feasible method for overcoming ground-wave jamming, but it believes propagation diversity can often beat sky-wave jamming. This technique takes advantage of the difference between ionospheric bending properties seen from Russia and those seen from other parts of the world at certain times of day.

Trend to integral ECCM design

Electronic counter-countermeasures (ECCM) in radar design are commonly referred to as AJ techniques (*Table II*). While many circuits in modern radars owe their existence to anti-jamming, the trend has been away from "hanging" these circuits on as appendages and toward integrating the ECCM capability into the overhall design.

Since most ECM takes the form of RF interference, many of the design approaches that give better performance in natural noise and in a high signal environment of friendly radars will also work against ECM. For example, *Table I*, which shows methods of general "RF interference reduction," is essentially a listing of the basic AJ methods. Actually, discrimination against jamming is always based on one or more of the five types of selectivity listed in *Table I:* spatial, frequency, amplitude, time, and signal. Some of the precepts for ECCM designs that were advocated by radar engineers interviewed by SPACE/AERONAUTICS are:

- *Use unpredictable operating parameters* — The more orderly a radar is in its operation, the easier it is to predict what it will do next, and consequently the simpler it is to confuse or deceive. Uncertainties in parameters like pulse length, PRF, frequency, modulation, and antenna characteristics make ECM more difficult.
- *Recognize your own signal*—Radars that fit this description are based on information theory. Their AJ capability depends on waveform, spectral, or statistical discrimination. The waveforms used in such radars normally have large bandwidths and time duration products.
- *Use trained operators*—A good operator is hard to confuse or deceive. Two operators observing two redundant radars, with each using more than one type of display (PPI, A-scan, etc.) are proof against a lot of ECM.
- *Operate at reduced performance levels when jammed*—No system should be designed that can't exchange resolution for vulnerability to jamming. A system that breaks down completely after a certain jamming level is reached, is poorly designed.
- *Come up with a good basic design*—Provide large dynamic range, a bandwidth just wide enough to receive your pulse, a stable transmitter, large main-to-side and main-to-back lobe ratios in the radiation pattern, etc.

FD is most effective in ECCM

Many of the advanced radars now being built are actually ECCM tools in themselves. Such techniques as pulse compression, noise correlation, pulse-to-pulse frequency or phase change, triangular FM-CW, waveform coding, and matched filters are forcing the ECM designer to apply much more sophisticated equipment.

Radars using these advanced techniques are relatively impervious to both continuous narrow-band swept wide-band noise, but are somewhat susceptiable to continuous wide-band noise. However, for jamming, even such noise must match the fine structure of the radar signal, or it will be rejected (unless the jamming signal has a decided power advantage).

Today's most effective ECCM technique is frequency diversity (FD). A family of long range ground-based (FD) radars has already been developed under a multimillion dollar RADC program—GE's FPS-24, Crosley's FPS-26, Westinghouse's FPS-27, Raytheon's FPS-28, and Sperry's FPS-35. Each of these radars operates in a different frequency band, within which it can rapidly shift frequency. Set up with over-lapping coverage to guard a defense sector, a group of such radars represents about the best type of ECCM yet developed.—End observers believe that USAF's interest in the TWT as an amplifier for high powered ECM systems has cooled off.

The high weight, low efficiency (about 10 per cent at high powers), and relatively high cost of the tube are the main reasons. Most current high power TWTs use electromagnets for focusing. These are quite heavy, especially at the lower frequencies (L-band). Permanent magnet period focusing looks promising for weight reduction and may solve the problem. It has worked out in low power tubes and is now being applied to higher-powered ones. Another technique is electrostatic focusing; however, it raises the power supply weight.

What is needed is a microwave amplifier with an average output between 0.8 and 1.5 kw, 50 per cent efficiency, and at least a 2:1 bandwidth. Efficiency is extremely important in airborne high power amplifiers, since primary power in aerospace vehicles is limited and also because of the cooling problem created by dissipated and unradiated energy.

WADD emphasizes that efficiency is extremely important where high powers are generated, since primary power in aerospace vehicles is limited, and also because of the problem of getting rid of dissipated (unradiated) energy.

Crossed-field forward-wave amplifiers (FWAs) might be one answer. Two M-type FWAs are known to be in development: one by Hughes Aircraft, another by CSF in France. The latter reportedly has a bandwidth of 25 per cent, an average power output, of six kilowatts, and an efficiency of 50 per cent at L-band.

Communications ECM runs a poor second to radar ECM. Part of the reason is that in World War II it was more profitable to eavesdrop on the enemy than to jam him. This situation might be drastically different in a future conflict—secure communications methods now are making it far more difficult to extract intelligence from intercepted signals.

Communications ECM has its own peculiar set of problems, such as:

- In communications, the energy from the defensive equipment is not necessarily beamed at you, as in radar, so quite often you have to look for the signal you want to counter.
- A communications jammer doesn't have the $1/R^2$ vs $1/R^4$ advantage, since signal transmission is over a one-way path. Quite often, too, the jammer is farther from the enemy's receiver than his transmitter.
- The propagation losses of a battlefield jammer might run as high as R^4-R^6 because of ground absorption, reflection, etc.
- Point-to-point communications are well-nigh impossible to detect, let alone to jam.
- Military communications transmitters use low powers, compared with radars. Also, the time constants in communications are in milliseconds rather than microseconds.
- Most military channels are quite narrow in bandwidth, so that large frequency blocks can be jammed with relatively simple equipment.

Jamming power versus bandwidth

The trend in the military has been to narrow-band circuits to get more channels within a given part of the spectrum. This trend would seem to make the jammer's job easier. The military is apparently hoping the enemy would be using the same parts of the spectrum and would be reluctant to jam over wide bandwidths for fear of self-interference. Military communications experts point out that they can always make up for the vulnerability of narrow-band channels by sending the same message simultaneously over several channels.

Dr. J. P. Costas, of GE's Heavy Military Electronics Dept., is an outspoken critics of this approach. As he sees it, an enemy with unlimited resources and determination can jam any communications system. The trick therefore is to force him—through good design—to expend an unreasonably large effort in jamming. "If the most efficient system design is assumed for a fixed data rate in each case," Costas explains, "the necessary power required to jam the circuit varies in direct proportion to system bandwidth. The broader the bandwidth, the more difficult it will be to jam the circuit. Conversely, the narrower the bandwidth, the easier it becomes to jam the circuit."*

Our experience with the Voice of America shows that the Russians are quite expert at communications jamming. Since 1950, they have been silencing about 90 per cent of VOA's broadcasts to countries behind the

* "Poisson, Shannon and the Radio Amateur;" *Proc. IRE*, Dec. '59.

more on next page

TYPICAL random-noise, sweep-through jamming system for L- to X-band.

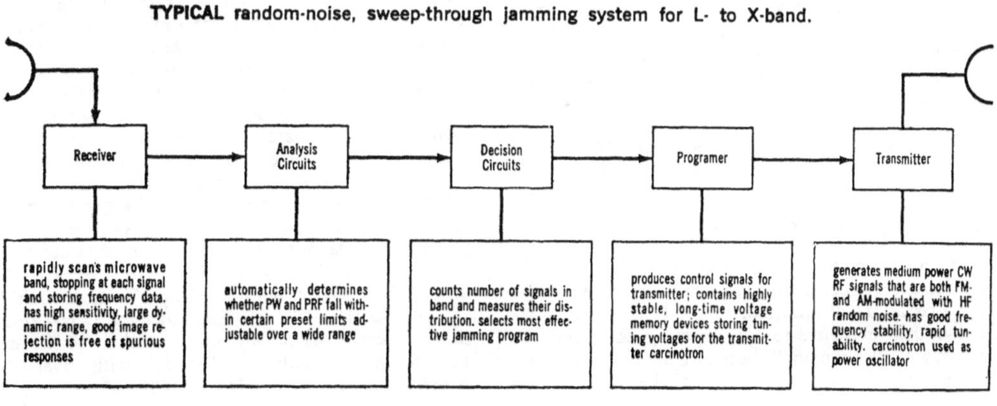

Table II: ECM vs ECCM

Countermeasure	AJ Counter-countermeasure
Low-PRF random pulse jam	pulse delay and comparison with pulse on next sweep and blanking
Wide-band noise Very-high-PRF random pulse jam	increased look time and integrating; antenna pattern with high sidelobe discrimination
CW jam	methods for avoiding saturation (IAGC and the like, followed by FTC and third detector)
Pulse repeater jam	PRF jitter, then as for random pulse jam
Short pulse jam	limiting at output of wide-band front end
Slow sweep-through jam	as for random narrow pulse jam
Fast sweep-through jam	as for wide-band noise
Decoys	discrimination on basis of reflectivity, mass, and size
Chaff	Doppler shift, acceleration gating
Range gate stealing	Doppler shift, narrow range gate, acceleration gating
Angular track breaking	simultaneous lobing techniques

Iron Curtain. Today they are estimated to have several thousand high-powered jamming transmitters (with outputs as high as ½ Mw) connected by a highly efficient intercept and control network (*Fig. 1*). Some of these transmitters send their signals over sky waves, others over ground waves. The Red jammers have amazing frequency agility—if VOA jumps to a new frequency, they are right on top of it in seconds.

VOA has studied several possibilities for piercing this electronic curtain, such as higher power, frequency jumping, "cuddle-up" (getting next to an important Russian broadcast station in frequency, so that, in jamming VOA, the Reds would also jam their own station), and propagation diversity. VOA admits it has no feasible method for overcoming ground-wave jamming, but it believes propagation diversity can often beat sky-wave jamming. This technique takes advantage of the difference between ionospheric bending properties seen from Russia and those seen from other parts of the world at certain times of day.

Trend to integral ECCM design

Electronic counter-countermeasures (ECCM) in radar design are commonly referred to as AJ techniques *(Table II)*. While many circuits in modern radars owe their existence to anti-jamming, the trend has been away from "hanging" these circuits on as appendages and toward integrating the ECCM capability into the overhall design.

Since most ECM takes the form of RF interference, many of the design approaches that give better performance in natural noise and in a high signal environment of friendly radars will also work against ECM. For example, *Table I*, which shows methods of general "RF interference reduction," is essentially a listing of the basic AJ methods. Actually, discrimination against jamming is always based on one or more of the five types of selectivity listed in *Table I:* spatial, frequency, amplitude, time, and signal. Some of the precepts for ECCM designs that were advocated by radar engineers interviewed by SPACE/AERONAUTICS are:

- *Use unpredictable operating parameters* — The more orderly a radar is in its operation, the easier it is to predict what it will do next, and consequently the simpler it is to confuse or deceive. Uncertainties in parameters like pulse length, PRF, frequency, modulation, and antenna characteristics make ECM more difficult.
- *Recognize your own signal*—Radars that fit this description are based on information theory. Their AJ capability depends on waveform, spectral, or statistical discrimination. The waveforms used in such radars normally have large bandwidths and time duration products.
- *Use trained operators*—A good operator is hard to confuse or deceive. Two operators observing two redundant radars, with each using more than one type of display (PPI, A-scan, etc.) are proof against a lot of ECM.
- *Operate at reduced performance levels when jammed*—No system should be designed that can't exchange resolution for vulnerability to jamming. A system that breaks down completely after a certain jamming level is reached, is poorly designed.
- *Come up with a good basic design*—Provide large dynamic range, a bandwidth just wide enough to receive your pulse, a stable transmitter, large main-to-side and main-to-back lobe ratios in the radiation pattern, etc.

FD is most effective in ECCM

Many of the advanced radars now being built are actually ECCM tools in themselves. Such techniques as pulse compression, noise correlation, pulse-to-pulse frequency or phase change, triangular FM-CW, waveform coding, and matched filters are forcing the ECM designer to apply much more sophisticated equipment.

Radars using these advanced techniques are relatively impervious to both continuous narrow-band swept wide-band noise, but are somewhat susceptiable to continuous wide-band noise. However, for jamming, even such noise must match the fine structure of the radar signal, or it will be rejected (unless the jamming signal has a decided power advantage).

Today's most effective ECCM technique is frequency diversity (FD). A family of long range ground-based (FD) radars has already been developed under a multimillion dollar RADC program—GE's FPS-24, Crosley's FPS-26, Westinghouse's FPS-27, Raytheon's FPS-28, and Sperry's FPS-35. Each of these radars operates in a different frequency band, within which it can rapidly shift frequency. Set up with over-lapping coverage to guard a defense sector, a group of such radars represents about the best type of ECCM yet developed.—End

Electronic Dogfight: ECM vs ECCM in Search Radar
P.R. Dax
Space/Aeronautics Magazine, Vol. 33, April 1960
Pages 134-137

2.A.3

- Wide-band jamming keeps down equipment weight and complexity
- Frequency shifting is good countermove to jamming
- Decoy discrimination difficult for radar

by Peter R. Dax, Fellow Engineer, Radar Equipment Engineering, Electronics Div., Westinghouse Electric Corp.

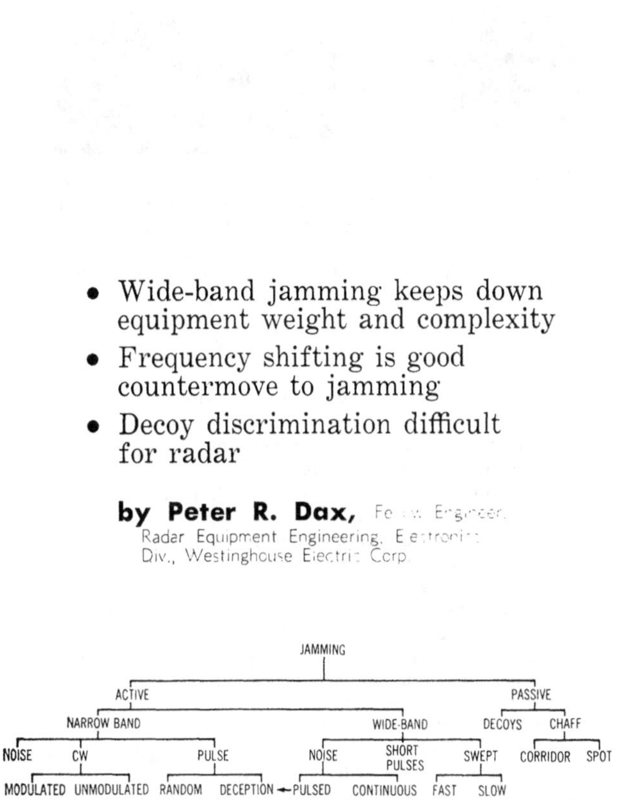

FIGURE 1: Basic jamming methods.

THE CONTEST between electronic countermeasures (ECM) and counter-countermeasures (ECCM) is a highly technical battle whose problems are not always fully understood by the actual contestants in the field. On the other hand, the designer in his lab often fails in understanding, too—when it comes to the practical problems of the contest as it is actually fought out in the field.

What with multiple jammers, multiple radars, and combinations of jam and anti-jam techniques, there's no doubt the situation is rather complex. Take, for example, the basic problems faced by a search radar in a jamming environment and by the airborne jammer that is trying to jam the radar.

The purpose of a radar is to detect targets and determine their position at the greatest possible range. It must report only on *real* targets; and in other words, it must not be deceived by *false* targets.

The purpose of the jammer is to keep the radar from locating a target of interest and possible certain other friendly targets (such as an air-to-surface missile fired at the radar). The jammer does this either by reducing the radar's ability to detect targets or by confusing the radar through the generation of false targets that are hard to tell from real ones. In the latter case, the jammer achieves its purpose by overloading the enemy's data processing system or CIC organization. At the same time, the jammer must try to avoid revealing his presence.

Figure 1 shows the basic jamming methods. In active jamming, the jammer is actually radiating electromagnetic energy. In passive jamming, reflecting material is used to re-radiate energy, generating pseudo-targets where no real ones exist.

* Electronics Div., Westinghouse Electric Corp., P. O. Box 1897, Baltimore 3, Md.

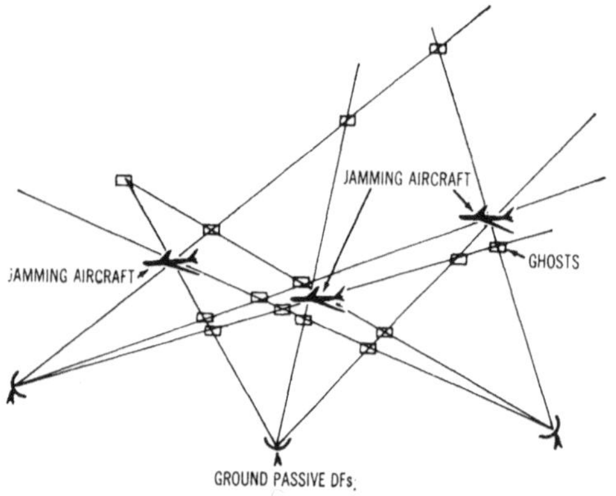

FIGURE 2: Passive direction finders can spot jamming aircraft by taking cross-bearings on their jamming signals. The big problem is ghosts (shown as squares whose centers are the points at which the bearing lines cross but where there are no targets). New designs have succeeded in overcoming the ghost problem.

FREQUENCY DIVERSITY is one of the trump cards in the hand of the radar designer. USAF is developing a network of search radars that operate in different frequency bands and each of which can quickly change frequency within its band. One of the radars in this group is this Westinghouse FPS-27.

Active jamming may be narrow- or wide-band. Combinations of jammers can, of course, mix wide-band noise with CW or active with passive jamming, and any form of jamming may be switched on and off intermittently.

A radar may fail to detect a target because of a poor signal-to-noise ratio or because circuit saturation makes it lose the target information. It can also be deceived by false targets generated by random pulses or by deliberate spoofing.

In practice, an increase in background noise will decrease the weak-target detection capability and increase the rate of false target reporting. The probability of detection of a strong target may remain relatively unaffected, however, providing the receiver is not driven into saturation. This saturation, which may take place anywhere in the receiver, will drive the circuits beyond their region of linear operation, and the target information will be lost.

AJ circuitry prevents saturation

One of the main functions of anti-jam (AJ) circuitry, therefore, is to prevent saturation by performing an AGC action that insures that at least the information on targets larger than the background jamming is preserved.

Overloading of a radar by false targets dilutes the radar's capabilities. The result is delays in target processing, which in turn means attacking vehicles will be able to concentrate the defense system more deeply—just as though there had been a loss in detection range.

Certain facts about active jamming must be clearly understood from the beginning:

- The radar receiver is extremely sensitive. It is therefore relatively easy for a jammer to generate RF power many times greater than the skin return from the target.
- The radar receiver is sensitive only to a very small range of frequencies. For effective jamming, an appreciable amount of power must fall within the passband of the receiver.
- Weight considerations limit the sophistication of the airborne jamming equipment and the amount of available jamming power.
- If the jammer manages to inject noise into the radar receiver, the result is equivalent to a degradation of the receiver's noise factor. Range performance is proportionally lowered, and no AJ circuitry can improve the situation (assuming the receiver has an optimum bandwidth for the radar pulse).

Active jamming most often is a matter of getting unwanted power into the radar's passband in excess of the power the radar gets from the wanted signal. Let's assume a search radar with a peak power of one megawatt and a 30-db antenna gain is illuminating a target with a radar cross-section of 10 m^2. In this average situation, the peak re-radiated power (P) from the target at a range of 200 km (108 nm) is:

$$P = P_t G\sigma / 4\pi R^2 = (10^6 W \times 10^3 \times 10 m^2)/[4\pi(200 \times 10^3)^2]$$
$$= 0.02 W,$$

where the symbols are those used in the standard radar equation.

Jammer power increase is lower

Admittedly, it is fairly easy for a jammer to transmit two W/mc, or 100 times the target skin return for our example, over a wide band of frequencies. However, as range decreases, the skin return (as seen from the radar) increases as the reciprocal of the fourth power of the range, while the jammer power increases only as

CARCINOTRON is by far the most widely used variable-frequency power generator of jamming signals in the UHF-to-microwave region. It can tune over a 1.5:1 bandwidth at rates of over 100 mc/usec. Peak power output is well over 100 kw. One form of sweep-through jammer could use a carcinotron modulated by noise to produce a random sweep. The Russians are known to have bought a large supply of carcinotrons in France.

the reciprocal of the square of the range:

$$P_{signal} = P_t G\sigma A/(4\pi R^2)^2,$$
$$P_{jammer} = P_j BA/4\pi R^2,$$

where A is the effective antenna area; B, the receiver bandwidth; and P_j, the jammer power per unit bandwidth.

From these equations, it is obvious that at a certain range the skin return is greater than the jamming background and the target becomes detectable. There are several methods for computing this so-called "self-screening range" (R_{sc}). For our example we can assume that the target will be seen when the power of the returned signal equals the jammer power:

$$(P_t G\sigma A/(4\pi R_{sc}^2)^2 = P_j BA/4\pi R_{sc}^2,$$
$$R_{sc}^2 = \tfrac{1}{4}\pi \times P_t/P_j \times G\sigma/B.$$

We find that R_{sc} equals 15 nm (assuming B equals 0.5 mc).

Our computation gives merely an order of magnitude for the effect of distance on received signal and jammer power. A more accurate formula would take into account the actual S/N required for detection and therefore the number of pulses integrated in a beamwidth (the jamming adds incoherently) and the variation of antenna gain with elevation. Furthermore P and σ are not independent: in practice, a large vehicle puts out more jamming power.

We can see that the first job of the jammer is to find the radar frequency and jam on this frequency. Conversely, the radar must avoid the jammer's frequency. A good anti-jamming radar above all must be able to shift frequency quickly and, if possible, over a wide band.

The ultimate in anti-jamming is reached when the frequency is changed on each pulse in a random manner. With the fastest possible reaction time, a narrow-band, or "spot", jammer can then only jam the receiver from the instant the radar pulse is received to the end of the pulse interval. The problem is further complicated for the jamming vehicle because it probably must jam several radars at once. The only solution is to use wideband jamming and accept the resulting power dilution.

Spot jamming, we have seen, is no good against radars with frequency "agility". However, there are older radars that don't have this agility, and there is also deception jamming, which the radar may not even realize is going on. In these cases, some form of countermeasures must be used against spot jamming.

Figure 1 shows that narrow-band jamming can be CW, pulse, or noise. Normally the output of the radar receiver consists of target returns competing with front-end noise (if we neglect interference due to clutter or weather). The target returns have certain predictable characteristics, like pulse length and, for a group of target pulses, beamwidth.

Theoretically, the receiving and processing circuitry can filter out signals that do not show the expected target characteristics. For instance, each received pulse that exceeds a certain threshold can be stored before display for the duration of one sweep and compared with the same range increment on the next sweep. If the pulse turns out to be isolated (i.e., random), it can be blanked out, so that it will not reach the display. Long pulses can be differentiated in a short-time-constant circuit and reduced to short pulses.

This theory for the elimination of random pulse jamming (which also covers accidental interference, as from another radar), holds true only at low repetition rates. If the rep rate is high compared with the PRF of the radar, the blanked periods may become too long and targets will be suppressed.

Special circuits help against jam

Combatting CW jamming is mainly a question of avoiding receiver saturation by using special circuits (of which IAGC is a classical example). If saturation can be avoided, the dc component of the detected output can be removed by a fast-time-constant circuit and the beat between wanted signal and CW jamming can be detected by a third detector rectifying the video. However, since the AGC action will reduce the gain in proportion to the jamming, both noise and target will be reduced in amplitude unless there is a corresponding automatic increase in video gain. The whole procedure still offers only a poor alternative to a shift in frequency, especially since fast-time-constant circuits can be foiled by modulated CW.

Deception jamming, usually in the form of pulse repeater jamming, consists of re-transmitting the radar pulse received by the jammer after a suitable delay and gating the transmission for a time corresponding to the beamwidth of the radar antenna. From the point of view of the radar, the received jamming looks exactly like a target, especially since it can be programed to have realistic motion.

Certain problems are associated with this form of jamming. The RF pulse, for one thing, cannot be delayed by more than a few microseconds at the present state of art. If there is any further delay, the pulse first must be converted to video and the delayed video signal used to modulate the transmitter.

If the jamming signal happens to be narrow-band, finding the correct frequency is a problem. One solution is to transmit a wide-band signal, such as a burst of wide-band noise. If the radar transmitter PRF is jittered, then the false target can only be generated at a range greater than the jammer range—if the pulses are delayed until they appear on the next sweep, they will not be correlated in range and therefore can be eliminated by the receiver as random.

If the radar doesn't shift frequency on a pulse-to-pulse basis, it may be jammed by a deception jammer and never know it, especially if the jammer is generating realistic tracks. The only solutions in such a case are frequent and radical frequency changes irrespective of the jamming situation or picture correlation with another radar.

Narrow-band jamming uses CW modulated by noise. Usually, the depth of modulation is such that there is a considerable amount of CW. The noise jamming cannot be eliminated, but anti-CW measures will help. Against all other forms of narrow-band jamming, a shift in frequency is the only really effective counter-measure.

Operator or complex gear needed

Narrow-band jamming requires an ECM operator or fairly complex equipment to determine the frequency of the radar and to track it when it shifts. In addition, there must be as many jamming circuits as there are radars to jam. For a given equipment weight and minimum complexity, therefore, a wide-band jammer appears to be best.

Wide bandwidth may be obtained by transmitting very short pulses, by sweeping a "narrow-band" jammer over a band of frequencies, or by amplifying wide-band noise in a wide-band traveling-wave-tube transmitter. The very-short-pulse method is hardly wideband enough to be included in this list, yet it is definitely out of the narrow-band category. A short pulse in the receiver passband has enough energy to make it ring for an appreciable period—on the order of the radar pulse length (τ). If the PRF of the jamming pulse is on the order of $1/\tau$, then the receiver is continually jammed.

This technique requires a high-powered jamming pulse. The receiver can cut down on its effect by limiting at some value greater than the maximum useful target signal. For maximum efficiency, the limiting should take place at a point in the receiver at which the bandwidth is on the same order as the jamming pulse—limiting should take place after a wide-band-width front end.

The simplest way to wide-band-jam is to sweep a narrow spectrum across the band. If the sweep rate is low, the effect is similar to that of random-pulse jamming. If the frequency is high, with the sweep rate greater than $1/\tau$, the problem is similar to that posed by very narrow pulses. If the sweep rate is random (as for a carcinotron modulated by noise), the effect is the same as that of wide-band noise.

Faced with a wide-band noise jammer, the radar cannot use frequency shift tactics. Since the jamming looks like receiver noise, there is nothing, once the optimum bandwidth has been reached in the receiver, that AJ circuitry can do short of trying to "burn through" by increasing the time that the radar looks in any particular direction and integrating. The best solution, therefore, is to detect at least those targets that are stronger than the noise and avoid overloading by false targets generated by random noise pulses. It can be achieved by some form of noise AGC that operates on the false-alarm count and keeps it constant.

The radar can also discriminate against the jammer on the basis of azimuth. However, the 30 db or so available from the antenna's directional characteristics may not be enough, particularly if the sidelobe level is only 15 dbn down.

Decoys and chaff jam passively

As *Figure 1* shows in the table, the two forms of passive jamming in current use are decoys and chaff. The most realistic decoys are small, aircraft-type vehicles that can be carried and launched by conventional bombers. They can be fitted with corner reflectors or other radar enhancement devices to give them a radar cross-section comparable to that of a much larger aircraft. In a tactical situation, these decoys would be launched by attackers outside the normal radar detection ranges to cause dispersion and overloading of the defenses when the real raid takes place.

Decoy discrimination is difficult, although it might be done on the basis of reflection characteristics or size. Since a decoy might conceivably carry a bomb, the best policy possible would be to destroy all unfriendly targets, including known decoys.

Chaff at present is used in spot and corridor forms. The spot variety consists of small bundles of strips of metallic foil that are released intermittently. Until scattered by the wind, these bundles look like normal targets and overload the defense systems. Soon after release, though, they lose the forward motion imparted to them by the carrying vehicle and drift with the wind while falling. Because of this drift, moving-target circuitry can spot them for what they really are. Even without MTI, a competent operator should be able to recognize them.

The theory behind corridor chaff is that the load planes in an attack can, by continuously releasing chaff, produce a long cloud, or corridor, through which the following aircraft can fly undetected. In practice, this technique has two drawbacks:

- The corridor pinpoints the releasing aircraft.
- It is very difficult to coordinate a long range attack so that the second wave of aircraft is sure to fly down the corridor before the chaff has been dispersed.

Ground DFs can spot ECM aircraft

Passive systems can be used for ECCM as well as ECM. Passive direction finders can pick up radiations from a jamming vehicle for triangulation by two or more ground receiving stations.

However, this approach is not as easy as it may first seem. When there are more than three or four jamming aircraft, passive DFs can become confused by the number of intersections of cross-bearings—some of which locate real targets while others locate only "ghosts" (*Fig. 2*). The ghosts increase in number as the square of the number of jammers, and they can be quite hard to get rid of. "Ghost" discrimination is helped by narrow beamwidths on the DFs, discriminating in elevations, and good data links between the triangulating stations (for passing on such correlation information as type of jamming).

The ground defenses can also use decoy transmitters to foil airborne countermeasures. These simple and cheap sets might look like a full radar station to an attacking vehicle and induce it to waste some of its limited power as it jams over a wide band of frequencies.—End

Counter-Countermeasures: Good Design Can Beat the ECCM Threat
M. Simpson
Space/Aeronautics Magazine, Vol. 33, April 1960
Pages 138-142

2.A.4

- Ground defense radars have clear advantage
- ECCM has led to broadening of frequency range
- Radar operator is important ECCM "device"

by Murray Simpson,
Assistant Vice President
and Technical Director, The W. L. Maxson Corp.*

IN ITS EARLIEST and most basic form, radar countermeasures equipment was designed primarily to obliterate the radar echo by increasing the receiver noise level until the radar return was no longer visible. This effect is readily achieved by generating a CW noise-modulated signal strong enough so that it arrives at the radar receiving antenna at a level higher than that of the "skin echo" from the target. The ECM transmitter has a power advantage, because the radar operates on a two-way basis—the received power is in inverse proportion to the fourth power of range (in accordance with the radar equation), while the power received at the radar from the ECM transmitter is inversely proportional to the square of range (in accordance with the beacon equation). This makes it practical to use relatively moderate ECM transmitter powers against high power radar systems.

* The W. L. Maxson Corp., 475 Tenth Ave., New York 18, N.Y.

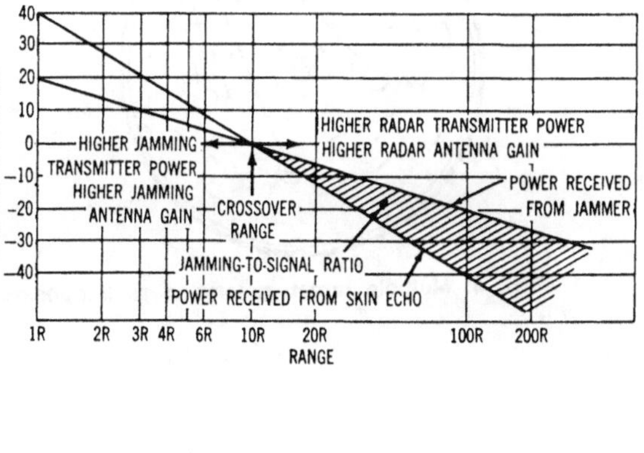

FIGURE 1: Radar-received power varies as $1/R^4$ for the skin echo and $1/R^2$ for the jamming signal. As the jammer moves in, the ratio of jamming power to signal power (J/S) is seen as the receiver decreases. If a radar cannot pick out a signal below the noise level, it can detect the approaching target only when J/S is less than 1 (i.e., after the crossover point). If the effective radiated jamming power is increased, the carrying vehicle is screened from detection at a closer range. Likewise the detection range of the radar is increased if the radar's radiated power is increased.

In addition to the noise jammer, many other types of ECM have been used that depend on various methods of confusing the radar operator. In general, these types have one thing in common: they do not attempt to saturate the radar receiver on a CW basis. Rather they provide signals similar to the normal skin echo return from the target but displaced in so as to make it extremely difficult or even impossible for the radar operator to distinguish the true target from the false signals.

Such systems have many important advantages. For one thing they can operate with relatively low-powered ECM transmitters. In certain cases, it is impossible for the radar operator to know that he is being deceived.

To find the amount of ECM power required for effective jamming, we first must know the power returned to the radar from its own signal. This is given by the well-known equation:

(1) $$P_{rr} = P_{tr}G_r^2\sigma\lambda^2/(4\pi)^3R^4,$$

where P_{rr} is power at radar receiver from skin echo; P_{tr}, power of radar transmitter; G_r, radar antenna gain; σ, target cross-section; λ, wavelength; and R, range.

This equation shows that the received power is inversely proportional to the fourth power of range. The power received from the ECM transmitter is given by the beacon, or one-way transmission, equation:

(2) $$P_{rj} = P_{tj}G_rG_j\lambda^2/(4\pi R)^2,$$

where P_{tj} is the power of the jamming transmitter; G_j, jammer antenna gain; and P_{rj}, power at the radar receiver from the jammer.

Greater range lowers power needs

We can see from *Equation 2* that the power received at the radar from the jammer is inversely proportional to the square of the range. To find the amount of ECM transmitter power needed to drown out the radar skin echo return, you simply divide *Equation* 2 by *Equation* 1 and introduce the proper jamming-to-signal ratio *(P_{rj}/P_{rr}):*

(3) $$P_{tj} = P_{rj}/P_{rr} \times P_{tr}G_r\sigma/4\pi R^2G_j,$$

Some very interesting conclusions can be drawn from *Equation 3*. Most important is that we now see that the amount of required ECM transmitter power is *inversely proportional to range squared.* So the greater the range of the target from the radar, the less ECM transmitter power is required for effective jamming. This is the opposite of normal communications and radar practice. Moreover, at some range the signal received from the jammer at the radar receiver falls below the level necessary for establishing the minimum jamming-to-signal ratio. Short of this so-called "cross-over" range, satisfactory jamming is impossible.

An important principle in ECM design is to make sure that the cross-over range provided by a particular design meets the tactical requirements. For example, assume an aircraft carrying a jammer must penetrate a radar defense system. The cross-over range for that particular jammer in operation against the radars of the defense net is 30 miles, and the ground defenses can shoot down the aircraft up to 15 miles range. The vulnerability of the aircraft is extremely high. The airborne jammer will not protect the aircraft on this mission.

Figure 1 shows the power received at the radar receiver from the jammer and the radar skin echo as a

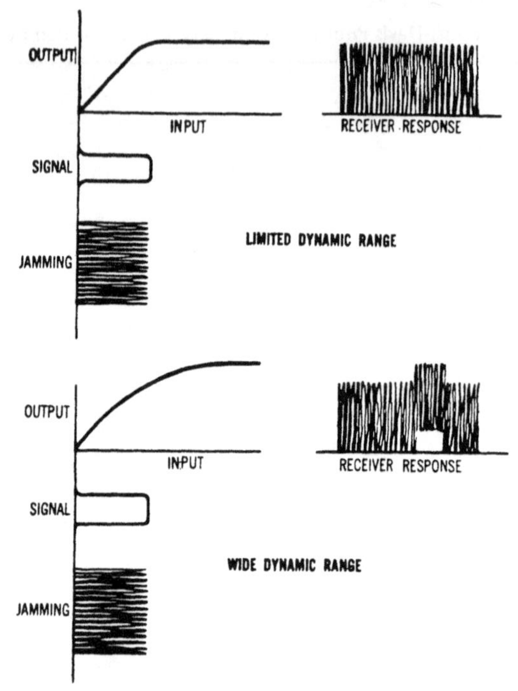

FIGURE 2: Effects of receiver saturation under jamming conditions.

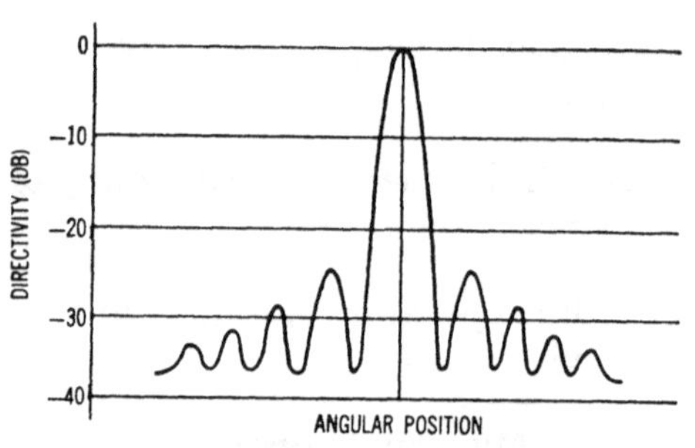

FIGURE 3: Typical radar antenna pattern.

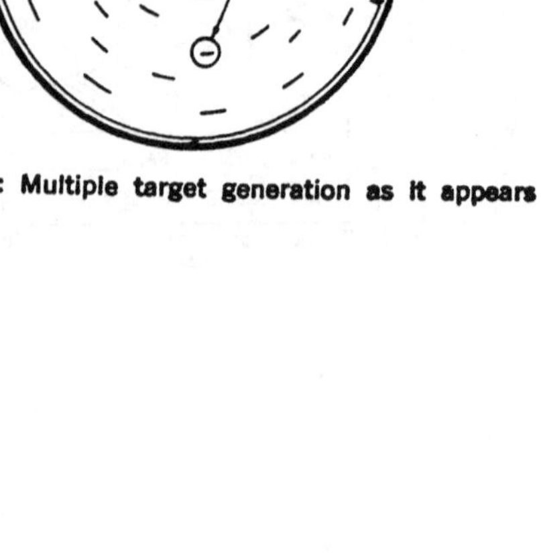

FIGURE 4: Multiple target generation as it appears on PPI.

ECM TEST facility at Rome Air Development Center has 20 different types of radars covering the entire usable radar spectrum. Included are the FPS-20 (foreground) and a number of duplicates of Russian radars. The facility primarily evaluates new ECM systems, components, and techniques.

function of units of range. The cross-over range is clearly seen where the two curves cross.

Noise jamming, as we've noted, has always been a basic form of ECM. It is extremely attractive because of its fundamental nature—it is identical to the normal receiver noise which always exists in a radar system. Therefore it cannot be distinguished from receiver noise and makes the counter-countermeasures problem extremely difficult.

Early radar receivers easy to jam

The limited dynamic range of early radar receivers made them extremely vulnerable to jamming. Against these receivers, the jamming signal could even be weaker than the skin echo. So long as it exceeded the receiver's dynamic range it would drive the receiver into saturation. Once saturation has occurred, no additional amount of input signal power would make the input signal detectable.

Figure 2 shows the effect of the response of a limited-dynamic-range receiver to a jamming signal equal to the skin echo response. The saturation in the receiver shrouds any information from the skin echo. *Figure 2* also shows a receiver with a wider dynamic range with similar values of skin echo and jamming signal. In this case, the skin echo is definitely visible.

Directional information is obtained in a radar system from a highly directional antenna. When a signal is received it is assumed that the direction of the target corresponds uniquely to the direction of the radar antenna beam.

Ideally, the radar antenna should not have any sidelobes, so that there will be no reception from directions other than that of the main beam. We know, however, that sidelobes can't be avoided. *Figure 3* shows a typical radar antenna pattern in one plane.

We can see from *Equation 1* that the normal radar signal is proportional to the square of the antenna gain. Since the sidelobes for a reasonably well-designed antenna are at least —20 db or lower, it is unlikely that the normal skin echo return, except for extremely close targets, will be picked up by the sidelobes. However, in a jamming situation, *Equation 2* shows, that power received by the radar is directly proportional to the radar antenna gain. Under these conditions, it is therefore definitely feasible for jamming signals to enter the sidelobes, provided the jammer power is 20 db or higher than the normal radar skin echo. In many cases, this jamming technique can be used quite effectively, since it destroys information over a very large azimuth on the radar PPI scope.

False targets produced to deceive

Another important form of jamming, designed primarily to confuse, generates a large number of false targets with characteristics very close to those of the actual target. When picked up by the radar, these false targets appear in various azimuth and range positions along with the true target, making it difficult, if not impossible, for the radar operator to tell for sure which is the right blip.

The multiple-target technique likewise takes advantage of radar antenna sidelobe imperfections. The jammer transmits many pulses per radar pulse period, and these pulses are made strong enough to enter the antenna's sidelobes. This effect is shown in figure four. Many degrees of sophistication can be added to this basic technique to maximize the confusion.

Chaff has been used since the earliest days of ECM to produce large numbers of false targets. It is still a very important form of ECM.

Since ECM has become a more or less standard technique in military operations, radar system designers are forced to consider the operation of their system in a countermeasures environment. To cope with this environment, several electronic counter-countermeasures (ECCM) techniques have been evolved.

Just as the ECM designer must understand details of the operation of the various radar sytems that he must operate against, so must the radar ECCM designer understand the various forms of ECM that his radar system is likely to encounter under tactical conditions. He must be prepared to operate his radar against several forms of ECM, since he cannot know in advance what particular technique may be used against him.

ECCM looks impossible at first

At first glance, it may seem that this task is impossible. However, further analysis shows that this is far from being the case. Indeed, in many areas of electronic warfare, radar has a very definite advantage. This is particularly true in the case of ground defense radar systems operating against airborne jammers. The ground radar usually is a high power unit of substantial size, whereas the airborne jammer must necessarily be limited in size and power capability and, at the same time, must be extremely versatile in frequency and modulation characteristics.

Radars have been designed in various parts of the frequency spectrum, ranging from VHF to very high microwaves. In the early days, specific frequency bands were assigned to certain types of radar. In many cases, these bands were relatively narrow and the radars were usually fixed-tuned to a particular frequency. This obviously made the ECM job relatively simple, since the ECM transmitter had to cover only the frequency bands of the operating radars.

As the importance of ECCM became better understood, a strong trend toward broadening the frequency range developed. New radar systems were designed to operate over a much wider part of the spectrum. Jammers were then forced to follow suit. They now had to cover the entire potential radar frequency band if they were to be effective. Moreover, if the radar is continuously monitoring the jamming signals, it can readily pick holes in the jamming spectrum and be shifted to a clear channel (*Fig. 5*). Using available state-of-the-art components and techniques, the designer can make this frequency shift quite rapid and even automatic. Today rapid tunability is one of the strongest ECCM design tools.

We have seen that a receiver with a limited dynamic range makes a radar system very susceptible to noise jamming, even though the radar may operate reasonably well in a non-ECM environment. Obviously extending the receiver's dynamic range is a very useful radar ECCM technique. It makes it possible to observe

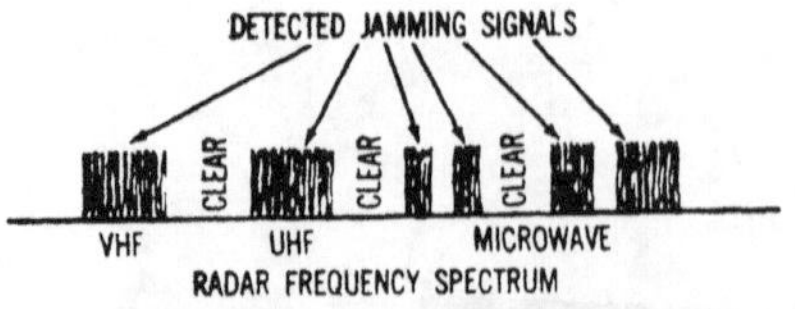

FIGURE 5: Continuous scanning of the spectrum is mandatory to enable the radar to choose clear channels within its operating frequency range.

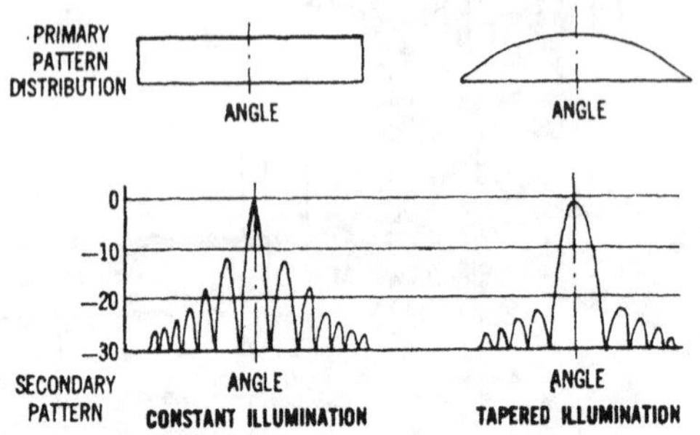

FIGURE 6: Variation in radar antenna beamwidth and sidelobe level resulting from the primary pattern distribution.

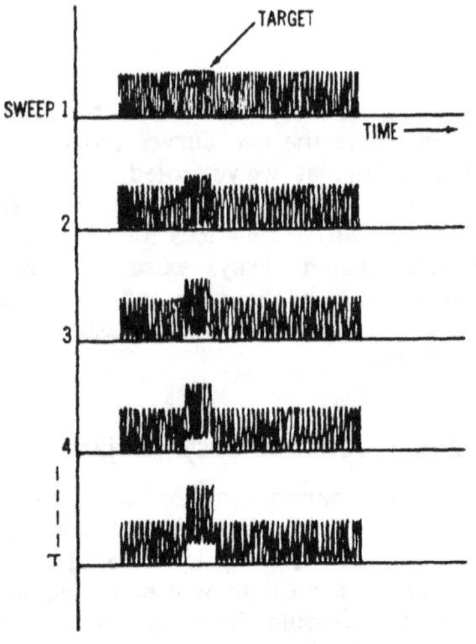

FIGURE 7: Improvement of signal detection as a result of linear integration of τ sweeps.

the radar skin echo even when it equals or is less than the jamming signal (*Fig. 2*). Radar receiver dynamic range may be increased by many techniques, like logarithmic receiver response and other non-linear receiver characteristics.

The most direct counter to sidelobe jamming is to improve the radar antenna and reduce the sidelobe levels. This technique cannot be depended upon completely, because of the physical limitations of radar design. The maximum gain of the radar antenna is obtained when the antenna is illuminated uniformly by the primary feed. However, in such a design, the first sidelobe is only 13 db down, in accordance with the *sin x/x* distribution (*Fig. 6*). A lower sidelobe level is obtained by a tapered distribution using a cosine or cosine-squared function. This approach cuts the sidelobe level to —20 db or less, but also broadens the main beam and decreases the antenna gain (*Fig. 6*). This tradeoff rapidly approaches a limit beyond which it is no longer practical to taper the illumination in return for reduced antenna gain.

Signal detectability is improved

As we have seen, one of the most common forms of jamming uses noise modulation. This has the effect of increasing the noise output of the receiver. Therefore, an obvious counter-countermeasure would be to improve signal detectability in the presence of noise. Needless to say, this technique is important even in the absence of jamming as a means for extending normal radar range. When there is ECM, it becomes an absolute necessity if you want to preserve standard radar performance.

The basic technique used to recover a radar signal in noise relies on the fact that the desired signal is video-coherent—it occurs on every radar pulse at the same point on the time scale for a fixed target. Noise, on the other hand, is fluctuating and non-coherent.

Figure 7 shows the appearance of signal and noise on successive radar sweeps. In a single sweep, the radar receiver can't possibly determine the difference between the signal and noise, if the two are comparable in value. However, if many looks are stored and linearly integrated, the video signal return will add on successive sweeps in a direct manner, while the noise will add in an RMS fashion. With integration, you can therefore "extract" the signal from the noise, provided the integration time is long enough.

The price paid for the signal-to-noise enhancement is time. The signal sensing circuits must dwell on a target for some time (the integration period) before the target can be reported.

Various sophisticated techniques for achieving this effect involve the use of correlation functions. If the integration period is long enough and the target does not change its characteristics in that period, very substantial improvements in signal detectability in noise can be achieved.

Figures 1 & 3 show that the required jamming power is directly proportional to the radar transmitter power for a fixed jamming-to-signal ratio. As the radar transmitter power is increased, the cross-over range moves out, making the target substantially more vulnerable. So the development of larger power radar transmitters greatly improves the ECCM capability of the radar.

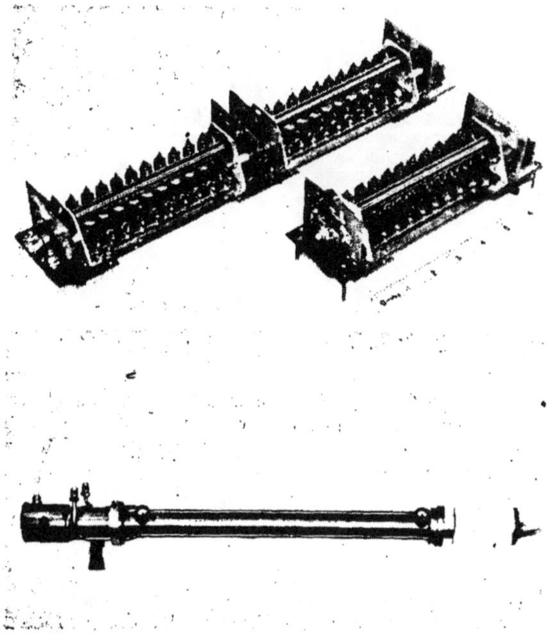

SPERRY AMPLIFIERS for automatic jamming. Distributed-amplifier units (top) serve as wide-band receiver amplifiers for the low frequency channels; 200-W X-band traveling-wave tube (bottom) is typical of the power amplifiers used to transmit jamming signals. This tube is less than 28 in. in length, weighs 11 lb, and provides 40 db minimum gain.

Large antennas for most future radars

Very large antennas, too, are practical on the ground. In fact, most future ground radars will use large antennas to get higher resolution and gain. As *Equation 3* shows, high gain also improves the ECCM capability of a radar. If gain is increased, the jammer must correspondingly increase his power if he wants to maintain his effectiveness. Since pencil-beam antennas have substantially larger gains than fan-beam types, they also have considerably greater ECCM capability for a given radiating area.

One particularly important ECCM "device" is the radar operator. Because of the tremendous impact of ECM on radar performance, today's operator must be thoroughly trained to recognize all possible forms of ECM and their effects on his particular radar. Above all, he must avoid the confusion which usually is the prime objective of ECM.

In view of the many new requirements heaped on the radar operator, a new occupational specialty should be created—the electronic warfare specialist. His prime function would be to understand and use ECCM techniques in the presence of jamming. Since these techniques are, in many cases, sophisticated and very specialized, he may often operate independently of the normal radar operator. His principal duty would be to make the information presented to the radar operator sufficiently accurate and clear so that the radar operator can perform his normal duties.—End

Philosophy of ECCM Utilization
S.L. Johnston
Electronic Warfare Magazine, Vol. 7, No. 9, May/June 1975
Pages 59-61

2.B.1

Stephen L. Johnston
U.S. Army Missile Command
Redstone Arsenal, AL 35809

In a recent issue of *Electronic Warfare*[1] Dr. Joseph Saloom posed some significant questions regarding the utilization of ECCM. His questions were both proper and timely. Indeed, it was very interesting to note at the 11th AOC International EW Symposium, several speakers mentioned ECCM in their talks. This is a welcome development that portends a turn from almost exclusive focus on offensive ECM in recent years.

Fortunately, ECCM has not been dead or even asleep; it was partially eclipsed in the public press by offensive ECM. Certainly offensive ECM played a much stronger role in the Southeast Asian conflict than ECCM. Although they have not attracted much attention, there have been a number of discussions in the open literature on ECCM. There have been several reviews in the open literature of ECM, but the author is unaware of any recent comprehensive treatise on ECCM—that is, other than the November/December 1974 edition of *EW* magazine. With so many open discussions of ECCM techniques taking place around the industry a survey of ECCM appears warranted.

Over ten years ago, Skolnik[2] devoted several pages to interference and ECCM. He cited several open literature references on a variety of ECCM techniques. In issue of *EW* magazine[3], the author discussed measures of ECCM performance, proposed the parameter ECCM Improvement Factor (EIF) and presented a list of approximately 150 ECCM types. Even so, Dr. Saloom's questions have not previously been addressed so they deserve consideration.

At the start, ECCM should be defined: "ECCM is that major subdivision of Electronic Warfare involving actions taken to insure our own effective use of electromagnetic radiations despite the enemy's use of countermeasures"[4]. Several points should be immediately apparent:

1) The purpose of ECCM is to combat or negate the effects of ECM.
2) While it is implied that ECCM is *electronic*, this definition does *not* limit ECCM to electronic means. Accordingly, physical destruction of a hostile jammer could be considered an ECCM!

Radar and radio are both the province of electronics engineers. Accordingly, both ECM and ECCM specialists tend to think in terms of electronics. Mr. George Steeg of the Office of the Secretary of Defense recently pointed out in a talk to an EW conference that EW need not be electronic, i.e. there are more ways of combating ECM than by purely electronic means. A weapon system engineer must consider *all* possibile means of countering ECM and must select the most cost effective method(s) whether they be electronic or not. In fact, there are some very effective non-electronic methods of countering ECM that are very simple and easy to implement. Tactics, deployment, and operational doctrines are non-electronic weapons, but are very important means of countering ECM.

3) ECCM is actually the approximate equivalent in a hostile environment of the benign environment EMC (Electromagnetic Compatability). In fact, a radar (or radio set) that is immune to unwanted friendly interference (thru EMC) is probably immune to hostile ECM and vice versa.

Where does ECCM start? Some would say that first comes the radar, then ECM then ECCM (then ECCCM, etc.). In the beginning, yes[3], but not today. In a modern radar, ECCM should be designed into the radar from the start. Technology does advance, resulting in increased effectiveness of ECM which must then be countered by ECCM advances; but it is only good engineering practice to design in both ECCM and EMC from the start. ECCM, when incorporated by a field change, can be expensive. Even as an automaker strives to minimize factory recalls, a good radar system engineer should design in ECCM.

There exist numerous ECM systems (pods, collections of black boxes, and ECM suites, etc.). While it is desirable that an ECM have general application, frequently this is not so, even if just different bands for a noise jammer. Some ECM's are designed specifically to exploit some known weakness in a hostile radar.

There are almost no "ECCM systems" as mentioned by Dr. Saloom. (The term *system* is meant to a complete separate working entity.) It should be recalled that there is an exception to every rule—except this one! The HIPAR (High Power Acquisition Radar) of the U.S. Army Nike Hercules Guided Missile System is actually an ECCM (to offset the effects of ECM against the original Nike Hercules acquisition radar—LOPAR).

An ECCM should be an integral portion of a radar system. Consequently ECCM's are usually achieved in fielded radar systems by the radar contractor. In a weapon system, ECCM characteristics must be closely related to overall weapon system performance. As the author pointed out in a recent article[5] increase of burn-through range could result in no improvement in the overall performance of the weapon system: i.e., increase in first engagement range, number of aircraft killed in a raid, etc.[6]

ECCM UTILIZATION

There are many ECM technology programs in the services (due to Vietnam emphasis) but a lesser number of ECCM technology programs. The RADC radar ECCM test bed and some ECOM communications ECCM programs are notable examples of ECCM technology programs. Within Army missile programs ECCM activities have been primarily centered within the various missile projects. Here potential ECM problems must face competing priorities, e.g. propulsion system problems. While ECCM proponents have an insatiable appetite for funds, on the whole, ECM problems generally have been addressed fairly well in Army missile systems. An ECCM advocate would have the project manager spend all his funds and attention on ECCM problems and none to propulsion, missile structure, etc.! The project manager must produce a balanced system.

ECCMs come in all shapes and sizes—from non-electronic means (e.g., an alternate guidance concept such as an inertial system, or the addition of an optical tracker to a handful of transistors) to a complete new radar (e.g., HIPAR). They are thus often hard to recognize and may not be identified as an ECCM.

The cheapest, simplest, most readily available and usually most effective ECCM is Homo The Sap (Homo Sapiens), the human operator. In communication, the human ear is an excellent very narrow band filter; a good operator can copy code in the midst of noise and false code signals. The human eye has excellent integration capbilities and can recognize many false signals on a radar display scope. In some cases a human operator can track better than an electronic tracker. Certainly a human can use his intelligence and new instruction to make decisions whereas a digital computer can only do what it was programmed to do. Continuous extensive training is required to retain operator proficiency at a high level.

Some ECCMs are introduced as multibenefit system changes. For example, a digital moving target indicator (DMTI) might provide improved performance in a benign environment, require less adjustments (be more stable), have increased reliability and be more flexible while providing superior resistance to ECM.

Some ECCMs that could be retrofitted, but although desirable, are very costly.

There has been a lack of funds to do all the ECCMs that some proponents would like. For these reasons in some cases ECCMs have been applied as Band Aids.

As a tube manufacturer, Dr. Saloom undoubtedly subscribes heartily to the British radar rule: "It is more blessed to transmit than to receive."[6] Certainly a very possible ECCM is to outshout the opposition by increasing the transmitter power output. This approach is often not cheap and may not be feasible where space, cooling and/or prime power are limited. It may even be inadvisable from an electronic security standpoint. Consultation of the ECCM techniques listed in Ref. 3 will show that there are many ways to combat jamming. Signal processing type ECCMs are generally easier to implement than an increase of power output.

It will be appreciated that there are a wider number of possible ECCMs for one ECM. The choice of which ECCM to use may be difficult. Certainly the topic is much broader than merely tubes, Dr. Saloom's principal interest.

Because of limited hostile engagement experience, ECCM has tended to receive less attention than offensive ECM. Consequently there has tended to develop the philosophy of not taking a medicine unless you have an illness. In some cases there is inadequate data on the effectiveness of a broad spectrum of ECMs and ECCMs. In fact, as Ref. 3 pointed out, there exists both a lack of performance data on ECCMs and a lack of ECCM performance criteria. This and the ever-changing threat characteristics make ECCM a very challenging and sometimes frustrating experience.

The specific tube ECCM considerations cited by Dr. Saloom are most interesting. It is highly possible that there may be a lack of appreciation for those techniques in ECCM circles. Certainly one can visualize their potentiality. But is the cure appropriate to the ailment? This can only be answered by a very careful weapon system study for one or more specific systems. Unfortunately, such a study would probably be very sensitive to threat characteristics.

Some suggestions for electronics components suppliers/circuit designers are offered:

- Learn about the effects of various types of ECM

- Become familiar with ECCM technology

- Be innovative. Conceive novel flexible unique methods of operation of tubes, new signal processing techniques, etc.

- Work closely with a radar/weapon design group to include your concepts in an overall systems analysis.

- If you invent it and it's worthwhile, it will ultimately be used, possibly as a surprise. Remember the radar designer is hungry for unique ECCMs. ■

Views and conclusions expressed herein are the primary responsibility of the author and should not be interpreted as representing the official opinion or policy of the Department of the Army.

References

1. ECCM—In a Fix (An Interview with Dr. Joseph Saloom), *EW* Magazine, Vol. 6, No. 5; Sept/Oct 1974, pp 48-49

2. Merrill I. Skolnik, *Introduction to Radar Systems*, McGraw Hill Book Co., New York, 1962, pp 554-567

3. Stephen L. Johnston, ECCM Improvement Factors, *EW* Magazine, May/June 1974, Vol. 6, No. 3, pp 41-46

4. *Dictionary of United States Army Terms:* AD Department of the Army Regulation AR 310-25, June 1972/ Change 1, 15 Jan. 1974.

5. Stephen L. Johnston, "Burnthrough Range—An Overrated Parameter", *EW* Magazine, Nov/Dec 1974, pg 49.

6. Stephen L. Johnston, Military Radar—Weapon System Analysis, presented 18 Sept 70 as lectures in the short course "Principles of Radar", Georgia Institute of Technology, Atlanta, Georgia. Published as lecture notes.

Radar ECCM: A European Approach
M.F. Radford
IEEE ELECTRO-76 Rec pt 30, pp. 30-6-1 to 5.
Pages 194-198

2.B.2

SUMMARY

The radar designer must optimise parameters for performance in ECM and avoid any constraints which could be exploited by ECM. Any ECM/ECCM strategy must take into account EMC, including enemy EMC.

Adaptivity improves search performance, but also reduces ECCM performance in other ways. While electronic scanning avoids some constraints, it imposes others, particularly in the frequency domain. It is argued that simple radars of good performance may be as cost-effective in ECM as more complex systems.

Some examples of recent developments are given.

INTRODUCTION

European work in the advanced radar field has in the past been based on relatively small national home markets, and this has tended to inhibit the development of large complex radars and encourage the investigation of simpler, but not necessarily less cost-effective systems. This paper summarises the arguments that have led some European manufacturers to produce families of military radar which although at first sight conventional have good current and also good potential future ECCM performance.

RADAR ECCM

The life of a radar system design is at least twenty-five years, ten years from research to development plus fifteen or more in service. The gestation period of a countermeasure is by comparison very short, a year or two in peacetime or only a few weeks in time of war. Thus it is not possible for the radar designer to know in advance what threats his system may be called upon to face. However, any fundamental constraint or weakness in the system is likely to be discovered and subsequently exploited by purpose-built ECM equipment, and such equipment may be ready for service almost as soon as the radar because of the disparity in development time.

The task of the radar designer is therefore not only to optimise the parameters for ECM resistance, but also to avoid inbuilt limitations which could restrict the choice of operational routines or provide openings for future ECM.

Threats

The ECM environment has been described by Meskill[1], and in a recent Aviation Week and Space Technology survey[2]. These publications give some indication of the wide variety of ECM threats which are currently possible. However, it is unlikely that any radar will be called upon to face all the possible ECM threats at maximum intensity at any one time. Radar designers are familiar with the restrictions imposed by compatibility requirements, and the same restrictions apply to ECM. The restrictions arise in two ways; first from a need for the ECM to avoid interference with friendly radar, communication and other requirements, and secondly from a need to operate intercept receivers in order to direct ECM activities. To give a few examples:

1. ECM may affect defensive radars on its own side.

2. Airborne ECM may reduce the sensitivity of homing and warning receivers.

3. Responsive jammers may respond to each other.

4. Airborne ECM may interfere with terrain avoidance radar.

Study of any list of ECM techniques will suggest many more such possible conflicts. It follows that any attack is likely to have a coordinated overall ECM plan, in which ECM is confined to predetermined limits in coverage, power, time and frequency, and in which ECM operators are given only limited freedom for individual initiative. The radar must be able to recognise and respond correctly to the ECM situation and this requires maximum operational flexibility as well as maximum ECM resistance.

Signal/Jamming Ratio

The most difficult radar task is the initial detection of the target. Once detected, the target is more easily recovered since the radar can take advantage of the target data it already holds.

Most current surveillance radars are optimised for maximum detection performance in clear conditions, that is, for maximum signal/thermal noise ratio. Future military radars need to be optimised for maximum detection performance in ECM, that is, for maximum signal/jamming ratio. It may be assumed that severe jamming will completely swamp thermal noise in the majority of systems.

The radar equation is unclassified, and it is easy to show that the signal/jamming ratio of a simple radar in search mode is given approximately by the formula

$$\frac{P\, B\, \sigma\, T\, S}{\Omega\, R^2\, L\, J}$$

The mean transmitted power P, the occupied bandwidth B over which the jamming is spread, the sidelobe protection S and the losses in transmission and processing L are all within the control of the radar designer. The remaining parameters, the echoing area σ of the target at range R, the effective radiated mean power J of the noise jammer (also at range R) and the time T taken to search the coverage solid angle Ω are all dictated by the requirement.

It is interesting to note that frequency does not appear explicitly, although several parameters are obviously frequency-dependent. Aperture, beamwidth and gain do not appear, but are taken into account in determining S. Receiver RF losses and noise factor do not appear as they do not affect signal/jamming ratio when the jamming is strong.

When the jammer is near the target, little or no sidelobe protection is available and detection range is drastically reduced. It is thus advantageous to have an adaptive radar in which extra energy can be allocated to the heavily jammed sectors at the expense of the rest of the coverage.

Signal/Chaff Ratio

The signal/chaff ratio is given by the formula

$$\frac{\sigma\, M}{C\, R^2\, \theta\, \phi\, r}$$

where C is the chaff density, θ and ϕ the azimuth and elevation beamwidths and r the range cell size. The improvement in signal/chaff ratio given by the MTI is M, where M is dependent upon the dwell time t. The exact relationship is determined by the type of MTI and the various assumptions made about the chaff doppler spectrum. The mean dwell time available in the search mode is given by

$$t = \frac{N\, T\, \theta\, \phi}{\Omega}$$

where N is the number of parallel receiving channels, i.e. the number of receiving beams if a multiple beam antenna is used.

Substitution of practical values shows that in order to obtain adequate signal/chaff performance the MTI must produce a substantial improvement. This implies a sufficiently large t for a number of pulses to be coherently processed, and this in turn implies several parallel receiving channels if the beams are to be reasonably narrow. If the chaff is confined to a limited sector an alternative solution is to have an adaptive radar in which extra dwell time can be allocated to the chaff at the expense of the rest of the coverage.

IS ADAPTIVITY WORTHWHILE ?

In both jamming and chaff, it has been shown that search performance can be improved by adaptive electronic control of the antenna beam. However, practical methods of electronic scanning penalise ECM resistance in other ways. Returning to the formulae derived for signal/jamming and signal/chaff ratio, consider the effect of various electronic scanning systems on each parameter in turn.

Power

Mean output power P is not significantly affected, since electronic scanning systems can be designed to cope with any output power currently available.

Bandwidth

Maximum available antenna bandwidth is significantly reduced. Current non-adaptive antennas[3] achieve bandwidths of 30%. Frequency-scanned antennas can be designed for various bandwidths from 5% to over 25%, while phase-scanned antennas have typical bandwidths of 10% to 15% depending upon the design of the phase shifters.

The instantaneous bandwidth of the non-adaptive antenna may again be 30%, thus it may be used in true simultaneous diversity systems. The instantaneous bandwidth of a frequency-scanned antenna is limited by its dispersion characteristics, and severe distortion is experienced when the bandwidth of the pulse becomes comparable with the frequency shift required to scan the beam through a beamwidth. However, the frequency-scanned antenna can be used to form a number of simultaneous beams having different radar frequencies. The instantaneous bandwidth of a phased array is the bandwidth which can be covered without altering the phase shifter settings, and this may be only 1% to 2% depending upon the beam-width and deflection angle.

Target Area

Target echoing area σ is at first sight independent of the antenna system. However, when the statistical fading characteristics of the target are considered, it is found that detection probability can be improved by frequency diversity. It has been shown above that frequency diversity is more readily obtained with non-adaptive systems. On the other hand, adaptive scanning permits dwell time to be extended when the results of the initial inspection are inconclusive, and this technique also improves detection performance.

Sidelobe Level

The sidelobe protection S of a non-adaptive antenna may be 35 dB for the first sidelobe and as much as 55 dB for the far sidelobes where site conditions permit. Frequency-scanned antennas normally have a cubic phase error which is generated by the reactance of the couplers along the dispersive line, and although this error can be compensated at midband it is difficult to eliminate it completely. At large beam deflection angles aperture-matching problems arise and a further degradation occurs. Thus, while a frequency-scanned antenna may achieve 32 dB first sidelobes at midband, the level may fall to 25 dB or worse at the ends of the band.

Phased arrays again have aperture-matching problems at large deflection angles, but their sidelobe performance is also limited by economic considerations. The high cost of the aperture makes it necessary to use a compromise amplitude distribution, and the desirability of using simple phase shifters with the minimum number of bits results in a further compromise with quantisation sidelobes. Thus the sidelobe protection of a cost-effective phased array may be as little as 25 dB for the near sidelobes. It is worth noting that when the jamming threat is constrained by ECM mutual compatibility considerations, a 10 dB difference in sidelobe protection leads to an overall 20 dB difference in signal/jamming ratio. This is because intercept sensitivity must increase 10 dB to detect the radar transmitting sidelobes, and this increase reduces by 10 dB the maximum tolerable output of an adjacent noise jammer. The improved sidelobe protection of the radar receiver contributes the second 10 dB.

Losses

The transmission losses are inevitably increased by electronic scanning. A frequency-scanned antenna uses a dispersive line in which an appreciable amount of energy is stored, and this storage results in a predictable heat loss. Waveguide delay lines are the least lossy but are rather bulky. With compact folded lines it is likely that a minimum additional loss of about 1 dB will be introduced.

Phased arrays have phase-shifter losses, and these are likely to be around 1.5 dB in a typical antenna of simple basic design. Losses may be higher if the design incorporates dielectric loading or lossy microstrip dipole radiators.

Taking all of these factors together, a fully adaptive phased array antenna may have a disadvantage of as much as 10 dB on basic ECCM parameter values, and this cancels out at least part of the potential advantage of adaptive dwell time.

Chaff

A non-adaptive antenna generally has no difficulty in providing a narrow azimuth beam θ, a small range cell r and an adequate dwell time t. Unless the antenna design is complicated by the incorporation of multiple beam reception in elevation, the elevation beamwidth ϕ tends to be rather large for good chaff control. The difficulty lies not so much in the actual beamwidth, which is likely to exceed the thickness of a chaff layer at maximum range, as in the possibility of two or more layers with different velocities occurring in the same resolution cell. This latter circumstance makes the MTI task considerably more difficult.

A frequency-scanned antenna can use frequency diplexing to form multiple beams and so has no dwell time problem. It has however, an instantaneous bandwidth restriction on its range cell size.

A phase-scanned antenna with a single beam must use its adaptivity to optimise dwell time against localised chaff. In widespread chaff it is at a considerable disadvantage and the provision of more than one search beam becomes desirable.

MTI

Two different types of processing are possible, continuous processing and batch processing. The former is traditionally associated with rotating antennas, the latter with step-scanned non-rotating systems. However, these differences are not fundamental, and it is possible for either antenna to use either system.

Continuous processing automatically imposes an amplitude weighting on the received pulse train, and enables returns to be accepted from well outside the nominal 3 dB beamwidth of the antenna. Batch processing tends to be less efficient in search and is less sensitive to targets half-way between two beam positions unless specially programmed to overcome this.

Taking all these factors together, there is no overwhelming reason for preferring any particular type of system so far as chaff suppression is concerned. In different scenarios, different systems may prove the most effective.

Constraints

The first section of this paper mentioned the need to avoid constraints. Each system, however, has constraints of a different form. The non-adaptive antenna has the angle/time constraints resulting from mechanical rotation, but is unconstrained in frequency. The phased array has no dwell time constraint, but its coverage is restricted to 90° in most designs. (The 'dome' is a notable exception.[4]) The phased array has some frequency constraints, but in the frequency-scanned array the frequency constraints become more serious. Frequency-scanned azimuth look-back in a rotating system is tolerable, but simple frequency-scanning in elevation is undesirable because of the vulnerability to narrow-band jamming covering attacks at a given elevation angle. Phased arrays are unable to generate simultaneous independently steered multiple beams, while frequency scanned arrays can do so quite easily.

Thus the adoption of electronic scanning does not in fact confer a greater overall degree of freedom, it merely exchanges one set of constraints for another. However, the electronically scanned radar gives more information to an intercept receiver, and automatically attracts attention as a worthwhile target.

Costs

Since adaptive radars are likely to be more expensive to develop, a smaller proportion of any given budget is available for ultimate production. Since their production costs are likely to be higher, it follows that the total number of adaptive radars which can be procured for a given total investment is substantially smaller than if a simpler design is chosen.

Thus, for the same investment, a simpler radar can be deployed in larger numbers, giving 'graceful degradation' under attack and giving an improved overall detection performance because of the better coverage overlap.

SOME ACHIEVEMENTS

Having discussed some of the arguments which have led to the continued pursuit of apparently orthodox rotating radars, it may be of interest to mention some recent achievements.

Transmitters

In order to permit the use of advanced coherent ECCM processing techniques, it has been necessary to design transmitters having unusually clean spectra. A typical high power transmitter has intra-spectral-line noise 120 dB below the peak pulse output.

Low Sidelobe Antennas

The objective of 35 dB sidelobe protection and 30% bandwidth has been reached. The antenna can handle high power and can be adapted to provide a difference channel for monopulse working.

Multiple Beam Antennas

Various ways of generating multiple beams without undue loss of sidelobe protection have been demonstrated. Note that multiple-beam antennas are particularly effective against deception jamming, since they always observe the true jammer position.

Signal Processing

A wide variety of digital and analogue techniques are available. Various CFAR systems are employed. Adaptive analogue processing[5] has been developed as well as the more usual digital processing. The latter includes FFT and FFT-derived systems, MTD and adaptive digital processing. Microprocessors offer useful benefits.

Data Handling and Control

Families of modular digital processors have been developed which can be used to perform a wide variety of data handling and control tasks. The basic modules are designed so that they can be reprogrammed or extended during the service life of the radar, enabling additional ECCM to be added as required.

CONCLUSION

For obvious reasons this has had to be a vague survey with nearly all the interesting arithmetic omitted. Nevertheless, it is hoped that two points have emerged.

First, that unless the application requires adaptivity for some special task such as high data rate tracking, a relatively simple radar may be as cost-effective in ECM as a more complex adaptive system.

Second, that European industry has some appreciation of the radar ECCM field, and has some interesting and worthwhile techniques to contribute.

The author is grateful to his colleagues for many valuable discussions, and to the Technical Director of GEC-Marconi Electronics for permission to present this paper.

REFERENCES

1. Why ECCM: The ECCM Environment W.D. Meskill. Paper presented in this Conference.

2. Special Report: Electronic Warfare Aviation Week and Space Technology January 27th, 1975.

3. Wideband Squintless Linear Arrays A. Rogers, Marconi Review, 4th Quarter, 1972.

4. Dome Radar - A New Phased Array System. P.M. Liebman, L. Schwartzman, A.E. Hylas, IEEE 1975 Radar Conference.

5. Some Advances in Pulse Doppler Processing. T. Hair, M.G. Cross. IEEE 1975 Radar Conference.

This paper includes work which has been carried out with the support of Procurement Executive, Ministry of Defence.

Guided Missile System ECM/ECCM
S.L. Johnston
Microwave Journal, Vol. 21, No. 9, September 1978
Pages 20-24

2.C.1

INTRODUCTION

A previous paper[1] defined electronic warfare, its three principal divisions, and described their effects. Principal emphasis in that paper was a systematic organization of radar electronic counter-countermeasures (ECCM) with extensive tables of nearly 200 types of ECCM. Although it was indicated that some ECCMs can only be used in certain types of radar, further consideration of application of ECCM techniques to various radar systems is warranted. Barton[2] presented an extensive review of radar technology. Initial military application of radar was aircraft detection and tracking. Although that use is still important, military radar applications now also include battlefield surveillance, guided missile systems, terrain avoidance systems for aircraft and satellite identification. As a class, guided missile systems utilize radar for a wide variety of purposes. Accordingly, this paper will use guided missile systems as a means for further treatment of application ECCM techniques to radar.

GUIDED MISSILE SYSTEMS

Most authors consider guided missile systems by application e.g. surface to air, anti-tank, etc. Actually, all guided missile systems utilize the same principal functional elements, although their embodiment may vary greatly in different applications. A generic guided missile system is shown in **Figure 1**. These functional elements will be recognized in the various guidance systems presented by Locke[3]. Missile navigation systems require measurement of position (distance), velocity, and angle. Basic navigation methods and navigational measure-methods were given in another paper.[4] Principal guidance methods include (alphabetically): beam rider, command, command-to-line-of sight, homing, and inertial. Each of these guidance methods has certain ECM/ECCM implications which will be addressed further in a later section. As a further means of illustrating radar ECM/ECCM, a radar guided air defense missile system will be considered. Although centered around microwave radar, this treatment is also extendable in concept to optical/IR/laser guided missile systems.

TARGET DETECTION/ ACQUISITION

Target detection/acquisition is usually accomplished in a "search" radar.

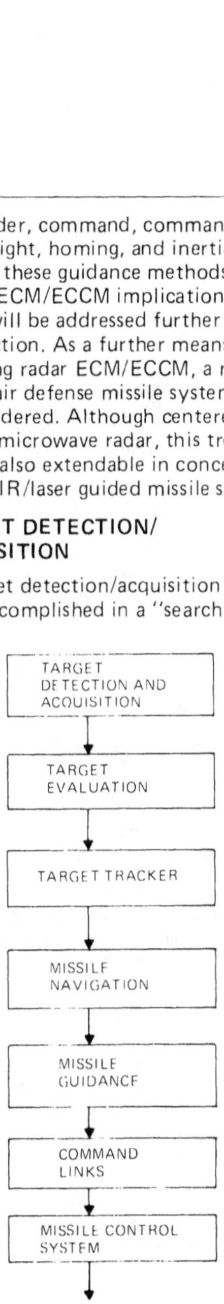

Fig. 1 Generic Guided Missile System.

This radar generally operates in L or S band. (Radar band designations of Reference 5.) It may be either 2D, i.e., provides range and azimuth, 3D, or range and velocity (range rate). The radar can also be a multifunction type by also providing target tracking and missile tracking. Radford,[6] however, pointed out the limitations of electronic scanning radars in an ECM environment. Additionally, the phased array radar generally simplifies hostile ECM since it performs search and track at the same frequency band, whereas mechanically scanned search systems usually perform search and track in different bands. Multi-radar systems have another advantage since a tracking radar may also be used for search/ acquisition if the latter radar is jammed.

As with nearly every radar, ECM against the search radar falls into two classes — denial and deception. Each of these may be either active, i.e., it generates electromagnetic energy, or it may be passive.[1]

Bittencourt[7] has a good general review of ECM techniques with limited discussion of appropriate ECCMs. Alred gave an excellent review of British World War II naval radar ECCM.[8] **Table I**, Search Radar ECM/ECCM Matrix, was evolved using Table VI of Reference 1 as a basis. Selected ECCMs from Tables I - V of Reference 1 were added in addition to some from Reference 7.

Several examples of application of ECCM to search radars may be cited, especially Johnson and Stoner,[9] who describe a number of ECCMs. The Marconi S-631 Radar[10,11], which uses back-to-back L and S radar combinations, is an excellent example of frequency diversity (diplexing). Lind[12] analyzes ECCM aspects of frequency agility. Dixon[13] covers general spread spectrum techniques. Skolnik[14] analyzed bistatic radar and pointed out several practical limitations in its employment. One of these is the form in which target location data is obtained, i.e., it is not in polar coordinates as in monostatic radar. Certainly bistatic radar provides inherent protection of the receiver(s) from anti-radiation missiles.

Radar antenna sidelobe characteristics are extremely important in standoff active denial ECM. The antenna of the previously cited Marconi dual radar was designed to have low sidelobes.[15] Rudge[16] and Anderson[17] are recent references on low sidelobe radar antennas. Sidelobe blanking[18,19] and sidelobe cancellers[20–26] are means to reduce the effective received sidelobe pattern.

A modern search radar makes use of digital processing to provide adaptivity in operation of the radar as exemplified by Di Nardo.[27] Norsworthy[28] described use of an electronic correlator for passive ranging. This is another ECCM against active denial ECM. Its primary use is against self screening targets, however.

Constant false alarm rate (CFAR) techniques[29–31] have received extensive attention as ECCM against active denial ECM. The main benefit from CFAR is the reduction of false detections (due to noise) reported to the radar target evaluation computer. Although this is important, it must be recognized that CFAR will not enhance detection of targets embedded in noise.

TABLE I

SEARCH RADAR ECM/ECCM MATRIX

	Generic ECM	Typical ECM	ECCM
A.	Active Denial	Jamming:	
		CW	frequency diversity, frequency agility, fast time constant, gain controls, pulse width discrimination, bistatic radar, triangulation, spread spectrum, low sidelobe antenna, passive correlation
		long pulse	all of the above plus integration
		spot noise	frequency diversity, frequency agility, integration, Dicke-Fix, CFAR, bistatic radar
		barrage noise	Dicke-Fix, CFAR, log FTC, automatic video noise level, local oscillator off, higher ERP, longer time on target
		medium and fast sweep jamming	Dicke-Fix, CFAR, variable IF Bandwidth, integration, pulse compression, frequency agility, guard band blanking
		impulse jamming	Dicke-Fix, CFAR
		short pulse jamming	integration, pulse compression
		sidelobe repeaters	PRF jitter and integration, pulse compression, sidelobe blanking, bistatic radar
B.	Passive Denial	Chaff	MTI, Doppler radar, improved resolution (narrow beamwidth, short pulse, pulse compression) log FTC, gain controls, low scan rate on search
		Radar absorbing materials	higher ERP, longer time on target
C.	Active Deception	Repeaters	complex radar waveform, spread spectrum technique, frequency agility, PRF jitter and integration, sidelobe blanking
		False target generator	PRF jitter, frequency agility, lobe sidelobe antenna, diplexing

Some CFAR systems are sensitive to the spectral distribution of the noise. A previous paper[32] pointed out that noise jammers do not necessarily produce white Gaussian noise. This was also shown by Boyd et al.[33] Turner et al.[34] recently treated the noise produced by jammers and introduced the parameter "noise quality".

The Dicke-Fix[31,35] is another widely used ECCM. Its chief application is against fast sweep jamming, e.g., generated by the Carcinotron (backward wave oscillator).[7] It has several forms.[1]

The human ear, because of its extreme sensitivity and ability to perform integration, can be used as an ECCM. In aural radar displays, the radar operator listens (either on a loudspeaker or headphones) to amplitude variations of radar video. This requires, however, that the search radar be used in a searchlight mode, i.e., non-scanning.

Van Vleck and Middleton[36] presented a theoretical comparison of the effectiveness of visual, aural and meter radar displays. Pew and Elkind[37] described experimental tests for use of aural displays in radar target identification, e.g., differentiation between a jet aircraft and propeller aircraft.

Aural displays are useful against both active and passive denial ECM. This technique has a low traffic capacity and is hence of little use in a high density target environment.

Moving Target Indicators (MTI) were developed during the last of World War II as a means of enabling radar to detect moving targets against a large background (30 dB) or from fixed or slowly moving "clutter" – trees, clouds, etc. Several MTI schemes including wind compensation were in development in Germany at the end of World War II.[38,39] MTI, when properly designed, can be effective against passive denial ECM (chaff). MTI originally used analog techniques but now uses digital techniques. Ellis[40] reviews digital MTI and indicates that it can also be effective against impulse jamming. The Lincoln Laboratory Moving Target Detector MTD[41] uses automatic adaptive thresholding and fully coherent processing as a digital replacement for the (analog) Dicke Fix.

TARGET TRACKING

Target tracking involves the "precision" determination (usually but not always) through continuous observation of a selected target (or targets). This is in contrast to the periodic search of a volume of space by a search radar. The target tracker provides measurement of target slant range and/or range rate, azimuth angle and elevation angle. Tracking radars generally operate in C or X bands. This is in part *(from page 22)* **ECM/ECCM**

TABLE II

TRACKING RADAR ECM/ECCM MATRIX

	Generic ECM	Typical ECM	ECCM
A.	Active Denial	Jamming (various)	frequency agility, frequency diversity (diplexing), higher ERP, pulse compression, angle track on jam
B.	Passive Denial	Chaff	MTI in angle track, Doppler radar, improved resolution (narrow bandwidth, short pulse, pulse compression)
		Radar Absorbing Materials	higher ERP, frequency diversity
C.	Active Deception	Repeaters	complex radar waveform, spread spectrum techniques, frequency agility, PRF jitter and integration
		Range and Velocity Gate Stealers	PRF jitter, leading edge tracking, acceleration limiting, Doppler/range rate checking
		Inverse gain modulation (scan modulation jamming)	monopulse, lobe on receive only, variable conical scan frequency, CONOPULSE
		Angle Gate Stealers (TWS Radars)	lobe on receive only
D.	Passive Deception	Decoys	target/decoy trajectory characteristics, multistatic radars, bistatic operations
		Chaff	higher resolution in range and angle, speedgate loop, MTI in angle tracking, leading/trailing edge range tracking
		forward chaff	ECCM guard gates/coast

due to the increased angular accuracy requirements and practical limitations on antenna diameter in mobile tracking radars (~ 3 m).

Most tracking radars use pulse waveform to facilitate range measurement. As with search radars, the magnetron in one of several forms was long used in tracking radars. In both search and tracking radars, advantages of fully coherent operation have encouraged use of MOPA (master oscillator-power amplifier) transmitters recently.

Angle measurement originally used conical scan.[4] This technique was originally developed for automatic guiding of astronomical telescopes.[4,42] Monopulse[4] is now extensively used although conical scan is still used in some systems (especially IR) due to its simplicity and cost advantage.

Table II, Tracking Radar ECM/ECCM Matrix, is similar to **Table I**. Since a tracking radar must in fact detect a target (with one notable exception) before it can track it, **Table I** also generally applies to tracking radars in the detection-lock-on mode.

False target generators would not generally be used as an ECM against tracking radars. Similarly CFAR, Dicke Fix, passive correlation and sidelobe blanking/cancellers are generally not employed in tracking radars.

The inherently narrower beamwidth of tracking radars generally nullifies standoff ECM, i.e., "self screen" ECM is used against tracking radars. Frequency diversity (diplexing) is a very useful ECCM against jamming. Its use forces the attacking aircraft to carry jammers, repeaters, etc., for two frequency bands. The HSA Flycatcher which uses an X/K_a band radar[43] is an excellent example.

Angle track-on-jam is the notable exception to the requirement for detection before track initiation. Most angle tracking radars can provide good angle tracking on noise (jamming) sources. The German World War II Wurzburg anti-aircraft fire control radar employed a track on jam ECCM known as Stendal-A.[38] As will be indicated later, track-on-jam requires auxiliary range tracking, e.g. by a separate range-only radar, laser range finder,[44] etc. for some missile navigation systems such as command guidance.

Moving Target Indicators in angle trackers are important ECCM against chaff. Most of the literature on MTI address (but fail to so state) its use in search radars since they consider its effect on the video (range) function. Sorkin and Steinberg[45] (USA, 1958) and Subramanian[46] (India, 1974) have addressed MTI in both coherent and noncoherent conically scanned radars.

Pulse compression is generally considered as a means to increase peak power without increase of average power or increase of range resolution without exceeding duty cycle limitations. Actually pulse compression was invented in Germany during World War II on a crash basis as an ECCM (then anti-jam) against Allied chaff.[47,38] In Germany it was called *kugelschale* (*landsknecht*). This used frequency modulation during the transmitted pulse. The intent was to put more energy on the target and less on the chaff by narrowing the radar resolution cell.

Monopulse angle tracking radars are not affected by inverse gain modulation repeaters as conical scan radars are. The usual conical scan radar performs conical scan on both transmit and receive thus disclosing the con scan frequency. This facilitates use of inverse gain modulation ECM. In lobe-on-receive-only (LORO), the conical frequency is not disclosed.

Sakamoto and Peebles[48] describe several hybrid monopulse-conical scan systems (CONOPULSE). It is interesting to note that Peebles invented Conopulse independently several years after others had invented it.[49] Sakamoto and Peebles have performed an extensive analysis of conopulse.

MISSILE NAVIGATION

As defined in Reference 4, navigation is the determination of the position, velocity, and acceleration vectors of the missile at any instant. In most radar guided missiles, the polar coordinate navigation method of Reference 4, slant range, azimuth and elevation angle is used. The navigation method used is further modified by the missile guidance method employed. ECM/ECCM considerations of missile navigation systems are essentially the same as those of the previously discussed tracking radars, since a tracking radar usually measures slant range, azimuth and elevation angles.

MISSILE GUIDANCE

As defined in Reference 4, guidance is the derivation of correcting signals to offset any errors in the flight path. As previously indicated, four types of guidance are used in air defense missile systems: beam riding, command, command to line of sight, and homing.[3,50] Each of these guidance methods has ECM/ECCM implications.

In beam riding, a radar tracks the

target in azimuth and elevation angle.[3,50] The air defense missile is launched into the beam of that target tracking radar. Receiver(s) aboard the missile sense its position with respect to the center of the radar beam. Steering commands then cause the missile to fly out the beam until it intercepts the target. As an historical note, beamriding was first used in some versions of the German World War II V-2 ballistic missile[51] for azimuth guidance.

From an ECM standpoint, the beamrider system has some good features. The system does not use range from radar to target in the guidance process (approximate target range at launch is required but only to determine when to launch, i.e., whether the missile could intercept the aircraft if launched then). The target tracking radar would still have to radiate the guidance beam. Due to geometry, the beamrider receiver is rather hard to jam. As an ECCM, various coding schemes have been derived to provide the receiver with angular references. The principal disadvantage to beamriding is not ECM related but rather that inherently it is a divergent system. Beam diameter increases as the missile leaves the radar.

In command guidance, two tracking radar beams are required: one to track the target and one to track the air defense missile[3,50]. A computer predicts the intercept point and generates steering command for the missile. Command guidance has several ECM disadvantages: a. It requires two tracking radar beams, b. Both radar beams must very accurately track in range as well as in azimuth and elevation, c. The command transmission link is a potential victim of ECM, and d. Command guidance is also a divergent system. Various ECCMs are available for command guidance in addition to those previously listed for tracking radars; e.g., the tracking radar beams may be in different frequency bands thus tempting the attack aircraft to try to jam on two bands rather than concentrate on one. Various ECM-free command transmission techniques such as those of Eckler[52] have been developed. Russian data transmission techniques[53] have also been identified.

Command-to-line-of-sight is a hybrid beam rider and command guidance method. The defense missile is tracked and continually guided into the center of the target tracking radar beam. Various methods which permit the use of only one tracking radar beam have been conceived.

CLOS has some of the ECCM features of beamriders, e.g., range tracking is not required, TOJ is possible and one tracking radar is permissible. As in command guidance, a command transmission link is required. It too is a divergent system.

There are three types of homing guidance[3,50] — active, semiactive and passive. In all three types a receiver in the air defense missile angle tracks electromagnetic energy from the target. In active homing, the defensive missile carries a transmitter which radiates energy to the target. In the semiactive homing system, an "illuminating radar" on the ground (for a ground based air defense system) tracks the target in angle only and provides the source of energy reflection from the target. In passive homing, the target itself provides the energy source, e.g., a jamming transmitter (HOJ), radar(s) on the attacking aircraft, IR radiation from the hot jet engine, or IR from the sun reflected by the attacking aircraft.

In all homing systems, range to target is required only for launching as in the beamrider. All homing systems are convergent. A very powerful ECCM of active and semiactive homing guidance missiles is their ability to use home-on-jam (HOJ). Monopulse radar need not be used in homing missiles since they inherently use LORO (lobe-on-receive-only) to eliminate the use of inverse modulation repeater ECM.

Many homing systems employ Doppler radar although that is not essential. Homing Doppler radars do permit use of very narrow band Doppler "speedgate" filters as described by Ivanoff.[50] Babanov et al,[54] and Sifonov[55] both with lengthy bibliographies describe extensive Russian work in interference elimination, e.g. ECCM. Some of this is applicable to radars. Track via missile[50] is actually a combination of command guidance and homing guidance, i.e., a sensor in the missile makes position measurements with respect to the target. These are then transmitted to the ground where missile steering commands are computed for subsequent transmission to the missile.

ECCM EFFICACY ASSESSMENT

A previous paper[1] indicated the necessity for total systems consideration in ECCM efficacy assessment. **Figure 1** clearly illustrates this point. If we now consider an air defense missile system which employs search radars in two frequency bands, a homing guidance missile (which could also use diplexing) and overlapping (mutually supporting) deployment of missile systems it will be recognized that ECCM efficacy assessment is much more than burnthrough range of a search radar. This is even more complex if a mixed deployment, i.e., radar directed guns and missile system, is utilized.

REFERENCES

1. Johnston, S.L., "Radar Electronic Counter-Countermeasures", *IEEE Trans. Aerospace Electron. Syst.*, Vol. AES-14, No. 1, Jan. 1978, pp. 109-117.
2. Barton, D.K., "Real World Radar Technology", *IEEE Int'l Radar Conf. Proc.*, April 1975, pp. 1-22.
3. Locke, A.S., *Guidance*, D. Van Nostrand Co., Princeton, NJ, 1955.
4. Johnston, S.L., "A Survey of Navigational Measurements Methods for Missile Guidance Systems", *IRE Trans. Instr. PGI-5*, June 1956, pp. 60-65.
5. *IEEE Standard Letter Designations for Radar Bands*, IEEE STD 521-1976, November 30, 1976.
6. Radford, M.F., "Radar ECCM: A European Approach", *IEEE Trans. Aerosp. and Electron. Syst.*, Vol. AES-14, No. 1, Jan 1978, pp. 194-198.
7. Bittencourt, R.P., *Electronic Warfare Technology*, US Naval Postgraduate thesis, June 1976, DDC AD A 028210.
8. Alred, R.V., "Naval Radar Anti-Jamming Techniques" JIEE, Vol. 93, pt. IIIA, No. 10, 1946, pp. 1593-1601.
9. Johnson, M.A., and D.C. Stoner, "ECCM from the Radar Designer's Viewpoint", *Microwave Journal*, March 1978, pp. 59-63.
10. Brookner, E., *Radar Technology*, Artech House, Dedham, MA, 1977.
11. Marconi S600 Series of Radars, Interavia, 1/1968, pp. 73-75.
12. Lind, G., "Analysis of ECCM Characteristics of Frequency Agile Radars," *Proc. Mil. MW Conf.*, London, England, Oct. 1978.
13. Dixon, R.C., *Spread Spectrum Techniques*, IEEE Press, NY, 1976.
14. Skolnik, M.I., "Analysis of Bistatic Radar", *IRE Trans. of Aerosp. and Nav. Electron.*, Vol. ANE-8, No. 1, Mar 1961, pp. 19-27.
15. Rogers, A. "Wideband Squintless Linear Arrays", *Marconi Rev.*, Vol. 35., No. 187, 4th Qtr. 1972, pp. 221-243.
16. Rudge, A.W., "Low Sidelobe Radar Antennas", Proc. Mil. MW Conf., London, England, Oct. 1978.
17. Anderson, L.G., "A Low Sidelobe Cosec^2 Antenna with Integrated IFF", Proc., Mil. MW Conf., London, England, Oct. 1978.
18. Dax, P.R., "Noise Jamming of Long Range Search Radar", *Microwaves*, Vol. 14, No. 9, 1975, pp. 54.
19. Aranciba, P.P., "A Sidelobe Blanking System Design and Demonstration", *Microwave Journal*, Vol. 21, No. 3, Mar. 1978, pp. 69-73.
20. Klemm, R., "Suppression of Jammers by Multiple Beam Signal Processing", *IEEE Int'l. Radar Conf.*, Apr. 1975, pp. 176-180.
21. Kretschner, F.F., "Effect of Cascading Sidelobe Canceller Loops", *IEEE Int'l Radar Conf.*, Apr. 1975, pp. 181-185.
22. Fielding, J.G., et al., "Adaptive Interference Cancellation in Radar Systems", *IEE Conf. Pub. No. 155*, Radar 77, Oct. 1977, pp. 212-217.
23. Wirth, W.D., "Radar Signal Processing with an Adaptive Receiving Array", *IEE Conf. Pub. No. 155*, Radar 77, Oct. 1977, pp. 218-221.

24. Turner, R.M., "Null Placement and Antenna Pattern Synthesis by Control of the Element Steering Phases of a Phased Array Radar", *IEE Conf. Pub. No. 155*, Radar 77, Oct. 77, pp. 222-225.
25. Dicken, L.W., "The Use of Null Steering in the Suppressing Main Beam Interference," *IEE Conf. Pub. No. 155*, Radar 77, Oct. 1977, pp. 226-231.
26. Trunk, G.V., "Automatic Detector for Suppression of Sidelobe Jamming", *IEE Conf. Pub. No. 155*, Radar 77, Oct. 77, pp. 280-284.
27. DiNardo, V., "Adaptivity Criteria of a Modern Air Traffic Control Radar," *Rev. Tech Selenia.*, Vol. 2, No. 4, 1975, pp. 17-32.
28. Norsworthy, K.H., "Electronic Correlator for Passive Ranging," Proc. Int'l. Countermeasures Conf., London, England, 1976, pp. 290-319.
29. Hansen, V.G., "Constant False Alarm Rate Processing in Search Radars", *Int. Conf. on Radar* – Present and Future, *IEE Conf. Pub. No. 105*, Oct. 1973, pp. 325-332.
30. Cope, R.W., "Tackle Noise Variations at Video", *Electronic Design Magazine*, Vol. 20, Sept. 1968, pp. 54-56.
31. Hansen, V.G., and Zottl, A.J., "The Detection Performance of the Siebert and Dicke-Fix CFAR Radar Detectors" *IEEE Trans. Aerosp. and Electron. Syst.*, Vol. AES-7, No. 4, Jul. 1971, pp. 706-709.
32. Johnston, S.L."Burnthrough Range, an Overrated Parameter", *EW Magazine*, Vol. 6, No. 6, Nov/Dec 1974, p. 49ff.
33. Boyd, J.A., D.B. Harris, D.D. King, H.W. Welch, Jr., *Electronic Countermeasures*, Ann Arbor, Univ. of Michigan, IST 71204 R 1961, DDC AD 357877, Section 14.26.
34. Turner, F.M., et al., "Noise Quality Optimizes Jammer Performance", *EW Magazine*, Vol. 9, No. 6, Nov/Dec 1977, p. 117ff.
35. Carpentier, M.H., *Radars: New Concepts*, Gordon and Breach, New York, 2nd Ed., 1968, p. 175.
36. Van Vleck, J.H., and D. Middletown, "A Theoretical Comparison of Visual, Aural, and Meter Reception of Pulsed Signals in the Presence of Noise", *J. Appl. Phys.*, Vol. 17, No. 11, Nov. 1946, pp. 940-971.
37. Pew, R.W., and J.I. Elkind, "Radar Target Identification by Aural Display", *IRE Nat. Conv. Rec.*, Pt 5, 1961, pp. 294-301.
38. O'Brien, R.S., and R.A. Soderman, "Intelligence Information on RCM Effectiveness in the E.T.O." *American British Laboratory Div. 15 Report 1045-MR-15*, June 16, 1945, ATI 62978.
39. Sard, R.D., "Survey of German Radar from the Countermeasures Point of View" *Radio Research Laboratory Report 411-95*, Sept. 4, 1944, ATI-15357.
40. Ellis, J.G., "Digital MTI, A New Tool for the Radar User", Marconi Rev. V. 36, No. 191, 4th Qtr., 1971, pp. 237-248.
41. Cartledge, L., and R.M. O'Donnell, "Description and Performance of the Moving Target Detector", *Rpt. FAA-RD-76-190*, 1977 (Lincoln Lab).
42. Whitford, A.E., and G.E. Kron, "Photoelectric Guiding of Astronomical Telescopes" *Rev. Sci. Inst.*, Vol. 8, March 1937, pp. 78-82.
43. Klaver, L., "Combined X/Ka Band Tracking Radar" *Proc. Mil. MW Conf.*, London, England, Oct. 1978.
44. Jelelanian, A.V., et al. "OADARS (Optical Aides to Detection and Ranging Systems)" *IEEE Electro 1976, Rec.*, Pt. 30, pp. 30-3-1 to 5.
45. Sorkin, C.S., and B.D. Steinberg, "Conical Scan Tracking in Ground Clutter with Coherent and Non-Coherent MTI", *IRE East Coast Conf. on Aero. and Nav. Electron. Proc.*, 1958, pp. 35-45.
46. Subramanian, A.K., "Design Considerations of a Coherent (On Receive) MTI System for Conically Scanned Tracking Radar", *Electro-Technol*, Vol. 18, No. 4, Dec. 1974, pp. 95-98.
47. Johnston, S.L., "Radar ECCM History" Submitted to NAFCON 1976 (Unpublished).
48. Sakamoto, H., P.Q. Peebles, Jr., "Conopulse Radar", *IEEE Trans.*, Vol. AES-14, No. 1, Jan. 1978, pp. 198-208.
49. Peebles, P.Z., private communication.
50. Ivanoff, Alex, "Improved Radar Designs Outwit Complex Threats", *Microwaves*, Apr. 1976, pp. 54-71.
51. Stublinger, E., O. Hirschler, private communication.
52. Eckler, A.R., "The construction of Missile Guidance Codes Resistant to Random Interference", *Bell Syst. Tech. J.*, Vol. 39, No. 4, Jul. 1960, pp. 973-994.
53. Khvorostenko, N.P., "The Noise Immunity of Optimum Higher Binary Codes" *Telecomm. and Radio Engr.*, Pt II, Vol. 18, No. 12, 1963, p. 1ff.
54. Babanov, Yu. N., et al., "Development of Methods for Protecting Radio Communications Systems from Impulse Noise", *Telecomm. and Radio Engr.* Pat. II, Vol. 22, No. 11, Nov. 1967, pp. 82-88.
55. Sifonov, V.I., "Development of the Theory of Interference Suppression in the USSR" *Radio Engr. and Electron. Phys.*, Vol. 2, No. 11, 1957, pp. 139-168.

Performance of Current Radar Systems in an EW Environment
J.R. Prior and N. Woodward
Proceedings of the Military Microwaves Conference, London, October 1978
Pages 39-50

2.C.2

ABSTRACT

In the previous paper an assessment has been given of the electronic warfare threat posed to NATO radar systems. It is the purpose of this paper to discuss the likely performance of current radar systems in this threat environment and indicate the trend for the future.

The paper will start by considering the more important ECCMs incorporated into radars. It will then consider the effects of EW against various types of radar systems. Long Range Early Warning (ADGE) radars, Airborne Early Warning, SAM Acquisition Tracking radars and Airborne intercept radars will be considered. For each of these systems typical present day characteristics will be defined, their vulnerability to the EW threat discussed and the benefits to be obtained from various ECCMs, both technical and operational, will be considered. An indication will be given of the future trends in the design of such systems where relevant to their operation in ECM and the role of the more modern anti-jamming aids arising from new technologies will be described.

1 INTRODUCTION

The previous paper has discussed the possible ECM threat to defence radar systems. This has stated not only the technical characteristics of the threat in terms of noise power densities and deception techniques but also the likely scenario in which these techniques will be used. It is the purpose of this paper to consider the various effects of ECM on typical radar systems, to consider the measures that can be used both technically and operationally to counter these ECM tactics and to indicate the extent to which reasonable operational performance can be maintained in the face of the ECM threat.

The threat derived in the earlier paper was comprehensive and considered noise, deception and passive countermeasures employed by aircraft in a number of tactical roles such as Stand-off-jammers, Escort jammers and self screening jamming. It should be noted that the extent of the threat to any radar system will depend strongly on the operational scenario. This paper will deal with a generalised attack scenario incorporating Stand-off and Escort jammers screening 'quiet', ie non-jamming attack aircraft on the one hand, and also with self-screening attack aircraft. The use of chaff as a passive countermeasure is also considered.

2 SUMMARY OF THE THREAT

By building upon the previous paper we can postulate the following threat scenario. Firstly, self-screening jamming aircraft employing power outputs of roughly 1 KW through aerials of 6 dB gain. These aircraft could spot jam with a noise bandwidth of 20 MHz or could employ broad band (10% of the radar frequency) jamming. Escort aircraft, of the same construction as the attack aircraft but outputting 3 or 4 times the power levels, could be part of the attack force. These aircraft would again use narrow band or broad band noise jamming depending on the density of radars to be countered. One might also envisage stand-off jamming aircraft. These could be transport aircraft dedicated to the jamming task hence could have sufficient space to employ roughly 10 jammer power generators radiating through 10 dB gain aerials. Such aircraft are postulated to employ broad band jamming only since they would operate at large stand-off ranges of for example 100 km. A number of such aircraft could be used to support an air attack. A final aspect of the threat is that concerned with chaff. Typically one can expect chaff clouds providing echoing areas of 10^{-6} M^2/M^3 to be deployed where the objective is to obscure the presence of a raid.

3 THE EFFECTS OF ECM ON RADARS

3.1 Surveillance Radars

Essentially surveillance systems can be degraded by three mechanisms. Firstly noise jamming from Stand-off jammers (SOJ), Escort Screening jammers (ESJ) or Self Screening jammers (SSJ) can be used to deny a measurement of the range of the attack aircraft. This tactic can conceal both the position and the magnitude of a raid. When deployed from a SOJ, with the attack aircraft quiet, complete concealment of the raid position can be achieved, in theory, until a late stage in the operation. Use of Escort or Self Screening jammers can have a similar effect in concealing the magnitude of the raid but a general indication of the possible attack directions can be obtained by a use of the jamming strobes. Widely deployed escort aircraft creating strobes from many directions only one of which supports the attack aircraft will have a major effect in creating confusion within the defence radar systems. Such confusion will have a significant effect on the reaction time of the defences. In the case of an early warning radar it will be necessary to obtain collateral information from unjammed systems or even to scramble aircraft to investigate each strobe direction. For a short range SAM system, collateral data may be obtainable by slewing the radar or an optical tracker to examine each direction in turn. In each of these cases a significant price in performance will result, either by ineffective use of fighters or by considerable lengthening of the SAM system reaction time.

The second mechanism arises from the use of deception techniques in which signals equivalent to those from genuine targets are introduced into the radar system at erroneous ranges or directions. These effects can be caused by the use of on-board equipment such as a repeater jamming pod. The use of staggered prf on the surveillance radar causes all false targets to be at ranges greater than the jamming aircraft but sufficient confusion can be created to deny accurate estimate of the raid size. If inverse gain is used the direction of attack can also be confused by creating false targets through the aerial sidelobes. This is achieved by generating deception pulses whose magnitude is inversely related to the incident radar signal. The radar is then unable to use jamming signal strength as an indicator of the direction of the jamming aircraft.

The third mechanism arises through the deployment of corridors of chaff to conceal the presence and magnitude of a raid. If many corridors can be laid then the true direction of attack can be denied to the defence system. The best anti chaff measures are often found to be in conflict with the best anti jamming measures. Together chaff and attack ECM can be very troublesome.

3.1.1 Surveillance Radar ECCM

All ECCMs built into the radar system exploit some feature which distinguishes the true radar return from the jamming signal. The distinguishing feature may be exploited by means of spatial resolution, spectral resolution or temporal resolution of the radar. A well designed ECM tactic therefore employs signals whose fundamental characteristics are such as to be indistinguishable from genuine radar signals or noise. If this policy is followed the radar designer can only attempt to dilute the effects of the jamming.

In the sections below the more significant ECCMs build into a modern radar are considered in terms of aerial features providing spatial selectivity, transmitter and receiver characteristics providing spectral and temporal selectivity and system features.

3.1.1.1 Aerial features

It will be shown below that for most practical purposes in the battle for good radar detection in the main beam, in the presence of chaff and noise jamming, the advantage lies with the ECM threat.

The most significant ECCM available to the radar is to achieve low sidelobes so that the noise jamming threat is minimised in the near and far sidelobe regions. This requirement leads to the use of planar array aerials or aerials with offset feeds to avoid aperture blocking since in both these cases a lower sidelobe level can be achieved. Second to this is the need to ensure a narrow beam in elevation and azimuth so that the radar is able to detect targets narrowly spaced from the jammers. The implication of this requirement is that, in systems where space and mobility considerations dictate the size of the aerial, advantages are to be gained from operating at higher frequencies. There are however two factors that tend to oppose this trend to higher frequencies. Firstly as the RF becomes higher so rain attenuation becomes more significant. In the EW content the more significant difficulty arises from the need to reject chaff and the use of an MTI radar for this purpose. As the azimuth beamwidth becomes narrower so the dwell time of the beam on the target becomes shorter. For a long range radar which requires a low prf it is not difficult to reach the situation where the number of radar pulses incident on the target is too small to allow effective MTI performance. Countering this by widening the azimuth beamwidth increases the resolution cell size and hence increases the chaff echo.

Although high gain and low sidelobes are undoubtedly the most important features they are supported in some circumstances by the additional features of sidelobe blanking and sidelobe cancellation. In the case of sidelobe blanking the objective is to reject pulsed signals entering through the aerial sidelobes. This is achieved by using an omnidirectional aerial and a receiver system which gives a gain just greater than that achieved in the main aerial sidelobes. Amplitude comparison of the main receiver signals with the omni-receiver signals is used to select only those returns that are greater as seen in the main receiver. Thus all sidelobe returns are rejected. Such a technique is an effective counter to the pulsed signals of inverse gain angle deception jamming mentioned above. It is however not of value against range denial noise screening. In this case some advantages can be obtained from the technique of sidelobe cancellation. Once again subsidiary aerials must be used but the objective now is to use an adaptive control loop so that the phase and amplitude of the noise signal received in the subsidiary aerial is adjusted to cancel the noise entering the main aerial. This system can achieve roughly 20 dB cancellation of noise signals arriving in the sidelobes from a single jammer but will not cancel main beam jamming or multiple signals in the sidelobes. To counter an extensive array of stand-off jammers many independent subsidiary aerials and cancellations loops are required. The problems of creating more than 3 or 4 loops have not been solved in operational equipments. An indication of the theoretical performance of a 3 or 4 loop system in the presence of many jammers is given in figure 1.

One favoured way of achieving sidelobe cancellation arises with the use of multi-element array aerials when the phase and amplitude of the signal received in each element is adjusted to form the receiver beam. By a technique known as adaptive null steering these systems can arrange to position aerial nulls in the direction of the jammers. The use of multi-element array techniques provides additional advantages since it can allow the beam to dwell preferentially in the direction of known targets thus allowing a greater coherent integration time on the target and hence improved performance in jamming. However in practical situations the advantages to be obtained are small compared to the other techniques mentioned above.

3.1.1.2 Transmitter and Receiver features

The single most powerful feature of the transmitter and receiver to counter the ECM threat described earlier as that of frequency diversity. The aim of the ECM force is to concentrate its jamming over as narrow a frequency band as possible, preferably by the use of responsive jamming, thus maximising the jamming power entering the receiver passband. It is essential to counter this capability. One powerful method is to use pulse to pulse frequency agility thereby defeating the responsive jammer and forcing the use of broad band jamming. Unfortunately this transmission is not compatible with doppler processing hence low level performance, chaff and rainfall rejection is degraded. A somewhat less optimal technique for use with short range high prf radars is to use burst to burst agility in which a number of pulses are transmitted on each RF during the beam dwell time. In this case doppler processing can still be used but only a slowly reacting responsive jammer can be defeated. Frequency spread can also be introduced in stacked beam radars by employing differing RFs on each elevation beam. The technique whilst not defeating the responsive self screening aircraft or an escort in close formation with the attack will force the stand-off jammers to use broad band jamming since they will not know which elevation beam, and hence frequency, is being used to track the attack force at any time. Typically within one radar one can expect a diversity of roughly 10% of the RF.

The deception jamming threat arising from on-board repeater jammers can be minimised by prf agility which denies the jammer the capability of creating pulse bursts like those from a true target at ranges shorter than that of the jammer. The uncorrelated random pulses arriving at shorter ranges than the target are effectively removed by a man initiating tracks at a PPI or by pattern recognition capabilities of the automatic plot extractor second detector.

Another use of a wide frequency spectrum within the radar transmission is that of pulse compression which can be of value in improving performance in chaff. The objective here is to reduce the radar resolution cell size to minimise the amount of chaff that competes with the target echo. This technique is particularly valuable where the frequency agility requirement and narrow azimuth beamwidth have prevented the use of MTI processing. Compression ratios of 10:1 have been achieved, producing typical effective pulse lengths of 0.1 μS. Even with this short pulse and with a narrow beam aerial chaff echoing aerials of 100 m^2 can be encountered.

As far as the receiver is concerned the objective is to select any feature of the jamming signal input that distinguishes it from a true target.

The first line of defence is to incorporate a constant false alarm rate circuit which rapidly reacts to changes in background noise level to adjust the detection threshold. Such circuits frequently operate by a moving window technique in which the threshold used at any range cell is derived from the signal level in a number of cells at slightly shorter and greater ranges. The technique can be so successful that jamming noise and target signals are removed from the radar display and a separate jamming level display is required to alert the operator to the existence and direction of jammers.

Increased resistance to jamming can also be obtained merely by operating the radar at a higher RF because if the jammer can be forced into broad band (10%) jamming the frequency spread to be covered is greater at higher frequency. Examination of the frequency utilisation of a large number of radars clearly shows a trend to the use of higher frequencies in more modern systems.

A final important ECCM incorporated into most radars is the Dicke Fix receiver. This utilises a wide band limiting amplifier followed by a narrow band amplifier to reject non gaussian noise signals such as swept narrow band noise or other signals generated from imperfect ECM sources.

3.1.1.3 System features

If all else fails there are still ECCMs that can be incorporated into the overall system both by technical means and also by operational means. Thus frequency diversity between many radars in an area can be used to increase the problems of the jammers particularly if the various radars can be netted. Should overlapping cover be available then netting of radar plots and jammer location by strobe triangulation become powerful additional ECCM that are incorporated into current systems.

In the case of mobile radars on the ground it may in some circumstances be possible to locate them in positions so that local site features obscure the principal direction of ECM attack while with airborne radars such as AIs it is advantageous to arrange that the sector of search is offset with respect to an airborne stand-off jammer force.

A final, and most important feature of any radar system is the incorporation of a man into the ECM control. A modern system can have many optional signal processing features and many of them whilst improving ECM performance carry a clear performance penalty. The proper selection of options can vary from sector to sector of the radar coverage. It is therefore important in many situations that a man should have a supervisory role in selecting these options according to his judgement of the nature of the ECM.

3.1.2 Summary - Surveillance Radar ECCM/ECM

Table 1 gives a summary of the relevance of the various surveillance radar ECCM techniques discussed and indicates their usefulness against the various tactical roles of ECM attack. This diagram illustrates the conflict between the ECCMs to Active and Passive ECM.

TABLE 1 - SUMMARY OF SURVEILLANCE RADAR ECM/ECCM INTERACTION

RADAR ECCM FEATURES	THREAT				
	SOJ	ESJ	SSJ	CHAFF	DECEPTION
AERIAL ECCM					
Low Sidelobes	/	/			/
Narrow elevation beam	/	/		/	
Narrow azimuth beam	/	/		X	
Adaptive Dwell	X	/	/		
Sidelobe Blanking					/
Sidelobe Cancellation	/	/			
TRANSMITTER					
Higher RF	/	/	/	X	
Chirp	/	/	/	/	/
RF agility pulse/pulse	NBN	NBN	NBN	X	
RF Agility burst/burst	NBN	NBN	NBN		
PRF agility					/
RF diversity in elevation	/				
RECEIVER					
CFAR	/	/	/	/	
Dicke Fix	NGN	NGN	NGN		
Pulse length discrimination	SN	SN	SN		/
SYSTEM					
Strobe triangulation		/	/		
RF diversity between systems	/	/			

NBN - effective against narrow band noise
NGN - effective against non gaussian noise
SN - effective against switched noise
X - counter productive
/ - useful

It should be emphasised that the foregoing brief review of surveillance radar ECCM options merely indicates some of the main tools available to the radar designer. The precise way in which these components are selected and combined will vary according to the operational scenario in which the equipment will be used.

3.2 The Effects of ECM on Tracking Radars

Many of the above ECCM features apply also to tracking radars however due to the different nature of the task of a tracking radar the ECM techniques which can be employed are rather different from the surveillance radar case. The aims of the jammer are to deny or confuse the measurement of range, angle or velocity of the attack aircraft by the use of noise or deception techniques. Although tracking radars can be degraded by Escort or Stand-off aircraft the emphasis below is on the effects of self screening jammers.

3.2.1 Tracking Radar ECCM

3.2.1.1 Attacks on range tracking

Target range can be denied to the radar by the use of noise jamming and by the use of the Range gate pull off deception technique (RGPO) in which a false echo generated on the jammer is progressively delayed to reduce the target tracking gate away from the true target echo so that an infinite J/S ratio can be obtained for angle deception purposes. The use of noise jamming is most effective against a missile system that does not have a Track-on-Jam or Home-on-Jam mode of operation. In other situations its effect is to create uncertainty as to whether the target is within cover of the weapon system. Short range gun systems increasingly rely on laser rangefinders to provide immunity to this type of ECM. Range gate pull-off can be countered by using prf jitter (to deny the possibility of forward pull-off) in combination with leading edge tracking techniques to reduce the probability of the range gate being seduced. In a leading edge gate the true pulse is preferentially selected by employing preferential weighting to the forward gate signal. This can lead to a 10 dB advantage to the true echo. In very high J/S situations leading edge tracking may not be effective however a well trained operator will rapidly recognise that pull-off is occurring.

3.2.1.2 Attacks on velocity tracking

A well designed velocity gate pull off system is capable of hiding the true target amongst a set of false targets at different doppler frequencies. The true target is not easily recognised by a man but the man is aware that deception is taking place. VGPO is usually used to create an infinite J/S situation in which effective angle deception can occur. Hence VGPO is best countered by switching to the Track-on-Jam mode of operation. Similar considerations apply to the velocity denial mode of a noise jammer.

3.2.1.3 Attacks on angle tracking

The main objective of the SSJ is to create angular errors; RGPO and VGPO if successful can assist in this by creating a situation in which the jamming to signal ratio is very large. Older types of tracking radars employed a conical scanning pencil beam and used the phase and modulation depth of the amplitude modulated echo from an off axis target to derive the angular error signal. Such systems are easily confused by amplitude modulated jamming hence modern tracking radars use monopulse transmitter and receiver systems. These are extremely resistant to angle deception modulation from a single jamming source.

There are several other techniques that are applicable to jamming of monopulse systems. For example jamming from several aircraft, ground reflection effects, artificial glint from two sources on the aircraft, cross polarisation jamming, and the use of active expendable jammers. These are all operationally or technically difficult to implement and generally do not form part of the current threat.

3.2.2 Chaff

A final significant countermeasure is that arising from cartridge dispensed chaff. Chaff dispensed from a high speed aircraft in this way rapidly blooms to create a large echoing area, but also rapidly slows to the point that is is rejected by an MTI system. The effectiveness of chaff used in this way depends on many detailed features of the radar design such as the range gate structure, the range servo control loop logic, the radar aerial beamwidth, the MTI system employed and the range and aspect angle of the aircraft with respect to the radar.

4 PERFORMANCE OF RADARS IN ECM

The extent to which ECM degrades the performance of a radar, and the degree to which the remaining performance is operationally useful, depends on the role of the radar. This section examines typical examples of a:

- Long Range Early Warning(LREW) radar
- SAM Surveillance radar
- SAM/AAA Tracking radar
- AI radar

to illustrate the factors involved.

4.1 LREW Radar

A modern LREW radar will use a peak transmitter power of a few megawatts at a frequency in the 3 GHz region. Its antenna assembly will consist of several narrow pencil beams stacked in elevation, rotating in azimuth at a rew rpm. The main beam gain will be high, about 45 dB, and attention will have to be given in the design to achieve low sidelobes. In the absence of jamming, its detection range will be of the order of 400 km.

Fig 2 shows the detection ranges which might be obtained against a number of different threats:-

- A single SOJ in a main beam would reduce the detection range against a quiet target to about 20 km. However this is an unlikely case because the target and SOJ are not likely to be in the same elevation beam.
- A much more likely case is several SOJs in the sidelobes, which reduce the detection range by more than 50%. Thus the main function of the radar, LREW, is seriously impaired.
- An SSJ will deny range measurement until it is at about 30 km range.
- If an Escort jammer could stay in the same beam as the target, it would deny range to below 15 km, but in the more likely case where it is in the near sidelobes, burn through would occur at about 150 km range.
- Used without support jamming from SOJs, the SSJ and ESJ give away their angular position and their true position may be obtainable by using collateral information from other radars. However, the addition of jamming strobes from SOJs would greatly confuse the total picture.

For long range operation the prf is limited and at long range the resolution cell is large but the number of pulses per dwell time is small. Hence MTI cannot be effectively employed and these radars are vulnerable to the effects of chaff.

Older types of LREW radars use fan beams in elevation (combined with a separate height-finding radar) and less attention was paid to sidelobe reduction. Elevation discrimination between an SOJ and a quiet target is not achieved and the performance in jamming is significantly worse, as illustrated in Fig 3.

4.2 SAM Surveillance

A short range SAM Surveillance radar might typically operate at 3 GHz and use a fan beam covering 30° in elevation by 3° in azimuth. To achieve low level cover a pulse doppler radar could be used. Fig 4 gives an indication of the performance of such a radar against a number of different threats. The boxes on the graph indicate the power levels of typical SSJ, ESJ and SOJ threats. It can be seen that

1 The SSJ, even when operating in 10% broad band jamming mode can provide screening to very short ranges.
2 The ESJ can similarly protect aircraft which are at the same range from the target radar and for which it can provide jamming protection through the near sidelobes.
3 The SOJ provides good protection for aircraft penetrating along its jamming strobe but even 10 SOJ aircraft are of limited effectiveness in protecting attackers widely spaced in azimuth from the jammers.
4 Against the chaff threat, assuming a chaff rejection of 60 dB due to the use of a pulse doppler radar, no significant difficulty exists.

4.3 SAM/AAA Tracking Radar

A typical modern tracking radar for use with short range gun or missile air defence systems might operate at 15 GHz with a coherent MTI transmission and use a staggered prf with a pulse length of 0.5 microsecs. A 40 dB gain aerial (1° pencil beam) and a monopulse angle measurement system would be used. The range tracking system would employ a split range gate with preferential weighting to the forward gate, and a range rate acceleration limit of 3 to 5 'g' would be applied. A boresight optical tracking system would be used to provide reversionary angle tracking while for a gun system a laser rangefinder would be incorporated.

Such a radar would be somewhat more resistant to range denial noise jamming than any associated surveillance radar since it operates at higher frequency and uses a high gain aerial. For a short-range weapon system the required maximum effective range for the radar need be only 10 to 20 km. Thus, as shown in Fig 5, the SOJs and ESJs in sidelobes produce no effective degradation. An SOJ in the main beam is only marginally effective, and in any case this geometry is difficult to achieve and maintain against a narrow beam radar. The radar postulated here has a maximum unjammed range of 40 to 50 km, ie it has been designed to operate in a jamming environment with a consequent excess of performance in the clear. An SSJ employing narrow band jamming from a single transmitter tube delays burn through until less than 1 km range, thus forcing the use of a laser rangefinder in a AAA system, or the Track-on-Jam mode in a missile system. In the broad band mode of jamming, burn through might occur at 2 to 3 km, thus leaving some capability in a AAA system without a laser.

With a split range gate width of 150 metres and an acceleration limit of 4 'g' it will take roughly 2 seconds to seduce the gate away from the target. Using a repeated and time delayed pulse which is at least 10 dB greater than the aircraft echo, RGPO could be achieved over the 2 second period. However an alert operator could observe this on an A-scope display and intervene manually to negate the effect. The problem with using chaff released by the target itself to try to pull-off the range gate is that because the chaff slows down very rapidly to the point where the MTI will remove its echo, it will be difficult to produce a high velocity, high echoing area chaff cloud for the necessary two seconds at a range of at least 100 metres from the aircraft. Considerably greater effectiveness can be expected from non-MTI tracking radars against crossing targets.

The use of monopulse tracking ensures a high degree of invulnerability to practical angle deception techniques.

The trend in modern radars is to use digital signal processing which offers the advantage of speed and stability in manipulating the signals for range gating, MTI etc. An important additional advantage which ensues is that of very rapid mode switching. Hence, if deception ECM can force the radar to lose track on the target signal, a digital radar will be able to re-acquire it in a timescale of milliseconds rather than seconds, as required by analogue or manual mode switching.

4.4 AI Radars

An AI radar enjoys the tactical freedom of a manned mobile platform and conceptually at least can manoeuvre to avoid the worst effects of the jamming. For example, a fighter which found itself at the same altitude as a SOJ force and its target would suffer the degradation shown in Fig 7(a). However, it has the option of rapidly changing altitude and its performance could then improve to that indicated by the line (b). It could also alter its direction of approach (or its patrol pattern) to search across rather than into the jammed sector, putting the SOJs into its far sidelobes. The disadvantage of the fighter arises from its short endurance and the limited detection range of the radar; it therefore requires initial direction from ground radars whose performance could also be seriously degraded.

Against an SSJ or ESJ threat the AI is denied range information until the fighter is within a few kilometres of the jammer. However the jamming strobe provides the direction of the target and methods of obtaining crude estimates of range from the rate of rotation of the sight line are available. Generally, only a 'within-range' indication is required for missile release, and at gun ranges, visual ranging is possible. Thus while self screening or Escort jamming may force the fighter to use less effective tactics, it will not prevent the fighter from closing the target.

Older types of AI using conscan angle tracking are vulnerable to deception ECM, but as with other tracking radars, the use of monopulse removes this vulnerability. Modern AIs employ doppler processing to achieve a look-down capability, and this also confers an invulnerability to chaff.

5 CONCLUSION

This Paper has reviewed the trends in radar design and performance arising from an increasing ECM threat. While conventional longer range surveillance radars have fundamental limitations and will suffer a degradation in performance, it has been shown that shorter range systems, by over-designing to cope with the ECM threat, by utilising latest technology and by incorporating (with the penalty of extra cost and complexity) opto-electronic devices, can retain a useful performance. The interaction between radar and ECM is one of continuing development and both new technologies and new concepts of operational employment need to be explored and exploited.

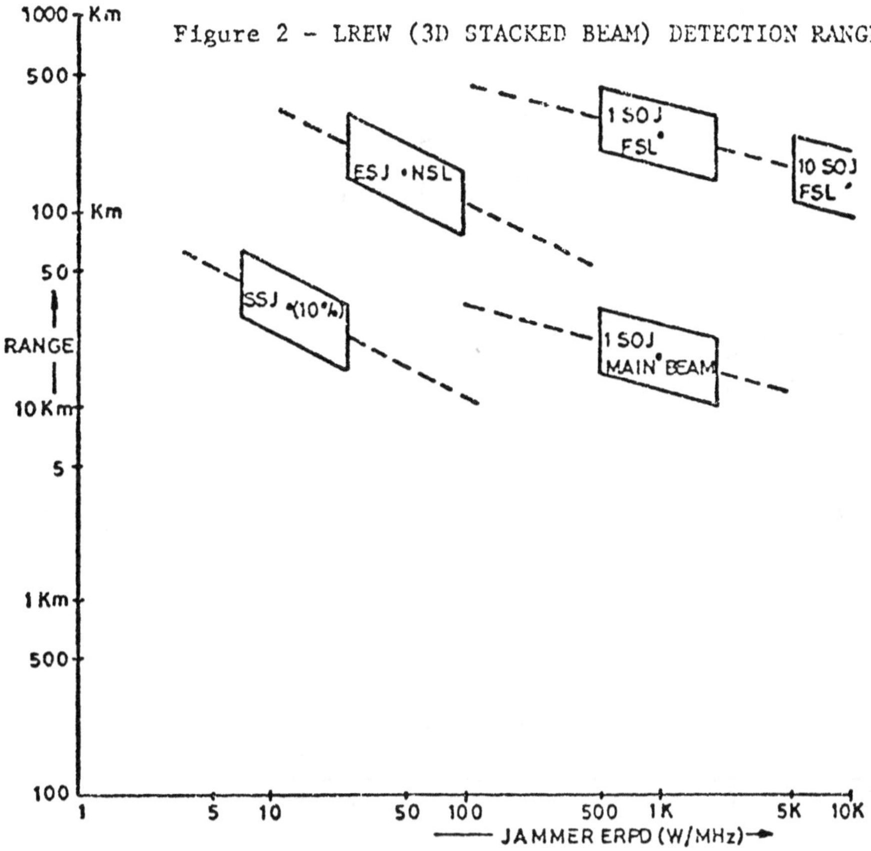

Figure 2 - LREW (3D STACKED BEAM) DETECTION RANGES

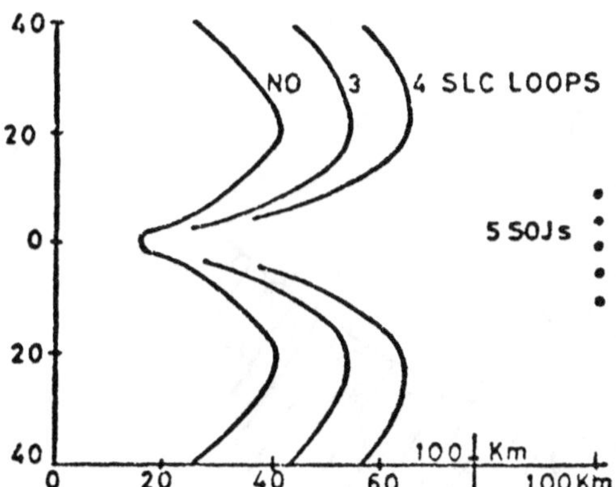

Figure 1 - SIDE LOBE CANCELLATION

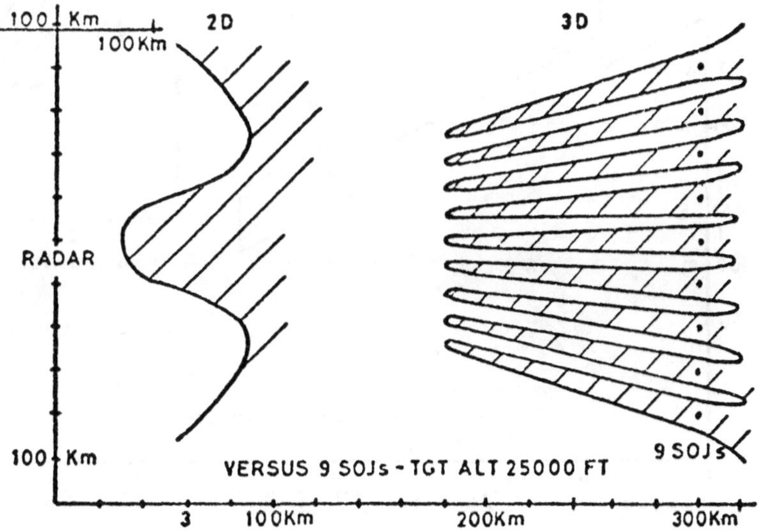

Figure 3 - RELATIVE PERFORMANCE OF 2D & 3D LREW

Figure 4 - SAM SURVEILLANCE RADAR DETECTION RANGES

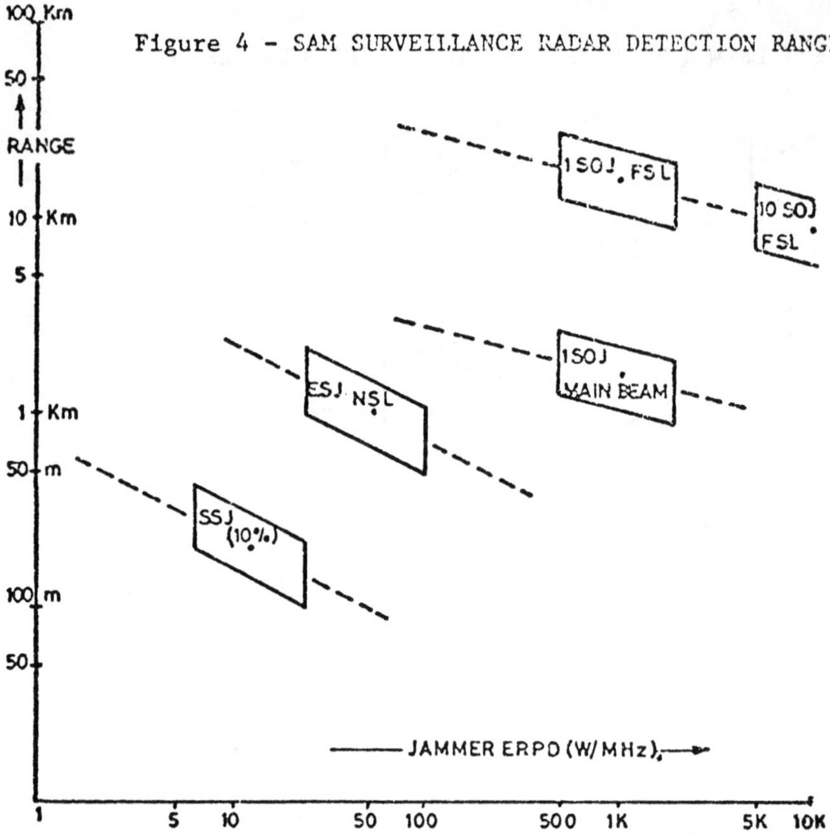

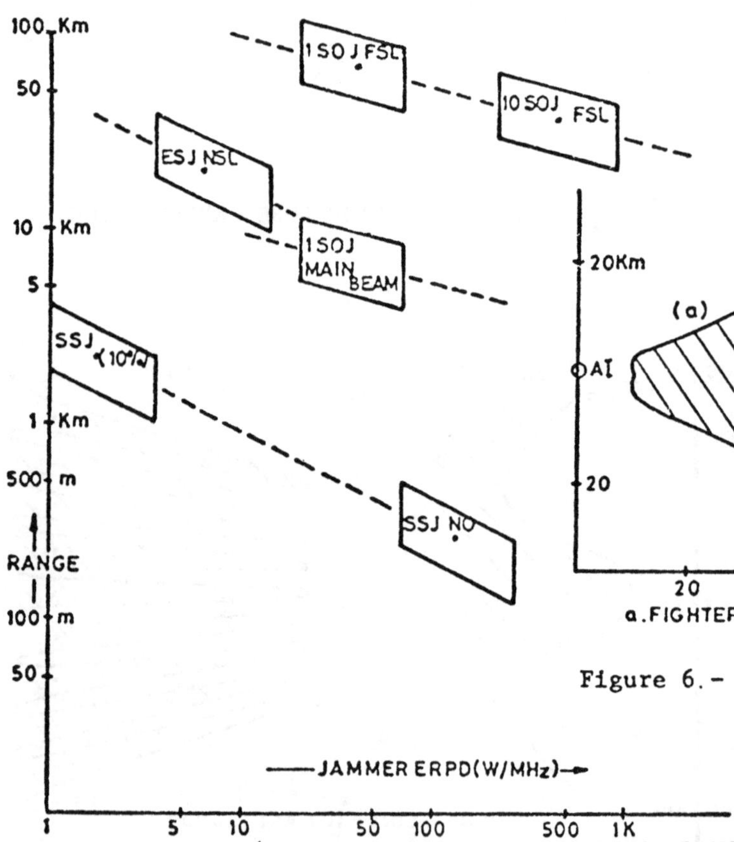

Figure 5 - SAM/AAA TRACKING RADAR DETECTION RANGES

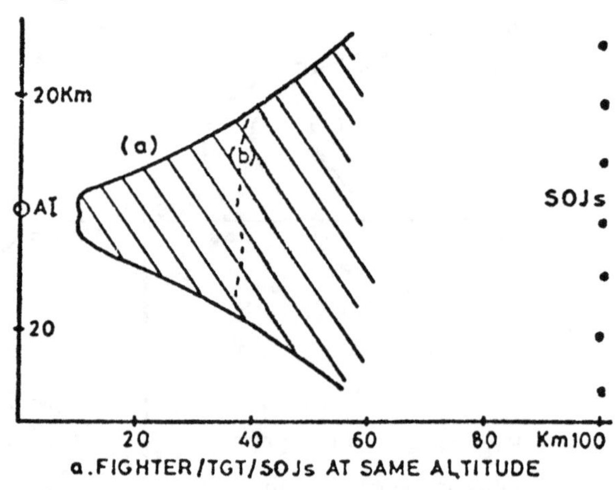

Figure 6.- AI RADAR DETECTION RANGES

ECCM from the Radar Designer's Viewpoint
M.A. Johnson and D.C. Stoner
IEEE ELECTRO-76 Record, Part 30, May 1976
Pages 30-5-1 – 30-5-11

2.C.3

INTRODUCTION

The previous paper by Steve Johnston might imply that for every possible mode of jamming, an appropriate counter has been invented and developed. Therefore, you all may be thinking there is little left to be concerned with. Apparently, all the radar designer need do is simply add sufficient counter techniques to satisfy the specified threat and ship the equipment.

Unfortunately, from the military customer's point-of-view (or should I say fortunately, from the continued engineering employment point-of-view), things are not quite that simple. There are four hurdles to cross before the Electronic Counter Counter Measures (ECCM) technique performance measured in the laboratory can be realized on an operating radar:

1. Many ECCM techniques require basic radar performance beyond that required in the absence of jamming. Improvement of the receiver stability, dynamic range, bandwidth rolloff or out-of-channel characteristics, and achievement of matched multiple receiver channels, for instance, may be necessary. Thus, it is usually necessary to review and possibly redesign the basic receiver in light of the ECCM techniques to be employed. In addition, some techniques make special demands on the transmitter and antenna.

2. When several ECCM techniques are to be employed, it may be found that they are incompatible with each other, so that they cannot be used simultaneously.

3. The presence of radar clutter of "friendly" interference may require counter techniques of their own which can work at cross purposes to the desired ECCM techniques.

4. One approach to the compatibility problem is to give the radar operator a jamming monitor and a set of switches so that he can choose those techniques which best counter the threat of the moment without using simultaneously incompatible techniques. There is a tendency with this approach, however, to give the operator so many choices that he is unable to find the right combination when he really needs it.

It is the intent of this paper to illustrate the kind of solution to these problems which has proven to be practical in the experience of the General Electric Company, and to show the general kind of thinking which led to that solution. Due to the restrictions of time allotted and security, the answers to all threats and all design problems cannot be covered. However, the kind of approach necessary can be illustrated.

DEFINITION OF AN ILLUSTRATIVE CASE

We shall consider a conventional 2-D air search radar which mechanically scans a fan beam in azimuth and radiates simple short pulses. The jamming and interference environment will include:

1. Multiple standoff jammers

2. Both narrow and wideband jamming

3. Continuous Wave (CW) and pulse interference.

4. Ground clutter

5. Weather clutter and/or chaff

Figure 1 illustrates an ECCM radar configuration we have evolved to meet this type of threat. It uses frequency domain, time domain, and spatial domain techniques combined to reduce the effect of the various kinds of jamming and interference, either singly or in combination. The techniques used are individually adaptive to the threat present to maximize radar performance with very little operator interaction required. Let us first briefly summarize its features, then detail them individually.

1. A frequency agile transmitter is adaptively programmed to the frequency of least interference.

2. Several auxiliary antennas and receivers are used to suppress sidelobe interference using both adaptive cancellation and short interval gating.

3. Ground clutter suppression is provided by appropriately tailored Moving Target Indicator (MTI) cancellation.

4. Chaff and weather clutter suppression is achieved by adaptively weighted cancellation.

5. A "clean" display presentation is provided by a Constant False Alarm Rate (CFAR) threshold utilizing split-window normalization.

6. Elimination of a synchronous pulse interference in the mainlobe is achieved by a pulse-to-pulse correlator.

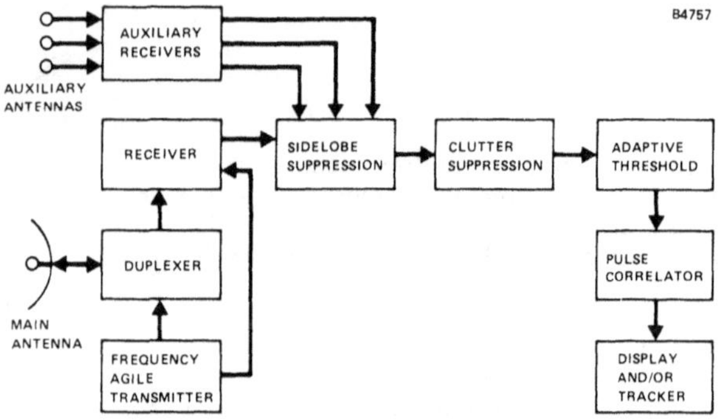

Fig. 1. ECCM Radar Configuration

DESCRIPTION OF ECCM TECHNIQUES

Transmitter

The basic technique applied here is frequency agility, appropriately programmed. An agility programmer is required to force a jammer to "spread" his energy over the entire agile bandwidth of the radar, thus reducing the effective jamming power. The agility programmer samples the interference prior to the next radar transmission (pre-look). This is done in such a way that the radar is not forced to change frequencies until the interference level is sufficient to degrade performance in a normal channel. The pre-look programmer is also very useful for the mainlobe jammer case in that spot jammers become less of a threat. It has been found that by examining the jammer spectrum and selecting a "hole" in this spectrum for the next radar transmission, a significant range improvement is obtainable. For typical parameters, the feature will also extend "burnthrough" ranges.

There are other benefits to be derived from the operation of this frequency agility over a wide bandwidth. If the frequency is varied over a reasonable percentage of the bandwidth in a systematic manner, it has the following advantages:

1. If the radar is used in an environment where a reflecting surface produces interference nulls in the elevation coverage pattern, the effects of a frequency agility will tend to fill in these nulls, hence eliminating the problem.

2. Pulse-to-pulse fluctuations of the target return caused by the frequency changes should increase the detection probability of the target.

In the "agile mode", the pre-look sampling of the environment occurs during the radar interpulse period (on a pulse-to-pulse basis) prior to each transmission. The next transmission then takes place at the frequency of least interference. In the MTI mode, the pre-look sampling is accomplished after every fourth pulse to maintain the MTI processing.

Receiver

The receiver uses dual conversion with a high dynamic range. Since the bandwidth at the receiver input will be broadband to accommodate the system frequency agility, the dual conversion assures that a jammer cannot transmit two frequencies separated by the first Intermediate Frequency (IF) and defeat the agility of the radar by producing a constant signal at the IF. The high dynamic range is essential to prevent a jammer of any realistic size from saturating the receiver and reducing system sensitivity.

A Sensitivity Time Control (STC) in the receiver reduces system sensitivity to avoid detections of birds, ground moving vehicles, etc. It also suppresses close-in ground clutter that may otherwise be acted upon by the sidelobe canceller system. By placing the STC actuators in front of the sidelobe cancellers, this problem was eliminated economically. Other solutions such as "sample-and-hold" techniques within the sidelobe canceller loop and directive auxiliary antennas are also viable solutions. One must take care that the former solution is not degraded in a "swept jamming" environment.

Since the Coherent Sidelobe Canceller (CSLC) used in this system represents considerable cost and capability, it is good strategy to use it only on jammers that are actually perturbing the used frequency of the radar. This is best accomplished by placing a filter matched to the radar transmission before each loop to reject out-of-band jamming. As an illustration: If a Radio Frequency ("RF") sidelobe canceller was used in the system which covered the entire agile bandwidth of the radar, it may devote all of its capabilities to the largest jammer in the agile band. If a smaller jammer is used to cover the transmission frequency, this CSLC arrangement may be effectively defeated. This, the loops preceded by the matched filter sees only the in-band jammer and simplification in the design of the auxiliary antenna system and the CSLC loops are achieved. To obtain maximum performance from the CSLC, each receiver (main and auxiliaries) must also be matched to each other.

Sidelobe Suppression

The sidelobe suppression system consists of a coherent sidelobe canceller (CSLC) array (three loops) to remove barrage, CW and swept jamming and a compatible sidelobe blanking system to remove repeater and deception jamming.

The CSLC system is used to counter high-duty cycle jamming received in the sidelobes of the main antenna.[1,2] As shown in Figure 2, an auxiliary antenna array is used to sample the jamming entering the sidelobes of the main antenna. This sampled signal is corrected in phase and amplitude to match that of the main antenna, and subtracted in a canceller system. The correlation between the jamming signal in the auxiliary antennas and the output signal to the summer are measured by the correlation mixer and the narrowband filter. This results in a weighted signal which is fed to the steering mixer and will amplitude weight and phase shift the auxiliary antenna signal also entering the steering mixer. These translated signals leaving the steering mixer are then subtracted from the main signal in the summing network. Cancellation thereby subtracts this array beam from the main antenna sidelobe pattern, forming a narrow null at the jammer location.

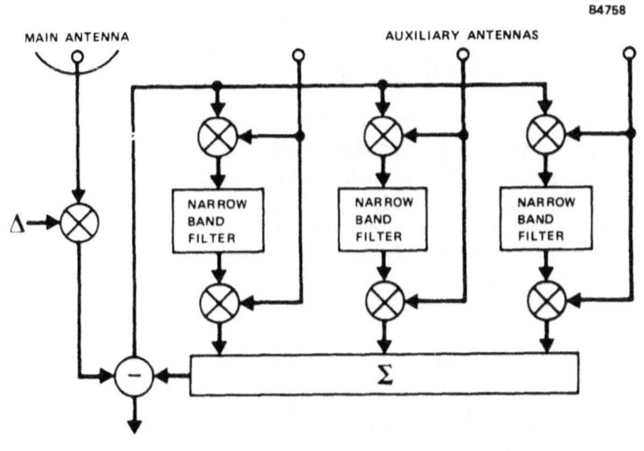

Fig. 2. 3-Loop Sidelobe Canceller

In addition to the interaction with a responsive sidelobe jamming threat, the CSLC must be configured, constrained and integrated into the radar system to insure it will perform its function without perturbing or degrading other system functions.

For example:

- It must not degrade noise figure or subclutter visibility in the normal environment where clutter may be present.

- It must perform in a multiple threat environment where Electronic Countermeasures (ECM), clutter and mutual interference may be present.

- It must be designed to not eliminate desired target signals.

Because this is a simple pulse system, the lockup time of the CSLC was made fast enough to achieve good cancellation without cancelling the signal.

The sidelobe blanker is used to remove low-duty cycle repeater of "spoofer" jammers entering the sidelobes of the main antenna in either a deception or saturation mode. The blanker utilizes two of the same auxiliary antenna and receiver systems as the CSLC to sample the signal, determine through logic processing the presence of these same signals in the sidelobes of the main antenna, and gate out the unwanted signals, as illustrated in Figure 3.

This logic is designed to allow compatible operation of the sidelobe blanking and the correlation sidelobe canceller in the presence of combined jamming.

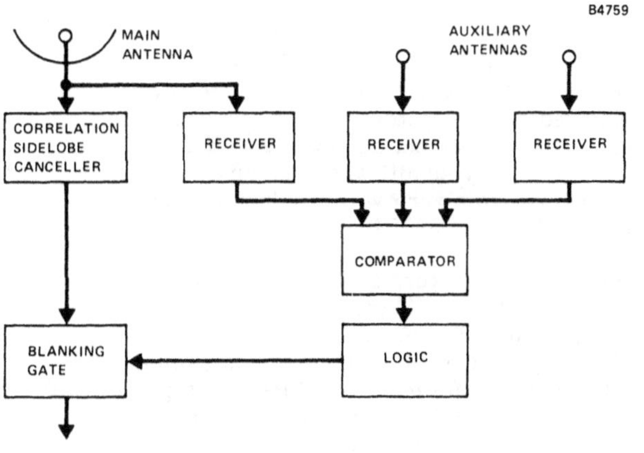

Fig. 3. Sidelobe Blanker

Clutter, Chaff and Weather Suppression

The MTI system consists of two filters, a fixed-filter function for stationary clutter with the filter centered at zero Doppler and an adaptive-notch filter function for nonstationary clutter. The adaptive system is used to place the zero of the filter at the clutter Doppler response. This is shown in Figure 4.

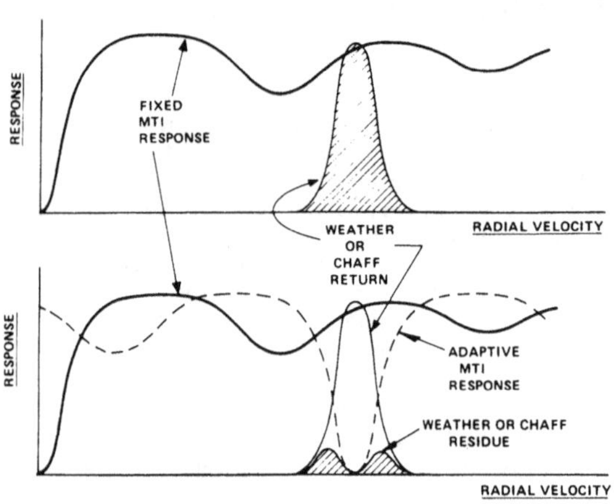

Fig. 4. MTI Response

The MTI function used for ground clutter is a conventional four-pulse MTI filter response with Pulse Repetition Period (PRP) stagger to remove the blind speeds. The system stability is adequate to support the improvement achievable with a four-pulse canceller. Any significant departure of the clutter spectrum from zero will result in poorer performance of the fixed-filter MTI.

To minimize the return from chaff and weather, the fixed-filter MTI is followed by the four-pulse adaptive MTI which is illustrated on Figure 5. The correlation loops placed at each tap output samples the output of the MTI filter and modifies the tap weights (phase and amplitude) to maximize the signal-to-interference ratio (SIR) at the sum output port. This configuration will maintain its performance in a continually varying clutter environment on a pulse-to-pulse basis.[3]

This clutter suppression scheme of a fixed and variable filter has worked well in a combination of mixed clutter spectra. The adaptive MTI varies its response to various clutter spectral shapes including both chaff and weather.

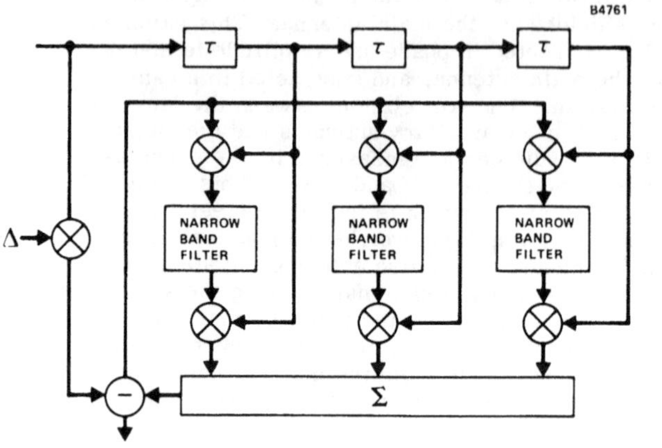

Fig. 5. Adaptive MTI

Adaptive Threshold and Correlator

Automatic thresholding for target detection is provided by a normalizer (a CFAR device) and a two-pulse correlator [Pulse Repetition Frequency (PRF) discriminator] to reduce nonsynchronous interference. A functional diagram of the range normalizer and correlator is shown in Figure 6.

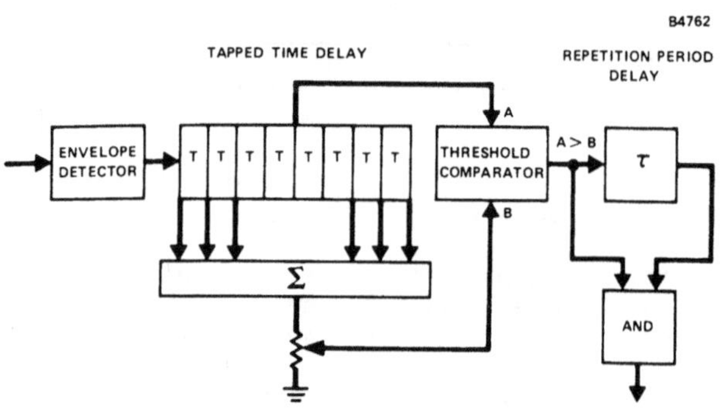

Fig. 6. Adaptive Threshold and Correlator

The range background estimator is probably the most cost-effective signal processing function one can implement. The adaptive threshold suppresses unwanted noise and other residual uniformly-distributed clutter while obtaining "super clutter visibility" on targets in the clutter region. This adaptive threshold is implemented as a range normalizer to provide CFAR in a noise environment. While its primary purpose is CFAR in a receiver noise background, it also provides CFAR in any reasonably homogeneous interference. This device adjusts a threshold proportional to the mean background level in some number of range cells preceding and following the range cell under examination. When the range cell under examination is sufficiently large with respect to this background (A > B), the normalizer gives a video output.[4] This output is then sent to the correlator, as shown in Figure 6. As stated before, the normalizer provides no subclutter visibility by itself, but it has proven an excellent adjustable threshold when as much interference as possible has been eliminated by the processor functions prior to its estimation processing. To prevent "capture" by a very large signal in a single range cell, the transfer characteristic preceding this device is a linear-logarithmic function.

This adaptive thresholding scheme is preferred over base clipping for an automatic detection system. With base clipping, if the detection level is set on the basis of receiver noise, then clutter returns over a large dynamic range will change the noise level and raise the false alarm probability drastically. This rise in the false alarms has an intolerable effect on automatic detection systems.

The two-pulse correlator following the normalizer serves several purpose. It essentially doubles the false-alarm rate of the adaptive threshold, thus decreasing the sensitivity required by the device alone. It also removes impulse jamming, even in the mainlobe, unless this interference has been synchronized to the radar PRF. The adaptive threshold and the correlator thus work together to reduce the false-alarm rate at the output of the radar system to a tolerable rate.

Other Design Considerations

The preceding portion of the paper has presented the selection and grouping of compatible techniques which enhance the sensor sensitivity to signal power while reducing sensitivity to clutter and jamming in a multiple threat environment. The incorporation of:

1. Frequency processing
2. Antenna pattern processing
3. Velocity processing
4. Time processing
5. Amplitude processing

have all been required to counter the environment.

It was important that these capabilities be "on-line" at all times and require minimum operator decisions to function properly. This not only insures the proper response to the environment for maximum detection, but at the same time reduces operator training requirements. The operator is no longer required to recognize the total environment and adapt the radar lineup to that situation. Consequently, he needs to devote little time to this. The requirement that these features be self-compatible is a necessary condition.

In the system design, the compatibility of the frequency agility with the philosophy of MTI operation must be addressed. For this system, the ground rules were to maintain MTI operation at all times, giving it up only when the jamming becomes large enough to render it useless. This requires holding a single carrier frequency for at least four consecutive pulses. When extreme jamming is sensed, this requirement is automatically relaxed. Once the jamming has been effectively countered, MTI operation will resume.

It must also be remembered that this was a simple pulse, long range search radar with the PRP being compatible with the processing. With the advent of long duration waveforms, faster response times, higher PRP's, and more degrees-of-freedom care must be taken to avoid signal cancellation, distortion, low-level deception jamming and an increase in noise carryover or throughnoise in the range sidelobes of large Signal-to-Noise Ratio (SNR) returns. These will all degrade the performance of the ECCM processor if not handled properly. Since this particular design has worked well for this system does not mean it is appropriate for all situations. As always, each system must be examined in light of its own specification.

A radar ECCM processing lineup has not been designed to insure immunity from most ECM tactics. There are several additional tactics that come to mind which can with high probability defeat this radar processor. However, these tactics require that the jammer teams utilize considerable resource expenditures. This means that our processing performs its function well; that of maximizing the ratio of cost to defeat the system relative to the system cost.

QUALITATIVE PERFORMANCE RESULTS

Specific improvement performance cannot be given due to security restrictions. However, the following Plan Position Indicator (PPI) photographs were selected to give an indication of the performance of individual ECCM fixes and their working compatibility.

The photographs were taken at the General Electric facility with targets of opportunity and externally applied jamming signals. While the total ECCM processor is on-line in all the photos (i. e., the MTI is usually removing ground clutter that is always present) only the major ECCM contributor is described for its performance. The before photos are with basic receiver and the after photos are with the ECCM processor.

CSLC Against Swept CW

Figure 7(a) shows the effect of a narrowband swept CW jammer radiating 1000 W of Effective Radiated Power (ERP) without the CSLC connected. Figure 7(b) shows processed video due to the CSLC system. The range rings are 50 nmi each.

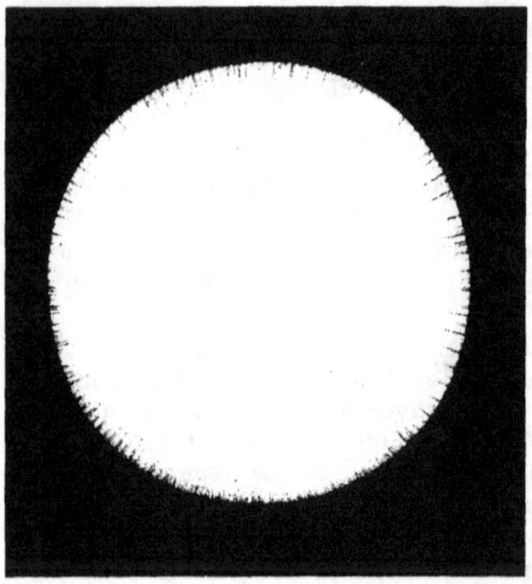

a. Canceller Off

b. Canceller On

Fig. 7. CSLC Performance with Swept CW

Sidelobe Blanker Operation

These photos were taken during flight tests at an operational site with controlled aircraft. The solid strobe line in the photograph is a result of instrumentation being used and would not normally be present.

Figure 8(a) shows the false target repeater in the false target mode of operation. The false targets are close to maximum range at 11 o'clock. The effectiveness of the sidelobe blanker is seen in Figure 8(b) where the true targets are visible and the false targets have been removed.

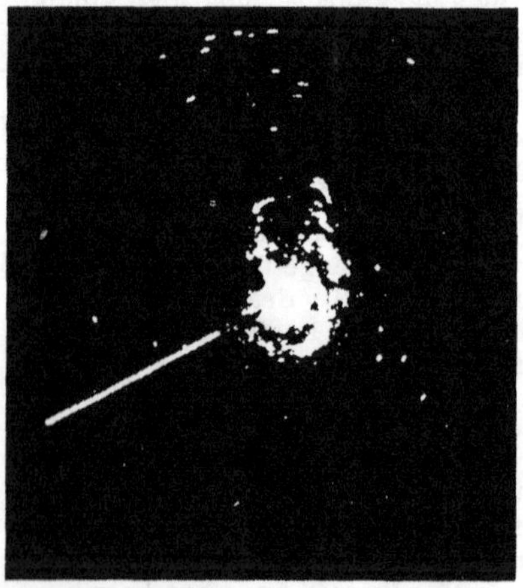

a. Sidelobe Blanker Off

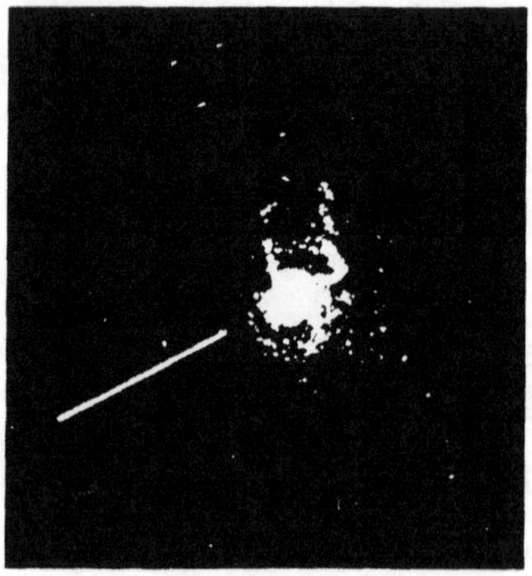

b. Sidelobe Blanker On

Fig. 8. Sidelobe Blanker Performance with False Targets

Sidelobe Blanker and CSLC Compatibility

Figure 9 demonstrates the compatibility of the two sidelobe suppression techniques. Figure 9(a) shows a high-duty cycle jammer obscuring targets from maximum range into a fixed range with both sidelobe suppression fixes off. Figure 9(b) has only with the sidelobe blanker on and demonstrates how it "breaks down" under this high-duty cycle jamming to prevent it from shutting down the radar continuously. At these higher duty rates, the CSLC should turn on and cancel the jamming as shown in Figure 9(c). Note that with only the CSLC on in Figure 9(c), there is a continuous ring around the PPI photo at close range. This is caused by the CSLC lockup time. (The CSLC is continuously locking and unlocking on this high-duty cycle repeater jammer.) It is during this time that the radar is still vulnerable to any low-duty cycle jamming. Figure 9(d) is taken with the sidelobe blanker and the CSLC both on. Note that the ring in Figure 9(c) is now gone. The sidelobe blanker is now blanking any low-duty cycle jamming during the CSLC lockup time and the CSLC turns on to cancel the high-duty cycle jamming. The MTI was off-line during this photo sequence.

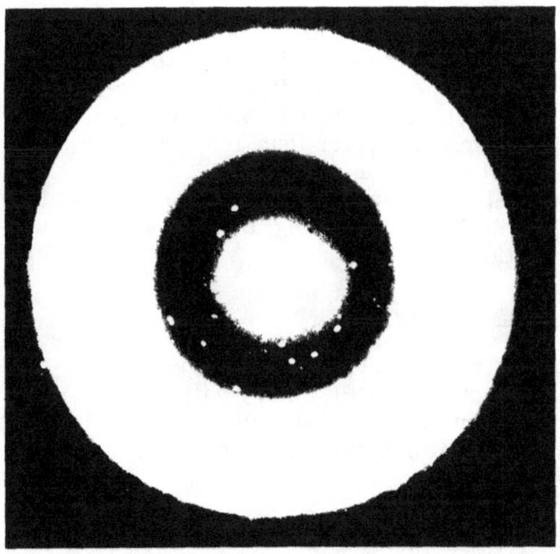

a. Both Blanker and CSLC Off

b. Only Sidelobe Blanker On

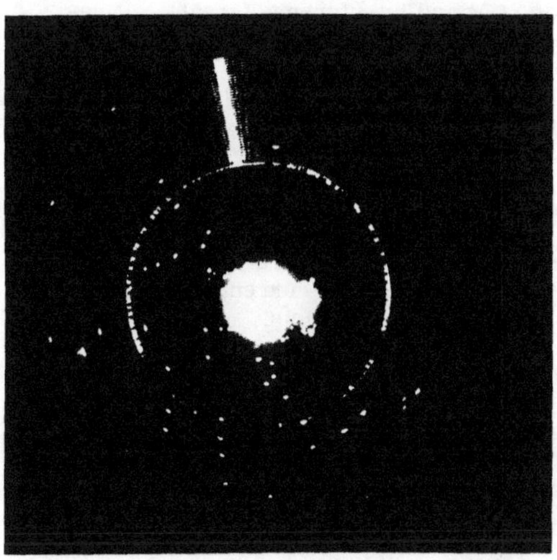

c. Only CSLC On

d. Both Blanker and CSLC On

Fig. 9. Combined Sidelobe Blanker and CSLC Performance with High-Duty Cycle Pulse Jammer

Clutter, Chaff and Weather Suppression

Figure 10 shows MTI and adaptive threshold performance. Figure 10(a) shows the close in ground clutter profile with no MTI or adaptive threshold. Figure 10(b) demonstrates the close in improvement achieved in this ground clutter. This photo has an elapsed exposure time to show the traceability of the targets.

a. MTI and Adaptive Threshold Off

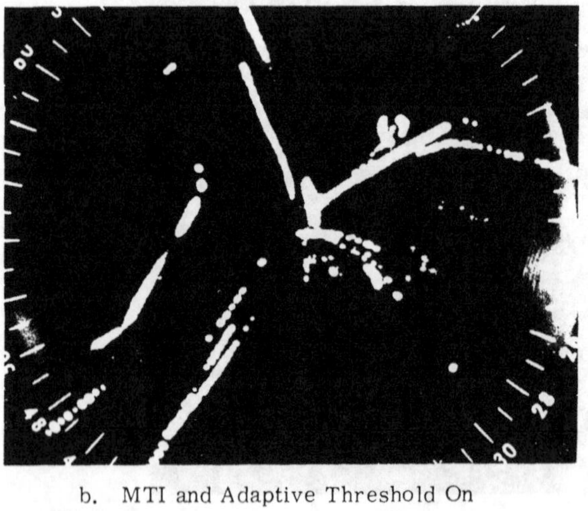

b. MTI and Adaptive Threshold On

Fig. 10. Ground Clutter Suppression Performance

Figure 11 shows the adaptive MTI working against a chaff drop at 100 nmi. The top photos show a chaff drop with a test target ring inserted for test purposes. Both the PPI and A-scope photos are shown for clarity. The bottom set of photos are with the adaptive MTI, on-line.

Compatibility of Sidelobe Suppression, Clutter Suppression and Adaptive Threshold Techniques

The three pictures on Figure 12 indicate the combined performance of the CSLC eliminating barrage jamming, the MTI eliminating ground clutter, the adaptive MTI eliminating a weather storm and the adaptive threshold action to maintain a false alarm rate to the PPI.

Figure 12(a) shows the profile of a weather storm out to approximately 60 nmi, where the ground clutter still exists at the site (shown previously in Figure 10(a)). Figure 12(b) is barrage noise jamming which is greater than 100,000 W of ERP from a standoff jammer at 100 nmi. Figure 12(c) gives an indication of the performance delivered by this compatible lineup operating to give a clear target presentation.

Pulse-to-Pulse Correlation

The effect of the correlator is shown in Figure 13. Figure 13(a) shows a PPI display with mutual interference present, and Figure 13(b) indicates the results of the correlator action to remove the interference.

CONCLUSION

The ECCM radar configuration we have described has proven to be highly effective in operational use. It provides reliable performance and requires little operator intervention to counter a wide range of ECM and interference threats. It has extended the basic radar capability to cope with these threats at a cost which is only a fraction of the unprotected radar cost. This has been achieved by selecting a relatively small number of inherently adaptive and complementary ECCM techniques which had been individually proven, and carefully integrating them into the complete system. In view of the currently expected ECM threat and the success of this counter approach, we believe that it should be designed into or added to all currently considered operational military radar systems.

This obviously is not the end of the story. Both the total ECM threat and the types of radars employed are gradually but continuously growing. Consequently, it will be necessary not only to develop and evaluate effective new ECCM techniques, but also to integrate them into complete operational radars.

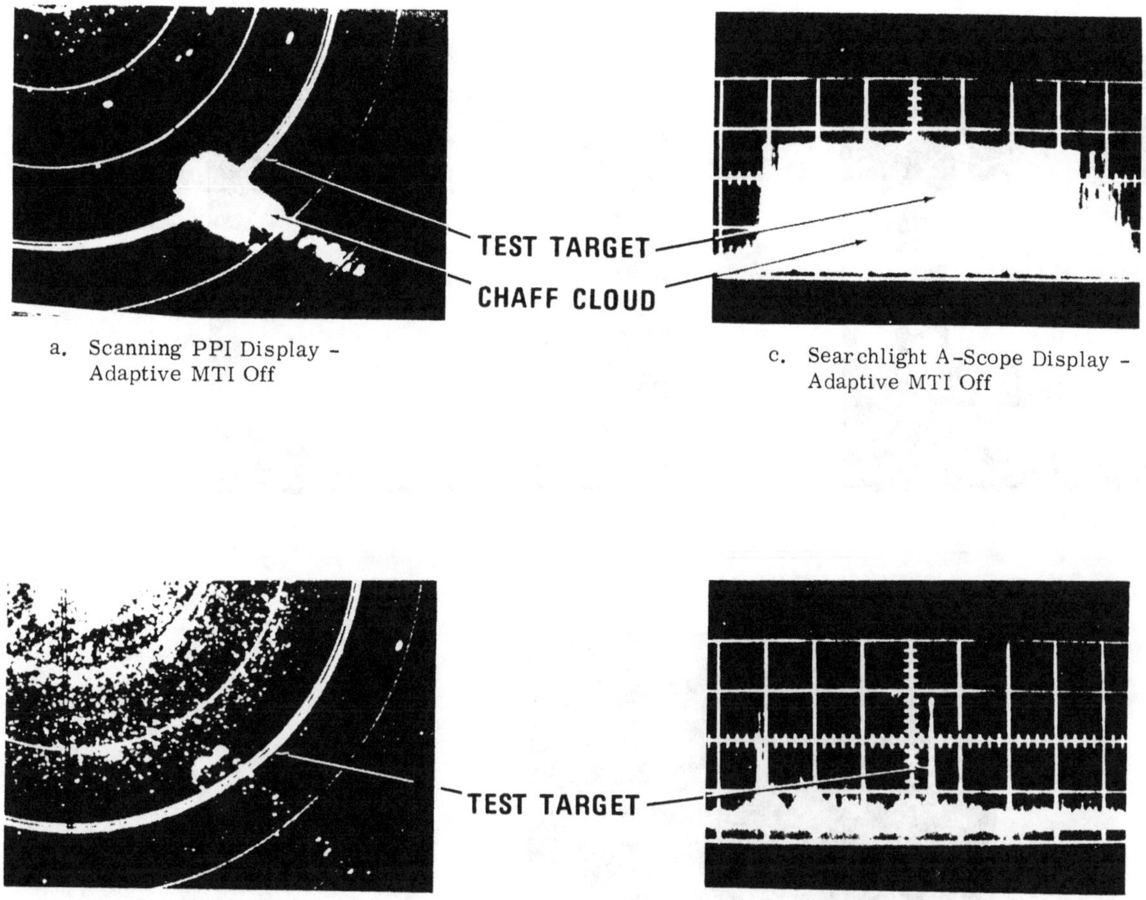

a. Scanning PPI Display - Adaptive MTI Off

c. Searchlight A-Scope Display - Adaptive MTI Off

b. Scanning PPI Display - Adaptive MTI On

d. Searchlight A-Scope Display - Adaptive MTI On

Fig. 11. Adaptive MTI Performance with Chaff

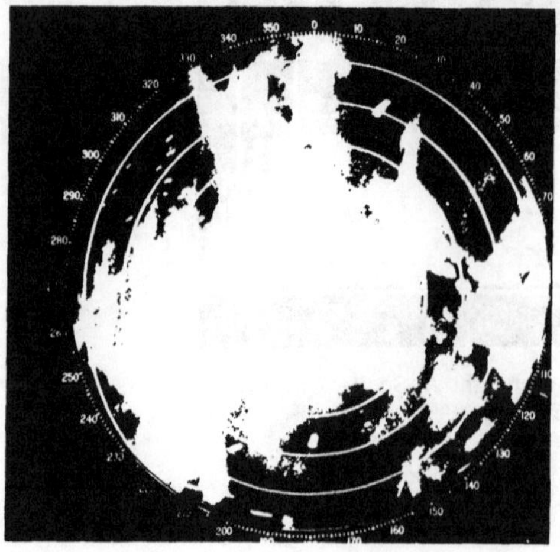

a. Weather and Ground Clutter

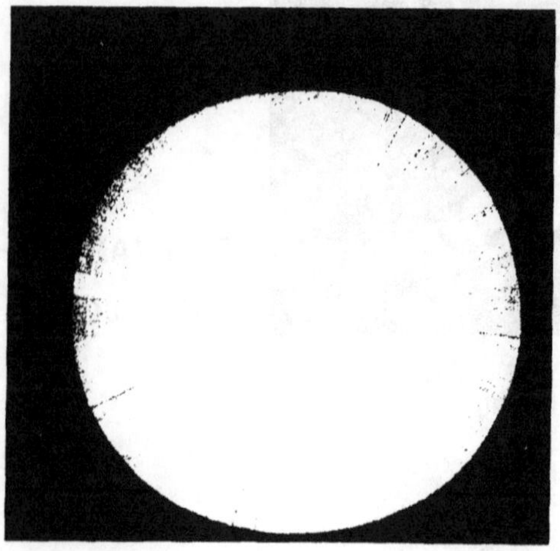

b. High-Power Barrage Noise Jamming

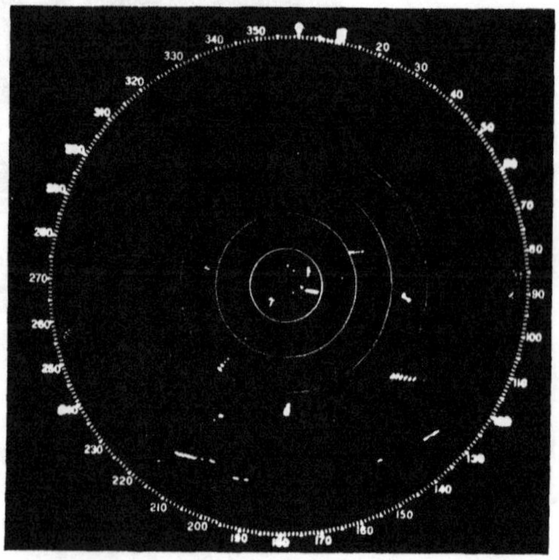

c. Sidelobe and Clutter Suppression Plus Adaptive Threshold

Fig. 12. Compatibility of Sidelobe Suppression, Clutter Suppression and Adaptive Threshold

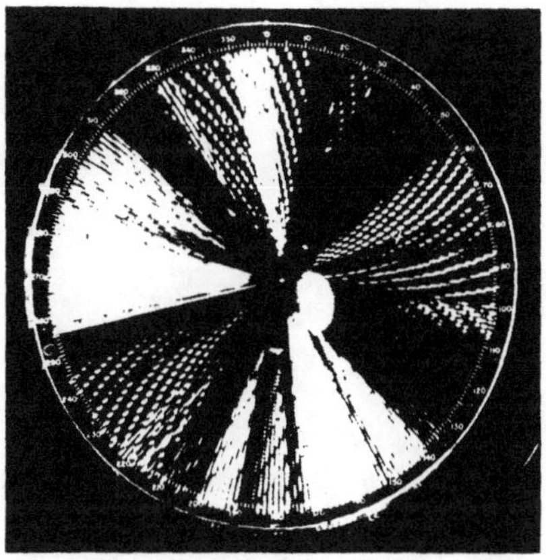

a. Correlator Off

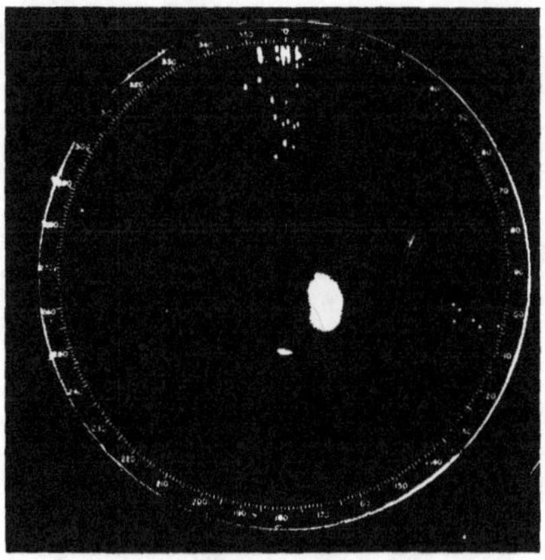

b. Correlator On

Fig. 13. Correlator Performance - With Mutual Interference

REFERENCES

1. Bernard L. Lewis and James P. Hansen, "Understanding and Optimizing Multiple Sidelobe Canceller Operation", NRL Report 7610, October 1973.

2. F. F. Kretschmer, "Effects of Cascading Sidelobe Canceller Loops", The Record of the IEEE 1975 International Radar Conference, April 1975.

3. Lloyd J. Spafford, "Optimum Radar Signal Processing in Clutter" (U), IEEE Transactions on Information Theory, Vol. IT 4, No. 5, September 1968, pp. 734-743.

4. L. W. Bauer, "Detection in the Sonar Environment" TIS No. R69EMH21, Class 1, GE Company, June 1969.

Chapter 3
ECCM in the Transmitter

Summary

These papers address various aspects of the transmitted signal-power, waveform, and frequency. Typical ECCMs in the transmitter are listed in Paper 1.1, Table I (page 14 of this volume.) Many of the techniques of this table are described in papers in the bibliography for this chapter.

Synopsis of Reprint Papers in Chapter 3

3.1 "Effects of ECCM Techniques on Transmitter Requirements"—Gardner

Some current ECCM techniques place special requirements on radar transmitters. These techniques include short burst high average power operation, wide tunable bandwidth (i.e., frequency agility), and wide signal bandwidth. Each of these is examined in this paper. The importance of amplitude and phase distortion as functions of frequency and time are emphasized. It is indicated that nonlinear considerations of high power transmitters are sometimes ignored or glossed over in the consideration of unique ECCM waveforms although they are well known. These effects can convert AM to PM, and generate harmonics and unwanted intermodulation distortion which can cause problems for the radar receiver designer. This paper was presented at the DDRE Radar ECCM Symposium in 1976.

3.2 "Analysis of ECCM Characteristics of Frequency Agile Radar Systems"—Bergkvist

Frequency agility is a very powerful ECCM technique in that it forces the jammer to spread its power over a large frequency region and thus lowers the effective jammer power at any single frequency. This paper describes methods of generating frequency agility and discusses requirements imposed on various parts of a radar system by the use of frequency agility. It is pointed out that frequency agility can cause violent variations in the antenna far sidelobe levels. It is interesting to note that this paper contains an extensive discussion of jammer requirements when used against a frequency agile radar. This paper was presented at the Military Microwaves Conference in 1978.

3.3 "Combined X/K_A Band Tracking Radar"—Klaver

This paper describes a dual-band tracking radar for a short range air defense system. The principal thrust is the use of the K_A-band to alleviate multipath effects in low altitude aircraft tracking. To that end, consideration is given to atmospheric effects such as rain and sea clutter. A short description of the radar is given along with its principles of operation and results of tracking tests. While multipath is important in the benign environment, a dual band radar can also be of great benefit in an ECM environment. It forces the enemy to employ jammers or repeaters in two bands thus causing the enemy aircraft to incur both weight, space, and prime power penalties in addition to reliability problems. If either jammer fails the dual band radar is "home free". This paper was presented at the Military Microwaves Conference in 1978.

Bibliography Topics for Chapter 3

Frequency Agility/Diversity
Pseudo-Noise
Pulse Compression
Transmitter PRF ECCMs
Various Transmitter ECCMs

BIBLIOGRAPHY FOR CHAPTER 3

FREQUENCY AGILITY/DIVERSITY

1. Books and General Bibliography

Barton, D.K., *Frequency Agility and Diversity: Radars, vol. VI*, Artech House Books, Dedham, MA, 1977.

Barton, D.K., "International Cummulative Index on Radar Systems," *IEEE Pub Cat No. JH 4675-5*, 1978 (extensive bibliography on frequency agility and diversity, p. 26).

2. Journal Articles

Dax, P.R., "The Statistics of Radar Video After Linear and Nonlinear Mixing," *Proc NEC*, vol. 15, 1959, pp. 850-9.

Leger, J., "Surveillance Radar: Theoretical Study of Frequency Diversity," *Rev Tech CFTH*, vol. no. 39, Dec. 1963, pp. 31-45 (in French, trans. abstract SA 65-1376).

Chiles, W.H., and H.G. Lafuse, "Sweeping Carrier Signals Through Interference," *Electronics Magazine*, vol. 37, no. 16, 18 May 1964, pp. 94-6.

Skolnik, M.I., "Analysis of a Proposed Two-Frequency Radar Waveform," *Naval Res Lab Rpt MR-2926,* Nov 1964, DDC AD A002942.

Calamia, M., "Improvement of Signal to Noise Ratio Obtainable in Multichannel Radar," *Alta Freq.,* vol. 34, no. 10, Oct. 65, pp. 697-706 (in Italian, trans. abstract SA 66-6555).

Shapovalov, V.I., "On the Construction of Signal Detection Characteristics of Frequency Diversity Radar Stations," *Problems in Radio Electronics,* Gen Engrg Ser 1965, no. 3, pp. 51-6 (in Russian).

Waters, P.L., "Frequency Diversity Performance of Ground Surveillance Radar," *IEE Electron Lett.,* vol. 1, no. 10, Dec 1965, pp. 282-3.

Kroszczniski, J., "Working of Multifrequency Radar Systems," *Problemy Elektron Telekommun Metody Wspolczesnej Radiol,* vol. 11, 1966, pp. 81-8 (in Polish, trans. abstract SA 68-6570).

Gilardini, A., "Variable Radiofrequency Radar," *Alta Freq.,* vol. 38, no. 9, Sept 1969, pp. 657-65 (in Italian, trans. abstract SA 70-18340).

Kleiber, P. and L. Reinhard, "Fire Control Radar for Ships," *Albieswech-Ber* (Switzerland), vol. 20, no. 3, Nov 1968, pp. 79-84, (in German, trans. abstract SA 69-5387, Uses frequency agility)

Chieco, G.A. and G. Scerch, "The Use of Amplitrons as Microwave Amplifiers in High Power Radar Systems with Frequency Agility," *Int. J. of Electron.,* vol. 26, no. 3, Mar 1969, pp. 237-252.

Kaplan, A., "Detection and Analysis of Frequency Hopping Radar Signals," *Electron Warfare Magazine,* vol. 1, no. 4, Conv 1969, pp. 13-15.

Nathanson, F.E. and J.P. Reilley, "Frequency Agility for Radar Target Detection and Tracking," *Applied Physics Lab Tech Dig.,* vol. 9, no. 6, July/Aug 1970, pp. 2-8.

"Radar Stations with Changing Carrier Frequency," *Ekspress-Informatsiya Ser. Radiolokatsiya,* Televideniye, Svyaz' 1970, no. 6, pp. 1-13 (in Russian).

Lind, G., "Clutter and Frequency Agility," *J. Inst. of Engrs* (India), vol. 52, no. 1, pt ET-1, Sept 1971, pp. 4-6.

Murtly, A.N. and J.P. Gupta, "Frequency Agile Radar Techniques," *J. Inst Engrs* (India), *E and TE,* vol. 52, no. 1, pt ET-1, Sept 1971, pp. 11-16.

Jones, P., E.A. Mickle, and G.D. Swetnam, "Homing Radar Tracking Accuracy Improvement with Frequency Diversity," *J. Spacecraft and Rockets,* vol. 8, no. 9, Sept 1971, pp. 983-9.

Fuller, J.B., "Subsystem Approach to Improvement in Radar Performance by Frequency Agility," *IEEE Conv of Electr and Electron Eng in Israel,* 8th Proc Tel-Aviv, vol. 3, Apr 30-May 3, 1973, pp. 1-9.

Schwarzmann, A., "Reducing Multipath Effects on Hand Held Radars," *RCA* (Princeton, N.J.) *Rpt TN-948,* Dec 1973 (uses frequency agility).

Gadre, M.N., A.K., Subramanian, and V.S. Taneji, "A Scheme for Continuously Tunable Stable Local Oscillator for Radar Systems," *Electro-Technol* (India), vol. 18, no. 1, Mar 1974, pp. 27-9.

Fuller, J.B., "Frequency-Agile Local Oscillator," *Microwave J.,* vol. 17, no. 5, May 1974, pp. 31ff.

Rohlss, J. and M.A. Vonk, "Spin-Tuned Magnetrons for Frequency-Agile Radar," *Philips Product Information,* no. 61, 25 June 1974.

Hull, J.F., "Frequency Diversity Radar," *US Patent No. 3,858,219,* 31 Dec 1974 (abstract SA 75-21126).

Marinos, P.N. and W.M. Waters, "A Proposed Coherent Integration Scheme for Use in Frequency Agile Radar Systems," *Proc IEEE Southeastcon Region III Conf.,* pt. II, Apr 1975, pp. 6E-3-1 thru 4.

Scherer, E.F., "Compact Multi-Frequency Stalo Sources in Frequency Agile Radar Systems," *Microwave J.,* vol. 19, no. 3, Mar 1976, pp. 41-44.

"Spin Tuned Magnetrons for Frequency Agile Radars," *Countermeasures Magazine,* Mar 1976, pp. 52-6.

Lob, C., "Introduction to Frequency Agile Radar Sub-Systems," *Proc Int Countermeasures Conf,* EW Institute Ltd, Los Altos, CA, London, Sept 8-10, 1976, pp. 329-343. Also issued as a Varian/Beverly Brochure; Fuller, J.B. and J.R. Martin.

Davies, N.G. and D.B. Wohlberg, "Spread Spectrum Techniques," *IEE Conf. Pub. 144,* "The Impact of New Technologies on Signal Processing," Sept 20-24 1976, pp. 144-154.

Weissglas, P. and S. Svensson, "Local Oscillators for Frequency Agile Systems," *Proc Int Countermeasures Conf.,* London, EW Institute Ltd, Los Altos, CA, Sept 8-10 1976, pp. 239-245. Also published in Microwave J., vol. 20, no. 1, Jan 1977, pp. 35-8.

Philips, "Frequency Agile Radar Design: Using the Philips Spin-Tuned Magnetron," *Philips Application Note PM 61395,* 8 Dec 1976.

Nourse, O., and M. Crossfield, "Quad-Pulse Radar—A Frequency Agile Radar with Doppler Capability," *IEE Conf Pub no. 155, Radar 77,* London, Oct 1977, pp. 186-90.

Bergkvist, B., "Analysis of ECCM Characteristics of Frequency Agile Radars," *Proc Mil Microwaves Conf.,* London, Oct 1978, pp. 51-6 (reprinted in this volume).

Pickering, A.H., "Electronic Tuning of Magnetrons," *Proc. Mil. Microwaves Conf.,* London, Oct 1978, pp. 98-102.

Lewis, M., "A Compact Agile Microwave Frequency Generator for Radar Applications," *Proc. Mil. Microwaves Conf.,* London, Oct 1978, pp. 234-237.

Knowles, R.M., "VCO's for Modern Frequency Agile Systems," *Proc. Mil. Microwaves Conf.,* London, Oct 1978, pp. 279-286.

Czyz, Z.H., "Coverage Diagrams for the Dual Frequency-Diversity Radar System Operating with Reflections from Ground," *Proc Int Conf on Radar,* Paris, 4-8 Dec 1978, pp. 71-5.

Petrocchi, G., S. Rampazzo and G. Rodriguez, "Anti-clutter and ECCM Design Criteria for a Low Coverage Radar," *Proc Int Conf on Radar,* 4-8 Dec 1978, pp. 194-200 (reprinted in this volume) (describes a method for making MTI compatable with frequency agility).

TRANSMITTER–ECCM: PSEUDO NOISE

1. Books and General Bibliography

Dixon, R.C., *Spread Spectrum Techniques,* IEEE Press, 1976.

Barton, D.K., "International Cumulative Index on Radar Systems, " *IEEE Pub Cat No. JH 4675-5,* 1978: extensive bibliography on Doppler and CW Radar pp. 22-24 (includes "noise" radar). See also measurement pp. 32-35.

Barton, D.K., *CW and Doppler Radar; Radars, vol. VII,* Artech House Books, Dedham, MA, 1978.

2. Journal Articles

Bourette, R., "A Proposed Technique for the Improvement of Radar Determination with Noise Radar," *Proc. IRE,* vol. 45, no. 12, p. 1744c; also 1958, p. 1652, p. 1757.

Fishbein, W., and O.N. Rittenbach, "Correlation Radar Using Pseudo-Random Modulation," *IRE Int Conv Rec,* vol. 9, pt. 5, 1961, pp. 259-77.

Reid, M.S., "Millimeter Wave Pseudorandom Coded Meteorological Radar," *IEEE Trans.,* vol. GE-7, no. 3, Jul 1969, pp. 146-56.

Carpentier, M.H., "Application to Radar of Pseudo-random Codes: Attempt at Synthesis," *Onde Elect,* vol. 49, no. 510, Sept 1969, pp. 836-837. (in French, trans. abstract SA 70-1816).

Smit, J.A., "Rudar—An Experimental Noise Radar System," *AGARD Conf Proc Advanced Radar Syst. CP - 66,* 1970, pp. 39-1 thru 17.

Bosc, H. and J.M. Colin, "RACINE Concepts for Radar, Principles and Experimental Models," *Elec Comm.* vol. 46, no. 4, 1971, pp. 277-282.

Smit, J.A., and W.B.S.M. Kneefel, "Rudar—An Experimental Noise Radar System," *Ingenieur* (Netherlands), vol. 83, no. 32, 13 Aug 1971, pp. 99-110.

Berrie, D.W., "The Effects of Pseudo-random Noise Upon Radar System Performance," *Ohio State Univ. Thesis,* 1974, Int Microfilms No. 74-24298. abstract SA 75-21121.

TCSF, "Correlation Distance Measuring Radar Device," *UK Patent No. 1380451,* 15 Jan 1975 (abstract SA75-30129).

Yamamoto, M., et al, "Spread Spectrum Range and Range Rate Equipment," *Rev Radio Res Lab* (Japan), vol. 21, no. 112, Jan 1975, pp. 1-17 (in Japanese, trans. abstract SA 76-17864).

Botnev, V.N., "Autocorrelation Reception of Pseudo-random Signals," *Telecomm. and Radio Engrg., Pt.2,* vol. 30, no. 6, Jun 1975, pp. 109-113.

Jordan, G.B., "Spread Spectrum Primer," *Electron Warfare Magazine,* vol. 7, no. 4, Jul/Aug 1975, pp. 69ff.

Salapski, G.D., "Spread Spectrum Transceiver Using Analog Programmable Analog Filters," *Electron Warfare Magazine,* vol. 7, no. 4, Jul/Aug 1975, pp. 75-9.

Forrest, J.R. and J.P. Meeson, "Solid State Microwave Noise Radar," *IEE Electron Lett,* vol. 12, no. 15, 22 Jul 1976, pp. 365-6, A76-39769.

Albanese, D.G., "Pseudo-Random Code Waveform Design Trade-Offs for CW Radar Applications," *IEE Conf Pub No 155 Radar 77,* London 1977, pp. 482-7.

Forrest, J.R., and J.P. Meeson, "Solid State Microwave Noise Radar," *IEE Conf Pub No155 Radar 77,* London, Oct 1977, pp. 531-4.

Albanese, D.F. and A.M. Klein, "Pseudorandom Code Waveform Design for CW Radar," *IEEE Trans,* vol. AEE-15, no. 1, Jan 1979, pp. 67-75.

PULSE COMPRESSION

1. Books and General Bibliography

Barton, D.K., "International Cumulative Index on Radar Systems," *IEEE Pub Cat No. JH 4675-5,* 1978: extensive bibliography on pulse compression: see signal design (general) p. 60, signal design (FM) p. 62, signal design (phase code) p. 63, signal design (pulse bursts and codes) p. 63.

Barton, D.K., *Pulse Compression; Radars, Vol. III,* Artech House Books, Dedham, MA, 1975.

2. Journal Articles

Hutmann, E., "A Procedure for Distance Measurement," *German Patent No. 768068,* Mar 22, 1940 (in German).

Darlington, S., "Pulse Transmission," *US Patent No. 2,678, 997,* May 18, 1954.

Huffman, D.A., "The Synthesis of Linear Sequential Coding Networks," *Proc. Third London Symp. on Information Theory,* Sept 1955, pp. 77-95.

Sakamoto, T., et al, "How Coded-Pulse Techniques Extend Radar Range," *Electronics Magazine,* vol. 36, no. 47, Nov 22, 1963, pp. 34-37.

Klassen, R.W., "Comparison of Frequency Shift Keyed and Phase Shift Keyed Pulse Compression Systems," *IEEE Int. Conv. Rec.,* vol. 14, pt 4, 1966, pp. 13-20.

Kroszczynski, J., "Pulse Compression in Radar," *Prace PIT,* vol. 17, no. 58, 1967, pp. 1-31 (in Polish, trans. abstract SA 68-19035; trans. DDC AD 864868).

Nicodemus, K.L., "Chirp Radar Technique for Suppressing Second Time Around Echoes," *US Patent No. 3,363,248,* Jan 9, 1968.

Sperry Rand Corp., "Radar Systems," *UK Patent No. 1125521,* 28 Aug 1968 (abstract SA 69-5406).

Shirman, Ya D., et al, "First Soviet Investigations of Radio-Pulse Shortening (Pulse Compression)," *Telecom and Radio Engr., Pt 2,* vol. 25, no. 3, March 1970, pp. 141-147.

Tournois, L, et al, "Pulse Compression Dispersive Filters," *I.E. (India) J, E. and T.E.,* vol. 51, May 1971, pp. 202-205.

Ragan, L.H., "Pulse Compression Radar Study," *FAA Rpt RD-73-192,* US Dept Transp., March 1974.

Gutleber, F.S., "Method and Apparatus for Adaptively Suppressing Unwanted Lobes in a Compressed Coded Radar Signal," *US Patent No. 3.889.199.* 22 May 1974. (abstract SA 74-42998).

Marconi Ltd., "Pulse Compression Radar System," *UK Patent No. 1379863,* 28 Jan 1975 (abstract SA 75-30127).

Gager, C.H. and R.B. Van Deusen, "Generation of Flexible Pulse Compression Waveforms with Computer Control," *IEE Conf Pub. No. 155, Radar 1977,* Oct 25-28, 1977, pp. 488-492.

TRANSMITTER ECCMs

1. Bibliography and Book

Luck, D.G., *Frequency Modulated Radar,* McGraw Hill, New York, 1949.

Barton, D.K., "International Cumulative Index on Radar Systems," *IEEE Pub Cat No. JH 4675-5:* extensive bibliography on transmitters, pp. 72-4.

2. Journal Articles

Clavier, A.G., "Signal-to-Noise Ratio Improvement in a PCM System," *Proc. IRE,* vol. 37, no. 4, Apr 1949, pp. 357-9.

Herrick. G.Q., "How VOA Combats Jamming," *Electronics Magazine,* vol. 22, Sept 1949, pp. 82-4.

Rosien, R.A., and R. Shavlach, "Bandwidth Conservation in Pulse Modulated Radars," *IRE Conv Rec.,* vol. 6, pt. 8, 1958, pp. 139-54.

Salati, O.M., and R.A. Rosien, "Technical Considerations in the Assignment of Operating Frequencies in a Communications (Radar) System," *IRE Nat Conv Rec,* vol. 7, pt. 8, 1959, pp. 93-113.

Roman, J., and F. Murphy, "Proposed Radar Spectrum Occupancy Design Standard," *Proc 8th Tri-Service Conf. on Electromag Compat.,* Oct 1962, pp. 737-45.

Stimson, R.G., "A Modulation Technique for Eliminating Interference in CW Radars," *Proc IEEE,* vol. 53, no. 6, June 1965, pp. 641-2c.

Hair, T. and W.O. Agar, "A Precision C.W. Ranging Technique," *Marconi Rev.,* vol. 29, 4th qtr, 1966, pp. 191-205.

Edwards, R.E., "Reduction of Spurious Output from Radar Magnetron Tubes," *IEEE Southeast Electromag. Compat. Symp. Rec.,* Atlanta, Oct 1969, pp. 340-53.

Wolf, F.H., "Cut Pulse-Radar Frequency Drift," *Electron. Design Magazine,* vol. 17, no. 25, Dec 6, 1969, pp. 80-2.

Teich, M.C., "Three Frequency Heterodyne System for Acquisition and Tracking of Radar and Communications Signals," *Appl Phys Lett,* vol. 15, no. 12, Dec 15, 1969, pp. 420-3.

CFTH, "Radar" *UK Patent No. 1,220,321* (France No. 862,199) 27 Jan 1971 (Use of Special Anti-Jamming Transmitted Waveform) (abstract SA71-23538).

Davis, D.E.N., "Two Frequency Secondary Radar Incorporating Passive Transponders," *IEE Electron Lett,* vol. 9, no. 25, Dec 13, 1973, pp. 592-3.

Saloom, J., "ECCM Tubes in a Fix," *Electron. Warfare Magazine,* vol. 6, no. 5, Sept/Oct 1974, pp. 48ff.

Ofverberg, M.J., "Multiple Target Resolution in CW-FM Tone Ranging System," *Univ of Cincinnati Thesis,* 1975, 314 pp., Univ Microfilm No. 75-21415 (abstract SA 76-21817).

Harmuth, H.F., "The Case for Nonsinusodial Carriers in Radar," *IEEE Electro-magnetic Compatibility Symp.,* Montreux, Switzerland, 20-22 May 1975, p. 549 (abstract only).

Gardner, W.L., "Effect of ECCM Techniques on Transmitter Requirements," *DDRE ECCM Symp.*, 1976 (reprinted in this volume).

Fleming, M.A., F.H. Mullins, and A.W.D. Watson, "Harmonic Detection Systems," *IEE Conf. Pub No. 155, Radar-77,* London, Oct 1977, pp. 552-4.

Klaver, J., "Combined X/K_A-Band Tracking Radar," *Proc Mil Microwaves Conf.*, London, Oct 1978, pp. 147-56. (reprinted in this volume).

Johnston, S.L., "Some Aspects of Millimeter Radar," *Proc Int Conf on Radar,* Paris, 4-8 Dec 1978, pp. 148-59 (use of higher frequencies).

TRANSMITTER PRF ECCMs

Javid, M., "Performance Characteristics of Baseline Scanning Ambiguity Filters for High PRF Radars," *PhD Thesis,* McGill Univ, Aug 1955.

Cohn, G.I., "Suppression of False Range Indications in High Pulse Repetition Rate Radars," *Proc NEC,* vol. 13, 1957, pp. 744-760.

Potter, N.S., "Range Ambiguity Resolution in High PRF Radars," *IRE Int Conv Rec.,* vol. 8, pt. 8, 1960, pp. 65-80.

Shenoy, R.P., "Target Resolution of a Doppler Radar with Statistically Variable PRF," *J. Inst Telecomm Engrs* (India), vol. 12, no. 10, Oct 1966, pp. 501-510.

Cooper, D.C., D. Greenhalgh, and J.B.G. Roberts, "Clutter Cancellation and Unambiguous Doppler Filtering in Jittered Pulse Radars," *Proc Int Conf on Radar,* Paris, 4-8 Dec 1978, pp. 32-40.

Isnard, J., J.C. Debuisser, and J.L. Bossenec, "Matched Filtering in Jittered PRF Radars," *Proc Int Conf on Radar,* Paris, 4-8 Dec 1978, pp. 317-22 (in French).

Effect of ECCM Techniques on Transmitter Requirements
W.L. Gardner
Proceedings of the DDRE Radar ECCM Symposium, 1976.
Pages 29-1 – 29-13
Reprinted by permission.

3.1

SUMMARY

With the evolvement of emphasis toward implementing advanced ECCM technologies there arises the question as to whether various techniques can be applied effectively to tactical radar systems now in existance. In particular the requirements placed on the transmitter become significantly important and must be examined in light of present capabilities. Discussed first are various transmitter modifications that can be made to implement short burst, high average power operation. Then the conditions imposed on transmitter tubes when required to operate at large tunable bandwidths for frequency diversity are considered. Finally, the distortions to signal fidelity that are introduced when transmitters are operated at wide signal bandwidths are examined.

1. SHORT BURST, HIGH AVERAGE POWER OPERATION

The purpose of this section is to examine the potential for implementing a short burst, high average power mode of ECCM operation to increase the radar energy on specified targets as a means of improving detection in an ECM environment. It is assumed for this technique to work that an increase in ECM noise has been detected as the radar scans past an aircraft radiating ECM for self-screening purposes. The radiated power might then be increased for a short period of time, nominally the beam dwell time on target, to improve the probability of detection.

There are three obvious ways and their combinations to increase the power output of a transmitter for short periods of time; (1) increase the peak power, (2) increase the pulse width, or (3) increase the pulse repetition rate. Of course, a number of potential problems can arise when such operation is attempted requiring a certain degree of equipment modification such as:

a. An increase in the power supply and modulator output capability to deliver the required voltage and/or current.

b. The alteration of the pulse forming network, pulse transformer, and in some cases heat transfer units to accomodate higher duty operation.

c. Changes in the RF plumbing to enable the transmission of increased peak power.

d. A redesign of the transmitter tube to meet increased power handling and heat dissipation requirements.

Of the above the first three modifications are usually straight forward, with low to moderate costs, and requiring little or no development effort. On the other hand any redesign of the transmitter tube can entail substantial amounts of money and development effort. Hence, it becomes important to examine the potential of transmitter tubes to operate at higher peak powers and/or higher duties under short burst conditions without modification. One can then examine the trade-offs of development costs versus enhanced short burst power capability and decide whether additional development is justified.

For the most part a tube's peak power cannot be increased to any great extent without significant design changes. Of the high power tube types usually found in tactical radars with ECCM capability, the coupled cavity TWT and Twystron are limited to an increase in peak power of usually less than 0.5 dB because of the sensitivity of beam voltage and current levels needed to maintain synchronism and maximum coupling of the beam to the RF circuit. High power klystrons being somewhat less sensitive to parameter change can generally handle a peak power increase on the order of 3 dB.

When increasing a tube's duty by either increasing the pulse width or pulse repetition rate or both the major consequence to the tube is the increase in the thermal losses that must be dissipated. For the most part limitations to thermal dissipation in a tube occur either in the collector where the electron beam is dissipated or in the RF interaction circuit where RF heating losses are maximum. In either case there are certain common factors effecting average power limitations. These are:

a. The maximum allowable temperature of the material being heated.

b. The power density input to the surface being heated.

c. The wall thickness between the surface being heated and the coolant.

d. The rate of heat transfer between the surface and the coolant, determined by the thermal conductivity of the material and the temperature gradient.

e. The mass flow and specific heat of the coolant.

Of the factors listed above the temperature at the surface of the material being heated has the most significant impact under short burst conditions. Operation above this temperature will substantially effect the material's evaporation rate, gas permeability, and mechanical strength. Essentially the surface temperature is determined by the difference between the input power to be dissipated and the rate of dissipation away from the surface. Initially as input power is increased the temperature gradient through the material, hence the rate of heat dissipation away from the surface, increases. This thermal inertia phenomenon moderates the temperature rise at the surface allowing the possibility for tube operation under short burst conditions. Furthermore, as can be perceived conceptually, when input power is increased the allowable time span of the burst pulse will necessarily decrease to sustain a surface temperature below critical levels.

An example of potential transmitter modifications for short burst operation such as have been discussed can be found in the RADC SEEK SCREEN Study. In this study the AN/TPS-43A radar with an arbitrarily suggested 1/30 second every 10 seconds burn-through mode was considered. A primary requirement of the study was that only minor design changes or "fixes" using existing technologies could be proposed. A summary of the findings can be found in Table 1. The term "temperature limited" where found indicates that either tube and/or equipment design limits will be exceeded should the transmitter be operated beyond the burst mode time constraints. Lines 1 and 2 list the radar and tube design parameters respectively. Lines 3, 4, and 5 list necessary modifications and cost estimates pertaining to an increase in the transmitter duty. It should be noted that the pulse widths given in lines 4 and 5 represent, respectively, a safe level of operation and a marginal level of operation, i.e. having no safety margin. Lines 6, 7, and 8 indicate the changes to transmitter operation created by the use of different tubes of the same family. The cost estimates, however, do not reflect modification costs implied by the higher duties listed for these tubes. It can easily be seen that increases in output power on the order of 3 dB for short time periods at nominal modification and cost can be obtained by just increasing the transmitter duty. Furthermore, by using the VA145F Twystron with minor modifications an increase on the order of 6 dB can be achieved though not without substantial changes to the transmitter. Going one step further, assuming one is not restricted in changing the transmitter an increase on the order of 10 dB might be achieved with modification to a Twystron such as the VA915. Of course such an increase in output power may not be justified in light of the considerable costs involved.

2. WIDE TUNABLE BANDWIDTH CONSIDERATIONS

Because noise jammers are in general average power limited, it is desirable to force them to operate broadband to dilute their signal energy, hence increasing the radar's signal to jammer advantage. If for reasons of complexity and cost a wide signal bandwidth is not practical, then a wide tunable bandwidth can also be very effective in forcing these jammers to operate broadband. However, for this to be true the radar should be tunable on a pulse to pulse basis or, as a minimum, on a beam dwell to beam dwell basis.

The only requirement placed on the transmitter for such operation is that the transmitter tube have a wide instantaneous bandwidth. Of course, to maintain maximum signal to jammer advantage tube bandwidths must be specified between the 1 dB points of the passband ensuring that at least 80% of the peak output power will always be transmitted. However, it is important to point out that due to the sensitivity of tube output to drive power, the tube's drive characteristic producing the 1 dB bandwidth output must be closely matched in the transmitter.

Of the high power tube types mentioned in the previous section the coupled cavity TWT and Twystron are inherently broadband having in general 8 to 13% 1 dB bandwidth capabilities. (Percentage bandwidth is nothing more than a normalization of the tube bandwidth by its center frequency). High power klystrons on the other hand usually have 3 to 5% 1 dB bandwidths with some broadband models achieving approximately 8% 1 dB bandwidths. Generally, the transmitters in tactical radars need not be modified to operate over a wide tuning range. For example, the AN/TPS-43A as was mentioned previously has a 2 MHz signal bandwidth while the transmitter tube has an 8% or 252 MHz 1 dB bandwidth capability making it easy to implement wide tunable bandwidth operation for ECCM.

3. WIDE SIGNAL BANDWIDTH CONSIDERATIONS

It is generally agreed that a wide signal bandwidth has a greater potential for multiple ECM threat improvement than any other single factor to be identified. All such improvements are due either directly or indirectly to the increase in time resolution implied by a wide signal bandwidth. However, as a consequence, the use of wide signal bandwidths increases the amount of transmitter distortion applied to the signal. Whether or not these distortion levels are acceptable will depend on the range resolution and sidelobe levels required.

a. Amplitude and Phase Distortions as a Function of Frequency

For any transmitter the amplitude and phase response as a function of frequency will contain distortions to the ideal response. As is well known, mathematical treatment of such distortions is accomplished using paired-echo analysis. Stated simply the band limited distortion function is expressed in terms of a Fourier series expansion that completely describes the function over the signal bandwidth. Each of the harmonically related Fourier components is associated with a pair of time sidelobes; one to either side of the main lobe. Each sidelobe has a magnitude that is one-half the corresponding Fourier component magnitude and is displaced in time from the main lobe by the reciprocal of the signal bandwidth times the corresponding Fourier component's harmonic integer.

Figure 1 as taken from Cook and Bernfield is a plot of the first paired-echo (which is usually the largest) in dB below the main lobe as a function of the Fourier component magnitude. Such a graph is many times utilized to specify maximum amplitude and phase distortion levels in a transmitter tube and can easily lead to costly over specification. For example, Figure 2 is a plot of the relative phase error in an actual transmitter tube. It has a maximum peak to peak ripple of approximately 28° implying from Figure 1 a sidelobe level of approximately 12 dB below the main lobe, which in many instances would be unacceptable. Yet when this phase error curve is applied to a simulated linear FM waveform, whose nondistorted correlation peak with Hamming weighting is shown in Figure 3a, the first sidelobe level is less than 25 dB below the main lobe as shown in Figure 3b. For many tactical radar applications this would be a more than satisfactory range sidelobe level. Hence to be realistic in evaluating spectrum distortions the whole tube distortion response should be measured and analyzed as above.

b. Amplitude and Phase Distortions as a Function of Time

Because of symmetries found in Fourier analysis the effect of respective time domain distortions on a signal will be to produce frequency sidebands in the signal spectrum. Time domain distortions can be either in the form of amplitude ripple across the pulse or in the form of time dependent changes in the transmitter tube time delay (electrical length) across the pulse. However, when the signal is either a pulse compression or coded waveform the paired sideband signals will generally become dispersed in time, and under certain conditions may be completely separable. The standard example for a sinusoidal time varying phase (time delay) ripple applied to a linear

FM pulse is shown in Figure 4. Not shown is the fact that each sidelobe contains a frequency shift from the main lobe of magnitude fm/2.

In the same manner as was discussed for spectral distortions, time domain distortions may also be expressed in terms of a Fourier series expansion describing the distortion function over the pulse width. Each Fourier component can then be associated with a pair of time sidelobes, and the same cautions as given in the previous section in specifying maximum distortion levels will apply to time domain distortions as well.

The sources of time domain distortion are generally derived from time dependent gains and delays in a high power transmitter chain. Typical sources are modulator ripple and droop characteristics, power supply ripple, and load mismatches from stage to stage and within modulator pulse forming networks. The total time domain distortion in a transmitter chain will not be merely the summation of all the distortions from these sources, but will, in general, also depend upon the interaction of the various transmitter components. This, of course, indicates that the only true means for determining these distortions is to measure them over the total transmitter chain.

In addition, some of the distortions will create amplitude modulation of the pulse compression, which in turn will be converted to phase modulation by the saturation properties of the high power amplifier stages. Some typical sources of transmitter chain phase modulation error are found in Table II. It must be remembered though that phase ripple and electrical length can vary significantly between tubes within a class, i.e. between various klystrons or TWTs, implying that actual values for these quantities must be used for individual systems rather than typical values.

c. Non-Linear Considerations

In the previous two sections only linear distortions have been considered. However, a high power transmitter tube operates in the non-linear region of its power characteristic. For an actual tube Figure 5 shows the power characteristics at low, midband, and high frequencies. As indicated the nominal operational input power level for this tube is well into the saturation region. For the same tube Figure 6 indicates the relative phase change at the output as a function of input power for several frequencies in the band. In general, all high power tubes will have similar types of power and phase characteristics that will vary in shape across the band. For radar pulses with constant amplitude the consequences of these nonlinearities on the signal will normally be negligible or easily handled. Furthermore, such other non-linear effects as AM to PM conversion, harmonic generation, and intermodulation distortion can be handled in a fairly straight forward manner.

Such non-linear properties of high power tubes are well known and tend to be glossed over or ignored when talking about utilizing various unique waveforms for ECCM purposes. In this respect extreme caution is advised. The effects of tube non-linearities on wide band signals that have rapid amplitude and/or phase changes are not known. What is meant by rapid changes is that significant signal energy is invested in switching times within the pulse train. The implication is that such waveforms as fast binary codes, polyphase codes, and burst pulse codes, could be seriously modified with consequences not predictable from linear analysis. Only the development of non-linear analytical techniques backed by experimental corroboration can predict the effects of tube non-linearities on such waveforms.

4. CONCLUSIONS

In the first section it was demonstrated that average power increases on the order of 3 dB could be obtained in a transmitter under short burst conditions with minor modification to the power supply, modulator, and perhaps cooling systems. Indicated also was the fact that substantial enhancement of average power may not be justifiable due to extensive and costly transmitter modifications. The second section dealt with the specification of transmitter tubes for wide tunable bandwidth operation where it was pointed out that the only necessary specification is the tube bandwidth between the 1 dB points of the passband. It was further noted that the tube drive characteristic producing the 1 dB bandwidth output should be reproduced in the transmitter. In the third section amplitude and phase distortions as a function of both frequency and time under wide signal bandwidth conditions were discussed. In particular it was shown that over specification can be a significant problem if the total transmitter chain distortion response is not analyzed completely. Also mentioned were various non-linearities created by operating the transmitter tube beyond saturation. The consequences of such non-linearities to wideband waveforms are not well known since requisite non-linear analytical techniques have not yet been developed.

REFERENCES

1. RADC private communication.

2. Allen, S., "Super Power Tubes for Radar and Accelerator Applications", Notes published by Varian Associates.

3. Cook C.E. and Bernfeld, M., Radar Signals, Academic Press (1967).

4. Cook, C.E., "Transmitter Phase Modulation Errors and Pulse Compression Waveform Distortion", The Microwave Journal (May 1963).

	P_{PK} MW	P_{AV} KW	PRF PPS	T μSEC	DUTY	P_{AV} ADVAN-TAGE dB	COMMENTS
NORMAL AN/TPS-43A	2.8	4.7	250	6.7	.00168	0	FREQUENCY RANGE 2.9-3.1 GHz
VA145E TWYSTRON RATING	3	5.25		7.0	.00175	+.48	117 KV, 80A BEAM
2X PRF	2.8	9.4	500	6.7	.00335	+3.0	TEMPERATURE LIMITED. REDUCE MODULATOR CHARGING CHOKE INDUCTANCE. $1K
RAISE T	2.8	7.0	250	10.0	.00250	+1.7	TEMPERATURE LIMITED. MODIFY MODULATOR $5K - $15K
RAISE T	2.8	9.1	250	13.0	.00325	+2.9	TEMPERATURE LIMITED. MODIFY MODULATOR. $5K - $15K
VA145 J TWYSTRON RATING	4.5	10.0		10.0	.0025	+3.2	FREQ RANGE APPROX SAME (3.015-3.205 MHz) 117 KV BEAM. DIFFERENT FOCUS COIL. SOMEWHAT TEMPERATURE LIMITED. $20K
VA145B TWYSTRON RATING	5	12.0		10.0	.0025	+4.1	LOWER FREQ RANGE (2.75-2.95 GHz). 125 kv, 92A BEAM. DIFFERENT FOCUS COIL. TEMPERATURE LIMITED. $20K
VA145F TWYSTRON RATING	8	17.0		10.0	.002	+5.5	HIGHER FREQ RANGE (3.1 - 3.6 GHz). $50K 155KV, 125A BEAM. DIFFERENT FOCUS COIL. REQUIRES SUBSTANTIAL TRANSMITTER CHANGES

Table I. RADC SEEK SCREEN Study results for burn-through mode implementation in the AN/TPS-43A radar.

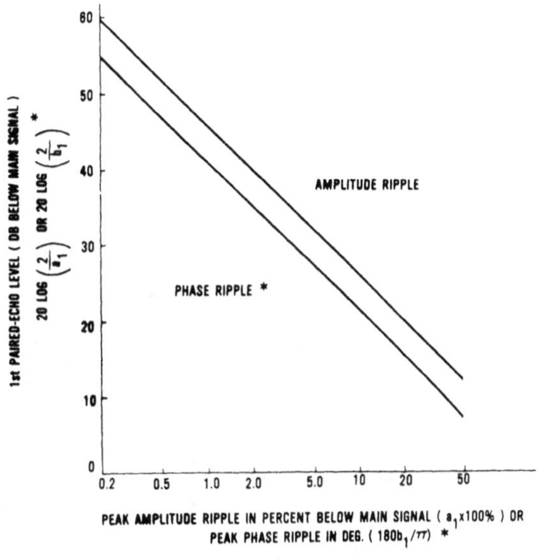

Figure 1. Amplitude of first paired-echo as a function of amplitude and phase distortion magnitude.

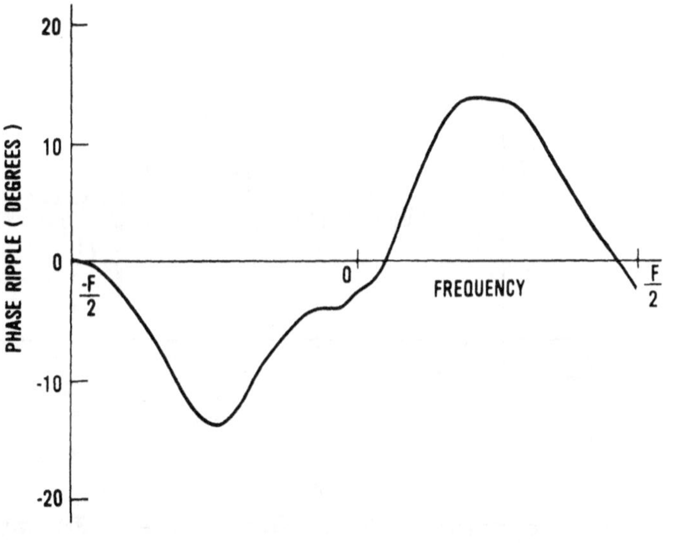

Figure 2. Phase ripple for a high power transmitter tube.

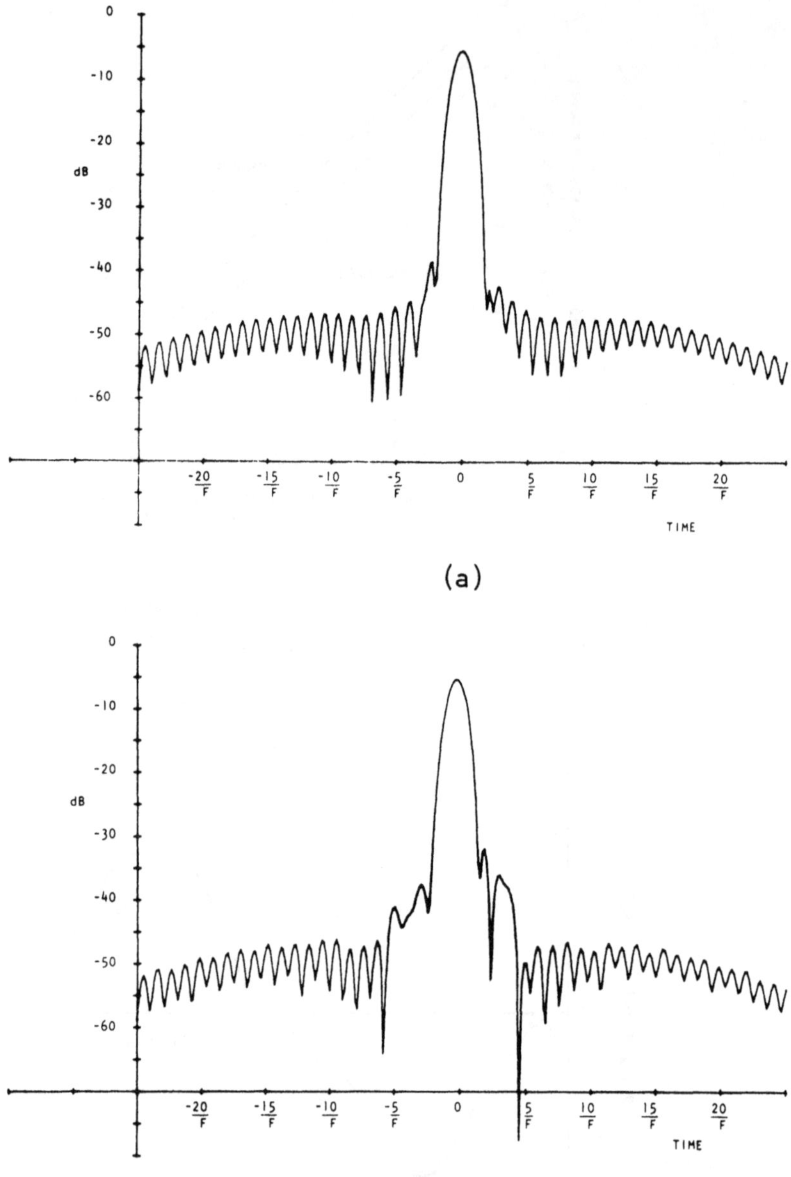

Figure 3. Correlation peak for a linear FM waveform with time-bandwidth product of 2500 and Hamming weighting. (a) No distortion. (b) Distortion effect due to phase ripple shown in Figure 2.

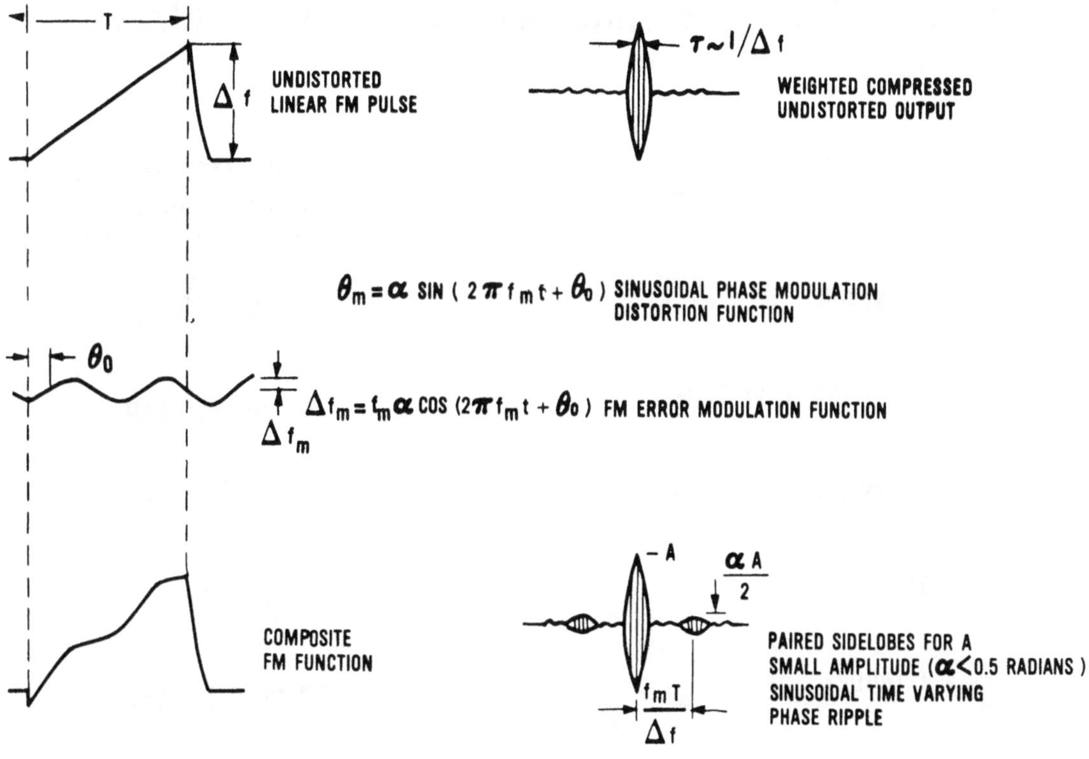

Figure 4. Induced paired-echoes from sinusoidal time modulation of linear FM.

TABLE II

TRANSMITTER TUBE TIME DEPENDENT PHASE RIPPLE

AMPLIFIER TYPE	APPROXIMATE MAGNITUDE OF PHASE RIPPLE $^{*}\alpha$
KLYSTRON	$\frac{1}{2}\left(\frac{\Delta E}{E}\right)\phi$
TRAVELING WAVE TUBE	$\frac{1}{3}\left(\frac{\Delta E}{E}\right)\phi$
CROSSED FIELD AMPLIFIERS	$\frac{1}{25}\left(\frac{\Delta E}{E}\right)\phi$
TRIODE	$50\left(\frac{\Delta I}{I}\right)$
TETRODE	$(1 \text{ to } 50)\left(\frac{\Delta I}{I}\right)$
AMPLITRON	$40\left(\frac{\Delta I}{I}\right)$

* ϕ IS THE ELECTRICAL LENGTH OF THE DEVICE, $\Delta E/E$ IS THE VOLTAGE RIPPLE RATIO, AND $\Delta I/I$ IS THE CURRENT RIPPLE RATIO.

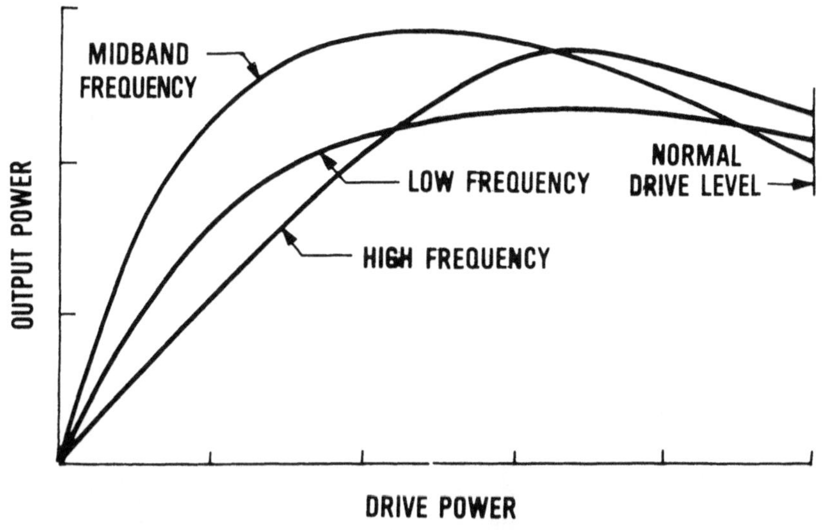

Figure 5. Power characteristic for a high power transmitter tube at low, midband, and high frequencies.

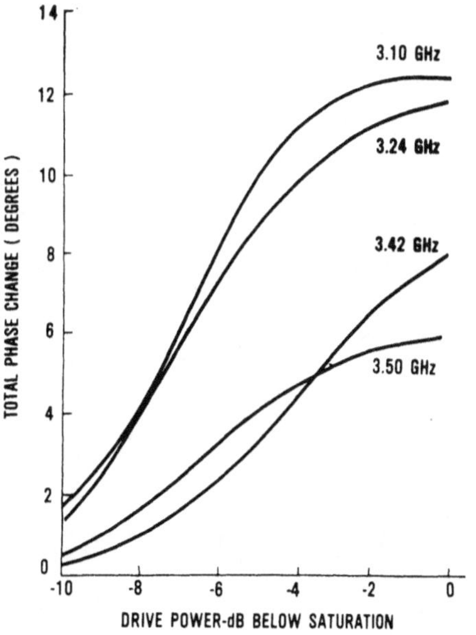

Figure 6. Relative phase change through a high power transmitter tube as a function of drive power for several frequencies in the band.

Analysis of ECCM Characteristics of Frequency Agile Radar Systems
B. Bergkvist
Proceedings of the Military Microwaves Conference, London, October 1978
pp. 51-57.

3.2

After an introductory résumé of frequency agility radar system properties the general ECCM characteristics of such systems are discussed. Frequency agile transmitters are presented and system implications for optimized ECCM design are pointed out. Principles for jamming a frequency agile radar are discussed, and the required jammer complexity for different levels of jamming ambition is presented. Finally the ECCM properties of mechanically tuned magnetrons are discussed in the light of modern signal processing and the conclusion is drawn, that well designed pulse generation patterns in combination with fast tunable magnetrons will have a substantial ECCM value even when confronted with sophisticated jammers.

1. GENERAL

Introducing a Frequency Agile transmitter into a radar system will give it a number of desirable qualities. Target detectability is enhanced by target decorrelation and the same mechanism also serves to reduce the glint error and thus to increase measurement accuracy. Target visibility in regions of distributed clutter is improved by clutter decorrelation and the constantly changing frequency of the radar avoids certain types of interference and eliminates multiple time around echoes. Definition of critical parameters as well as estimates of obtainable improvements by Frequency Agility for these tasks are well described in available open litterature (Ref 4). The ECCM properties of frequency agile systems are however much less treated (Ref 6,7) even if the improvement of the jamming resistance originally was the main goal for the introduction of frequency agility. Thus little guidance to the radar designer on how to select critical system parameters is available.

In this paper, some thoughts are presented dealing with system design of frequency agile systems optimized for maximum ECCM performance. Before that, however, the basic principle of how Frequency Agility achieves its jamming resistance is discussed and some methods to implement agile transmitters are presented. Requirements on the radar receiver in order to take maximum advantage of an agile transmission are pointed out. Finally, different methods of jamming agile systems are discussed. It is shown, that advanced jammers incorporating frequency measuring receivers, signal processing and tuneable transmitters, will be able to degrade the jamming resistance of such agility radars, which do not include a sufficient degree of randomness in their frequency choices. The ability to achieve such behaviour with mechanically tuned transmitters, such as magnetrons, is discussed, and it is shown that a high degree of unpredictability can be designed into these systems provided their tuning rates are high enough.

2. INTRODUCTION

All Radars can be jammed. However, the more power an enemy has to dedicate to jam a radar station in order to achieve a certain degradation of its performance, the more jamming resistant the radar is considered to be.

The jamming effort required depends, among other things, on how well the parameters of the radar are known to the jammer. If little information is available, the jammer has to spread its jamming power over a number of possible parameter choices, which will dilute the jamming.

One important radar parameter is the radar carrier frequency. If the radar transmits a waveform with a comparatively narrow bandwidth but with a carrier frequency, which varies from pulse to pulse, frequency agility (FA) is said to be used. The ECCM mechanism of frequency agility is to deny the jammer information on the instantaneous radar carrier frequency and thus force it to jam over a much broader band than the passband of the radar receiver.

3. METHODS OF GENERATING FREQUENCY AGILITY

Frequency agility can for instance be generated as illustrated by fig. 1. A noise generator controls the input voltage level to a voltage controlled oscillator (VCO). Amplitude variations of the noise source are thus transformed into stochastic frequency variations of the oscillator. By using the VCO output as input to a pulsed RF power amplifier, RF pulses with random carrier frequency are generated. The noise generator and the VCO can, with no loss of generality, be substituted by a bank of oscillators and randomly controlled switches to select a carrier frequency provided that the number of oscillators are spaced no further than one pulse bandwidth apart. (Fig 2).

The comparatively high costs of RF power amplifiers have led to the development of a number of fast tuneable RF oscillators of the magnetron type. Only mechanical tuning has as yet been useable to generate attractive bandwidths. Basically two methods are dominating, namely rotating and reciprocating tuning.

Rotating tuners can be designed for very high tuning speeds due to the small power required to keep the tuner in motion. Another factor which makes this principle more apt to fast tuning is the ability of the tuner to generate many modulation cycles by any single turn of the rotating device.

Reciprocating tuners have considerably lower tuning speeds but can more easily be supplied with additional center frequency control. Both types of magnetrons normally generate periodic frequency variations with time.

4. RADAR SYSTEM REQUIREMENTS

A frequency agile transmitter is a powerful tool to improve the jamming resistance of a radar, but other radar design parameters are also very important in order to take full advantage of the agility. Frequency agile systems only use narrow bandwidths (normally the reciprocal of the pulsewidth) for reception. To take advantage of this fact, the receiver has to be fast and accurately tuned to the selected carrier frequency and to have high selectivity during the listening period. Many practical environments, especially those where sea and multipath are involved, are characterized by a jamming power density, which varies rapidly with frequency. Single side band mixers with high rejection of the image frequency are then important to incorporate in the receiver design. Special attention shall be paid to the bandwidth of the IF channel where not only the 3 dB but also the 40 dB bandwidth is of importance. The linearity of the receiver before reducing bandwidth is of course vital. High IF selectivity is rather useless if it is implemented after a saturated RF amplifier!

Another important consideration is the CFAR properties of the radar's receiver, i e the ability of the receiver to keep the background at an approximately constant level. In practical environments, multipath effects in combination with often unintentional deficiencies of the jammer due to inefficient modulation schemes or improper installation of the jamming antennas in the carrier, generate jamming power densities which are highly variable in both frequency and time. The input jamming power to the radar is further modulated by the radar antenna pattern, which especially in the far sidelobe region often exhibits violent variations due to the carrier frequency change (sidelobe "fanning"). A good radar receiver has to be capable of taking advantage of this variability - to take good care of receptions occuring when the jamming input power is low. In order to obtain such properties, care has to be exercized both regarding the instantaneous gain of the receiver and on the integrator design. Excellent video level control can be achieved by the use of Dicke fix receivers or by fast gain control, IAGC. The use of a Dicke fix receiver is however somewhat adventurous. One drawback is the requirement for a locally homogeneous jamming background (the same power density in the guard band as in the final pass band is a prerequisite for good CFAR performance),another drawback the inherent susceptibility of this receiver to specially designed guard band jammers. Therefore well designed IAGC amplifiers are generally to be preferred.

In search radars a number of pulses are normally combined before thresholding, video integration. This method will generate a substantial increase in sensitivity, if the signal to noise ratio of successive pulses are of the same order of magnitude. For an agile system operating in a jammed environment, the signal to noise ratio may vary orders of magnitude between different radar sweeps, and in these cases video integration may very well only degrade the sensitivity. This degradation can become excessive if nonparametric detectors are used. The final integrator choice however has to be a compromise between sensitivity to single echoes and susceptibility to pulsed interference.
The conclusion of the above is, that a Frequency Agile transmitter is a good start to make a jamming resistant radar, but that the less glamorous receiver, integrator and threshold design is very important if full use of the agility is to be obtained.

5. HOW TO JAM A FREQUENCY AGILE RADAR

The most unsophisticated form of jamming is to use broadband barrage jamming with a bandwidth, which is equal to the agile bandwidth of the transmitter plus margins to compensate for bandwidth estimation errors and jammer imperfections. This kind of jamming has very low efficiency. Depending on the agility bandwidth and the pulse width, a FA radar has in the order of 10 to 1000 transmission channels. Thus every channel will experience just 1/10 to 1/1000 (-10 to -30 dB) of the jammer's total strength. The possible jamming impact on a radar by such a mission has to be carefully weighted against other effects of broadband jamming, e g effects on own forces.

It is assumed above, that the frequency channels of the radar are adjacent which not necessariely has to be the case when dealing with radars using RF power amplifiers with a limited choice of selectable oscillators. Significant improvement regarding power density and ability to evade interference with own forces can then be achieved by switching to spot jamming of the actual frequencies. A slightly more complex jammer has to be used and intelligence on actual frequencies has to be made available, or to be measured, which complicates the jamming equipment.

A first degree of sophistication is to measure the instantaneous frequency of the frequency agile radar and after that, tune all available power to that frequency. There is one important drawback with this method. The jammer can not tune his high efficiency jammer before the radar pulse is received, and thus targets at shorter range, and even the echo from the jammer carrier, will not be concealed by narrow band jamming. The option of the jammer to operate close to the radar is not always compatible with survivability expectations for jammers. To operate at very long ranges from the radar is neither possible because of the requirement to be within the unambiguous range of the radar. If the radar has tuned to a new frequency when the jammer is tuned to the original one, the radar pass-band will be unjammed and only jamming attenuated by the selectivity of the receiver will affect the performance.

The tuned jammer of above requires in excess of a fast tunable frequency source a receiver, which can measure frequency of the relevant pulses. In tactical environments the pulse density within the interesting frequency range may be very high, and therefore the task to sort out pulses from one agile radar is normally much more complicated than the frequency measurement. The pulse sorting problem is aggravated if the ECM receiver can not rely on some unique property of the frequency agile transmission, such as pulse interval or pulse width. Pulse amplitude sorting, which for fixed frequency radars often is an efficient selection tool, is hardly useful at all for Frequency Agile radars due to both environmental and sidelobe effects. Thus definition of pulse pattern and pulse width selection have impact on the jamming susceptibility for radar systems. Another important design consideration is not to synchronize an agile transmission to other more recognizable radiation sources.

The next degree of sophistication of the jammer is to include means for frequency prediction into the jammer's signal processing. Such prediction is, as indicated above, highly desirable even for the fast tuned jammer for pulse sorting purposes, and if narrowband jamming for protection of targets at ranges closer to the radar than the jammer is desired, frequency prediction is of course mandatory. Frequency prediction is only possible, if some not completely random frequency distribution is used by the radar and of course only interesting if the number of possible transmission frequencies is too large to make spot jamming efficient. For mechanically tuned agile magnetrons, the inertia of the fast tuning mechanism will generate a comparatively strong correlation between frequency and time, the actual tuning pattern can be traced. In order to track the frequency tuning curve of a magnetron a number of consecutive transmissions have to be measured. From these data, estimators (descramblers) could be designed based on e g linear prediction methods or phase locked loops, possibly biased by tuning rate estimates from Fourier Analysis of available data (not necessarily evenly sampled). Before however the conclusion is drawn, that mechanically tuned Frequency Agile transmitters are obsolete in the era of digital signal processing, the consequences of a successful magnetron modelling by an ECM processor on the jamming susceptibility has to be analysed.

Assume that the actual radar is at a range R from the jammer. Further assume, that this range is exactly known to the jammer as well as the exact modulation pattern of the agility source in the radar i.e. both the sorting and prediction problems have been solved. How well is it possible then in this rather artificial scenario for the jammer to tune his power to the correct frequency?
The jammer will be informed that a pulse is transmitted from the radar at first when the pulse reaches the ECM receiver. If the radar does not use a fixed pulse repetition pattern, the exact time of transmission can not be estimated by the jammer. If the transmission instant is completely unknown, the jammer has to use an instantaneous bandwidth equal to the frequency excursion possible in the time interval between transmission by the radar and reception by the ECM receiver. As an example consider a fast tuned magnetron, having a sinusoidal tuning pattern with a tuning rate of 1000 Hz and frequency deviation of $\pm$ 225 MHz. Assume further that the jammer is at 30 km range from the radar, i e the delay time between transmission and reception is 100 μs. During this time, the frequency of the magnetron may change by as much as 140 MHz which is also the bandwidth required by the jammer for effective jamming. Thus, in this example a power density gain from the jammers point of view of 450/140 = 3.2 (5 dB) can be achieved, which is significant - but not spectacular - considering the amount of hardware required and the tough assumptions made (known range, the sorting problem solved, exactly predictable magnetron tuning). The conclusion of this simple example is therefore, that transmitters with high tuning rates and with randomized pulse intervals will remain useful even when attacked by the most sophisticated jammers. Lower tuning rates as well as less unpredictable pulse patterns will decrease the ECCM value of the agility but to what degree has to be analysed for every specific case.

REFERENSES

1.Thourel, L; "Les Radars de Veille Modernes", Masson et Cie. Paris 1965

2.Gustafsson, B.G and Ås B-O; "System properties of jumping frequency radars", Philips Telecom Rev 25, No 1, July 64, pp 70-76

3.Lind G, "Reduction of radar tracking errors with frequency agility", IEEE Trans AES-4, No 3, May 68, pp 410-416

4.Barton D.K; "Radars Vol 6. Frequency Agility and Diversity", Artech House Inc 1977. (Includes reprints of ref 2 and 3).

5.Barton D.K; "Simple Procedures for Radar Detection Calculations", IEEE Trans AES-5, No 5, Sept 69, pp 837-846

6.International Cumulative Index on Radar Systems IEEE 1978

7.Johnston S.L: Radar Electronic Counter-Countermeasures, IEEE Trans AES-14, No 1, Jan 78, pp 109-117.

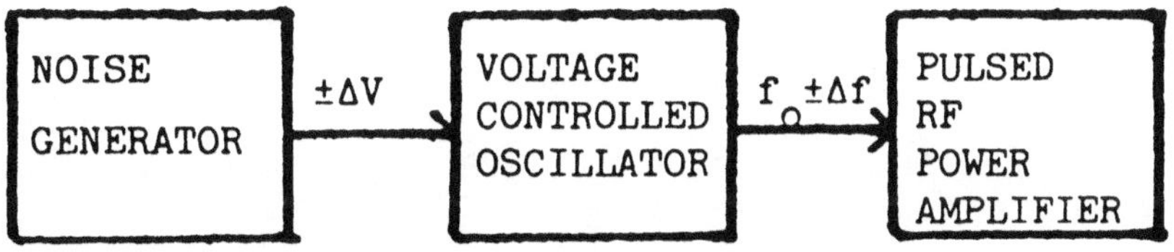

Fig. 1 General Frequency Agile Transmitter

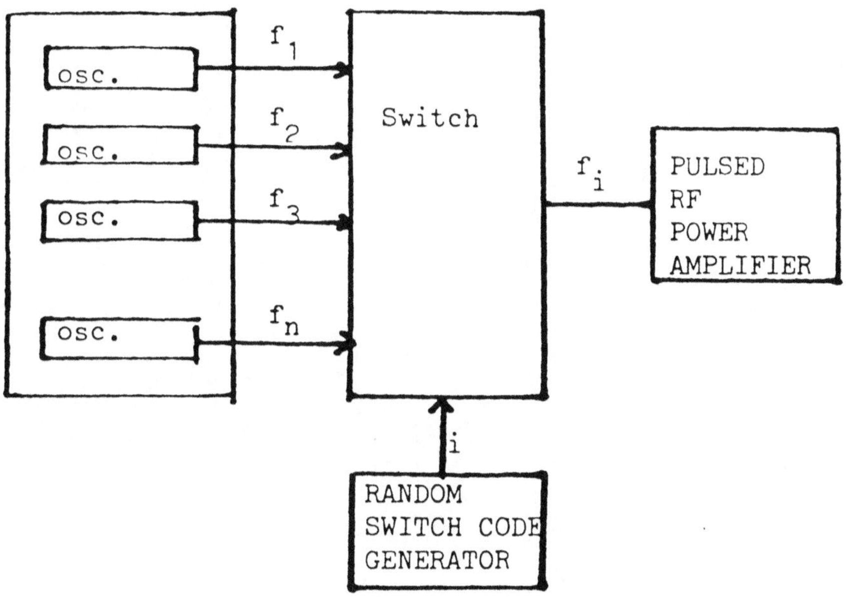

Fig. 2 Frequency Agile Transmitter with oscillator bank

Combined X/K_a-Band Tracking Radar
L. Klaver
Proceedings of the Military Microwaves Conference, London, October 1978
Pages 147-156

3.3

ABSTRACT

First a short historical survey will be given how SIGNAAL arrived at the present solution to track low-flying targets through a combined X/Ka-band system, thus adequately suppressing the image effect.

Attention will be paid to the effects of atmospheric conditions, including rain and sea clutter.

After six years of evaluation and testing, the final configurations are in full serial production.

In the system, target detection is performed by a normal search radar, after which the acquisitionprocess is executed with the X-band tracking radar. As soon as video quality permits, target tracking is performed in the Ka-band.

This paper deals with the practical problems which are experienced when tracking low-flying target by radar and the solutions devised by SIGNAAL. The mathematical problems of the multipath phenomenon have been thoroughly investigated by several well-known authors during the last ten years; hence the theory will not be elaborated upon here.

History

In early 1971, during our activities in the field of short-range air defence, SIGNAAL was faced with multipath propagation phenomena. It became quite evident during test runs serving to check our X-band tracking radars, using very low-flying targets, both over land and over sea (< 100 ft).

The author is with Hollandse Signaalapparaten B.V.
Hengelo, The Netherlands.

It appeared that inaccuracies caused by forward scatter did not permit an optimum exploitation of the maximum effective range of medium to short-range guns and missiles.

We investigated a wide range of possible solutions including frequency control, electro-optics and various methods of video processing.
Electro-optical solutions for the sole reason of image suppression were rejected because of their non-all-weather performance.
Various methods of processing the amplitude and phase relationship between real target and image were also rejected because we did not believe an adequate "closed-loop" processing for tracking radars was feasible.
This is so because:

- the phenomenon is very complex, since specular and diffuse reflections are involved,
- the occurrence of images is seldom predictable and its position varies continuously, since it is a function of target altitude, antenna height and surface roughness, all three of which are never constant,
- for fire control systems establishing track at the earliest possible time is a primary requirement, thus processing time is at a premium.

Therefore, SIGNAAL studied possibilities to prevent the problem from reaching the antenna in the first place.
The simplest solution is to use very small beamwidths in order to prevent the energy from hitting the surface at all, thus avoiding interfering reflections completely. But apart from the fact that small beamwidths would only lead to long acquisition times (because of the required volume search patterns), they also require large antenna dimensions at frequencies normally used for fire control radars.
The next logical step was to realise that a very narrow beamwidth could also be obtained by choosing a higher frequency.
The choice of frequency for the narrow beam radar was between the Ku- and Ka-bands. Ka was selected because of its even narrower beamwidth possibilities than Ku-band, the avoidance of atmospheric absorption at around 20 GHz, and ready availability of military componentry. Furthermore, actual trials have shown that the Ku-band frequency was still sensitive to lobing.

In order to simplify the antenna structure and to avoid all parallax compensation problems, the Ka-band antenna was integrated into the primary tracking radar (Fig. 1).

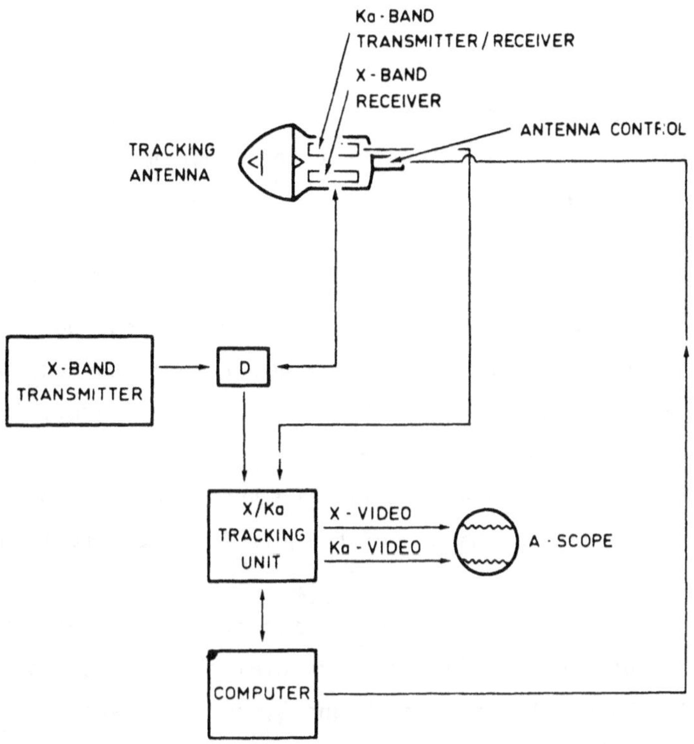

FIG. 1 COMBINED X/Ka-BAND TRACKING RADAR

Patents for this process have been obtained in a number of countries, including the US.

This integration philosophy thus allowed:

- target acquisition by the wide beam X-band radar, thus ensuring fast reaction,
- rejection of the false image by means of the Ka-band radar,
- possible use of the various combined modes for range and angle tracking,
- avoidance of both the physical size and weight problems of having two radar antennas, and the associated parallax error and alignment problems as well.

The solution proposed and the accuracies thus obtained convinced military authorities that the SIGNAAL configuration was the answer to the image problem.
Since 1974 serial production of this integrated X/Ka-band tracking radar system has been in progress, and a large number of units are operational by now.
For security reasons I am not allowed, however, to provide more details on the customers who have purchased this radar system and on the configuration in question.

Trials

By the end of 1972 a prototype of this integrated X/Ka-band tracking radar was available for trials. The extensive trial program was executed both over land and sea, under various weather conditions.

I would like to give you a general idea of the tracking performance by showing you a segment of film obtained from a monitor during trials. The pictures shown on the monitor originate from a camera mounted on top of the tracking antenna. It shows a 16 mrad beamwidth Ka-band radar, integrated with a 64 mrad X-band radar performing tracking trials over sea with a very low-flying LEAR-jet target. One specific run shows the aircraft at an altitude of 20 ft, thus representing the proper heights for a tactical environment.

For security reasons no tracking accuracy figures can be mentioned at this open conference, but the detailed data are available on request.
For an example, please refer to Fig. 2 and 3, showing part of an actual recording (TV as well as radar) of a target flying at an altitude of approximately 30 ft over sea. The antenna height during the trials was approximately 20 ft.

FIG 2 Ka-BAND TRACKING, FLYING ALTITUDE 30 FT
TV-CAMERA RECORDING

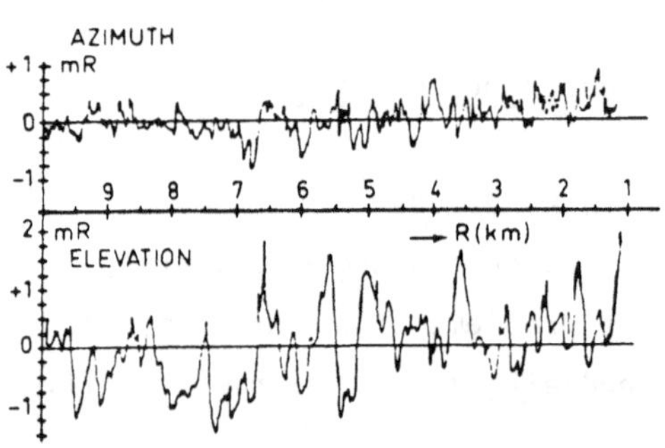

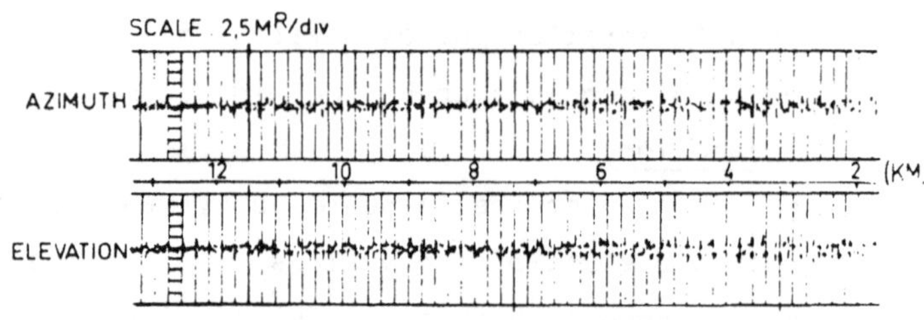

FIG 3 Ka-BAND TRACKING, FLYING ALTITUDE 30 FT
RADAR RECORDING

Sea Surface Roughness

Apart from the narrow beamwidth of the Ka-band tracking system, through which only a small area of the sea surface is illuminated (and consequently little energy can be reflected), the surface roughness of the sea is also an important factor as it is a direct function of the radar wavelength. Therefore the roughness factor $(\delta h/\lambda)$ is tabulated (Fig. 4) for radar frequencies in the X-band (9 GHz, $\lambda = 3.3$ cm) and the Ka-band (35 GHz, $\lambda = 8.6$ mm).

		$\delta h/\lambda$	
SEA STATE	δh (mm)	9 GHz	35 GHz
0	0-6	0-0.18	0-0.7
1	20-75	0.6-2.3	2.3-8.6
2	105-220	3.2-6.7	12-75
3	250-350	7.6-10.5	29-40
4	465-525	14-16	53-60
5	610-915	18.5-28	70-105
6	990-1680	30-51	114-193

SURFACE ROUGHNESS FACTOR ($\delta h/\lambda$)

Fig. 4

The significant heights $\overline{H}\ 1/3 = 4\delta_h$ for each sea state have been derived from a US publication [1].

As can be seen from Fig. 4, there is a marked difference in the apparent sea status (surface roughness) for the two frequency bands.

For instance, in sea state 2 the Ka-band radar experiences a very rough surface which approximately equals sea state 4 or 5 for the X-band radar, while sea state 3 for the Ka-band radar equals sea state 6 for the X-band radar.

Even at sea state 0 there is a fair chance that the Ka-band radar encounters a certain roughness of the sea surface.

In general this means that the Ka-band radar always experiences a rougher surface than lower frequency radars, which implies that the value of the specular reflection coefficient for Ka-band frequencies is smaller than for lower frequencies.

Multiplication of the roughness factor and the grazing angle results in the rms roughness, which is given in [2] . See Fig. 5.

For a roughness of 0.08 (specular reflection coefficient = 0.6), the grazing angle of Ka-band frequencies can be as small as 3.2 mrad in sea state 2, whereas the same value for X-band frequencies is only reached at sea state 5.

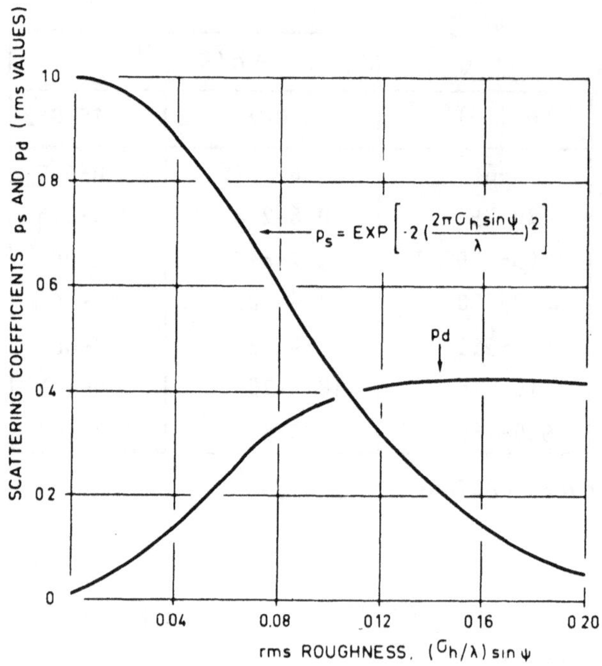

FIG 5 SPECULAR AND DIFFUSE SCATTERING COEFFICIENTS VERSUS ROUGHNESS (AFTER [2])

Sea Clutter

ENVIRONMENT: SEA STATE

WAVE HEIGHT : 191 CM

129 CM

ANTENNA HEIGHT 12 M

BEAM POINTED AT SEA SURFACE

MEASURED REFLECTIVITY

- X-BAND : IN ACCORDANCE WITH THEORY
- Ka-BAND : FAR LESS THAN PREDICTED
- VISIBLE UP TO 1 KM (A-SCOPE)

The measured sea clutter reflectivity was well in accordance with the theoretical values for X-band frequencies, but at ranges up to 1 km, the Ka-band measuring results were approx. 4 - 5 dB less then expected from theory. Beyond the 1 km range the clutter was no longer visible on the A-scope. The values measured have been compared with values computed from:

$$A_c = \tfrac{1}{2}\sqrt{2} \cdot \theta_h \cdot \frac{\tau c}{2} \cdot R \cdot \sigma_o \qquad [3] \text{ page } 234 - 236$$

Rain Clutter

RAIN CLUTTER (Ka-BAND)

ENVIRONMENT : 3 mm/hr
40 KN
ABOVE SEA
ELEVATION 15°

MEASURED ATTENUATION WITH RESPECT TO X-BAND

. UP/DOWNWIND : 0.1 dB/KM] (TWO-WAY)
. CROSSWIND : 1.7 dB/KM]

MEASURED CLUTTER SPECTRUM:
IN ACCORDANCE WITH THEORY

MEASURED REFLECTIVITY:
5-10 dB LESS THAN PREDICTED BY THEORY

A number of measurements on the reflectivity and attenuation of rain clutter have been performed.
The rain was measured at the same place and time for both radar systems.
It was found that the measured results and the theory agree for the X-band, based on:

$$V_c = \frac{\pi}{8 \ln 2} \cdot \theta_h \cdot \theta_v \cdot \frac{\tau c}{2} \cdot R^2 \cdot \sigma o$$ [3] page 60

and $$\sigma_o = 6.10^{-14} \cdot r^{1.6} \cdot \lambda^{-4}$$ [4] page 138

and attenuation figures of 5 page 544.

For the Ka-band system, the situation was different; the Ka-performance in rain was markedly better than theory predicted.
This experience is in accordance with the findings in [6] , which state that: "Some tracking has been performed with an intervening heavy rain squall, demonstrating that the Ka-band radar has significant capability to see through heavy rain".

As an example, with the following conditions:
- rainfall rate 3 mm/hr
- wind speed 40 knots
- elevation 15°

the Ka-band radar suffered an extra loss of only 0.1 dB/2 km, as compared with the X-band radar for upwind and downwind situations, and 1.7 dB/2 km

extra loss for a crosswind situation, though one would have expected ([5], page 544) an extra attenuation of 2 dB/2 km.
As a result of the various measurements it was found that the reflectivity of rain clutter for Ka-band frequencies was on the order of 5 - 10 dB less than predicted by theory.

Short Description of the Radar

The X-band monopulse cluster is mounted at the focal point of the auxiliary reflector of a cassegrain antenna, while the Ka-band cluster is mounted at the focal point of the main reflector.
The ratio of energy levels received in the horns is a measure of the deviation of the target position with respect to the beam axis.

The Ka-band transmitter, as well as the r.f.-receiver and i.f.-amplifiers of both radars, are mounted at the back of the antenna. The i.f.-amplifiers contain three normal monopulse receiving channels.

The ΔE and ΔB receivers are connected to the angular tracking circuits. The output voltages of these units drive the tracking antenna in azimuth and elevation.
The Σ receiver output controls the range tracking circuit, which continuously corrects the position of the range gate with respect to the target video.
Doppler filters containing phase-sensitive detectors are incorporated for Pulse Doppler Tracking (PDT).
The two X and Ka-tracking video signals are processed in the tracking unit which includes angular and range tracking circuits as well as a logic circuit to select either one of these signals for target tracking.
Both tracking signals are displayed on an A-scope.

Operation

The principles of operation of the combined radar system are as follows:
The X-band radar acquires the target.

Target take-over is always effected with the aid of the X-band radar, ensuring the fast reaction of the system.
When the Ka-band tracking radar has obtained sufficient echo data from the target, the system switches to automatic Ka-band tracking.

The range, azimuth and elevation of the target are determined by the Ka-band radar. The error voltages of the X-band radar are replaced by the error voltages of the Ka-band radar.
The switch-over time between "lock-on" by the X-band radar and "tracking" by the Ka-band radar was measured in the reaction time trials.
During 98 runs at various altitudes (down to 30 ft) and at ranges between 15 and 18 km, the measured time lapse between lock-on (X-band) and tracking (Ka) was negligibly small.
Automatic switch-over is possible even at X-band elevation errors of up to 10 mrad.

Conclusion

The proof of an idea is in how well it works in practice. The method of rejection of images using high frequency narrow-beam radar integrated with a primary tracker works. SIGNAAL has proven its integrated X/Ka-band tracking radar performance against many types of low-flying targets during actual trials.
This tracking method is an integral part of SIGNAAL's radar fire control system product line, which is in actual service now.

References

[1] Title and Data unknown
Origin: John Hopkins University
Applied Physics Laboratory
Silver Spring, Maryland

[2] I. Beckmann and A. Spizaichino, "The scattering of EM waves from rough surfaces".
New York: Pergamon 1963.

[3] F.E. Nathanson, "Radar Design Principles".
Mc. Graw-Hill 1969.

[4] Barton and Ward, "Handbook of Radar Measurements".
Mc. Graw-Hill.

[5] M.I. Skolnik, "Introduction to Radar Systems".
Mc. Graw-Hill 1962.

[6] D.D. Howard, "Investigation and Application of Radar Techniques for Low-Altitude Target Tracking".
US Naval Research Laboratory, presented at RADAR "77 in London, (25 - 28 October 1977).

Chapter 4
ECCM in the Antenna

Summary

The papers presented in this chapter treat various ECCM techniques in the antenna. A list of ECCM techniques in the antenna is given in Paper 1.1, Table II (page 14 of this volume.) Owing to their current popularity, sidelobe blankers and cancellers will be considered separately in Chapter 5.

Synopsis of Reprint Papers in Chapter 4

4.1 "Low Sidelobe Radar Antennas"—Rudge

Several papers in Chapter 2 point out the importance of antenna sidelobes in the presence of standoff jammers and false target deception ECM. This very recent paper reviews the state-of-the-art in low-sidelobe radar antennas. Some examples are treated in more detail. Predicted and measured performance data are presented. Both co-polarized and cross-polarized patterns are given. The latter is extremely important as an ECCM against cross polarization ECM. This paper was presented at the Military Microwaves Conference in 1978.

4.2 "Effect of Fresnel Zone Blockage on Very Low Sidelobe Antennas"—Fante

As in most endeavors, extremely severe requirements incur penalities for their achievement. In radar, it is a matter of an exchange of constraints as Radford indicated in Paper 2.B.2. Very low sidelobe antennas are very attractive ECCMs for search radars, but they require careful attention to obstacles in the Fresnel zone. This paper presents an approximate method for estimating the azimuthal sidelobe level due to partial blockage by obstacles in the Fresnel zone and applies it to problems that are typical as well as practical. As one application, a tactical military commander could greatly deteriorate the antenna sidelobe characteristics of a radar by a poor choice of site.

An additional influence on very low sidelobe antennas not discussed in this paper is the effect of the antenna *vertical* radiation pattern. The vertical pattern determines illumination of objects in the Fresnel zone. Reduction of this illumination may require sacrifice in low angle radar coverage however.

4.3 "Tolerance Errors for Very Low Sidelobe Antennas"—Fante

Close control of the manufacturing tolerances in the surface of a radar antenna is necessary to achieve 50 dB antenna sidelobes. This paper analyzes the effect of surface tolerance on antenna sidelobe level. As an illustrative example, it is shown that an S-band parabolic cylinder antenna with a 225 ft^2 aperture requires a tolerance of 4.87×10^{-3} inches for a probability of 0.01 that sidelobes beyond 7 degrees will exceed -50 dB.

4.4 "Antenna Sidelobe Reduction Using Spatial Filters"—Mailloux and Schell

Reduction of antenna sidelobes on existing radars usually requires replacement of the antenna. This paper presents an approach that could be retrofitted onto existing antennas. A multilayer structure is placed in front of the antenna to function as a wave filter. These spatial filters are configured as parallel layers of dielectrics or metal gratings extending across the entire antenna aperture. Experimental results are shown at X-band for both a horn antenna and an array antenna.

It is proposed that this spatial filter be incorporated as a part of the antenna radome and thus serve to protect the antenna as well as reducing undesired sidelobe radiation. This technique is still experimental. This paper was presented at the DDRE Radar ECCM Symposium in 1976.

4.5 "Conopulse Radar"—Sakamoto and Peebles

Military angle tracking radars orginally employed conical scan techniques [1], however such radars can be affected by inverse amplitude modulation ECM. Monopulse radar is an effective ECCM in that it has no target scan modulation. Monopulse is more complex than conscan however. Another ECCM against inverse scan ECM is lobe on receive only (LORO). This can be readily implemented in a CW radar which uses separate antennas for receiving and transmitting. Use of LORO in radars which have only one antenna can present problems however. A hybrid conical scan-monopulse radar called *conopulse* is described in this paper. It has the advantage of monopulse freedom from inverse scan ECM and requires less receiver equipment than monopulse. Its disadvantages are its lower data rate than monopulse and its somewhat more complexity than conical scan. This paper contains an extensive analysis of two forms of conopulse.

References

[1] Johnston, S.L.,"A Survey of Navigational Measurements Methods for Missile Guidance Systems," *IRE Trans.*, vol. PGI-5, June 1956, pp. 60-5.

Bibliography Topics for Chapter 4

Angle Measurement, General
Antennas, General
Antennas, Low Sidelobe
Bistatic Radar
Clutter Fences
Monopulse

BIBLIOGRAPHY FOR CHAPTER 4

ANGLE MEASUREMENT, GENERAL

Sklar, J.R., and F.C. Schweppe, "On the Angular Resolution of Multiple Targets," *Proc. IEEE*, vol. 52, no. 9, Sept 1964 pp. 1044-5 c.

Gutkin, L.S., "Comparison of Actual and Optimum Direction-Finding Accuracy, Part 1, Continuous Signals," *Telecomm and Radio Engrg, Pt.2,* vol. 20, no. 6, June 1965, pp. 107-117.

Gutkin, L.S., "Comparison of Actual and Optimum Direction-Finding Accuracy, Part 2, Pulsed Signal," *Telecomm Radio Engrg, Pt. 2,* vol. 21, no. 3, 1966, pp. 114-8.

Shirokov, V.V. and V.A. Loginov, "The Effect of Noise and Signal Fluctuation on Angle-Measuring Systems," *Radio Engrg and Electron Physics,* vol. 11, no. 8, Aug 1966, pp. 1192-1200.

Samsonenko, S.V., "Comparative Characteristics of Methods of Measuring Angular Coordinates in the Presence of Noise," *Telecomm and Radio Engr, Pt. 2,* vol. 21, no. 8, Aug 1966, pp. 128-133.

Ostrovityanov, R.V., "On the Problems of Angular Noise," *Radio. Engrg. and Electron. Phys,* vol. 11, no. 4, Apr 1966, pp. 507-515.

Yurchenko, Yu. S., "Investigation of Noise Immunity of Tracking Radar Direction Finders in Presence of External Noise," *Radio Engrg and Electron Phys.,* vol. 14, no. 10, Oct 1969, pp. 1641-3.

Basalov, F.A. and R.V. Ostrovitianov, "Effect of Signal Polarization on Angular Noise Characteristics," *Radio Engrg. and Electron. Phys.,* vol. 12, no. 12, Dec. 1967, pp. 2048-2050, IA68A33500.

Andreyev, F.M., and G.F. Bogdanov, "Statistics of Bearing Errors Caused by Interfering Reflections," *Radio Engrg and Electron Phys,* vol. 18, no. 9, Sept 1973, pp. 1429-32.

Dobykin, V.D., "Direction-finding Accuracy of Ground or Seaborne Radiating Sources Using Low-flying Objects," *Izv VUZ Radioelektron,* vol. 17, no. 10, Oct 1974, pp. 29-35 (in Russian, trans. abstract SA 75-9726).

Glazov, Ge. N., Gr. N. Glazov, and N.P. Krasnenko, "Bearing Determination Accuracy Using Antenna Arrays," *Izv VUZ Radioelektron,* vol. 17, no. 10, Oct 1974, pp. 46-9 (in Russian, trans. abstract SA 75-9727).

Brennan, L.E., and I.S. Reed, "Adaptive Angle Tracking," *Technology Service Corp Rpt TSC-PD-119-4,* Feb 1975 (angle tracking with phased array antenna in external noise environment).

Mingalev, G.K., and V.M. Chilikin, "Analysis of the Fluctuation Characteristics of a Direction Finder," *Izv VUZ Radioelektron,* vol. 19, no. 1, Jan 1976, pp. 123-5, (in Russian, trans. abstract SA 76-21838).

Sakamoto, H., and P.Z. Peebles, Jr., "Conopulse Radar, U. of Tenn. Dissertation, Univ. Microfilms, 1976, 76-11088, 199p.

Sakamoto, H., and P.Z. Peebles, Jr., "Conopulse Radar, a New Tracking Technique," *IEEE Reg. 3 Conf. Rec,* SOUTHEASTCON–76, Clemson, S.C., Apr 5-7, 1976, pp. 130-2.

Sakamoto, H., and P.Z. Peebles, Jr., "Accuracy of Conopulse Tracking Radar," *IEEE Reg. 3 Conf. Rec,* SOUTHEASTCON–76, Clemson, S.C., Apr 5-7, 1976, pp. 133-4.

Wilcox, R.E., "The Relative Performance of Con-scan, Cosro and Two-axis Monopulse in Air-to-Ground Ranging," *IEEE NAECON Rec.,* 1976, pp. 525-32.

Sakamoto, H., and P.Z. Peebles, Jr.,"Conopulse Radar," *IEEE Trans.,* vol. AES-14, no. 1, Jan 1978, pp. 199-208 (reprinted in this volume).

Evans, J.D., "Aperture Sampling Techniques for Precision Direction Finding," *IEEE ELECTRO-78 Conv. Rec.,* vol. 28, May 1978, pp. 28/3 - 1 thru 8.

ANTENNAS, GENERAL

1. Books and General Bibliography

Love, A.W., *Reflector Antennas,* IEEE Press, 1978.

Barton, D.K., "International Cumulative Index on Radar Systems," *IEEE Pub. Cat. No. JH 4675-5,* 1978: extensive bibliographies on antennas (General) p. 1; antennas (gain and pattern considerations) pp. 1-2; antennas (lenses) pp. 2-3; antennas (radomes p. 3; antennas (reflector systems) pp. 3-4; antennas (scanners) pp. 4-5; arrays (general) pp. 5-7; arrays (gain, pattern and accuracy considerations) pp. 7-8; arrays (phase shifter, frequency scanning and control) pp. 8-9; array radar systems pp. 9-10.

2. Journal Articles

Bahret, W.F., "Technique for Amplitude Modulating a Van Atta Radar Reflector," *Proc. IRE,* vol. 49, no. 11, Nov 1961, p. 1692c.

Bradley, G.A., "How to Determine Antenna Pattern Loss," *MicroWaves Magazine,* vol. 6, no. 1, January 1967, pp. 38-41.

Radford, M.F., "Horn Antenna Having Flural Convergent Waveguide Paths," *US Patent No. 3,438,040,* April 8, 1969 (abstract SA 70-4669)

Holtum, A.G., Jr., "Parabolic Reflector with Dual Cross-polarized Feeds of Different Frequencies," *US Patent No. 3,438,041,* April 8, 1969 (abstract SA 70-4693).

Braccini, A.A., "Dish Reflector with Adjustable Subreflector," *US Patent No. 3,438,045,* April 8, 1969 (abstract SA 70-4673).

Rice, P.L., W.I. Thompson, and J.L. Noble, "Idealized Pencil Beam Antenna Patterns Used in Interference Studies," *IEEE Trans.,* vol. COM-18, no. 1, Feb 1970, pp. 27-32.

Potekhin, V.A., and I.Z. Klimov, "On Selecting the Polarization Characteristics of a Radar Antenna," *Rzh-Radiotekhnika,* no. 7, July 1971, Abstract no. 7B24 (in Russian, trans. abstract JPRS 55283).

Turner, E.M., "The Constant Index Lens Antenna," *IEEE NAECON—73 Record,* pp. 46-51.

Pozdnyak, S.I., Yu D. Ermolaev, and V.M. Yanyshev, "Optimum Control of a Radioelectronic System by Polarization Parameters in the Presence of Noise," *Radio Eng. and Electron Phys.,* vol. 18, no. 11, Nov 1973, pp. 1762-1764.

Brunner, A. and E. Kress, "Monopulse Antennas with Doubly Curved Reflectors," *Proc. 4th European Microwave Conf.,* 1974, pp. 357-361.

Marconi Ltd, "Volume Scanning Radar," *UK Patent No. 1377202,* 11 Dec 1974 (abstract SA 75-21123).

Nathanson, F.E., "adaptive Circular Polarization," *IEEE Int. Radar Conf. Rec.,* Apr 1975, pp. 221-225.

Cleveland, F.H., "Universal Radar Antenna Design Charts," *Proc. IEEE Reg. V Conf. Digest on Electr. Engrg. for the Decade,* 14-16 Apr 1976, pp. 282-5.

Bega, L.J., "System Design Philosophy of a 3-D Antenna Radar Application," *Proc. Int. Conf. on Radar,* Paris, 4-8 Dec 1978, pp. 351-6.

LOW SIDELOBE ANTENNAS

Jones, E.M.T., "Low Sidelobes in Pencil Beam Antennas," *IRE Conv. Rec.,* 1953, Pt. 2, pp. 64-67.

Blacksmith, P.A., et al, "Method of Reducing Far-out Sidelobes," *AFCRL Rpt TR 57-115,* July 1957, DDC AD 133707.

Taylor, T.T., "Design of Circular Apertures for Narrow Beamwidth and Low Sidelobes," *IEEE Trans.,* vol. AP-8, no. 1, Jan 1960, pp. 17-22.

Peters, L., Jr., R.C. Rudduck, and L. Du, "RFI Reduction by Control of Antenna Sidelobes," *IEEE Trans.,* vol. EMC-6, no. 1, Jan 1964, pp. 1-11.

Gray, C.L., "Estimating the Effect of Feed Support Member Blocking on Antenna Gain and Sidelobe Level," *Microwave J.,* vol. 7, no. 3, March 1964, pp. 88-91.

Johnson, R.C., "Reducing Wide-angle Sidelobes in Radar Antennas," *Microwave J.,* vol. 9, no. 8, Aug 1966, p. 48ff.

Ruze, J., "Feed Support Blockage Loss in Parabolic Antennas," *Microwave J.,* vol. 11, no. 12, Dec 1968, pp. 76-80.

Hatcher, R.B., "Graphs Estimate Feed Blockage," *Microwave J.,* vol. 14, no. 7, Jul 1971, pp. 31-32.

Lopez, A.R., "Line-source Excitation for Maximum Aperture Efficiency with Given Sidelobe Level," *IEEE Trans.,* vol. AP-19, no. 4, July 1971, pp. 530-2c.

Yur'Yev, A.N., "Minimization of Sidelobe Radiation Level of Antennas with a Circular Aperture," *Radio Eng. and Electron Phys.,* vol. 16, no. 7, July 1971, pp. 1106-1119.

Rogers, A., "Wideband Squintless Linear Arrays," *Marconi Rev.,* vol. 35, no. 187, 4th Q, 1972, pp. 221-243.

Dahlsjo, O., "A Low Sidelobe Cassegrain Antenna," *IEE Conf. Pub. 105, Radar—Present and Future,* London, Oct 23-25, 1973, pp. 408-413.

Josefsson, K *et al.,* "A Stripline Flat-Plate Antenna with Low Sidelobes," *IEEE Int. Symp. on Ant and Prop.,* 1974, p. 282-5.

Elliott, R.S., "Design of Line Source Antennas for Narrow Beamwidth and Asymmetric Low Sidelobes," *IEEE Trans.,* vol. AP-23, no. 1, Jan 1975, pp. 100-107.

Mayhan, J.T. and A.J. Simmons, "A Low Sidelobe K_A-Band Antenna-radome Study," *IEEE Trans.,* vol. AP-23, no. 4, July 1975, pp. 579-72c.

Nishida, S., and Wagatsuma, "Sidelobe-level Reduction of the Yagi-ude Array;" pt II: "Horizontally Polarized Antenna," *Rec. Electro. and Comm. Engr. Conver.*, Sazione Tohoku Univ, Japan, vol. 44, no. 3, Aug 1975, pp. 7-10 (in Japanese, trans. abstract SA 76-9706).(For Pt I see *ibid*, p. 1-6).

Gladman, B.R., "A Theoretical Study of Low Sidelobe Antenna Arrays," Ph.d. thesis, Univ. of London, September 1975.

Green, T.J., G.J. Halford, and D. Woller, "Antenna Requirements for Naval Radars," *Proc. Military Microwaves Conf.*, London, Oct 1975, p. 244. (late paper).

Goto, N., and D.K. Cheng, "Phase-shifter Thinning and Sidelobe Reduction for Large Phased Arrays," *IEEE Trans.*, vol. AP-24, no. 2, March 1976, pp. 139-143.

Hansen, R.C., "A One-parameter Circular Aperture Distribution with Narrow Beamwidth and Low Sidelobes," *IEEE Trans.*, vol. AP-24, no. 4, July 1976, pp. 477-480.

Kilgus, C.C., "A Technique for the Synthesis of Low-sidelobe Shaped Beam Radiation Patterns," *IEEE Int. Symp. AP-5*, Oct 15, 1976, pp. 472-5.

Mailloux, R.J., and A.C. Schell, "Antenna Sidelobe Reduction Using Spatial Filters," *DDRE ECCM Conf.*, 1976 (reprinted in this volume).

Fante, R.L., "Design of Parabolic Cylinder Reflector Systems with Low Sidelobes," *RADC TR 77-204*, June 1977; DDC AD A045619.

Fante, R.L., "Tolerance Errors for Very Low Sidelobe Antennas," *RADC TR-77-360*, October 1977 (reprinted in this volume). DDC AD A049259.

Fante, R.L., "Effect of Fresnel Zone Blockage on Very Low Sidelobe Antennas," *RADC TR-77-361*, Oct 1977 (reprinted in this volume). DDC AD A051188.

Fante, R.L., "Beam Deflection Caused by Antenna Phase Errors," *RADC TR-78-41*, Feb 1978, DDC A053209.

Fante, R.L., "On Reducing the Field Strength in the Fresnel Zone of an Antenna," *RADC TR-78-130*, May 1978, DDC AD A059225.

Cassel, E., and G. Lind, "Considerations on the Location of Antennas Aboard Warships," *Proc. Int. Nav. Technol. Expo 78, (Interavia)*, June 6-8, 1978, Rotterdam, the Netherlands, pp. 88-97.

Rudge, A.W., and P.R. Foster, "Low Sidelobe Antennas," *Proc. Military Microwaves Conference*, London. October 25-27, 1978, p. 245, abstract only, (reprinted in this volume).

BISTATIC RADAR

Harris, D.B., "A Forward Transmission Echo-ranging System," *Proc. IRE*, vol. 37, no. 7, July 1949, pp. 767-770.

Siegel, K.M., "Bistatic Radars and Forward Scattering," *NAECON Rec.*, 1958, pp. 286-290.

Easton, R.L., and J.J. Fleming, "The Navy Space Surveillance System," *Proc. IRE*, vol. 48, no. 4, Apr 1960, pp. 663-9.

Skolnik, M.I., "Analysis of Bistatic Radar," *IRE Trans.*, vol. ANE-8, no. 1, March 1961, pp. 19-27.

Peters, L., "Passive Bistatic Radar Enhancement Devices," *Proc. IEE*, vol. 109-C, no. 15, monogr. 455E, March 1962, pp. 1-10.

Matsua, M., "Effective Coverage of Bistatic Radar," *J. Inst. Commun. Engr.*, Japan, vol. 49, no. 4, April 1966, pp. 771-772, (in Japanese, trans. abstract SA 67-7510).

Davis, D.E.N., "Use of Bistatic Radar Techniques to Improve Resolution in the Vertical Plane," *IEE Electron. Lett.*, vol. 4, no. 9, May 3, 1968, pp. 170-171.

Yamane, K., Y. Yuba, and M. Matsuo, "Unified Analysis of Effective Coverage of Scatter Communication and Bistatic Radar Systems," *Electron. and Commun. in Japan*, vol. 53, no. 10, October 1970, pp. 105-112.

Hoisington, D.B., and C.E. Carroll, "Improved Sweep Waveform Generator for Bistatic Radar," *US Naval Postgraduate School Report, NPS-52, Hs 72081*, August 1975.

Milne, K., "Trends in Array Signal Processing," *IEE Conf. Pub.*, no. 144, 20-24 Sept 1976, pp. 121-134.

Lorti, D.C., and M. Balser, "Simulated Performance of a Tactical Bistatic Radar," *EASCON-77 Proc.*, Sept 1977, pp. 4-4A thru N.

Milne, K., "Principles and Concepts of Multistatic Surveillance Radars," *IEE Conf. Pub. 155*, Radar-77, Oct 1977, pp. 46-52.

Ewing, E.F., "The Applicability of Bistatic Radar to Short Range Surveillance," *IEE Conf. Pub. 155*, Radar-77, Oct 1977, pp. 53-58.

Mrstik, A.V., "Multistatic-radar Binomial Detection," *IEEE Trans.*, vol. AES-14, no. 1, Jan 1978, pp. 103-8.

Fawcette, J., "Bistatic Radar May Find a Sanctuary in Space," *Electronic Warfare/Defense Electronics Magazine*, vol. 10, no. 1, Jan 1978, p. 84ff.

Ewing, E.F., and L.W. Dicken, "Some Applications of Bistatic and Multistatic Radars," *Proc. Int. Conf. on Radar*, Paris, France, Dec 4-8, 1978, pp. 222-31.

CLUTTER FENCES

Preikschat, F.K., "Screening Fences for Ground Reflection Reduction," *Microwave J.*, vol. 7, no. 8, Aug 1964, pp. 46-50 (reprinted in Barton, D.K., ed., *Radars, Vol. 4,* Artech House, Dedham, MA, 1975.)

Becker, J.E., and J.C. Sureau, "Control of Radar Site Environment by Use of Fences," *IEEE Trans.,* vol. AP-14, no. 6, Nov 1966, pp. 768-773 (reprinted in Barton, D.K., ed., *Radars, Vol. 5,* Artech House, Dedham, MA, 1975.)

Albersheim, W.J., "Elevation Tracking Through Clutter Fences," *IEEE Trans.,* vol. AES-3, no. 6, Nov 1967, pp. 366-373 (Sup) (reprinted in Barton, D.K. ed., *Radars, Vol. 4,* Artech House, Dedham, MA, 1975.

Boithias, L., "Screening Effect of the Terrain for Various Propagation Methods," *AGARD Conf. Proc. No. 127: Propagation Effects on Frequency Sharing,* 7-11 May 1973, pp. S3/ - 1 thru 9 (in French, trans. abstract SA 74-3194).

MONOPULSE

1. Books and General Bibliography

Barton, D.K., ed., *Radars, Vol. I: Monopulse Radar,* Artech House, Dedham, MA, 1974.

Barton, D.K., "International Cumulative Index on Radar Systems," *IEEE Pub. Cat. No. JH 4675-5,* 1978: extensive bibliography on monopulse, pp. 35-37. See also measurement, pp. 32-5, and tracking radar, pp. 71, 2.

2. Journal Articles

Franklin, S.B., C.L. Hilbers, W.E. Kosydar, "A Wideband Two Channel Monopulse Technique," *IEEE 9th Mil. Electron. Conf. Rec.,* 1965, pp. 174-176.

Vakin, S.A., B.Kh. Krivitskii, and L.N. Shustov, "Direction Finding Characteristics of Single Pulse Automatic Tracking Systems," *Soviet Radio Engineering,* vol. 8, no. 5, Sept-Oct 1965, pp. 408-17.

Chadwick. G.G., and J.P. Shelton, "Two-channel Monopulse Techniques—Theory and Practice," *IEEE 9th Mil. Electron. Conf. Rec.,* 1965, pp. 177-181.

Riggs, R.F., "The Angular Accuracy of Monopulse Radar in the Presence of Clutter," *Proc. Nat. Electron. Conf.,* vol. 22, 1966, pp. 237-242.

Wojtowicz, H., "Monopulse Radar," *Problemy Elektron Telekomun Metody Wspolczesnej,* no. 11, 1966, pp. 128-142, (in Polish, trans. abstract SA 68-6571).

Noblit, R.S., "Reliability Without Redundancy from a Radar Monopulse Receiver," *Microwaves Magazine,* vol. 6, no. 12, Dec 1967, pp. 57-60.

Elia, A.D., and R.F. Schmidt, "Monopulse System with an Electronic Scanner," *US Patent No. 3,438,044,* 8 April 1969 (abstract SA 70-4672).

Yurchenko, Yu. S., "Effect of Fluctuation Noise on Monopulse Tracking Radar," *Izv VUZ Radioelektron,* vol. 14, no. 8, Aug 1971, pp. 894-901 (in Russian, trans. abstract SA 71-37998, JPRS 55283).

Yurchenko, Yu. S., "The Effect of Extraneous Fluctuations Noise on Monopulse Radar," *Izv VUZ Radioelecktron,* vol. 15, no. 4, Apr 1972, pp. 487-91 (in Russian, trans. abstract SA 72-36868).

Cooper, D.C., and B.A. Wyndham, "A Comparison of Monopulse Techniques for Direction Finding in Interference," *IEE Conf. Pub. 105, Radar Present and Future,* London, 23-25 Oct 1972, pp. 154-159.

Rajeshwari, B.V., and O.P.N. Calla, "A Brief Survey of Monopulse Techniques," *Inst. Telecomm. Engr. J.,* vol. 18, no. 2, Dec 1972, pp. 576-581.

Hollandse Signaalapparaten, "Monopulse Radar," *UK Patent No. 1300105,* Dec. 20, 1972, (abstract SA 73-21192).

Scholtz, R.A., "On the Passive Resolution of Broadcasting Targets with a Monopulse Antenna," *IEEE Trans.,* vol. AES-9, no. 5, Sept 1973, pp. 643-648.

Brunner, A., and E. Kress, "Monopulse Antennas with Doubly Curved Reflectors," *Proc. 4th European Microwaves Conf.,* Montreux, Switzerland, Sept 1974, pp. 357-61 (includes treatment of cross polizarization).

Gonzalez, D.G., and G.E. Pollon, "Cross-polizaration Limitation on Accuracy of Monopulse Reflector Antennas," *Proc. Int. Conf. on Radar,* Paris, 4-8 Dec 1978, pp. 333-6.

Low Sidelobe Radar Antennas
A.W. Rudge and P.R. Foster
Proceedings of the Military Microwaves Conference, London, October 1978

4.1

ABSTRACT

The paper briefly reviews the state-of-the-art in high-gain low sidelobe antennas suitable for radar applications. The relationship between electromagnetic aperture theory and the simplified scalar analyses based upon Fourier transform techniques is shown. The paper proceeds to discuss the relative merits of reflector, lens and array antennas in these applications. The point is made that very low sidelobe performance is not only the prerogative of antenna arrays but can also be achieved with offset reflector systems.

INTRODUCTION

The desirability of low sidelobe levels, as a protection against jamming in surveillance and tracking radars, is well known. For example, Dax (1) has expressed the opinion that for long range radars to avoid jamming, the level of all antenna sidelobes must be suppressed to better than -40dB and desirably to better than -60dB. Certainly there are many applications in which the peak sidelobes in the vicinity of the main beam are required to be at least -30dB down, with a rapid decay to below the -50dB at wide angles. Unless it is unavoidable, it is generally not advisable to specify the sidelobe requirements of an antenna by a single level for all directions in space (2). The price which must be paid for the specified sidelobe performance, in terms of the reduction in antenna efficiency, is a maximum in this case. For the purposes of this paper the -30dB to -50dB template referred to above will be referred to as a low side-lobe characteristic. Antennas having a sidelobe performance significantly better than this will be termed 'very low sidelobe antennas'. The Dax requirements certainly fall into this latter category.

* The authors are with the RF Technology Centre, ERA, Cleeve Road, Leatherhead, Surrey.

The comments made here will be generally concerned with high gain radar antennas, although for reflector-antenna variants the performance of the low gain primary-feed is also of interest. To simplify the discussion, we can crudely divide the angular space around the antenna into three regions:

- the near-in region, defining a cone of angles about the antenna boresight
- the back-lobe region defining a similar cone, about the direction 180 degrees from the boresight
- the far-out or wide-angle region defining all angles between the two cones.

BASIC THEORY

In the near-in region the antenna sidelobe performance is dominated by the diffraction characteristics of the radiating aperture. For any aperture-type antenna, the sidelobe performance in all regions can be accurately predicted using the well established vector-potential methods, providing the tangential electric fields in the aperture plane are known. For unblocked apertures, and to a lesser extent for blocked apertures, even a comparatively crude approximation for the tangential electric field distributions can provide quite good predictions for the near-in sidelobes. The normalised co-polar (E_{pn}) and cross-polar (E_{qn}) far-fields radiated from an aperture can be expressed exactly in terms of a conventional spherical co-ordinate system (r, ψ, Φ) as (3). (See Fig 1).

$$\begin{bmatrix} E_{pn}(\psi,\Phi) \\ E_{qn}(\psi,\Phi) \end{bmatrix} = \frac{1+\cos\psi}{2} \begin{bmatrix} 1-t^2\cos 2\Phi & t^2\sin 2\Phi \\ t^2\sin 2\Phi & 1+t^2\cos 2\Phi \end{bmatrix} \begin{bmatrix} F_p(\psi,\Phi) \\ F_q(\psi,\Phi) \end{bmatrix} \tag{1}$$

where F_p and F_q are the spatial Fourier transforms of the co-polar and cross-polar tangential electric-field distribution ($\overline{E}_a$) in the projected aperture-plane of the antenna, and $t = \tan\psi/2$. In equa.1 the polarisation of the radiated fields is defined in terms of Ludwig's third definition (4) which corresponds to the field components measured using the conventional antenna-range techniques described in Silver's text (5). Assuming a principal polarisation aligned with the y-axis of the antenna aperture, the Fourier transforms can be expressed as

$$F_p(\psi,\Phi) = FT\left\langle \overline{E}_a . \overline{a}_y \right\rangle \tag{2a}$$

$$F_q(\psi,\Phi) = FT\left\langle \overline{E}_a . \overline{a}_x \right\rangle \tag{2b}$$

where FT denotes the transform operation and $\overline{a}_x$, $\overline{a}_y$, are the unit vectors in the x- and y- directions respectively.

From equa. 1 it can be seen that when ψ is small the relationships reduce to

$$E_{pn}(\psi,\Phi) \simeq F_p(\psi,\Phi) \tag{3a}$$

$$E_{qn}(\psi,\Phi) \simeq F_q(\psi,\Phi) \tag{3b}$$

Hence the co-polar and cross-polar near-in sidelobes are given by the scalar Fourier transforms of the x- and y- directed components of the antenna aperture-plane field distributions.

APERTURE DISTRIBUTIONS AND SIDELOBE LEVELS

For unblocked apertures the extensive literature on scalar aperture distributions and Fourier transforms can be employed to good effect to examine the principal parameters governing the near-in sidelobe performance (2,6,7). For example, Table 1 illustrates the relationships between peak sidelobe levels, antenna half-power beamwidth and aperture efficiency for a circular aperture of radius 'a' with an aperture field distribution function derived by Hansen (7). The table also shows the level of the field distribution at the edge of the aperture relative to that at the centre (edge taper). It must be stressed, however, that the resulting sidelobe levels are dependent upon the overall 'shape' of the aperture distribution function and not merely by the value of the field at the edge of the reflector. The Hansen distribution is only one of many which have been reported in the literature, but the table serves to illustrate the fundamental fact that there is a price to be paid for low sidelobe performance in terms of lower efficiency and broader beamwidths for any radiating aperture of a given size.

Clearly the sidelobe performance resulting from an idealised aperture distribution may be some way from the case of a practical antenna. However, the Fourier transform relationship does provide a convenient basis, upon which studies of the fundamental characteristics can be performed. For example, amplitude and phase errors can be introduced into the distribution functions to examine the sensitivity of a given distribution to tolerances. Sidelobe performance is particularly sensitive to phase-errors in the aperture distribution and the literature in this area is extensive(2). To produce significant degradation in the near-in region, periodic phase error distributions must have relatively low spatial periods. For randomly distributed errors large correlation intervals are necessary to affect the near-in region. To maintain the wide-angle sidelobes to within 3dB of the 'zero error' level, Dragone and Hogg (8) have indicated that aperture tolerances must be maintained to within the order of $\lambda/100$ peak.

Although idealised aperture distribution functions can be employed to investigate many of the fundamental factors governing antenna sidelobe performance, the results must be interpreted with some caution in attempting to predict the performance, or optimise the design of practical low-sidelobe antennas. For low sidelobe reflector antennas, the actual field distribution achieved in the projected-aperture plane of the reflector is a function of both the primary-feed characteristics and the reflector geometry. In view of the sensitivity of the sidelobe performance to the aperture illumination characteristics there is little point in assuming a particular aperture illumination function which may not be realizable by a practical feed and reflector configuration. In addition the sidelobe performance of any reflector antenna will be strongly influenced by primary-feed spillover and aperture-blockage effects, and for axisymmetric antenna systems this will be a dominant factor in defining the sidelobe radiation characteristics.

For offset reflector systems or lenses the blockage effects can be avoided but the aperture illumination taper still cannot be independently specified without due consideration of the primary-feed and reflector, or lens characteristics. This is not to infer, however, that offset reflector or lens systems cannot produce low sidelobe characteristics.

For antenna arrays the required aperture illumination distribution can be specified more freely and in this case the scalar theory can be applied more directly. However, the effects of mutual coupling, tolerance errors and quantisation of the continuous aperture distribution must all be accounted for ultimately.

ARRAY ANTENNAS

In principal it might be thought that an array antenna is the obvious choice when low sidelobe performance is sought. The aperture need not suffer from blockage and the freedom to adjust the excitation coefficients of the array elements should permit desired illumination characteristics to be achieved. Before jumping to this conclusion too readily, however, it is worthwhile bearing in mind some of the practicalities involved.

For small arrays (ie linear dimensions of less than 20 wavelengths) the quantisation effects introduce a degradation in the anticipated sidelobe performance for a given taper distribution. Josefsson (9) reported that for a circular array of 52 elements (approximately 7 x 7) a Chebyshev distribution designed for -35dB sidelobes, only realised levels of the order of -30dB. A theoretical distribution for -40dB sidelobes was found necessary to achieve levels of -35dB. The consequences of this need for increased tapering are a further reduction in aperture efficiency and the requirement for a larger array to provide a specified beamwidth. For comparison Louapre and Fujioka (10) have described a 49 element <u>linear</u> array which achieved the design goal of -35dB sidelobe performance.

For larger arrays illumination tapering can also be achieved by use of an element density taper, in which the number of uniformly excited radiating elements are decreased toward the edges of the array. Density tapering must be random to avoid additional grating-lobe problems. Although the beamwidth of such an array is essentially equal to that of a filled illumination-tapered array, the wide-angle sidelobe performance is significantly degraded. A computed example for a 50 wavelength aperture indicated wide-angle sidelobes of -32dB, or approximately 25dB worse than the equivalent illumination-tapered filled array (2).

Although spillover and backlobe problems are absent from most array configurations, grating lobes which can occur in both the near-in and wide-angle regions must be avoided for low sidelobe applications. This requirement governs the maximum inter-element spacing which can be used in the array design. If the array is to be electronically scanned then this maximum spacing is further reduced. If the maximum scan-angle is not greater than $\pm 45°$ an inter-element spacing of the order of 0.6λ will be required. Since array thinning techniques degrade the sidelobe performance, something like 15,000 elements could be required to produce a 1 degree pencil beam with a -35dB sidelobe specification.

Mutual coupling effects between the closely spaced array elements can significantly modify the illumination taper and this must be allowed for by use of numerical techniques during the design of the array(11, 12). Edge effects can be reduced by the use of dummy elements at the array edge.

Problems can also arise due to 'reverse coupling', in which power is reflected within the feed element and transferred via the feed network to other radiating elements. This can result in significant sidelobe degradation which is strongly frequency dependent. The use of directional couplers in the feed network eliminates this problem but the implications can be significant in terms of feed hardware for a large array.

The avoidance of phase and amplitude errors through the feed network of a large array are a major difficulty if low-sidelobe performance is to be avoided. This can only be achieved by maintaining very accurate tolerances throughout the feed network and clearly this is an exacting and costly task for a large array. As a consequence of the illumination weighting sidelobe performance is likely to be more sensitive to the phase of the centrally located array elements. Attenuation losses through the feed network are also likely to be significant, values of between 1 and 3dB being common in the literature. The efficiencies quoted for array antennas are often aperture-efficiency only. When the feed network losses are included overall efficiencies of the order of 20-30% are not uncommon in the published data.

In summary array techniques can be employed to provide low sidelobe performance but the practical problems and cost implications are significant. From published data near-in sidelobe levels of the order of -25dB are readily and frequently achieved, while levels of -40dB are rare. It is certain that low sidelobe and very low sidelobe performance can be realised with these antennas but for high gain applications the present state-of-the-art suggests that the cost will be high and that very low sidelobe performance will only be maintained over relatively narrow bandwidths.

AXISYMMETRIC REFLECTOR AND LENS ANTENNAS

Of the various high-gain antenna systems which can be employed for radar systems, the parabolic reflector antenna and its variants have been clearly the most popular in the past. The preference for the parabolic reflector antenna has been soundly based on its performance and cost-effective characteristics. However, in the light of the growing jamming threat, the question of the performance limitations of these systems demands investigation.

Axisymmetric reflector antennas, of either the prime-focus or Cassegrain varieties, suffer from a major drawback with regard to sidelobe performance. The location of the feed system, or sub-reflector, with their supporting struts in the aperture of the antenna, results in blockage and scattering effects which seriously degrade the sidelobe performance. Near-in sidelobe levels of these antennas are typically in the range -15 to -30dB and, as a consequence of the blockage, the fall-off in sidelobe level is often comparatively slow. Increasing the taper of the aperture illumination produces little benefit, since blockage effects tend to dominate at the -30dB level and overall near-in sidelobe performance is rarely better than -28dB. Wide-angle radiation from axisymmetric reflector antennas can be reduced by the use of hoods or 'blinders' about the reflector periphery. Fig 2 indicates the order of improvement which can be realised with these techniques (13).

The backlobe radiation from a reflector antenna is very dependent upon the edge illumination of the reflector. The radiation level can also be strongly influenced by the use of hoods or blinders around the reflector rim. Knop (14) has derived the following approximate formula for the front-to-back ratio (R) of an axisymmetric reflector without hoods or blinders.

$$R = G + T + K - G_F \quad dB$$

where all parameters are in decibels and G is the reflector gain, T is the edge taper, K is a factor dependent on F/D and G_F is the feed gain. K has a value of approximately 3dB for an F/d of 0.25, decreasing to 1dB at F/D = 0.5 and less than 0.5dB for F/D ratios greater than 1.

Blockage effects on the near-in sidelobes can be reduced by twist-reflector Cassegrain techniques (15) at the price of constraining the antenna to a single linear polarisation. Many of the future high-performance radar systems are likely to require a full dual-polarised capability and thus the twist-reflector approach provides only a limited solution.

Lens techniques, which offer freedom from aperture blockage are generally not well-suited to large radar antenna applications. When compared to a large reflector antenna, the equivalent lens tends to be massive and to suffer from structural limitations. For physically small systems however, particularly at the high microwave or millimetre-wave frequencies, the lens antenna could find applications as a low sidelobe device. Yamada and Takano (16) have described a shaped dielectric lens antenna operating at 19.5GHz with excellent wide-angle sidelobe performance. After extracting what they identify as range-effects, the authors claim a wide-angle performance close to the -70dB level (-30dBi). The near-in sidelobes of this antenna are relatively high, however, at -20dB.

OFFSET REFLECTOR ANTENNAS

The offset parabolic reflector is a somewhat more complicated structure to deal with both structurally and analytically. Nevertheless its potential advantages with regard to low sidelobe performance are significant. The single-reflector offset geometry is illustrated in Fig 3. Since the configuration avoids aperture-blocking near-in sidelobe levels are dictated by reflector diffraction effects. Dragone and Hogg (17) have computed near-in sidelobe performance improvements of the order of 10dB over equivalent (blocked) axisymmetric reflectors. Rudge (3) has shown predicted and measured data for offset antennas with aperture efficiencies approaching 70% and peak sidelobes below -30dB. One example of a number of these designs with sidelobe peaks in the range -32 to -38dB is shown in Fig 4.

These results indicate that a good efficiency-sidelobe compromise is realisable with offset systems. Computer-aided design methods, based upon Physical-optics techniques for the reflector and combined with accurate vector-radiation models for the primary-feed, are now developed to the point where fine optimisation of offset reflector antennas on the computer is both practical and realistic (18). Employing these techniques performance in the near-in region can be predicted to good accuracy for low sidelobe applications.

The reflector-diffraction contribution to the far-field sidelobes decays rapidly with angle from the boresight and the wide angle radiation can be largely attributed to primary-feed spillover. In low sidelobe designs this can be dealt with by a combination of primary-feed design and hooding techniques similar to those employed for axisymmetric reflector antennas.

The single offset reflector does have a major drawback when used as a radar antenna. The asymmetrical geometry of the reflector results in a low level depolarisation of the signal received or transmitted by the antenna. This does not constitute a significant power loss and in a linearly-polarised 'sum' mode of operation the cross-polar lobes are essentially contained within the co-polar envelope of the main beam. In a monopulse tracking-mode, however, the co-polar radiated field has a minimum on the antenna boresight but the cross-polar radiation has a peak value at this point. The tracking-mode radiation-patterns are illustrated in Fig 5. The polarisation of radar returns from the target are effectively random and thus the relative phase and magnitude of the co-polar and cross-polar signals incident upon the receiving antenna aperture will be time-varying. When these two components are summed in the tracking channel the angular position of the antenna boresight, as defined by the null in the difference-mode radiation-pattern, becomes dependent upon the relative phases and magnitudes of the vector components in the radar return. The symptom of this condition is termed 'boresight-jitter' and an uncertainty exists regarding the true location of the target. The angular range of this uncertainty can be related to the relative magnitude of the offset reflector antenna cross-polarised field and it represents a major constraint upon the use of the single offset reflector in a precision tracking role.

Two solutions to this problem have been developed. The first is the use of a dual offset reflector configuration in which the asymmetric sub-reflector is adjusted such that it cancels the cross-polarisation introduced by the asymmetry of the main offset reflector. This technique is satisfactory providing the dual-reflector structural configuration is appropriate and that the sub-reflector dimensions are larger than about 20 wavelengths. Diffraction effects at small subreflectors tend to interfere with the cancellation process (18). The second, and more recent technique is based upon the design of a primary-feed antenna which essentially compensates for the reflector depolarisation. Several forms of this feed concept are under development and one has been recently reported in the literature (19). When optimised both the dual reflector and primary-feed methods essentially result in an unblocked aperture antenna with pure polarisation properties. These configurations have excellent potential both as low sidelobe antennas and as precision radar trackers.

CONCLUSIONS

Low sidelobe radar antennas are clearly feasible at the present state-of-the art. However, axisymmetric (blocked) reflector systems are generally unsuitable and at best marginal, with regard to the near-in sidelobe performance. Very low sidelobe performance is probably unattainable with these configurations. Lens antennas exhibit the important advantage of an unblocked aperture, but these devices are not well suited structurally for large radar antennas.

Array antennas can satisfy low sidelobe specifications with the aid of good design practice but the requirements on tolerances are severe. Large arrays can probably meet 'very low' sidelobe specifications but the tolerances are very severe and at the present state-of-the-art the bandwidth over which this performance can be maintained is likely to be very narrow. The cost of such an antenna is likely to be very high, particularly if any form of electronic scanning is implied. No examples of very low sidelobe performance, as defined here, have appeared in the literature to the author's knowledge.

Offset reflector antennas in either their single or dual-reflector configuration appear to offer excellent potential for an acceptable compromise in terms of performance and cost. The ultimate sidelobe performance which can be achieved with these antennas has yet to be determined, in theory at least there is no fundamental reason why offset reflector antennas cannot equal or even exceed the best available sidelobe performance of arrays.

REFERENCES

1. Dax P 'Noise jamming of long range search radars'. Microwaves 14, Sept 1975 p52.

2. Foster P R and Rudge A W 'Low sidelobe antenna study.' Final Report. Part 1 Literature Survey and Review. MoD Contract. Rept No: BR56799 ERA Leatherhead Oct 1977.

3. Rudge A W 'Multiple-beam antennas: offset reflectors with offset feeds.' IEEE Trans Ant & Prop AP-23 May 1975 p 317.

4. Ludwig A C 'The definition of cross polarisation'' IEEE Trans Ant and Prop AP-21 Jan 1973 pp116.

5. Silver S 'Microwave antenna theory and design' McGraw-Hill 1949.

6. Taylor T T 'Design of circular aperture for narrow beamwidth and low sidelobes' Trans IEEE on Ant and Prop AP-8 Jan 1960 p17.

7. Hansen R C 'A one-parameter circular aperture distribution with narrow beamwidth and low sidelobes' Trans IEEE Ant & Prop AP-24 July 1976 p477.

8. Dragone C and Hogg D C 'Wide-angle radiation due to rough phase fronts' BSTJ 42 Sept 1963 p2285.

9. Josefsson K et al 'A stripline flat-plate antenna with low sidelobes' IEEE Int. Symp on Ant and Prop. 1974 p282.

10. Louapre M E and Fujioka J K 'A high efficiency electrically-scanned phased array for space-borne radiometric applications' Proc IEEE 56 Nov 1968 p 2011

11. Alexander D K and Sichelstein B A 'A dual beam antenna - A unique waveguide phased array with independently steered beams' Phased Array Antennas Artech House, 1972. p357.

12. Gladman B R 'A theoretical study of low sidelobe antenna arrays' PhD Thesis Univ of London Sept 1975

13. Blacksmith P A 'Method for reducing far-out sidelobes' AFCRL Rept AFCRL-TR-57-115 July 1957

14. Knop C M 'On the front to back ratio of a parabolic dish antenna' Trans IEEE on Ant and Prop AP-24 Jan 1976 p109.

15. Dahlsjo O 'A low sidelobe Cassegrain antenna' IEE Conf. Proc. Radar Present and Future London 1973

16. Yamada Y and Takano T 'Dielectric lens antenna with shaped surfaces' Proc Int Symp on Ant and Prop (ISAP) Sendai Japan Aug 1978 p239.

17. Dragone C and Hogg D C 'The radiation pattern and impedance for offset and symmetrical near-field Cassegrainian and Gregorian antennas' IEEE Trans Ant and Prop AP-22 May 1974 p472.

18. Rudge A W and Adatia N A 'Offset parabolic reflector antennas: A review' Proc IEEE Dec 1978

19. Rudge A W and Adatia N A 'Primary feeds for boresight-jitter compensation of offset reflector radar antennas' Proc of IEEE Int Conf Radar 77 London Oct 1977 p409.

ACKNOWLEDGEMENT

This work has been carried out with the support of Procurement Executive, Ministry of Defence.

TABLE I: Hansen one-parameter circular distribution

1st sidelobe (dB)	Edge taper (dB)	Beamwidth $(\times 2a/\lambda)$	G/Go*
-17.57	0	1.0000	1.0000
-20.00	-4.49	1.0483	0.9786
-25.00	-12.35	1.1408	0.8711
-30.00	-19.29	1.2252	0.7595
-35.00	-25.78	1.3025	0.6683
-40.00	-31.98	1.3741	0.5964
-45.00	-38.00	1.4407	0.5390
-50.00	-43.89	1.5038	0.4923

* Go is the forward gain of a uniformly illuminated circular aperture.

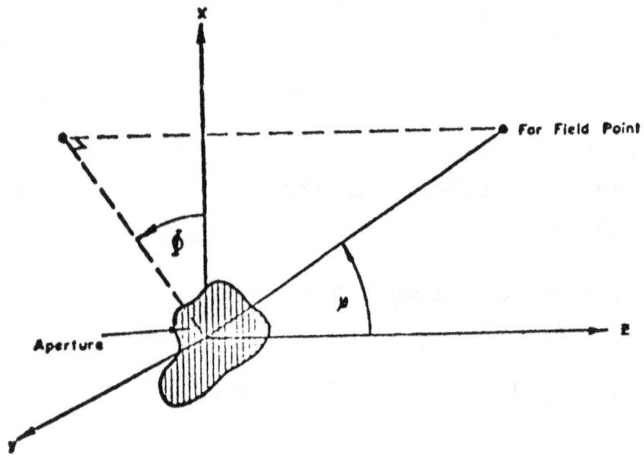

FIG 1 Aperture geometry

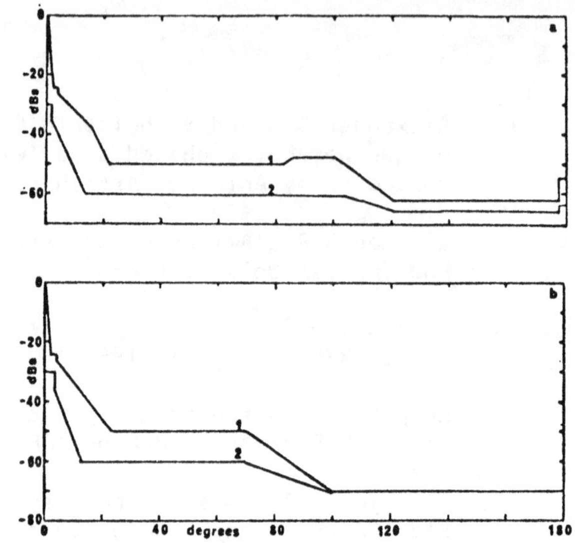

FIG 2 Radiation envelopes for front fed 12ft dia. paraboloid at 11GHz. 1 co-polar, 2 cross-polar (a) without blinders, (b) with blinders

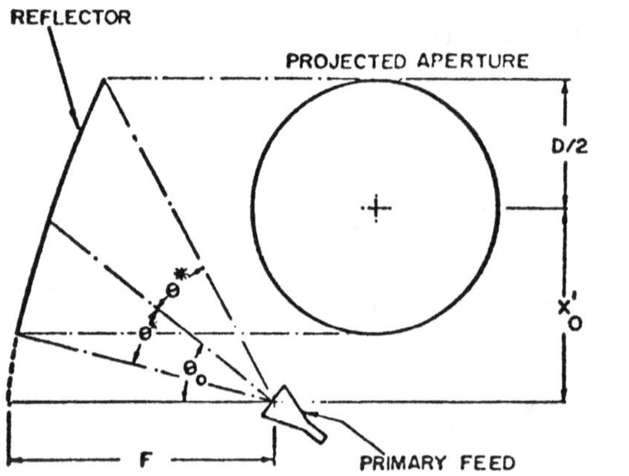

FIG 3 Offset reflector geometry

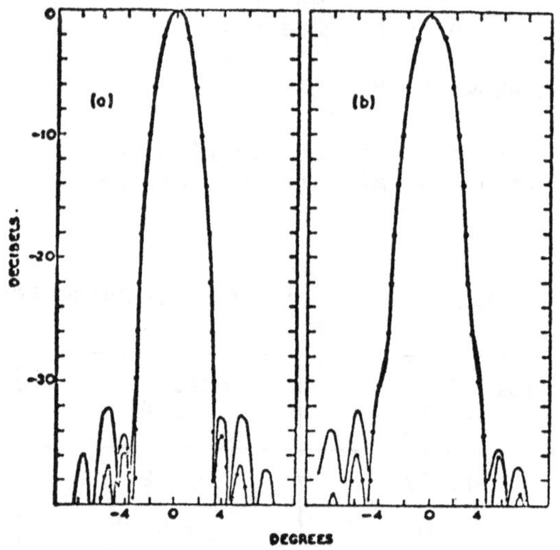

FIG 4 Radiation for offset antenna $F=22.5\lambda$ $\theta_o=44$ $\theta^*=30°$ fed by 2.8λ diam corrugated horn (a) $\Phi = 90°$ (b) $\Phi = 0°$

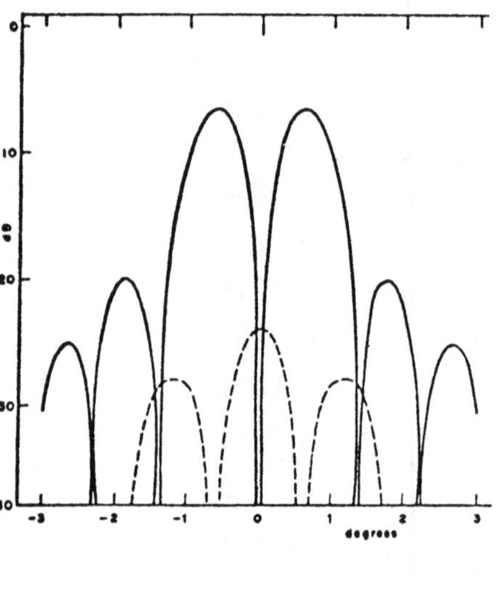

FIG 5 Offset antenna monopulse difference pattern
co-polar ————
cross-polar - - - -

Effect of Fresnel Zone Blockage on Very Low Sidelobe Antennas
R.L. Fante
RADC TR 77-361, October 1977, DDC AD A051188
Reprinted by permission.

4.2

1. INTRODUCTION

At present the United States Air Force is interested in the development of radars with -50 dB azimuthal sidelobes. In order to maintain such low sidelobes, unfortunately, it is necessary to find a radar site which is free from all main-beam blockage within quite a distance from the radar. In this paper, we will present an approximate method for estimating the effect of partial-beam blockage by an obstacle (such as a tree) located in the Fresnel zone of the antenna.

2. FORMAL THEORY

2.1 Field Scattered by the Obstacle

Consider an obstacle in the Fresnel zone of a radiating antenna, as shown in Figure 1. The far-electric field scattered by this obstacle can be written as[1]

(Received for publication 26 October 1977)

1. Morse, P., and Fishback, H. (1953) Methods of Theoretical Physics, vol. 2, McGraw-Hill, New York.

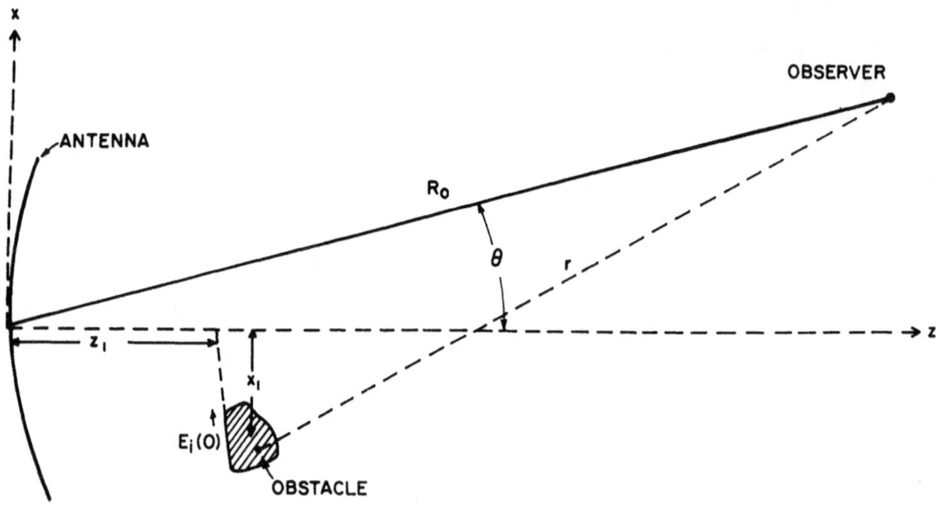

Figure 1. Obstacle in the Fresnel Zone of an Antenna

$$\underline{E}_s(\underline{x}) = \frac{k^2}{4\pi} \iiint_V (\epsilon - 1)\, \underset{\sim}{G}_o \cdot \underline{E}\, \frac{e^{-ikr}}{r} d^3 x'$$

$$= \frac{k^2 e^{-i\psi_o}}{4\pi R_o} \iiint_V (\epsilon - 1)\, \underset{\sim}{G}_o \cdot \underline{E}\, e^{ik(x' \sin\theta\cos\phi + y' \sin\theta \sin\phi + z' \cos\theta)}$$

$$dx'\, dy'\, dz' \quad , \qquad (1)$$

where

k = the signal wave number,
ϵ = the relative (complex) permittivity of the obstacle,
V = the volume of the obstacle,
$\underline{E}$ = the electric field inside the obstacle,
$\underset{\sim}{G}_o = \underset{\sim}{I} - \hat{v}\hat{v}$,
$\underset{\sim}{I}$ = the unit dyad,
$\hat{v} = \hat{x} \sin\theta \cos\phi + \hat{y} \sin\theta \sin\phi + \hat{z} \cos\theta$,
θ = the polar angle,
ϕ = the azimuthal angle,
$\hat{x}$, $\hat{y}$, $\hat{z}$ are unit vectors, and
$\psi_o = k(R_o - x_1 \sin\theta\cos\phi - y_1 \sin\theta\sin\phi - z_1 \cos\theta)$.

Let us now assume that the antenna is a planar, rectangular aperture lying in the z = o plane. We further assume that the antenna is linearly polarized, and excited with an electric field distribution $\hat{x}\, e_o(x, y)$. Then the Fresnel-zone field, near the z-axis, in the plane $z = z_1$ can be approximated as

$$E_i(z_1) = \hat{x}\, \alpha(x, y)\, e^{-ikz_1} \quad , \tag{2}$$

where $\alpha(x, y)$ is plotted[2] for various aperture-field distributions. For the case of a square aperture with a cosine distribution in the x-direction and a cosine squared taper in the y-direction, the values of the relative axial field $\alpha_o \equiv \alpha(0, 0)/e_o(0, 0)$ are given in Figure 2, assuming the linear aperture dimension is 15 ft and the signal frequency 3.35 GHz. We observe that there is no significant dropoff in the field strength until z_1 is approximately 200 ft; consequently, for $z_1 \lesssim 200$ ft we can use the approximation

$$\alpha(x, y) \simeq e_o(x, y) \quad . \tag{3}$$

Next we need to calculate the field inside the obstacle. For mathematical simplicity, we shall assume that the obstacle is a homogeneous rectangular structure oriented as shown in Figure 3. We also assume that the transverse dimensions of the obstacle (that is, its x and y dimensions) are considerably larger than the signal wavelength. In this case, we may approximate the field transmitted into the obstacle by that which would be transmitted into a planar slab, of infinite extent in the x-y directions. Upon using the aforementioned assumptions, along with Eq. (2), and ignoring the effect of the obstacle on the aperture distribution, we find it straightforward to show that the electric field inside the obstacle is given by

$$\underline{E} \simeq \hat{x}\, \alpha(x, y) \left[A\, e^{-ik_2 z'} + B\, e^{ik_2 z'} \right] e^{-ikz_1} \quad , \tag{4}$$

where

$$z = z_1 + z' \quad , \quad k_2 = k\,\epsilon^{1/2}$$

2. Air Force Handbook, Electromagnetic Radiation Hazards, Air Force Communications Service (E-1 Standard) TO 31Z-10-4, 1 Aug. 1966.

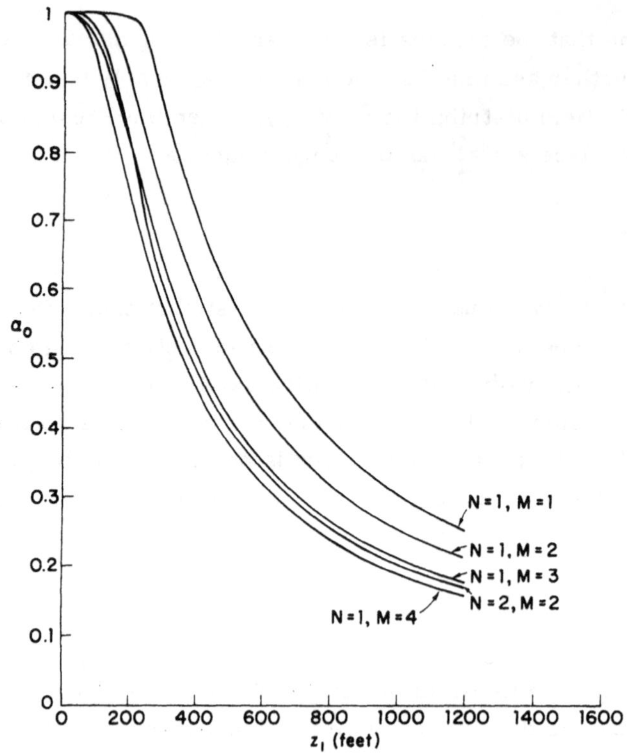

Figure 2. Values of α_o for a Square Aperture With $2L_1 = 2L_2 = 15$ ft and a taper $\cos^N(\pi x/2L_1)$ $\cos^M(\pi y/2L_1)$. The frequency is 3.35 GHz

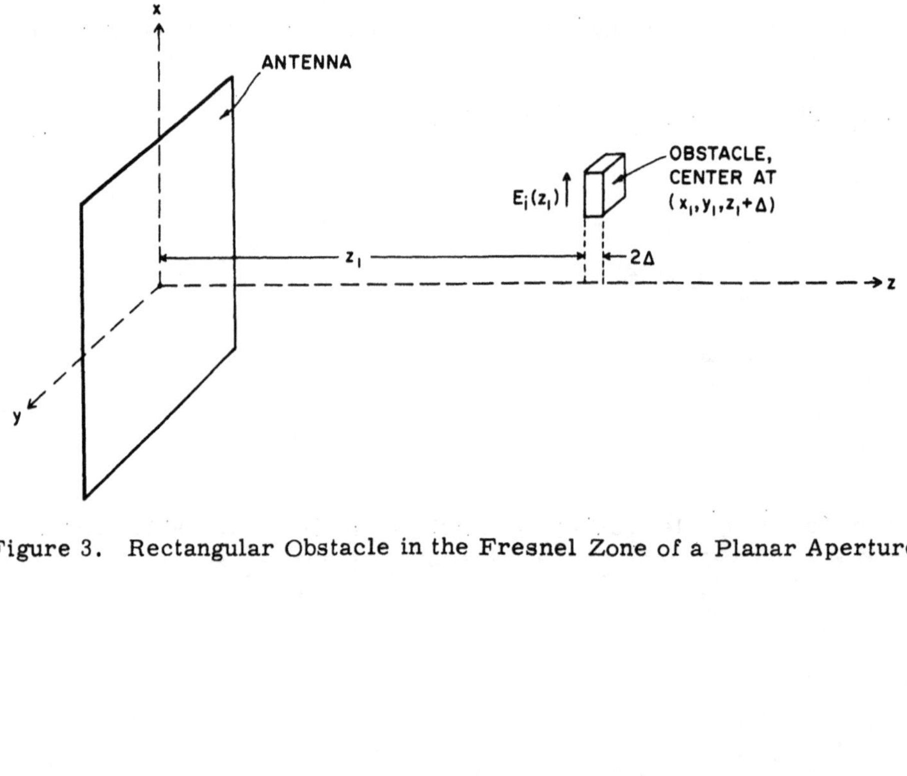

Figure 3. Rectangular Obstacle in the Fresnel Zone of a Planar Aperture

and

$$A = \frac{2(\epsilon^{1/2} + 1)\exp(i\,4\,k_2\,\Delta)}{(\epsilon^{1/2} + 1)^2 \exp(i\,4\,k_2\,\Delta) - (\epsilon^{1/2} - 1)^2} \quad , \tag{5}$$

$$B = \frac{2(\epsilon^{1/2} - 1)}{(\epsilon^{1/2} + 1)^2 \exp(i\,4\,k_2\,\Delta) - (\epsilon^{1/2} - 1)^2} \quad . \tag{6}$$

If we recall that $x = x_1 + x'$ and $y = y_1 + y'$ and then substitute Eq. (4) into Eq. (1), we obtain

$$\underline{E}_s = \frac{k^2 F e^{-i\psi_1}}{4\pi R_o} (\underset{\sim}{G}_o \cdot \hat{x}) \iint_{S_o} dx'\, dy'\, \alpha(x_1 + x', y_1 + y')$$

$$e^{i k \sin\theta (x' \cos\phi + y' \sin\phi)} \quad , \tag{7}$$

where S_o = transverse surface area of obstacle,

$$\psi_1 = k(R_o - x_1 \sin\theta \cos\phi - y_1 \sin\theta \sin\phi + 2z_1 \sin^2 \frac{\theta}{2}) \quad ,$$

$$F \equiv (\epsilon - 1) \int_0^{2\Delta} dz'\, e^{i k z' \cos\theta} \left[A\, e^{-i k_2 z'} + B\, e^{i k_2 z'} \right] \tag{8a}$$

$$\simeq \frac{-2\, e^{i\gamma}}{i k} \left[2\, i \sin\gamma - \frac{b(1 - e^{i\beta})}{a - b\, e^{i\beta}} e^{i\gamma} \right] \quad , \tag{8b}$$

and

$$\gamma = k\,\Delta(\epsilon^{1/2} + 1) \quad ,$$

$$\beta = 4\,k\,\Delta\,\epsilon^{1/2} \quad ,$$

$$a = (\epsilon^{1/2} - 1)^2 \quad ,$$

$$b = (\epsilon^{1/2} + 1)^2 \quad .$$

Note that Eq. (8a) is valid for all values of θ, but Eq. (8b) is specialized to the case when θ is small, so that $\cos\theta \simeq 1$.

It is interesting to observe from Eq. (8) that when $|\epsilon| \to \infty$, $F \to (2i/k)$. Consequently, when $|\epsilon| = |\epsilon_r - i\sigma/\omega\epsilon_o|$ is very large, Eq. (7) reduces to

$$\underline{E}_s \to \frac{ik\, e^{-i\psi_1}}{2\pi R_o} (\underset{\sim}{G}_o \cdot \hat{x}) \iint_{S_o} dx'\, dy'\, \alpha(x_1 + x', y_1 + y')$$

$$e^{ik \sin\theta (x' \cos\phi + y' \sin\phi)} \quad . \tag{9}$$

Finally, if the obstacle is within 200 ft of the antenna, we may use the approximation in Eq. (3) to get

$$E_s \to \frac{ik\, e^{-i\psi_1}}{2\pi R_o} (\underset{\sim}{G}_o \cdot \hat{x}) \iint_{S_o} dx'\, dy'\, e_o(x_1 + x', y_1 + y')$$

$$e^{ik \sin\theta (x' \cos\phi + y' \sin\phi)} \quad . \tag{10}$$

Equation (10) is the standard result for the effect of near-field blockage by a perfectly conducting obstacle. In the opposite limit when ϵ is near unity, we find $F \simeq 2\Delta(\epsilon - 1)$, provided $k\Delta|\epsilon - 1| \ll 1$.

2.2 Direct Field of the Antenna

The far field of the antenna in the absence of any blockage can also be written in the same form as Eq. (7). From Silver[3] we have

$$\underline{E}_D = \frac{ik\, e^{-ikR_o} (\underset{\sim}{G}_1 \cdot \hat{x})}{4\pi R_o} \iint_{S_A} dx'\, dy'\, e_o(x', y')\, e^{ik \sin\theta (x' \cos\phi + y' \sin\phi)} \quad , \tag{11}$$

where $\underset{\sim}{G}_1 = (1 + \hat{z} \cdot \hat{v})\underset{\sim}{I} - \hat{v}\hat{v} - \hat{z}\hat{v}$, and S_A is the area of the antenna aperture. For values of $\theta < 15^o$, we can approximate $\underset{\sim}{G}_1 \cdot \hat{x}$ by $2\hat{x} - 2\hat{z} \sin\theta \cos\phi$.

---

3. Silver, S. (1965) Microwave Antenna Theory and Design, Dover, New York, p. 162.

2.3 Total Radiated Field

The total field radiated when an obstacle is located in the Fresnel zone of an antenna is obtained by combining Eqs. (7) and (11), provided it is acceptable to assume that the presence of the obstacle does not alter the aperture field distribution. Let us, therefore, combine Eqs. (7) and (11) and assume that $\theta < 15^{\circ}$, so that $\underset{\sim}{G}_1 \cdot \hat{x} \simeq 2\hat{x} - 2\hat{z} \sin\theta \cos\phi$ and $\underset{\sim}{G}_o \cdot \hat{x} \simeq \hat{x} - \hat{z} \sin\theta \cos\phi$. We obtain

$$\underline{E}_T = \frac{ike^{-ikR_o}}{2\pi R_o} \hat{s} \left[\int_{-L_1}^{L_1} dx' \int_{-L_2}^{L_2} dy' \, e_o(x', y') \, e^{ik \sin\theta (x' \cos\phi + y' \sin\phi)} \right.$$

$$- \frac{ikF}{2} e^{i\psi_2} \int_{-X_o}^{X_o} dx' \int_{-Y_o}^{Y_o} dy' \, \alpha(x_1 + x', y_1 + y')$$

$$\left. e^{ik \sin\theta (x' \cos\phi + y' \sin\phi)} \right] , \qquad (12)$$

where $\hat{s} = \hat{x} - \hat{z} \sin\theta \cos\phi$, $\psi_2 = k(x_1 \sin\theta \cos\phi + y_1 \sin\theta \sin\phi - 2z_1 \sin^2 \frac{\theta}{2})$, the aperture antenna has dimensions $2L_1$ and $2L_2$ in the x and y directions, respectively, and the obstacle has transverse dimensions $2X_o$ and $2Y_o$. Note that when $z_1 \lesssim 200$ ft, we can replace $\alpha(x_1 + x', y_1 + y')$ by $e_o(x_1 + x', y_1 + y')$.

In studying Eq. (12) further, we will assume that $z_1 > 200$ ft, because it is unlikely that a radar would be placed in a site with obstacles closer than that distance. We shall also assume that the obstacle is the principal source of the sidelobes. When we make this latter assumption, it is easy to see from Eq. (12) that the relative (that is, relative to the axial power density) sidelobe power density S is as follows:

$$S = \frac{\frac{k^2|F|^2}{4} \left| \int_{-X_o}^{X_o} dx' \int_{-Y_o}^{Y_o} dy' \, \alpha(x_1 + x', y_1 + y') \, e^{ik \sin\theta (x' \cos\phi + y' \sin\phi)} \right|^2}{\left| \int_{-L_1}^{L_1} dx' \int_{-L_2}^{L_2} dy' \, e_o(x', y') \right|^2} \qquad (13)$$

For the case of a square aperture with $e_o = \cos(\pi x/2L_1)\cos^2(\pi y/2L_1)$, we can rewrite Eq. (13) as

$$S = \left(\frac{\pi k}{8L_1^2}\right)^2 |F|^2 \left| \int_{-X_o}^{X_o} dx' \int_{-Y_o}^{Y_o} dy' \, \alpha(x_1 + x', y_1 + y') \, e^{ik \sin\theta(x' \cos\phi + y' \sin\phi)} \right|^2 \tag{14}$$

In much of the Fresnel zone ($0.3 \lesssim p \leq 2$, where $p \equiv \lambda z_1/4L_1^2$) we have found that $\alpha(x, y)$ can be approximated by the analytic form

$$\alpha(x, y) = \alpha_o \exp\left\{- m_1 \left(\frac{x}{L_1}\right)^2 - m_2 \left(\frac{y}{L_1}\right)^2\right\} , \tag{15}$$

where the values of m_1 and m_2 depend on the aperture taper and p. Values of m for different aperture field tapers are presented in Table 1. Values of α_o are presented in Figure 2.

Table 1. Values of m for Different Aperture Field Tapers (Values of m versus $p = \lambda z_1/rL_1^2$, where $2L_1$ = aperture dimension and λ = wavelength

Cosine Taper		Cosine-squared Taper		Cosine-cubed Taper		Cosine-fourth Taper	
p	m	p	m	p	m	p	m
0.5	0.9	0.3	1.43	0.3	1.15	0.3	0.975
1.0	0.25	0.5	0.605	0.5	0.45	0.5	0.38
2.0	0.059	1.0	C.155	1.0	0.125	1.0	0.1
		2.0	0.040	2.0	0.035	2.0	0.025

When the obstacle area is considerably smaller that the aperture area, it is possible to simplify Eq. (14) by realizing that α varies rather slowly over the extent of the obstacle. Consequently, we can approximate $\alpha(x_1 + x', y_1 + y')$ by $\alpha(x_1, y_1)$. If this is done and the integrals in Eq. (14) are then evaluated we find, after using Eq. (15), that

$$S(\theta, \phi) = \left(\frac{\pi k \alpha_o}{2}\right)^2 |F|^2 \left(\frac{X_o Y_o}{L_1^2}\right)^2 \exp\left\{-2m_1 \left(\frac{X_1}{L_1}\right)^2 - 2m_2\left(\frac{y_1}{L_1}\right)^2\right.$$

$$\cdot \operatorname{sinc}^2 (kX_o \sin\theta \cos\phi) \operatorname{sinc}^2 (kY_o \sin\theta \sin\phi) \quad , \tag{16}$$

where x_1 and y_1 are the coordinates of the center of the obstacle, we shall be primarily concerned with the azimuthal sidelobes. Consequently, we can set $\phi = \pi/2$ in Eq. (16). Also, for θ well outside of the main beam, the envelope of the function $\operatorname{sinc}^2 (kY_o \sin\theta)$ is $(kY_o \sin\theta)^{-2}$. Therefore, the magnitude of the azimuthal sidelobe at θ is

$$|S(\phi) = \pi/2| = \frac{\left(\frac{\pi k \alpha_o X_o}{2L_1}\right)^2 |F|^2 \exp\left[-2m_1 \left(\frac{X_1}{L_1}\right)^2 - 2m_2 \left(\frac{y_1}{L_1}\right)^2\right]}{(kL_1 \sin\theta)^2} . \tag{17}$$

3. NUMERICAL EXAMPLES

3.1 Fence-like Obstacle

As a first example, let us consider the situation in Figure 4. The obstacle is assumed to be a metal fence or a very dense forest at a distance z_1 of 800 ft from the antenna. For this case $|F|^2 \simeq 4/k^2$, $\alpha_o \simeq 0.31$ and $p \simeq 1$. Because we are unable to perform the x- integration in Eq. (14), we will approximate $\alpha(x_1 + x')$ by $\alpha(x_1)$, where $x_1 = -2L_1$, (see Figure 4). If we assume the obstacle is infinitely long in the y - direction, we obtain from Eq. (14)

$$S\left(\phi = \frac{\pi}{2}\right) = \frac{0.094}{m_2} \exp\left[\frac{-(kL_1 \sin\theta)^2}{2m_2}\right] . \tag{18}$$

From Table 1, we find $m_2 = 0.155$, so that for θ outside the main beam, the side-lobe level is smaller than -400 dB, which is clearly no problem.

3.2 Finite Obstacle

We now consider the same geometry as in Figure 4, except with the obstacle width, $2Y_o$, smaller than the aperture width. A pictoral illustration of the projection

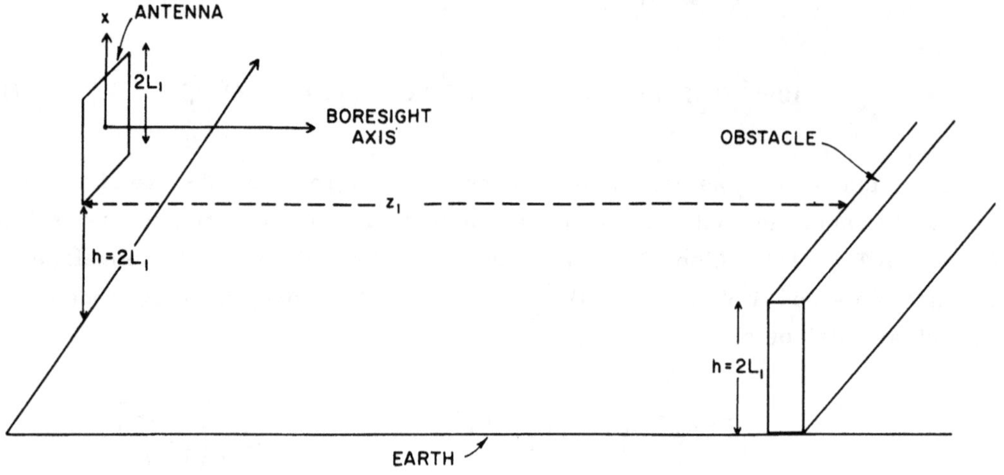

Figure 4. Fence-like Obstacle in the Fresnel Zone of an Antenna

of the obstacle onto the aperture is shown in Figure 5. In this case, we may use the approximate formula in Eq. (17) to obtain

$$\left|S\left(\phi = \frac{\pi}{2}\right)\right| = \frac{4.7 \times 10^{-6} \left(\frac{k|F|}{2}\right)^2}{\sin^2 \theta} . \tag{19}$$

If the obstacle is metal or dense wood, such as a tree trunk

$$|F|^2 \simeq 4/k^2 .$$

However, if the obstacle is the branches and leaves of a tree, it is difficult to specify $|F|$, although it is expected that $|F|^2 < 4/k^2$. In order to be pessimistic,* we can assume

$$|F|^2 \simeq 4/k^2 ,$$

*Note that for appropriate combinations of ϵ and Δ we find that $|F|^2 > 4/k^2$, but generally $|F|^2 \leq 4/k^2$ otherwise.

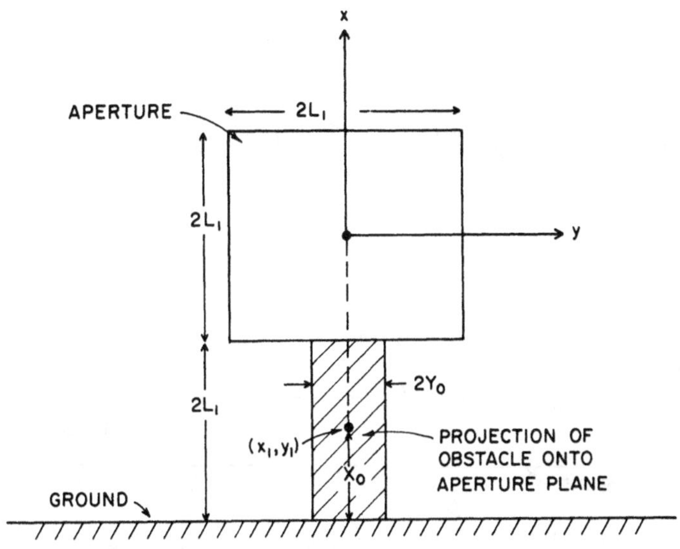

Figure 5. Projection of the Obstacle onto the Aperture Plane

so that

$$|S(\phi = \pi/2)| \simeq 4.7 \times 10^{-6}/\sin^2 \theta \quad .$$

Consequently, for $\theta = 10^\circ$ the sidelobe level produced by the obstacle will be -38.1 dB. This is unacceptable, because we require -50 dB sidelobes. However, when the main beam is elevated, we expect that the azimuthal sidelobe level will decrease. In order to study the effect of beam elevation, we realize that for a given beam elevation the value, x_1, of the location of the obstacle center is

$$x_1 = -2L_1 - 800 \tan \beta_o \text{ ft} \quad , \tag{20}$$

as is clear from Figure 6. For this case we get, in place of Eq. (19), the result

$$\left|S\left(\phi = \frac{\pi}{2}\right)\right| \simeq \frac{3.5 \times 10^{-5} \left(\frac{k|F|}{2}\right)^2}{\sin^2 \theta} \exp\left[-2(1 + 53.3 \tan \beta_o)^2\right] \tag{21}$$

If we again assume that $|F|^2 \simeq 4/k^2$, we obtain the results in Table 2.

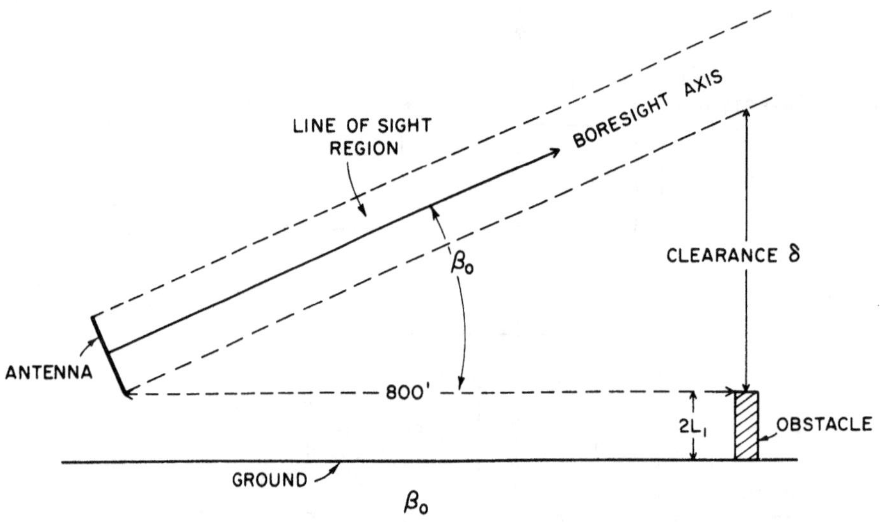

Figure 6. Effect of a Beam Elevation

Table 2. Azimuthal Sidelobe Levels for the Obstacle in Figure 5

β_o (degrees)	Clearance δ (ft)	$\lvert S(\theta = 6^o)\rvert_{dB}$	$\lvert S(\theta = 10^o)\rvert_{dB}$
0	0	-33.7	-38.1
0.5	7	-43.6	-48.0
1	14	-57.3	-61.7
1.5	21	-74.8	-79.2

4. CONCLUSIONS

We have found that for high (complex) permittivity obstacles of transverse dimensions $2X_o$ and $2Y_o$ centered at (x_1, y_1) relative to the aperture center, the relative azimuthal sidelobe level is

$$\left|S\left(\phi = \frac{\pi}{2}\right)\right| \simeq \frac{\left(\frac{\alpha_o X_o}{L_1}\right)^2 \exp\left[-2m_1\left(\frac{x_1}{L_1}\right)^2 - 2m_2\left(\frac{y_1}{L_1}\right)^2\right]}{(2L_1 \sin\theta/\lambda)^2}$$

where $2L_1$ is the linear aperture dimension, λ is the wavelength, and α_o and m are given in Figure 2 and Table 1. By using the above result, we have shown that even when there is not actual aperture blockage, the effect of obstacles could lead to sidelobe levels greater than -50 dB. In general, it is not always sufficient that the entire aperture have a clear line of sight; this conclusion is illustrated pictorially in Figure 7.

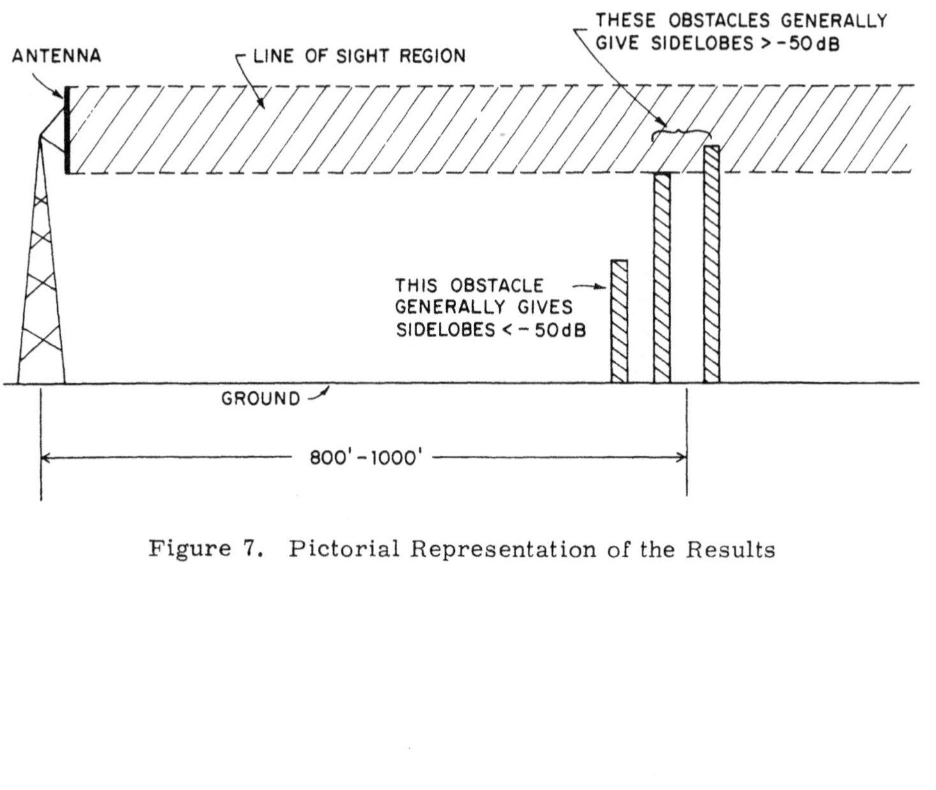

Figure 7. Pictorial Representation of the Results

Tolerance Errors for Very Low Sidelobe Antennas
R.L. Fante
RADC TR 77-360, October 1977, DDC AD A049259
Reprinted by permission.

4.3

Tolerance Errors for Very Low Sidelobe Antennas

1. INTRODUCTION

In a previous report[1] we considered the design of a parabolic cylinder with very low sidelobes. That report was devoted to the illumination required to achieve -50 dB sidelobes; this report is devoted to calculating the manufacturing tolerances required for the reflector. The method employed in making these calculations is an extension of the analysis due to Ruze.[2]

2. NUMERICAL CALCULATIONS

The average sidelobe power density at (θ, ϕ) due to the surface errors is given by the second term in Eq. (A19) in the Appendix, and is

$$\langle I(\theta, \phi)\rangle_S = \pi a^2 \sigma^2 e^{-\sigma^2} g(0) \exp\left[-\frac{k_o^2 a^2 \sin^2\theta}{4}\right] , \qquad (1)$$

(Received for publication 26 October 1977)

1. Fante, R. (1977) Design of Parabolic Cylinder Reflector Systems With Low Sidelobes, RADC-TR-77-204.

2. Ruze, J. (1966) Antenna tolerance theory - a review, Proc IEEE 54:633-640.

where a is the surface correlation length, $\sigma = 2\,k_o\,\delta\,\cos^2(\theta/2)$, $k_o = 2\pi/\lambda$, δ = surface bump height, (θ, ϕ) are shown in Figure 1, and

$$g(0) = \iint_A |e_o(\underline{\rho})|^2\, d^2\rho \quad . \tag{2}$$

In Eq. (2), $e_o(\underline{\rho})$ is the antenna field distribution and A denotes the antenna surface. We note that Eq. (1) is valid only if $\sigma \ll 1$.

The main-beam power density is given by the first term in (A19) evaluated at $\theta = \phi = 0$, and is

$$\langle I\rangle_M = e^{-\sigma^2}\, G_o(0, 0) \quad , \tag{3}$$

where

$$G_o(0, 0) = \left| \iint_A e_o(\underline{\rho})\, d^2\rho \right|^2 \quad . \tag{4}$$

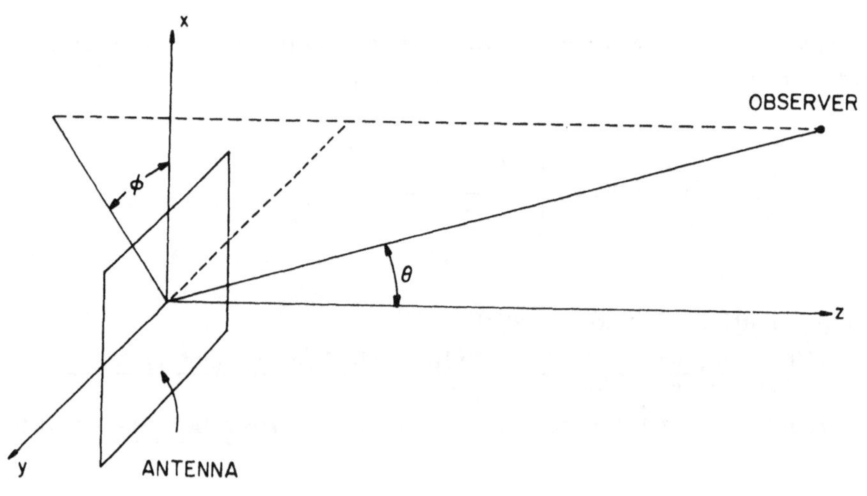

Figure 1. Assumed Geometry

By using Eqs. (1) and (3) it is easy to see that the relative average sidelobe level at (θ, ϕ) is

$$\langle S(\theta, \phi)\rangle = \frac{\langle I\rangle_s}{\langle I\rangle_M} = \frac{\pi a^2 \sigma^2 \exp\left(-\frac{k_o^2 a^2 \sin^2\theta}{4}\right)}{\gamma}, \tag{5}$$

where

$$\gamma = \frac{\left|\iint_A e_o(\underline{\rho})\, d^2\rho\right|^2}{\iint_A |e_o(\underline{\rho})|^2\, d^2\rho} .$$

For a rectangular aperture with excitation $(-L_1 \leq x \leq L_1,\ -L_2 \leq y \leq L_2)$

$$e_o = \cos^n\left(\frac{\pi x}{2L_1}\right) \cos^m\left(\frac{\pi y}{2L_2}\right) \tag{6}$$

we can write

$$\gamma = A\,\beta_x\,\beta_y\ , \tag{7}$$

where

$$\beta_x = \frac{\left[\int_{-L_1}^{L_1} dx \cos^n\left(\frac{\pi x}{2L_1}\right)\right]^2}{2L_1 \int_{-L_1}^{L_1} dx \cos^{2n}\left(\frac{\pi x}{2L_1}\right)}, \tag{8a}$$

$$\beta_y = \frac{\left[\int_{-L_2}^{L_2} dy \cos^m\left(\frac{\pi y}{2L_2}\right)\right]^2}{2L_2 \int_{-L_2}^{L_2} dy \cos^{2m}\left(\frac{\pi y}{2L_2}\right)}, \tag{8b}$$

and $A = 4L_1L_2$ is the area of the aperture. Values of β for n = 0 through 6 are presented in Table 1.

Table 1. Values of β for n = 0–6

n	β
0	1
1	0.81
2	0.667
3	0.575
4	0.515
5	0.468
6	0.433

For a given value $\theta = \theta_o$ it is evident from Eq. (5) that the relative sidelobe level will be a maximum (that is, worst case) if the correlation length

$$a = \frac{2}{k_o \sin\theta} \quad . \tag{9}$$

When a has this value, the average sidelobe level is

$$\langle S \rangle_{MAX} = \frac{16\,\pi\, e^{-1}\, \delta^2 \cos^4\left(\frac{\theta_o}{2}\right)}{\beta_x\, \beta_y\, A \sin^2\theta_o} \quad . \tag{10}$$

Furthermore, it is also clear from Eq. (5) that for all values $\theta > \theta_o$ the average sidelobe level will be smaller than $\langle S \rangle_{MAX}$. Consequently, Eq. (10) is the maximum average sidelobe level for all $\theta \geq \theta_o$.

In the Appendix we also show that the probability density, satisfied by I, and consequently $S = I/\langle I \rangle_M$, is

$$p(S) = \langle S \rangle^{-1} \exp\left\{-\frac{S}{\langle S \rangle}\right\} \quad . \tag{11}$$

Therefore, the probability that the sidelobe level will exceed a specified sidelobe level S_o is

$$P(S > S_o) = \frac{1}{\langle S \rangle} \int_{S_o}^{\infty} \exp\left[-\frac{S}{\langle S \rangle}\right] dS$$

$$= \exp\left\{-\frac{S_o}{\langle S \rangle}\right\} \quad . \tag{12}$$

When a is given by Eq. (9), so that we have the worst case, the probability that S exceeds S_o at $\theta = \theta_o$ is

$$P(S > S_o) = \exp\left\{-\frac{S_o A \beta_x \beta_y \sin^2 \theta_o}{18.5\, \delta^2 \cos^4 \frac{\theta_o}{2}}\right\} \quad . \tag{13}$$

The probability that $S > S_o$ for $\theta > \theta_o$ is smaller than the value given by Eq. (13). Therefore, if $P(S > S_o)$ is acceptable at $\theta = \theta_o$, it will also be acceptable for all values $\theta > \theta_o$.

We now apply Eq. (13) to a radar equivalent to the parabolic cylinder considered in reference[1]. We will set $A = 225\ ft^2$ and assume that e_o can be approximated by $\cos^4 (\pi x/2L_1) \cos (\pi y/2L_2)$, so that $\beta_x = 0.515$ and $\beta_y = 0.81$. The signal frequency is chosen as 3.35 GHz, and we require that for all $\theta \geq 7^o$ the sidelobe level be smaller than -50 dB, with a probability P_o of exceeding this value. For this case Eq. (13) becomes

$$P(S > -50\ dB) = \exp\left\{-\frac{8.8 \times 10^{-6}}{(\delta/\lambda)^2}\right\} \quad . \tag{14}$$

The required value of δ so that the sidelobe level exceeds $S_o = 10^{-5}$ with a value P_o for all $\theta \geq 7^o$ is

$$\frac{\delta}{\lambda} = \left[\frac{-8.8}{\ell n\, P_o}\right]^{1/2} \times 10^{-3} \quad . \tag{15}$$

Values of δ for various values of P_o are given in Table 2.

The equivalent results for the tolerance required so that the sidelobe level exceeds $S_o = 10^{-5}$ with a value P_o for all $\theta \geq 10^o$ are given in Table 3.

Table 2. Tolerance δ Required so That There is a Probability P_o of Sidelobe Exceeding -50 dB for all $\theta \geq 7^o$

P_o	δ (inches)
0.005	4.55×10^{-3}
0.01	4.87×10^{-3}
0.03	5.57×10^{-3}
0.08	6.6×10^{-3}
0.15	7.58×10^{-3}
0.30	9.53×10^{-3}

Table 3. Tolerance δ Required so That There is a Probability P_o of the Sidelobe Exceeding -50 dB for all $\theta \geq 10^o$

P_o	δ (inches)
0.005	6.46×10^{-3}
0.01	6.95×10^{-3}
0.03	7.97×10^{-3}
0.08	9.38×10^{-3}
0.15	1.08×10^{-2}
0.30	1.36×10^{-2}

Antenna Sidelobe Reduction Using Spatial Filters
R.J. Mailloux and A.C. Schell
Proceedings of the DDRE Radar ECCM Symposium, 1976
Pages 40-1 – 40-11
Reprinted by permission.

4.4

SUMMARY

(U) A new technique has been developed for the suppression of sidelobes of existing antennas. A multilayer structure is placed in front of the antenna to function as a wave filter. Radiation from the antenna is minimally attenuated in directions near the main beam, while sidelobes at angles away from the main beam are suppressed. These spatial filters are configured of parallel layers of dielectrics or metal gratings extending across the entire antenna aperture. Procedures for the synthesis of desired spatial characteristics are drawn from analogous methods used for frequency filters.

(U) In this paper the synthesis of dielectric wave filters with Chebyshev bandpass characteristics is described. Results of the initial tests of an experimental model are presented. The development of this type of spatial filter is at an early stage, and considerably more work is necessary for both the assessment of the filter properties and for the successful applications to equipment.

1. INTRODUCTION

Low antenna sidelobe levels are a desirable attribute of ECM-resistant radar and communications systems. For many applications one of the best ECCM features is a sidelobe level substantially below the range that is common to current systems. In order to reduce the vulnerability of existing systems that do not have very low sidelobes, it is often necessary to completely redesign the antenna. However, a new technology has been developed to provide an option that in certain cases can upgrade the ECCM capability of a radar or communications system without requiring antenna replacement. This technology is called spatial filtering.

Spatial filters are structures that are placed in front of an existing antenna to provide minimally attenuated transmission in the angular region near the main beam while suppressing radiation in other directions. Spatial filters consist of several parallel layers of uniform dielectric or metallic gratings. The reflection coefficients of the layers and the interlayer spacings are chosen to produce the desired angular filter characteristic. Tradeoffs can be made among the frequency bandwidth, angular range of transmission and filter characteristics by varying the physical parameters of the filter.

The principles of layered dielectric frequency domain filters are well established [1-5], and insofar as possible the techniques for analysis and synthesis have been extended to the spatial domain. The fundamental difference between synthesis in the frequency domain and in the spatial domain arises because the transmission coefficients of layers that have a high dielectric constant are strongly frequency dependent but relatively invariant with the angle of incidence. If a wave from a medium having a low dielectric constant is incident on a medium of high dielectric constant, then for any angle of incidence the wave propagation direction in the latter is almost perpendicular to the interface.

This property necessitated a fundamental change in filter design from the frequency domain transformers synthesized by Collins and others [1-5], which consist of various dielectric layers sandwiched together. The spatial domain filters reported in this paper consist of quarter-wave sections of dielectric separated by half-wave or full-wave air spaces. The methods that Collin outlined in two basic studies [1,6], were adapted to their synthesis. The end result is a technique for the synthesis of Chebyshev bandpass characteristics for the transmission response over an angular region.

2. ANALYSIS

Collin used the wave matrix formalism to derive convenient expressions for the transmission properties of layered impedance sections,[1] and to describe the spatial properties of abutting dielectric layers.[6] The same formalism will be used here to derive properties of the stratefied dielectric filter. Consider the basic filter section shown in Fig. 1. The incident and reflected waves in medium 1 are a_1 and b_1 respectively. The incident electric fields are assumed to be either parallel-polarized or perpendicular - polarized, with no cross-polarized components excited. The input - output parameters of the section are related by

$$\begin{pmatrix} a_1 \\ b_1 \end{pmatrix} = \begin{pmatrix} A_{11} & A_{12} \\ A_{21} & A_{22} \end{pmatrix} \begin{pmatrix} a_2 \\ b_2 \end{pmatrix}$$

The parameters A_{mn} can be related in terms of the conventional scattering matrix. The wave matrix of a cascade of networks is the product of the wave matrices of each network. A parameter of particular importance is

$$A_{11} = \left.\frac{a_1}{a_2}\right|_{b_2=0}$$

which is the inverse of the filter transmission coefficient. For the case of near-broadside incidence on a quarter-wave dielectric slab, the wave matrix is approximately

$$[A] = \begin{pmatrix} A_{11} & A_{12} \\ A_{21} & A_{22} \end{pmatrix} = \frac{1}{S_{12}} \begin{pmatrix} e^{jk_zS} & -S_{11}e^{jk_zS} \\ S_{11}e^{-jk_zS} & -e^{-jk_zS} \end{pmatrix}$$

where

$$S_{11} = \frac{\epsilon - 1}{\epsilon + 1}, \quad S_{12} = \frac{-2j\sqrt{\epsilon}}{\epsilon + 1} \quad \text{and} \quad k_z = k_0 \cos\theta.$$

The angle θ is the polar angle from broadside in air.

The wave matrix of a filter comprised of a number of such sections is obtained by multiplying in sequential order the matrices of the sections, as

$$[A]_{filter} = \prod_{n=1}^{N} [A(s_n, \epsilon_n)]$$

This procedure is used to derive the filter properties for parallel and perpendicular polarizations and arbitrary angles of incidence.[7]

3. SYNTHESIS

The procedure for filter synthesis depends on the properties of the polynomial expression for the power loss ratio, defined by Collin[1] as the power ratio associated with the inverse of the filter transmission coefficient:

The power loss rates of a filter consisting of quarter wave sections of transmission line can be expressed as an even polynomial of degree 2n, where n is the number of sections in the filter.

The difference between the design for spatial filters and the work of Collin and others is the inclusion of the air spaces S_j between the dielectric layers. The electrical path length through the dielectric and the air-dielectric interface characteristics do not vary appreciably with the angle of incidence. It is the variation with angle of the paths through the air spaces that causes the filter properties.

The synthesis procedure is to set the power loss ratio of an m section filter equal to unity plus a Chebyshev polynomial of order 2 m. The form of A_{11} that is used is the approximate version, stated earlier, that neglects the angular variation of the interface reflection and the path through the dielectric. Thus the synthesis procedure does not distinguish between the E-plane and the H-plane variation. The dielectric slabs are considered to be lumped reactive components, and the filter synthesis consists of choosing the magnitudes of these reactances.[7]

A representative result of the synthesis procedure is shown in Figure 2. The filter consists of four layers with an interlayer separation of 0.5λ. The pass band extends to $\pm 11.5^\circ$ from broadside. The two inner layers of dielectric have a permittivity of $\epsilon_2 = 15.14$, while the outer dielectric layers have an $\epsilon_1 = 3.08$. The transmission characteristics for this filter have been calculated using the accurate formulation of the wave matrix, and the results are shown in Figures 3a and 3b. The filter reflection for the transmitted polarization and the cross polarization are given. The u-axis ($u = \sin\theta\cos\phi$) is the H-plane, while the v-axis ($v = \sin\theta\sin\phi$) is the E-plane.

The transmission characteristics of a four layer filter with λ interlayer spacings is shown in Figures 4a and 4b. The dielectric constants are the same as given before. Comparison of this with the previous case shows that for the given dielectric layers much sharper angular cutoff is possible with the larger spacing. However, the λ-spacing results in a second passband occurring near $\theta = 60^\circ$. The filter behavior for layer scan angles ($\theta > 60^\circ$) is not plotted because of the complexity of the curves and the generally unsatisfactory filter performance for the larger angular range.

4. EXPERIMENTAL RESULTS

A four-section dielectric layer spatial filter has been constructed. The filter consists of two outer layers of $\epsilon = 3$ dielectric and two inner layers of $\epsilon = 15$ dielectric material, each layer having a thickness of a quarter wavelength in the material at 11GHz. The spacing between layers is a free-space wavelength at 11GHz. Each layer is approximately 25X75 cm in planar extent, and styrofoam is used between layers for stability. The theoretical transmission characteristics are described by Figure 4.

Preliminary tests of this filter have been made using a horn and an array as sources. In Figure 5 are shown the radiation patterns of the horn alone and the horn with the filter positioned in front of the horn aperture. The filter has substantially attenuated the horn radiation at angles near ± 30 degrees. In Figure 6 are shown the radiation patterns of the array alone and the array with the filter. This array has a set of grating lobes that can be easily distinguished. Note that the filter attenuated those lobes falling within the stop-band of the filter, but did not attenuate the grating lobe at $\theta = 60^{\circ}$, which is within the second pass-band of the filter.

5. CONCLUSIONS

The suppression of antenna sidelobes is possible using spatial filters placed in front of the antenna aperture. These spatial filters do not require modification to, or even knowledge of, the details of the antenna design and construction. A spatial filter consists of a number of planar reflecting layers separated by air spaces. These layers are uniform in the transverse direction, and may consist of dielectric or wire grids. They function as a series of reactances that produce a very small net reflection of near-broadside waves but strongly reflect waves travelling at angles far from broadside. The shape of the spatial filter pass band and stop band can be designed as desired, and a procedure for Chebyshev filter synthesis procedures has been adapted from analogous procedures for waveguide filters. Experimental results have confirmed the basic properties of these filters although much remains to be done to understand all aspects of this device.

There are a number of limitations to this type of spatial filter. The filter must cover the antenna aperture, and should extend somewhat beyond the aperture edge in order to contain all sidelobe energy. If a narrow angular pass band is desired, the filter must be thick. If additionally there is to be a single pass band, the filter must contain a sufficient number of layers (typically, one per half-wavelength of extent in the broadside direction). A narrow angular pass band implies a narrow frequency bandwidth of operation. This type of filter tends not to be well suited to suppressing sidelobes resulting from aperture blockage, because these sidelobes lie near the main beam, and a very thick filter is needed. However, a spatial filter can be very effective in suppressing grating lobes or other wide-angle sidelobes without redesigning the basic antenna. A spatial filter can be incorporated as a part of the antenna radome and thus can serve to protect the antenna as well as reducing undesired sidelobe radiation.

6. REFERENCES

1. Collin, R.E., Theory and design of wideband multisection quarter-wave transformers, Proc. IRE, Vol 43, No 2, pp 179-185, Feb 1955

2. Cohn, S.B., Optimum design of stepped transmission-line transformers, IRE Trans. MTT, Vol M-TT-3, No 2, pp 16-21, Apr. 1955

3. Riblet, H.J., General synthesis of quarter-wave impedance transformers, IRE Trans. MTT, Vol MTT-5, No 1, pp 36-43, Jan. 1957

4. Young, L. Tables for cascaded homogeneous quarter-wave transformers, IRE Trans. MTT, Vol MTT-7, No 2, pp 233-244, Apr. 1959

5. Young, L. Stepped-impedance transformers and filter prototypes, IRE Trans. MTT, Vol MTT-10, No 5, pp 339-359, Sept. 1962

6. Collin, R.E. Field Theory of Guided Waves, McGraw Hill, pp 79-93

7. Mailloux, R.J., Synthesis of spatial filters with Chebyshev characteristics, IEEE Trans. Ant & Prop, Vol AP-24, No 2, pp 174-181, March 1976

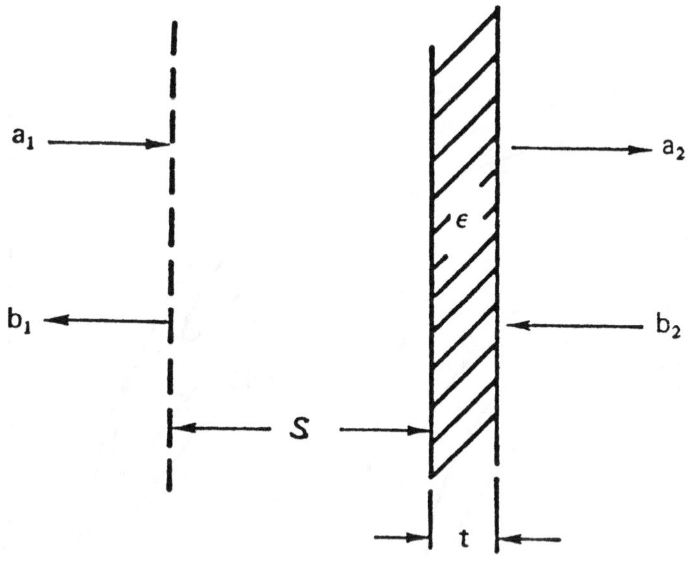

Fig. 1 The Basic Filter Section

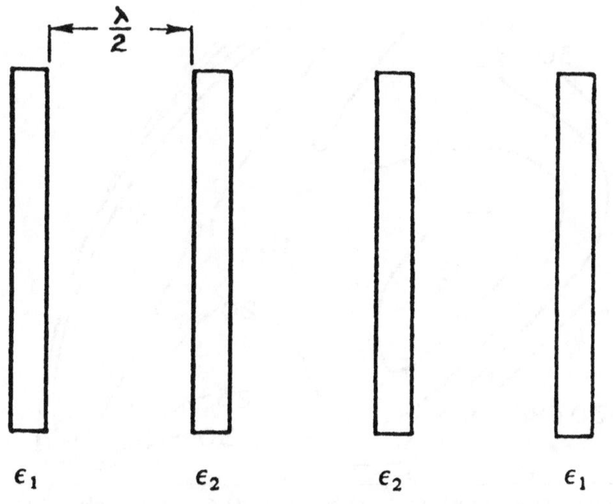

Fig. 2 A Typical Spatial Filter

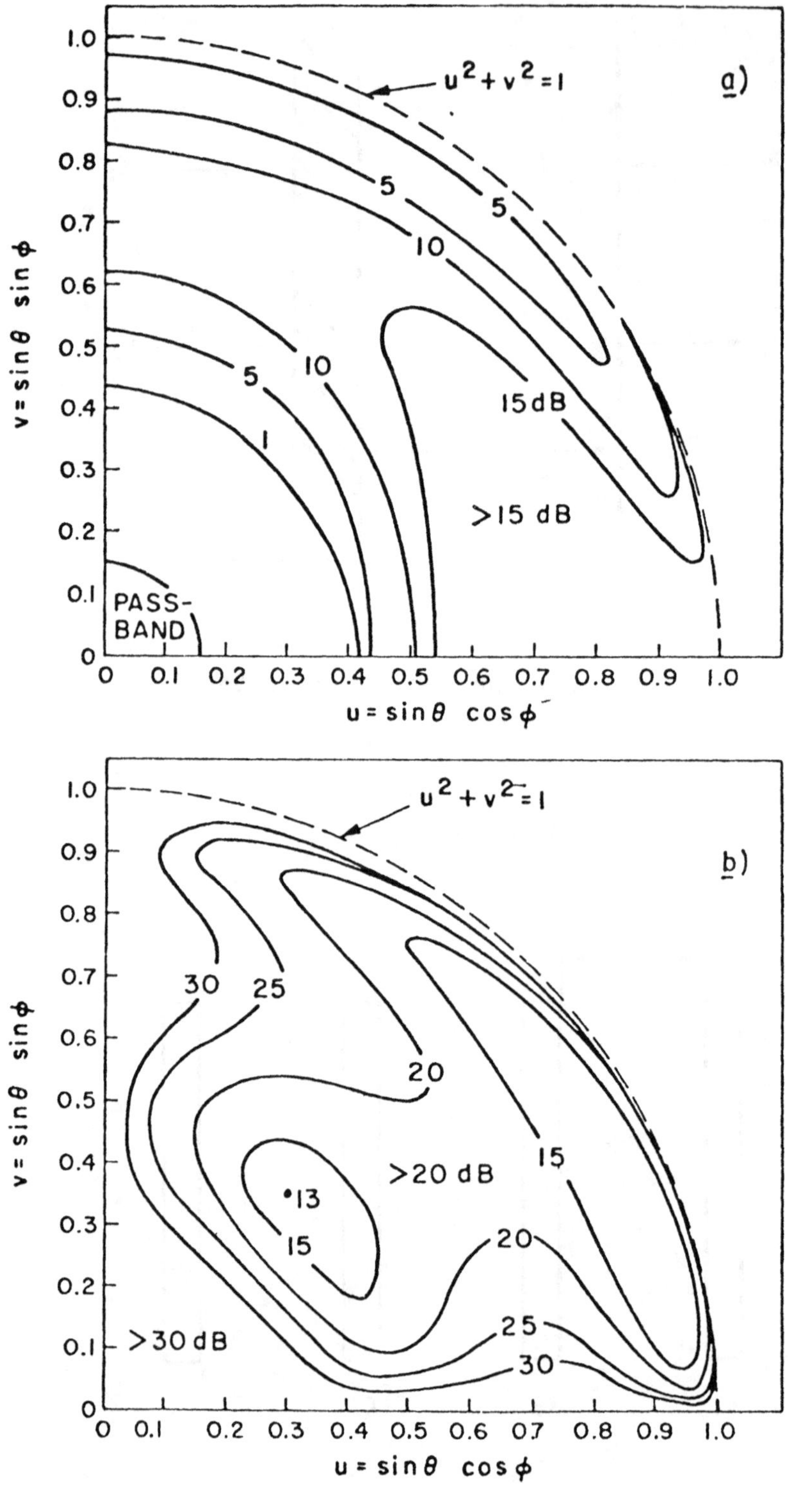

Fig. 3. Transmission Characteristics of a Four-Layer Chebyshev Filter (ϵ_1 = 3.08, ϵ_2 = 15.14, S = 0.5 λ) (a) Filter Rejection for the Transmitted Polarization, (b) Isolation between the transmitted polarization and cross polarization.

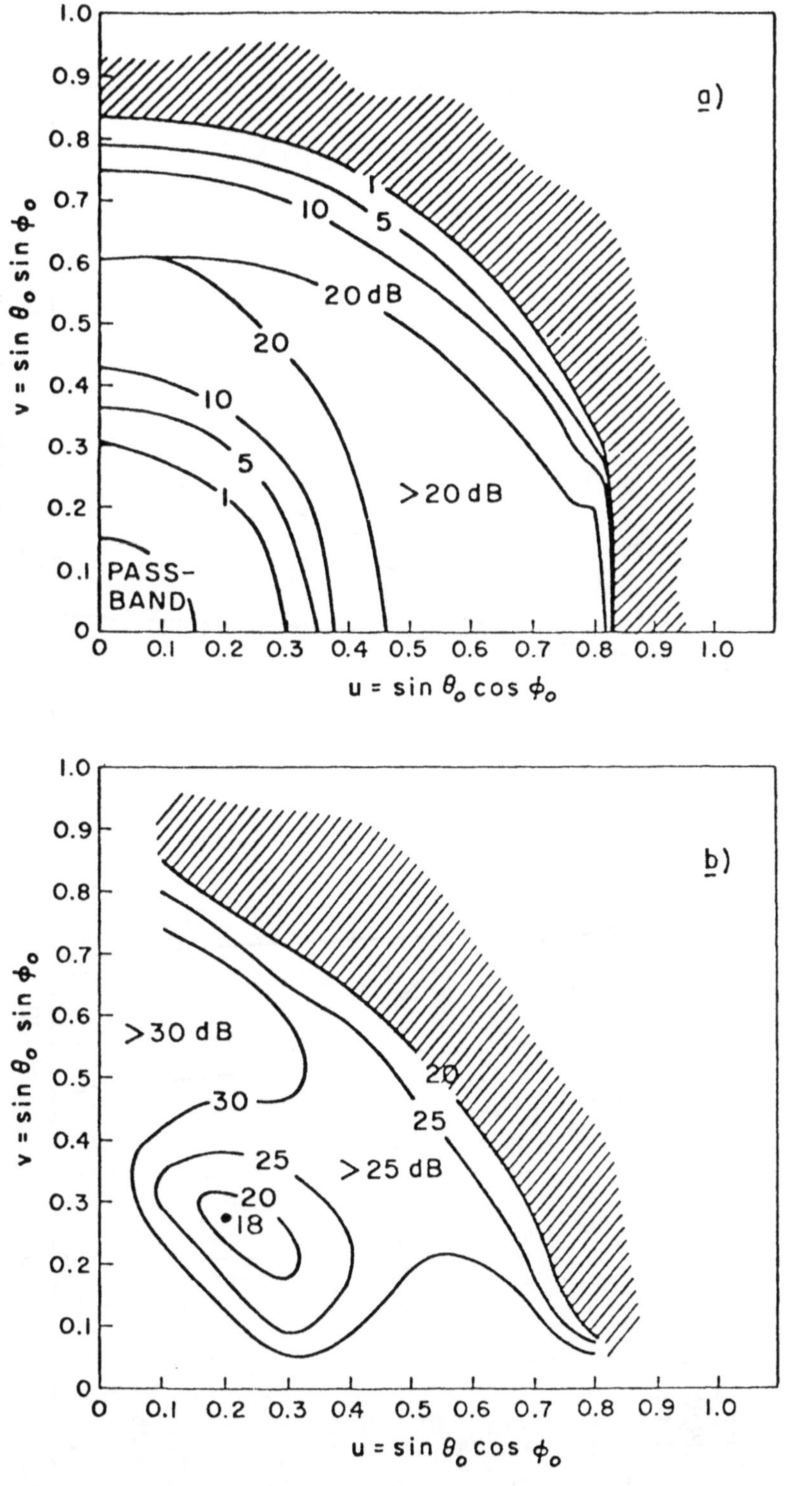

Fig. 4. Transmission Characteristics of a Four-Layer Chebyshev Filter ($\epsilon_1 = 3.08$, $\epsilon_2 = 15.14$, $S = \lambda$) (a) Filter Rejection for the Transmitted Polarization; (b) Isolation between the transmitted polarization and cross polarization

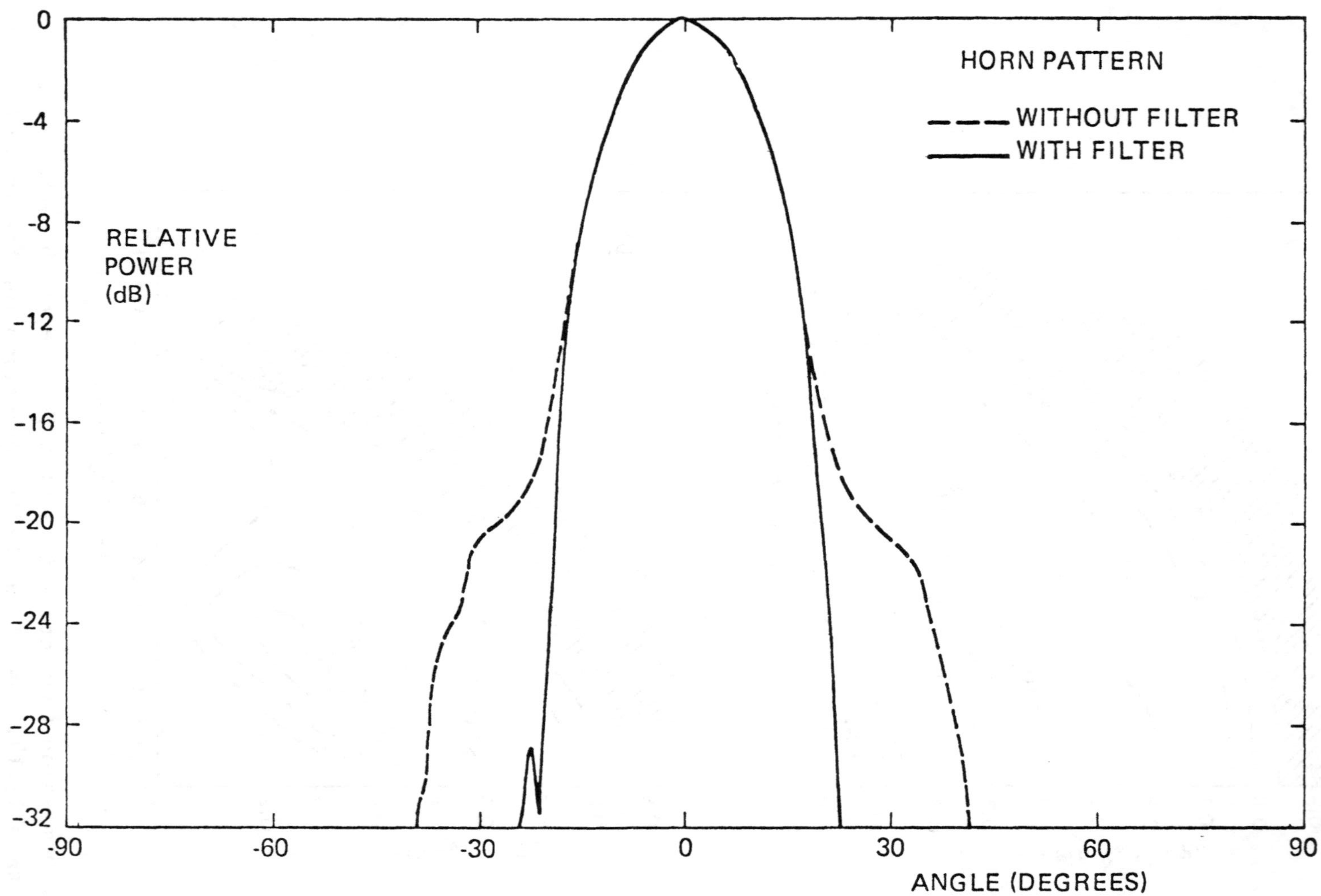

Fig. 5 Experimental radiation patterns of a horn (dotted line) and a horn with a four layer spatial filter (solid line).

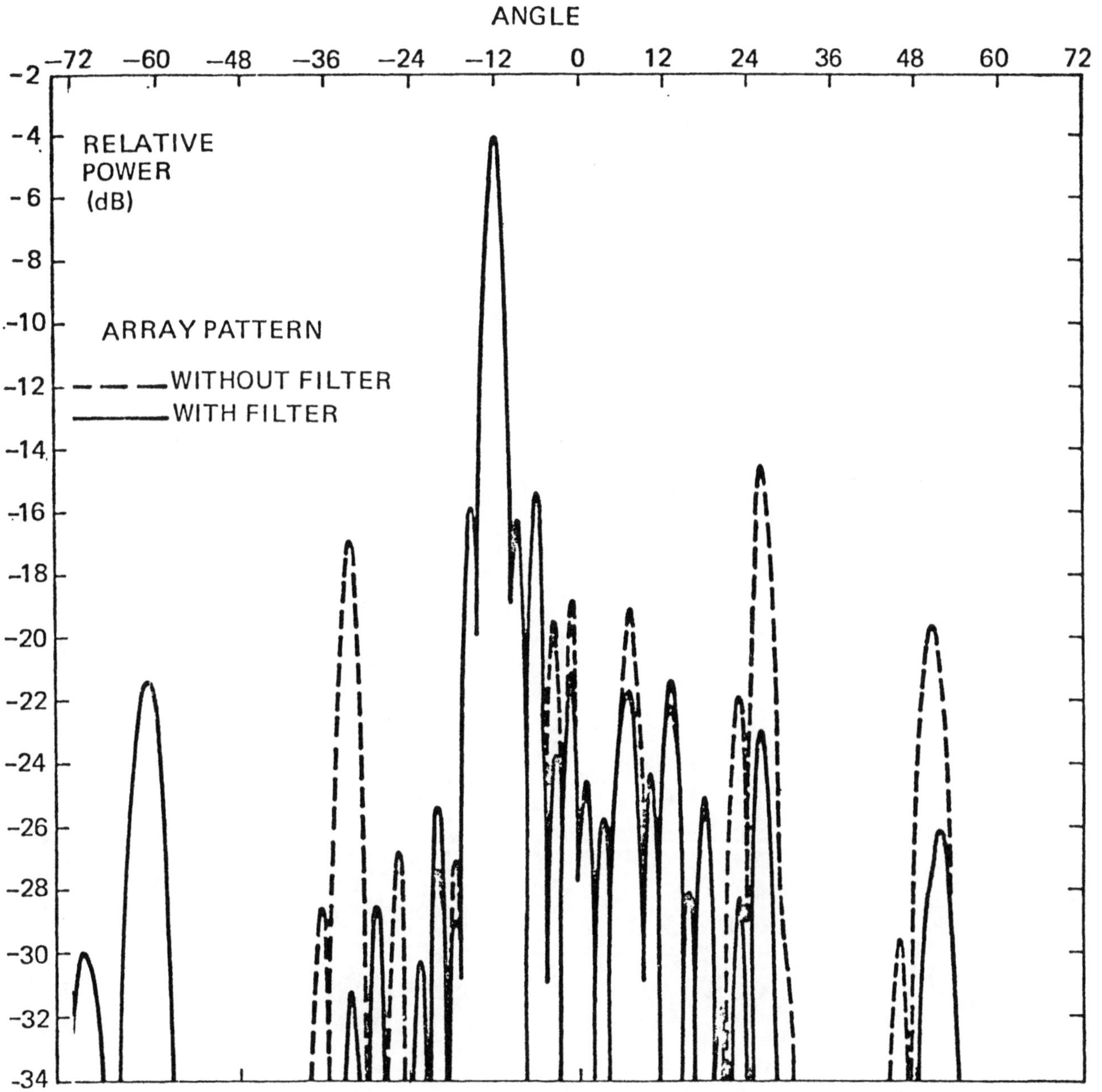

Fig 6. Experimental radiation patterns of an array with grating lobes (dotted line) and the same array with a four-layer spatial filter (solid line).

Conopulse Radar
H. Sakamoto and P.Z. Peebles, Jr.
IEEE Transactions on Aerospace and Electronic Systems, Vol. AES-14, No. 1, January 1978, pp. 199-208. Correction Vol. *AES-14, No. 4,* July 1978, p. 673c.

4.5

Abstract

The methods usually adopted for implementing an angle-tracking radar are monopulse and conical scan. A monopulse system is free from target-signal amplitude-fluctuation effects, in principle, but usually requires three channels of receiver equipment. A conical scan system, on the other hand, requires only a one-channel receiver, but is subject to target-signal amplitude-fluctuation effects. This paper describes and analyzes a hybrid system called *conopulse* that requires only two receiver channels and is free of target-signal amplitude-fluctuation effects. Part of the signal processing in a conopulse system uses *conical* scan techniques and part uses *monopulse* methods. The principal advantages of conopulse are that it has the monopulse freedom from angle errors caused by target fluctuations and that it requires less receiver equipment than monopulse to achieve this freedom. Its main disadvantages are its lower data rate and the practical problems associated with the required antenna feed, which is somewhat similar to but more complicated than conical scan.

Manuscript received March 24, 1977, revised September 16, 1977.

Authors' addresses: H. Sakamoto, Research and Planning Division, Bureau of Equipment, Defense Agency, Yokahama, Japan; P.Z. Peebles, Department of Electrical Engineering, University of Tennessee, Knoxville, TN 37916.

0018-9251/78/0100-0199 $00.75 1978 IEEE

Two principal methods historically have been available to a radar system engineer interested in designing an angle-tracking radar. The older method called conical scan [1, p. 166] has the main advantage of simplicity. It requires a relatively simple antenna and only one channel of signal processing equipment in the receiver. One of the main disadvantages of conical scan is its susceptibility to target signal amplitude fluctuations (hereafter called simply target fluctuations or amplitude fluctuations). Target fluctuations add to the effective noise level and degrade tracking accuracy.

The other tracking method called monopulse [1-6] is not, in principle, susceptible to target fluctuation effects. However, it has the disadvantage of requiring three receiver channels in its usual implementation, and a conceptually more complex antenna structure than in conical scan.

This paper describes and analyzes an alternative tracking method called *conopulse* that is a sort of hybrid *conical* scan-*monopulse* system. It has independence from the target fluctuation of monopulse but requires fewer receiver channels (conopulse requires only two channels).

The conopulse concept described herein is not considered to be new.[1] The principal contributions made in this paper consist of tutorial discussions of both theoretical and practical aspects of the concept as well as analysis that includes some information about the nonlinear nature of the angle-error voltage response of the system.

The following sections of the paper describe the conopulse concept both conceptually and with respect to possible implementations (Section II.). Two forms of conopulse are analyzed to determine the system's angle-error voltage response function (Section III and IV.). Finally, the most important conclusions are summarized and discussed (Section V.).

II. The Conopulse Concept

The key idea in conopulse is that *two* independent receiving beams (antenna patterns) are scanned in space about the tracking or boresight axis instead of one, as in conical

[1] The conopulse concept described herein was conceived by P.Z. Peebles, Jr., in 1973. Subsequent discussions of the technique with Dean Howard of the U.S. Naval Research Laboratory revealed that Cook Research Laboratories (CRL) had done work for the U.S. Air Force on a similar concept in 1958; apparently, the CRL concept was called konopulse. Subsequent discussions with E.H. Beneke at CRL, who worked on the project, verified that a prototype system was built around the original ideas of Robert MacAskill. Limited tests of the prototype were conducted, but very little analysis of the system was performed. Results of the work were not available to the authors due to Air Force restrictions placed on the reports generated. Others have pointed out to the authors that Bakut et al. [7], have described a tracking method called "scanning with compensation" that is essentially the same as conopulse. However, little information is shed on practical implementation or on the nonlinear nature of the system's angle error voltage response.

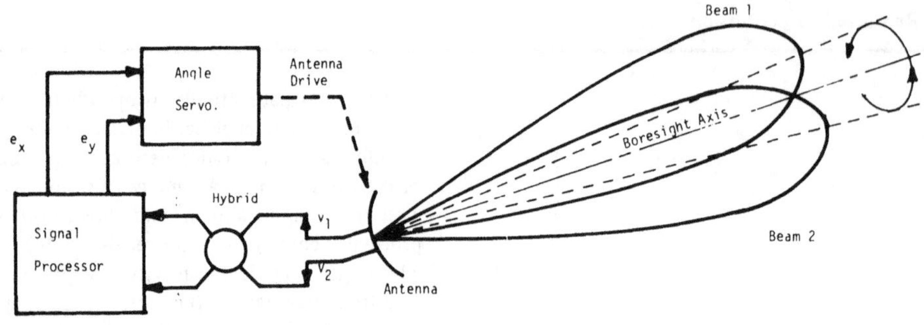

Fig. 1. A general conopulse receiving system.

Fig. 2. Functional block diagram for a monopulse-conical scan (MOCO) implementation of a conopulse system.

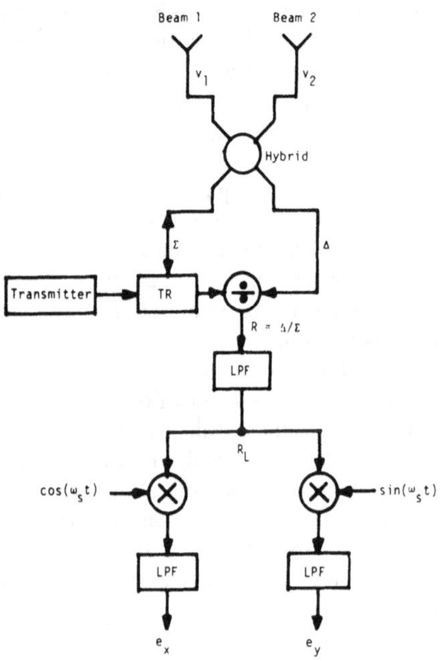

scan. Fig. 1 illustrates the general conopulse system. The target return signals ν_1 and ν_2 emerging from the respective beams are modulated both by target fluctuations and by beam scanning. Fluctuations affect ν_1 and ν_2 equally by the same factor. Modulation factors in ν_1 and ν_2 due to scanning are identical to that achieved when the two beams are considered separately to be conical scan beams. If the two patterns are identical except that one lags the other by one half scan period, then the two scan modulation factors would be identical except for a relative timing of one half scan period.

Signals ν_1 and ν_2 each possess sufficient information to obtain full-angle tracking by conical scan processing. However, by use of *both* outputs, there is also sufficient information to obtain independence from target fluctuations. The basic idea is to form the ratio of two signals that are each functions of ν_1 and ν_2 so that the common modulation factor due to fluctuations cancels. The cancellation step takes place in the signal processor, which, for reasons of convenience that will become clearer later, is provided with sum Σ and difference Δ signals derived from a microwave hybrid network.

Signal processing in the receiver is done partly by monopulse methods and partly by conical scan methods. The end results of processing are two error voltages to drive the angle tracking servomechanisms; the voltages are proportional to space-orthogonal target-angle errors from boresight. According to the usual custom we call these the elevation and azimuth angle error signals. As will be shown, conopulse ultimately derives these voltages using conical scan methods, so that conopulse requires a number of pulse transmissions to develop angle data. Therefore, conopulse cannot produce single-pulse angle estimates as true monopulse can in principle.

Two basic methods exist for conopulse implementation. They differ according to which form of signal processing is done at intermediate frequency (IF) first. In one case, monopulse normalization is done first. This form of implementation is called Monopulse-Conical scan (MOCO). If conical-scan operations are done initially, a Conical scan-Monopulse (COMO) system results.

MOCO System Description

Fig. 2 gives a block diagram of the functions to be performed in a MOCO system. The sum Σ and difference Δ of the two beam output voltages ν_1 and ν_2 are first formed in a hybrid network. The ratio Δ/Σ is next formed in a manner identical to an ordinary monopulse receiver, which is the process of normalization in ordinary monopulse (see Appendix A). It is in this ratio that target amplitude fluctuation effects are removed. If the ratio is formed by instantaneous automatic gain control (IAGC), wideband or rapid target fluctuations are removed. If a slower form of gain control is used, only slow target fluctuations may be removed.

The signal R is a train of video pulses occurring at the radar pulse rate. This pulse train will possess amplitude modulation caused by the scanning of the antenna beams. In practical stiuations, the radar pulse rate is much larger

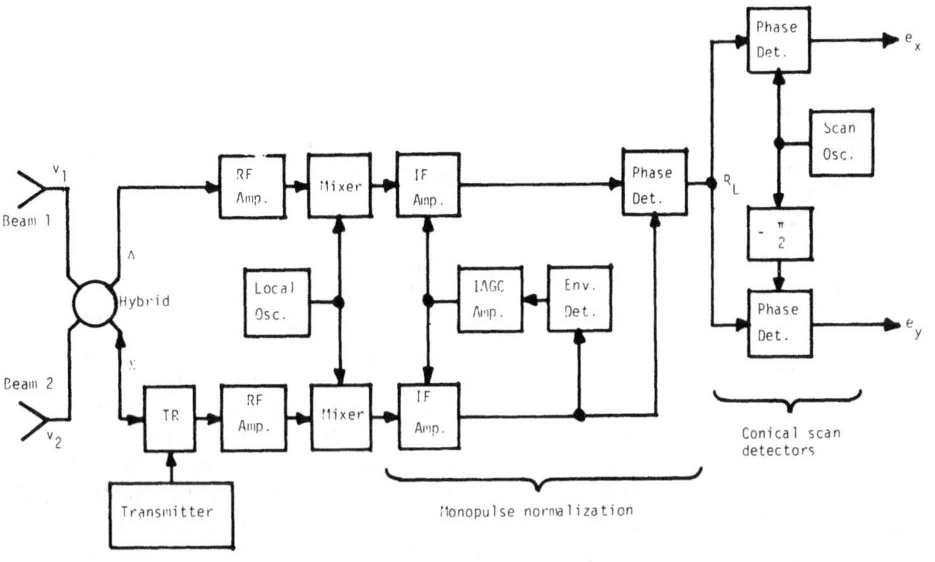

Fig. 3. Block diagram of a possible practical conopulse system of the MOCO type.

than the scan rate, so that a low-pass filter (LPF) can be used to smooth out the pulsed waveform and obtain a voltage R_L proportional to the beam-scanning modulation.

The voltage R_L is almost periodic[2] and is finally processed by two conical scan detectors (the product device and LPF) to develop angle error signals. These detectors are nothing more than devices that produce voltages proportional to the sine and cosine coefficients of the fundamental or scan frequency term in the Fourier series expansion of R_L.

A Practical MOCO System

Although the main concern of this paper is to discuss conopulse in theoretical and conceptual terms, some brief comments on the implementation of a practical system are in order. The two areas of concern are the antenna and the signal processor.

The problem of how to realize two independent scanning beams has not, as yet, been adequately solved and remains an open challenge to antenna designers. If the system is to have a single linear polarization, then it is clear that the two beams should nutate as opposed to rotate because nutation preserves polarization. Beam rotation would rotate polarization, and differences in target cross section due to polarization would cause undesirable target amplitude fluctuations at the scan rate. Although such fluctuations are theoretically removed by the monopulse normalization step in Fig. 2, the removal is not complete in practice. The result can be a serious angle-error voltage component in either e_x, e_y, or both. A further disadvantage of beam rotation would be that the polarization-induced fluctuations could limit the usable dynamic range of the monopulse normalization network (the IAGC amplifier, see Appendix A).

If the system has a dual polarization, such as circular, it would seem that either nutation or rotation of beams may be possible. In the latter case, one implementation might involve two circular horns, with circular polarizers, rotated about the boresight axis of a paraboloid and excited from a dual, concentric, coaxial rotary joint. Of course, such a rotary joint may limit the power capability of the system.

The signal processor for the MOCO system is readily established. The block diagram of one possible design is shown in Fig. 3. It implements the functions specified in Fig. 2. For simplicity, the angle- and range-tracking functions and the associated gating operations have been omitted. A discussion of how the monopulse normalization operations produce a signal proportional to Δ/Σ is included in Appendix A.[3]

COMO System Description

The COMO system signal processor functional block diagram is illustrated in Fig. 4. As in the MOCO system, sum and difference signals Σ and Δ are generated. These signals are first processed to remove carrier phase, preserving the correct sense of the difference signal. Coherent detectors (product followed by LPF) are illustrated to perform this task. However, it is not really necessary that

[2] R_L is truly periodic only for a target that is stationary relative to the radar. Since the radar is presumed to be tracking in range and in angle, the approximation that R_L is periodic is adequate for nearly all cases of practical interest and is used throughout this paper without further comment.

[3] An alternative normalization method involving logarithmic amplifiers may also be possible. See M.I. Skolnik, ed., *Radar Handbook*. New York: McGraw-Hill, 1970, pp. 21-30, and [4, p. 57].

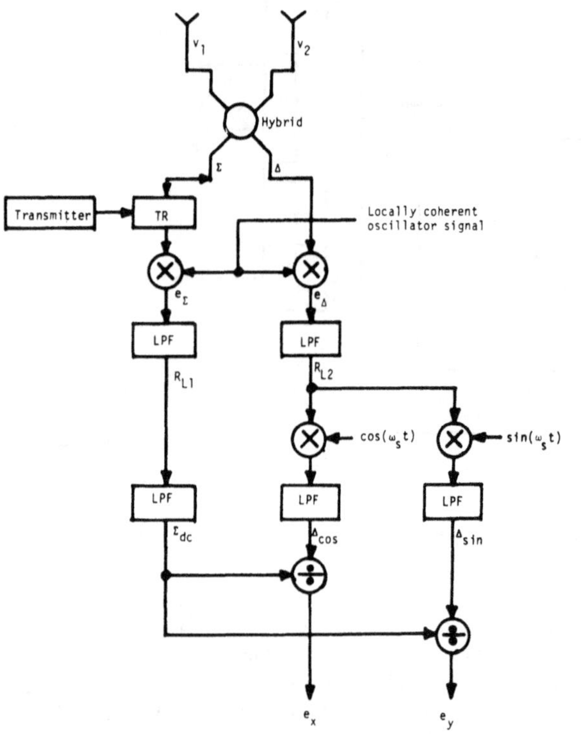

Fig. 4. Functional block diagram for a conical scan-monopulse (COMO) implementation of a conopulse system.

the system be coherent, and therefore, in the practical implementation to be described later, a noncoherent approach is demonstrated. The product outputs e_Σ and e_Δ are video pulse trains amplitude modulated by both target fluctuation and motion of the beams. These signals are next low-pass filtered to produce voltages R_{L1} and R_{L2}, respectively, that are proportional to the total modulation present on the signals Σ and Δ, provided only that the significant spectral extent of the total modulation does not exceed half the radar pulse rate.[4] We assume this condition satisfied. Thus, the maximum bandwidth of the low-pass filters that generate R_{L1} and R_{L2} is half the radar pulse rate.

Since both signals R_{L1} and R_{L2} have two factors, one due to target fluctuations (the same for both R_{L1} and R_{L2}) and one due to scan modulation (different for R_{L1} and R_{L2}), and since the scan modulations can be considered approximately periodic and thereby represented by Fourier series, then R_{L1} and R_{L2} each may be approximated as a dc term plus a sum of sinusoids all modulated (multiplied) by target fluctuation. With these considerations in mind, it can be shown by considering the spectra of R_{L1} and R_{L2} that the system of Fig. 4 completely eliminates target fluctuation if the spectral extent denoted B, of the fluctuation does not exceed the beam scan frequency f_s.[5] The bandwidth of the low-pass filters (acting on R_{L1} and at the outputs of the conical scan demodulator product devices) should not exceed $f_s - B$.

Analysis shows that if $B < f_s/2$ and the low-pass filters have bandwidths of B then Σ_{dc}, Δ_{cos}, and Δ_{sin} all fluctuate equally, and fluctuation is entirely eliminated by the normalization (ratio) operation in Fig. 4. However, if $f_s/2 \leqslant B < f_s$, the filter bandwidths must not exceed $f_s - B$, and part of the fluctuation elimination is due to filtering action of the low-pass filters and part is due to cancellation by normalization. For $B > f_s$, the COMO system cannot offer complete cancellation and some fluctuation noise will show up on e_x and e_y; the noise level will depend on the spectral form of the fluctuation.

From the above discussion it is clear that the COMO system is capable of removing only narrowband target fluctuations having a maximum spectral extent equal to the beam scan frequency. However, from earlier discussions it was found that the MOCO system can remove wideband fluctuations having a maximum spectral extent of half the radar pulse rate. Thus for many applications, the MOCO realization may be preferred.

A Practical COMO System

One possible practical COMO system block diagram is illustrated in Fig. 5. Again for simplicity, various range and angle tracking functions are omitted. When compared with Fig. 3, it is found that the principal difference in the two systems hinges around the location of the monopulse normalizing gain-control loops.

III. MOCO System Analysis

The following paragraphs analyze the MOCO system illustrated in Fig. 2. The goal is to derive expressions for the angle error signals e_x and e_y. The analysis begins by defining the geometry of space angles.

Space Angles

Fig. 6 illustrates angular locations of the target and beams in space with respect to the boresight axis at the origin. The x (reference) axis is the azimuth error axis and the y axis is the elevation error axis. The target is located by its offset angle θ_T and its rotation angle ϕ_T. The maximum or nose of beam one is squinted off boresight by a constant angle θ_q, and it therefore traces a circular path with time. Its angular frequency of rotation (scan) is denoted $\omega_s = 2\pi f_s$, and its angular position with time is $\phi_1 = \omega_s t$. Beam two is also squinted by an amount

[4] This condition, through application of the sampling theorem, allows the recovery of the total modulation function.

[5] If the beams are identical, theoretically it can be shown that the spectral extent can be twice as large as the scan frequency. In practice, however, beams are never identical and the theoretical situation probably should not be relied upon.

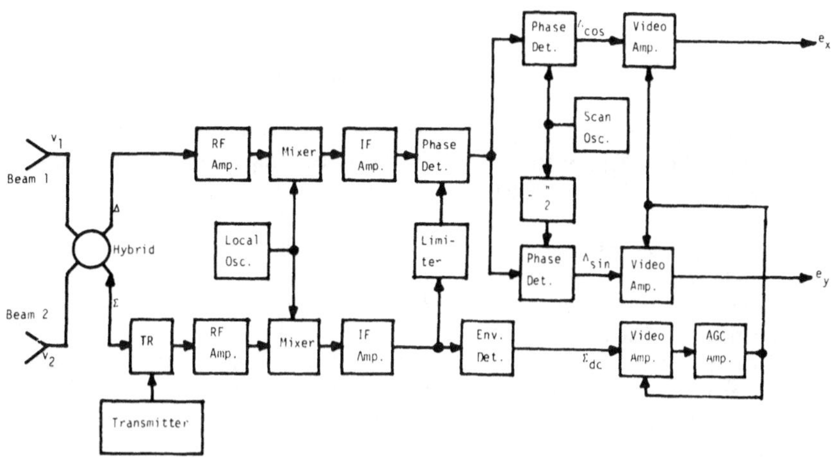

Fig. 5. Block diagram of a possible practical conopulse system of the COMO type.

θ_q, but its scan position is 180° from that of beam one. Thus $\phi_2 = \omega_s t + \pi$. The instantaneous target angles relative to the beam positions are denoted θ_1 and θ_2, as shown, and are given by

$$\theta_1^2 = \theta_q^2 + \theta_T^2 - 2\theta_q\theta_T \cos(\omega_s t - \phi_T) \tag{1}$$

$$\theta_2^2 = \theta_q^2 + \theta_T^2 + 2\theta_q\theta_T \cos(\omega_s t - \phi_T). \tag{2}$$

Sum and Difference Signals

By assuming complex waveforms, the transmitter output signal can be represented by $F(t)\exp(j\omega_0 t + j\theta_0)$, where ω_0 is the radar angular frequency, θ_0 is an arbitrary phase angle, and $F(t)$ represents a periodic train of video pulses, each with duration τ. The period between pulses is denoted T_R.

For simplicity, assume that the one-way voltage patterns of the beams are rotationally symmetric, so that they become functions of only the offset angles of the target from the beam maxima. If these patterns are denoted $G_1(\theta)$ and $G_2(\theta)$, where θ is the angle from the beam's maximum, it is found that the signals Σ and Δ in Fig. 2 are:

$$\Sigma = K[G_1(\theta_1) + G_2(\theta_2)]^2 F(t - \tau_T)\exp[j\omega_0 t + j\theta_0 + j\beta(t)] \tag{3}$$

$$\Delta = K[G_1(\theta_1) + G_2(\theta_2)][G_1(\theta_1) - G_2(\theta_2)]F(t - \tau_T) \cdot \exp[j\omega_0 t + j\theta_0 + j\beta(t)]. \tag{4}$$

Here $\beta(t)$ is a phase angle associated with the target, τ_T is the two-way time delay to the target, and K is a nonnegative real proportionality factor that is determined by both target cross-section fluctuations[6] and factors in the range

[6] Target phase fluctuations are assumed to be part of $\beta(t)$.

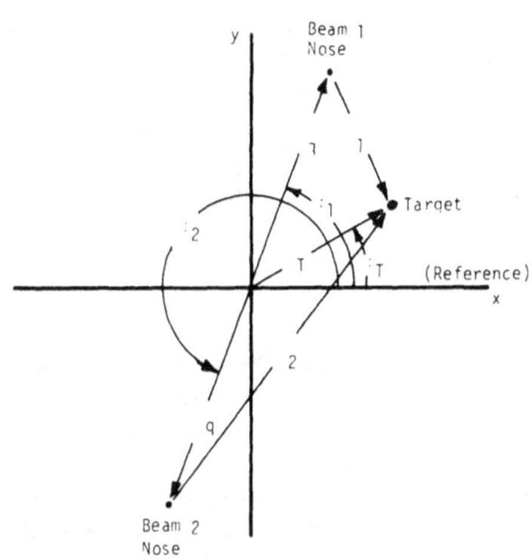

Fig. 6. Geometry of angles that locate beam and target positions relative to the radar boresight axis.

equation. In deriving (3) and (4) it was assumed that the transmitter radiates through the sum pattern $G_1(\theta_1) + G_2(\theta_2)$.

Signal Processing

By forming the ratio $R = \Delta/\Sigma$ through the use of ordinary monopulse normalization methods, an amplitude-modulated pulse waveform is produced:

$$R = \Delta/\Sigma = \{[G_1(\theta_1) - G_2(\theta_2)]/[G_1(\theta_1) + G_2(\theta_2)]\} \cdot \{\sum_{k=-\infty}^{\infty} \text{rect}[(t - \tau_T - kT_R)/\tau]\}, \tag{5}$$

where we define

$$\text{rect}(t/\tau) = 1, |t| < \tau/2$$
$$= 0, |t| > \tau/2. \qquad (6)$$

The second braced factor in (5) represents the normalized train of rectangular received pulses[7] (envelope of RF pulses), while the beam-dependent factor is the modulation function due to scanning.

We have assumed that the target is either stationary or slowly moving with respect to the radar, so τ_T may be taken as approximately constant and the second braced term in (5) is a periodic function with period T_R. The Fourier transform of R then consists of replica spectra of the modulation function centered, in general, at frequencies $kf_R = k/T_R$, $k = 0, \pm 1, \pm 2, \cdots$. In other words, it is just the spectrum of a naturally sampled pulse amplitude-modulated signal. It is well known that for such signals a band-limited modulation function can be recovered at the output of a low-pass filter if the "sampling" frequency, analogous to f_R here, is not less than twice the spectral extent of the modulation function [8]. This modulation function approximately satisfies these conditions, since its bandwidth will not be greatly larger than f_s, while for most radar systems $f_R \gg f_s$. Thus, if the bandwidth of the low-pass filter in Fig. 2 is about $f_R/2$, R_L will be given approximately by

$$R_L = (\tau/T_R)\{[G_1(\theta_1) - G_2(\theta_2)]/[G_1(\theta_1) + G_2(\theta_2)]\}. \qquad (7)$$

Now because R_L is approximately periodic, with fundamental angular frequency ω_s, the product and low-pass filter operations of Fig. 2 function as coherent detectors operating on the term in the Fourier series having angular frequency ω_s. Thus, the outputs e_x and e_y will be proportional to the "cosine" and "sine" Fourier series coefficients corresponding to the angular frequency ω_s. More precisely, if $T_s = 1/f_s$, then

$$\left.\begin{matrix} e_x \\ \\ e_y \end{matrix}\right\} = 1/T_s \int_{-T_s/2}^{T_s/2} R_L(t) \left\{\begin{matrix} \cos(\omega_s t) \\ \\ \sin(\omega_s t) \end{matrix}\right\} dt. \qquad (8)$$

This result will be called the exact solution for e_x and ϵ_y. In general, it requires computation by digital computer after specific patterns have been specified.

In order to obtain closed-form approximations for e_x and e_y, some method of approximating arbitrary patterns is required. One approach has been developed. For clarity of presentation, however, the details have been relegated to Appendix B. Applying (51) and (52) from Appendix B to (7) gives

$$(T_R/\tau)R_L \approx -\sum_{\substack{r=1 \\ r\ \text{odd}}}^{\infty} A_r Y^r \Big/ \sum_{\substack{r=0 \\ r\ \text{even}}}^{\infty} A_r Y^r. \qquad (9)$$

[7] For arbitrarily shaped pulses, the second braced factor in (5) is replaced by $|F(t - \tau_T)/F(0)|^2$.

Here A_r is a coefficient defined by (50) and Y is time dependent, being defined by (45). We now consider only two special cases of target angle.

The conical scan detectors of Fig. 2 produce voltages with amplitudes proportional to the amplitudes of the Fourier series components of R_L that have frequency f_s: in other words, the frequency of Y. For $\theta_q < \theta_B/2$ and a target near the boresight, as in a tracking situation, $|Y| \ll 1$ and a good approximation for R_L is

$$R_L \approx (-\tau A_1/T_R A_0)\, Y. \qquad (10)$$

For conditions of larger beam squint and target offset angles, it is assumed that terms in Y of order five or larger are negligible. With this assumption:

$$R_L \approx (-\tau/T_R)(A_1 Y + A_3 Y^3)/(A_0 + A_2 Y^2 + A_4 Y^4). \qquad (11)$$

Now the exact behavior of the coefficients A_r is difficult to determine except for specific pattern assumptions. However, if A_0 and A_1 are approximated by the leading term in their series, it is readily found that (10) reduces to

$$R_L \approx (-\tau/T_R)(c_2 d_1^2/c_0)(\theta_q \theta_T/\theta_B^2) \cos(\omega_s t - \phi_T), \qquad (12)$$

where the coefficients c_0, c_2, and d_1 are defined by (39) and (41) in Appendix B. Finally, angle-error signals become

$$\left.\begin{matrix} e_x \\ \\ e_y \end{matrix}\right\} \approx (-\tau/T_R)(c_2 d_1^2/2c_0)\, \theta_q \theta_T/\theta_B^2 \left\{\begin{matrix} \cos(\phi_T) \\ \\ \sin(\phi_T) \end{matrix}\right\} \qquad (13)$$

for the case of small target offset angles.

For the intermediate case of target angle, (11) can be further approximated by its Maclaurin expansion in the variable Y and neglecting terms of order five or higher. After algebraic reduction,

$$\begin{aligned} R_L \approxeq\ & (-\tau/T_R)\{(A_1/A_0) - (3/4)[2\theta_q \theta_T/\theta_B^2]^2 \\ & [(A_1 A_2/A_0^2) - (A_3/A_0)]\}\, (2\theta_q \theta_T/\theta_B^2) \cos(\omega_s t - \phi_T) \\ & \cdot - (\tau/4T_R)(A_1 A_2/A_0^2 - A_3/A_0)(2\theta_q \theta_T/\theta_B^2)^3 \\ & \cdot \cos(3\omega_s t - 3\phi_T), \end{aligned} \qquad (14)$$

so, at the outputs of the conical scan detectors,

$$\left.\begin{matrix} e_x \\ \\ e_y \end{matrix}\right\} = -(\tau/T_R)[A_1/A_0 - (3/4)(2\theta_q \theta_T/\theta_B^2)^2 (A_1 A_2/A_0^2 - A_3/A_0)]\, \theta_q \theta_T/\theta_B^2 \left\{\begin{matrix} \cos(\phi_T) \\ \\ \sin(\phi_T) \end{matrix}\right\}. \qquad (15)$$

Further reduction of (15) is possible. The procedure used is to approximate A_0 by its series terms up to fourth order and expand either $1/A_0$ or $1/A_0^2$ into a Maclaurin's series.

Upon forming appropriate products, dropping higher order terms, and defining coefficients a_n by

$$a_n = c_{2n} d_1^{2n}/(2n)!, \quad n = 1, 2, 3, 4, \tag{16}$$

we obtain

$$\left.\begin{matrix} e_x \\ e_y \end{matrix}\right\} = (-\tau/T_R)\{a_1/a_0 - (a_1^2/a_0^2 - 2a_2/a_0)X + (a_1^3/a_0^3 - 3a_1a_2/a_0^2 + 3a_3/a_0)X^2 - (a_1^4/a_0^4 - 4a_1^2a_2/a_0^3 + 4a_1a_3/a_0^2 + 2a_2^2/a_0^2 - 4a_4/a_0)X^3 - (3/4)(2\theta_q\theta_T/\theta_B^2)^2[a_1a_2/a_0^2 - a_3/a_0 - (2a_1^2a_2/a_0^3 - 4a_1a_3/a_0^2 - 2a_2^2/a_0^2 + 4a_4/a_0)X]\}\,\theta_q\theta_T/\theta_B^2 \begin{cases} \cos(\phi_T) \\ \sin(\phi_T) \end{cases} \tag{17}$$

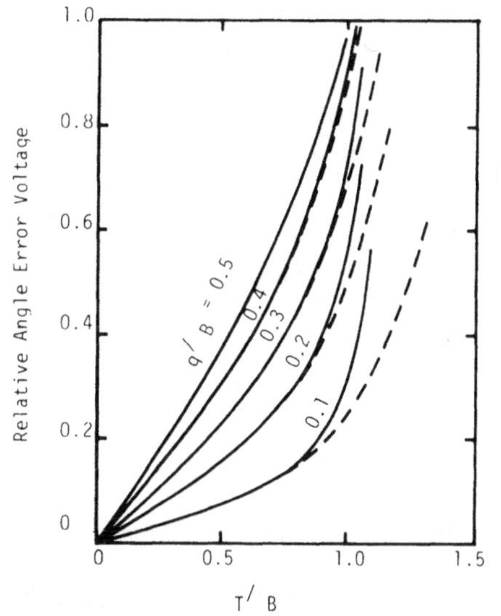

Fig. 7. Angle error voltage for a MOCO form of conopulse radar when beam patterns have the form $2J_1(f)/f$.

Example of Bessel Function Pattern

The three methods described above for obtaining angle-error voltage may be illustrated for the pattern defined by (37) and (36) of Appendix B. The exact method uses (7) with (8). Fig. 7 (solid lines) is the result of a calculation of (8) by digital computer. The ordinate is the value of $e_x T_R/\tau$ when $\phi_T = 0$ or the value of $e_y T_R/\tau$ when $\phi_T = \pi/2$.

The coefficients a_n required in the approximation (17) applicable to intermediate target offset angles are found from (16) using (39):

$$a_n = (-1)^n (1.02\pi)^{2n}/2^{2n+1}\, n!\,(n+1)!, \quad n = 0, 1, 2, 3, 4. \tag{18}$$

Calculation of (17) produces the dashed curves in Fig. 7. It is seen that the intermediate approximation gives nearly the same values as the Fourier series (exact) method when $\theta_T/\theta_B < 0.8$ and $\theta_q/\theta_B < 0.5$. Similarly, by direct calculation, the small-angle linear approximation (13) can be found to produce nearly exact results when $\theta_T/\theta_B < 0.3$ and $\theta_q/\theta_B < 0.3$. Of course, these conclusions are strictly valid only for the assumed pattern but should not be greatly different for other patterns.

IV. COMO System Analysis

Because the COMO system of Fig. 4 may not be as attractive as the MOCO system for some applications, its analysis will be brief.

Signals

After carrier-phase removal, the video signals e_Σ and and e_Δ are

$$e_\Sigma = K[G_1(\theta_1) + G_2(\theta_2)]^2 F(t - \tau_T) \tag{19}$$

$$e_\Delta = K[G_1(\theta_1) + G_2(\theta_2)][G_1(\theta_1) - G_2(\theta_2)]\,F(t - \tau_T). \tag{20}$$

These waveforms are each pulse trains (period T_R) amplitude modulated by beam scanning, which is an approximately periodic function T_s and target fluctuation contained in K. If $T_R \ll T_s$ and target fluctuation is slow, the low-pass filter output signals are

$$R_{L1} = F(0)(\tau/T_R)K[G_1(\theta_1) + G_2(\theta_2)]^2 \tag{21}$$

$$R_{L2} = F(0)(\tau/T_R)K[G_1(\theta_1) - G_2(\theta_2)][G_1(\theta_1) + G_2(\theta_2)] \tag{22}$$

where $F(0)$ is the amplitude of the pulses (assumed rectangular) of duration τ. The bandwidth of each low-pass filter is assumed to be approximately $1/2T_R$.

Next, assume that the spectral extent of target fluctuation does not exceed $1/2T_s$. Then the signal R_{L1}, after passing through a low-pass filter of bandwidth $1/2T_s$, will become $F(0)\tau K/T_R$ times the dc component of $[G_1(\theta_1) + G_2(\theta_2)]^2$. Now let us consider only the case of the small target offset angle θ_T. From (51), (45), and (50), the dc component of $[G_1(\theta_1) + G_2(\theta_2)]^2$ is approximately $(2A_0)^2 \approx 4c_0^2$. Hence,

$$\Sigma_{dc} \approx F(0)\tau 4c_0^2 K/T_R. \tag{23}$$

The action of the conical scan detectors is to pick out

the sine and cosine components of the Fourier series representation of R_{L2} that correspond to the angular frequency ω_s. Thus, the only term in a Fourier series expansion of R_{L2} that will contribute to the formation of $\Delta_{\cos}$ or $\Delta_{\sin}$ is of the form $[F(0)\ \tau K/T_R]\ \alpha \cos(\omega_s t - \phi_T)$, where α is the amplitude of the term in the Fourier series expansion of $[G_1(\theta_1) + G_2(\theta_2)][G_1(\theta_1) - G_2(\theta_2)]$ that has frequency f_s. Forming the product of (51) and (52) and taking a first-order approximation for the resulting series gives

$$\alpha \approx -4c_0c_2d_1^2\theta_q\theta_T/\theta_B^2. \tag{24}$$

The conical scan detector outputs are now seen to be

$$\Delta_{\cos} = -(F(0)\tau 2c_0c_2d_1^2K/T_R)(\theta_q\theta_T/\theta_B^2)\cos(\phi_T) \tag{25}$$

$$\Delta_{\sin} = -(F(0)\ \tau 2c_0c_2d_1^2K/T_R)(\theta_q\theta_T/\theta_B^2)\sin(\phi_T), \tag{26}$$

if each low-pass filter has a bandwidth of $1/2T_s$ and the spectral extent of target fluctuation does not exceed $1/2T_s$.

Finally, forming the ratios indicated in Fig. 4 yields the angle-error signals

$$\left.\begin{matrix} e_x \\ \\ e_y \end{matrix}\right\} \approx -(c_2d_1^2/2c_0)\,\theta_q\theta_T/\theta_B^2 \left\{\begin{matrix} \cos(\phi_T) \\ \\ \sin(\phi_T) \end{matrix}\right. . \tag{27}$$

These signals are free of slow target fluctuations.

More general expressions for angle-error voltages have been developed [9] that are analogous to (17) for the MOCO system. However, we note that (27) is equivalent to the more general expression when θ_T is small, as would be the case when tracking with small error. Consequently, (27) is adequate for most practical situations.

V. Summary and Discussion

In this paper a radar-angle-tracking technique called conopulse has been described and analyzed to determine the angle-error voltages. It is a sort of hybrid system where signal processing is done partly by monopulse and partly by conical scan techniques. The main advantage of the conopulse tracking technique is its independence from target cross-section fluctuations. Its main disadvantage lies in the difficulty of realizing an antenna that can provide the required scanning of two simultaneous beams.

Two conopulse system implementations were described. In one, called the MOCO system illustrated in Fig. 2, rapid target fluctuations could be eliminated, in principle. The angle-error voltages for targets near the boresight axis are given by (13). For somewhat larger target error angles the error voltages are given by (17). For still larger error angles, (8) applies.

The second form of conopulse, called the COMO system, is somewhat less attractive than the MOCO system because it is only capable of reducing slow fluctuations in target signal strength. It is illustrated in Fig. 4, and its small angle-error voltages are given by (27). Comparing (27) with (13) we see that the MOCO system outputs differ from those of the COMO system by only the scale constant τ/T_R. Thus, the two systems give the same noise-free error-voltage performance. The factor τ/T_R is of no consequence as it represents only a gain constant. In fact, if a sample-hold circuit and equalizing filter [see 8, p. 314] are inserted at the divider output in Fig. 2, the error voltages of the MOCO system become identical to those of the COMO system given by (27).

Finally, we note that Johnson [10] has recently discussed the performance of conopulse (called scan with compensation in [10]) in a jamming environment.

Appendix A

Consider Fig. 8. The ratio Δ/Σ of two pulsed-carrier signals Σ and Δ is formed using two identical gain-controlled intermediate-frequency (IF) amplifiers of voltage gain A, a phase detector, an envelope detector, and a baseband or video amplifier. Now

$$e_1 = A\Delta \tag{28}$$

$$e_2 = A\Sigma. \tag{29}$$

The envelope detector, video amplifier, and one IF amplifier form a wide-band automatic gain-controlled amplifier loop. The loop is assumed so fast as to be nearly instantaneous, thereby providing instantaneous automatic gain control (IAGC). The loop acts to force e_2 to have a fixed-pulse magnitude γ. Then $|e_2| = \gamma = A|\Sigma|$, so

$$A = \gamma/|\Sigma| \tag{30}$$

$$e_2 = \gamma\Sigma/|\Sigma| \tag{31}$$

$$e_1 = \gamma\Delta/|\Sigma|. \tag{32}$$

If e_1 and e_2 are expressed in exponential form, the action of the phase detector is to produce a voltage $R = e_1e_2^*$, where e_2^* is the complex conjugate of e_2. Thus,

$$R = \gamma^2\Delta\Sigma^*/|\Sigma|^2 = \gamma^2\Delta/\Sigma \tag{33}$$

and R is proportional to Δ/Σ.

If Δ and Σ are real video (pulsed) signals, the phase detector of Fig. 8 becomes a product device. As before $|e_2| = \gamma = A|\Sigma|$, so $A = \gamma/|\Sigma|$. Now, however, $e_1 = \gamma\Delta/|\Sigma| = (\gamma\Delta/\Sigma)(\Sigma/|\Sigma|) = (\gamma\Delta/\Sigma)\ \mathrm{sgn}(\Sigma)$ and $e_2 = \gamma\Sigma/|\Sigma| = \gamma\mathrm{sgn}(\Sigma)$, where $\mathrm{sgn}(\Sigma)$ represents the sign of Σ. The product is

$$R = e_1e_2 = \gamma^2\Delta/\Sigma. \tag{34}$$

In the special case where Σ is a nonnegative quantity, the product and envelope detector can be eliminated. Now $e_2 = A\Sigma = \gamma$ and $A = \gamma/\Sigma$. Thus, $e_1 = A\Delta = \gamma\Delta/\Sigma$ and e_1 becomes proportional to Δ/Σ.

Appendix B

It is convenient to write an arbitrary[8] pattern $G(\theta)$ in terms of a function f of θ instead of θ. In other words, $G(\theta) = G[f(\theta)]$. For example, the pattern of a uniformly illuminated circular aperture is

$$G(\theta) = 2J_1[(1.02\pi/\theta_B)\sin(\theta)]/(1.02\pi/\theta_B)\sin(\theta), \tag{35}$$

where $J_1(\cdot)$ is the Bessel function of the first kind of order one, and θ_B is the pattern's 3-dB beamwidth. Defining

$$f = 1.02\pi/\theta_B \sin(\theta), \tag{36}$$

allows the convenient representation

$$G(f) = 2J_1(f)/f. \tag{37}$$

The approximation of an arbitrary pattern is accomplished by using a Maclaurin series expansion:

$$G(\theta) = G[f(\theta)] = \sum_{n=0}^{\infty} [c_{2n}/(2n)!]\,[f(\theta)]^{2n} \tag{38}$$

where

$$c_{2n} = d^{2n}G(f)/df^{2n}|_{f=0}. \tag{39}$$

For many narrow-beam systems the important values of θ are small enough for $f(\theta)$ to be approximated well by the first term of its Maclaurin's series expansion:

$$f(\theta) \approx d_1(\theta/\theta_B) \tag{40}$$

where

$$d_1 = \theta_B[df(\theta)/d\theta]|_{\theta=0}. \tag{41}$$

With this approximation (38) may be written as

$$G(\theta) \approx \sum_{n=0}^{\infty} [c_{2n}d_1^{2n}/(2n)!]\,(\theta/\theta_B)^{2n}. \tag{42}$$

To approximate the response of beam one we write (1) as

$$(\theta_1/\theta_B)^2 = (\theta_q/\theta_B)^2 + (\theta_T/\theta_B)^2 - (2\theta_q\theta_T/\theta_B^2)\cos(\omega_s t - \phi_T) = X - Y \tag{43}$$

where X and Y are defined as

$$X = (\theta_q/\theta_B)^2 + (\theta_T/\theta_B)^2 \tag{44}$$

$$Y = (2\theta_q\theta_T/\theta_B^2)\cos(\omega_s t - \phi_T). \tag{45}$$

[8] All patterns are assumed to be rotationally symmetric (pencil beams) but may have an arbitrary shape.

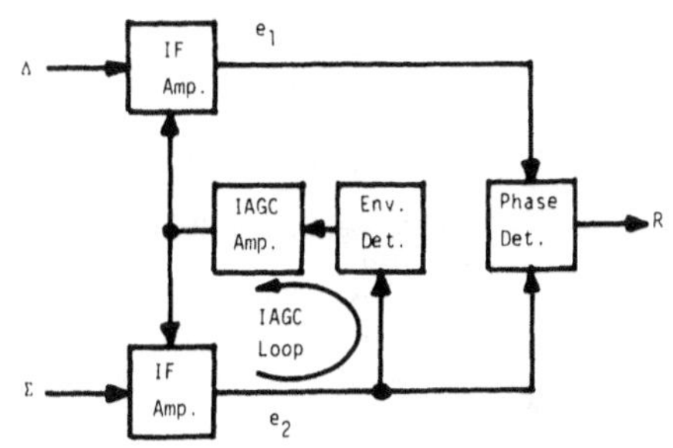

Fig. 8. Block diagram of monopulse normalizing networks.

Thus, using the binomial expansion theorem, we can write

$$(\theta_1/\theta_B)^{2n} = (X - Y)^n = \sum_{r=0}^{n} (-1)^r \binom{n}{r} X^{n-r}\, Y^r, \tag{46}$$

where

$$\binom{n}{r} = n!/r!(n-r)!. \tag{47}$$

Equation (38) can now be written as

$$G_1(\theta_1) \approx \sum_{n=0}^{\infty} [c_{2n}d_1^{2n}/(2n)!] \sum_{r=0}^{n} (-1)^r \binom{n}{r} X^{n-r}\, Y^r, \tag{48}$$

or equivalently,

$$G_1(\theta_1) \approx \sum_{r=0}^{\infty} (-1)^r \Big\{\sum_{n=r}^{\infty} [c_{2n}d_1^{2n}/(2n)!] \binom{n}{r} X^{n-r}\Big\}\, Y^r. \tag{49}$$

For $G_2(\theta_2)$ the same expression is obtained but without the $(-1)^r$ factor in each term; that is, the sign of Y changes.

If coefficients A_r, $r = 0, 1, 2, \cdots$, are defined by

$$A_r = \sum_{n=r}^{\infty} [c_{2n}d_1^{2n}/(2n)!] \binom{n}{r} X^{n-r}, \tag{50}$$

then by substitution of appropriate expressions developed above, we arrive at the approximations

$$G_1(\theta_1) + G_2(\theta_2) \approx 2 \sum_{\substack{r=0\\ r\ \text{even}}}^{\infty} A_r Y^r \tag{51}$$

and

$$G_1(\theta_1) - G_2(\theta_2) \approx -2 \sum_{\substack{r=1\\ r\ \text{odd}}}^{\infty} A_r Y^r. \tag{52}$$

Acknowledgments

The authors express their thanks to E.H. Beneke and Dean Howard for helpful comments regarding this paper. Thanks are also extended to D.K. Barton and the reviewers whose comments and suggestions have materially improved the paper.

References

[1] M.I. Skolnik, *Introduction to Radar Systems*. New York: McGraw-Hill, 1962.
[2] D.K. Barton, *Radar System Analysis*. Englewood Cliffs, N.J.: Prentice-Hall, 1964.
[3] R.M. Page, "Monopluse radar," *IRE Conv. Rec.*, vol. 8, pp. 132-134, 1955.
[4] D.R. Rhodes, *Introduction to Monopulse*. New York: McGraw-Hill, 1959.
[5] P.Z. Peebles, Jr., "Signal processors and accuracy of three beam monopulse tracking radar, *IEEE Trans. Aerosp. Electron. Syst.*, vol. AES-5, pp. 52-57, Jan. 1969.
[6] P.Z. Peebles, Jr., and R.S. Berkowitz, "Multiple target monopulse radar processing techniques, *IEEE Trans. Aerosp. Electron. Syst.*, vol. AES-4, pp. 845-854, Nov. 1968.
[7] P.A. Bakut, et. al., *Questions of the Statistical Theory of Radar*, vol. II, Defense Documentation Center report AD 645775, June 28, 1966.
[8] P.Z. Peebles, Jr., *Communication System Principles*, Reading, Mass.: Addison-Wesley, 1976.
[9] H. Sakamoto, "Analysis of conopulse radar systems," Ph.D. dissertation, University of Tennessee, Dec. 1975.
[10] G.R. Johnson, "Jamming passive lobing radars," *Elec. World*, pp. 75-86, Mar./Apr. 1977.

Corrections to "Conopulse Radar"[1]

Two errors in the above paper have been brought to our attention by Mr. Gilbert R. Johnson. The ninth line of the third paragraph starting on p. 201 should read "... fluctuations at twice the ... ." Reference [10] should read "... *Elec. Warfare*, pp. 75-86, ..."

Several other corrections that should be made are: (1) parentheses should be placed around the factor $1/T_s$ in (8), around the factor $\theta_q \theta_T/\theta_B^2$ in (15), (17) and (27), around the factor $1.02\pi/\theta_B$ in (36); (2) the product dot on the third line of (14) should be on the second line; (3) the third line of text following (20) should have T_s replaced by (period T_s).

[1] H. Sakamoto and P.Z. Peebles, Jr., *IEEE Trans. Aerospace and Electronic Systems*, vol. AES-14, Jan. 1978, pp. 199-208.

Hajime Sakamoto was born in Yokohama, Japan, on January 23, 1938. He received the B.S., M.S. degrees in electrical engineering in 1963 and 1965, respectively, from Waseda University, Tokyo, Japan. He received his Ph.D. degree in electrical engineering in 1975 from the University of Tennessee.

In 1965 he entered the Defense Agency, Japan, and his work has involved the research and development of communication systems. He is now a staff official of the Research and Development Planning Division, Bureau of Equipment, Defense Agency, Japan.

Dr. Sakamoto is a member of Eta Kappa Nu and Phi Kappa Phi.

Peyton Z. Peebles, Jr., was born in Columbus, Ga., on September 10, 1934. He holds the B.S., M.S., and Ph.D. degrees from Evansville College (1957), Drexel Institute of Technology (1963), and the University of Pennsylvania (1967), respectively, all in electrical engineering. From 1957 to 1969 he worked for RCA on various radar design and system problems. He attended the University of Pennsylvania for two years on a David Sarnoff fellowship awarded by RCA. Since 1969 he has been with the University of Tennessee, except during 1976-77 when he was visiting professor of electrical engineering at the University of Hawaii, Honolulu. He is now professor of Electrical Engineering at the University of Tennessee where his research interests primarily involve radar and communication systems.

Dr. Peebles is a member of Tau Beta Pi, Eta Kappa Nu, Sigma Pi Sigma, and Phi Beta Chi.

Chapter 5
ECCM in the Antenna: Sidelobe Blanker/Canceller

Summary

The papers in this chapter treat two special aspects of ECCM in the antenna. As indicated in Chapter 4, sidelobe blanker and sidelobe canceller are treated separately due to their current importance.

Papers in Chapter 2 have cited the importance of antenna sidelobes in an ECM environment and the general principles of sidelobe blankers and cancellers. Sidelobe cancellers are an alternative/companion to low sidelobe antennas of Chapter 4. As noted in Paper 1.1, a very large number of papers have been published on sidelobe cancellers including entire issues of an *IEEE Transactions.* Several of these papers are survey papers. Accordingly, this chapter will include only representative papers on these topics. One of these papers is a "classic", another has had only limited distribution previously, while the last paper addresses important aspects which may be overlooked by others.

Synopsis of Reprint Papers in Chapter 5

A. Sidelobe Blankers

5.A.1 "A Sidelobe Blanking System Design and Demonstration" — Arancibia

This is a recent paper which is representative of the literature. Basic theory and principles of sidelobe blanking and the design and implementation of working circuits are presented together with examples of the displayed signals and results. It will be noted that the SLB is intended for use against low duty cycle, e.g., 0.01 (pulsed) ECM, and not noise. This paper shows results for only one jammer. Typically several jammers at the same azimuth or at different azimuths may be encountered. As indicated, failure to suppress main beam jamming may make detection of targets in the direction of the jammer more difficult.

B. Sidelobe Cancellers

5.B.1 "Adaptive Arrays and Sidelobe Cancellers; A Perspective" — Chapman

This survey paper presents the history of sidelobe canceller work at the Syracuse Research Corporation from 1959. The theory of operation, implementation, limitations, and conclusions serve as an excellent introduction to this topic.

5.B.2 "Adaptive Arrays" — Applebaum

This is one of the earliest papers on this subject and was originally published in 1966. The concepts of adaptive arrays presented here are actually a generalization of the coherent sidelobe cancellation techniques. There are many practical considerations regarding sidelobe cancellers which are not presented by Applebaum. Some of these are given in the next two papers in this chapter.

5.B.3 "Understanding and Optimizing Multiple Sidelobe Canceller Operation" — Lewis and Hansen

This paper describes a simple technique for analyzing the operation of multiple sidelobe canceller loops on multiple jammers. That technique is a simple graphical method which requires very little mathematics yet gives its user a feel for canceller operation that is missing in the conventional mathematical approach. Examples of the application of this technique which are given demonstrate the fallacy of the common belief that *n* canceller loops can adequately cancel the signals from *n* jammers when the jammer geometry is variable and the canceller geometry is fixed. (Authors' summary)

5.B.4 "Problems of Adaptive Sidelobe Suppression" Ries and Krücker

Decorrelation effects on adaptive sidelobe suppression are investigated. Sources of decorrelation treated include the antenna configuration, mismatch between the receiving channels, target signal characteristics, bandwidth, and correlator offset errors in the adaptive control loops. Results of a special experimental system are included. The term "processing gain" has now been defined as the ECCM Improvement Factor (see Papers 1.1 and 9.2). In one illustrative example, bandlimited white noise is assumed.

The authors indicate that the "processing gain" of an adaptive sidelobe canceller is affected by bandwidth of the system. This as stated, in turn, is affected by the interfering noise spectrum. Results are given only for white noise. Tong and Steichen (Paper 7.3) in their examination of CFAR performance in an ECM environment consider the effects of amplitude modulated noise and non-Rayleigh noise. Turner *et al.* (Paper 7.4) provide experimental data which shows that some noise jammers are non-Gaussian. Johnson and Stoner (Paper 2.C.3) present a valid illustration of the performance of a coherent sidelobe canceller.

Bibliography Topics for Chapter 5

Sidelobe Blankers
Sidelobe Cancellers

BIBLIOGRAPHY FOR CHAPTER 5

SIDELOBE BLANKERS

Lawson, J.L., "Pulsed Interference Suppression," *MIT Radiation Lab Rpt R-826,* Oct 24, 1945, DDC ATI 24688.

Engels, R., "An Impulse Interference Blanker for Communications Receivers," *IRE Trans.,* vol. CS-4, no. 3, Oct 1956, pp. 22-26.

Kurland, H., "Interference Blanker for Normal and MTI Video," *Proc. NEC,* vol. 13, 1957, pp. 738-43.

Maisel, L., "Performance of Sidelobe Blanking Systems," *IEEE Trans.,* vol. AES-4, no. 2, March 1968, pp. 174-180.

Finn, H.M., R.S. Johnson, and P.Z. Peebles, Jr., "Fluctuating Target Detection in Clutter Using Sidelobe Blanking Logic," *IEEE Trans.,* vol. AES-7, no. 1, Jan 1971, pp. 147-159.

Marcum, J.I., "Interference Suppression," *US Patent No. 4,010,469,* Mar 1, 1977.

Arancibia, P.O., "A Sidelobe Blanking System Design and Demonstration," *Microwave J.,* vol. 21, no. 3, March 1978, pp. 69-73 (reprinted in this volume).

SIDELOBE CANCELLERS

Ludbrook, R.J., "Side-lobe Suppression in S.S.R., (Secondary Surveillance Radar)," *Marconi Instruments,* vol. 9, no. 5, March 1964, pp. 121-126.

Croney, J. and P.R. Wallis, "A Side-lobe Suppression System for Primary Radar," *The Radio and Electronic Engr.,* vol. 28, no. 4, Oct 1964, pp. 247-260.

Howells, P.W., "Intermediate Frequency Side-lobe Canceller," *US Patent No. 3,202,990,* Aug 24, 1965.

Widrow, B., P.E. Mantey, L.J. Griffiths, and B.B. Goode, "Adaptive Antenna Systems," *Proc. IEEE,* vol. 55, no. 12, Dec 1967, pp. 2143-2159.

Griffiths, L.J., "Signal Extraction Using Real-time Adaptation of Linear Multi-channel Filter," Stanford Univ Center for Systems Res., *Syst. Theory Lab Tech. Rpt. No. 6788-1,* DDC AD 836761, Feb 1968.

Lacoss, R.T., "Adaptive Combining of Wideband Array Data for Optimal Reception," *IEEE Trans.,* vol. GE-6, no. 2, May 1968, pp. 78-86.

Crotty, R.E. and C.G. Goss, "Sidelobe Response Reducing System," *US Patent No. 3,412,405,* Nov 19, 1968, (abstract SA 69-8023).

Mantey, P.E., and L.J. Griffiths, "Iterative Least-squares Algorithms for Signal Extraction," *Proc. 2nd Hawaii Int. Conf. on Systems Sciences,* 1969, pp. 767-770.

Howard, D.D., "Sidelobe Cancelling System for Array Type Target Detectors," *US Patent No. 3,435,453,* March 25, 1969(abstract SA 69-21150).

Mermoz, H., "Elimination of Jammers by Optimum Treatment of Antenna Arrays," *Ann. Telecommun.,* vol. 24, no. 7-8, July 1969, pp. 282-293 (in French, trans. abstract SA 60-4675).

Drabowitch, S., "Signal Processing Antennas," *E.I.* (India), vol. 51, May 1971, pp. 144-152.

Marconi Co., Ltd., "Radar System," *UK Patent No. 1,236,501,* 23 June 1971, abstract SA 71-34554).

Baird, C.A., Jr. and J.T. Rickard, "Recursive Estimation of Array Processing," *Proc. 5th Asilomar Conf. on Circuits and Systems,* Nov 8-10, 1971, pp. 509-513.

Baird, C.A., Jr., and C.L. Zahm, "Performance Criteria for Array Processing," *IEEE Conf. on Decision and Control Proc.,* 1971, pp. 564-565.

Waddle, F.M., "A Preliminary Investigation of a Multiple Channel Sidelobe Canceller," *US Army Missile Command Rpt. RS-TR-72-2,* Sept 1972, DDC AD 753571.

Brennan, L.E. and I.S. Reed, "Theory of Adaptive Radar," *IEEE Trans.,* vol. AES-9, no. 2, Mar 1973, pp. 237-252.

Zahm, C.L., "Application of Adaptive Arrays to Suppress Strong Jammers in the Presence of Weak Signals," *IEEE Trans.,* vol. AES-9, no. 2, March 1973, pp. 260-271.

Riegler, R.L., and R.T. Compton, Jr., "An Adaptive Array for Interference Rejection," *Proc. IEEE,* vol. 61, no. 6, June 1973, pp. 748-758.

Brook-Footitt, E., "Equipment for Suppressing Reflected Secondary Radar Signals," *IEE Conf. Pub. No. 105, Radar—Present and Future,* London 23-25 Oct 1973, pp. 38-42.

Lewis, B.L. and J.P. Hansen, "Understanding and Optimizing Multiple Sidelobe Canceller Operation," *Naval Res. Lab. Rpt. 7610,* Oct 25, 1973 (reprinted in this volume). DDC AD 914536.

Chapman, D.J, "ECM/ECCM Studies, Vol 3: Adaptive Array Technique Study," *Syracuse Univ. Res. Corp. Rpt. SURC-TR-73-186-vol. 3,* Feb 1974, DDC AD 919655.

Krücker, K. and H.W. Bock, "Theoretische and Experimentelle Untersuchungen zur Signalverarbeitung in eim Adaptiven Raumlichen Filter," *Forschungsbericht Nr 3-74,* FHP-Werthoven (in German).

Brennan, L.E., I.S. Reed, and P.Swerling, "Adaptive Arrays," *Microwave J.,* vol. 17, no. 5, May 1974, pp. 43ff.

Staudaher, F.M., "An Experimental Evaluation of Adaptive Radar," *IEEE Int. APS Symp. Digest,* June 1974, p. 423 (abstract only).

Marchand, N., "Evaluation of Lateral Displacement of SLS Antennas," *IEEE Trans.,* vol. AP-22, no. 4, Jul 1974, pp. 546-550.

Reed, I.S., J.D. Mallett, and L.E. Brennan, "Rapid Convergence Rate in Adaptive Arrays," *IEEE Trans.,* vol. AES-10, no. 6, Nov 1974, pp. 853-863.

Marchand, N. and D. Riva, "Laterally Displaced ISLS Antenna for Tactical Radar," *J. of the Inst. of Nav.,* vol. 21, no. 4, Winter 1974-1975, pp. 326-332.

Applebaum, S.P., "Adaptive Arrays," *Syracuse Univ. Res Corp Rpt TR 66-001R,* Mar 1975.

Klemm, R., "Suppression of Jammers by Multiple Beam Signal Processing," *IEEE Int. Radar Conf. Proc.,* Apr 1975, pp. 176-180.

Krestschmer, F.F., "Effects of Cascading Sidelobe Canceller Loops," *IEEE Int. Radar Conf. Proc.,* April 1975, pp. 181-185.

Ekholdt, R., "Adaptive Interference Cancellation in Nyquist Rate Scanned Arrays," *IEEE Trans.,* vol. AES-11, no. 3, May 1975, pp. 359-362.

Lau, C.K., and C.T. Leondes, "Optimal Adaptive Control of an Array of Passive Sensors," *IEEE Trans.,* vol. AES-11, no. 4, July 1975, pp. 608-11.

Widrow, B., *et al,* "Adaptive Noise Cancelling: Principles and Applications," *Proc. IEEE,* vol. 63, no. 12, Dec 1975, pp. 1692-1716.

Bergsagel, E. and C. Hittle, "Coherent Sidelobe Canceller Analysis Study," *US Army Office of Missile Electronic Warfare Report ECOM-5708,* Dec 1975.

Gabriel, W.E., "Adaptive Antenna Arrays—An Introduction," *Proc. IEEE,* vol. 64, no. 2, Feb 1976, pp. 239-273.

Ries, G. and K. Krücker, "Problems of Adaptive Sidelobe Suppression," *AGARD Conf. Proc. 197,* 14-16 June 1976, pp. 23-1 to 12 (reprinted in this volume).

Nitzberg, R., "Interference/Clutter Rejection Obtained when Adapting a Fraction of the Array Elements," *EASCON Rec. 1976,* pp. 62A-D.

Gabriel, W.F., "Special Issue on Adaptive Antennas," *IEEE Trans.,* vol. AP-24, no. 5, Sept 1976, pp. 573-574.

Howells, P.W., "Explorations in Fixed and Adaptive Resolution at GE and SURC," *IEEE Trans.,* vol. 24, no. 5, Sept 1976, pp. 575-584.

Applebaum, S.P., "Adaptive Arrays," *IEEE Trans.,* vol. AP-24, no. 5, Sept 1976, pp. 585-598 (reprinted in this volume).

Compton, R.T., Jr., "Adaptive Arrays for Communication Systems: An Overview of Research at the Ohio State University," *IEEE Trans.,* vol. AP-24, no. 5, Sept 1976, pp. 599-607.

Brennan, L.E., J.D. Mallett, and I.S. Reed, "Adaptive Arrays in Airborne MTI Radar," *IEEE Trans.,* vol. AP-24, no. 5, Sept 1976, pp. 607-615.

Widrow, B., and J.M. McCool, "A Comparison of Adaptive Algorithms Based on the Method of Steepest Descent and Random Search," *IEEE Trans.,* vol. AP-24, no. 5, Sept 1976, pp. 615-637.

Baird, C.A. and G.G. Rassweiler, "Adaptive Sidelobe Nulling Using Digitally Controlled Phase-shifters," *IEEE Trans.,* vol. AP-24, no. 5, Sept 1976, pp. 638-649.

Applebaum, S.P. and D.J. Chapman, "Adaptive Arrays with Main Beam Constraints," *IEEE Trans.,* vol. AP-24, no. 5, Sept 1976, pp. 650-662.

Takao, K., M. Fujita, and T. Nishi, "An Adaptive Array Under Directional Constraint," *IEEE Trans.,* vol. AP-24, no. 5, Sept 1976, pp. 662-669.

White, W.D., "Cascade Preprocessors for Adaptive Antennas," *IEEE Trans.,* vol. AP-24, no. 5, Sept 1976, pp. 670-684.

Chapman, D.J., "Partial Adaptivity for the Large Array," *IEEE Trans.,* vol. AP-24, no. 5, Sept 1976, pp. 685-696.

Compton, R.T., Jr., "An Experimental Four-element Adaptive Array," *IEEE Trans.,* vol. AP-24, no. 5, Sept 1976, pp. 697-706.

Griffiths, L.J., "Time Domain Adaptive Beamforming of HF Backscatter Systems," *IEEE Trans.,* vol. AP-24, no. 5, Sept 1976, pp. 707-720.

Washburn, T.W. and L.E. Sweeney, Jr., "An Adaptive Beamforming Capability for HF Backscatter Radar," *IEEE Trans.*, vol. AP-24, no. 5, Sept 1976, pp. 721-732.

Taheri, S.H., and B.D. Steinberg, "Tolerances in Self-cohering Antenna Arrays of Arbitrary Geometry," *IEEE Trans.*, vol. AP-24, no. 5, Sept 1976, pp. 733-739.

Brennan, L.E. and J.D. Mallett, "Efficient Simulation of External Noise Incident on Arrays," *IEEE Trans.*, vol. AP-24, no. 5, Sept 1976, pp. 740-741.

Wirth, W.D., "Suboptimal Suppression of Directional Noise by a Sensor Array Before Beamforming," *IEEE Trans.*, vol. AP-24, no. 5, Sept 1976, pp. 741-744.

Gobert, J., "Adaptive Beam Forming," *IEEE Trans.*, vol. AP-24, no. 5, Sept 1976, pp. 744-749.

Rassweiler, G.G., *et al*, "A Miniaturized Lightweight Wideband Null Steerer," *IEEE Trans.*, vol. AP-24, no. 5, Sept 1976, pp. 749-754.

Leavitt, M.K., "A Phase Adaptive Algorithm," *IEEE Trans.*, vol. AP-24, no. 5, Sept 1976, pp. 754-756.

Thompson, P.A., "Adaptation by Direct Phase-shift Adjustment in Narrow-band Adaptive Antenna Systems," *IEEE Trans.*, vol. AP-24, no. 6, Sept 1976, pp. 756 -760.

Sensiper, S., "Suggested Signal Generator-Simulator for Adaptive Antenna Test," *IEEE Trans.*, vol. AP-24, no. 6, Sept 1976, pp. 760-764.

Chapman, D.J., "Adaptive Arrays in Sidelobe Cancellers— A Perspective," *Microwave J.*, Aug 1977, pp. 43-6. (Reprinted in this volume.)

Fielding, J.G., *et al*, "Adaptive Interference Cancellation Radar Systems," *IEE Conf. Pub. No. 155*, Radar 77, London, Oct 25-28, 1977, pp. 212-217.

Wirth, W.D., "Radar Signal Processing with an Active Receiving Array," *IEE Conf. Pub. No. 155*, Radar 77, London, Oct 25-28, 1977, pp. 218-221.

Turner, R.M., "Null Placement and Antenna Pattern Synthesis by Control of the Element Steering Phases of a Phased Array," *IEE Conf. Pub. No. 155*, Radar 77, London, Oct 25-28, 1977, pp. 222-225.

Dicken, L.W., "The Use of Null Steering in Suppressing Main Beam Interference," *IEE Conf. Pub. No. 155*, London, Oct 25-28, 1977, pp. 226-231.

Trunk, G.V., "Automatic Detector for Suppression of Sidelobe Jamming," *IEE Conf. PUb. No. 155*, Radar 77, London, Oct 1977, pp. 280-284.

Kretschmer, F.F., Jr. and B.L. Lewis, "A Digital Open-Loop Adaptive Processor," *IEEE Trans.*, vol. AES-14, no. 1, Jan 1978, pp. 165-71.

Kretschmer, F.F., Jr. and B.L. Lewis, "An Improved Algorithm for Adaptive Processing," *IEEE Trans.*, vol. AES-14, no. 1, Jan 1978, pp. 172-177.

White, W.D., "Adaptive Cascade Networks for Null Steering Antenna," *Military Microwaves Conf.*, London, 25-27 Oct 1978, pp. 357-365.

Raymond, G. and D.S. Hicks, "Adaptive and Sidelobe Cancellers for Communications and Radar Applications, " *Military Microwaves Conf.*, London, 25-27 Oct 1978, pp. 366-378.

Bock. H.W., "Digital Signal Processing in the Adaptive Sidelobe Canceller," *Int. Conf. on Radar*, Paris, 3-8 Dec 1978 (late paper).

Turner, R.M., A.L. Poirier, and P.J. Wilkins, "A Steering-phase Control Architecture and its use for Null Steering in a Phased-array Radar," *Proc. Int. Conf. on Radar*, Paris, 3-8 Dec 1978, pp. 441-449.

Rodgers, W.E. and R.T. Compton, "Adaptive Array Bandwidth with Tapped Delay-line Processing," *IEEE Trans.*, vol. AES-15, no. 1, Jan 1979, pp. 21-28.

Vural, A.M., "Effects of Perturbations on the Performance of Optimum/Adaptive Arrays," *IEEE Trans.*, vol. AES-15, no. 1, Jan 1979, pp. 76-87.

A Sidelobe Blanking System Design and Demonstration
P.O. Arancibia
Microwave Journal, Vol. 21, No. 3, March 1978
Pages 69-73

5.A.1

INTRODUCTION

Antennas for search radars which continuously scan through 360 degrees in azimuth usually are characterized by a high-gain main lobe and many minor lobes. It is important that the minor lobe gain be very much less that of the relatively narrow main lobe in order to:

- have an antenna with high main lobe gain
- reduce the possibility of detecting a target on a sidelobe
- reduce the susceptibility of the radar to interfering signals arriving through the side lobes.

The field in the main lobe and the first few side lobes of a radar antenna are determined chiefly by the amplitude and phase distributions of the field in the aperture. However, the rest of the minor lobes in a 360 degree pattern are influenced by many other factors, such as spillover and back radiation from the primary feed horn, edge currents on the reflector, scattering off of the primary feed and supporting structures and leakage through mesh type reflectors. All of these factors contribute to make the sidelobes in a practical radar antenna non-negligible.

One method of reducing the effects of sidelobe signals is the sidelobe blanking technique. Sidelobe blanking systems are useful in preventing acquisition of strong targets in the antenna sidelobes and also in rejecting pulsed interference originating in the sidelobes.

With an effective sidelobe blanker one expects that only signals received through the main beam of the antenna will be presented on the display. This is achieved by comparing signals received by the rotating antenna of the radar with those received simultaneously from an omnidirectional antenna. A comparator is used that blanks the video output whenever the signal received in the omnidirectional channel is stronger than the signal in the main channel.

The objective of this work was to eliminate from the display in the PPI of the AN/UPS-1 and AN/SPS-12 radar sets any low duty cycle signal coming through the sidelobes of the radar antenna, using the sidelobe blanking (SLB) technique.

BASIC PRINCIPLE

The basic operation of the SLB system is shown in **Figure 1** where the detected outputs of two channels, identical except for antenna patterns, are compared. Typical relative antenna patterns for the two channels are shown in **Figure 2**. When the target or pulsed interference is in the sidelobes of the acquisition antenna, the output from channel "B" will be greater than that of channel "A", provided level "a" is greater than level "b". Thus, the threshold difference can be used to blank channel "A" before the final detection circuits or displays, thereby preventing display of the unwanted

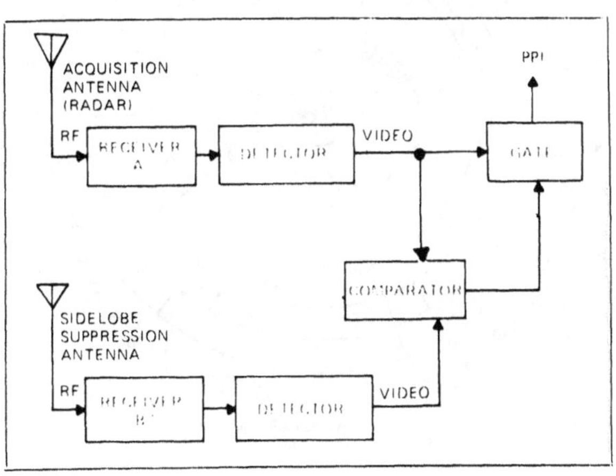

Fig. 1 Basic Sidelobe Blanking System.

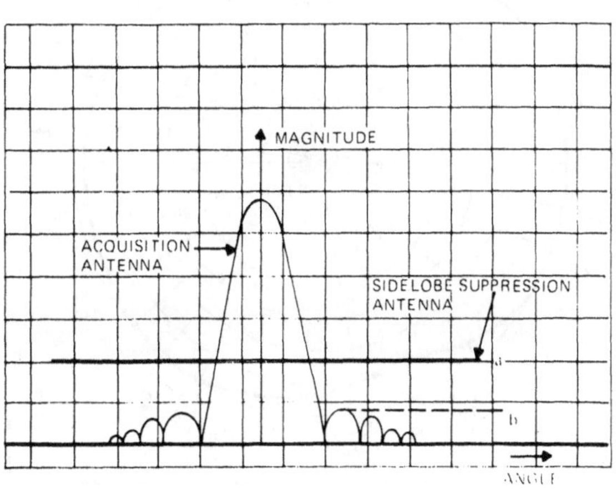

Fig. 2 Idealized Antenna Pattern.

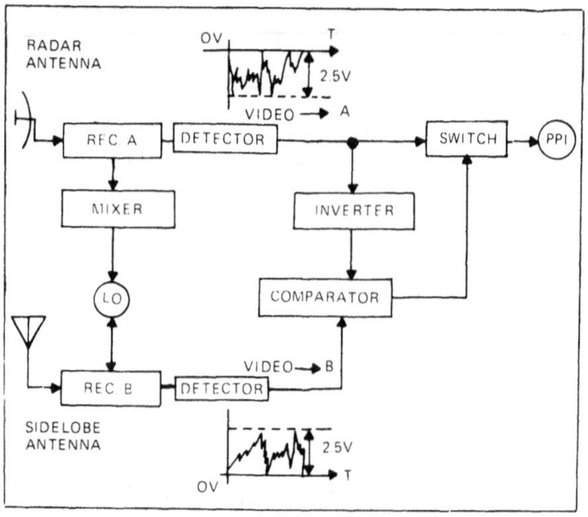

Fig. 3 Basic Sidelobe Blanking System for AN/UPS-1 Radar.

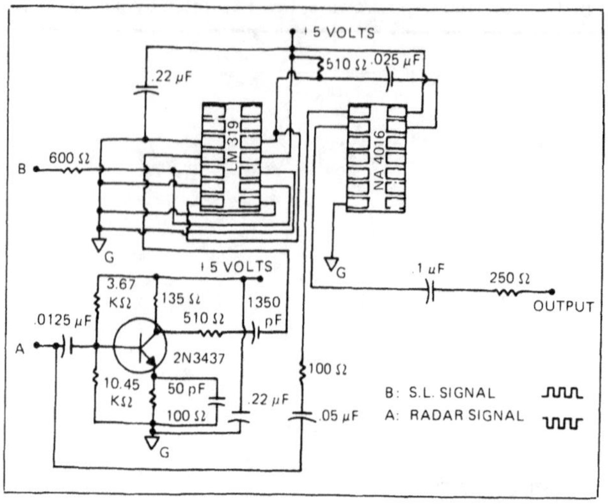

Fig. 4 Comparator Circuit for Blanker System.

sidelobe targets or interference. The gate in **Figure 1** is normally closed, but when signal |B| ≥ |A|, the switch must be opened to prevent output to the display.

COMPARATOR AND BLANKER SYSTEM DESIGN

To investigate the sidelobe blanking technique using the AN/UPS-1 it was necessary to design a circuit to compare the amplitude of the video pulse (0 to 2.5 volts) signals received by the main radar antenna with the video pulses (also 0 to 2.5 volts) received by the sidelobe blanking auxiliary antenna (**Figure 3**). Once this comparison is done, the system must be able to blank the output before the final detection circuits and displays whenever the signal coming from the sidelobe antenna is greater than the main signal coming from the radar antenna. (A suitable comparator is shown in **Figure** 4. This contains an inverter for the main lobe signal, which in the radar had a polarity opposite to the blanking signal). In this way, at the output of the analog switch we have the original radar signal when the radar signal is greater than the sidelobe signal, but no signal when the sidelobe signal is greater.

ANTENNA PATTERN MEASUREMENT

It was found desirable for the sidelobe blanking antenna to have a relative wide bandwidth with omnidirectional radiation and horizontal polarization (the same as the radar). A log-periodic antenna structure was designed for this requirement and its E-plane pattern, taken at 1320 MHz, is shown in **Figure 5.**

For measuring the patterns of the radar and omnidirectional antennas, a transmitting antenna was mounted at 250 meters from the radar antenna at the same elevation (30 meters). A hori-

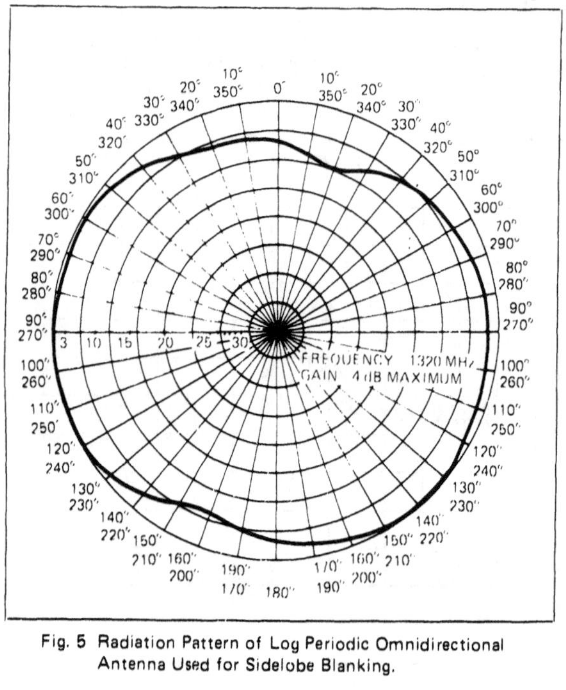

Fig. 5 Radiation Pattern of Log Periodic Omnidirectional Antenna Used for Sidelobe Blanking.

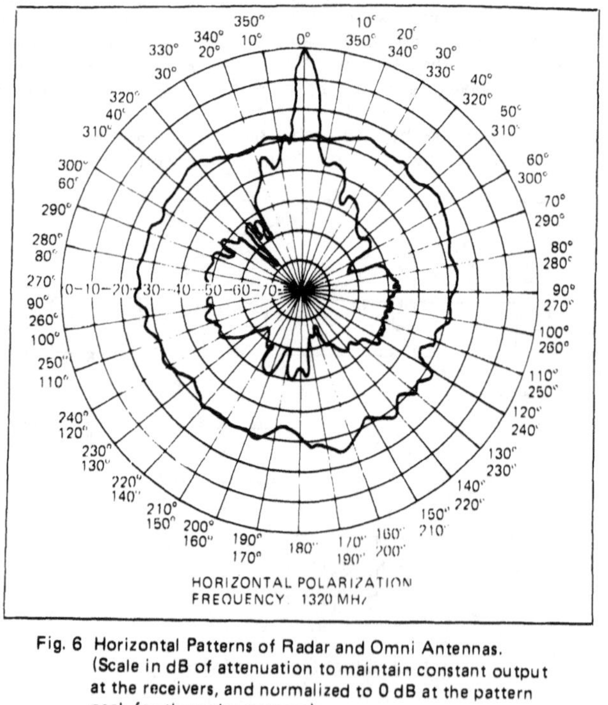

Fig. 6 Horizontal Patterns of Radar and Omni Antennas. (Scale in dB of attenuation to maintain constant output at the receivers, and normalized to 0 dB at the pattern peak for the radar antenna)

zontally polarized echo box antenna (AT-43/SPS-12) with a reflector behind it was used. The transmitter output was 1.5 watts at 1320 MHz.

The change in signal strength in dB with antenna rotation for both receivers, radar and sidelobe, was determined for every two degrees of azimuthal rotation using a continuous attenuator installed at the bottom of the transmission line between each antenna and its receiver. The attenuation in dB necessary to maintain constant power at the output of each of the two similar receivers was so measured and the antenna patterns obtained for both the radar and log-periodic omnidirectional antenna used for blanking are plotted in **Figure 6**. The sensitivity of both the radar and auxiliary receiver for a 2 to 1 (S+N)/N ratio were almost identical (106 and 105 dBm respectively).

SIDELOBE BLANKING TEST

The blanking system effectiveness was demonstrated by "jamming" the radar using the transmitter formerly employed in the antenna pattern measurements, first with a 10 μsec wide pulse signal at a prf of 3 kHz, (duty cycle = 0.03) 100 mW power, and then with a 17 μsec wide pulse signal at a prf of 4 kHz (duty cycle = 0.068) 100 mW output power.

The horizontally polarized echo box antenna AT-43/SPS-12 with a gain of 2 dB over an isotropic antenna was

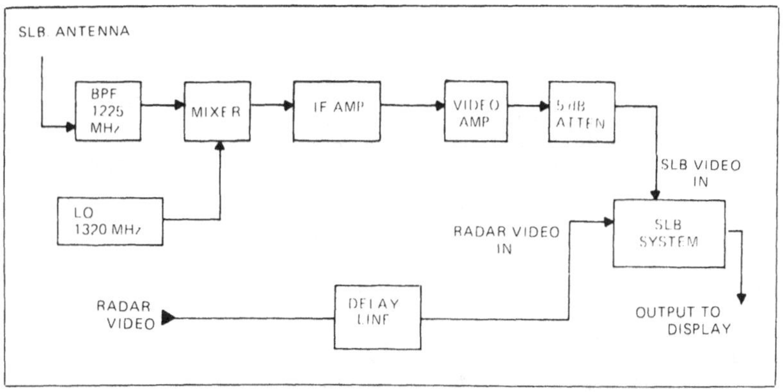

Fig. 7 Experimental Set-up for Test and Measurements with Pulse Jamming.

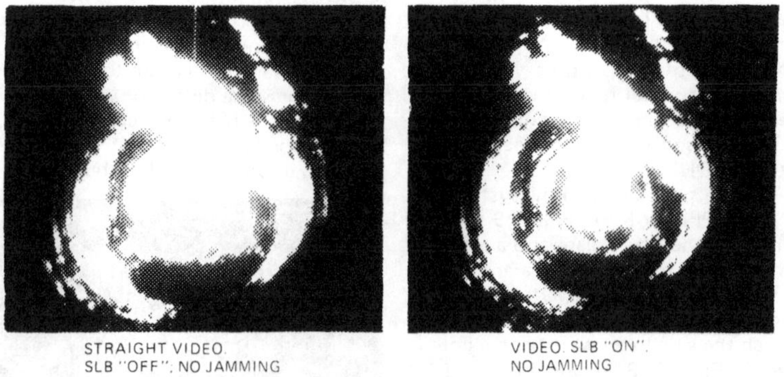

Fig. 8 Sidelobe Blanking System Sensitivity Test
(Radar AN/SPS-12 Video Cursor at 000°, Range 5 miles).

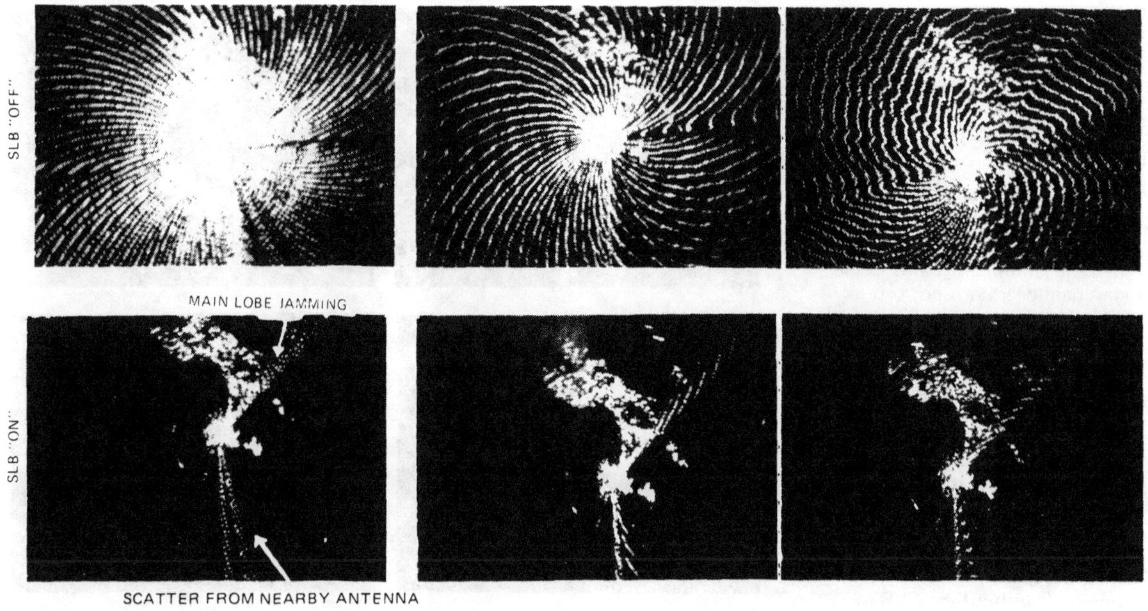

Fig. 9 Radar AN/SPS-12 Video PPI Displays with Pulse Jamming, High Pass Video "ON". (10 μsec jamming pulses at 3 kHz. The radar prf was changed by 5 percent between each pair of pictures).

mounted at the same elevation and about 250 meters from the radar and sidelobe antenna and was used for the jamming purpose. The output of the sidelobe blanker system (**Figure 7**) was fed into a PPI (SPA-8A) for display purposes. In order to compensate for the long line used to feed the radar signal into the blanking system, together with the other system delays (in normal usage both antennas would be close together), it was necessary to delay the sidelobe signal by 2.5 μsec.

For each of the two jamming signals the video pattern of the radar receiver was changed by varying the radar transmitter prf by ± 5 percent. The results are shown in **Figures 8** through **11**. All pictures were taken on the 50 mile range scale. In **Figure 8** true North is indicated by the cursor position. Note that these two pictures, with no jamming, show that there is no noticeable change in sensitivity incurred as the sidelobe suppression channel is switched "ON".

With the sidelobe blanking system "ON" the PPI photos in **Figure 9** show jamming in the main beam and a reflection produced by one antenna around the SPS-12 radar antenna. Other jamming signals entering the radar antenna through the side lobes have been eliminated. In **Figures, 9, 10** and **11**, the radar prf was changed by 5 percent between each pair of pictures.

CONCLUSIONS

The sidelobe blanking system designed is demonstrably effective in preventing low duty cycle signals from entering through the sidelobes of the radar antenna without compromising performance in detecting targets in the main lobe.

The results can be summarized as follows:

1. When the jamming signal is in the radar main-beam it is not suppressed. It is then more difficult to detect targets in the direction of the jammer.
2. When the jamming signal arrives through the main antenna side lobes it is suppressed.
3. A reduction in probability of detection when using the SLB with the jammer in the side lobes was not observed in the tests made for duty cycles used. Detection probability was greatly reduced by the "jamming" without the SLB in operation.
4. The direction of the jamming signal is easily determined with SLB turned on, as seen in **Figure 9**.

SUMMARY

The whole purpose of the sidelobe blanker system can be stated as the ability to produce the maximum percent sidelobe blanking with the least loss of system sensitivity. This was achieved by processing the SLB antenna signal in the same manner as the signal in the radar main channel. When the SLB signal level exceeds the level in the main beam, as it would for targets in the main antenna sidelobes, "target detection" is blanked. When the main beam signal is greater, the detected target in the main lobe is displayed.

The criteria for evaluating performance of the sidelobe blanker can be sensitivity reduction and percent blanking. Sensitivity reduction is defined as the increase in signal-to-noise ratio necessary for a given target to achieve the same detection probability as without the blanker. Percent blanking is defined as the percentage of sidelobe signals uniformly distributed in angle which are prevented from being processed as targets.

In general, comparison of the main signal and blanker antenna outputs with equal channel gains would not result in the required blanking due to transmission line losses. For this reason, normally the sidelobe blanker channel gain must be increased so that sidelobe signals produce a larger response at the blanker channel output than the normal radar antenna channel and thereby provide the required blanking.

Although not observable in these tests, the presence of the sidelobe blanker theoretically reduces the probability of detection of a target in the main beam, because there is a finite probability at any given time that thermal noise in the blanker channel output will have a larger amplitude than signal plus noise at the radar receiver antenna output. This could result in blanking of a signal which should suc-

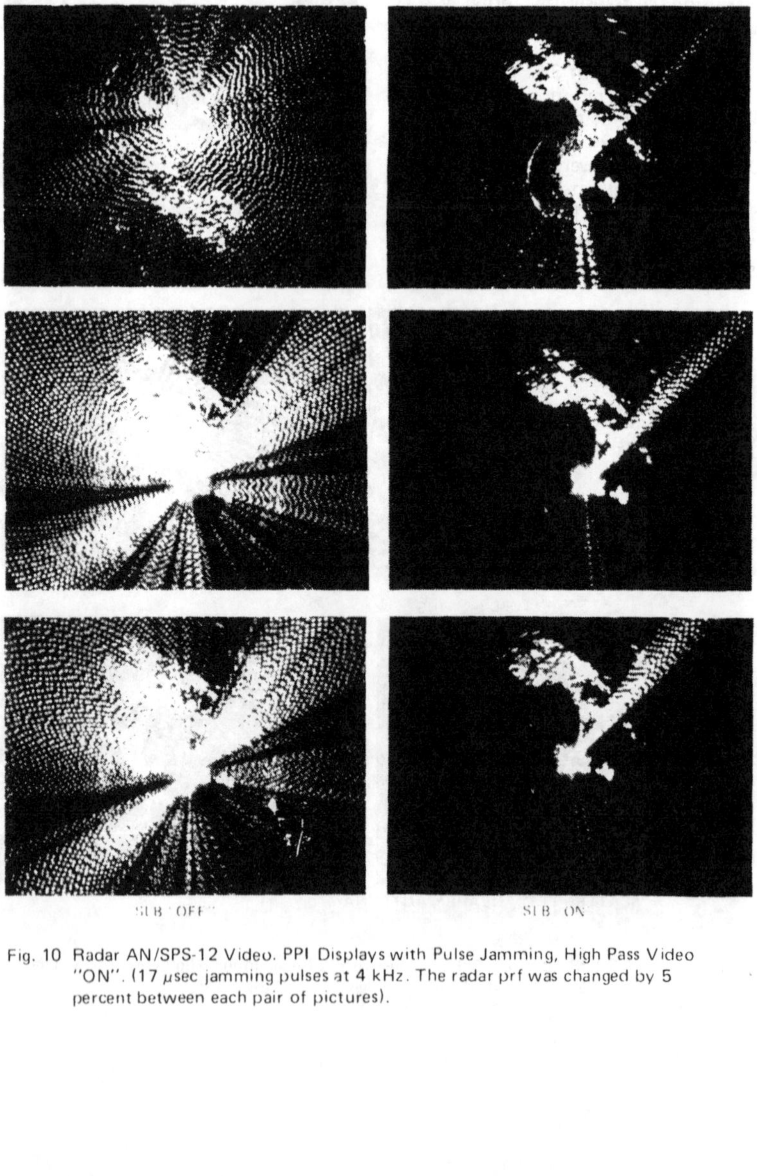

Fig. 10 Radar AN/SPS-12 Video. PPI Displays with Pulse Jamming, High Pass Video "ON". (17 μsec jamming pulses at 4 kHz. The radar prf was changed by 5 percent between each pair of pictures).

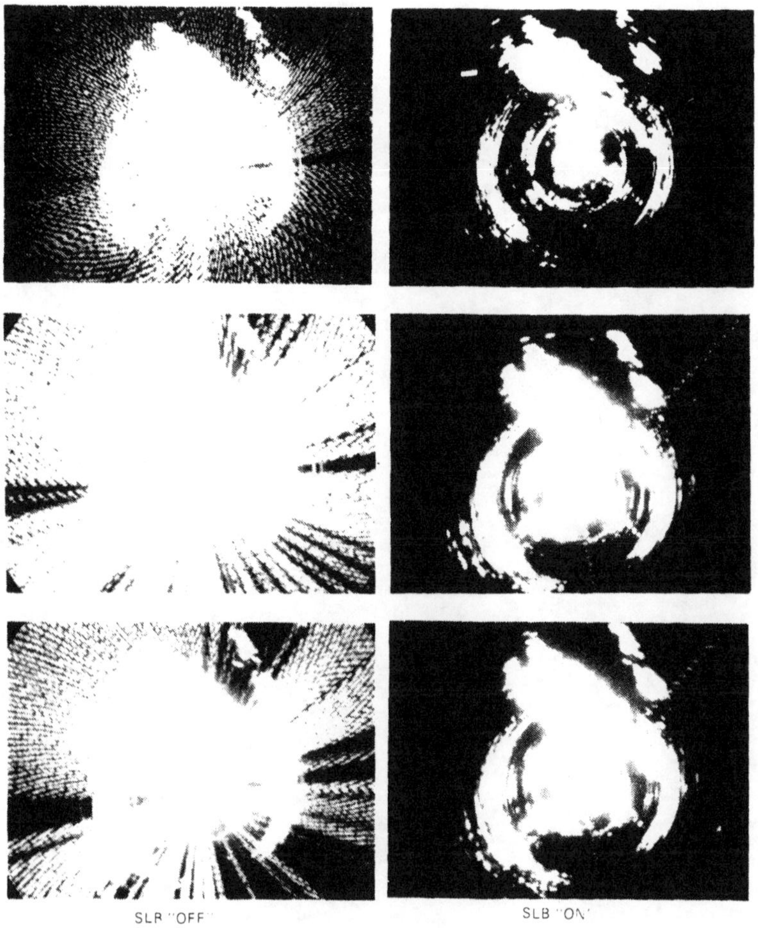

Fig. 11 Radar AN/SPS-12 Normal Video, PPI Displays with Pulse Jamming, High Pass Video "OFF". (17 μsec jamming pulses at 4 kHz. The radar prf was changed by 5 percent between each pair of pictures.

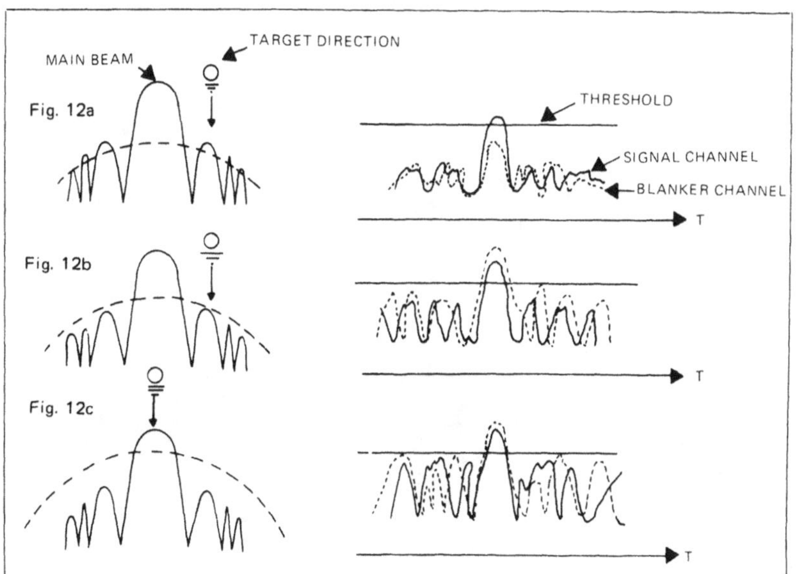

Fig. 12 Sensitivity Reduction.

cessfully exceed the false-alarm-rate threshold in a radar without SLB. Sensitivity reduction is specifically defined as the increase in target average signal-to-noise ratio for a given target necessary to achieve the same detection probability as would be obtained with no blanker in the system.

This effect of the sidelobe blanker on system sensitivity can be better explained using **Figure 12. Figure 12a** shows a sidelobe target producing a larger signal out of the receiver radar antenna than out of the blanker channel. This results if the blanker channel gain is too low and results in a sidelobe target whichis not suppressed. **Figure 12b** applies to the same sidelobe target which is successfully blanked due to an increase in blanker channel gain.

Figure 12c shows a small target (low SNR) which is blanked even though it is in the main beam direction because blanker channel noise exceeds signal plus noise in the receiver radar antenna. It is this phenomenon that reduces detection probability and must be accounted for in reassessing system detection performance with the sidelobe blanker.

An alternate measure of system sensitivity reduction can be made in terms of system fractional range reduction. Specifically, the system range reduction can be defined as that percent reduction in range for a given target necessary to achieve the same probability of detection as would be obtained without SLB, (with the same probability of false alarm). In this case, system reduction according to the following equation. Percent range reduction is equal to

$$100\,(10^{\frac{\Delta\bar{x}}{40}} - 1)$$

where $\Delta\bar{x}$ is the sensitivity reduction in dB. For a specified detection probability the target signal to noise ratio, SNR, must increase as blanker gain increases. Sensitivity loss is measured as the increase in target SNR to maintain 95% probability of detection.

ACKNOWLEDGEMENT

The author wishes to thank Professor David B. Hoisington of the U.S. Naval Postgraduate School for his guidance in this work.

BIBLIOGRAPHY

1. Jasik, H., "Antenna Engineering Handbook", MacGraw-Hill, 1961.
2. Barton, D. K., "Radar Principles", Sec. 25 of *Electronic Engineers Handbook*, edited by D. S. Fink, McGraw Hill, N.Y., 1975.
3. Maisel, Louis, "Performance of Sidelobe Blanking System", IEEE Trans. Aerospace and Electronic Systems, March 1967.

Adaptive Arrays and Sidelobe Cancellers; A Perspective
D.J. Chapman
Microwave Journal, Vol. 20, No. 8, August 1977
Pages 43-46

5.B.1

Modern radar and communications systems must be designed to cope with directional interference, whether intentional or not, in order to satisfy the needs of the users. This requirement has led to increasing interest in adaptive antenna techniques in recent years. As with many modern technologies, research and development has progressed down several parallel paths to attain today's state-of-the-art. This paper will trace and attempt to place into perspective the findings obtained at Syracuse Research Corporation* over the past decade.

BACKGROUND

The beginnings of this branch of the research into adaptive techniques can be traced to the original patent[1] for the multiple sidelobe canceller (MSLC) granted to Paul Howells (then with General Electric in Syracuse) in 1959. This device was initially developed as a "fix" for air defense radars which were particularly vulnerable to sidelobe jamming. Paul and Sid Applebaum continued their research upon joining SRC leading to Sid's now classic paper[2] on Adaptive Arrays, first published in 1964. This paper considered the extension of the concept of sidelobe cancellation to the point where all elements of a phased array antenna were made adaptive.

Interest in sidelobe cancellation techniques was then extended to the ballistic missile radar application, and in 1967 SRC entered into an experimental program to prove the feasibility of the technique using the HAPDAR (Hard Point Demonstration Array Radar) phased array at the White Sands Missile Range, N.M. From then until 1972 (when HAPDAR was shut down) two generations of 16 loop sidelobe cancellers were built and successfully tested, culminating in 1972 in the successful demonstration of a 16-element linear fully adaptive array operating in real-time at a 15 MHz instantaneous bandwidth.

Since that time, research into partially and fully adaptive arrays has been carried out for the Ballistic Missile Defense Advanced Technology Center in Huntsville, Alabama under contracts F30602-72-C-0075 and DAHC60-73-C-0044. This work has consisted of studies and simulations as well as the evolution of a new generation of processor hardware.

THEORY OF OPERATION

To better understand the capabilities and limitations of the technique of self-adapting arrays, it is necessary to delve somewhat into the fundamentals of the process. Throughout the remainder of this paper, the terms fully adaptive and partially adaptive array and sidelobe canceller will be used. The definitions of these terms, although somewhat arbitrary, are as follows: A fully adaptive array is one in which the feed coefficient of each element in the array is adaptively controlled. A partially adaptive array is one in which approximately 10 to 50% of the elements are controlled adaptively. The array in which less than 10% of the elements are adaptively controlled is considered a multiple sidelobe canceller. Note that the MSLC technique is also applied to reflector type antennas as well. One can then generalize by considering the percentage of the degrees of freedom of the antenna far field pattern which can be adaptively controlled rather than the percentage of elements controlled.

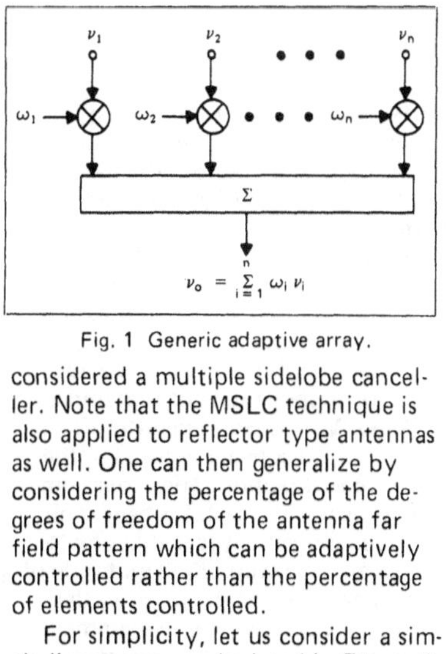

Fig. 1 Generic adaptive array.

For simplicity, let us consider a simple linear array as depicted in **Figure 1**. Signals received in each element channel are weighted by a complex coefficient and summed together to form the output of the array. Assume initially that these signals consist of one or more sources of directional interference (jamming, perhaps) as well as ordinary receiver thermal noise. Consider a vector V whose elements are the voltages in each channel induced by these sources of noise. We form a matrix M according to the equation

$$M = E(V^{*t}) \qquad (1)$$

That is, the conjugate of the complex vector V is multiplied by the transpose of the vector and the resulting matrix is the averaged (expected) value over time.

*Formerly known as Syracuse University Research Corporation.

The result is known as the covariance matrix and is fundamental to all of the adaptive processor theory. It contains all of the information about the interference environment. The diagonal terms are a measure of the power in each channel. The off-diagonal terms contain information about the directions of arrival of the various interference signals.

To compute the set of adaptive coefficients or weights we need to specify the direction in which we wish to steer the main beam. We define a vector S which can be thought of as a vector of the voltages induced in our array from a simple plane wave arriving from the desired steering direction. Then, from Applebaum's original paper, the set of weights which yields the best signal-to-interference ratio for a signal arriving from the desired direction is given by

$$W_{opt} = \mu M^{-1} S^* \qquad (2)$$

or that one must invert the covariance matrix and multiply it by the conjugate of the steering vector to yield a vector proportional to the theoretically optimum set of weights or feed coefficients.

To obtain an intuitive feel for what physically happens in an adaptive array, it is possible to show that, for the case of one jamming source, the optimum weight vector can be factored into two components. With reference to **Figure 2a**, one component results in a simple sin(x)/x beam pointed in the desired scan direction. The other component results in a sin(x)/x beam pointed at the jammer and scaled in amplitude to match exactly the amplitude of the original beam at the point in the sidelobe structure where the jammer is received. The first beam is referred to as the quiescent beam and the latter as the cancellation or perturbation beam. When the cancellation beam is subtracted from the quiescent beam, the composite result exhibits a simple null in the direction of the jammer, as depicted in **Figure 2b**. The condition for finite bandwidth and multiple sources of interference can be explained in a similar although much more complicated fashion by considering the eigenanalysis of the matrix M.[3]

IMPLEMENTATION

There are at least three major methods of implementing processors to control an adaptive array and several variations to each. The particular choice usually depends upon the number of adaptive channels, the bandwidth, and the resources available. Three techniques currently being used or studied will be outlined here. These are the Sample Matrix Inversion technique, the correlation loop, and the Modified Random Search algorithm.

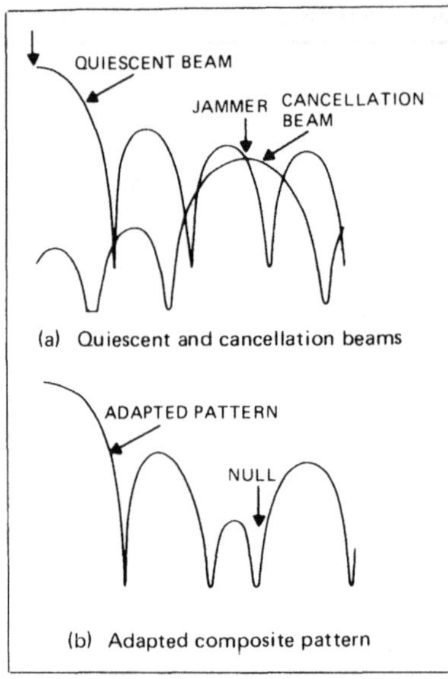

Fig. 2 Single jammer adaptive array patterns.

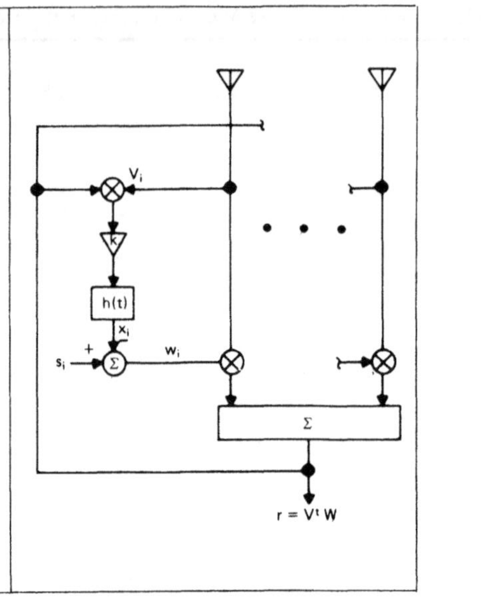

Fig. 3 Howells-Applebaum loop.

The sample matrix inversion technique, treated recently in a paper[4] by Brennan *et al.*, was first used for adaptive seismic arrays. It is a direct application of the fundamental control equation (see **Equation 1**). Although the calculation of the covariance matrix involves an infinite average, it has been shown that a finite average over about 2N samples (where N is the number of adaptive channels) yields a very acceptable result. This result puts a very nice upper bound on the number of arithmetic operations required to arrive at the solution and is the principal advantage of this approach. It is a digital algorithm and is not amenable to wideband analog techniques as are some of the other algorithms. It is also an open loop process which implies a requirement for relatively high arithmetic precision in order to get good agreement with theoretical performance predictions.

Probably the most popular algorithm for radar work is the correlation loop processor. This usually takes the form of the Howells-Applebaum maximum signal-to-noise or the Widrow Least-Mean-Squares circuit. These algorithms, which differ only in the method in which they are steered, are based upon a calculus technique known as the Method of Steepest Descent. If one considers the residue power output of the adaptive array for an arbitrary set of weights, this power describes a surface in weight space which is concave and quadratic in behavior. This means there is only one minimum at the bottom of the "bowl" described by this function and the coordinates of that minimum are the theoretical optimum weights. This algorithm calls for an initial guess as to the weights to be used. At the point on the surface corresponding to the initial weights, the gradient (direction of steepest slope) is computed. A small step is taken in that direction (along the negative gradient). The whole process is then repeated until the bottom of the bowl is reached. The process is tolerant of arithmetic errors (or equivalent nonlinearity in analog components) since a step in a less than optimum direction just prolongs the process slightly. The choice of step size is equivalent to adjusting the servo loop gain.

Figure 3 shows a typical implementation of the Howells-Applebaum loop. The residue is fed back and correlated (multiplied and filtered) with each of the element signals. The output of the filters, which is proportional to the gradient, is subtracted from the steering signal to form the adaptive weight vector. Practical implementations would also include receiver components such as amplifiers, mixers, and couplers. The entire process could, of course, also be done digitally.

Figure 4 depicts one form of the Widrow LMS processor. The principal difference between this and the previous loop is that the steering vector is not injected. For this development, it was assumed that the direction of arrival was not known but that a replica of the desired signal (called a pilot signal) was available. This is compared with the received signal and the two are matched in a least mean squares sense. The result of this is that a beam is retrodirectively formed in the direc-

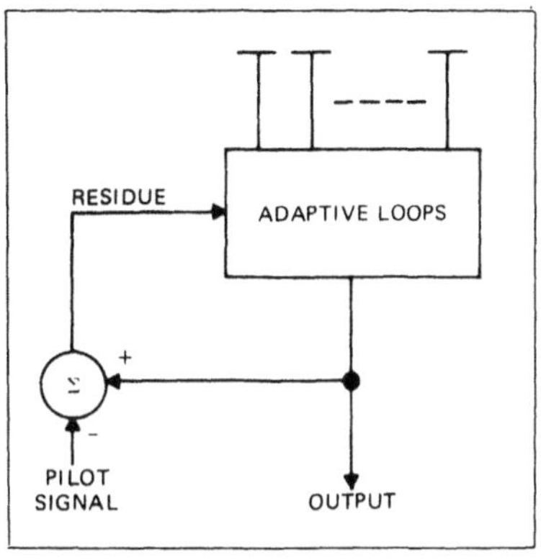

Fig. 4 Widrow LMS (least mean square) algorithm.

tion of the desired signal and nulls are formed in the direction(s) of signals which are not correlated with the pilot signal. This technique is very useful in communications systems.

The third technique appears in many different versions under slightly different names, but will be termed here the Modified Random Search technique. The concept is extremely simple. An initial guess for the weights is made and the residue power is noted. An arbitrary small change is made in the weight vector and the residue is again noted. If it decreased, the same change increment is repeated. If not, a new increment is tried. This method is probably the slowest to converge of the three.[5] However inelegant it may seem, this algorithm possesses several advantages. It is extremely simple to implement in either analog or digital form. Furthermore, it is tolerant of even the most extreme nonlinearities. A good example of its application is in an array in which only the phase of the elements is controlled. The main variant in the different versions of this algorithm is the philosophy used to choose the next weight increment, particularly when an improvement in residue is noted.

LIMITATIONS

As with any other technique, there are very definite limitations to the application of adaptive array techniques. In this section, several of the major limiting effects will be discussed, particularly with application to sidelobe cancellation. Rules of thumb, evolved from years of experience with MSLC systems, will be included where applicable.

One of the first things one must define when considering adaptivity as an ECCM adjunct to a radar system is the severity of the expected interference environment. The MSLC system is sensitive to the number of independent sources of directional noise. The bandwidth of an array is frequently measured in terms of the product (BT) of the bandwidth and the time of propagation across the aperture of the array. For a BT product which is orders of magnitude less than unity, one needs as a minimum, one auxiliary element and loop for each independent source of interference. This can be understood in terms of the null-steering concept, where each loop is considered to control one null. For wideband systems, a simple null is no longer adequate since, for signals impinging upon the array at angles other than normal to the array, the signal appears to span an arc in angle. Multiple closely spaced nulls are required in this case. Typically, then, the number of loops might be specified at up to three times the number of expected jamming sources.

One of the most dominant sources of less than ideal performance in an MSLC system is decorrelation of the jamming signal in the main channel with respect to that same signal in any of the auxiliary channels. If the tolerances in amplitude and phase ripple in the receiver components are not held low, decorrelation will occur. The most common source of decorrelation in either array or dish antenna systems is the frequency response of the main antenna in the sidelobe region. This phenomenon can cause an apparent mismatch in bandpass response between an otherwise clean main channel and an auxiliary channel which uses a relatively simple antenna element. Unfortunately, this is an aspect of antenna performance which is not usually controlled.

A problem which is particularly evident in reflector type antennas with linear polarization is the significant sidelobe response to energy in the orthogonal polarization. Since the well-designed jammer will transmit with random polarization, the sidelobe canceller with auxiliary antenna elements all polarized to match the main antenna primary polarization will not perform well in those regions of the sidelobe structure where cross-polarized response is high. It is necessary in this case to include additional auxiliary channels which have antenna elements responsive to cross-polarized energy. Phased array antennas are usually much less susceptible to this effect.

Finally, multipath effects can adversely affect the operation of a sidelobe canceller system. If the reflected path from a jammer to the radar is sufficiently longer than the direct path (difference in time of arrival exceeds the inverse of the processing bandwidth), the two signals will be essentially uncorrelated and will appear to the MSLC as two distinct sources. If the array of auxiliary elements has sufficient degrees of freedom to cope with this additional apparent jammer, the performance will not suffer. If not, more channels will again have to be added (or the auxiliary array geometry changed) to cope with this problem.

First order adaptive array theory assumes that the desired signal is very small compared with the interference power. There is no inherent property of the adaptive processor which allows it to distinguish between desired signal and undesired interference. In a radar application, time constants were originally adjusted so that short duration target return pulses would not cause a measurable response from the MSLC circuits. However, certain applications requiring large MSLC systems, the use of pulse compression, and the presence of extended clutter or large discrete targets all contribute to a tendency for the MSLC system to attempt to cancel within the main beam. Three methods have been investigated[6] in an attempt to control this effect. They attack the problem in the time, frequency, and angle domains, respectively. First, a technique called "weight holding" makes use of the fact that most strong clutter returns occur early in a range sweep. The loops are opened and the weights frozen during this period. They are then allowed to continue adapting during the remainder of the PRI. In the frequency domain, a technique called offset correlation can be employed. Since even spot jamming usually extends well beyond the radar bandpass limits, this technique employs only that energy which is outside the radar instantaneous bandpass, filtering out the jamming (and clutter!) within the bandpass of the radar receiver. Both techniques have drawbacks. The weight holding system could be vulnerable to a properly timed blinking jammer and the offset correlation scheme suffers somewhat in performance for having to process a wider bandwidth.

The spatial domain techniques can be looked at as a means of blinding the auxiliary array to energy arriving from the direction of the main beam. The use of steerable nulls in the auxiliary element patterns is one obvious method. There are also several ways in which the weight vector can be processed to remove components in the direction of the main beam. One way which has been tested at the HAPDAR site is to inject a pseudopilot signal into each auxiliary array channel similar to the Widrow LMS algorithm. The pilot tone is offset in frequency from the desired receiver signal and is filtered out later. These techniques require precise knowledge of the signal direction since the angular zone of protection is very narrow. Some research and testing has been carried out[6] with respect to multiple constraints. The reader is directed to the cited references for further information in this area.

CONCLUSIONS

Extensive experimentation by SRC and others has shown that adaptive array techniques, when properly applied, can greatly extend the usefulness of a radar or communications receiver in an environment of directional interference. Particularly in the case of MSLC, the installation is frequently a retrofit to an existing system. The optimum situation is to consider ECCM a design requirement right from the inception of a system (just like MTI) rather than after one is committed to a design which may not be particularly compatible with adaptive array techniques.

It has been demonstrated experimentally that MSLC techniques can be applied in real time at instantaneous bandwidths exceeding 100 MHz. Current development programs are directed toward large systems with several hundred degrees of freedom. Hybrid digital/analog processor development is making use of the extraordinary advances in digital integrated circuitry coupled with the cost-effective use of analog devices for handling wideband signals. A study was also conducted[7] into the use of partial adaptivity for the large ballistic missile defense radar array. These results, which are reported in the open literature, show that certain configurations have promise of yielding useful performance but are sensitive to array element errors.

Finally, there are many areas of related research being carried out by others, including two-dimensional adaptation in both the frequency and spatial domains for airborne MTI applications[8] and for Over-the-Horizon radar.[9] The interested reader is directed to the September 1976 issue of the *IEEE Transactions on Antennas and Propagation* (Vol. AP-24, #5) for an excellent cross-section of the current work in the area of adaptive arrays.

REFERENCES

1. Howells, P.W., "Intermediate Frequency Sidelobe Canceller," US Patent 3,202,990, filed May 4, 1959 and granted August 24, 1965.
2. Applebaum, S.P., "Adaptive Arrays," *IEEE Trans. Antennas and Propagation,* vol. AP-24, no. 5, September 1976, p. 585.
3. Gabriel, W.F., "Adaptive Arrays - An Introduction," *Proc. IEEE,* vol. 64, no. 2, February 1976, pp. 239-272.
4. Reed, I.S., *et al.,* "Rapid Convergence Rate in Adaptive Arrays," *IEEE Trans. Aerospace and Electronic Systems,* vol. AES-10, no. 6, p. 853.
5. Widrow, B. and J.M. McCool, "A Comparison of Adaptive Algorithms Based on the Methods of Steepest Descent and Random Search," *IEEE Trans. Antennas and Propagation,* vol. AP-24, no. 5, September 1976, p. 615.
6. Applebaum, S.P. and D.J. Chapman, "Adaptive Arrays with Main Beam Constraints," *IEEE Trans. Antennas and Propagation,* vol. AP-24, no. 5, September 1976, p. 650.
7. Chapman, D.J., "Partial Adaptivity for the Large Array," *IEEE Trans. Antennas and Propagation,* vol. AP-24, no. 5, September 1976, p. 685.
8. Brennan, L.E., J.D. Mallett, and I.S. Reed, "Adaptive Arrays in Airborne MTI Radar," *IEEE Trans. Antennas and Propagation,* vol. AP-24, no. 5, September 1976, p. 607.
9. Griffiths, L.J.,"Time-Domain Adaptive Beamforming of HF Backscatter Radar Signals," *IEEE Trans. Antennas and Propagation,* vol. AP-24, no. 5, September 1976, p. 707.

Adaptive Arrays
S.P. Applebaum
IEEE Transactions on Antennas and Propagation, Vol. AP-24, No. 5, September 1976
Pages 585-598

5.B.2

Abstract—**A method for adaptively optimizing the signal-to-noise ratio of an array antenna is presented. Optimum element weights are derived for a prescribed environment and a given signal direction. The derivation is extended to the optimization of a "generalized" signal-to-noise ratio which permits specification of preferred weights for the normal quiescent environment. The relation of the adaptive array to sidelobe cancellation is shown, and a real-time adaptive implementation is discussed. For illustration, the performance of an adaptive linear array is presented for various jammer configurations.**

I. Introduction

Array antennas consisting of many controllable radiating elements are very versatile sensors. The pattern of the array can be steered by applying linear phase weighting across the array and can be shaped by amplitude and phase weighting the outputs of the array elements. Most arrays are built with fixed weights designed to produce a pattern that is a compromise between resolution, gain, and low sidelobes. The versatility of the array antenna, however, invites the use of more sophisticated techniques for array weighting. Particularly attractive are adaptive schemes that can sense and respond to a time-varying environment. In this report, we show how adaptive techniques can be applied to an antenna array to reduce its susceptibility to jamming or interference of any kind. We begin with a derivation of the "control law" for the array weights that will maximize the signal-to-noise ratio (SNR) of the array output in the presence of any spatial configuration of noise sources.

II. Signal-to-Noise Optimization

It is well known that a uniformly weighted array gives the maximum SNR when the noise contributions from the element channels have equal power and are uncorrelated. These conditions are approximately valid when receiver noise and uniformly distributed sky noise are the predominant noise contributions. (They pertain exactly in linear halfwave space array antennas.) However, when there is directional interference from other in-band transmitters, jammers, or natural phenomena, the noise out of the element channels will be correlated, and uniform weighting will not optimize the SNR.

Manuscript received April 20, 1976. This paper is reprinted from Syracuse University Research Corporation, Syracuse, NY, Technical Report SURC TR 66-001, August 1966 (revised March 1975). This study was performed under the sponsorship of the Advanced Projects Agency, Ballistic Missile Defense Office (Project Defender), on Contract AF 30 (602)-3523, ARPA Order 561, Program Code 4720.

The author was with the Syracuse University Research Corporation, Syracuse, NY. He is now with the General Electric Company, Syracuse, NY 13201.

Editor's Note: This paper was first published as a technical report. Its circulation was somewhat limited, consequently, this classic report is being reprinted in this issue with the kind permission of the Syracuse University Research Corporation.

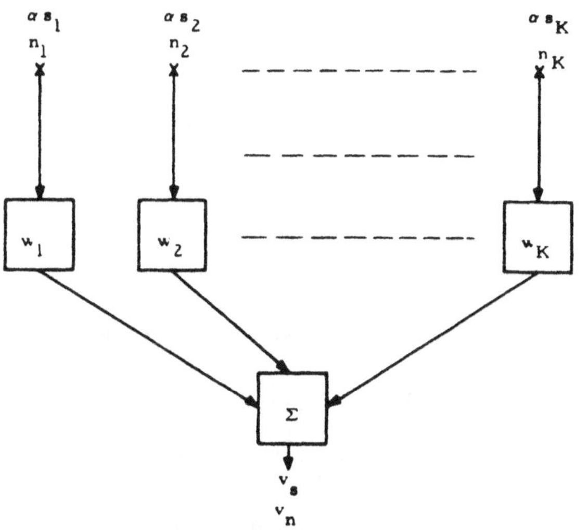

Fig. 2-1. Functional representation of "optimum" linear coherent combiner.

In this report, we consider the problem of determining the array weights that will maximize the SNR for any type of noise environment. The problem may be viewed as that of finding an optimum coherent combiner for K channels as shown in Fig. 2-1. We assume in this discussion that all signals have "bandpass" frequency spectra. All signals are represented by their complex envelopes. These are assumed to modulate a common carrier reference that never appears explicitly.

Each of the K channels contains a noise component whose complex envelope is denoted by n_k. The envelope power in the kth channel is denoted by μ_{kk} and the covariance of n_k and n_l by

$$\mu_{kl} = E(n_k{}^* n_l) \tag{2-1}$$

where the asterisk (*) denotes the complex conjugate. We note that

$$\mu_{lk} = E(n_l{}^* n_k) = \mu_{kl}{}^*. \tag{2-2}$$

When the K channels represent the output of the element of an array antenna, the covariance terms μ_{kl} are determined by receiver noise and the spatial distribution of all noise sources "seen" by the antenna. Here we assume that the covariances are known. The desired signal, when it occurs, is assumed to be present in the K channels in proportion to the known complex numbers s_k. The signal in channel k is represented as αs_k, where α defines the level and time variation of the signal. In a linear array antenna with equally spaced elements, the s_k are determined by the direction of the desired signal. Thus, if the desired direction

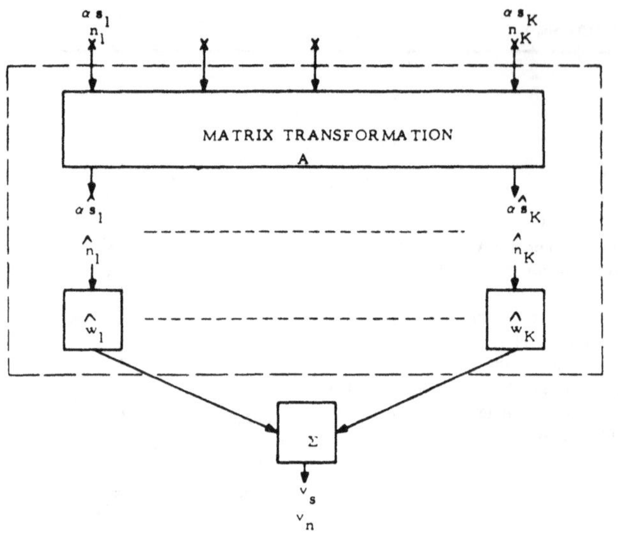

Fig. 2-2. Functional representation of combiner that is equivalent to one shown in Fig. 2-1.

is θ rad from mechanical boresight, we have

$$s_k = \exp\left(j\frac{2\pi kd}{\lambda}\sin\theta\right) \tag{2-3}$$

where d is the element spacing and λ is the wavelength.

The problem is to choose the weights w_k (see Fig. 2-1), so that the SNR at the output of the combiner is maximized. We will show that the optimum weights are determined by

$$MW = \mu S^* \tag{2-4}$$

where

$$M = [\mu_{kl}] = \text{covariance matrix of the noise outputs} \tag{2-5}$$

$$W = \begin{bmatrix} w_1 \\ \vdots \\ w_K \end{bmatrix} \tag{2-6}$$

$$S = \begin{bmatrix} s_1 \\ \vdots \\ s_K \end{bmatrix} \tag{2-7}$$

are column vectors and μ is an arbitrary constant.

To derive (2-4), we start by writing expressions for the signal and noise outputs of the combiner. The signal output is

$$v_s = \alpha \sum_{k=1}^{K} w_k s_k. \tag{2-8}$$

This may be conveniently expressed in matrix notation as

$$v_s = \alpha W_t S = \alpha S_t W \tag{2-9}$$

where the subscript t means transposed. Similarly, the noise output may be expressed as

$$v_n = W_t N = N_t W \tag{2-10}$$

where

$$N = \begin{bmatrix} n_1 \\ \vdots \\ n_K \end{bmatrix}. \tag{2-11}$$

The expected output noise power will be

$$\begin{aligned} P_n &= E\{|v_n|^2\} = E\{|W_t N|^2\} \\ &= E\{(W_t N)^*(N_t W)\} \\ &= E\{W_t^* N^* N_t W\}. \end{aligned} \tag{2-12}$$

The expectation operator E will affect only the noise terms; hence,

$$P_n = W_t^* E\{N^* N_t\} W \tag{2-13}$$

$$= W_t^* M W \tag{2-14}$$

where

$$M = E\{N^* N_t\} = [\mu_{kl}]. \tag{2-15}$$

As noted earlier, M is the covariance matrix of the noise components. If the noise components are uncorrelated, the matrix M will be a diagonal matrix. In general, however, M may have nonzero entries in any position. Since $\mu_{kl} = \mu_{lk}^*$, the matrix M is Hermitian, that is, $M_t = M^*$. It is also positive definite since the output noise power P_n is greater than zero whenever $W \neq 0$.

Since M is a positive definite Hermitian matrix, it can be "diagonalized" by a nonsingular coordinate transformation. This means that a transformation exists that will transform the problem into one in which all channels have equal power noise components that are uncorrelated.

Assume that the transformation matrix is A and consider the block diagram shown in Fig. 2-2. After the transformation shown there, the signal and noise components, respectively, become

$$\hat{S} = AS \tag{2-16}$$

and

$$\hat{N} = AN. \tag{2-17}$$

A caret (ˆ) has been used to denote quantities after the transformation. The transformation matrix A followed by a combiner, as shown in Fig. 2-2, may be made equivalent to the combiner of Fig. 2-1 by properly relating the weights used in each case. If the outputs of the transformation matrix A are combined with the weights $\hat{W}_k$, the output signal will be

$$v_s = \alpha \hat{W}_t \hat{S} \tag{2-18}$$

$$= \alpha \hat{W}_t A S. \tag{2-19}$$

Similarly, the output noise will be

$$v_n = \hat{W}_t \hat{N} \tag{2-20}$$

$$= \hat{W}_t A N. \tag{2-21}$$

Comparing (2-19) and (2-21) with (2-9) and (2-10), we note that combining the channels after the transformation matrix A with the weight vector $\hat{W}$ is equivalent to using the weight vector $A_t \hat{W}$ without the transformation matrix. Thus, for equivalent outputs we have

$$W = A_t \hat{W}. \tag{2-22}$$

The output noise power expressed in terms of the quantities in Fig. 2-2 is

$$P_n = E\{|\hat{W}_t \hat{N}|^2\} \tag{2-23}$$

$$= E\{\hat{W}_t^* \hat{N}^* \hat{N}_t \hat{W}\} \tag{2-24}$$

$$= \hat{W}_t^* E\{\hat{N}^* \hat{N}_t\} \hat{W}. \tag{2-25}$$

Since the transformation matrix A decorrelates the noise components and equalizes their powers, the covariance matrix of the noise components after A is simply the identity matrix of order K; thus,

$$E\{\hat{N}^* \hat{N}_t\} = 1_K. \tag{2-26}$$

Using (2-26) in (2-25), we obtain

$$P_n = \hat{W}_t^* \hat{W} = \|\hat{W}\|^2. \tag{2-27}$$

If the configurations of Figs. 2-1 and 2-2 are equivalent, we may write, from (2-14) and (2-22),

$$P_n = W_t^* M W$$

$$= \hat{W}_t^* A^* M A_t \hat{W}. \tag{2-28}$$

Comparing this with (2-27), we see that

$$A^* M A_t = 1_K \tag{2-29}$$

and, therefore,

$$M = (A_t A^*)^{-1}. \tag{2-30}$$

Equation (2-29) expresses the fact that the transformation matrix A diagonalizes the matrix M. It is well known that the optimum choice for the weighting vector $\hat{W}$ in Fig. 2-2, where the noise components $\hat{n}_k$ have equal power and are uncorrelated, is given by

$$\hat{W}_{opt} = \mu \hat{S}^* \tag{2-31}$$

where μ is an arbitrary constant.

To show that (2-31) is the optimum, apply the Cauchy–Schwartz inequality to (2-18). This yields

$$|v_s|^2 \le |\alpha|^2 \|\hat{W}\|^2 \|\hat{S}\|^2 \tag{2-32}$$

where

$$\|\hat{S}\|^2 = \hat{S}_t^* \hat{S} \tag{2-33}$$

and

$$\|\hat{W}\|^2 = \hat{W}_t^* \hat{W}. \tag{2-34}$$

From (2-27) we have $P_n = \|\hat{W}\|^2$; thus, dividing both sides of (2-32) by P_n, we obtain

$$\frac{|v_s|^2}{P_n} \le |\alpha|^2 \|\hat{S}\|^2. \tag{2-35}$$

This puts an upper bound on the SNR. However, if we substitute $\hat{W} = \mu \hat{S}^*$ into (2-18) and (2-27), we obtain

$$v_s = \alpha \mu \hat{S}_t^* \hat{S} = \alpha \mu \|\hat{S}\|^2 \tag{2-36}$$

and

$$P_n = |\mu|^2 \|\hat{S}\|^2. \tag{2-37}$$

Dividing the magnitude square of (2-36) by (2-37), we have

$$\frac{|v_s|^2}{P_n} = |\alpha|^2 \|\hat{S}\|^2. \tag{2-38}$$

By (2-35), this is the maximum possible value of the SNR. Thus, we have shown that $\mu \hat{S}^*$ is the optimum value for $\hat{W}$. The optimum value of W may now be obtained from (2-22):

$$W_{opt} = A_t \hat{W}_{opt} = A_t \mu \hat{S}^* \tag{2-39}$$

or using (2-16),

$$W_{opt} = \mu A_t A^* S^* \tag{2-40}$$

and, finally, from (2-30),

$$W_{opt} = \mu M^{-1} S^*. \tag{2-41}$$

Thus the optimum weight vector W_{opt} for the combiner in Fig. 2-1 is the value of W that satisfies the equation

$$M W = \mu S^*. \tag{2-42}$$

The SNR corresponding to the optimum weighting can be obtained by substituting (2-16) into (2-38). The result is

$$\left(\frac{S}{N}\right)_{opt} = |\alpha|^2 S_t M^{-1} S^*. \tag{2-43}$$

It is important to emphasize that the solution, or control-law, expressed by (2-42) maximizes the SNR. In Section 6, we show how the optimum weights may be determined adaptively in real time.

III. Generalized Signal-to-Noise Optimization

If an array is designed so that the weights are adaptively controlled to satisfy (2-42), then the array designer is forced to accept the SNR as the governing criterion for all noise environments, including the normal "quiescent" environment (no jamming). In most applications, however, array designers are willing to compromise on SNR in order to obtain some control of the pattern shape, particularly sidelobe levels. For example, Dolph–Chebyshev weights can be used to obtain 30 dB sidelobes with less than 2 dB loss in SNR. It is apparent, then, that we need a criterion for the control of array weights that will give more flexibility in beam shaping.

A tractable approach can be developed as follows. Suppose that, in the normal quiescent environment, the most desirable array weights are given by the weight vector W_q. We assume that W_q represents an optimum compromise between gain, sidelobes, etc. Let the covariance matrix in the normal quiescent environment be M_q, and define the column vector T by the equation

$$M_q W_q = \mu T^* \tag{3-1}$$

where μ is a normalizing constant.

The point of view we wish to present is that in choosing W_q as being optimum, we have, in effect, decided to optimize to an equivalent signal vector T instead of the actual signal vector S. If the environment changes so that the covariance matrix becomes M, then from the results of the previous section, to continue optimizing on T, the weights should be adjusted so that

$$MW = \mu T^* = M_q W_q. \tag{3-2}$$

From the results of the previous section, we also know that controlling the array so as to always satisfy (3-2) is equivalent to maximizing the ratio

$$\frac{|W_t T|^2}{W_t^* M W}. \tag{3-3}$$

This is a more general criterion than maximizing the SNR, which it includes as a special case when $T = \alpha S$. We refer to T as a generalized signal vector and the ratio, (3-3), as the generalized signal-to-noise ratio (GSN). The GSN is a measure of how close the weights are to the generalized signal vector T in the "dot" on "inner" product sense, relative to output noise. Since the "level" of T is arbitrary, the numerical value of the GSN by itself has no significance. It is only significant in comparisons in which the vector T is held fixed.

Comparison of the GSN of a fixed array with that of an adaptive array is of interest. Suppose, then, that the weight vector is held fixed at W_q for all noise environments. Then the GSN will be

$$(\text{GSN})_f = \frac{|(W_q)_t T|^2}{(W_q)_t^* M W_q}. \tag{3-4}$$

Using (3-1), this may be written as

$$(\text{GSN})_f = \frac{1}{|\mu|^2}\left(\frac{|(W_q)_t M_q^* W_q^*|^2}{(W_q)_t^* M W_q}\right). \tag{3-5}$$

For an adaptive array controlled by (3-2), we have

$$(\text{GSN})_a = \frac{|W_t T|^2}{W_t^* M W} = \frac{1}{|\mu|^2}\left(\frac{|W_t M^* W^*|^2}{W_t^* M W}\right) \tag{3-6}$$

$$= \frac{1}{|\mu|^2}(W_t^* M W) \tag{3-7}$$

$$= \frac{1}{|\mu|^2}((W_q)_t^* M_q M^{-1} M_q W_q). \tag{3-8}$$

In the normal quiescent environment where $W = W_q$ and $M = M_q$, (3-5) and (3-8) both reduce to

$$(\text{GSN})_q = \frac{1}{|\mu|^2}(W_q)_t^* M_q W_q. \tag{3-9}$$

In order to put (3-6) and (3-8) into more meaningful forms, we note that the covariance matrix M differs from M_q because of the presence of additional noise sources (jamming). If these are statistically independent of the noise sources in the quiescent environment, we may write

$$M = M_q + M_j \tag{3-10}$$

where M_j is the covariance matrix of the noise components due to the additional noise sources.

Using (3-10) and (3-11) in (3-6) and (3-9), we can obtain the following expressions:

$$(\text{GSN})_f = \frac{(\text{GSN})_q}{1 + \dfrac{(W_q)_t^* M_j W_q}{(W_q)_t^* M_q W_q}} \tag{3-11}$$

and

$$(\text{GSN})_a = \frac{(\text{GSN})_q}{1 + \dfrac{(W_q)_t^* M_q M^{-1} M_j W_q}{(W_q)_t^* M_q M^{-1} M_q W_q}}. \tag{3-12}$$

The ratio in the denominator of (3-11) is the ratio of the jammer noise output to the quiescent noise output when the weights are fixed at W_q. Designating this ratio by $(J/N)_q$, we may write (3-11) and (3-12) as

$$(\text{GSN})_f = \frac{(\text{GSN})_q}{1 + (J/N)_q} \tag{3-13}$$

and

$$(\text{GSN})_a = \frac{(\text{GSN})_q}{1 + \Gamma(J/N)_q} \tag{3-14}$$

$$= (\text{GSN})_q\{1 - \gamma(J/N)_q\} \tag{3-15}$$

where

$$\Gamma = \frac{\gamma}{1 - \gamma(J/N)_q} \tag{3-16}$$

and

$$\gamma = \frac{(W_q)_t^* M_q M^{-1} M_j W_q}{(W_q)_t^* M_j W_q}. \tag{3-17}$$

When the weights are fixed, we note from (3-13) that the GSN varies almost as the inverse of the jammer-to-noise ratio. From (3-14), however, we see that the effect of the jammer on the GSN with adaptive control is reduced by the factor Γ. This factor Γ is analogous to the jammer cancellation ratio that appears in analyses of sidelobe cancellation; it is a figure-of-merit that describes the performance of an adaptive array in a specific noise environment.

The output noise power of an adaptive array controlled according to (3-2) will not remain constant as the noise environment varies. This may require using an AGC following the array combiner to maintain a constant noise level.

Without the AGC, the change in the noise level will be

$$\Delta P_n = W_t{}^*MW - (W_q)_t{}^*M_qW_q. \quad (3\text{-}18)$$

Substituting $W = M^{-1}M_qW_q$ from (3-3), we get

$$\Delta P_n = (M^{-1}M_qW_q)_t{}^*MM^{-1}M_qW_q - (W_q)_t{}^*M_qW_q \quad (3\text{-}19)$$

$$= (W_q)_t{}^*M_qM^{-1}M_qW_q \quad (3\text{-}20)$$

$$= -(W_q)_t{}^*M_qM^{-1}M_jW_q. \quad (3\text{-}21)$$

Since M_q, M_j, and M are positive definite Hermitian matrices, M^{-1} and, hence, $M_qM^{-1}M_j$ are also positive definite Hermitian matrices. This implies that, because of the minus sign in (3-21), the output noise level will decrease when additional noise sources are added to the environment.

It is interesting to observe that the control law given by (3-2) can also be developed from another point of view. Assume, again, that the weight vector W_q represents an optimum compromise between gain, sidelobes, and other factors in the quiescent environment. When the environment changes due to the introduction of jamming noise sources, we recognize that the weight vector W_q may no longer be optimum. How should the weights be changed so that the effects of the jamming are reduced? We want a criterion which reduces the jamming effects and still recognizes the optimality of W_q in the quiescent environment. A suitable criterion is to choose a weight vector W that minimizes the quantity η, where

$$\eta = W_t{}^*M_jW + (W - W_q)_t{}^*M_q(W - W_q). \quad (3\text{-}22)$$

The first term in (3-22) represents the output noise power due to the jamming sources. The second term is a measure of the deviation of the weight vector W from the quiescent optimum W_q. The deviation is measured with respect to a metric based on the quiescent covariance matrix M_q. In so doing, the importance of minimizing the deviation, $\|W - W_q\|$, is emphasized when the noise introduced by the jamming is small compared to the quiescent noise; conversely, when the jamming power is large compared to the quiescent noise, minimizing the jamming power will receive greater weight.

The expression for η, (3-22), can be manipulated into the form

$$\eta = (W_q)_t{}^*(M_qM^{-1}M_j)W_q + (MW - M_qW_q)_t{}^*M^{-1}(MW - M_qW_q). \quad (3\text{-}23)$$

It is easy to verify that (3-23) does reduce to (3-22). Now, as has already been remarked, M^{-1} and $M_qM^{-1}M_j$ are positive definite Hermitian matrices. This implies that both terms in (3-23) are nonnegative. Since the first term is fixed, η will have its minimum value when the second term is minimized. The minimum value of the second term is 0, and this occurs if, and only if,

$$MW = M_qW_q. \quad (3\text{-}24)$$

This is the same as (3-2). Note that in the quiescent environment M reduces to M_q and W is then equal to W_q as it should be. It is also interesting to observe that the value of η is the negative of ΔP_n (see (3-21)) when (3-24) is satisfied.

IV. Linear Adaptive Array

To illustrate the concepts and results of the previous section, we consider here a linear, uniformly spaced array that is weighted in accordance with (3-2). We assume first that in the quiescent environment the noise output of the element channels have equal powers and are uncorrelated. The covariance matrix in the quiescent environment would then be a diagonal matrix of the form

$$M_q = \begin{bmatrix} p_q & & & 0 \\ & p_q & & \\ & & \ddots & \\ 0 & & & p_q \end{bmatrix} \quad (4\text{-}1)$$

$$= p_q 1_K \quad (4\text{-}2)$$

where

p_q noise power output of each element,
K number of array elements,
1_K identity matrix of order K.

Now let the desired weight vector in the quiescent environment for a signal in the direction θ_s from mechanical boresight be

$$W_q = \begin{bmatrix} a_1 \\ a_2 \exp -j\beta_s \\ a_3 \exp -j2\beta_s \\ \vdots \\ a_K \exp -j(K-1)\beta_s \end{bmatrix} \quad (4\text{-}3)$$

In (4-3), the a_l are amplitude weights (real numbers) and

$$\beta_s = \frac{2\pi d}{\lambda} \sin \theta_s. \quad (4\text{-}4)$$

If the amplitude weights are all equal, the resultant patterns will be of the form $(\sin Kx)/(\sin x)$. In general, however, they will not be equal. The pattern obtained with any weight vector W_q will be

$$G_q(\beta) = \sum_{l=1}^{K} a_l \exp j(l-1)(\beta - \beta_s) \quad (4\text{-}5)$$

where

$$\beta = \frac{2\pi d}{\lambda} \sin \theta. \quad (4\text{-}6)$$

This can be expressed in matrix notation as

$$G_q(\beta) = B_tW_q \quad (4\text{-}7)$$

where

$$B = \begin{bmatrix} 1 \\ \exp j\beta \\ \exp j2\beta \\ \vdots \\ \exp j(K-1)\beta \end{bmatrix}. \quad (4\text{-}8)$$

Since the weights on the array are controlled in accordance with (3-2), the control law for the weights W in any noise environment will be

$$MW = M_q W_q = p_q W_q \tag{4-9}$$

or

$$W = p_q M^{-1} W_q. \tag{4-10}$$

The noise environment we wish to study is that caused by a single jammer added to the quiescent environment. Let the jammer be located at the angle θ_j from mechanical boresight. Then if the jamming signal in the first element channel is $J(t)$, the jamming signal in the lth channel, assuming narrow bandwidth, will be $J(t) \exp j(l - 1)\beta_j$, where

$$\beta_j = \frac{2\pi d}{\lambda} \sin \theta_j. \tag{4-11}$$

The covariance of the jamming signals in the kth and lth channels, therefore, will be $p_j \exp -j(k - l)\beta_j$, where p_j is the envelope jamming power in each channel. The covariance matrix due to the jamming signal, therefore, will be

$$M_j = p_j(\exp[-j(k - l)\beta_j]). \tag{4-12}$$

This is a Hermitian matrix of order K in which all terms on the same diagonal are equal. Because of its simple structure, it can be conveniently expressed as

$$M_j = p_j H^* U H \tag{4-13}$$

where H is the diagonal matrix

$$H = \begin{bmatrix} 1 & & & 0 \\ & \exp j\beta_j & & \\ & & \exp j2\beta_j & \\ 0 & & & \exp j(K-1)\beta_j \end{bmatrix} \tag{4-14}$$

and U is a $K \times K$ matrix of ones.

The covariance matrix for the total noise environment, quiescent plus jammer, is just the sum of the two covariance matrices

$$M = M_q + M_j \tag{4-15}$$

$$= p_q 1_K + p_j H^* U H. \tag{4-16}$$

To obtain the weight vector W from (4-10), we require the inverse of M. Using the property $H^* = H^{-1}$, it is easy to verify that

$$M^{-1} = \frac{1}{p_q}\left\{1_K - \left(\frac{p_j}{p_j + Kp_j}\right) H^* U H\right\}. \tag{4-17}$$

Using (4-17) in (4-10),

$$W = \left\{1_K - \left(\frac{p_j}{p_q + Kp_j}\right) H^* U H\right\} W_q \tag{4-18}$$

$$= W_q - \left(\frac{p_j}{p_q + Kp_j}\right) H^* U H W_q. \tag{4-19}$$

From the definitions of W_q and H we have

$$HW_q = \begin{bmatrix} a_1 \\ a_2 \exp j(\beta_j - \beta_s) \\ \vdots \\ a_K \exp j(K-1)(\beta_j - \beta_s) \end{bmatrix} \tag{4-20}$$

therefore,

$$UHW_q = G_q(\beta_j) \begin{bmatrix} 1 \\ 1 \\ 1 \\ \vdots \\ 1 \end{bmatrix} \tag{4-21}$$

and finally,

$$H^* U H W_q = G_q(\beta_j) B_j^* \tag{4-22}$$

where

$$B_j = \begin{bmatrix} 1 \\ \exp j\beta_j \\ \vdots \\ \exp j(K-1)\beta_j \end{bmatrix}. \tag{4-23}$$

Substituting (4-22) in (4-19), we get

$$W = W_q - \left(\frac{p_j}{p_q + Kp_j}\right) G_q(\beta_j) B_j^*. \tag{4-24}$$

The pattern obtained with the weight vector W may now be expressed as

$$G(\beta) = B_t W \tag{4-25}$$

$$= B_t W_q - \left(\frac{p_j}{p_q + Kp_j}\right) G_q(\beta_j) B_t B_j^*. \tag{4-26}$$

From (4-7), however,

$$B_t W_q = G_q(\beta) \tag{4-27}$$

and it is easy to show that

$$B_t B_j^* = C(\beta - \beta_j) \tag{4-28}$$

where

$$C(x) = \exp j\left\{\frac{(K-1)x}{2}\right\} \frac{\sin Kx/2}{\sin x/2} \tag{4-29}$$

thus, substituting (4-27) and (4-28) into (4-26) gives

$$G(\beta) = G_q(\beta) - \left(\frac{p_j}{p_q + Kp_j}\right) G_q(\beta_j) C(\beta - \beta_j). \tag{4-30}$$

This says that the pattern of the adaptively controlled linear array in the presence of a jammer consists of two parts. The first is the quiescent pattern $G_q(\beta)$, and the second, which is subtracted from the first, is a $(\sin Kx)/(\sin x)$ shaped beam centered on the jammer. This is shown in Fig. 4-1. The gain of the array in the direction of the jammer will be

$$G(\beta_j) = G_q(\beta_j) - \left(\frac{p_j}{p_q + Kp_j}\right) G_q(\beta_j) C(0). \tag{4-31}$$

Since $C(0) = K$, (4-31) reduces to

$$G(\beta_j) = \left(\frac{p_q}{p_q + Kp_j}\right) G_q(\beta_j). \tag{4-32}$$

If the array weights remained fixed at W_q in the presence of the jammer, the gain in the direction of the jammer would be $G_q(\beta_j)$. Hence, the adaptive control reduces the gain in

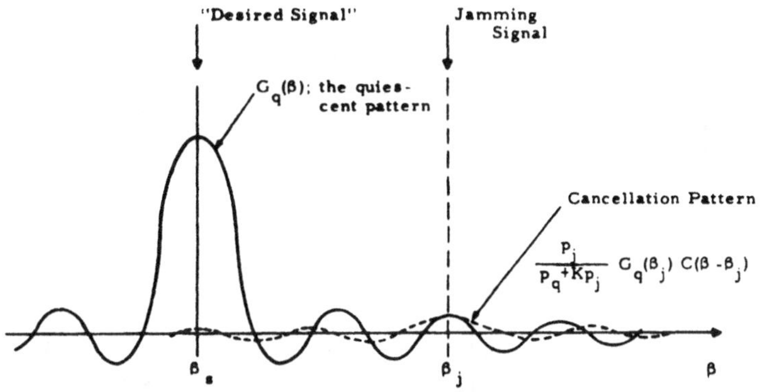

Fig. 4-1. Sketch of pattern of adaptively controlled array. (Resultant pattern is difference between quiescent and cancellation pattern.)

the direction of the jammer by the factor

$$\frac{G(\beta_j)}{G_q(\beta_j)} = \frac{p_j}{p_q + Kp_j} = \frac{1}{1 + \dfrac{Kp_j}{p_q}}. \tag{4-33}$$

The ratio in the denominator of (4-33), Kp_j/p_q, is the jammer-to-noise ratio in the "cancellation" beam, $C(\beta - \beta_j)$. Since $G(\beta)$ is a voltage gain pattern, it is tempting to conclude that the performance of the array against the jammer power had been improved by the square of the inverse of Equation (4-33). This, however, would not be correct. The proper measure of the improvement against the jammer is the cancellation ratio Γ defined by (3-16) and (3-17), which are repeated for convenience:

$$\Gamma = \frac{\gamma}{1 - \gamma(J/N)_q} \tag{3-16}$$

and

$$\gamma = \frac{(W_q)t^* M_q M^{-1} M_j W_q}{(W_q)t^* M_j W_q}. \tag{3-17}$$

To evaluate (3-17) in the present context, we note that

$$M_q M^{-1} = 1_K - \left(\frac{1}{p_q + Kp_j}\right) M_j \tag{4-34}$$

and

$$M_j M_j = Kp_j M_j \tag{4-35}$$

therefore,

$$M_q M^{-1} M_j = M_j - \left(\frac{Kp_j}{p_q + Kp_j}\right) M_j \tag{4-36}$$

$$= \left(\frac{p_q}{p_q + Kp_j}\right) M_j. \tag{4-37}$$

Using (4-37) in (3-17), we see that

$$\gamma = \frac{1}{1 + \dfrac{Kp_j}{p_j}}. \tag{4-38}$$

Now the maximum possible value of $(J/N)_q$ is Kp_j/p_q, and this will occur only if the jammer is at the peak of the main beam of a $(\sin Kx)/(\sin x)$ quiescent beam. In that case, we obtain $\Gamma = 1$, indicating no improvement of performance against the jammer, which is to be expected. The adaptive control will be most effective against jammers in the sidelobe region of the quiescent pattern. In that case, we will have $(J/N)_q \ll Kp_j/p_q$, so that $\gamma(J/N)_q \ll 1$ and, hence, $\Gamma \approx \gamma$. Thus, for a jammer in the sidelobe region, the adaptive control will "cancel" the jammer power by approximately the jammer-to-noise ratio in the cancellation beam.

It is interesting to note that this result is the same as the reduction in voltage gain of the pattern in the direction of the jammer and that it is almost independent of the choice of W_q and, hence, of the shape of the quiescent beam.

V. Sidelobe Cancellation

The concepts presented in this report are actually a generalization of the coherent sidelobe cancellation techniques. For the sake of completeness and as a second illustration of the application of the concepts, we shall show that sidelobe cancellation may be viewed as a special case of an adaptive array.

A sidelobe cancellation system is shown in Fig. 5-1. It consists of a main, high gain antenna whose output is designated as channel "o" and K auxiliary antennas. The auxiliary antenna gains are designed to approximate the average sidelobe level of the main antenna gain pattern. The amount of desired target signal received by the auxiliaries is negligible compared to the target signal in the main channel. The purpose of the auxiliaries is to provide independent replicas of jamming signals in the sidelobes of the main pattern for cancellation.

To apply the concepts of the adaptive array we assume that the outputs of the channels are weighted and summed. The problem is to find an appropriate control law for the weights. The first step is to select a T vector. In this case, since the signal collected by the auxiliaries is negligible and the main antenna has a carefully designed pattern, we choose the $K + 1$ column vector

$$T = \begin{bmatrix} 1 \\ 0 \\ 0 \\ \vdots \\ 0 \end{bmatrix}. \tag{5-1}$$

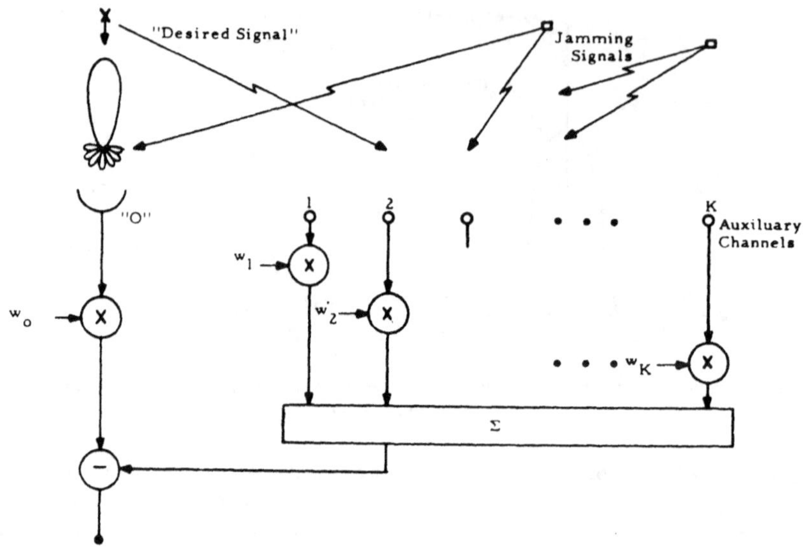

Fig. 5-1. Sidelobe cancellation system.

The optimum control law would then be

$$M'W' = \mu T \tag{5-2}$$

where M' is the $K + 1$ by $K + 1$ covariance matrix of all the channels and W' is the $K + 1$ column vector of all the weights.

Now let M be the K by K covariance matrix of the auxiliary channels only and W be the K column vector of the auxiliary channel weights. Equation (5-2) now may be partitioned as follows:

$$\begin{bmatrix} p_0 & \Lambda_t^* \\ \Lambda & M \end{bmatrix} \begin{bmatrix} w_0 \\ W \end{bmatrix} = \begin{bmatrix} \mu \\ 0 \end{bmatrix}. \tag{5-3}$$

In (5-2) we have set

$$p_0 = \mu_{00}' \text{ of } M' = \text{noise power output of the main channel} \tag{5-4}$$

and

$$\Lambda = \begin{bmatrix} \mu_{1o}' \\ \mu_{2o}' \\ \vdots \\ \mu_{Ko}' \end{bmatrix}. \tag{5-5}$$

Note that μ_{lo}' is the cross-correlation of the output of the lth auxiliary with the output of the main channel. Equation (5-3) may be written as two separate equations: a scalar equation,

$$p_o w_o + \Lambda_t^* W = \mu \tag{5-6}$$

and a matrix equation,

$$MW = -w_0 \Lambda. \tag{5-7}$$

The control law expressed by (5-2) can be implemented using $K + 1$ control loops to control the $K + 1$ weights $w_o, w_1, \cdots, w_k$. However, because of the unique form of the T vector, it is possible to achieve an optimum combiner with only K control loops. To do this, we observe that if the weight vector W' is optimum for a given noise environment (a given M'), then any multiple of W' will also be optimum since the figure of merit, GSN, is a ratio in which the level of W' does not matter. This means, in effect, that we may let the parameter μ in (5-2) vary freely. Thus, instead of fixing μ and trying to control w_o to satisfy (5-6), we can fix w_o at a nonzero value if we are sure that no solution to (5-2) will require w_o to be 0. However, since the inner product $W_t'T = w_o$, the GSN will never be optimized with $w_o = 0$. With w_o fixed at $\hat{w}_o$, we only need control loops for the auxiliaries, and the control law is given by (5-7) with $w_o = \hat{w}_o$,

$$MW = -\hat{w}_o \Lambda. \tag{5-8}$$

Since the inner product $W_t'T = \hat{w}_o$ is fixed, and the control law, (5-8), optimizes the ratio of the inner product to the output noise power, it must clearly be minimizing the output power. Thus, (5-8) is the sidelobe cancellation control law for minimizing the output noise power. Equation (5-8) has been obtained using this criterion in previous studies of sidelobe cancellation.

VI. Implementation

The control loops for an adaptive array can be implemented using the same circuitry as is used for coherent sidelobe cancellation. The basic idea is shown in the functional block diagram in Fig. 6-1. The notation used there gives the complex envelope of signals on phase coherent carriers. The signal in each channel u_k is multiplied by the weight w_k, and then the weighted signals are summed. The multiplication and summation occur on carrier frequencies at IF.

The weights w_k are derived by correlating u_k with the sum signal Σ, substracting the correlation from the desired vector component t_k^*, and then using a high gain amplifier.

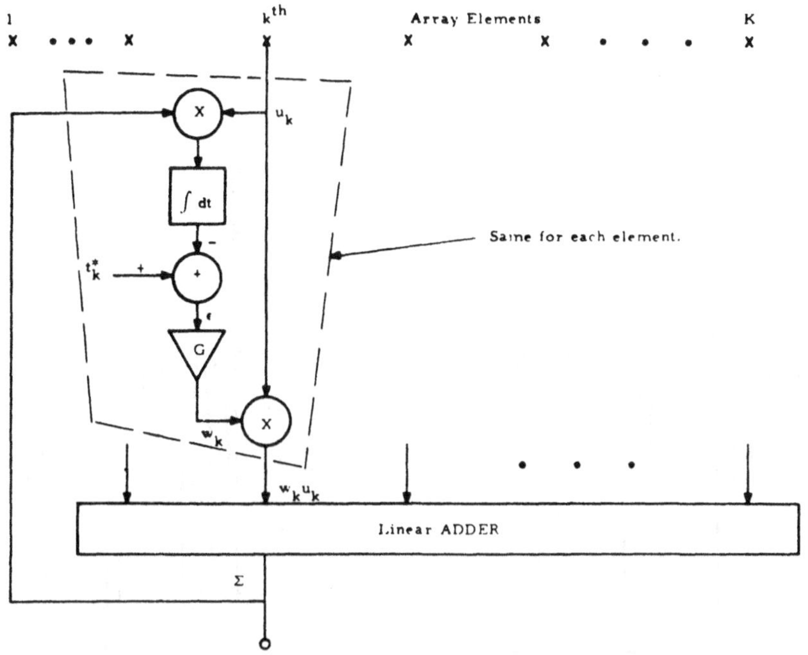

Fig. 6-1. Functional block diagram of implementation of element control loop for adaptive array.

For each w_k, we have

$$w_k = G\left\{t_k^* - u_k^* \sum_{l=1}^{K} w_l u_l\right\} \qquad (6\text{-}1)$$

or

$$\sum_{l=1}^{K} w_l\left(u_k^* u_l + \frac{\delta_{lk}}{G}\right) = t_k^* \qquad \text{for } k = 1,\cdots,K \quad (6\text{-}2)$$

where

$$\delta_{lk} = \begin{cases}1, & \text{when } k = l\\ 0, & \text{all other } k.\end{cases}$$

Recalling that $u_k^* u_l$, is an element of the covariance matrix M, the K equations represented by (6-2) may be written as

$$\left\{M + \frac{1_K}{G}\right\} W = T^*. \qquad (6\text{-}3)$$

This differs from the optimum control law, (3-2), by the addition of a term inversely proportional to gain. This term introduces an error analogous to the "servo error" of a type 0 servo. Its effect can be made negligible with sufficiently high gain.

The preceding discussion demonstrates that the steady-state solution of the control loops is essentially the desired one. The dynamic behavior of the control loops is determined by the "integrators" shown in Fig. 6-1. In practice these are high Q, single-pole circuits. The differential equations describing the dynamic behavior of the loops can be shown to be

$$\tau\frac{dw_k}{dt} + w_k = G\left\{t_k^* - u_k^* \sum_{l=1}^{K} w_l u_l\right\} \qquad (6\text{-}4)$$

where τ is the time constant of the "integrator" circuits.

The loops are designed so that the weights w_k will vary slowly compared to the bandwidth of the signals u_l. Hence, the weights will be uncorrelated with u_l. Thus, if we apply the expectation operator to both sides of (6-4) we get

$$\tau\frac{d\bar{w}_k}{dt} + \bar{w}_k = G\{t_k^* - \sum \bar{w}_l\overline{u_k^* u_l}\}. \qquad (6\text{-}5)$$

This equation determines the dynamic behavior of the expected value of w_k. The K equations obtained from (6-5) may be represented in matrix form as

$$\tau\frac{d\bar{W}}{dt} = -(GM + 1_K)\bar{W} + GT^*. \qquad (6\text{-}6)$$

It can be shown easily that matrix equations of this form are stable if the matrix $GM + 1_K$ has only positive eigenvalues. Since M is a positive definite Hermitian matrix, $GM + 1_K$ is also a positive definite Hermitian matrix. It, therefore, has only positive eigenvalues. Thus, the implementation shown in Fig. 6-1 is stable under all conditions for the assumptions made. In practice, second-order effects, neglected in the preceding analysis, may make the loops unstable if the loop gains are allowed to become excessive.

VII. Computer Simulation

Figs. 7-1 through 7-9 show the results of the computer simulation of an adaptive array. The array used in these studies was a 21-element linear array with half-wavelength spacing. The servo error discussed previously was neglected, and it was assumed that the array weights were determined by the control law

$$MW = T^*. \qquad (7\text{-}1)$$

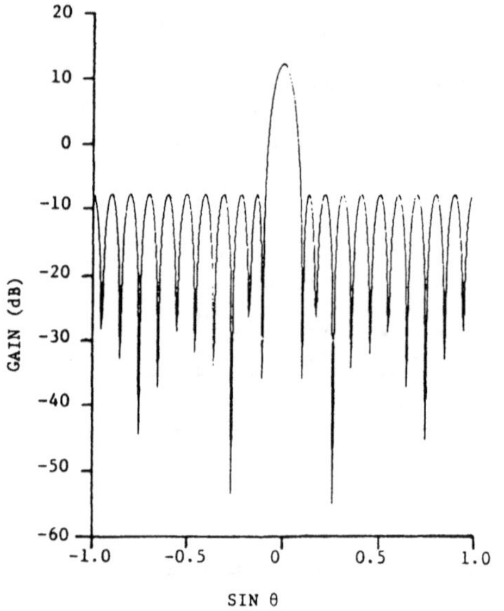

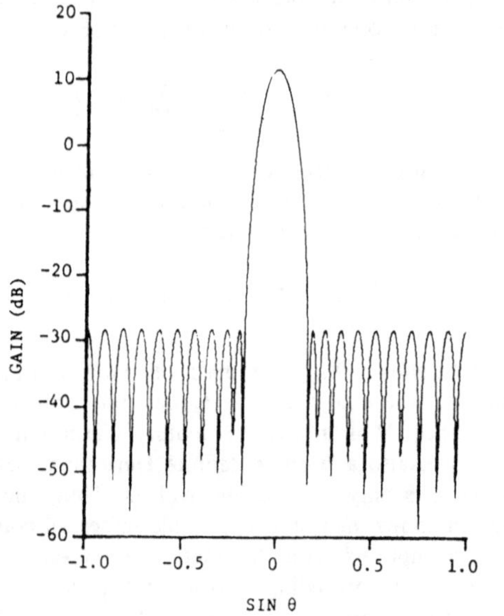

Fig. 7-1. Quiescent patterns (no jammers present). (a) −20 dB sidelobes. (b) −40 dB sidelobes.

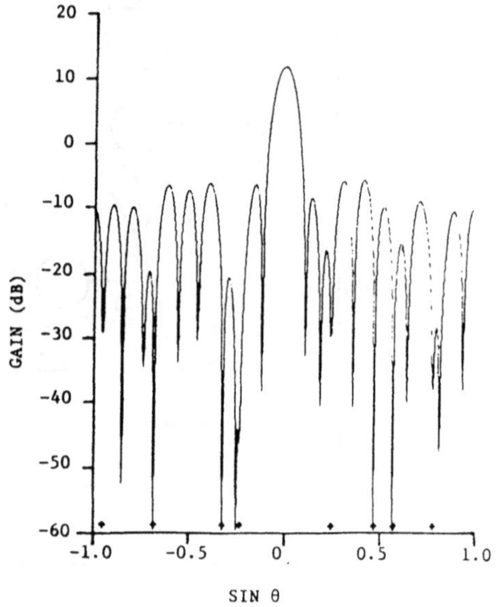

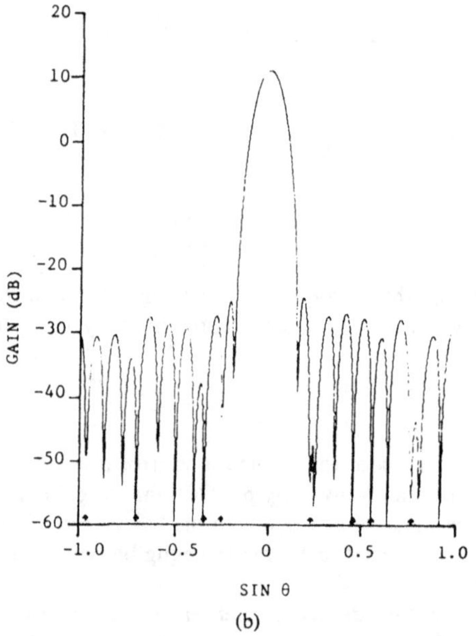

Fig. 7-2. Patterns with eight "randomly" located narrowband jammers. Nulls are produced at locations of jammers. Nulls below −60 dB are plotted as 60 dB. (a) −20 dB sidelobes (quiescent). (b) −40 dB sidelobes (quiescent).

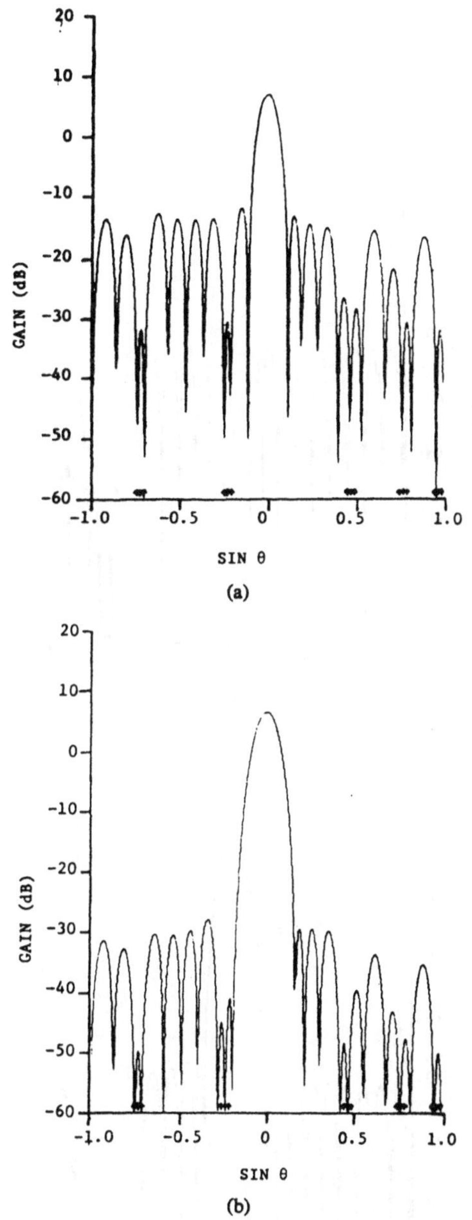

Fig. 7-3. Jammer configurations consist of five clusters of three each. Each cluster is about half a sidelobe width wide. Jammer-to-noise ratio in each element is 0 dB. (a) −20 dB sidelobes (quiescent). (b) −40 dB sidelobes (quiescent).

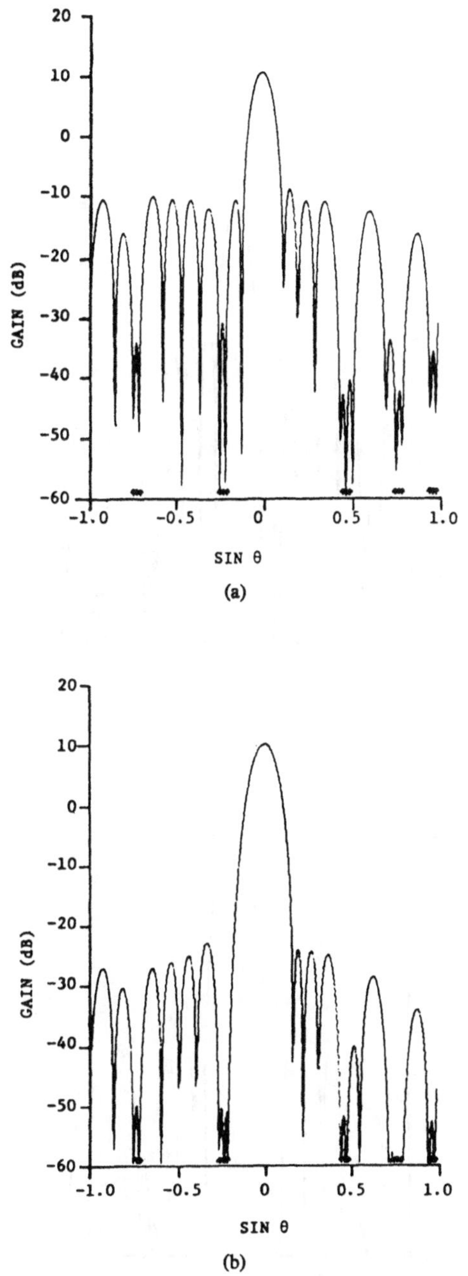

Fig. 7-4. Same as Fig. 7-3 with 15 dB jammer-to-noise ratio.

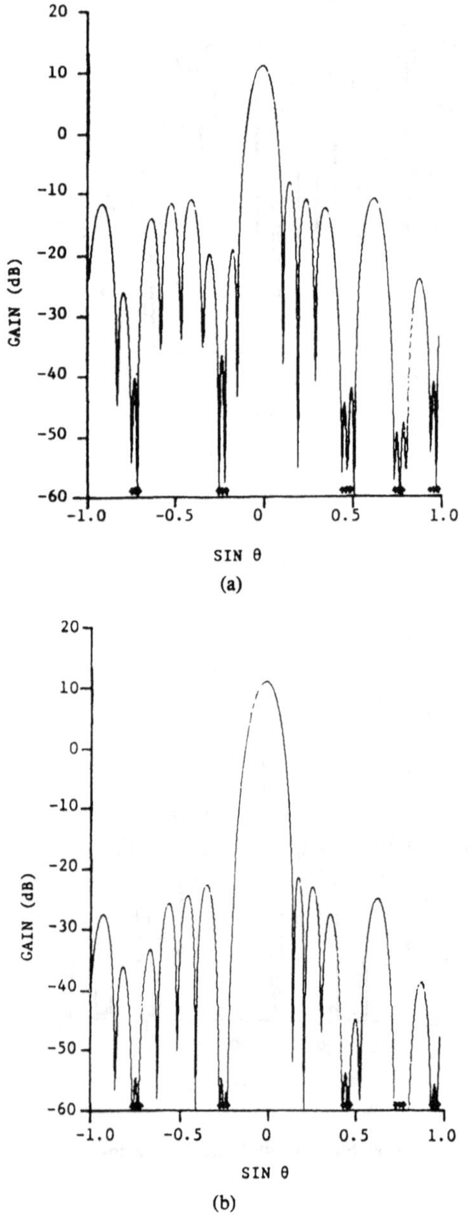

Fig. 7-5. Same as Fig. 7-3 with 30 dB jammer-to-noise ratio.

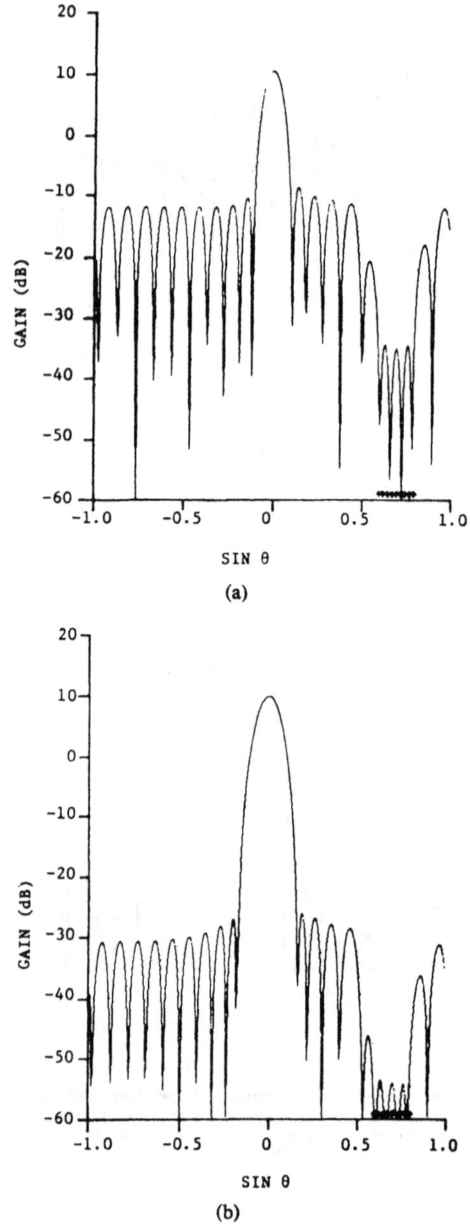

Fig. 7-6. Nine jammers are clustered and span almost two sidelobes.

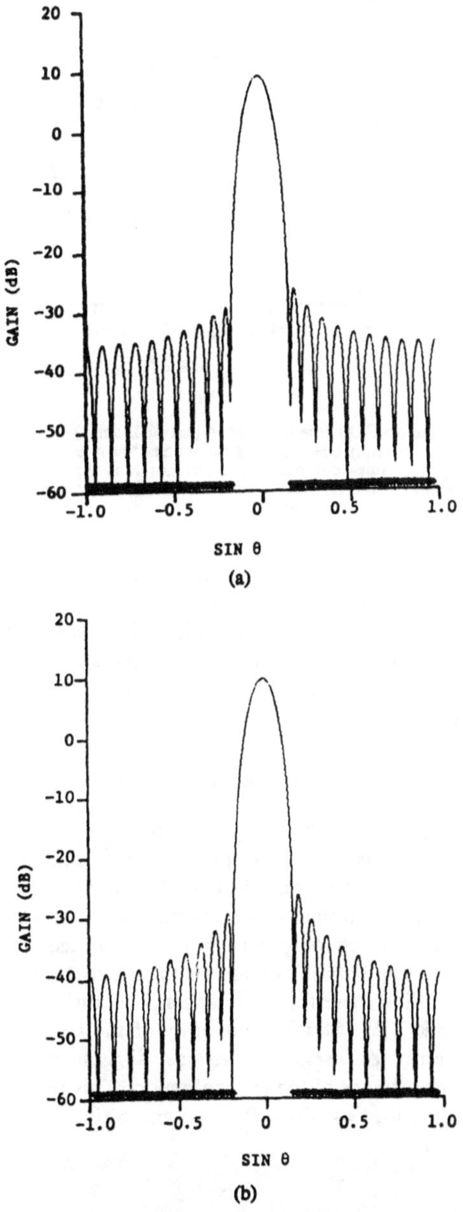

Fig. 7-7. 91 jammers span entire sidelobe region. Two resulting patterns are almost identical, indicating that under severe conditions choice of quiescent pattern has little effect. (a) −20 dB sidelobes (quiescent). (b) −40 dB sidelobes (quiescent).

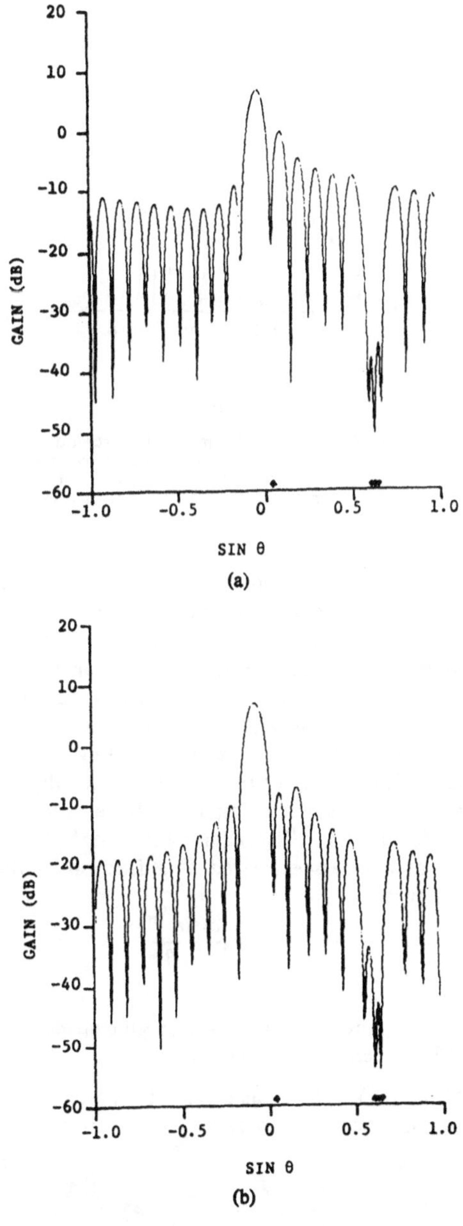

Fig. 7-8. Four jammers; one cluster of three and one in main beam. One in main beam apparently forces beam to move so as to put jammer in its first null. (a) −20 dB sidelobes (quiescent). (b) −40 dB sidelobes (quiescent).

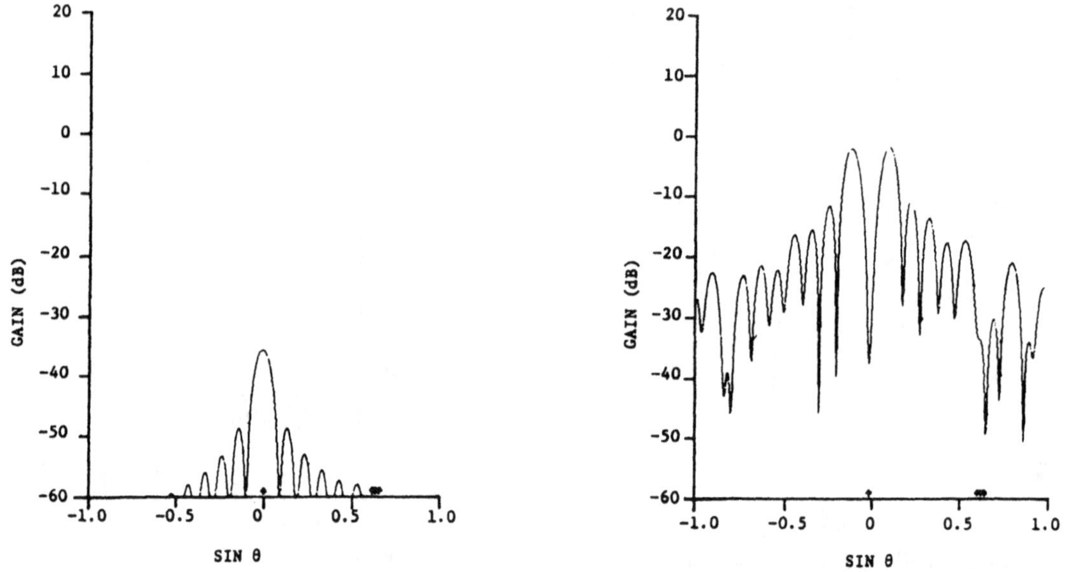

(a) (b)

Fig. 7-9. (a) One jammer is at boresight and cluster of three is in sidelobe. Quiescent patter is (sin Kx)/(sin x) beam. Shape of main beam is unaltered by presence of jammer since cancellation beam has same shape. Main effect is drastic lowering of level of pattern. (b) −30 dB Dolph–Chebyshev quiescent pattern. Deep notch is produced in center of main beam. This occurs because main beam of quiescent pattern is broader than main beam of (sin Kx)/(sin x) "cancellation" beam. There is much less effect on pattern level than in first case.

Various vectors T were used to give different quiescent beams. The quiescent beams studied were a (sin Kx)/(sin x) beam and Dolph–Chebyshev beams with 20, 30, and 40 dB sidelobes. The matrix M was computed for various spatial configurations of jammers, and narrow bandwidth was assumed. Wide bandwidth can be simulated by using narrowband calculations and spreading the jammers spatially. Thus, many of the figures show configurations of clustered jammers. Each cluster may be regarded as a single wideband jammer. The equivalent bandwidth Δf of a cluster of jammers expressed as a fraction of the center frequency is

$$\frac{\Delta f}{f_c} = \frac{\Delta(\sin\theta)}{\sin\theta_c} \tag{7-2}$$

where

$\Delta(\sin\theta)$ spread of the cluster in $\sin\theta$ space,
$\sin\theta_c$ center of the cluster.

The results are shown as plots of the pattern of the array in the presence of jammers. The abscissa is $\sin\theta$ for $0 \le \theta \le \pi/2$ from $-1 \le \sin\theta \le 1$. The desired target direction corresponds to the center of the plots. Except for Fig. 7-9, only 20 and 40 dB Dolph–Chebyshev patterns are shown. For ease of comparison, two plots are given in each figure with the same jammer environment assumed in both cases.

Acknowledgment

This report has been reprinted because of the recurring interest in this study. Except for format changes in some figures, the text corresponds to that in previous editions. The antenna gain patterns in Section 7 were replotted on a Calcomp plotter (originally they were plotted using a line printer). The author extends his thanks to D. J. Chapman for handling these format changes. The author also wishes to thank J. W. Stauffer, presently of the General Electric Company, for useful discussions, suggestions, and help with the artwork of the original report. This report has also been identified as SPL TR 69-76 and SPL TR 66-1.

Understanding and Optimizing Multiple Sidelobe Canceler Operation
B.L. Lewis and J.P. Hansen
NRL Report 7610, DDC AD 914536, October 25, 1973
Reprinted by permission.

5.B.3

INTRODUCTION

The conventional mathematical technique of analyzing the operation of multiple side-lobe cancelers on multiple jammers does not provide a clear insight into the actual processes involved, nor does it automatically suggest methods for optimizing the canceler performance in practical situations. As a consequence, a simple vector technique for analyzing such loops has been developed that provides a lucid picture of the processes involved and offers help in optimization.

This technique allows an analyst to determine easily what the residue of multiple loops will be under any given geometry of jammers and auxiliary antennas through a complete 360° scan of the radar antenna if desired. It then allows him to determine the effect of repositioning the auxiliary antennas or tells him that he must add more antennas and loops to obtain the desired performance.

The technique is based on the fact that side-lobe canceler loops are linear, and when operating in groups the operation of any loop is independent of the operation of all other loops to a first approximation. This allows an analyst to pick any loop to start his analysis and to ignore the operation of all others while he completes his analysis of the first loop. In this way he can analyze one loop after another, with each successive loop operating on the residue of those already analyzed, and thus determine an initial approximation of the final residue produced by the group operating together. The approximation can be made much more accurate by repeating the process, using this initial result as the radar input to the cancelers.

With practice, the analyst can obtain a feel for the problem that will allow him to design an optimum arrangement of auxiliary antennas to protect adequately a radar from any given number of jammers.

This technique has been tested experimentally and found to be valid and valuable. It reveals that the common belief that n loops can adequately cancel the signals from n jammers is false when the jammer geometry is variable and the auxiliary antennas are omnidirectional and fixed in space. It explains why poor cancellation can be obtained from a given system with certain jammer geometries and good cancellation can be obtained with other jammer geometries without invoking multipath effects as an excuse for the poor performance.

It is hoped that, because the technique is so easy to use, it will prove valuable both to engineers designing systems and to customers buying them.

BACKGROUND

Side-lobe canceler loops are usually composed of an auxiliary omnidirectional antenna and a feedback loop that uses a sample of the jamming environment to subtract jamming signals from a radar receiver. In this process, it is assumed that the spacing between the omnidirectional antenna and the radar antenna is small compared to the autocorrelation distance of any jamming signal to be considered. In this way, any jamming signal picked up by the auxiliary antenna can be considered to be correlated with its counterpart picked up by the radar antenna. This permits the relative radio-frequency phase and amplitude of the auxiliary antennas' signal to be specified with respect to its counterpart picked up by the radar antenna, independently of the spectral characteristics of the jamming signal.

DEMONSTRATION OF VECTOR TECHNIQUE

The vector method of analyzing the operation of multiple canceler loops on the signals from multiple jammers can be demonstrated as follows.

Figure 1 shows a scanning serach radar antenna with two omnidirectional auxiliary antennas placed symmetrically about the radar antenna. The two auxiliary antennas are shown driving two coherent side-lobe canceler loops. The phase center of the radar antenna is the chosen reference point for defining the phase angle of all signals incident at the three antennas. The wavefronts from two jammers are shown incident upon the antennas from two different directions. One signal is represented by a circle on the line and the other is represented by a square.

The average rms magnitudes and the phases of the two jamming signals received at the auxiliary antennas with respect to their counterparts received by the radar antenna can be graphically represented by vectors as illustrated in Fig. 2. It is obvious that "circle" signals will lag and "square" signals will lead their radar counterparts in phase at auxiliary antenna 1, since the wavefront of the circle has not arrived at antenna 1, whereas that of square has already passed antenna 1 when the wavefronts arrive at the radar antenna phase center. Similarly, circle will lead and square will lag their counterparts in the radar antenna in auxiliary 2 as illustrated.

Radar antenna scan can be accounted for by controlling the magnitudes of the vectors in the radar antenna and inverting one or the other as they move from side lobe to side lobe. In Fig. 2, the two jamming signals were arbitrarily set equal in magnitude at the auxiliary antennas. They could just as easily have been made unequal. However, their resultants in the radar antenna were made unequal to demonstrate the directive nature of the radar antenna.

The signals picked up by the radar antenna are frequency translated by ω_1 from a local oscillator and pass through adder 1 and adder 2. The output of adder 2 enters multiplier 2 where it is multiplied by the jamming signals picked up by auxiliary antenna 1. The difference frequencies formed in this process are passed through a narrowband amplifier centered on ω_1. The multiplier and narrowband amplifier act like a correlator and produce outputs only when the inputs to the multiplier are correlated.

Mathematically, this operation can be demonstrated by letting circle be $A \cos(\omega t + \theta)$ in auxiliary 1 and $B \cos \omega t$ in the radar antenna. The output of adder 2 will then be

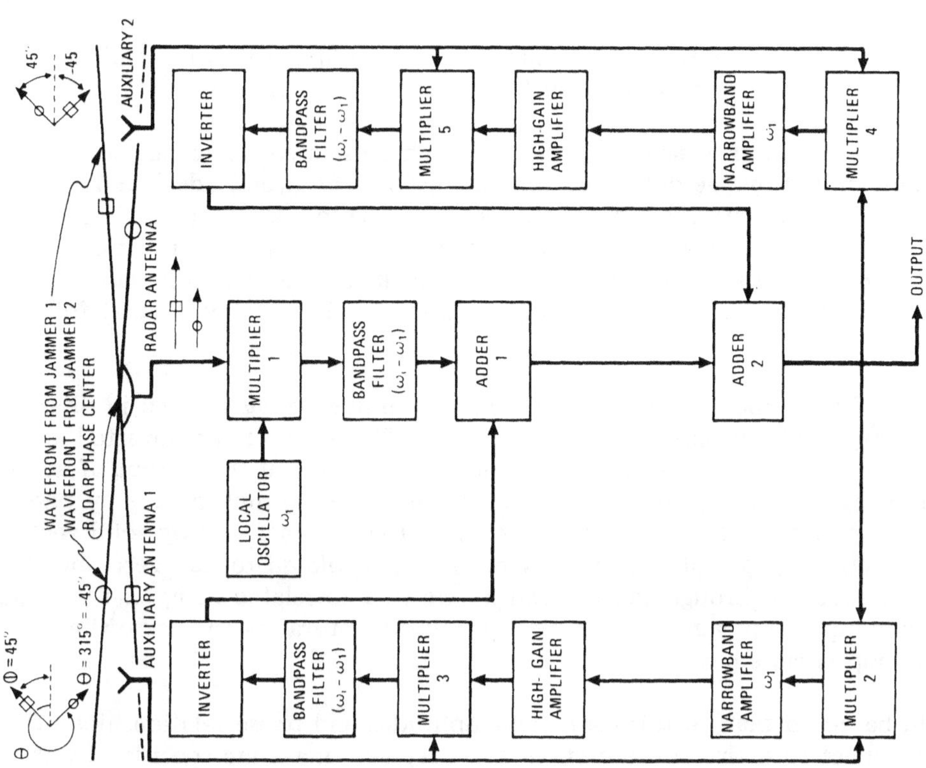

Fig. 2—Definition of phase and amplitude of the vectors at the antennas

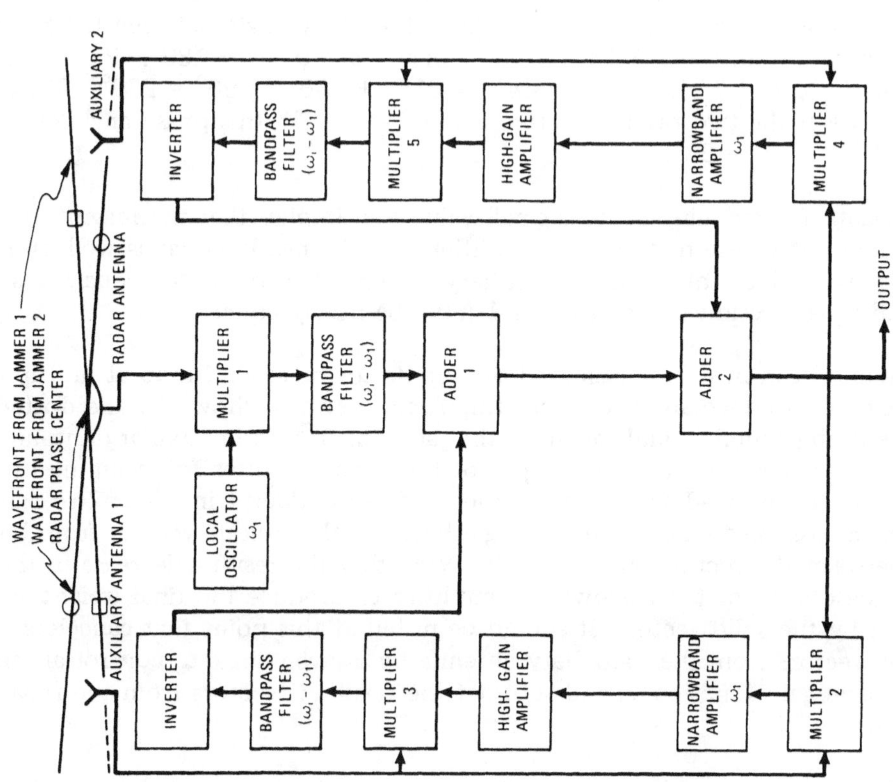

Fig. 1—Starting point in applying the vector technique

$KB \cos(\omega - \omega_1)t$, and the output of the narrowband amplifier following multiplier 2 will be $GKAB \cos(\omega_1 t + \theta)$, where G is a gain function greater than unity.

The output of the narrowband amplifier for each correlated signal can be represented vectorially as illustrated by the dotted vectors in Fig. 3. The magnitude k and phase α of the resultant can also be portrayed graphically since all vectors are correlated at this point. This initial resultant will be labeled F_1^1 and has been called a weighting function. F_1^1 enters multiplier 3 where it multiplies the jamming signals picked up on auxiliary antenna 1. The resultant lower sideband is passed by a bandpass filter closely matched to that following multiplier 1.

The result of this process will be the auxiliary jamming signals rotated in phase by $-\alpha$ and having magnitudes proportional to k (Fig. 4). These vectors are inverted and added to the radar input to adder 1 to form the output of canceler 1 at the output of adder 1 as illustrated in Fig. 5. These vectors now become the input to multiplier 2 by passing adder 2, and the new output of the narrowband amplifier is portrayed by solid vectors. Note that the resulting phase of F_1, the new resultant, is rotated toward zero and its magnitude is greatly reduced through cancellation of the two correlated components. This cancellation permits high loop gain to be employed without saturating the narrowband amplifier when the loop is closed.

It should be noted that, in this case, the amplitudes and phase relationships of the residues with respect to their counterparts in auxiliary antenna 1 are specified by the necessity for reducing F_1 by the amount required by the loop gain. If this gain is high, the phase of square signals out of the narrowband amplifier $(\phi - \gamma)$ must approach 180° with respect to circle whose phase will be $\theta - \delta$. Thus, the phase angle between the residue vectors must be such that $\phi - \gamma + 180° = \theta - \delta$ or $\gamma - \delta = \phi - \theta + 180°$. Putting in the assumed values for ϕ and θ yields $\gamma - \delta = 45° - 315° + 180° = -90° = 270°$. Since the phase angle between the two residue vectors is $360° - (\gamma - \delta)$, this phase angle can be seen to be $360° - 270° = 90°$.

The necessity for reducing F_1 to a small value also implies that the magnitudes of circle and square out of the narrowband amplifier must be nearly equal as well as nearly 180° out of phase. Since the signals in auxiliary antenna 1 were chosen equal, this means that circle and square residues out of adder 1 must be nearly equal.

These residues can now be considered the inputs to the second loop at multiplier 4 (Fig. 6). They will produce an initial weighting function F_2^1 as shown by the dotted vectors. This weighting function multiplied by the jamming signals in auxiliary antenna 2 will produce the vectors shown at the output of the bandpass filter following multiplier 5. These vectors are inverted and added to the residue of adder 1 in adder 2 to cancel the residue in the assumed case, since the angle between the residue vectors from loop 1 equals that between the vectors on antenna 2. Note that the residue decrease in the second loop reduces the output of the narrowband amplifier to produce the final weighting function F_2 shown by the solid vector. It should be noted at this point that cancelers can only rotate all vectors from their auxiliary antenna by a given angle α, control all magnitudes by the same gain function k, and subtract the resultant vectors from the radar antenna's signals.

The utility and power of this technique can now be illustrated by showing that two loops cannot always cancel two jamming signals. This is done in Fig. 7 where the phase angle of one jamming signal picked up on one auxiliary leads in phase by 90° its

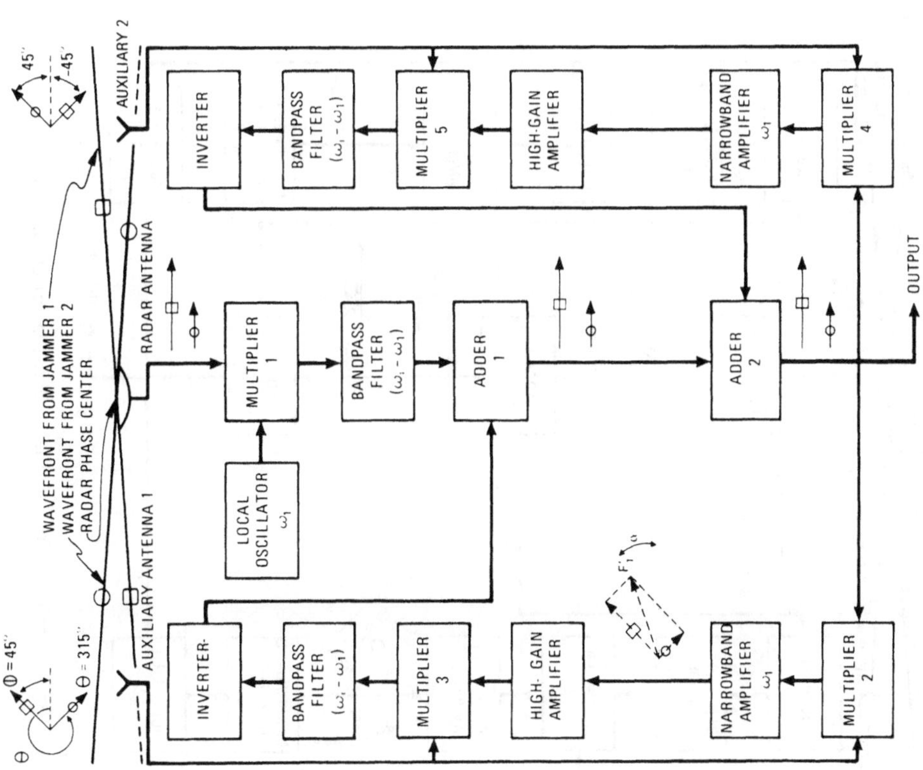

Fig. 3—Derivation of initial weighting function in loop 1

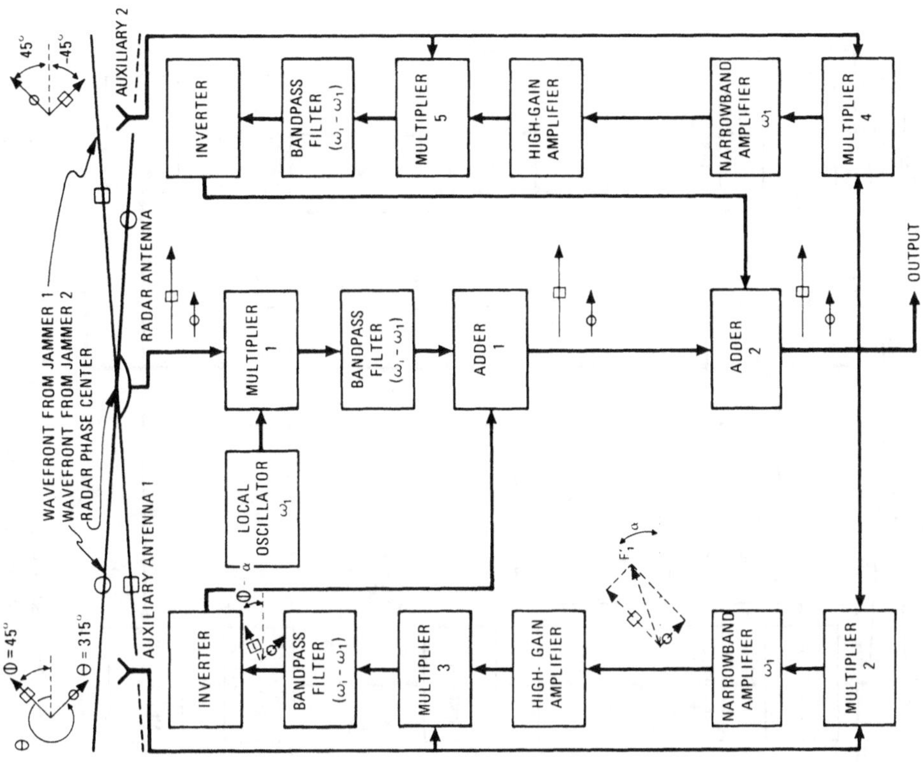

Fig. 4—Translation of auxiliary signals

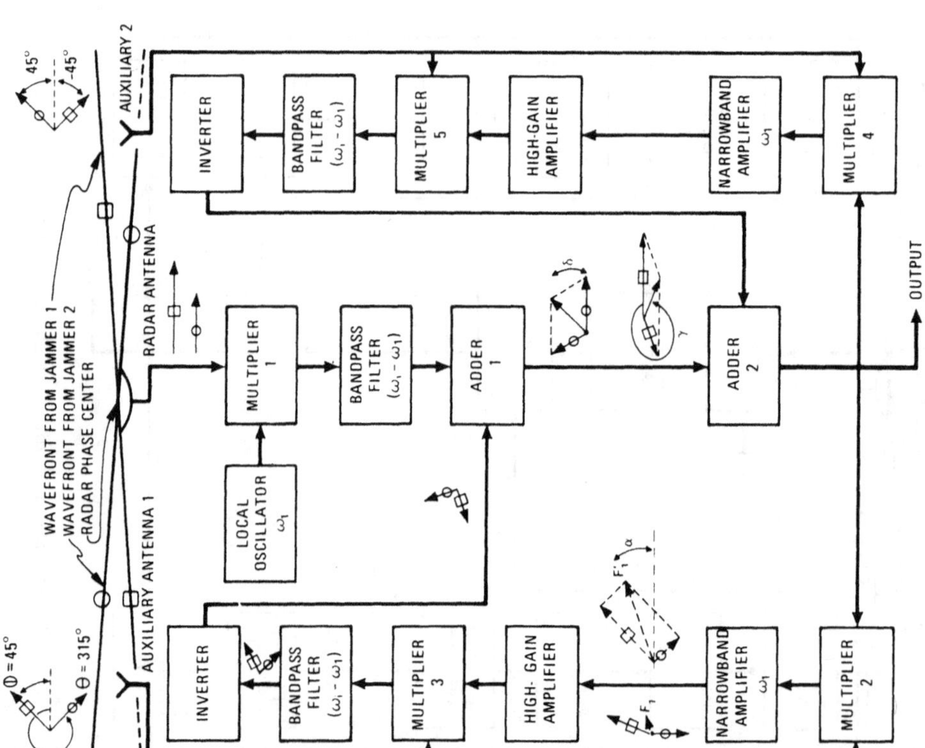

Fig. 5—Derivation of final weighting function in loop 1

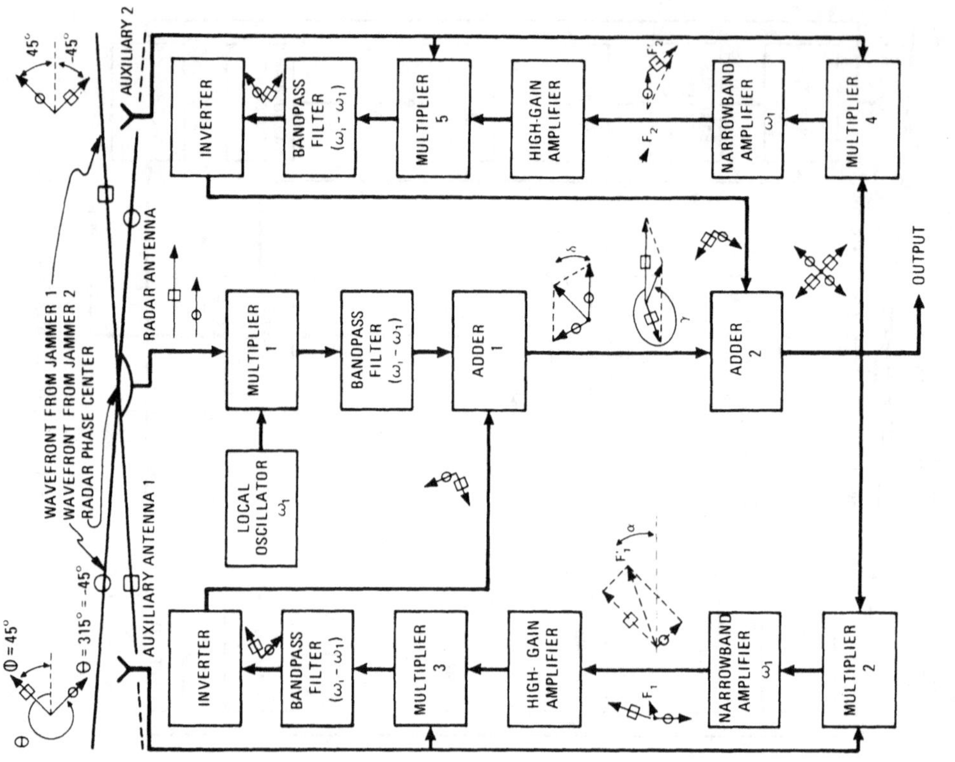

Fig. 6—Demonstration of vector technique for analyzing multiple canceler operation on the signals from multiple jammers

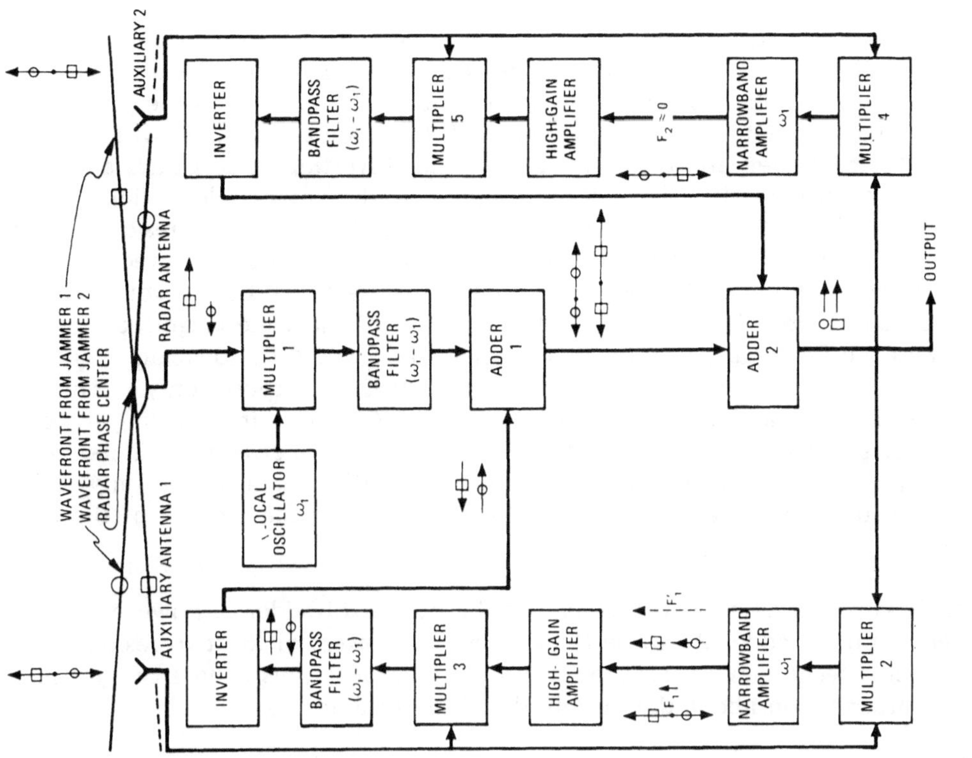

Fig. 8—Effect of phase inversion of one of the radar components in Fig. 7

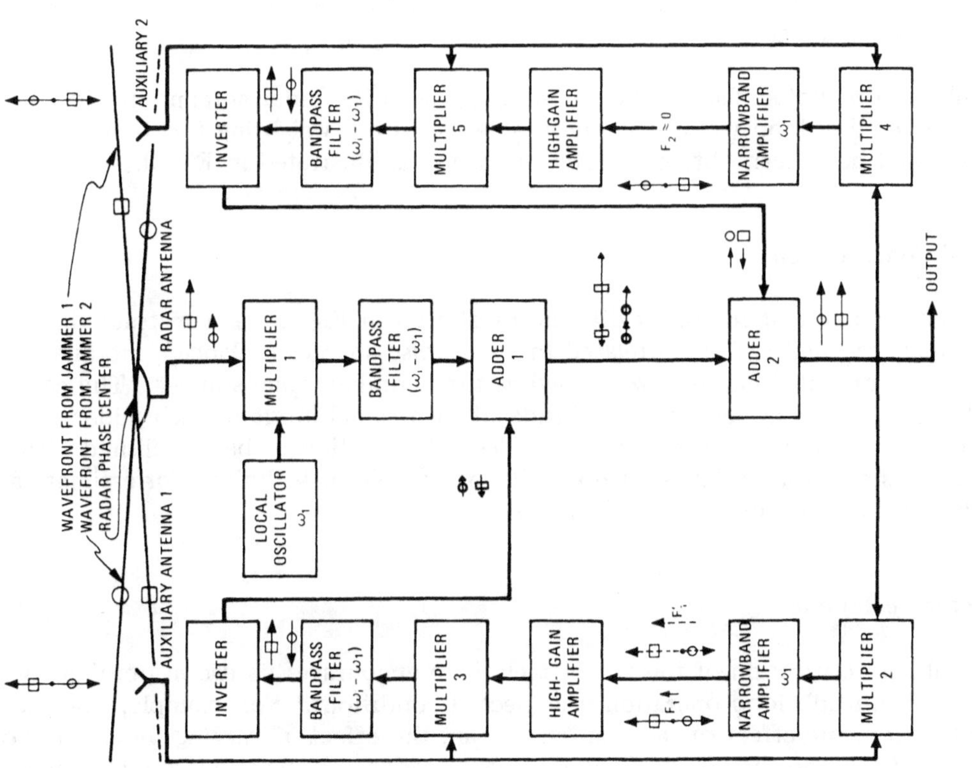

Fig. 7—Vector technique used to show a condition in which two cancelers offer poor cancellation of the signals from two jammers caused by antenna scan

counterpart in the radar and the other lags its counterpart by 90°. With symmetrical auxiliary spacing, the inverse will hold true on the other auxiliary. In this case, the initial weighting function F_1^1 has a phase angle of 90°. This subtracts from the phase angles of the signals in auxiliary 1 to produce a square with 0° phase and a circle with 180° phase with respect to their counterparts in the radar antenna. Inverting these signals and adding them to the radar signals in adder 1 results in square canceling and circle adding. This produces a final weighting function F_1 such that with high loop gain, the residue consists of nearly equal parts of square and circle with magnitudes intermediate between square and circle in the radar antenna (the loop adds circle to circle and subtracts square from square). The second loop will form a weighting function of the same magnitude but opposite polarity as the first, but it cannot be shown since it would be so small. (Note that function F_1 is shown larger than actual size for demonstration purposes only.) This will translate the signals from auxiliary 2 so that they aid the first loop in canceling square and adding to circle in adder 2. In trying to help loop 1, loop 2 effectively increases its gain but cannot significantly change the residue, since infinite gain would only make the residue of square and circle equal and opposite in phase at the outputs of the narrowband amplifiers.

Since circle is not correlated with square at the output of adder 2, however, the jammer power output of adder 2 will be the sum of the powers of the two residues. In this way, the vector technique reveals that two loops cannot always handle two jammers and it provides information about the magnitude of the residue.

It should be noted that radar antenna scan might invert the sense of one of the vectors in the radar antenna in the situation illustrated in Fig. 7. If this happens, better cancellation will be obtained as illustrated in Fig. 8. Again, in this case, the second loop cannot contribute significantly.

It should also be noted that, if the jamming signals in the radar antenna in Fig. 7 had been equal, neither loop would have been able to form a weighting function, and no cancellation would have been obtained. This condition is illustrated in Fig. 9.

Experimental Confirmation

An analog simulator of the block diagram illustrated in Fig. 1 was constructed. The antennas were adders, and two uncorrelated thermal noise sources (balanced, modulating, variable-phase, uncorrelated carriers) were used as the jamming signal sources. The carrier phase and the total signal average rms amplitude of each signal input to each adder simulating the antennas were variable to permit any desired condition to be established. The operation of this simulator confirmed the validity of the vector technique for estimating multiple canceler operation on multiple jammers.

Uses of Vector Technique

The simulator verification of the vector technique encourages its use in obtaining an understanding of multiple-loop operation in expected conditions. For example, the effect of the jammer location, the effect of radar antenna scan, the effect of varying the spacing of the auxiliary antennas, and the effect of rotation of the radar antenna about a point not located at its phase center can be determined. This knowledge can then be used to specify how many and what locations should be filled with auxiliary antennas in order to optimize the canceler operation. The effect of radar antenna scan about its phase center with symmetrical auxiliary antenna locations is analyzed in the next section as an example.

EFFECT OF RADAR ANTENNA SCAN

Radar antenna scan about its phase center will cause the relative magnitudes of the signals from two jammers at different locations to vary independently and will invert the sense of each jamming signal when the jammer moves from one side lobe to the next. An idea of the effect of this variation on multiple cancelers can be obtained as follows.

FIgure 10 illustrates one case in which one jammer (circle) is near a null of the radar antenna pattern and the signals on the auxiliary antennas are in phase. In this case, F_1^1 is produced by square alone. However, F_1^1 multiplies both square and circle from auxiliary antenna 1. The inverter thus provides two signals to adder 1 while the radar only provides one. The output of adder 1 becoming the new input to multiplier 2 produces an F_1 that is the difference of the two correlating components as illustrated. With high loop gain, the output of adder 1 will approach equal values of circle and square with a total power equal to half the power of the single component in the radar antenna. The half-power condition is evident by virtue of the residue being composed of two uncorrelated vectors, each having one-half the magnitude of the vector in the radar antenna. Each component thus has one-fourth of the power, which results in their sum having one-half the power. The output of adder 1 is now the input to multiplier 4 in the second loop. However, since these vectors are nearly equal and opposite, the second loop cannot form more than a token weighting function. As a consequence, it cannot contribute significantly to adder 2 and the output of adder 2 will remain that out of adder 1 with loop 1 gain increased by 2. In this case, the two loops will provide only a 3-dB cancellation ratio.

Figure 11 illustrates the action of the loop when the signals on the auxiliary antennas are in phase and equal in magnitude, while the signals in the radar antenna are 180° out of phase due to the jammers' being on side lobes oppositely sensed but equal in magnitude. Note that this condition prevents both loops from developing a weighting function, and no cancellation is obtained.

Figure 12 demonstrates the equal-amplitude, equal-phase case in which good cancellation is obtained by one loop, and the other does nothing but double the gain of the first.

The trend that falls between equality of signals and one signal missing in the radar antenna is illustrated for the in-phase condition in Fig. 13 and the out-of-phase condition in Fig. 14. In the in-phase case, Fig. 13, the residue of each signal has a magnitude equal to one-half the difference in magnitude of the signals in the radar antenna, and the second loop does little. In the out-of-phase case, Fig. 14, the residue of each signal has a magnitude equal to the magnitude of the smaller signal plus one-half the difference between the two signal amplitudes in the radar channel. In this case, also, the second loop does little.

It should be noted that jammer geometries that allow one jammer to enter a null on the radar antenna while the other jammer is near a side-lobe peak are bad cases if the signals from each jammer on the two auxiliaries are in phase. It should also be noted that jammer geometries in which the jamming signals arrive at both auxiliary antennas with 90° lags and leads are also bad. Good jammer geometries are those in which the jamming signals in the auxiliary antennas lead and lag by 45° in one auxiliary and lag and lead by 45° in the other. The latter case is illustrated in Fig. 6 and Figs. 15 and 16 which illustrate

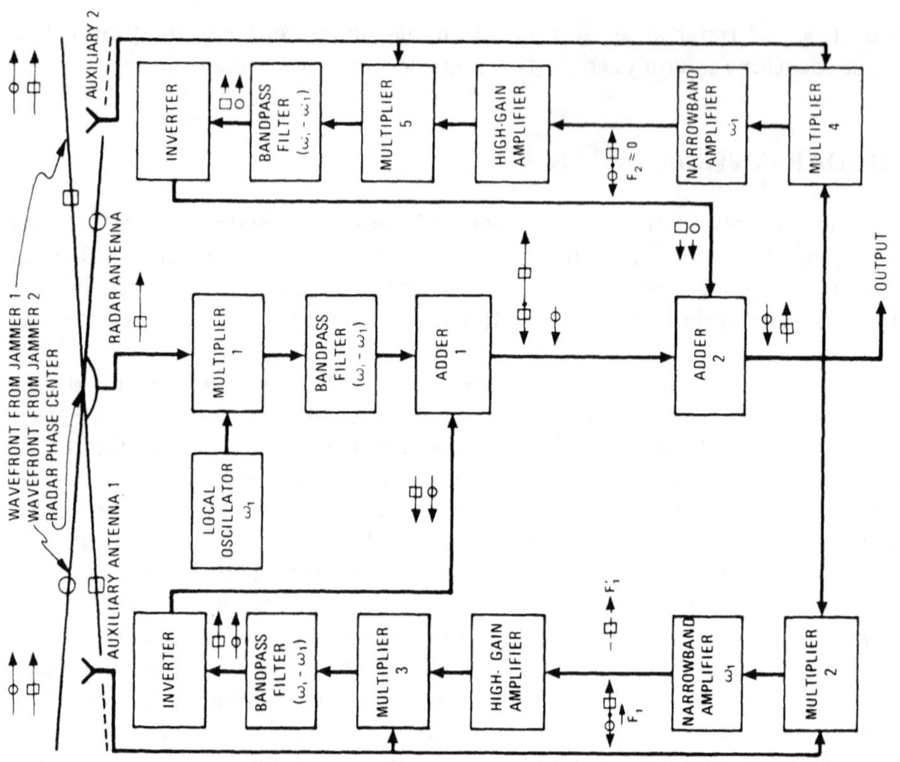

Fig. 10—One jammer in a radar antenna pattern null

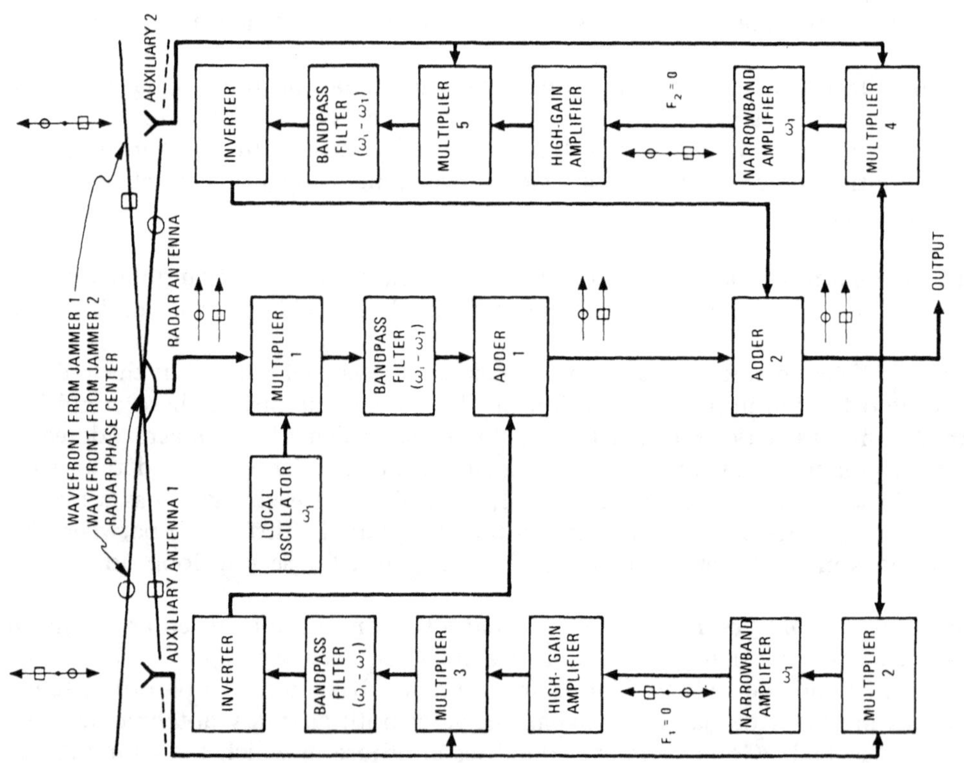

Fig. 9—The case of Fig. 7 with equal radar components

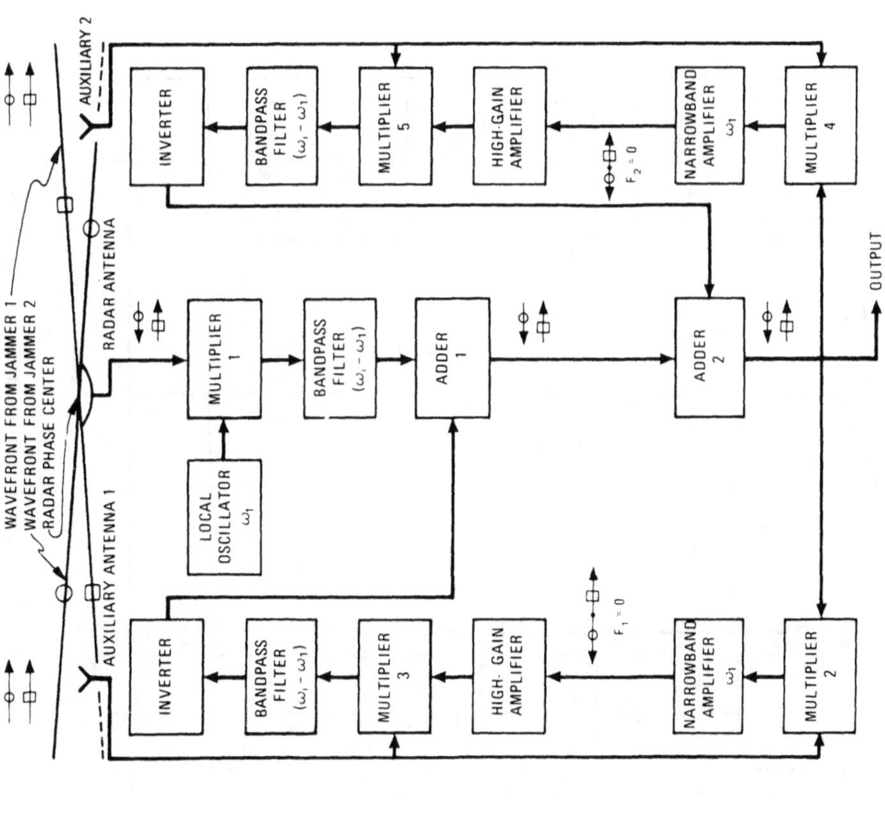

Fig. 11—Equal vectors with radar components 180° out of phase

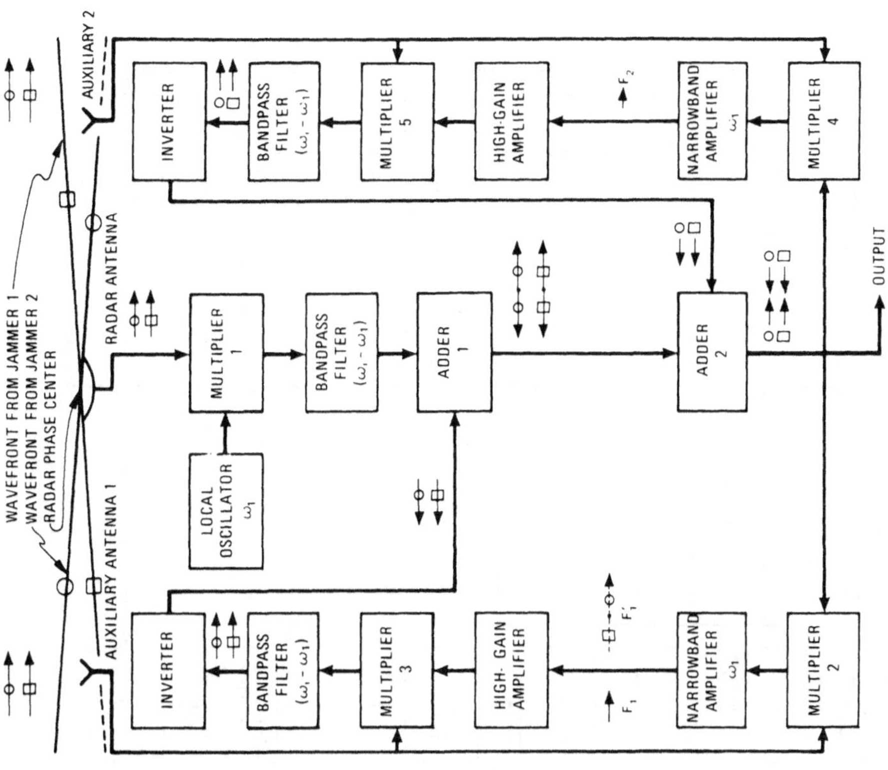

Fig. 12—Equal-amplitude components all in phase

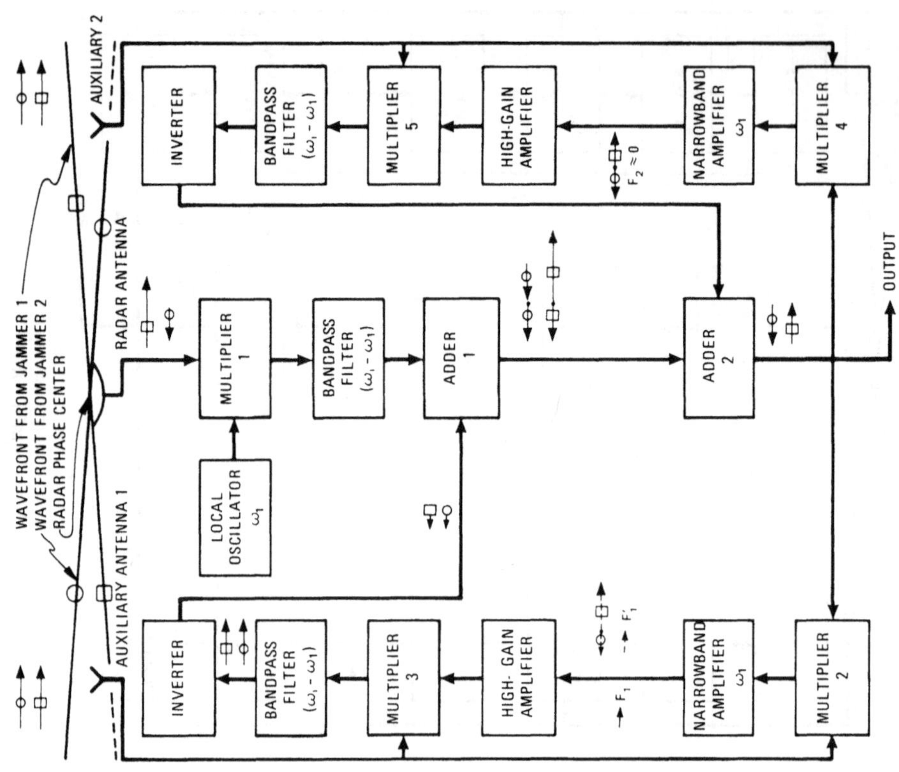

Fig. 14—Unequal amplitudes, 180° out of phase

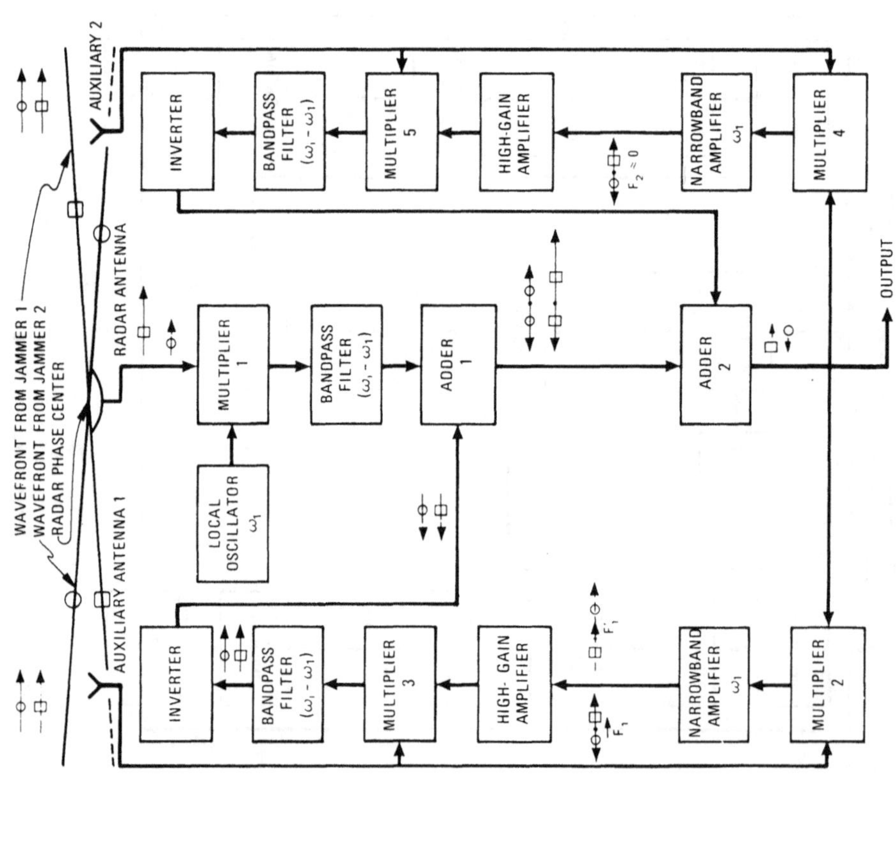

Fig. 13—Unequal radar components, in phase

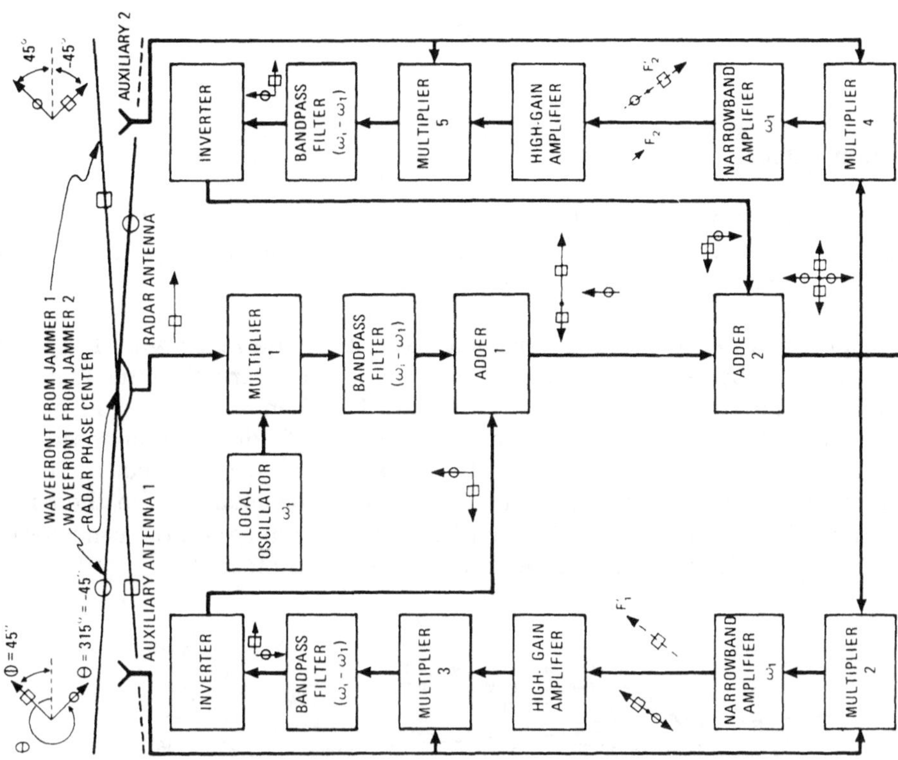

Fig. 16—Best-case auxiliary phasing, one jammer in radar antenna null

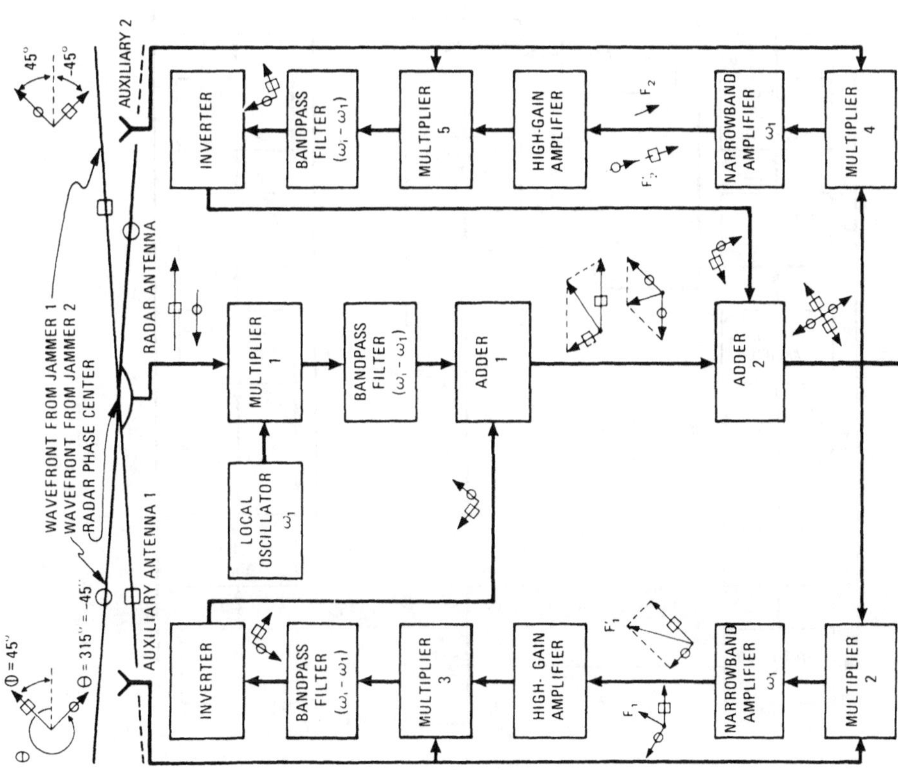

Fig. 15—Best-case auxiliary phasing, one vector flipped by radar antenna

typical variations that could occur due to radar scan and show good cancellation in all cases.

Loop Interaction Effects

A case in which the operation of one loop allows another loop to function is illustrated in Fig. 17. In this case, the order of loop analysis would be important or two successive analyses would be required.

Figure 17 portrays two equal and zero-phase signals in the radar with 180° and 90° relationships in the auxiliaries. If the analysis starts with auxiliary antenna 1, it is obvious that no weighting function can be developed by loop 1 and an analyst would proceed to loop 2. Loop 2 would be found to produce a residue out of adder 2 composed of equal parts of circle and square with phase angles of +45° and -45° with respect to their counterparts in the radar antenna. The residue magnitudes would be smaller than their radar counterparts by $(1/2)^{1/2}$. Loop 2 would then be said to produce a 3-dB cancellation ratio (ratio of power out of adder 2 with canceler loops open and closed).

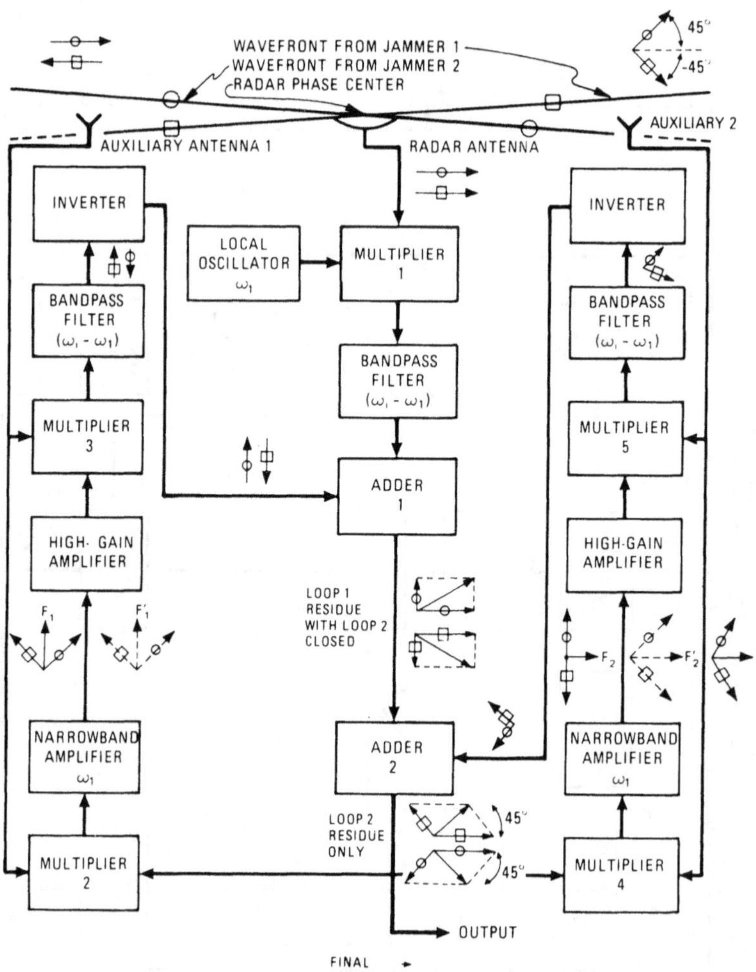

Fig. 17—Demonstration of loop interaction

At this point, the analyst should note that loop 1 develops a weighting function F_1^1. Loop 1 attempts to cancel the vertical components of the residue of loop 2. In so doing, it reduces the magnitudes and phase angles of the two components of the residue. The reduction in the phase angles is now opposed by loop 2 developing a new weighting function F_2^{11} that attempts to hold the phase angles at +45° and −45°. Stability is reached with loop 1 canceling the vertical components of the residue of loop 2 as loop 2 cancels the horizontal radar components. The cancellation ratio of both loops operating in this case would only be limited by the gain of the loops.

Loop interaction, when it exists, is evident as soon as the first approximation to the multiple loop residue is obtained. The angles and relative magnitudes of the residue components can be compared to the angles and relative magnitudes of the signals on the auxiliary antennas to determine if any loop or loops can further reduce the residue.

Analysis of Three-Loop Systems

A three-loop system can be obtained by adding a third adder following adder 2 and a third auxiliary antenna driving a third canceler that transmits inputs to adder 3. The addition of a third auxiliary antenna permits an extra sample of the jamming environment to be obtained. However, its effectiveness is still dependent upon the assumed jammer geometry and the assumed locations of the auxiliary antennas. For example, in the case illustrated in Fig. 6, the addition of a third loop would not significantly improve jamming since two loops are adequate. However, in Fig. 7, the third loop would be extremely useful if it were positioned to receive both jamming signals in phase with their radar counterparts, and nearly any phases other than those shown in Fig. 7 would yield some cancellation.

It should be noted that the best location for auxiliary antenna 3 is that in which the phases of the jamming signals are such that the loop can rotate them to be out of phase with the residues of the first two loops. In Fig. 7, this would require both the jamming signals on auxiliary 3 to be in or 180° out of phase with their radar counterparts. In Fig. 10, it would require one of the jamming signals on auxiliary 3 to be in phase with its radar counterpart and the other to be 180° out of phase.

It should be noted that the good cancellation obtained in Fig. 1, 15, and 16 would not have been obtained if auxiliary 2 had received a circle with lag and a square with lead, since loop 2 could not have rotated them to have each component out of phase with its counterpart in the residue of loop 1.

More than three loops can be obtained by adding more loops as done to obtain three and the analysis can be continued loop by loop. In all cases, the optimum position for the next auxiliary antenna can be determined by determining the phases of the residues it has to work on.

Triple Jammer Analysis

Three jammers and multiple loops can be easily accommodated by drawing a third phase front through the radar phase center and denoting it by a different symbol. The analysis with three jammers then proceeds as it did with two. However, three vectors contribute to the weighting functions and one, two, or three vectors may appear in the residue.

It is important to remember that, regardless of the number of loops and jammers, each loop can only rotate all of the vectors on its auxiliary antenna by the same angle, control all of their magnitudes by the same gain constant, and subtract the resultant vectors from the radar signals.

CONCLUSION

The vector technique of analyzing multiple side-lobe canceler loop operation on multiple jammers is a valid check of performance and permits optimum auxiliary antenna locations to be determined for any given jammer geometry. With practice, many loops and many jammers can be analyzed in a short time. In addition, the use of this technique and the simulator results reveal that the popular belief that n loops can effectively cancel n jammers in all condition is completely false when the auxiliary antennas are omnidirectional so that they cannot resolve the various jammers in angle.

Problems of Adaptive Sidelobe Suppression
G. Reis and K. Krücker
AGARD CP-197, New Deivces, Techniques and Systems in Radar, June 14-17, 1976
Pages 23-1 – 23-12

5.B.4

SUMMARY

In this paper we present some investigations dealing with decorrelation effects in a suboptimal spatial filter for adaptive interference suppression. Degradation caused by decorrelation is partly determined by the antenna configuration, partly by mismatch between the receiving channels. Additional degradation caused by the target signal and by correlator offset errors in the adaptive control loops are discussed, too. Finally the paper gives some results of a special experimental system.

1. INTRODUCTION

This paper deals with particular problems in adaptive spatial filtering for coherent sidelobe suppression. The advantages of this technique for radar applications in ECM environments are evident and should only be mentioned.

Vector-matrix notation is used throughout this paper. Column vectors are represented by underlined lower case letters, for example $\underline{d}$. The asterisk $\underline{d}^*$ is used to denote conjugate transposition. Matrices are denoted by underlined upper case letters, e.g. $\underline{Q}$. Expected values are denoted by overlined letters, e.g. $\overline{\underline{n}\,\underline{n}^*}$.

Starting with the problem of detection of target signals in spatially distributed interferences, some optimality criteria, for example the maximum signal to noise ratio S/N or the likelihood ratio test, lead to an optimum linear array configuration according to Fig. 1. Target and gaussian noise signals s and n of each array element, given by the complex signal vector $\underline{x}$, are multiplied by the complex weighting vector $\underline{k}_{opt}$. Compared with a conventionally steered array with a weighting vector equal to the target signal vector $\underline{s}$, the target signal vector has to be transformed by $\underline{Q}_{opt}^{-1}$ with $\underline{Q}_{opt} = \overline{\underline{n}\,\underline{n}^*}$, the hermitian noise covariance matrix. From this weighting an array pattern results with a main beam in target direction and minima for angles of interference. For broad-band signals the optimum weighting vector is a function of frequency but for narrow-band conditions all weighting elements may be approximated by frequency independent factors without severe degradation in noise suppression. Further reduction of the complexity of this configuration, especially in the matrix conversion process, results in a configuration as shown in Fig. 2, the sidelobe canceller /1/. Assuming a very small S/N at the isotropic auxiliar-array elements compared to the small S/N at the high gain main antenna, steering of the auxiliar array is neglected. For radar applications this assumption gives no severe degradation of the noise suppression. The weighting vector is determined for minimum noise at the output, which results in a processing gain or cancellation ratio g, defined as the ratio of output noise without and with auxiliary array.

$$g = \frac{\overline{n_0 n_0^*}}{\overline{|n_0 - \underline{k}^* \underline{n}|^2}} \qquad (1)$$

In the following sections degradations of the processing gain caused by the suboptimality of the configuration according to Fig. 1 as well as by component errors are discussed. Convergence problems of the adaptively controlled weighting vector are not considered. Finally this paper gives a description of an existing experimental system with an auxiliar array of 6 elements and its performance.

2. EFFECTS OF THE TARGET SIGNAL

As mentioned in the introduction the contribution of the target signal $\underline{s}$ in the auxiliar array to the output signal and from this the steering of the auxiliar array may be neglected as far as the number of the quasi isotropic auxiliar elements N is small compared to the directivity of the main antenna. Further, the effect of $\underline{s}$ on the processing gain according to eq. (1) has to be considered. The weighting vector $\underline{k}$ from Fig. 2 is given by:

$$\begin{aligned} \underline{k} &= (\underline{Q} + \overline{\underline{s}\,\underline{s}^*})^{-1} (\overline{\underline{n}\, n_0^*} + \overline{\underline{s}\, s_0^*}) \\ &\approx \underline{Q}^{-1} (\overline{\underline{n}\, n_0^*} + \overline{\underline{s}\, s_0^*}) \end{aligned} \qquad (2)$$

The approximation proves to be reasonable because of the much lower S/N in the auxiliar array compared to the main antenna S/N. The second term of $\underline{k}$ leads to a degradation of the processing gain g_o achievable without signal. From eq. (1) results for a configuration with only one auxiliar element, i.e. N = 1:

$$g = \frac{g_0}{1 + g_0 \left(\frac{|\overline{s s_0^*}|^2}{\overline{nn^*}\;\overline{n_0 n_0^*}} + 2\,\mathrm{Re}\left\{ \frac{(\overline{n n_0^*} - \overline{n^* n_0})\,\overline{s s_0^*}}{\overline{nn^*}\;\overline{n_0 n_0^*}} \right\} \right)} \tag{3}$$

According to this equation, effects of target signal have to be considered only for very large g_o. For targets in $\Theta = 0$ direction with complete correlation of s and s_o the maximum achievable gain is limited to:

$$g_{max} = \frac{\overline{n n^*}\;\overline{n_0 n_0^*}}{\overline{s s^*}\;\overline{s_0 s_0^*}} \tag{4}$$

Because of low average S/N over long integration periods the resulting g_{max} is sufficient. Low g_{max} for interference directions corresponding to minima of the main antenna pattern with high S/N of the main antenna are not significant. Low g_{max} may be problematic for short effective integration time constants, that means approximately equal to the pulse width of the target signal, when the higher pulse S/N has to be considered. But under these conditions additional effects are caused by the short-time correlation between interference and target signal, which was neglected in eq. (2).

The maximum output S/N of the suboptimal configuration according to Fig. 2 is limited by g_{max} to N/S in the auxiliar array element.

3. BANDWIDTH EFFECTS

The processing gain g may be derived from eq. (1) with the $\underline{k}$ from Fig. 2 for negligible signal effects. From this, the achievable g depends primarily on the correlation of the interference signals n_ν in the auxiliar array /2/. Starting with an n_ν given by only one quasi monofrequent interference with $\overline{n_\nu n_\nu^*} = \overline{nn^*}$ equal for all auxiliar elements ν and additional receiver noise power j in each channel, $\underline{Q}$ and g come out to:

$$Q = \overline{n\underline{n}^*} + j\,\underline{I}$$

$$g = \frac{\overline{n_0 n_0^*}}{\overline{n_0 n_0^*} - \overline{n_0 n^*}\,\frac{\overline{nn^*}N}{j + \overline{nn^*}N}} \tag{5}$$

Because of the uncorrelated receiver noise, g is limited, compared to $g \rightarrow \infty$ for $j \rightarrow 0$.

Additional gain reduction results from the signal bandwidth, which deforms the dyadic $\underline{Q} = \underline{n}\underline{n}^*$ to the Toeplitz matrix $\underline{Q} = \overline{\underline{n}\underline{n}^*}$. No simple expression for the inverted $\underline{Q}$ and from this for g can be derived except for an auxiliar array with only one element, i.e. N = 1 with:

$$g = \frac{1}{1 - \left|\frac{r_1}{q_{00}}\right|^2}$$

$$\text{with: } \frac{r_1}{q_{00}} = R_{01} = \frac{\int_{-\infty}^{\infty} S(f)\,|H(f)|^2\, e^{i\pi \frac{d}{\lambda/2} \sin\theta \frac{f}{f_0}}\, df}{\int_{-\infty}^{\infty} S(f)\,|H(f)|^2\, df} \qquad j \rightarrow 0 \tag{6}$$

From this g depends on the pre-processing filter function H (f) and the power spectrum S (f) of the interference process as well as on the element distance d and Θ, the direction of the interference. For white noise spectrum results:

1. rectangular filter of bandwidth B

$$g = \frac{1}{1 - \mathrm{sinc}^2\left(\frac{\pi}{2}\frac{B}{f_0}\frac{d}{\lambda/2}\sin\theta\right)} \tag{7}$$

2. sinc-filter of 3 dB bandwidth B

$$g = \frac{1}{1 - \left(1 - 0.35\,\frac{\pi}{2}\frac{B}{f_0}\frac{d}{\lambda/2}\sin\theta\right)^2} \tag{8}$$

Approximations of these equations reasonable for g ≥ 10 dB shows Fig. 3 where the bandwidth of both filters is optimized for a rectangular pulsed target signal of pulse length 1/1.37 B. According to this figure high processing gain g may be obtained by filter functions with strong sideband attenuation. The resulting pattern of the antenna configuration has a deep minimum for the center frequency only, and from this interference suppression for far-off side bands, for example in the spectrum according to filter function 2 proves to be insufficient.

The processing gain may be increased by pattern minimums of higher order. These are a result of additional degrees of freedom when the number of discrete interference directions is less than the number N of elements in the auxiliar array. For only one interference direction and N = 2 g is given by:

$$g = \frac{1}{1 - \frac{|R_{01}|^2 + |R_{02}|^2 - 2 R_{01} R_{02}^* R_{12}}{1 - |R_{12}|^2}} \tag{9}$$

Results for a rectangular filter function and $d_2 = 2d_1 = 2d$ shows Fig. 3, again as an approximation for g ≫ 10 dB and $j \to 0$. Because of the broadened pattern minimum g increases with the forth power of decreasing bandwidth compared to the second power for N = 1. High processing gain may be realized only neglecting receiver noise. On the other side it should be mentioned that the large time constant in the transition from the small g according to N = 1 to g according to N = 2, which is caused by one vanishing eigenvalue in the covariance matrix, proves to be problematic.

4. EFFECTS OF THE MAIN ANTENNA PATTERN

The degradation of the processing gain by bandwidth as discussed in the previous section is caused by frequency dependent phase angles at the interference signals of all antenna elements. An additional degradation results from the frequency dependent amplitude of the interference signal in the high-gain main antenna, given by its frequency dependent sidelobe pattern. For a main antenna of dimenskon D in direction of the auxiliar array, the pattern in the sidelobe region may be approximated by:

$$A = \frac{\sin\left(\pi \frac{D}{\lambda_0} \sin\theta \frac{f}{f_0}\right)}{\left(\pi \frac{D}{\lambda_0} \sin\theta \frac{f}{f_0}\right)^l} \qquad l \geq 1/2 \tag{10}$$

This pattern amplitude has to be considered as a frequency dependent weighting function for the spectrum of the interference signal in the main antenna. For the quasi isotropic elements of the auxiliar array the corresponding weighting function is constant within the bandwidth.

With A taken into consideration processing gain for a configuration with one auxiliar element, i.e. N = 1 may be computed according to eq. (6). Fig. 4 shows some results for only one interference direction, rectangular filter functions, and negligible receiver noise. For large distances d between the main antenna phase centre and the auxiliar element, compared to the main antenna dimension D, the phase effects as discussed in the previous section dominate, and amplitude effects of the main antenna pattern A may be neglected. The influence of A is also negligible for small d/D as far as the interference direction coincides with a sidelobe peak in A, i.e. $D/\lambda_0 \cdot \sin\Theta$ = 1.5, 2.5, 3.5... Severe gain degradations are observed for all other interference directions with A variations over the bandwidth. Unter these conditions no considerable increase of g may be obtained for d ≦ D/2, that means when the auxiliar element is within the main-antenna aperture.

5. DECORRELATION EFFECTS DUE TO RECEIVER MISMATCH

As already mentioned, influences which decorrelate the interference process on the way from the noise source to the weighting filter $\underline{k}$ degrade the processing gain of the canceller. Besides the effects discussed in the previous sections, mismatch in the receiver channels has to be considered. It should be mentioned that the optimum array processor Fig. 1 is sensitive to decorrelation in the same way as the sidelobe canceller.

We concentrate our attention in this chapter on amplitude response mismatch between main and auxiliary channels, deriving an upper limit for the cancellation ratio due to decorrelation in the channels of the receiver front-end. To derive an expression for the processing gain in the presence of decorrelating prefilters, the quadratic form $\underline{r}^* \underline{Q}^{-1} \underline{r}$ which results in eq. (1) for the optimum $\underline{k}$ according to Fig. 2, has to be evaluated. The crosscorrelations $r_\nu^*(o) = \overline{x_0(t) x_\nu^*(t)}$ and $q_{\nu\mu}(o) = \overline{x_\nu(t) x_\mu^*(t)}$ are components of the correlation vector $\underline{r}^*$ and the covariance matrix $\underline{Q}$. x_0, x_ν, x_μ can be regarded as new signal processes after convolving the original element signals $x'_{0,\nu,\mu}(t)$ with the impulse response of the corresponding decorrelating prefilter. With transfer functions $H_\nu(\omega)$, $H_\mu(\omega)$, the crosscorrelation of x_ν and x_μ is given by /3/:

$$q_{\nu\mu}(0) = \int_\omega S_{\nu\mu}(\omega)\, H_\nu(\omega)\, H_\mu^*(\omega)\, \frac{d\omega}{2\pi} \tag{11}$$

$S_{\nu\mu}(\omega)$ is the cross-spectral density of the input processes $x'_\nu(t)$, $x'_\mu(t)$. With $x'_\nu(t) = x'(t-(\nu-1)\delta)$, where δ is the propagation delay of the process between two adjacent elements for a given interference direction Θ, we get by Fourier-transform:

$$\begin{aligned} S_{\nu\mu}(\omega) &= \int q_{x'}(\alpha - (\nu-\mu)\delta)\, e^{-i\omega\alpha}\, d\alpha\; e^{-i\omega_0(\nu-\mu)\delta} \\ &= S_{x'}(\omega)\, e^{-i\omega(\nu-\mu)\delta}\, e^{-i\omega_0(\nu-\mu)\delta} \end{aligned} \tag{12}$$

with the power spectrum $S_{x'}(\omega)$ of $x'(t)$. From eq. (11) and eq. (12) we get the general form of the resulting correlation of two processes which are transformed by different transfer functions.

$$q_{\nu\mu}(0) = \int_{\omega} S_{x'}(\omega) H_\nu(\omega) H_\mu^*(\omega) e^{-i\omega(\nu-\mu)\delta} \frac{d\omega}{2\pi} e^{-i\omega_0(\nu-\mu)\delta}$$

$$r_\nu^*(0) = \int_{\omega} S_{x'}(\omega) H_m(\omega) H_\nu^*(\omega) e^{i\omega\nu\delta} \frac{d\omega}{2\pi} e^{i\omega_0\nu\delta} \qquad (13)$$

We apply eq. (13) in two examples, thus giving an impression of the required tracking tolerances. For simplicity the interference process is described by a bandlimited white noise process, $S_{x'}(\omega)$ = const., and the phase responses of all prefilter transfer functions are assumed to be linear and equal. Fig. 5 illustrates the cancellation ratio g as a function of relative bandwidth·array dimension for a linear decreasing and a cosine ripple amplitude response of the main channel.

a) $H_0 = H_m(\omega) = (H_{m0} + a - 2a \frac{|\omega|}{\Delta W/2})\, rect \frac{\omega}{\Delta W}$

b) $H_0 = H_m(\omega) = (H_{m0} + a \cos 2\pi n \frac{\omega}{\Delta W/2})\, rect \frac{\omega}{\Delta W}$

$H_\nu = H_a(\omega) = H_{a0}\, rect \frac{\omega}{\Delta W}$ for $\nu = 1 \dots N$

The corresponding formulas for g when $\delta = 0$ ($\Theta = 0$) - i.e. no additional bandwidth effects due to the antenna configuration - with g independent from the number of elements N, are given in Fig. 5, too. It is important to note that the processing gain is limited to 25-30 dB due to a mismatch of about 0.5 dB im amplitude response.

6. OFFSET-ERROR PROBLEM OF LOW PASS CORRELATOR CONTROL LOOPS

In changing interference environment the optimum filter weighting $\underline{k}^* = \underline{r}^* \underline{Q}^{-1}$ should be controlled in an adaptive way to guarantee optimum performance. Updating and matrix inversion methods or Gradient-following techniques could be applied. A well known solution is given by adaptive control loops which realize the Steepest-Descent algorithm in a continuous manner. /1/, /4/, /5/. Fig. 6 shows the principle network for adaptively controlled filter weights.
A lot of literature (see reference list in /6/) has been published dealing with the performance of the control loops with regard to dynamic and convergence problems.

We consider in this chapter the influence of an additive correlator error (correlator offset or drift) for a low pass version of the canceller system. Adding an offset vector $\underline{\varepsilon}^* = \underline{c}^* \underline{\Lambda}$ to the outputs of the correlator multipliers (see Fig. 6) results in a modified filter weighting:

$$\underline{k}^* = (\underline{r} + \underline{\varepsilon})^* \underline{Q}^{-1} \qquad (14)$$

with amplitude vector $\underline{c}^* = (c_1 \dots c_N)$ and diagonal matrix of the phase error $\underline{\Lambda}$ = diag ($e^{-i\vartheta_\nu}$). With $\underline{k}^*$ from eq. (14) we get the cancellation ratio eq. (1):

$$g = \frac{|n_0|^2}{|n_0|^2 - \underline{r}^* \underline{Q}^{-1} \underline{r} + \underline{\varepsilon}^* \underline{Q}^{-1} \underline{\varepsilon}} \qquad (15)$$

Eq. (15) shows that for an unavoidable offset error $\underline{\varepsilon}^*$ a severe reduction in processing gain occurs when $\underline{Q}$ becomes singular.
A simplified analysis of offset error influence for a single element canceller, neglecting bandwidth effects, $\Theta = 0$, gives a lower bound estimate for offset requirements. $\underline{\varepsilon}^* \underline{Q}^{-1} \underline{\varepsilon}$ reduces to c_1^2/P_j, with the interference power P_j at the correlator input. From eq. (15) we get with $|n_0|^2 - \underline{r}^* \underline{Q}^{-1} \underline{r}$, neglecting isotropic system noise:

$$g = \frac{1}{c_1^2/P_j^2} \qquad (16)$$

The following table shows that for a given P_i the offset requirements become very critical, even for a relatively modest cancellation ratio.

P_i/dBm	g/dB: 20	g/dB: 30
- 10	0.01 mV	3.2 µV
0	0.1 mV	32 µV

Table 1) Tolerable offset error for a given cancellation ratio.

7. SINGLE-SIDEBAND MODULATION WEIGHTING

In order to avoid correlator offset problems it seems reasonable to use IF-Bandpass correlator loops, /7/ with a bandpass version of the optimum weighting vector

$$\underline{k}^* = \underline{r}^* \, \underline{Q}^{-1} \, e^{-i\omega_1 t} \tag{17}$$

with $\omega_1 \geq \Delta\omega$, the interference process bandwidth, to prevent spectrum fold-over. However, the IF-loops are more critical with regard to control loop stability than the low-pass version /8/. Instability is caused by time delay in the loop network and by the higher-order filters, necessary to attenuate the unused sidebands. In practical cases, the loop gain and therefore the processing gain is stability-limited. Instability caused by additional filters can be avoided by considering the weighting process (see Fig. 6)

$$z = \underline{k}^* \underline{x} \tag{18}$$

carried out by the analytic signals $\underline{k} = \underline{w} + i\hat{\underline{w}}$ and $\underline{x} = \underline{b} + i\,\hat{\underline{b}}$ ($\hat{w}$, $\hat{b}$ are Hilbert-transformed processes). The real part of z is given by

$$Re\{z\} = \underline{w}^* \underline{b} + \hat{\underline{w}}^* \hat{\underline{b}} \tag{19}$$

The operation described by eq. (19) cancels the unused frequency components and is realized by two Hilbert-transformers and multipliers per array element. This procedure introduced here for the purpose of simultaneous weighting and filtering is known as Single Sideband Modulation and eliminates the need of additional sideband filters. Compared to the in-phase and quadrature channel low pass correlator, the number of correlators is reduced by a factor of two. Fig. 7 shows the discussed modification of the IF-canceller system.

8. EXPERIMENTAL RESULTS

In order to study some fundamental properties and problems of adaptive spatial filtering, an experimental system with an auxiliar array of 6 elements and loops implementing the results of chapter 7 has been built (see Fig.7). Fig. 8 shows some theoretical and experimental values of processing gain as a function of $B/f_0 \; d/\lambda_0/2 \sin\Theta$ for a single interference direction.

Decorrelation effects limit the cancellation ratio to a value of about 28 dB. For a relative bandwidth > 5%, bandwidth effects predominate. Increased gain, using additional spatial samples (N = 2) can be recognized. A qualitative impression of the benefits of the canceller system is given by the pictures of Fig. 9. Two target signals with a bandwidth ratio of about 4 are immersed in the main channel interference. With processing gain corresponding to the bandwidth of the process, target signals appear in a reduced noise background.

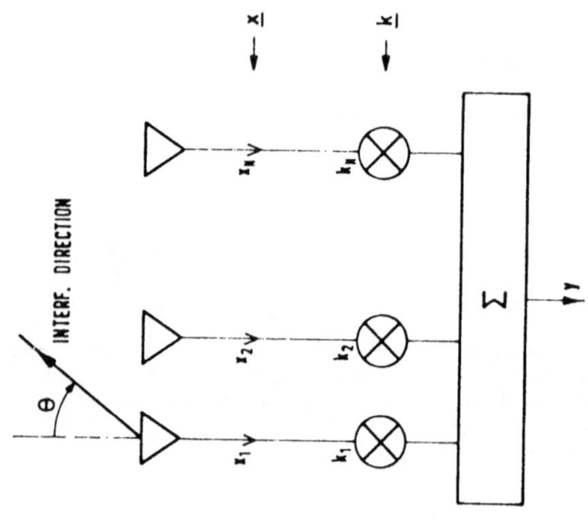

$$\underline{x}(\omega) = \begin{pmatrix} x_1 \\ \vdots \\ x_N \end{pmatrix} = \underline{s} + \underline{n}$$

$$\underline{k}_{opt}(\omega) = \begin{pmatrix} k_1 \\ \vdots \\ k_N \end{pmatrix} = \underline{Q}_{opt}^{-1}\,\underline{s}\ ; \quad \underline{Q}_{opt}(\omega) = \overline{\underline{n}\,\underline{n}^*}$$

$$y(\omega) = \underline{k}_{opt}^*\ \underline{x}$$

FIG. 1 OPTIMAL ARRAY

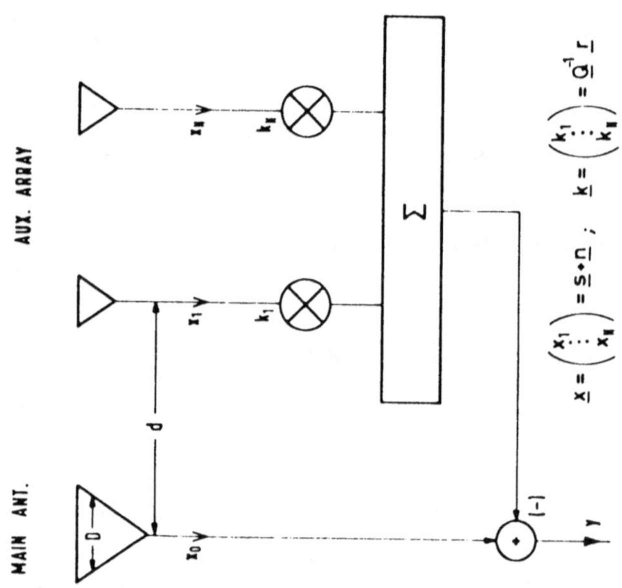

$$\underline{x} = \begin{pmatrix} x_1 \\ \vdots \\ x_N \end{pmatrix} = \underline{s} + \underline{n}\ ; \quad \underline{k} = \begin{pmatrix} k_1 \\ \vdots \\ k_N \end{pmatrix} = \underline{Q}^{-1}\,\underline{r}$$

$$\underline{Q} = \overline{\underline{n}\,\underline{n}^*}\ ; \quad \underline{r} = \overline{\underline{n}\,n_0^*}$$

$$y = x_0 - \underline{k}^*\,\underline{x}$$

FIG. 2 SUBOPTIMAL CONFIGURATION WITH A HIGH-GAIN MAIN ANTENNA

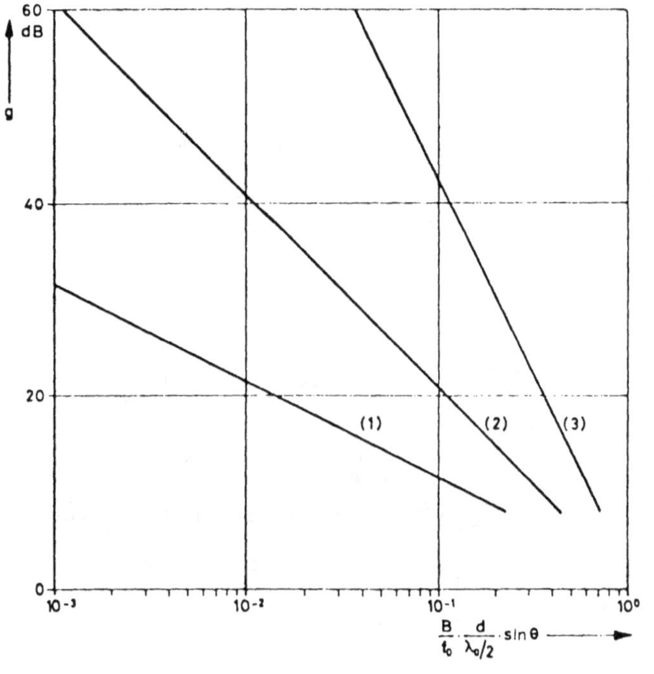

(1) ONE ELEMENT, sinc - FILTER

$$g = \frac{1,43}{\frac{0,66\,B}{f_0} \cdot \frac{d}{\lambda_0/2} \sin\theta}$$

(2) ONE ELEMENT, rect. - FILTER

$$g = \frac{1,21}{\left(\frac{B}{f_0} \cdot \frac{d}{\lambda_0/2} \sin\theta\right)^2}$$

(3) TWO ELEMENTS, rect - FILTER

$$g = \frac{1,84}{\left(\frac{B}{f_0} \cdot \frac{d}{\lambda_0/2} \sin\theta\right)^4}$$

FIG. 3 PROCESSING GAIN AS A FUNCTION OF BANDWIDTH FOR DIFFERENT FILTERS

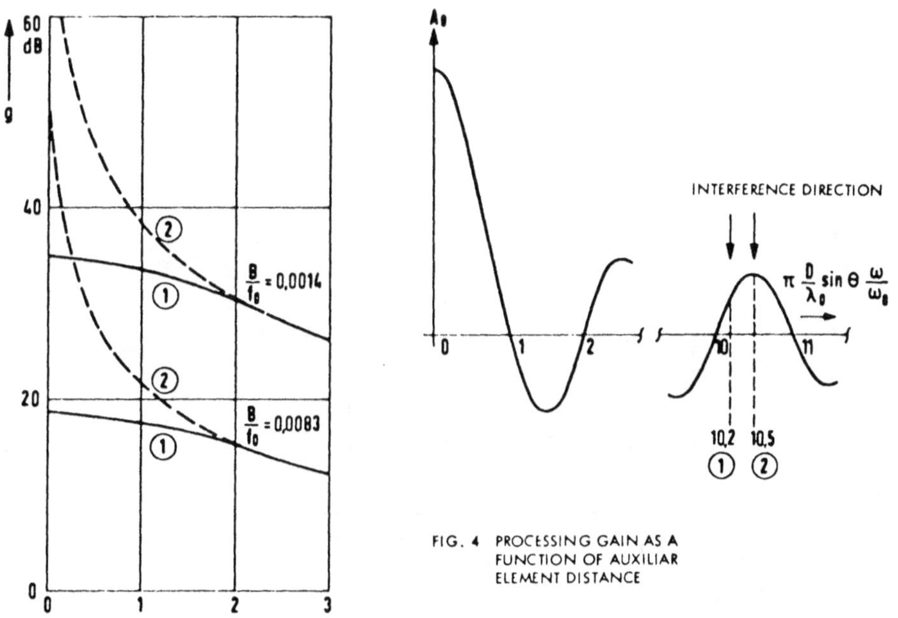

FIG. 4 PROCESSING GAIN AS A FUNCTION OF AUXILIAR ELEMENT DISTANCE

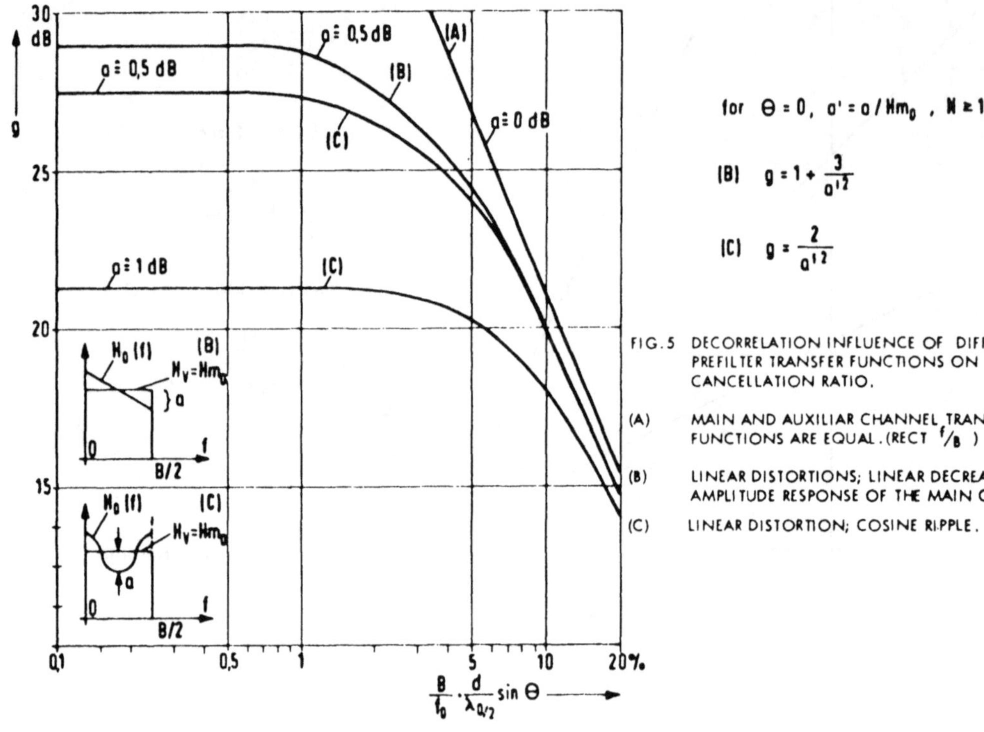

FIG. 5 DECORRELATION INFLUENCE OF DIFFERENT PREFILTER TRANSFER FUNCTIONS ON THE CANCELLATION RATIO.

(A) MAIN AND AUXILIAR CHANNEL TRANSFER FUNCTIONS ARE EQUAL. (RECT f/B)

(B) LINEAR DISTORTIONS; LINEAR DECREASING AMPLITUDE RESPONSE OF THE MAIN CHANNEL.

(C) LINEAR DISTORTION; COSINE RIPPLE.

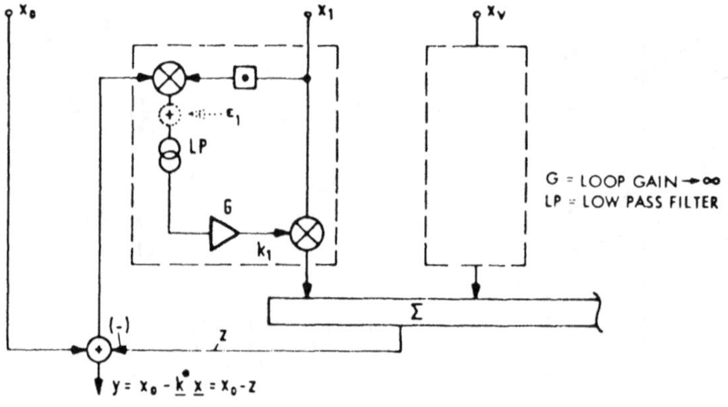

FIG. 6 REALISATION OF THE FILTERVECTOR $\underline{k}^* = \underline{r}^* Q^{-1}$ BY ADAPTIVE CONTROL LOOPS (APPLEBAUM / BRENNAN)

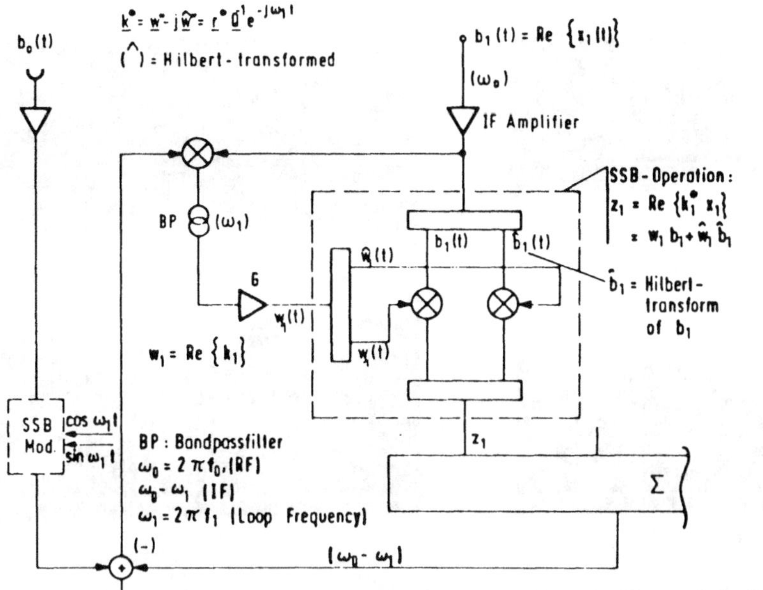

FIG. 7 INTERMEDIATE FREQUENCY ADAPTIVE CONTROL LOOPS WITH SINGLE-SIDEBAND WEIGHTING FILTER.

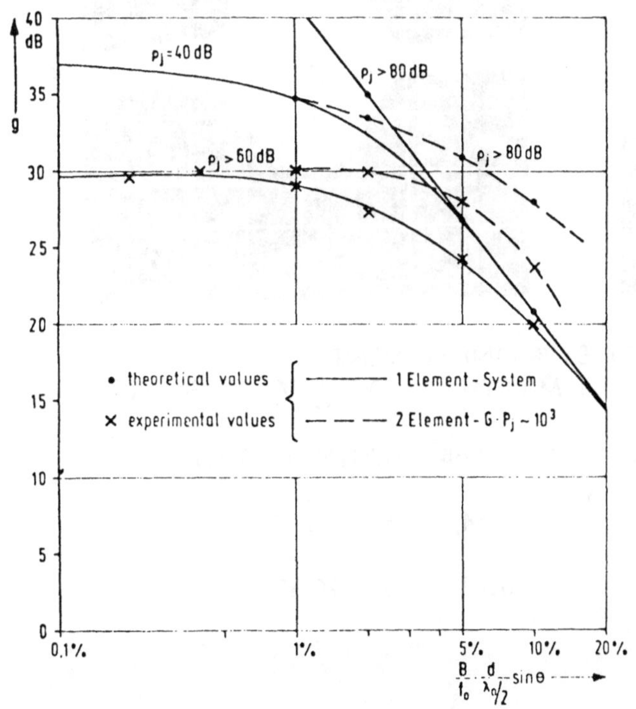

EXPERIMENTAL CONDITIONS:

PREFILTER TRANSFER FUNCTION: APPROXIMATELY RECT $\frac{\omega - \omega_0}{\Delta W} \cdot e^{-j(\omega - \omega_0)t_0}$

SINGLE NOISE SOURCE;

$p_j = \frac{P_j}{P_n}$ = JAMMER TO SYSTEM NOISE POWER RATIO, MEASURED AT AN ARRAY ELEMENT.

FIG. 8 INTERFERENCE SUPPRESSION g/dB AS A FUNCTION OF BANDWIDTH B, ELEMENT DISTANCE d AND JAMMER LOCATION θ FOR A ONE AND TWO ELEMENT IF CANCELLER WITH SSB WEIGHTING.

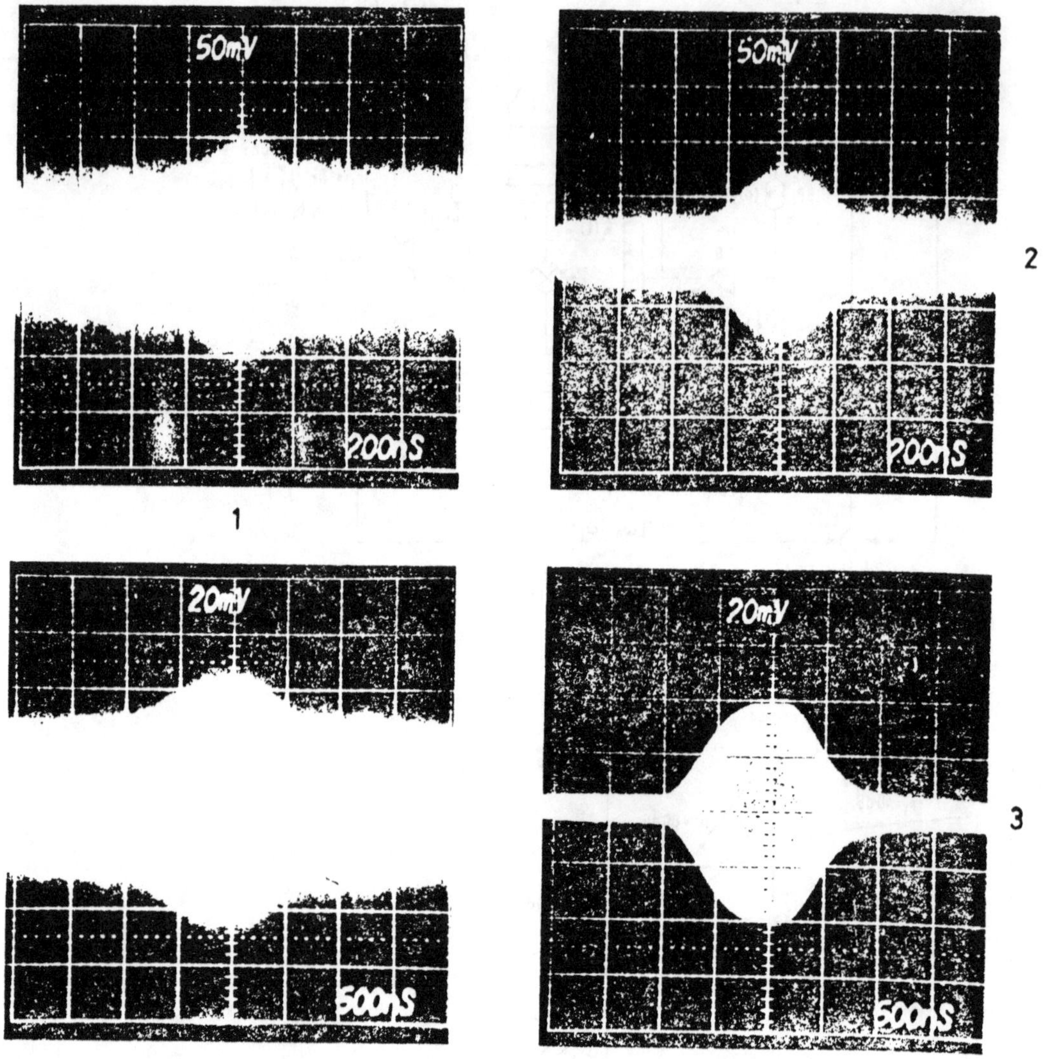

FIG. 9 IMPROVEMENT IN S/N BY A ONE ELEMENT SIDELOBE CANCELER SYSTEM WITH $d = 10\,\lambda_o/2$ AND $\Theta = 90°$.

(1) MAIN CHANNEL S/N = - 3 dB (WITHOUT SLC)

(2) WITH SLC: $\frac{B}{f_o} = 0.04$, $g \sim 8$ dB

(3) WITH SLC: $\frac{B}{f_o} = 0.01$, $g \sim 20$ dB

DISCUSSION

J FREEDMAN: Would you please indicate the results obtained for the case of m independent noise sources and n loops where m < n?

K KRUCKER: As long as m = n - 1, processing is similar to that discussed here for m = 1, n = 2 depending on the spatial distribution of the interferences.

R VOLES: Have you estimated the additional number of degrees of freedom taken by a source which is amplitude modulated at a rate faster than can be followed by the adaptive loops?

K KRUCKER: No, because we think that, in that case, the performance is not a function of additional degrees of freedom but only a function of the time behaviour of the control loops. When the interference pulse width is small compared with the canceller learning time constant, performance of the SLC is degraded. This statement is confirmed by our experiments.

K MILNE: Is there an optimum spacing for auxiliary antennas? Wide spacings are required to obtain closely-spaced nulls to deal with jamming sources with small angular separations but, on the other hand, close element spacing is desirable to achieve broadband operation. Is the optimum perhaps a random distribution of spacings?

K KRUCKER: We considered several element positions, especially different distances between main and auxiliary elements. The results are highly dependent on the interference conditions. In general, it is difficult to give an optimum configuration.

J S YU: Could you please comment on why your elements in the auxiliary channel are not fixed in their spacing to improve the processing gain within the narrow band of your main channel?

K KRUCKER: In order to get sufficient interference suppression it is necessary to have a small product of bandwidth x propagation delay. Even with narrow radar bandwidth, the distance between auxiliary elements and main antenna phase centre has to be small.

P BRADSELL: Would you please explain the advantage of the system using upper and lower sideband separation to prevent spectrum folding, over a simpler system using real and imaginary components of the signal?

K KRUCKER: A prime problem of an analog baseband SLC system is the offset error of the correlator multiplier. In the IF system only one sideband is processed and the other one must be suppressed. Foldover is avoided by a properly chosen frequency relationship. The advantage is that no offset error occurs.

G GALATI: Have you done any analytical evaluations of SLC with more than one auxiliary antenna and/or more than one interference? Have you taken account of the internal (white) noise?

K KRUCKER: As seen from Fig 3, analytical evaluations of a two-element system have been performed. Further analytical evaluations under narrow-band conditions have been done for a six-element SLC with up to three interference directions. In this case, uncorrelated (system) noise has been taken into account.

REFERENCES

/1/ Applebaum S.P. "Adaptive Arrays" Syracuse University Research Corporation Report No. SPL 769 June 30, 1964

/2/ Ries G., Söntgerath W. "Untersuchungen an analogen Regelschleifenschaltungen für adaptive Antennengruppen" Forschungsbericht 2-74, FHP Werthhoven

/3/ Papoulis A. "Probability Random Variables and Stochastic Processes" McGraw-Hill Book Company

/4/ Widrow B. et al. "Adaptive Antenna Systems" Proc. IEEE Vol. 55 No. 12 Dec. 1967 p. 2143-2159

/5/ Brennan L.E. "Theory of Adaptive Radar" IEEE Transact. on Aerospace and Electronic Systems, Vol. AES-9 No.2 March 1973

/6/ Rees F.L. "Adaptive Beam Forming and Filtering" TSC Course, Oct. 1973, Silver Spring, Maryland, USA

/7) Howells P.W. "Intermediate Frequency Sidelobe Canceller" US Patent No. 3,202,990

/8/ Krücker K., Bock H.W. "Theoretische und experimentelle Untersuchungen zur Signalverarbeitung in einem adaptiven räumlichen Filter" Forschungsbericht Nr. 3-74, FHP Werthhoven

Chapter 6
ECCM in the Receiver (General)

Summary

Papers address various ECCM techniques in the receiver including general receiver techniques, digital MTI, and MTI in tracking radars.

A list of ECCM techniques in the receiver is given in Paper 1.1, Table III (on p. 15 of this volume.)

Receiver Signal Processor ECCM techniques are treated in Chapter 7. Some receiver ECCM techniques are included in Papers 2.A.1, 2.A.2, 2.A.3, and 2.A.4.

Many of the techniques listed in Paper 1.1, Table III are described in papers in the bibliography at the end of this chapter. The extensive bibliography of receiver detection/noise reduction includes a number of Russian papers on noise, detection in noise, and noise reduction. Most of these papers do not appear in any of the general bibliographies listed. These Russian papers reflect the extensive work done in this area by the Soviets.

Synopsis of Papers in Chapter 6

6.1 "Naval Radar Anti-Jamming Techniques" – Alred

This paper is the most detailed ECCM paper of the World War II era. Most of the paper pertains to search (acquisition) radars. Extensive photographs show visual effects of denial ECM (jamming) on radar displays—both deflection modulated, e.g., A-Scope, and intensity modulated, e.g., PPI. Numerous anti-jam techniques (ECCM) for the radar receiver are described. "Paralysis" is known in the U.S. as blocking or desensitization, (Paper 6.3). The author indicated that careful attention to details of design yielded improvements of 30 dB in performance over earlier designs.

6.2 "Designing Radar Receivers to Overcome Jamming" – Galenian

This paper is similar to Papers 6.1 and 2.A.1. It provides many general guidelines for proper radar receiver design considerations. It is a "must" for receiver design engineers. Since this paper contains no references, the reader should consult the bibliography at the end of this chapter for more detailed treatments of the various topics in this paper.

6.3 "Desensitization in Transistorized PM/FM-Receivers" – Hansen

Desensitization is sometimes called blocking, although the latter term tends to imply a binary situation; i.e., a signal is either passed as is or is cut off completely. Actually a desired signal may be attenuated (often greatly, but not completely) by desensitization. Desensitization is the action of a strong signal which is out of the receiver pass band in reducing the receiver output for a desired in-band signal. In desensitization, generally cross modulation does not occur; i.e., modulation on the undesired out of band signal does not appear in the receiver output.

Desensitization is generally not a problem in vacuum tube receivers, but it may be a very severe problem in transistorized receivers. It is seldom mentioned in works on radar receivers. This paper addresses communications receivers, but the work is applicable to radar receivers. The principal source of desensitization is shown to be noise figure degradation because of noise conversion. A simple expression is derived for noise figure level increase as a function of interfering signal level. This is then converted into overall receiver blocking level for both RF stage and/or mixer stage blocking. Experimental measurements on practical receivers support the theory. A method for desensitization measurement is described.

6.4 "Tackle Noise Variations at Video" – Cope

Several techniques for obtaining constant false alarm rate (CFAR) are described. Originally, true CFAR processing had to be accomplished prior to signal detection. At IF, these include the Dicke-Fix, logarithmic receiver and automatic gain control (AGC). Some limitations of these techniques are given. A technique for performing CFAR processing at video is described. A video CFAR using a piecewise linear logarithmic amplifier is described which demonstrated preservation of CFAR action to within 1 dB over nearly 60 dB change of input level dynamic range.

6.5 *"Digital MTI, A New Tool for the Radar User"*—Ellis

This British paper is different from most MTI papers (either analog or digital) in that it addresses the effects of noise jamming in MTI systems. The moving clutter problem can be readily extended into falling/drifting chaff. Doppler compensation (for moving clutter) is often not treated by other MTI authors. The author's terms "phase channel" and "quadrature channel" are generally called I (in phase) and Q (quadrature), respectively, in U.S. works.

6.6 *"Conical Scan Tracking in Ground Clutter with Non-coherent and Coherent MTI"*—Sorkin and Steinberg

Practically all MTI papers have search radars as their intended application. This is one of two papers known to this author which treats the application of MTI to conical scan tracking radars. A first order investigation of the angle tracking loop using a conical scanning antenna with both coherent and non-coherent MTI signal processing is presented. The application considered is an airborne radar tracking a target in a ground clutter environment. This is extendable to a ground radar tracking an airborne target in a chaff environment. Interestingly, this paper is twenty years old.

6.7 *"Design Considerations of a Coherent (on receive) MTI System for a Conically Scanned Tracking Radar"*—Subramanian

This paper is complementary to the previous paper. It deals with aspects of design and integration of a coherent MTI system to a conically scanned tracking radar to improve its capability for detection and tracking under a clutter environment. The special considerations cited in this paper should be addressed by the designer of an MTI system for a conical scan tracking radar; e.g., the necessity for preservation of conscan modulation on the target in the MTI processor.

Bibliography Topics for Chapter 6

A F C
A G C
Analog MTI
CFAR
Detection/Noise Reduction
Digital MTI
Displays
General Receiver ECCM
Integration
Logarithmic Receiver
Receiver Front End
Tracking Receivers
Various Receiver ECCM Techniques

BIBLIOGRAPHY FOR CHAPTER 6

RECEIVER AUTOMATIC FREQUENCY CONTROL (AFC)

1. Books

Shakhgil'dyan, V.V., and A.A. Lyakhovkin, *Phase Automatic Frequency Control Systems,* Moscow: "Svyaz' " 1971 (in Russian).

2. Journal Articles

Tikhonov, V.I., "Phase-lock Automatic Frequency Control Operation in the Presence of Noise," *Automation and Remote Control,* vol. 21, no. 3, Mar 1960, pp. 209-214.

Rodionov, Ya. G., "Optimum Passband of a Filter in the Reception of Frequency-modulated Signals with Follow-up Tuning," *Radio Engrg.,* vol. 15, no. 9, Sept 1960, pp. 63-73.

Isabeau, J.G., "Single Pulse AFC Systems," *IRE Nat. Conv. Rec.,* vol. 9, pt. 5, 1961, pp. 278-284.

Rodionov, Ya. G., "On the Effect of Weak Harmonic Interference on FM Reception with Automatic Tracking," *Radio Engrg.,* vol. 16, no. 11, Nov 1961, pp. 33-38.

Tikhonov, V.I. and K.B. Chelyshev, "Brief Communications: Statistical Dynamics of Phase-synchronized AFC," *Radio Engrg. and Electron. Phys.,* vol. 8, no. 2, 1963, pp. 287-290.

Pervachev, S.V., "Lock-in Range of Phase-synchronized AFC," *Radio Engrg. and Electron. Phys.,* vol. 8, no. 2, 1963, pp. 291-294.

Zhuravlev, A.G., "Operation of a Phase-activated AFC System in the Presence of Harmonic Interference," *Telecomm. Radio Engrg., Pt. 2,* vol. 18, no. 9, Sept 1963, pp. 31-39.

Belousova, N.V., and V.L. Lebeduv, "Interruption of Tracking in Automatic Frequency Control System," *Telecomm. Radio Engrg., pt. 2,* vol. 18, no. 10, Oct 1963, pp. 33-41.

Shakhgil'dyan, V.V., "Filtering of Discrete Noise in a Nonlinear Phase-sensitive Automatic Frequency-control System," *Telecomm. Radio Engrg., Pt. 2,* vol. 19, no. 3, Mar 1964, pp. 95-99.

Shakhgil'dyan V.V. and A.A. Lyakhovkin, "Filtering of Monochromatic Signal by a Phase AFC System," *Telecomm. Radio Engrg, Pt. 1,* vol. 18, no. 4, Apr 1964, pp. 9-15.

Lyakhovkin, A.A., and V.V. Shakhgil'dyna, "Pull-in Range of an Inertial Phase-synchronized AFC System," *Telecomm. Radio Engrg., pt. 2,* vol. 19, no. 9, Sept 1964, pp. 88-94.

Zhuravlev, A.G., I.T. Alekeyevskiy, and V.V. Revyakin, "Operation of Phase AFC System with Modulated Signals in the Presence of Noise," *Telecomm. Radio Engrg., pt. 1,* vol. 19, no. 2, Feb 1965, pp. 11-17.

Shakhgil'dyan, V.V., "Influence of a Phase-modulated Reference Signal and Noise on a Phase AFC System," *Telecomm. Radio Engrg. pt. 1,* vol. 19, no. 6, June 1965, pp. 15-23.

Shaklhgil'dyan, V.V. and A.A. Lyakhovkin, "Choice of the Low-pass Filter in a Phase Automatic Frequency Control Circuit to Minimize the Frequency Variance of the Locked Oscillator," *Telecomm. Radio Engrg., pt. 1,* vol. 19, no. 6, June 1965, pp. 88-92.

Shaklhgil'kyan, V.V. and Yu. F. Ignatov, "Determination of the Region of Synchronous Operation of a Phase AFC System," *Telecomm. Radio Engrg., pt. 1,* vol. 19, no. 10, Oct 1965, pp. 25-28.

Shakhgil'dyan, V.V., "Determination of the Capture Band of a Phase-lock AFC System when the Reference Signal is Phase Modulated," *Radio Engrg. and Electron. Phys.,* vol. 10, no. 10, Oct 1965, pp. 1506-1510.

Shakhgil'dyan, V.V. and Yu. F. Ignatov, "The Effect of Noise on the Accuracy of a Phase-locked AFC System," *Telecomm. Radio Engrg. pt. 1,* vol. 20, no. 3, Mar 1966, pp. 32-37.

Shakhgil'dyan, V.V., "Investigation of Statistical Properties of a Phase AFC System with an Arbitrary Four-pole in the Feedback Loop," *Radio Engrg. and Electron. Phys.,* vol. 11, no. 9, Sept 1966, pp. 1348-1354.

Shakhgil'dyan, V.V., "Probability of Synchronization Failure in a Phase Automatic Control System," *Radio Engrg. and Electron. Phys.,* vol. 11, no. 10, Oct 1966, pp. 1535-1540.

Shakhgil'dyan, V.V., "Effect of Narrow-band Noise on the Performance of Phase Automatic-frequency Control Systems," *Automation and Remote Control,* vol. 27, no. 12, Dec 1966, pp. 2122-2126.

Shakhgil'dyan, V.V., "Pull-in Range of a Phase-locked Automatic Frequency Control System with Inertia in the Presence of Fluctuating Interference," *Telecomm. Radio Engrg.,* pt. 2, vol. 22, no. 2, Feb 1967, pp. 73-8.

Tsybul'ko, L. Ye., "Investigation of a Phase-locked AFC System by Digital-computer simulation," *Telecomm. and Radio Engrg., pt. 2,* vol. 25, no. 3, Mar 1970, pp. 100-104.

Antsibor, N.M., M.I. Zhodzishskiy, and A.V. Kurbatov, "False Pull-in Selectivity in a PAFC System with a Pulsed Phase Detector," *Telecomm. and Radio Engrg.,* pt. 1, vol. 24, no. 11, Nov 1970, pp. 34-40.

Kapranov, M.V. and T.I. Kurochkina, "The Pull-in Range and the Filtering Properties of a Phase-Lock AFC System Containing Non-Linear Attenuation," *Radio Engrg. and Electron. Phys.* vol. 19, no. 6, June 1974, pp. 38-44.

Blanchard, A., "Interference in Phase-locked Loops," *IEEE Trans.,* vol. AES-10, no. 5, Sept 1974, pp. 686-97.

Samoilo, K.A., and T.S. Fedosova, "Criterion of Disappearance of Cyclic-mode Oscillations in a Phase-Locked AFC System," *Izv VUZ Radioelektron.,* vol. 18, no. 11, Nov 1975, pp. 69-71 (in Russian, trans. SA 76-12454).

RECEIVER: AUTOMATIC GAIN CONTROL (AGC)

1. Books

Tartakovskiy, G.P., *Dynamics of Automatic Gain Control Systems,* Moscow, Gosenergoizdat, 1957 (in Russian).

Lukoshkin, A.P., "Radar Amplifiers with Wide Range Of Input Signals" Moscow, 1964 (in Russian, partial trans. JPRS 52891).

2. Journal Articles

Krivitskiy, B. KH., "Pulse Transmission by an Amplifier Having High Speed Automatic Amplification Control," *Radio Engrg.,* vol. 13, no. 4, 1958, pp. 93-105.

Shirokov, V.V. and V.G. Repin, "The Influence of Interference on an Automatic Gain Control System," *Radio Engrg.,* vol. 14, no. 4, 1959, pp. 88-98.

Shirokov, V.V., "Effect of Signal Fluctuations on a Receiver with AGC," *Radio Engrg. and Electron. Phys.,* vol. 6, no. 9, 1961, pp. 1288-1295.

Skirokov, V.V., "Interference Effects on a Two-channel Receiver Installation with AGC," *Telecomm.,* no. 1, Jan 1961, pp. 16-24.

Manske, R.A., "Detection Probability for Systems with Instantaneous Automatic Gain Control," *9th IEEE MIL Electron. Conf. Rec.,* 1965, pp. 28-31.

Yanishevskiy, A.A., "Instantaneous Automatic Gain Control Circuits," *Telecomm. and Radio Engrg.,* pt. 2, vol. 20, no. 6, June 1965, pp. 81-5.

Krivitskiy, B. Kh., "Dynamic Properties of Automatic Gain Control Systems in Consideration of Limitations," *Izv Vuzov SSSR, Ser. Radioelektronika,* vol. 10, no. 3, 1967, pp. 254-63 (in Russian).

Grigor'yev, L.N., "Action of Random Process on AGC System with Nonlinear Characteristics," *Voprosy Radioelektronika, Ser. Obshcheteknicheskaya,* no. 9, 1969, pp. 83-7 (in Russian).

Grigor'yev, L.N., "Action of Multiplicative Stationary Interference on AGC Systems," *Voprosy Radioelektronika,* Ser. Obshchetekhnicheskaya, *no. 11, 1970,* pp. 100-4 (in Russian).

Cochrane, W.A., and W.G. Briscoe, "Design and Operation of Automatic Gain Control (AGC) for AREST Receiver," *US Army Missile Res. Dev. and Engrg.* Lab. Rpt. RE-75-8, Aug 1974. DDC AD A005153.

Olenberger, C.F., "Effects of Automatic Gain Control on Phased-locked Loop Behavior in the Presence of Interference," *IEEE Trans.,* vol. AES-12, no. 6, Nov 1976, pp. 803-5c.

ANALOG MOVING TARGET INDICATOR (MTI)

1. Books and General Bibliography

Bibliography of Radiation Laboratory Literature on MTI as of Feb 6, 1945, MIT Radiation Lab., DDC ATI 155094.

Vishin, G.M., *Moving Target Selection,* Moscow: Voyenizdat, 1966 (in Russian, trans. AD 826182).

Barton, D.K., "International Cumulative Index on Radar Systems," *IEEE Pub. Cat. no. JH 4675-5,* 1978: extensive bibliography on MTI, pp. 37-39. See also Doppler radar pp. 22-4 and signal processing pp. 64-8.

Schleher, D.C., ed., *MTI Radar,* Artech House, Dedham, MA, 1978.

2. Journal Articles

Middleton, D. and P.J. Sutro, "Analysis for a Possible A/J System Against Window (Chaff)," *Radio Res. Lab., Harvard Univ., Rpt. 411-128,* Dec 9, 1944, DDC ATI 15322 (Analyzes use of a video differentiator as an ECCM against long chaff clouds).

Campanella, M.J. and M.C. Johnson, "Evaluation of the C73405 Radechon Storage Tube Using RF Signal Separation," *RCA Missile and Surface Radar Div., Rpt. TR 1600,* Sept 15, 1953.

Fishbein, W., "Area MTI Kit for Radar Set AN/MPS-501," *Signal Corp Eng. Lab. Tech. memo M-1549,* Dec 8, 1953.

Sorkin, C.S., and B.D. Steinberg, "Conical Scan Techniques in Ground Clutter with Coherent and Noncoherent MTI," *IRE East Coast Conf. on Aero. and Nav. Electron., Proc.,* 1958, pp. 35-45 (reprinted in this volume).

Rees, F.L. and G.F. Thomas, "Frequency Domain Approach to Subclutter Visibility," *IRE WESCON Conv. Rec.,* vol. 3, pt. 5, 1959, pp. 89-101.

d'Obrenan, G. VDB., "SDS Type Radar for Detection of Moving Objects at Ground Level," *L'Onde Electr.,* vol. 44, 1964, pp. 1032-1036 (in French).

Skolnik, M.I. "Analysis of Proposed Two-frequency Radar Waveform," *Naval Res. Lab. Rpt. MR-2926,* Nov 1964, DDC AD AOO 2942.

Torby, A., "Detectability of Signal in Noise When Using Range Gated MTI Radar," *Ericsson Tech. (Sweden),* vol. 21, no. 2, 1965, pp. 241-268.

Hair, T., "An Introduction to the Theory of Pulse Doppler Radar," *Marconi Rev.,* vol. 29, 4th qtr., 1966, pp. 175-190.

Harrison, R.N., "Selective Moving Target Indication," *Industrial Electron.,* vol. 5, no. 4, Apr 1967, pp. 146-148.

Brueggemann, R.P. and R.B. Rutledge, "Digital Simulation of Clutter Rejection MTI Radars," *IEEE Electromag. Compat. Symp. Rec.,* July 18-20, 1967, pp. 281-294.

Gomzin, V.N., "Correlation Coefficient of Doppler Signals in a Two-channel System," *Radio Engrg. and Electron. Phys.,* vol. 13, no. 9, Sept 1968, pp. 1477-1480.

Goldberg, R.M., "Evading Radar Detection is a Cinch (against MTI radar) for Aircraft Equipped with EASY," *Electronic Design Magazine,* vol. 14, no. 26, Nov 22, 1968, pp. 86-87.

Kroszczynski, J., "Efficiency of the Two-frequency MTI System," *Przemysl Inst. Telekommun.,* vol. 20, no. 68, 1970, pp. 1-5 (in Polish, trans. abstract SA 71-31474).

"Radar Clutter Reduction Techniques," *Shipping World Shipbuilder,* vol. 163, no. 3843, Mar 1970, pp. 450, 3.

Gomzin, V.N., "The Problem of Constructing a Two-channel Direction-tracking System with Selection of Moving Targets," *K. Voprosu O. Postroyenii Dvukhkanal'noy Sistemy ASN S SDPS,* Leningrad, 1970, pp. 1-14, Viniti No. 3592-71 (in Russian, trans. FTD ID(RS) I - 0322 - 76).

Gorin, I.I., "A Device for Detecting a Doppler Signal Against a Background of Interference," Moscow, Othrytiya, Izobreteniya, Promyshiennyye Obraztsy, Tovarnyye Znaki No. 32, 1970, *Soviet Patent No. 284064* (in Russian, trans. abstract JPRS 53117, p. 79).

Taylor, J.W., Jr., "MTI Radar System Utilizing Unique Patterns of Interpulse Period Choices to Simplify Weighting Coefficients for Individual Radar Echo Pulses," *US Patent no. 3566402,* Feb 23,1971, abstract SA 71-16087.

Mann, D., "The Detection of Moving Targets in Clutter with Doppler Filters," *NTZ,* vol. 24, no. 2, Feb 1971, pp. 72-76 (in German, trans. abstract EI 71-74858).

Galvin, A.A., "Radar Processor Having Adaptive Clutter Rejection," *US Patent no. 3576564,* Apr 27, 1971, abstract SA 71-23185.

Grasso, G., and P.F. Guarguaglini, "New Techniques in MTI Receivers," *IE* (India), *JE*, and *TE*, pt. ET-3, no. 10, vol. 51, May 1971, pp. 159-164.

Hayre, H.S., "Doppler Tracking Systems," *IE* (India), *JE*, and *TE*, pt. ET-3, no. 10, May 1971, pp. 169-172.

Shenoy, R.P., "Signal Processing Techniques for Clutter Rejection," *IE* (India), *JE*, and *TE*, pt. ET-3, no. 10, vol. 51, May 1971, pp. 173-179.

Shrader, W.W., "Modern Moving Target Indicator Radar," *IE* (India), *JE*, and *TE*, pt. ET-3, no. 10, vol. 51, May 1971, pp. 180-188.

Houston, W., "Improvements in or Relating to Radars," *UK Patent no. 1236501,* June 23, 1971.

Haykin, S. and J.F. Boulter, "Response of a Recursive MTI Filter to Inputs with Staggered PRF," *IEE Electron. Lett.,* vol. 7, no. 14, July 15, 1971, pp. 415-416.

Hartley-Smith, A., "Ground and Angel Clutter in Radar Systems: The Two Beam Antenna Solution," *GEC J Sci Technol.,* vol. 39, no. 4, 1972, pp. 173-180.

Gomzin, V.N., "On the Problems of Constructing a two-channel Direction Tracking Systems with Selection of Moving Targets," *Radiotekhnika i Elektronika,* vol. 17, no. 4, Apr 1972, pp. 903ff (in Russian trans. abstract JPRS 57112).

Rajendran, S., "Techniques for Clutter Suppression in Radar Systems," *IE* (India), *JE*, and *TE*, vol. 17, no. 4, Jul-Aug 1973, pp. 93-118.

Popov, D.I., "Efficiency of the Coherent-pulse Method of Signal Reception in the Presence of Passive Noise (Chaff)," *Telecomm. Radio Engrg., pt. 2,* vol. 28, no. 11, 1973, pp. 92-94.

Kailasam, T., and M. Venkatesan, "A Solid State MTI Canceller Using RGF Canceller," *Electro-Technol. (India),* vol. 17, no. 6, Nov/Dec 1973, pp. 181-9.(Author states this is applicable to search or tracking to reduce ground reflections or clutter.)

Ramakrishna, S., and P.R. Mahapatra, "A Method for Optimizing Radar Signal Waveforms for Clutter Rejection," *IE* (India), *JE*, and *TE*, vol. 20, no. 8, Aug 1974, pp. 394-398.

Subramanian, A.K., "Design Considerations of a Coherent (On Receive) MTI System for a Conically Scanned Tracking Radar," *Electro-Technol. (India),* vol. 18, no. 4, Dec 1974, pp. 95-98 (reprinted in this volume).

Zuech, D.I., "Radar Moving Target Tracking System," *US Patent no. 3858209,* 31 Dec 1974, abstract SA 75-21125.

Telefunken, "Radar System for Moving Targets," *UK Patent no. 1380040,* West Germany patent no. 18966, Jan 8, 1975 abstract SA 75-30128.

Konecny, J., "Analysis of the Velocity Characteristics of MTI Filters," *TELSA Electron.* (Czechoslovakia), vol. 8, no. 3, Sept 1975, pp. 67-72 (in Polish, trans. SA 76-17865).

Vnotchenko, S.L., "Reducing the Power of Passive Interference in the Forward Hemisphere of a Moving Radar," *Telecomm. and Radio Engrg., pt. 2,* vol. 30, no. 10, Oct 1975, pp. 66-9.

Bates, C.N., "New Advances in Doppler Radar," *Electron. Electro-optic Infrared Countermeasures,* vol. 2, no. 4, Apr 1976, pp. 34-5.

Kroszczynski, J., "Dual Frequency MTI System," *Radio and Electron. Engrg.,* vol. 46, no. 7, July 1976, p. 332.

CONSTANT FALSE ALARM RATE (CFAR)

1. General Bibliography

Barton, D.K., "International Cumulative Index on Radar Systems," *IEEE Pub. Cat. no. JH 4675-5,* 1978: extensive bibliography on CFAR, pp. 10-11, see also Detection pp. 18-22.

2. Journal Articles

Lamb, J.F., "A Noise-silencing I.F. Circuit for Superhet Receivers," *QST Magazine,* vol. 20, no. 2, Feb 1936, p. 11ff.

Siforov, V.I., "Effect of Interference on Radio Reception of Pulse Signals," *Radio Tekhnika,* vol. 1, no. 4, Apr 1946, pp. 3-19 (in Russian, trans. abstract SA 45-2724).

Schukin, A.N., "A Method of Preventing Impulse Interference with Radio Reception," *Bull ACAD SCI Ser. Phys. (USSR),* vol. 10, no. 1, 1946, pp. 49-56 (in Russian, trans. abstract SA 46-2935).

Utkina, G.I., and P.S. Davydov, "Evaluation of Effectiveness of Certain False Alarm Stabilization Systems in Nonstationary Gaussian Noise," pp. 116-126 in *Theory and Technology of Radar,* A.G. Saybel, ed., no. II, Moscow, Mashenostroyeniye, 1967 (in Russian).

Carpentier, M.H., *Radars: New Concepts,* NY: Gordon and Breech Science Pub., 2nd ed., 1968, p. 175.

Cope, R., "Tackle Noise Variations at Video," *Electronic Design Magazine,* vol. 20, Sept 26, 1968, pp. 54-56 (reprinted in this volume).

Bogdanovich, V.A. and V.N. Prokof'ev, "Rank-Quantitization Detector of Fluctuating Signals in Noise with Unknown Distribution," *Izv VUZ Radioelektronika,* vol. 14, no. 5, May 1971, pp. 522-526 (in Russian, abstract SA 71-27577) (also trans: AD BO14837).

Hansen, V.G., and A.J. Zottl, "The Detection Performance of the Siebert and Dicke-fix CFAR Radar Detectors," *IEEE Trans.,* vol. AES-7, no. 4, Jul 1971, pp. 706-709c.

Lomakin, D.V., "Efficiency of Detection Systems with Arbitrary Noise Distribution," *Izv VUZ Radioelektron,* vol. 14, no. 8, 1971, pp. 861-7 (in Russian, trans. abstract SA 71-37667).

Nitzberg, R., "Constant-false-alarm-rate Signal Processors for Several Types of Interference," *IEEE Trans.,* vol. AES-8, no. 1, Jan 1972, pp. 27-34.

Tong, P.S., "Greatest-of-N Distribution-free CFAR," *Technol. Service Corp. Intern. memo TSC-W4-30,* June 10, 1974.

Grieve, P.G., "A Unified Approach to the Construction of Constant False Alarm Probability Radar Detectors," *UCLA Thesis, University Microfilms No. 75-14055,* 1975. Abstract SA 76-17852.

Aristova, O.G., "Statistical Characteristics of Signal Detection for a Fluctuating Detection Threshold Level," *Telecomm. and Radio Engrg., pt. 2,* vol. 30, no. 6, June 1975, pp. 127-9.

Tong, P.S., and P.E. Steichen, "Performance of CFAR Devices in ECM Environments," *DDRE ECCM Conf.,* 1976 (reprinted in this volume).

Woerrlein, H.H., "CFAR-detector Derived from Correlation Processing in Adjacent Range Bins," *IEE Conf. Pub. no. 155,* Radar 77, London, 1977, pp. 266-70.

Cole, L.G., and Pin-wei Chen, "Constant False Alarm Detector for a Pulse Radar in a Maritime Environment," *IEEE NAECON Rec. 1978,* pp. 1110-1113.

Nitzberg, R., "Analysis of the Arithmetic Mean CFAR Normalizer for Fluctuating Targets," *IEEE Trans.,* vol. AES-14, no. 1, Jan 1978, pp. 44-47.

RECEIVERS: DETECTION/NOISE REDUCTION

1. Books and Bibliography

Momot, Ye. G., *Synchronous Reception Problems and Technology,* Moscow: Svyaz'izdat, 1961 (in Russian).

Kantor, L. Ya., *Methods of Improving Noise Immunity of FM Receivers,* Moscow: Svyaz'izdat, 1967 (in Russian).

Teterich, N.M., *Noise Generators and Measurement of Noise Characteristics,* Moscow: Energiya, 1968 (in Russian).

Rodinov, Ya. G., *FM Reception with Back Control,* Moscow: Soviet Radio 1972 (in Russian).

Vinitskiy, A.S., and A.G. Zyuko, eds., *Noise-Immune FM and PM Methods* (Collection of articles) Moscow: Soviet Radio 1972 (in Russian).

Stratonovish, R.L., *Principles of Adaptive Reception,* Moscow: Soviet Radio, 1973 (in Russian).

Barton, D.K., "International Cumulative Index on Radar Systems," *IEEE Pub. Cat. no. JH 4675-5,* 1978: extensive bibliographies on Detection (General) pp. 18-20, Detection (Automatic) p. 21, Detection (Sequential) pp. 21-22.

2. Journal Articles

Siforov, V.I., "Effect of Interference on Radio Reception of Pulse Signals," *Radiotekhnika,* vol. 1, Apr 1946, pp. 3-19 (in Russian, trans. abstract SA 45-2724).

McLaughlin, J.L.A., "Hetrodyne Eliminator," *Electronics Magazine,* vol. 22, Mar 1949, pp. 83-5.

Gutkin, L.S., "Effect of Interference on Super-regenerators, Effect of Interference Pulses," *Radio Tekhnika,* vol. 4, no. 1, 1949, pp. 61-76 (in Russian, trans. abstract SA 49-2256).

Moses, R.C., "Impulse Noise Limiters," *Radio-Electronic Engrg.,* vol. 23, June 1951, p. 16ff.

Smith, R.A., "The Relative Advantages of Coherent and Incoherent Detectors: A Study of their Output Noise Spectra under Various Conditions," *Proc IEE,* vol. 98, pt. III, *Radio and Commun. Engr. Monograph no. 6,* 1951, pp. 401-6.

Kaden, H., "The Compensation of Interference in Carrier Receivers by Opposition Cancellation of a Parallel Receiver," *Frequencz,* vol. 10, Mar 1956, pp. 76-87 (see Skolnik intro to radar, p. 569, ref. #63).

Gorbachev, A.A., "A Compensation Method for the Reduction of Non-linear Noise," *Radiotekhnika,* vol. 11, no. 4, 1956, pp. 67-74 (in Russian, trans. abstract SA 57-292).

Borg, H., "On the Technical Possibilities of Reducing Noise in the Reception of Weaker Signals," *Ann. Telecomm.,* vol. 11, no. 5, May 1956, pp. 90-110 (in French, trans. abstract SA 57-442).

Linder, I.W., and P. Swerling, "Performance of the Double-threshold Radar Receiver in the Presence of Interference," *Rand Report RM-1719,* 28 May 1956, DDC AD 115366.

Corbachev, A.A., "An Experimental Study of the Limiting of Pulse Noises with Spectrum Conversion and Shifting Threshold," *Radio Engr.,* vol. 12, no. 6, June 1957, pp. 87-93.

Livshits, A.R., "The Probability of n-Coincidence," *Radio Engrg., and Electron. Phys.*, vol. 2, no. 8, Aug 1957, pp. 1-6.

Gorbachev, A.A., "The Suppression of Pulse Interference by Non-linear Transformation of the Form of the Frequency Spectrum," *Radio Engrg.*, vol. 13, no. 1, 1958, pp. 72-80.

Tartakovskii, G.P., and Yu. M. Sergienko, "The Influence of a Sequence of Pulses Modulated by a Random Process on an Inert Pulse Detector," *Radio Engrg.*, vol. 13, no. 1, 1958, pp. 81-9.

Sidorov, V.M., "Action of Weak Pulse Interference on Frequency-modulation Receivers," *Radio Engrg.*, vol. 13, no. 3, 1958, pp. 26-46.

Troude, J., "Elimination of Background Noise and Jamming in Radar; An Application of Correlation," *Revue H.F.*, vol. 4, no. 2, 1958, pp. 25-31 (in French, trans. abstract SA 58-5608).

Medbedev, Iu. I., "An Experimental Investigation of the Extrapolation Method of Suppressing Pulse Interference in Broadcast Reception," *Radio Engrg.*, vol. 13, no. 6, 1958, pp. 66-74.

Tiberio, V., "Problems Connected with Echo Filtering in Radar," *Ricerca Sci.*, vol. 28, Dec 1958, pp. 2464-81.

Gudzenko, L.I., "On Periodic Non-stationary Processes," *Radio Engrg. and Electron. Phys.*, vol. 4, no. 6, 1959, pp. 220-4.

Boldyrev, N.M., "A Study of the Change in Spectrum of a Rectangular High Frequency Pulse in Passage Through a Tuned Pi-section Filter and Amplitude Limiter," *Radio Engrg.*, vol. 14, no. 8, Aug 1959, pp. 14-26.

Clark, D.L., "Detection of Signals in Non-Gaussian Noise," *MIT Lincoln Lab. Group Rpt. 47.36*, 15 Dec 1959, DDC AD 232332.

Blokh, E.L., and A.A. Kharkevich, "Certain Properties of Communications Systems with Fading Type Interference," *Radio Engrg.*, vol. 15, no. 9, Sept 1960, pp. 1-9.

Kharkevich, A.A., "Recognition of Continuous Signals," *Radio Engrg.*, vol. 15, no. 9, Oct 1960, pp. 12-15.

McLauchlan, E.C., "Noise Suppression in Pulse Receivers," *Australian J. Phys.*, vol. 13, no. 4, Dec 1960, pp. 750-2.

Petrovich, N.T., "Transmission of Signals by Pulse-Parameter Comparison Method," *Radio Engrg.*, vol. 16, no. 1, 1961, pp. 1-9.

Kharkevich, A.A., "Brief Communications: Asymptotic Expressions for Transmission Rates with High Reliability," *Telecomm. Radio Engrg., pt. 2*, vol. 17, no. 1, Jan 1961, pp. 76-80.

Gorbachev, A.A., "Impulse-noise Limiter with Variable Threshold," *Radio Engrg.*, vol. 16, no. 2, Feb 1961, pp. 61-72.

Davies, L.E., "A Method of Characterizing the Effect of Non-Gaussian Noise Upon Signal Reception by Linear or Non-linear Receivers (and Related Correspondence)," Weapons Systems Laboratory, Stanford Research Inst., *Tech Note 18A*, Mar 1961, revised. Apr 1961 correspondence.

Tsybakov, B.S., "Brief Communications: Shannon Scheme for Gaussian Information with a Uniform Spectrum and for a Channel with Fluctuation Noise," *Radio Engrg. and Electron. Phys.*, vol. 6, no. 4, Apr 1961, pp. 571-3.

Gorbachev, A.A., and M.I. Vinogradov, "On the Application of the Signal Extrapolation Method under Conditions of Pulse Interference Suppression," *Radio Engr.*, vol. 16, no. 8, 1961, pp. 49-55.

Buni'movich, V.I., and V.A. Morozov, "Brief Communications: Estimate of Frequency of Narrow-band Signal Against a White Noise Background," *Radio Engrg. and Electron. Phys.*, vol. 6, no. 9, Sept 1961, pp. 1403-7.

Teplyakov, I.M., "Signal-to-noise Ratio of a Receiver with a Noise-limiter Circuit in the Presence of Fluctuation Noise," *Radio Engrg.*, vol. 17, no. 4, Apr 1962, pp. 76-9.

Tikhonov, V.I., and V.T. Goryainov, "Effect of Normal Noise on Limiter," *Telecomm.*, no. 11, 1961, pp. 10-22.

Krivetskiy, A.A., "Method of Reducing Pulsed Noise in Radio-telegraph Reception," *Telecomm.*, no. 11, 1961, pp. 22-6.

Shabalin, A.M., "Method of Suppressing High-frequency Pulsed Interference in a Receiver by Intercepting High-frequency Wide-band Channel and Using Aperiodic Trap Filters," *Trudy GPI*, vol. 18, no, 2, 1962, pp. 74-90 (in Russian).

Sergiyevskiy, B.D., "Response of Receiver with Square-law Detector to Waves Modulated by Phase or Frequency Fluctuations," *Radio Engrg. and Electron. Phys.*, vol. 7, no. 5, May 1962, pp. 741-51.

Gorbachev, A.A., "On a Method of Signal Extrapolation by Power Series," *Telecomm. Radio Engrg., pt. 2*, vol. 17, no. 7, Jul 1962, pp. 9-12.

Winker, G., "The Effect of Averaging on the Operation of a Correlation Detector," *Arch. Elek. Ubertragung*, vol. 18, no. 7, Jul 1962, pp. 347-9 (in German).

Blokh, E.L., and A.A. Kharkevich, "Parasitic Modulation Caused by Small Additive Noise," *Telecomm. and Radio Engrg., pt. 2*, vol. 17, no. 11, Nov 1962, pp. 3-13.

Babanov, Yu. N., "Method of Increasing the Selectivity of Receivers for Two AM Signals with Superimposed Frequency Spectra," *Telecomm. Radio Engrg., pt. 2,* vol. 17, no. 12, Dec 1962, pp. 45-48.

Gorbachev, A.A., "Increased Noise Immunity in the Reception of AM Signals by Use of Non-linear Feedback in the High-frequency Section of the Receiver," *Telecomm. and Radio Engr., pt. 2,* vol. 18, no. 2, Feb 1963, pp. 36-41.

Levin, B.R., and Ya. A. Fomin, "Approximate Determination of the Distribution Function of Fade Lengths below Threshold Level for the Envelope of the Sum of a Deterministic Signal and Gaussian Stationary Noise," *Telecomm. and Radio Engrg., pt. 2,* vol. 18, no. 5, May 1963, pp. 20-7.

Bobnev, M.P., "Assessment of the Noise Immunity of the Automatic Pulse Rangefinder in the Signal-searching Mode of Operation," *Telecomm. and Radio Engrg. pt. 2,* vol. 18, no. 5, May 1963, pp. 57-65.

Babanov, Yu. N., "Experimental Investigation of a Method for Improving the Selectivity of Radio Receivers for Overlapping AM Signal Spectra," *Telecomm. and Radio Engrg., pt. 2,* vol. 18, no. 6, June 1963, pp. 27-30.

Ageyev, D.V., and Yu. N. Babanov, "A Method for Improving the Selectivity of Radio Reception when the Frequency Spectra of the Desired and Interfering AM Signals Overlap," *Telecomm. Radio Engrg., pt. 1,* vol. 18, no. 9, Sept 1963, pp. 8-16.

Sidorov, V.M., "Process of Transition to Reception of an Interfering Station in an FM Receiver," *Telecomm. Radio Engrg. pt. 2,* vol. 18, no. 11, 1963, pp. 33-43.

Shabalin, A.M., "On Noise Immunity of WLN System," *Trudy GPI,* vol. 20, no. 5, 1964, pp. 36-40 (in Russian).

Ageyev, D.V., and Yu. N. Babanov, "Reception of Amplitude-modulated Radio Signals in the Presence of Interference Signal with Overlapping Frequency Spectrum," *Radio Engrg. and Electron. Phys.,* vol. 9, no. 6, June 1964, pp. 944-8.

Babanov, Yu. N., "Methods of Reception of Amplitude Modulated Radio Signals in the Presence of Interfering Signals with Overlapping Spectrum," *Radio Engrg. and Electron. Phys.,* vol. 9, no. 7, June 1964, pp. 948-55.

Bykov, V.L., "Threshold in Frequency Modulation and Methods of Reducing It," *Telecomm. Radio Engrg., pt. 1,* vol. 18, no. 12, Dec 1964, pp. 17-26.

Levin, G.A., and A.M.Bonch-Bruyevich, "Design Methods for Detectors with Self-optimizing Threshold Level in Radio Communication System Receiving Equipment," *Telecomm. Radio Engrg., pt. 1,* vol. 19, no. 5, 1965, pp. 5-10.

Zelelin, A.P., "Use of a System of Resonators Operating in Parallel for the Suppression of Impulsive Interference," *Telecomm. Radio Engrg., pt. 1,* vol. 19, no. 5, May 1965, pp. 11-6.

Bugayev, V.A. "Multichannel Sequential Detection Theory with a Linearly Increasing Threshold," *Radio. Engrg. and Electron. Phys.,* vol. 10, no. 10, Oct 1965, pp. 1511-1515.

Gold'berg, A.P., "Characteristics of Pulse Noise Suppression Systems," *Telecomm. Radio Engrg., pt. 1,* vol. 20, no. 2, Feb 1966, pp. 22-31.

Andronov, I.S., "Noise Immunity of Coherent-summing Diversity Reception," *Telecomm. Radio Engrg., pt. 2,* vol. 21, no. 3, Mar 1966, pp. 100-5.

Stratonovich, R.L., and Yu. G. Sosulin, "Optimum Reception of Signals in Non-Gaussian Noise," *Radio Engrg. and Electron. Phys.,* vol. 11, no. 4, Apr 1966, pp. 497-507.

Shumlyanskiy, I.I., and V.A. Lobodzinskiy, "The Use of Amplitude Limiters in Frequency-selective Systems," *Telecomm. Radio Engrg., pt. 2,* vol. 21, no. 4, 1966, pp. 87-92.

George, T.S., "Output of a Tuned Circuit when Rapidly Swept by a Band Limited Noise Spectrum," *Proc. IEEE,* vol. 54, no. 6, June 1966, pp. 901-2c.

Finn, H.M., and R.S. Johnson, "Efficient Sequential Detection in the Presence of Strong Localized Interference," *RCA Rev.,* vol. 27, no. 3, Sept 1966, pp. 392-407.

Zelenin, A.P. and N.N. Glazkov, "Selectivity of Parallel-Operating Resonators," *Telecomm. and Radio Engrg., pt. 2,* vol. 22, no. 6, June 1967, pp. 75-79.

Ottinger, R.W., "The Effect of Hard Limiting in the Presence of Large Out-of-Band Interfering Signals," *IEEE Trans.,* vol. AES-3, no. 5, Sept 1967, pp. 830-834c.

Golovnya, Ya. P., "Analysis of Sources of Pulse Radio Interference," *Telecomm. and Radio Engrg., pt. 2,* vol. 22, no. 11, Nov 1967, pp. 140-2.

Banti, I., and L. Massotti, "Experiments in Improving the Signal-Noise Ratio by Means of Matched Filters," *Alta Freq.,* vol. 37, no. 1, Jan 1968, pp. 69-70 (in Italian, trans. abstract SA 68-19029).

Gorbachev, A.A., and V.I. Danilov, "Impulse-noise Limiter with Tracking Threshold in Receivers of Noise-like Signals," *Telecomm. Radio Engrg., pt. 2,* vol. 23, no. 7, July 1968, pp. 92-95.

Zachepitskii, A.A., and yu. I. Pakhomov, "Immunity of Radar Receivers Employing Pulse Interference Limiting," *Izv VUZ Radioelektronika,* vol. 12, no. 2, 1969, pp. 136-143 (in Russian, trans. abstract SA 69-24405).

Pauk, S.M., Ye, G. Servinskiy, "Effect of a Key on the Interference Resistance of a Receiver of Discrete Information with Multi-positional Frequency Manipulation under Conditions of Real Pre-detector Integration," Kiev, V sb *Inform.-izmerit. sistemy. Tr Seminara* (Measuring System Information. Works of a Seminar—collection of works), vol. 1, 1969, pp. 52-63, from *Rzh Radiolekhnika,* no. 4, 1970, (in Russian).

Bauer, L.W., "Detection in the Sonar Environment," *Phd. thesis,* Brooklyn Polytechnic Institute, 1969, Univ. Microfilms 69-20315.

Kamlet, S.G., "The Interference-free Feature of a System of Optimal Reception in the Presence of Frequency Separation," *Rzh - Radioteknika,* no. 4, 1970 Abstract No. 4A67 (in Russian, trans. abstract JPRS #53235).

Korado, V.A., "Two-channel Detection of Random Signals on Background of Interference of Unknown Intensity," *Izv an SSSR Ser. Tekhniches-kaya Kibernetika,* no. 4, 1970, pp. 152-9 (in Russian).

Krivosheev, V.I., "Optimum Extraction of Useful Information from a Mixture of AM Signal, AM Interference and Random Noise," *Radio Engr. and Electron. Phys.,* vol. 15, no. 5, May 1970, pp. 872-7.

Korado, V.A., "Optimum Detection of Signals in the Presence of Noise with Unknown Parameters with an Upper Bound on the False Alarm Probability," *Radio Engrg. and Electron. Phys.,* vol. 15, no. 7, Jul 1970, pp. 1215-22.

Terent'yev, A.S., "Measurement of the Parameters of a Pulse Signal in the Presence of Reflections," *Radio Engrg. and Electron. Phys.,* vol. 15, no. 7, 1970, pp. 1239-48.

Bykhovskiy, M.A., and V.S. Rabinovich, "Effect of Incomplete Separation of the Components of Composite Signals upon the Noise Immunity of Coherent Communications Systems," *Telecomm. and Radio Engrg., pt. 2,* vol. 25, no. 11, Nov 1970, pp. 77-80.

Orlov, I. Ya., "Optimum Radio Reception of an AM Signal when the Radio System Operates Nonlinearly," *Telecomm. and Radio Engrg., pt. 2,* vol. 25, no. 11, Nov 1970, pp. 81-83.

Chechulina, L.A., "Noise Immunity of Coherent Reception of a Spectrum-limited FM Signal," *Telecomm. and Radio Engrg., pt. 1,* vol. 24, no. 11, Nov 1970, pp. 28-33.

Ovarov, V.P. and Yu. V. Kon'kov, "Quantitative Evaluation of Interference Resistance of a Radar Detection System," Trans Tsentr Nauch Issled Inta Ekon i Ekspl Vodnogo trans, No 84, 71, pp. 46-tr (in Russian).

Levin, B.R., and Yu. S. Shinakov, "Bayesian System for the Simultaneous Discrimination of Several Signals and Estimation of their Parameters," *Telecomm. and Radio Engr., pt. 2,* vol. 26, no. 4, Apr 1971, pp. 67-70.

Troitskiy, Ye. V., "Classification of Nonstationary Gaussian Sets," *Telecomm. Radio Engr., pt. 2,* vol. 26, no. 4, Apr 1971, pp. 71-4.

Shakhgil'dyan, V.V. and M.S. Lokhvitskiy, "Generation of the Conditional Probability Density of a Mixture of Signal and Noise, Using a 'Learning' Sample with Interdependent Values," *Telecomm. Radio Engrg., pt. 2,* vol. 26, no. 4, Apr 1971, pp. 74-6.

Tartakovskiy, G.P., and V.G. Repin, "Statistical Synthesis of Adaptive Systems for Checking Hypotheses with Estimation of Distribution Parameters Related to these Hypotheses," *Izv VUZ Radioelektron.,* vol. 14, no. 4, Apr 1971, pp. 8-12 (in Russian).

Babanov, Yu. N., S.G. Manukyan, and M.A. Lavrov, "A Device for Suppressing Regular Interference in Reception of Wide-band Frequency-modulated Signals," *Otkrytia Izobretniya Promyshlennyye Obraztsy Tovarnyye Znaki,* May 1971 (in Russian, trans. abstract JPRS 54995).

Guarguaglini, P.F. and F. Marcoz, "A Range Sampled Detector," *I.E. India E. & T.E.,* vol. 51, May 1-3, 1971, pp. 165-169.

Gatkin, N., "Detection of Signals in Reverberation Noise," *Izv VUZ Radioelektron.,* vol. 14, no. 7, July 1971, pp. 758-761 (in Russian, trans. JPRS 54995, SA 71-37661).

Bogdanovich, V.A., "Detecting Coherent Pulse Packets in Noise of Unknown Powers," *Izv VUZ Radioelektron,* vol. 14, no. 7, July 1971, pp. 762-8 (in Russian, trans. abstract SA 71-37662, JPRS 54995).

Paulenko, K.V., "Detection Characteristics of Multiplicative Processing of Coherent Signals," *Izv VUZ Radioelektron.,* vol. 14, no. 8, 1971, pp. 843-51 (in Russian).

Gusinskaya, G.V., "Estimation of Noisy Radio Signals by the Method of Spectral Analysis," *Izv VUZ Radioelektron.,* vol. 14, no. 8, Aug 1971, pp. 951-3 (in Russian, trans. abstract SA 71-37674; JPRS 55283).

Zayezdnyi, A.M. Ye I. Plotkin, and Yu. A. Cherkasskiy, "Signal Processing Using their Structural Properties," *Telecomm. and Radio Engr., pt. 2,* vol. 26, no. 9, Sept 1971, pp. 61-5.

Khadzhi, B.A., "Measurement of the Time of Arrival of a Sinusodial Radio Signal Received in the Presence of Noise-like Impulse Interference," *Telecomm. Radio Engrg., pt. 2,* vol. 29, no. 9, Sept 1971, pp. 69-72.

Orlov, I. Ya., and E.V. Silin, "Extracting the Envelope of an AM Signal from its Mixture with a High-power

Unwanted Oscillation for Nonlinear Operating Conditions of the Amplifying Stages of the Receiver," *Telecomm. and Radio Engrg., pt. 2,* vol. 26, no. 9, Sept 1971, pp. 88-91.

Smol'vaninov, V.M., and L.I. Filoppov, "Noise Immunity of a Noncoherent Receiver," *Radio Engrg. and Electron. Phys.,* vol. 16, no. 9, Sept 1971, pp. 1504-8.

Kaszerman, P., "Frequency of Pulse Coincidence Given Radars of Different Pulsewidths and PRF's," *IEEE Trans.,* vol. AES-7, Sept 1971, pp. 1013-4c.

Kazakov, V.A., "Potential Noise Immunity of Radio Receivers Subjected to White Noise and Markov Noise," *Radio Engrg. and Electron. Phys.,* vol. 16, no. 10, Oct 1971, pp. 1763-5.

Osawa, H., N. Morinaga, and T. Namekawa, "Output Signal-to-noise-ratios in AM Correlation Systems and their Improvement," *Electron. and Commun. in Japan,* vol. 54A, no. 10, Oct 1971, pp. 1-8.

Bogdanovich, V.A., and V.N. Prokof'yev, "Optimum Recognition of Double-frequency Radar Signals with 'Interfering' Parameters," *Radio Engrg. and Electron. Phys.,* vol. 16, no. 12, Dec 1971, pp. 2022-31.

Shabalin, A.M., and B.D. Zabegalov, "Wide-band-limiter-trap System for Pulsed Interference Detection," *Trudy GPI,* vol. 27, no. 15, 1971, pp. 82-5 (in Russian).

Babanov, Yu.N. and S.G. Manukyan, "Optimum Reception of FM Signals in White Noise with Narrow Band AM Interference," *Izv VUZ Radioelektron,* vol. 14, no. 12, Dec. 1971, pp. 1502-1508 (in Russian, trans. abstract SA 72-14389).

Ruzhin, A.L., and S.V. Samsonenko, "A Detection Procedure Invariant with Respect to the Intensities of Signal and Noise," *Radio Engrg. and Electron. Phys.,* vol. 17, no. 1, Jan 1972, pp. 129-32.

Goryunov, M.V., "WLN Systems in which Maltuned Pulsed Interference is Standardized Independently of its Input Amplitude and Duration," *Proc. Gor'kiy Polytechnic Inst. imeni AA Zhdanov,* vol. 28, no. 7, 1972, pp. 5-8 (in Russian).

Gerlitzki, W., "Procedures Used in Detection Systems for Detecting Targets and Processing Target Data — Outgrowths of More Recent Results," *Ulm wehrtecknik,* no. 7, 1972, pp. 278-83 (in German, trans. abstract SA 72-33257).

Goryainov, V.T. and V.N. Smirnov, "Noise Immunity in the Reception of Binary Radio Signals in Correlated Noise," *Izv VUZ Radioelektron,* vol. 15, no. 9, Sept 1972, pp. 1077-1087 (in Russian, trans. abstract SA 73-13718).

Sviridenko, S.S., "Immunity of Synchronized Communications Systems Assuming Random Mistracking," *Izv VUZ Radioelektron,* vol. 15, no. 9, Sept 1972, pp. 1095-1100 (in Russian, trans. abstract SA 73-13719).

Kuzman, B.I., "A Model of Pused Interference in Sampled Data Systems," *Izv VUZ Radioelektron,* vol. 16, no. 1, Jan 1973, pp. 54-57 (in Russian, trans. abstract SA 73-20786).

Sidorov, Yu.Ye., "Optimum Detec-Algorithm for the Detection of Radar Signals in Noise of Unkown Power," *Izv VUZ Radioelektron,* vol. 16, no. 2, Feb 1973, pp. 101-102 (in Russian, trans. abstract SA 73-21182).

Golovanov, V.G., "Accuracy of Measurement of Phase of Signal for Normal *A Priori* Distribution of Frequency against the Background of White Noise," *Radio Engrg. and Electron. Phys.,* vol. 18, no. 11, Nov 1973, pp. 1764-6.

Pozdnyak, S.I., and V.V. Kuznetsov, "Estimation of Polarization Parameters of Signal against the Background of Randomly Polarized Noise," *Radio Engr. and Electron. Phys.,* vol. 18, no. 11, Nov 1973, pp. 1766-76.

Danilenko, A.I., "Nonlinear Processing of Orthogonal Signals," *Izv VUZ Radioelektron,* vol. 17, no. 2, Feb 1974, pp. 40-4 (in Russian, trans. abstract SA 74-23912).

Sidorov, Yu.Ye. and A.D. Shiskin, "Rank Law of Detection of Radar Signals in Unknown Noise," *Izv VUZ Radioelektron,* vol. 17, no. 2, Feb 1974, pp. 91-94 (in Russian, trans. abstract SA 74-23913).

Sosulin, Yu.G., and V.V. Kostrov, "Optimum and Quasi-Optimum Detection of Signals in the Presence of Arbitrary Interference by the Whitening Filter Method," *Radio Engrg. and Electron. Phys.,* vol. 19, no. 6, June 1974, pp. 45-52.

Sokolov, O.L., "Noise Immunity of a Detector which Analyzes the Difference in Flucutation Over-shoot of Additive Noise at Polar Levels," *Telecomm. and Radio Engrg., pt. 2,* vol. 29, no. 5, May 1974, pp. 105-7.

Kosova, A.L., "Noise Immunity of a Communication System with Phase-Shift Keying in the Presence of a Complex Non-linearity," *Telecomm. and Radio Engrg., pt. 2,* vol. 29, no. 5, May 1974, pp. 109-11.

Bobrova, Ye.P., and A.V. Larin, "Effectiveness of a Single-channel 'Whitening' Filter," *Izv VUZ Radioelektron,* vol. 17, no. 8, Aug 1974, pp. 83-5 (in Russian, trans. abstract SA 75-5958).

Galfi, II, "Examination of the Detection Properties of Radars, Application of Adaptive Procedures to Obtain

Optimum Detection Properties," *Hiradastechnika,* vol. 25, no. 9, Sept 1974, pp. 271-4 (in Hungarian, trans. abstract SA 75-3664).

Kazakov, V.A., "Noise Immunity of Optimal Receivers Operating in Conditions of White Noise and Signal Interference," *Izv VUZ Radioelektron,* vol. 17, no. 10, Oct 1974, pp. 16-22 (in Russian, trans. abstract SA 75-9368).

Simonskaya, S.F., "Influence of Noise Jamming on the Accuracy of a Non-servo-type Correlation Meter for Measuring the Time Shift of Random Signals," *Radiotekhnika Kharkov,* USSR, no. 29, 1974. pp. 55-9 (in Russian, trans. abstract SA 75-25397).

Daskalova, Kh., and T. Dzhubrailov, "Construction of an Optimum Receiver in the Existence of an Additive Noise and Intersymbol Interference," *Izv Inst. Tekh. Kibern* (Bulgaria), vol. 18, Nov 1974, pp. 115-28 (in Russian, trans. SA 75-5960).

Velkov, F., "Study of an Adaptive Receiver According to the Gradient Method," *Izv Inst. Tekh. Kibern* (Bulgaria), vol. 18, 1974, pp. 97-113 (in Russian, trans. abstract SA 75-5959).

Lowery, C.A., "Susceptibility of Binary DPSK to Periodic FM Interference," Thesis, 1974, Univ. Microfilms No. 74-21573, abstract SA 75-16286.

Sviridenko, S.S., "Effects of Random Time and Frequency Disturbances on the Noise Immunity of Binary Communication Systems," *Telecommun. J.*, (Switzerland), Engl. ed., vol. 41, no. 12, Dec 1974, pp. 741-744.

Beletski, H.Ya, "Synthesis of Target Azimuth and Detection Algorithms in the Presence of Internal Noise and Non-Gaussian Interference," *Izv VUZ Radioelektron,* vol. 17, no. 12, Dec 1974, pp. 52-5 (in Russian, trans. abstract SA 75-16708).

Bajic, B., "On the Noise Suppression by Means of Correlation Signal Processing," *Electroteknika Zagreb* (Yugoslavia), vol. 18, no. 1, 1975, pp. 18-22 (in Croatian, trans. abstract SA 75-20743).

Pantev, N., and V. Stanchev, "Optimal Filtration in the Presence of Time Correlated Noise," *Tekh. Misul,* Bulgarian, trans. abstract SA 75-3011).

Wolter, H., "Noise Removing and Utilization of Noise or of Other Attendant Exciters as a Means of Increasing Signal Pulses," *Frequenz,* vol. 29, no. 1, Jan 1975, pp. 26-8 (in German, trans. abstract SA 75-20747).

Richter, S.L., "Envelope Threshold Detection in a Class of Non-Gaussian Interference," *IEEE Trans.,* vol. AES-11, no. 1, Jan 1975, pp. 77-85.

Chabdárov, Sh.M., "Optimum Properties of Linear Processing for the Detection of Signals in Certain Types of Noise," *Izv VUZ Radioelektron,* vol. 18, no. 4, Apr 1975, pp. 123, 125 (in Russian, trans. abstract SA 75-29876).

Vasil'yev, K.K. and V.I. Kramuschenko, "The Asymptotic Efficiency of Certain Algorithms of Signal Detection in Noise with Unknown Power," *Radio Engrg. and Eletron. Phys.,* vol. 20, no. 5, May 1975, pp. 134-6.

Abramovich, Yu.I., A.D. Grinyaev, and O.V. Tsyganov, "Detection Characteristics of Signals Masked by Noise and Sidelobes of Strong Target Returns," *Izv VUZ Radioelektron,* vol. 18, no. 5, May 1975, pp. 120-123 (in Russian, trans. abstract SA 75-29877).

Kozolov, A.I., Yu. M. Demidov, and V.B. Ustinovich, "The Polarization Selection of Fluctuating Reflected Signals," *Radio Engrg. and Electron. Phys.,* vol. 20, no. 5, May 1975, pp. 138-9.

Akimov, P.S., V.S. Yefremov, and A.N. Kubasov, "Sequential Detection of Signals Using Statistics Based on a Sum of Ranks," *Radio Engrg. and Electron. Phys.,* vol. 20, no. 6, June 1975, pp. 119-23.

Rodikov, V.Ye. and S.I. Chibosov, "Estimation of the Effectiveness of Two Stage Detection on Non-Coherent Signal when the Detection Stages are Statistically Independent," *Radio Engrg. and Electron. Phys.,* vol. 20, no. 6, Jun 1975, pp. 123-4.

Rodikov, V.Ye., and S.I. Chibosov, "Two Stage Detection of Fast Fluctuation Signals," *Radio Engrg. and Electron. Phys.,* vol. 20, no. 6, Jun 1975, pp. 125-7.

Biletov, M.V., I.S. Vassenshtein, and V.V. Ryl'skiy, "Detection of Radar Signals Subject to Log Normal Intensity Fluctuations," *Telecomm. and Radio Engrg., pt. 2,* vol. 30, no. 6, June 1975, pp. 124-5.

Abramovich, Yu., A.O. Grinyaev, and O.V. Tsyganov, "Efficiency of an Adaptive Threshold Device Operating in Conditions of Pulsed Interference," *Izv Vuz Radioelektron,* vol. 18, no. 8, Aug 1975, pp. 74-9 (in Russian, trans. abstract SA 76-12452).

Fal'ko, A.K., "The Problems of Constructing Optimum Communications Systems in the Presence of Concentrated Interference, Part 1," *Telecomm. Radio Engrg., pt. 2,* vol. 30, no. 9, Sept 1975, pp. 56-62.

Nesteruk, V.F., and N.I. Perfir'eva, "Minimax Signal Reception Problem Under Conditions of Partial Indeterminacy of *A Priori* Probabilities and Loss Matrix," *Telecomm. Radio Engrg., pt. 2,* vol. 30, no. 9, Sept 1975, pp. 62-5.

Gol'dberg, N.I., "Synthesis of the Statistical Multivariate Probability Density Function of a Signal," *Telecomm. Radio Engrg., pt. 2,* vol. 30, no. 9, Sept 1975, pp. 66-70.

Siemens, AG., "Detecting Weak Signals in the Presence of Noise," *UK Patent no. 1, 406,294,* Sept 1975, *West Germany Pat. no. 2148650,* abstract SA 76-17867.

Siemens, AG., "Interference Suppression Circuit for Radar Receivers," *UK Patent no. 1,407,467, West Germany Patent no. 2155074,* 24 Sept 1975, abstract SA 76-21831.

Turin, G.L., "Minimax Strategies for Matched-filter Detection," *IEEE Trans.,* vol. COM-23, no. 11, Nov 1975, pp. 1370-1.

Pluzhnikov, A.D., "Protection for FM Radar from Narrowband interference," *Izv VUZ Radioelektron,* vol. 19, no. 5, May 1976, pp. 76-8 (in Russian, trans. abstract SA 76-39076). (Indicates 80 dB noise suppression in the receiver.)

Kondaurov, M.T., and B.I. Kuz/min, "Advantages of Adaptive Reception of Broadband Signals in Radio Links," *Izv VUZ Radioelektron,* vol. 19, no. 7, July 1976, pp. 134-6 (in Russian, trans. abstract SA 76-48285).

Levin, B.R. and Yu.S. Shinalov, "Jointly Optimum Algorithms of Signal Detection and Parameter Estimation, a Review," *Radio Engrg. and Electron. Phys.,* vol. 22, no. 11, Nov 1977, pp 1-15.

DIGITAL MOVING TARGET INDICATOR (DMTI)

1. Bibliography

Barton, D.K., "International Cumulative Index on Radar Systems," *IEEE Pub. Cat. no. JH 4675-5,* 1978: extensive bibliography on MTI, pp. 37-39. Also see Doppler radar pp. 22-4 and signal processing pp. 64-8.

2. Journal Articles

Hughes Aircraft "Surveillance Radar," *UK Patent no. 1,144,778,* 12 Mar 1969, abstract SA 69-21147.

Cionini, A., and I. Varano, "Analyses of a Digital Processor of Radar Echoes," *ALTA Frequenza,* vol. 38, no. 10, Oct 1969, pp. 758-767 (in Italian, trans. abstract SA 70-14985).

Likharev, V.A., A.S. Kartaskhin, and E.K. Lebedev, "Methods of Ranging and Selection of Moving Targets," *Izv VUZ Radioelek.,* vol. 13, no. 2, 1970, pp. 192-205 (in Russian, trans: JPRS 51136 and SA 70-32294).

"Radar Clutter Reduction Techniques-digital MTI Cancellers," *Radar and Electronics Magazine,* Spring 1970, pp. 15-6.

Nifontov, Yu. A., and V.A. Likharev, "Analysis of a Digital Moving Target Selection System," *Radio Eng. and Electron. Phys.,* vol. 15, no. 7, July 1970, pp. 1207-1214.

Reiss, A., "A Digital MTI Canceller," *AGARD Conf. Proc.,* no. 66 on Advanced Radar Systems, Nov 1970, pp. 27-1 thru (a)-9.

Bradsell, P., "The Plessey Digital MTI System," *AGARD Conf. Proc.,* no. 66 on Advanced Radar Systems, Nov 1970, p. 27(a)ff.

Gold, B., and C.E. Muehe, "Digital Signal Processing for Range-gated Pulse Doppler Radars," *AGARD Conf. Proc.,* no. 66 on Advanced Radar Systems, Nov 1970, pp. 31-1 thru 9.

Eschner, A., "Clutter Rejection in Radar Systems," *UK Patent no. 1224531,* 10 Mar 1971, abstract SA 71-23186.

Linder, R.A., and J.W. Taylor, Jr., "Digital Techniques Advance Tactical Radar," *Westinghouse Engineer,* vol. 31, no. 5, Sept 1971, pp. 147-155.

Gray, G.A., and F.E. Nathanson, "Digital Processing of Radar Signals," *IEEE EASCON Rec.,* 1971, pp. 208-215.

Lerner, V., "Effect of Amplitude Quantization in Digital Systems of Selection of Moving Targets has on the Suppression of Reflections by Local Objects," *Radio Electon. Phys.,* vol. 16, no. 10, Oct 1971, pp. 1753-1756.

Corbin, A.B., and M.P. Toohey, "Digital MTI Processors," *IEEE NAECON Rec. 1973,* pp. 279-87.

Ellis, J.G., "Digital MTI, A New Tool for the Radar User," *Marconi Rev.,* vol. 36, no. 191, 4th Q, 1973, pp. 237-248 (reprinted in this volume).

Ewell, G.W., "Design of Digital MTI Radar Processors," Thesis, abstract SA 75-16710, Georgia Institute of Technology, 1974, Univ. Microfilm No. 74-21 566.

Long, R.J., "Clutter Suppression by Means of Digital MTI as Applied to Precision Approach Radar," *MITRE Tech. Rpt. 2851,* (ESD-TR-74-304), Dec 1974.

Lerner, V. Ye., "Algorithms for Digital Selection of Moving Targets for a Varying Repetition Period," *Radio Engrg. and Electron. Phys.,* vol. 20, no. 4, April 1975, pp. 61-6.

Cartledge, L, and R.M. O'Donnell, "Description and Performance Evaluation of the Moving Target Detector," *Lincoln Lab Rpt. ATC-69,* FAA Rpt RD-190, 8 Mar 1977 (major portion reprinted in this volume).

O'Donnell, R.M., and L. Cartledge, "Comparison of the Performance of the Moving Target Detector and the Radar Video Digitizer," *FAA Rpt. FAA-RD-76-191* (MIT LL Rpt. ATC-20), 26 April 1977.

RADAR DISPLAYS

VanVleck, J.R., and D. Middleton, "A Theoretical Comparison of Visual and Meter Reception of Pulsed Signals in the Presence of Noise," *J. Appl. Phys.,* vol. 17, no. 11, Nov 1946, pp. 940-971.

Orman, L.M., "Pipology," *Coast Artillery J.,* vol. 90, no. 3, May-June 1947, pp. 31-6.

Gerdov. M.M., "Some Peculiarities of the Visual Indication of Radar Signals," *Radio Engr.,* vol. 14, no. 4, Apr 1959, pp. 54-60.

Potter, N.S., "A General Theory of Signal-to-noise Improvement with Application to the Visual Detection of Weak Signals," *IRE Int. Conv. Rec.,* vol. 8, pt. 4, 1960, pp. 167-81.

Pew, R.W., and J.I. Elkind, "Radar Target Identification by Aural Display," *IRE Int. Conv. Rec.,* vol. 9, pt. 5, 1961, pp. 294-301.

Katz, L., "PPI Interference Prediction," *IEEE Trans.,* vol. EMC-7, no. 2, June 1965, pp. 119-124.

Turner, F.M., "A Note on Deflection vs. Intensity-modulated CRT's as anti-jam Radar Displays," *Stanford Research Institute,* Nov 1970; also published as "Deflection vs. Intensity-modulation as Anti-jam Radar Displays," *Electronic Warfare Magazine,* vol. 3, Winter 1971, pp. 6-7.

RECEIVER: GENERAL

1. General Bibliography

Barton, D.K., "International Cumulative Index on Radar Systems," *IEEE Cat. Pub. no. JH-4675-5,* 1978: Receivers, pp. 55-6 (extensive bibliography).

2. Journal Articles

Bell, P.R., and F.M. Ashbrook, "Some General Microwave Anti-jam Design Considerations and Performance of a Special Receiver," *MIT Radiation Lab Div 14 Rpt.,* Feb 1944 DDC, ATI 44070.

Sifonov, V.I., "Development of the Theory of Interference Suppression in the USSR," *Radio Engrg. and Electron. Phys.,* vol. 2, no. 11, 1957, pp. 139-68 (contains 149 references).

Galenian, J.C., "Designing Radar Receivers to Overcome Jamming," *Electronics Magazine,* vol. 36, May 17, 1963, pp. 50-4 (reprinted in this volume).

Babanov, Yu. N., *et al,* "Development of Methods for Protecting Radio Communications Systems from Impulse Noise," *Telecomm. and Radio Engrg., pt. 2,* vol. 22, no. 11, Nov 1967, pp. 82-8 (contains 92 references, mostly to Russian authors).

INTEGRATION

Corum, C.E., "Signal-to-noise Improvement by Pulse Integration," *Nav. Res. Lab. Interim Rpt. 3699,* July 1950.

Cunningham, W.J., J.C. May, and J.G. Skalnik, "Integration Noise Reducer for Radar," *Electronics Magazine,* vol. 23, Sept 1950, pp. 76-78.

Fano, R.M., "On the Signal-to-noise Ratio in Correlation Detectors," *MIT/RLE TR no. 186,* Feb 1951 DDC (ATI 103-043).

Freedman, J., and J. Margolin, "Signal-to-noise Improvement Through Integration in a Delay-line Filter System," *MIT/LL TR no. 22,* 13 May 1953.

Miller, K.S., and R.I. Bernstein, "An Analysis of Coherent Integration and its Application to Signal Detection," *IRE Trans.,* vol. IT-3, no. 4, Dec 1957, pp. 237-248.

Ellis, R.H., and W.C. Rohlman, "Amplitude Modulated Video Integrator," *IRE Int. Conv. Rec.,* vol. 7, pt. 5, 1959, pp. 263-271.

Meinke, H., and K. Rihaczek, "Memory Processes to Improve the Signal-to-noise Ratio of almost Periodic Signals, Particularly in Radar Applications," *NTZ,* vol. 12, no. 4, April 1959, pp. 176-180 (in German, trans. abstract SA 59-6784).

Brookner, E., and J. Flink, "Coherent Enhancer for Pulse Radar Applications," *IRE Int. Conv. Rec.,* vol. 8, pt. 8, 1960, pp. 240-253.

Cooper, D.C., and J.W.R. Griffiths, "Video Integration in Radar and Sonar Systems," *J. Brit. IRE,* vol. 21, no. 5, May 1961, pp. 421-433.

Wilcox, R.C., "Method of Radar Return Integration Using a Dual-gun, Write-read, Recording Storage Tube," *US Patent no. 3,417,335,* 17 Dec 1968.

Trunk, G.V., "Detection Results for Scanning Radars Employing Feedback Integration," *IEEE Trans.,* vol. AES-6, no. 4, July 1970, pp. 522-527.

Minguzzi, B., and G. Picari, "Experimental Analysis of Two Types of Digital Extractors," *Alta Freq.,* vol. 39, no. 11, Nov 1970, pp. 933-942.

Lezin, Yu.S., "Exponentially Weighted Non-coherent Integration of Pulse Signals in the Presence of Gaussian Noise and Random Impulse Noise," *Radiotekhnika,* vol. 26, no. 10, 1971, pp. 31-7 (in Russian, trans. abstract SA 72-36539).

Hansen, V.G., "Post-detection Integration Loss for Logarithmic Detectors," *IEEE Trans.,* vol. AES-8, no. 3, May 1972, pp. 386-388. also 1974, p. 168c.

Glover, K.M., "A Precision Digital Radar Video Integratorm," *Bull Am. Met. Soc.,* vol. 53, no. 7, July 1972, p. 693ff.

Urkowitz, H., "Closed-form Expressions for Non-coherent Radar Integration Gain and Collapsing Loss," *IEEE Trans.,* vol. AES-9, no. 5, Sept 1973, pp. 781-783.

LOGARITHMIC RECEIVER

1. Book

Krylov, G.M., and A.V. Vishiyevskaya, *Planning of Logarithmic Amplifier with Continuous Signal Detection,* Moscow: Energiya, 1970 (in Russian).

2. Journal Articles

Croney, J., "Simple Logarithmic Receiver," *Proc. IRE,* vol. 39, no. 7, July 1951, pp. 807-813.

Rozenstein, S., "Design of a Logarithmic Receiver," *Proc. IEE,* vol. 102-B, Paper 1733-R, Jan 1955, pp. 69-74.

Volkov, V.M., "Transistor Logarithmic Amplifiers," Kiev, *Tekhnika,* 1955 (in Russian).

Weiss, M., and S.R. Sixbey, "Target Discrimator for Countermeasures," *Electronics Magazine,* vol. 28, no. 8, Aug 1955, pp. 118-120.

Croney, J., "Clutter on Radar Displays-reduction by Logarithmic Receivers," *Wireless Engr.,* vol. 33, no. 4, April 1956, pp. 83-96.

Vilesov, L.D., E.D. Lapchik, A.P. Lukoshkin, Yu. Ye. Monakhov, and A.A. Chude, "The Use of a Logarithmic Amplifier in Designing an Alternate-period Compensator for Passive Jamming (Chaff)," *Trudy Leningradskiy Institut Aviatsionnoqo Privorostroyeniya,* no. 55, 1958, pp. 208-12 (in Russian); Trans DDC AD A051472).

Green, B.A., Jr., "Radar Detection Probability with Logarithmic Detectors," *IRE Trans.,* vol. IT-4, no. 1, March 1958, pp. 50-52.

Hansen, V.G., "Studies of Logarithmic Radar Receiver Using Pulse-length Discrimination," *IEEE Trans.,* vol. AES-1, no. 3, Dec 1965, pp. 246-53.

Cope, R.W., "Logarithmic Receiver Detects Amplitude Modulation Linearly," *Microwave Journal,* Nov 1967, pp. 94-95.

Brown, A.J., and A. Haeberle, "Reduce Radar Display Clutter," *Electronic Design Magazine,* vol. 17, no. 6, Mar 1969, pp. 204-207.

Bykov, V.V., "Effect of External Noise on the Characteristics of an Automatic Range Finder with a Logarithmic Receiver," *Telecomm. Radio Engr., pt. 2,* vol. 25, no. 12, Dec 1970, pp. 102-107.

Roberts, J.A., "Logarithmic Detection with Post Detection Filtering," *IEEE Trans.,* vol. AES-14, no. 1, Jan 1978, pp. 35-43.

RECEIVER: FRONT END

Babanov, Yu. N., "Analysis of Signal Distortion in Special Converter Stages of a Highly Selective Receiving Equipment," *Telecomm. Radio Engrg., pt. 2,* vol. 18, no. 7, Jul 1963, pp. 54-7.

Reynolds, J., "Intermodulation in Transistors," *IEEE Trans.,* vol. VC-2, no. 1, Sept 1963, pp. 88-92.

Nitzberg, R., "Spurious Frequency Rejections," *IEEE Trans.,* vol. EMC-6, no. 1, Jan 1964, pp. 33-36.

Bridges, J.E., and R.A. Zalewski, "Orthogonal Detection to Reduce Common Channel Interference," *Proc. IEEE,* vol. 52, no. 9, Sept 1964, pp. 1022-1028.

Hansen, F., "Desensitization in Transistorized Receivers," *Proc. 18th Annual Conf. of IEEE Vehicular Technology,* New York, 1967, pp. 78-86 (reprinted in this volume).

Merriaman, R.H., "Quadruple-balance Mixer," *US Patent no. 3,515,993,* June 2, 1970.

Muzzi, A.G., "Amplitude Modulated Intensifier Method and Apparatus," *US Patent no. 3,628,155,* Dec 14, 1971.

Moore, R., "Designing Communications Receivers for Good Strong-signal Performance," *Ham Radio Magazine,* Feb 1973, pp. 6-15.

Maksimov, M.V., "Effect of Noise on Single-balance and Double-balance Converters in Continuous-wave Radar Stations," *Telecomm. Radio Engrg., pt. 2,* vol. 28, no. 5, May 1973, pp. 69-73.

RECEIVERS: VARIOUS ECCMs

1. Books and General Bibliographies

Lawson, J.L., and G.E. Uhlenbeck, *Threshold Signals,* Rad Lab Series, vol. 24, McGraw-Hill, 1950 (Chapter 12—Threshold signals in electronic interference; IAGC pp. 302-6, 345-6, log amp pp. 306-8, pulse shape descrim. p. 363, aural detect pp. 252-6, CW rejection filters pp. 346-7. Mouse under rug p. 248, STC p. 311, FTC p. 313, DBB p. 302, sidelobe blanker p. 360).

Barton, D.K., and H.R. Ward, *Handbook of Radar Measurement,* Prentice Hall, 1969 (Monopulse ECCM pp. 215-216, conical scan ECCMs p. 221, search radar azimuth errors p. 226, range measurement errors pp. 234-5, Doppler measurement errors, p. 238).

Barton, D.K., "International Cumulative Index on Radar Systems," *IEEE Pub. Cat. no. JH 4675-5,* 1978: extensive bibliography on Navigation, pp. 39-42 (some items have ECM/ECCM implications).

2. Journal Articles

Stone, A.M., and J.L. Lawson, "Infinite Rejection Filters," *J. Appl. Phys.,* vol. 18, 1947, pp. 691ff.

Sivaswamy, R., and R. Ramasubba, "Pulse Length Discriminator," *J. Inst. Telecomm. Engrs.* (India), vol. 11, no. 11, Nov 1965, pp. 392-5.

Picardi, F., and R. Ragonese, "A Delay Line Anti-Interference Circuit for Radar Systems," *Alta Freq.*, vol. 37, no. 12, Dec 1968, pp. 1147-53 (in Italian, trans. abstract SA 69-11692).

Blythe, J.H., "Separation of Radars on Common Frequencies by Pulse Repetition Frequency Discrimination," *G.E.C. J. Sci. and Technol.*, vol. 37, no. 4, 1970, pp. 157-162.

Halpern, H.M., and R.P. Perry, "Digital Matched Filters Using Fast Fourier Transform Techniques," *Proc. IEEE*, EASCON, 1971, pp. 222-30 (contains discussion of Kirbar ECCM fix.

Siemens, AG., "Radar Range Measuring Device," *UK Patent No. 1377032*, Dec. 27, 1972, abstract SA 75-21122.

Schmidt, R.O., "On Separating Interleaved Pulse Trains," *IEEE Trans.*, vol. AES-10, no. 1, Jan 1974, pp. 162-6.

TRACKING RECEIVER

1. Books

Vinitskiy, A.S., *Modulated Filters and FM Tracking Reception*, Moscow: Soviet Radio, 1969 (in Russian).

2. Journal Articles

Zelenin, A.P., "Theory of Tracking-tuned FM Reception," *Telecomm. Radio Engrg., pt. 2*, vol. 18, no. 12, Dec 1963, pp. 11-17.

Kantor, L. Ya., "The Effect of Fluctuation Noise on a Tracking FM Receiver," *Telecomm. Radio Engrg., pt. 1*, vol. 19, no. 4, Apr 1965, pp. 19-26.

Afanas'yev, Yu. A., and L. Ya Kantor, "Control Circuit Compensation in an FM Receiver with a Tracking Filter," *Telecomm. Radio Engrg., pt. 2*, vol. 21, no. 5, May 1966, pp. 87-93.

Belous, A.V., "Application of the Method of Equivalent Nonlinear Systems to the Analysis of the Noise Immunity of FM Tracking Receivers," *Telecomm. Radio Engrg., pt. 2*, vol. 28, no. 5, May 1973, pp. 73-78.

Jamuar, S.S. and S.K. Mullick, "Noise-free Analysis of a Tracking Filter FM Demodulator," *IEEE Trans.*, vol. AES-15, no. 1, Jan 1979, pp. 58-66.

Naval Radar Anti-Jamming Techniques
R.V. Alred
Proceedings of the IEE. Vol. 93m, Part IIIa, No. 10, 1946
Pages 1593-1601

6.1

SUMMARY

The use of radar in the 1939–45 War was quickly followed by the development of methods of "jamming" radar equipments in order to reduce their operational value. Specimen photographs are shown in the paper of the appearances of various types of jamming signal on radar displays.

Investigations of the effects of jamming led to the establishment of design practices which resulted in appreciable reduction of the general vulnerability of equipments to jamming; details are given of these practices, and also of a number of naval anti-jamming circuits developed for dealing with specific types of jamming.

(1) INTRODUCTION

A light cruiser may have as many as ten radar equipments, and a battleship as many as twenty-five, for the control of the ship's armament, and for warning of the approach of enemy aircraft or surface vessels; similar widespread use of radar is made by the other fighting services. It is not surprising, therefore, that much effort has been expended in attempts to reduce the effectiveness of radar by means of "jamming"; this may be defined as the deliberate radiation or re-radiation of electromagnetic waves with the object of preventing or impairing the reception of radio or radar signals.

Radar jamming can conveniently be divided into two categories—"electronic" and "non-electronic." Electronic jamming signals are modulated electromagnetic waves produced by a radio transmitter; amplitude-modulated transmitters have been used, with modulation frequencies as large as 1 Mc/s, and frequency-modulated transmitters, with sweep frequencies of 100 c/s and deviations of $\pm$ 5 Mc/s. Such jamming affects only those radar equipments for which the jamming carrier or side-bands occur within the radar receiver acceptance band. The strongest jamming signals are received when the radar aerial is directed at the jammer; with the aerial on other bearings, the strength of the jamming signal is dependent on the radiation pattern of the radar aerial. The jamming signals are generally emitted continuously, and tend to obscure echo-signals from targets at all ranges from the radar equipment. If a target-signal can be detected in the jamming pattern, the apparent range of the target, obtained in the normal manner, is nearly as accurate as would be obtained in the absence of jamming; the apparent bearing of the target, however, may be unreliable if the jammer is on a different bearing to the target.

Non-electronic jamming is produced by the use of large numbers of reflecting objects, which give corresponding echo-signals, confusing the radar displays; if the echo-signals occur at the same range and bearing as a target-signal, and are of greater amplitude, the latter is obscured. The most common reflectors are strips of aluminium foil or of graphite-coated paper, known as "window," which are cut approximately to a length resonant at the frequency to be jammed; they are generally dropped in self-opening packets, which may contain as many as 10 000 strips. Such strips are effective within a frequency band of about $\pm$ 5% either side of the frequency at which they resonate; long streamers of foil, known as "rope," affect equipments of frequencies within a larger band.

It is obvious that the principal anti-jamming measure is the spread of the frequencies of equipments over the widest possible range, in order to reduce the number of equipments which can be affected with a limited jamming effort. Not only should different equipments be designed to operate on widely-separated frequencies, but also the frequencies of equipments of the same type (i.e. designed for the same nominal frequency), should be spread over the largest range permitted by the electromagnetic characteristics of the components. Since changes in these characteristics, with variation of frequency, generally occur at a rate proportional to the nominal design frequency, the higher this frequency, the larger, in Mc/s, is the possible spread of frequencies of equipments of the same type.

Although the spread of frequencies reduces the probability of any given radar equipment and jammer being of the same frequency, it does not prevent an equipment receiving jamming which is of the same frequency; hence it is necessary that equipments be designed so that they are affected as little as possible by jamming which enters the receiver, and that if possible, means be provided for removing the jamming signals with the smallest possible loss of target-echo signals. These aspects are discussed in Sections 3 and 4 of the paper.

(2) VISUAL EFFECTS OF JAMMING ON RADAR DISPLAYS

(2.1) Radar Displays

Present-day radar displays can be divided into two basic types, deflection-modulation displays and intensity-modulation displays. With deflection-modulation displays, the cathode-ray-tube (c.r.t.) trace occurs substantially in the same position for each radar pulse transmission; distances along the trace are related to the ranges of possible targets, and the target-echo signals appear as deflections normal to the trace. The simplest type of such a display is one in which the distance along the trace and the target-range are linearly related; a photograph of such a display is shown in Fig. 1(*a*). The length of the trace shown corresponds to a target-range of 35 miles, and in addition to local echo-signals extending to about 4 miles, discrete echo-signals are visible at about 10, 12 and 30 miles respectively. Such a display is generally used with a radar equipment having an aerial beam which is not continually changing in direction.

When the aerial beam scans in bearing or elevation—or in both—intensity-modulation displays are more frequently used. The co-ordinates of the c.r.t. screen are made to correspond to two of the three position co-ordinates—range, bearing and elevation—the trace being positioned on the screen accordingly; as the aerial beam scans, so the position of the trace on the screen varies. Target-signals produce increased intensity of the C.R.T. trace at the corresponding points on the screen, and by using a screen with an afterglow, a planar presentation is obtained of the positions of targets in the scanned volume of space. The simplest example of such a display is a Plan Position Indication (P.P.I.) display, in which the c.r.t. trace rotates in synchronism with the radar aerial beam, distances along the trace from the centre of the screen corresponding linearly with target-ranges. A photograph

* Radio Section paper. † Admiralty Signal Establishment.

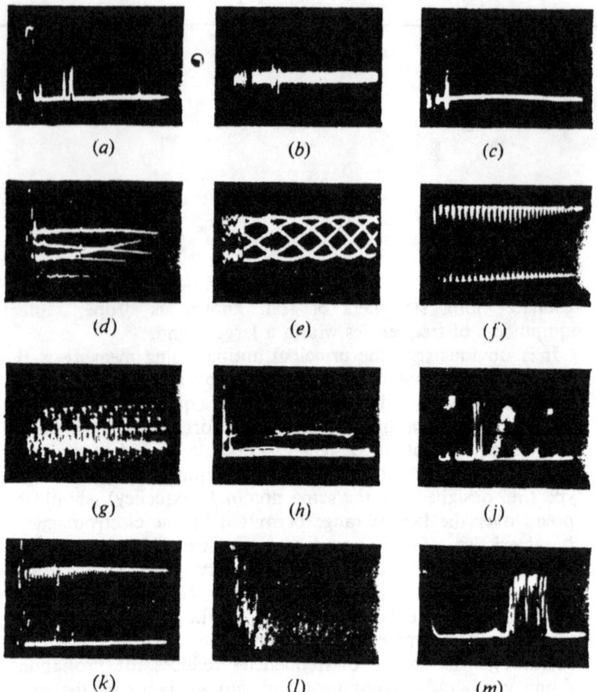

Fig. 1.—Appearances of various types of jamming on a deflection-modulation display. (The ranges represented and the target-signals are not the same for all the photographs.)

(*a*) No jamming. (*b*) C.W., weak. (*c*) C.W., very strong. (*d*) A.M.C.W., 400 c/s. (*e*) A.M.C.W., 4 kc/s. (*f*) A.M.C.W., 100 kc/s. (*g*) A.M.C.W., 20 kc/s + 100 kc/s. (*h*) Speech-modulation. (*j*) F.M.C.W., ± 5 Mc/s at 100 c/s. (*k*) Pulse-modulation, 100 kc/s. (*l*) Noise-modulation. (*m*) Window.

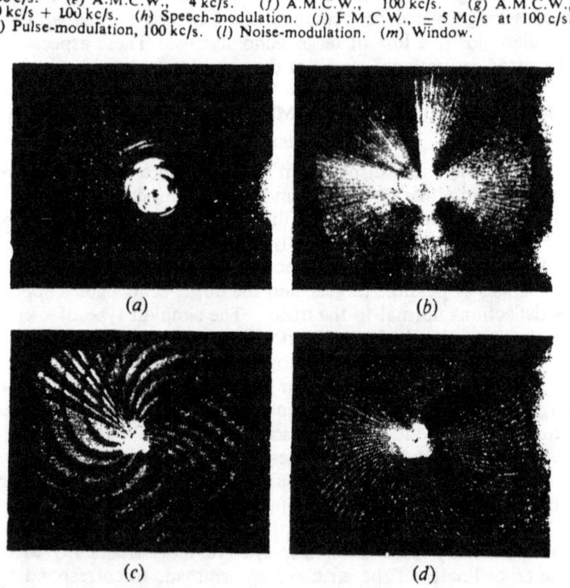

Fig. 2.—Appearances of jamming on an intensity-modulation display. (The ranges represented and the target-signals are the same for all the photographs.)

(*a*) No jamming. (*b*) A.M.C.W., 4 kc/s. (*c*) and (*d*) A.M.C.W., of frequency about 4 kc/s and nearly a multiple of the radar pulse-repetition frequency.

of such a display is shown in Fig. 2(*a*); the radius of the picture corresponds to a range of 100 miles, and in addition to local echo-signals, two discrete echo-signals can be seen at ranges of 45 and 50 miles respectively. Each echo-signal extends over an arc of about 35° corresponding to the beam-width of the radar aerial radiation pattern.

Most radar equipments are operated at the limit of sensitivity which is set by receiver "noise," such noise causing appropriate variations of the trace. These variations produce an irregular baseline on a deflection-modulation display, and a mottled background on an intensity-modulation display; these effects can be seen in Figs. 1(*a*) and 2(*a*).

(2.2) Appearances on Deflection-Modulation Displays

The general appearances of electronic jamming on a deflection-modulation display depend on the strength of the jamming. Weak jamming produces an apparent increase in the noise signals on the display, but in general target-signals are still clearly visible. With jamming of moderate strength, each trace is deflected in correspondence with the modulation of the jamming; and, owing to the persistence of vision, a number of traces appear to be superimposed, producing continually-changing patterns on the screen. If the jamming modulation is simple, target-signals can generally be detected superimposed on the jamming patterns; if the modulation is complex, however, the resulting patterns on the screen may obscure all target-signals. Very strong jamming may produce the complete suppression, as distinct from obscuration, of all target-signals and noise, so that the traces appear blank.

Photographs of various types of jamming on a deflection-modulation display associated with a radar equipment of 500 c/s pulse-repetition frequency are shown in Fig. 1. Still photographs do not adequately represent the picture as seen by a radar operator, since in operation the jamming patterns are continually changing and jumping about; nevertheless, the photographs give some indication of the visual effects of jamming. In each photograph, distances along the trace correspond approximately linearly with target-ranges, but the total ranges represented, and the target-signals, are not the same for all the photographs. Similar effects are obtained on other deflection-modulation displays in which distances along the trace and target-ranges are not related linearly, the patterns being distorted accordingly in the transverse direction.

Fig. 1(*a*) shows the display in the absence of jamming, target-signals being visible, as already stated, at distances along the trace corresponding to ranges of 10, 12 and 30 miles respectively. Fig. 1(*b*) shows the signals from the same targets in the presence of weak c.w. jamming; there is a small increase in noise level, and the target-signals show both above and below the trace. With an increased strength of jamming, the appearance on the display is very similar, the amplitudes of target-signals and noise being reduced owing to overloading of the receiving equipment; with very strong jamming, the trace appears as in Fig. 1(*c*). The appearance of target-signals both above and below the trace is due to the fact that, with successive radar pulse transmissions, the target-echo signals may be varying amounts in or out of phase with the jamming signals received at the same times; on rectification, addition or subtraction of the signals occurs accordingly. After rectification, the d.c. component corresponding to the c.w. jamming signal is removed by a condenser-coupling between the receiver and display, with the result that target-signals appear both positive-going and negative-going with respect to the trace.

With amplitude-modulated continuous-wave (a.m.c.w.) jamming, each trace is deflected and distorted in correspondence with the modulation of the jamming; the larger the modulation frequency, the greater is the distortion. Owing to the persistence of vision, a number of traces appear superimposed, and with a jamming modulation frequency of 400 c/s, a tram-line effect is produced, as shown in Fig. 1(*d*); with a modulation frequency of 4 kc/s a criss-cross pattern is obtained, as shown in Fig. 1(*e*). On both of these photographs, some target-signals can be seen superimposed on the jamming patterns. Jamming of 100-kc/s modulation frequency gives the appearance of a set of railings, as shown in Fig. 1(*f*), and jamming modulated at both 20 and 100 kc/s may give the complex appearance of Fig. 1(*g*); detection of target-signals in the presence of jamming of such high modulation frequencies is normally very difficult.

With speech-modulated jamming from a communications transmitter, the trace appears to jump about in an erratic manner, in correspondence with the speech, producing an effect such as is shown in Fig. 1(*h*); target-signals can often be seen superimposed on the jamming, or momentarily during pauses in the speech, as deflections of the trace.

The appearance of frequency-modulated jamming depends on the magnitude of the deviation of the frequency sweep. If the deviation is smaller than the band-width of the receiver, the effects of the jamming are similar to those of amplitude-modulated jamming. If the deviation is larger than the band-width of the receiver, the appearance is that of a number of humps, as shown in Fig. 1(*j*), moving across the screen; the humps occur as the jamming carrier sweeps through the acceptance band of the receiver. The latter type of jamming is a distraction, but should not, in general, prevent the observation of target-signals.

If pulse-modulated jamming is used, the pulses being widely spaced, the effect is very similar to that of large-deviation frequency-modulated jamming, already shown in Fig. 1(*j*). If the pulses are closely spaced, they give a double-trace effect, as shown in Fig. 1(*k*); with suitable design of the receiver, target-signals can generally be seen on the baseline, and also superimposed on the peaks of the jamming signals, if these are not of sufficient strength to saturate the receiver.

Noise-modulated jamming may be obtained by suitably amplifying the output of a diode, and applying the amplified signals to the modulating valve of the jammer. With such jamming, there is an apparent increase in noise amplitude on the display with no corresponding increases in target-signal amplitudes, as is illustrated in Fig. 1(*l*). Only those target-signals which are of amplitude greater than the jamming signals can be seen.

Window jamming produces a number of echo-signals on the trace as shown in Fig. 1(*m*); as the window is scattered by the wind, the corresponding echo-signals gradually spread in range and bearing, and decrease in amplitude. The time from appearance to disappearance of such echo-signals is generally between five and thirty minutes, depending on the height from which the window is dropped.

(2.3) Appearances on Intensity-Modulation Displays

On intensity-modulation displays, the appearances of most types of electronic jamming are very similar. This is due to the fact that the maximum intensity of the trace on such a display generally corresponds to a signal strength of only a few times the mean amplitude of the receiver noise; consequently with jamming of any amplitude greater than this, and of any complex modulation, each trace is practically of uniform intensity. As the aerial beam scans, the intensities of successive traces vary in accordance with the radiation pattern of the radar aerial, producing a corresponding pattern on the c.r.t. screen; for a P.P.I. display the pattern is star-shaped. When the jamming is of simple modulation, each trace is modulated in intensity at the jamming modulation frequency, and if this frequency is a multiple, or nearly a multiple, of the radar pulse-repetition frequency, a fingerprint appearance is superimposed on the previous pattern.

Fig. 2(*a*) shows the P.P.I. picture of a radar equipment of 35° beam-width, the echo-signals being from land in the neighbourhood of the equipment. Fig. 2(*b*) shows the corresponding picture in the presence of amplitude-modulated jamming of about 4-kc/s modulation frequency, the modulation frequency not being a multiple of the radar pulse-repetition frequency. The bearing of the jammer is about 310°, and the jamming causes increased intensity of the traces in a number of sectors; target-signals are completely obscured over the arc 270° to 360°, though over the arc 360° to 10° the target-signals present in Fig. 2(*a*) can just be discerned through the jamming. The complete suppression of all signals in the arc 345° to 360° was found to be due to overloading of the P.P.I. power supplies, the time-constants of which caused the visible effects to occur after the radar aerial beam had passed through the bearing of the jammer.

When the jamming modulation frequency is nearly a multiple of the radar pulse-repetition frequency, the pattern of Fig. 2(*b*) is modulated as already explained, having an appearance such as that shown in Fig. 2(*c*), or in Fig. 2(*d*); these photographs were obtained with jamming of approximately the same modulation frequency as that shown in Fig. 2(*b*). With jamming of larger modulation frequency than 4 kc/s, the lines of the fingerprint pattern, measured radially, are more narrow and more closely spaced.

Window jamming is more effective on intensity-modulation displays than on deflection-modulation displays, owing to the small signal amplitude required to give full intensity of the trace on the former displays; provided the jamming signals are of greater amplitude than this, a target-signal is obscured whatever its strength.

(3) ANTI-JAMMING DESIGN PRACTICES

(3.1) To Combat Off-frequency Jamming

As already mentioned, the principal anti-jamming measure is the spreading of the operating frequencies of equipments over the widest possible range. Towards the end of the 1939–45 War, naval radar equipments of various types operated on frequencies between 40 and 10 000 Mc/s; also the frequencies of equipments of the same types were spread over bands of the order of $\pm$ 2% of the nominal operating frequencies.

In addition to reducing the probability of experiencing jamming, such a large spread of frequencies gave a greater probability that any jamming encountered would not be exactly of the same frequency as any of the radar equipments, unless the enemy maintained an efficient monitoring system. In order that full advantage could be taken of any such difference of frequencies, the pass-bands of the later naval radar receivers were made the minimum width necessary for satisfactory target-signal reproduction. For a warning equipment the band-width in Mc/s, at -3 db, was made equal to the reciprocal of the transmitter pulse-duration in micro-seconds; for a gunnery equipment, in which greater accuracy of range-measurement was required, the band-width was made equal to twice this amount. Moreover, since jamming more than 40 db stronger than a target-echo was to be expected, the band-widths of receivers, at -40 db, were made as small as possible, being less than three times the corresponding widths at -3 db. To illustrate this, the frequency-response curves of an early and a modern radar receiver, both for use with a 2-microsec transmitter pulse, are shown in Fig. 3.

With such receivers, considerable reduction of the effects of jamming was obtained when the jamming was not exactly of the same frequency as the radar equipment; the improvement could be increased by slight change of local oscillator frequency, but the latter entailed some reduction in the amplitude of a received echo-signal if the transmitter was not correspondingly re-tuned.

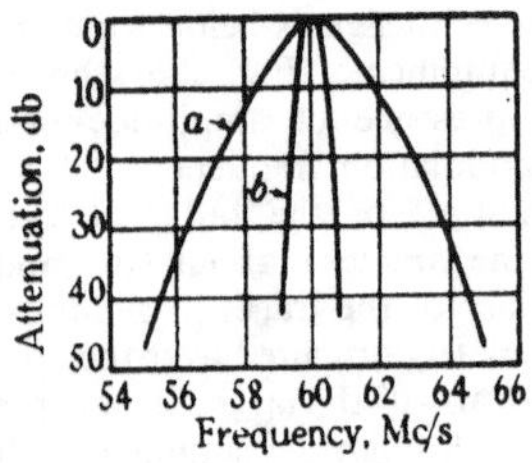

Fig. 3.—Frequency responses of radar receivers.
(*a*) Early receiver. (*b*) Modern receiver.

(3.2) **To Combat Amplitude-Modulated Jamming**

As jamming of the same frequency as the radar equipment would be accepted by a radar receiver, it was necessary to prevent such jamming overloading the receiver and display equipment, with consequent loss of target-signals. The principal causes of such overloading have been found to be:—

(*a*) The biasing back of valves owing to grid-current flow through high-impedance grid-leaks.

(*b*) The use of suppressor bias on the later i.f. valves.

(*c*) The use of a negative-going second detector with d.c. coupling to the succeeding cathode-follower or v.f. amplifying valve.

(*d*) The reduction of anode and screen voltages owing to increased currents through decoupling circuits.

(*e*) Changes in the supply and bias voltages owing to the increased currents required (c.w. jamming caused an increase from 50 to 120 mA in the h.t. current of a six-stage i.f. amplifier).

Such effects can be considerably reduced by careful design of receivers, and by the provision of adequate power supplies. Modifications were made to the early naval receivers for this purpose, and all new receivers are specified to accept amplitude-modulated jamming, of, say, 400-c/s modulation frequency, 30% depth, and of strength 40 db above the mean noise power, without causing the disappearance, on a deflection-modulation display, of a target-signal of strength equal to the mean noise power; the echo is detectable superimposed on the jamming as already shown in Fig. 1(*d*). For this measurement, it is presumed that no anti-jamming circuit is included specifically to remove such jamming.

A number of devices have been developed for reducing any remaining overloading, which is generally due to valve saturation. One method is a "back-biasing" circuit, the principle of which* is to provide automatically an additional bias to one or more i.f. valves to off-set any increase in the mean amplitude of the input signals, such as is caused by a continuous jamming signal; any target-signal modulation then occurs on the normal grid-voltage range of the valve and is amplified in the normal manner. A variety of circuits of this type have been developed by the different services; details of the circuit used in the early naval 3 000-Mc/s radar receivers, developed by H. E. Kettle, are given in Section 4.1.

Back-biasing circuits require an increase in mean valve current to provide the added bias, and the valve must normally be operated on the non-linear portion of its characteristic; a compromise between stage-gain and efficiency in biasing-back has therefore to be made. In addition, circuits have to be chosen with time-constants such that the biasing voltage follows the highest possible jamming modulation frequency, yet does not come into operation on a target-signal. Owing to these conflicting requirements, jamming of modulation frequencies less than a few kilocycles per second only, can be satisfactorily removed, although with jamming of higher modulation frequencies, some improvement is obtained in the visibility of target-signals against the jamming pattern on a deflection-modulation display.

An alternative to the use of back-biasing circuits is a system in which the receiver gain is reduced with consequent increase in the strength of the jamming signal necessary to cause overloading, the jamming modulation frequencies are removed after detection by means of a video-frequency filter, and the overall gain is restored to normal by an additional video-frequency amplifier. Such a system, proposed and developed by A. W. Ross, superseded back-biasing in later naval radar equipments; by its use, higher modulation frequencies can be removed with greater efficiency than by the use of back-biasing circuits. Circuit and performance details of one of the filter units are given in Section 4.2.

(3.3) **To Combat Pulse-Modulated Jamming**

With pulse-modulated jamming, or with wide-sweep frequency-modulated jamming, a number of humps are produced on the c.r.t. screen; if the repetition frequency of the pulses is a multiple of the radar pulse-repetition frequency, the humps appear stationary on the screen, but in general they appear to be moving across it. Target-signals and noise can be seen on the base-line during the gaps in the jamming, provided the amplification of the receiver is not reduced by the presence of the jamming signals.

Such reduction of amplification, persisting during gaps in the jamming, is known as paralysis, and has been found to be caused principally by circuits of long time-constant associated with the amplifying valves, or with the power supplies of the equipment. It can be reduced considerably by careful attention to details of design; by small modifications to early naval receivers, improvements of 30 db in performance against pulse-modulated jamming were obtained; the photographs of Fig. 4 illustrate this. New naval receivers are specified to accept unity mark/space pulses of duration five times the duration of the radar pulse, and of strength 60 db above the noise power, without loss on deflection-modulation displays of target-signals of strength equal to the mean noise power; the echo is observed on the base-line, through the jamming, as in Figs. 1(*k*) and 4(*b*). Many receivers meet the specification for all strengths of such jamming.

(*a*)

(*b*)

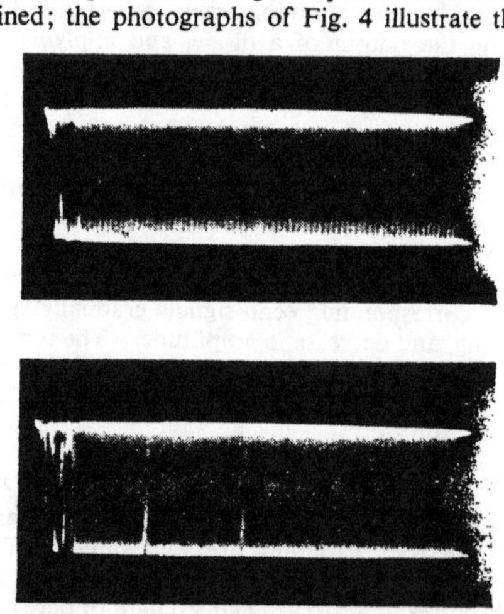

Fig. 4.—Improvement in target-signal visibility obtained by removing the causes of paralysis. Pulse jamming, 100 kc/s, 60 db above noise power.
(*a*) Original receiver. (*b*) Modified receiver.

In the same way that observation through a resolving shutter is

* Blumlein, A. D.: "Improvements in or relating to apparatus for the reception of modulated oscillations": British Patent Specification No. 587562.

improved by increasing the size of the open sectors, the observation of target-signals through pulse jamming is improved on intensity-modulation displays by reducing the duration of the jamming signals at video frequency by differentiation. Condenser-resistance circuits were used originally, but a delay-line differentiation unit was developed for later naval equipments. The principle of this* is to use a delay-line giving a delay of about half the duration of the radar pulse, with subtraction of the delayed from the undelayed signals, so that all jamming signals are thereby reduced in duration with little reduction in the amplitude of target-signals. Circuit details of the unit are given in Section 4.3; delay-line differentiation has been found preferable to condenser-resistance differentiation, particularly for radar equipments of pulse-duration smaller than one microsecond.

(3.4) To Combat Noise-Modulated Jamming

The problem of obtaining improvement in the detection of a signal in noise-modulated jamming is the same as that of obtaining improvement in the detection of a correspondingly smaller signal in receiver noise. Target-signals of amplitude greater than the jamming may be lost, however, owing to saturation of the C.R.T. screen by the jamming; in order to prevent this, an optional automatic gain-control system was developed, to maintain the jamming signal voltage at the display substantially the same as the receiver noise voltage in the absence of jamming. The circuit and performance details of this system are given in Section 4.4.

(3.5) To Combat Non-Electronic Jamming

In order to obscure echo-signals from targets in a large area by means of non-electronic jamming, widespread scattering of reflectors is required; since each reflector gives a returned echo-signal of duration equal to the radar pulse-duration, the effectiveness of the jamming can be reduced by reduction of the radar pulse-duration. An optional short pulse was provided in most naval radar equipments for this purpose. Further, as the reflectors scatter in the wind, a resulting composite echo-signal has a more gradual rise and decay than the echo-signal from a sharply-defined object; advantage can be taken of this in order to obtain improved visibility of target-signals in window jamming signals, by the use of a differentiation circuit with a time-constant approximately equal to the radar pulse-duration.

(4) ANTI-JAMMING CIRCUITS

(4.1) Back-Biasing Circuit

The principle on which back-biasing circuits are based is the operation of one or more i.f. valves on the curved portion of their characteristics, so that an increase in the mean amplitude of the input signals, such as is caused by a continuous jamming signal, gives corresponding increases in the mean valve currents; these increases are used to provide added bias to the valves. Overloading of the valves is thereby reduced, and also an improvement in the ratio of target signal to jamming signal is obtained, as is illustrated in Fig. 5. Fig. 5(*a*) shows the operating conditions of each i.f. valve for a normal target-signal, and Fig. 5(*b*) shows the corresponding operating conditions in the presence of a c.w. jamming signal, in phase with the target-signal. In addition to the improvement in the ratio of signal-strength to jamming-strength given by the use of the circuit, it may also appear from Fig. 5 that there should be increased target-signal amplification in the presence of jamming, since the modu-

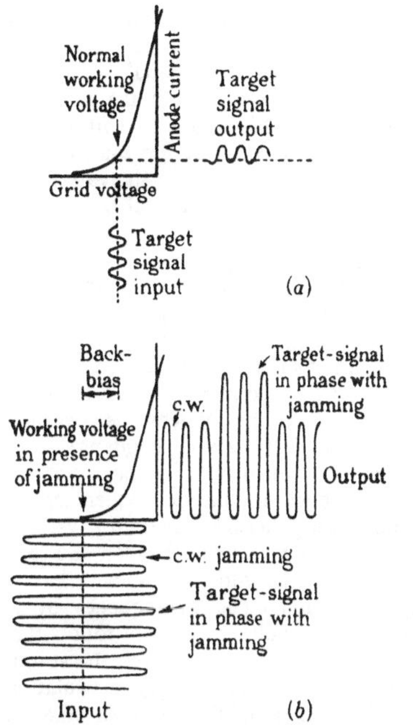

Fig. 5.—Reduction of jamming signals by back-biasing.
(*a*) Amplification of small target-signal alone.
(*b*) Amplification of small target-signal and jamming.

lation of the jamming envelope by the target-signal occurs further up the valve characteristic. In general this is not so, however, since only the fundamental components of the bursts of current are passed to the following valve by the tuned circuit.

Details of the back-biasing circuit fitted in the early naval 3 000-Mc/s radar equipments, are shown in Fig. 6; circuits were

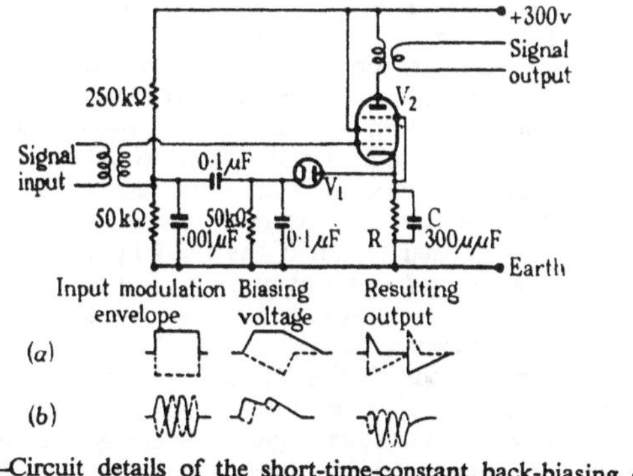

Fig. 6.—Circuit details of the short-time-constant back-biasing circuit, and its operation on the target-signal modulation-envelope for (*a*) on-frequency jamming, and (*b*) off-frequency jamming.

permanently incorporated in the last three i.f. stages (at 60 Mc/s) of the receivers. The principal biasing action on the i.f. amplifying valve V_2 is obtained from the cathode-resistor R, the cathode-condenser C being necessary to prevent the back-biasing action occurring on a target-signal. The remaining circuit components are included to reduce paralysis effects after a large ground-wave signal, as described later.

In order to obtain good biasing efficiency, resistor R is required to have a large value, 10 000 ohms being used, and in order that

* Parker, E., and Keyston, J. E.: "Improvements in and relating to gating circuits for the reception of pulse signals": British Patent Application No. 23362/46.

the stage-gain is not appreciably reduced from normal, the grid-voltage of the valve is raised to + 50 volts; the mean current of the valve in the absence of a signal is therefore about 5 mA, the operating point being just on the straight portion of the valve characteristic. This means that little back-biasing occurs on small jamming signals, but the biasing efficiency increases with the strength of the jamming; increased visibility of target-signals results, but complete removal of jamming signals is not achieved.

The cathode-condenser C, required to prevent the back-biasing action from occurring on a target-signal, limits the highest jamming modulation frequency which can be followed by the biasing voltage. For a 0·5-microsec transmitter pulse, assuming the impedance of the valve to be about 200 ohms, this ideally requires a condenser valve of 0·0025 μF; the discharge time-constant of the condenser is governed by the cathode resistance, and with the above condenser, the circuit would follow a jamming modulation frequency of about 50 kc/s.

In order that higher modulation frequencies could be followed, a smaller condenser, 300 $\mu\mu$F was used. With this value, back-biasing occurs on a target-signal when the jamming is of the same frequency as the radar equipment, as is illustrated in Fig. 6(*a*); if the jamming is off-frequency by such an amount that the back-biasing voltage does not follow the beat-frequency of jamming and target-signal, however, the voltage approximates to the peaks of the beat-frequency modulation, as shown in Fig. 6(*b*), and normal amplification then takes place. With the 300-$\mu\mu$F condenser, the back-biasing circuit follows jamming of modulation frequencies less than 400 kc/s; on deflection-modulation displays it gives improvements in the ratio of target-signal to jamming of about 3 db if the jamming is of the same frequency as the radar equipment, and improvements of about 12 db if the jamming is 1 Mc/s or more off-frequency. There is little improvement in the visibility of target-signals on intensity-modulation displays, however, since the jamming signals are not completely removed.

One trouble experienced with all back-biasing circuits is that they tend to come into operation in the presence of a strong signal, such as the ground-wave; the increased bias persists after the strong signal ceases, and causes temporarily reduced amplification, or paralysis, for a time which may be equivalent to 5 000 yd of target-range. In order to reduce the duration of this, the diode V_1, with a circuit of long time-constant in its cathode, was introduced, its action being two-fold. Firstly, in the presence of a strong signal, there is increased cathode current, and the cathode voltage of valve V_2 rises, with about 0·06-microsec time-constant, until it has risen by about 10 V; the diode V_1 then passes nearly as much current as valve V_2, and further rise of voltage is governed by the cathode circuit of the diode V_1 (50-microsec time-constant). Thus further rise of voltage of the cathode of valve V_2 is severely limited. Secondly, when the strong signal ceases, valve V_2 is non-conducting owing to its increased cathode voltage, giving the paralysis already mentioned. The increased bias voltage is still across the diode V_1, however, and therefore assists the discharge of the condenser C, and the return of the cathode voltage of valve V_2 to normal.

Because of this diode-limiting, valve V_2 passes grid-current with strong jamming signals, and when this happens, a second back-biasing action takes place in the grid potential-dividing chain. This is of about 50-microsec time-constant, and the grid biases back to some mean voltage dependent on the strength of the jamming, the short-time-constant cathode circuit then following the fluctuations about the mean level. The cathode of the diode V_1 is a.c. connected to the grid circuit of valve V_2, in order that it will follow changes in the mean grid voltage, and so restore the cathode voltage to the appropriate value.

(4.2) **Video-Frequency Filter Unit**

The principle of the video-frequency filter unit for reducing the effects of amplitude-modulated jamming is the reduction of the receiver amplification with consequent increase in the strength of the jamming signal necessary to cause overloading, the removal of the jamming modulation frequencies after detection by means of a v.f. filter, and the restoration of the overall amplification to normal by an additional v.f. amplifier.

The v.f. modulation components of a 0·5-microsec pulse extend substantially to about 2 Mc/s; in the presence of a strong jamming signal these components are considerably reduced in magnitude by the action of the second detector, the output consisting principally of the jamming modulation components, together with components due to beats between the jamming and radar-pulse frequencies. If the jamming is of small modulation frequency and is of the same carrier frequency as the radar equipment, these beat-frequency components correspond to the normal pulse-components; jamming of modulation frequencies less than about 600 kc/s can therefore be removed, together with the corresponding low-frequency beats of the jamming and target-signal, with a small reduction in the ratio of signal/noise energy. If the jamming is off-frequency more than 600 kc/s, however, the beat frequencies are greater than 600 kc/s, and the jamming components can then be removed with negligible reduction in the ratio of signal/noise energy. Thus a v.f. filter can be likened to an i.f. rejection circuit which is automatically tuned to correspond to the frequency of the jamming; when the jamming is of the same frequency as the radar equipment there is some reduction in the ratio of signal/noise energy, but when it is appreciably off-frequency the reduction in the ratio of signal/noise energy is negligible. The v.f. filter, of course, needs no tuning.

From the above, it is obvious that unless optional filters of various cut-off frequencies can be provided, a compromise must be accepted between the largest jamming modulation frequency which can be removed, and the resulting reduction in the ratio of signal/noise energy when the jamming is of the same frequency as the radar equipment. In naval applications, only one filter was, in general, fitted with each receiver; if the band-width at 3 db of the receiver was equal to the reciprocal of the radar pulse-duration, the filter cut-off frequency was made about three-eighths of this band-width; if the band-width was twice the reciprocal, the filter cut-off frequency was made about a quarter of the band-width.

Three-section filters were used in most applications, and in order to prevent the filter ringing on the receipt of a strong signal, such as the ground-wave, the sections were damped, with consequent reduction in the sharpness of cut-off. The first two sections were high-pass circuits of the desired cut-off frequency, and the third was a band-stop circuit tuned to about half this value, thereby giving a more abrupt cut-off. A specimen frequency-response curve of a filter for use with a 0·5-microsec pulse, is shown in Fig. 7; in the absence of the third section

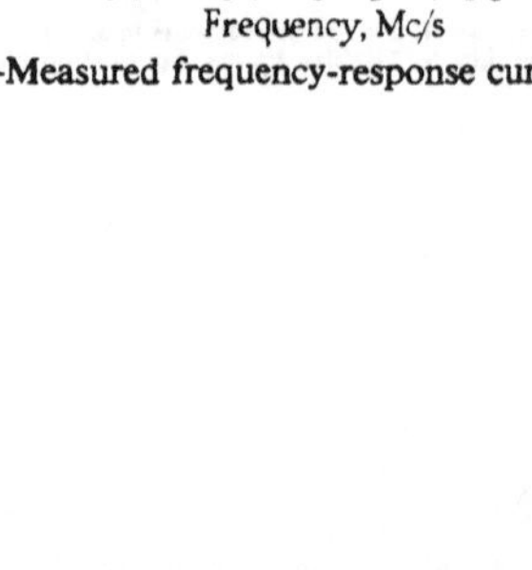

Fig. 7.—Measured frequency-response curve of filter.

of the filter, the response would be as shown by the dotted curve. For the measurement of such response curves, either a signal generator of less than 0·01% harmonic content must be used, or a harmonic-rejection circuit must be inserted between the signal generator and the filter unit.

As already stated, the filter is followed by a v.f. amplifier so that the i.f. amplifier can be operated at reduced amplification; the v.f. amplifier also compensates for loss of signal amplitude in the filter due to removal of some of the components. In the presence of jamming signals the output target-signal modulations from the filter are due to beat-frequencies, and hence the amplifier should handle both positive-going and negative-going signals; its band-width should also be sufficient to pass the highest beat-frequency of jamming and target-signals which may occur.

The circuit details of a complete v.f. filter unit are shown in Fig. 8; this is the unit of which the frequency response is shown in Fig. 7. If the jamming is of the same frequency as the target-echo, there is considerable distortion of the output pulse, as shown in Fig. 8(*a*); this latter distortion also occurs in the absence of jamming, and hence the use of the filter unit is made optional. If, however, the jamming is off-frequency so that the beats fall within the pass-band of the filter, a target-signal is passed through the filter with little distortion, as shown in Fig. 8(*b*). In order to produce a more normal picture, the signals after amplification, are rectified and smoothed; the effects on the pulse shapes are indicated in Figs. 8(*a*) and 8(*b*), the final output pulse being of duration about twice that of the input pulse, with a more gradual rise and decay. A full-wave rectifying circuit is used, fed by a pulse transformer, the low-frequency response of which is unimportant, since it is preceded by the high-pass filter. A phase-splitting valve with condenser-coupling was tried as an alternative, but trouble was experienced when responses from ground extending to a considerable range were received, owing to the charging of the coupling condensers, with subsequent paralysis and trace distortion.

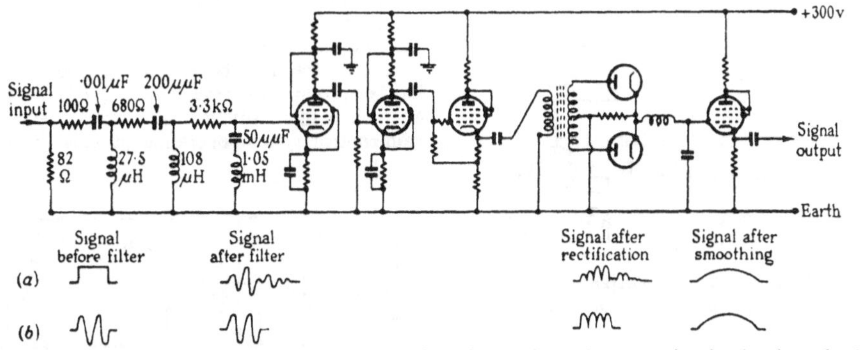

Fig. 8.—Circuit details of a video-frequency filter unit, and the effect of the unit on the target-signal pulse-shape for (*a*) on-frequency jamming, and (*b*) off-frequency jamming.

The filter units cause delays of the video signals, by amounts varying from about 1 microsec for a 1-Mc/s cut-off filter, to 10 microsec for a 25-kc/s cut-off filter; for accurate ranging equipments, a compensating delay-line was therefore provided, which was switched out of circuit when the filter unit was switched in.

Since the detector is operated at a low level when the filter is in circuit, it has a square-law response for small signals, and the output noise voltage therefore fluctuates, and is modulated, in correspondence with the strength and modulation of the jamming. An automatic gain-control (a.g.c.) system, such as is described in Section 4.4, was therefore provided in the later equipments to keep the peak output noise voltage from the filter unit of constant value for different strengths of jamming; such a system was essential for the satisfactory operation of the unit with intensity-modulation displays. The non-linearity of the detector also causes the production of harmonics of the jamming modulation frequencies; since these are of amplitude considerably less than the amplitude of the fundamental components, however, they are in general sufficiently attenuated by the filter, and the anti-jamming performance of the filter is substantially as indicated by its frequency response. For example, for the filter of which the response is shown in Fig. 7, jamming containing components of frequency 600 kc/s and power 20 db above noise power, or of frequency 200 kc/s and power 50 db above noise power, is reduced to less than the noise power. With a 0·5-microsec pulse there is a decrease in the ratio of signal/noise energy of about 4 db if the jamming is of the same frequency as the radar equipment, but there is negligible decrease if the jamming is off-frequency by 1 Mc/s or more.

Photographs illustrating the use of a filter unit, with an automatic gain-control system, to combat amplitude-modulated jamming are shown in Fig. 9; the lengthening of target-signals

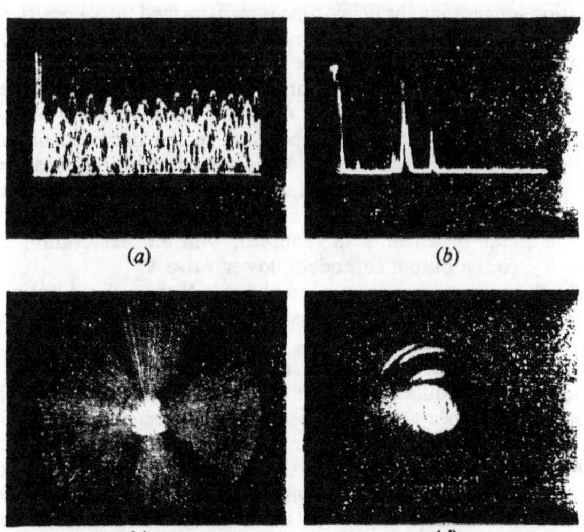

Fig. 9.—Removal of jamming signals by a video-frequency filter unit.

(*a*) and (*b*) A.M.C.W., 400 c/s + 4 kc/s with carrier frequencies 100 kc/s apart.
(*c*) and (*d*) A.M.C.W., 4 kc/s.

(*a*) Normal operation. (*b*) Filter in circuit.
(*c*) Normal operation. (*d*) Filter and a.g.c. in circuit.

by the filter unit can be seen by comparison of Figs. 9(*d*) and 2(*a*), although the target-echoes are not quite the same in the two cases. The filter is equally effective against speech-modulated and frequency-modulated jamming; in the latter case, there is the additional advantage that there is no reduction in the ratio of the peak signal energy to the noise energy, since with a sweeping jammer frequency, the jamming is periodically off-frequency by the requisite amount.

(4.3) Delay-line Differentiation Unit

Differentiation assists in combating pulse-modulated jamming, by reducing the duration of the jamming pulses at video frequency, so that target-signals and noise can be observed more easily in the gaps between the jamming pulses. The circuit details of the delay-line differentiation unit, developed for the later naval radar equipments, are shown in Fig. 10. Positive-going video signals

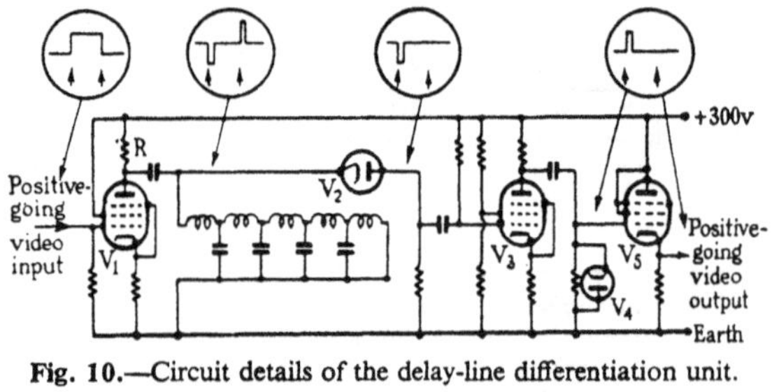

Fig. 10.—Circuit details of the delay-line differentiation unit.
The separation of the arrows within the circles corresponds to the duration of the original jamming signal.

are fed to the control-grid of valve V_1, the anode resistor R of which has a value equal to the characteristic impedance of the delay-line. Signals passed by the valve first appear across resistor R, and travel along the delay-line, being reflected with reversal of phase at the short-circuited end of the line; the total delay is twice the delay-time of the line, and since the latter is matched to the resistor R, further reflection at the input end is avoided. Any input signal of constant amplitude and of duration longer than the delay-time is transformed into a positive-going signal together with a negative-going signal, each of duration equal to the delay-time, separated by the duration of the original signal; the resultant voltage is the sum of these. This is illustrated on the circuit diagram, the separation of the arrows within the circles corresponding to the duration of the original jamming signal. Valve V_2 is a diode passing negative-going signals, corresponding to the leading edges of input signals; these negative-going signals are amplified and inverted in valve V_3, the anode of which is a.c. coupled, with d.c. restoration by valve V_4, to the output cathode-follower valve V_5.

The delay-time is made equal to about half the pulse-duration, so that the amplitude of a target-signal rises to its maximum value before cancellation occurs; the peak amplitude of signals passed by the unit is therefore unaltered by the cancellation. Cancellation produces a change in noise structure, but with noise from a linear or from a square-law receiver, where the difference of the peak and mean signals is much greater than that of the mean and minimum signals, there is little change in the peak amplitude of the noise signals. The ratio of the amplitude of a target-signal to that of receiver noise is then relatively unaffected by the action of the delay-line, yet jamming signals are reduced in duration to an amount equal to the delay of the delay-line, with consequent improvement in the visibility of target-signals through the jamming.

On deflection-modulation displays, differentiation gives negligible improvement in the detectability of a target-signal in pulse-modulated jamming, and on intensity-modulation displays the improvement is small—of the order of 2 to 4 db; the principal advantage of differentiation, however, is that it gives considerable reduction of the operational strain of detecting and tracking target-signals in the jamming.

(4.4) Automatic Gain-Control System

An efficient automatic gain-control system is required for keeping the mean output signal of a radar receiver, or of a v.f. filter unit, at a constant voltage, irrespective of large variations in the amplitude of any received jamming signal, such as the variations caused by rotation of the radar aerial beam through the bearing of the jammer.

In designing such systems for radar equipments, trouble is experienced owing to the fact that ground responses may extend for an appreciable portion of the repetition cycle, and so affect the control voltage. This could be completely eliminated by deriving the control voltage from signals and noise in a strobed portion of the trace beyond the ground-response signals, but it has been found that severe limiting of the signals is satisfactory and much more simple. The circuit details of an automatic gain-control system operating on this principle are shown in Fig. 11.

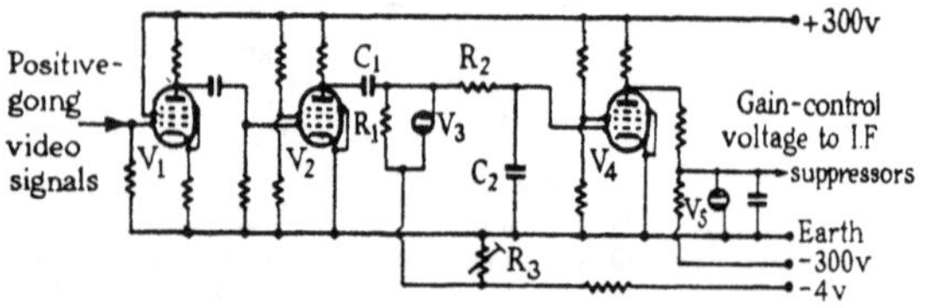

Fig. 11.—Circuit details of the automatic gain-control system.

Positive-going input signals and noise from the second detector of the radar receiver are amplified and inverted in valve V_1 to give a peak noise deviation of about 2 volts negative. Valve V_2 is non-conducting at this voltage and all signals are correspondingly limited by the valve; they are amplified by valve V_2 and then differentiated by the circuit C_1R_1, with d.c. restoration by valve V_3, to reduce the effects of ground-response signals. The resulting signals are smoothed by a long-time-constant circuit R_2C_2 before application to the control-grid of a further amplifying valve V_4. This has a large anode resistor of 47 000 ohms, and gives an amplification of about 40 db; the gain-control voltage for application to the suppressors of two or more i.f. valves (early in the receiver) is taken from a potential divider between the anode of valve V_4 and a negative h.t. voltage. Valve V_5 is a safety diode to prevent a positive voltage being applied to the i.f. valves in the event of circuit breakdown.

Measurements on this circuit have shown that it maintains the receiver noise-output power within 3 db of the value pre-set by the variable resistor R_3, for variations of 70 db in noise-input to the controlled stages, and for variations in the duration of ground-response signals from zero to a quarter of the radar pulse-repetition period; its use with a v.f. filter unit has already been illustrated in Figs. 9(*c*) and 9(*d*).

(5) CONCLUSION

The general principles followed for reducing the vulnerability of naval radar equipments to jamming have been described, together with a number of anti-jamming circuits for removing specific types of jamming. Of these circuits, that which proved most successful in the 1939–45 War was the v.f. filter unit. Its

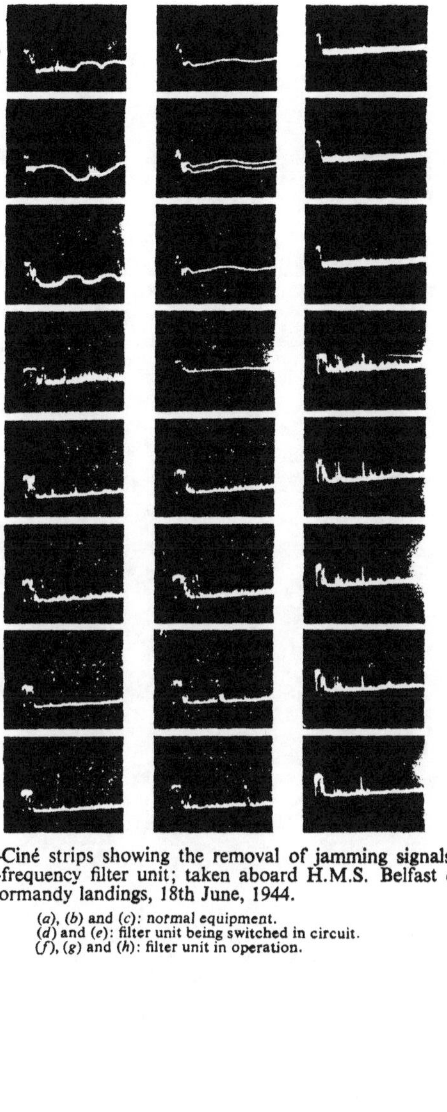

Fig. 12.—Ciné strips showing the removal of jamming signals by a video-frequency filter unit; taken aboard H.M.S. Belfast during the Normandy landings, 18th June, 1944.

(*a*), (*b*) and (*c*): normal equipment.
(*d*) and (*e*): filter unit being switched in circuit.
(*f*), (*g*) and (*h*): filter unit in operation.

use against jamming signals experienced during the Normandy landings is shown in Fig. 12; this comprises three strips from a ciné film taken by the author aboard H.M.S. Belfast on the 18th June, 1944. Frames (*a*), (*b*) and (*c*) of each strip show the jamming signals on the normal equipment, which looked more complex on the C.R.T. screen than they do on the film, owing to the apparent superposition on the screen of a number of traces; frames (*f*), (*g*) and (*h*) show the traces with the filter unit in operation, the establishment of stable operation on switching occupying about a tenth of a second, corresponding to the frames (*d*) and (*e*).

One of the most important effects of jamming, and one not previously mentioned in this paper, was its psychological effect on radar operators; it caused them considerable consternation when first encountered. To assist in preventing this, a training film "Radar Anti-jamming" was made, in conjunction with the Royal Naval Film Section; this showed the appearances of the various types of jamming, and the operation of the devices for overcoming them. Once the operators appreciated that they could do something to annul the effects of jamming they faced it with renewed confidence, and much of the operational success finally achieved was due to the intelligence and skill they displayed in the application of the various devices.

(6) ACKNOWLEDGMENTS

Close collaboration was maintained between the various fighting services and government scientific establishments in the examination of the effects of jamming, and in the development of appropriate countermeasures. Constant interchange of ideas took place, and the present report deals in detail only with those aspects in which members of Admiralty Signal Establishment made a major contribution. The author wishes to acknowledge the co-operation and assistance of many members of the establishment in the work described; he is particularly indebted to Mr. A. W. Ross, M.A., who was responsible for the direction of early naval anti-jamming development.

The author wishes to thank the Captain Superintendent for granting facilities for the preparation of this paper, and the Board of Admiralty for granting permission for its publication.

Designing Radar Receivers to Overcome Jamming
J.C. Galenian
Electronics Magazine, Vol. 36, May 17, 1963
Pages 50-54

6.2

RADAR TRANSMITTERS radiate energy over a wide range of frequencies, above and below the nominal operating frequency. This energy may be received by an adjacent radar set because of the large power output of the transmitter and the sensitivity of the victim receiver. Adjacent radar sets interfere with each other even though they operate on different frequency bands. This interference, often called "rabbits", appears as random pulses moving across an A-scope or as a series of moving bright dots on a pri or B-scope. The tendency of the rabbits to move will depend on the degree to which the pulse repetition frequency (prf) of the interfering radar is similar to that of the victim radar. If the two prf's are identical, then the interfering pulse will remain stationary on the indicator and may have the appearance of a legitimate target echo. This type of interference is not too troublesome to an A-scope indicator but may cause a confusing display on a ppi or B-scope display.

In addition to interference caused by the radar transmitter, the modulator generates large amounts of radiated energy over a wide range of frequencies. This also may cause interference to nearby radar and communications equipment.

Intentional interference or jamming may take many widely differing forms. The simplest form of jamming consists of an unmodulated carrier wave at the operating frequency of the radar set. This is continuous-wave (c-w) jamming as shown in Fig. 1A. The carrier may be amplitude modulated with random noise, or other waveforms. This is termed amplitude-modulated c-w (amcw) jamming.

The carrier frequency may be varied in step with a sawtooth or other waveform. This is f-m or swept-frequency jamming, Fig. 1B. The carrier may be pulsed on and off with pulses that may be either syncronized or not synchronized with the radar prf. This is known as pulse jamming. Nonsynchronous pulse interference is shown in Fig. 1C.

A jammer may receive the radar's transmitted pulse and re-transmit one or more replicas of it back to the radar. The jamming pulses may appear to be legitimate target echoes on the radar indicators and may cause confusion as to the number and location of true targets. This is known as repeater jamming. Passive reflectors, such as strips of aluminum foil or chaff, may also cause spurious targets to appear on the radar indicators.

Atmospheric phenomena, such as lightning and aurora borealis, spark type interference from nearby rotating machinery, engines, generators, neon signs and the like may also enter the radar receiver.

RECEIVER CHARACTERISTICS—Undesired signals may affect the radar receiver in several respects. The undesired signal may cause saturation or overload of the receiver, masking or hiding of desired signals or blocking of the receiver. The undesired signal may, effectively, be lengthened in its time dimension by the impulse response of the receiver. Spurious signals may appear in the output of the receiver.

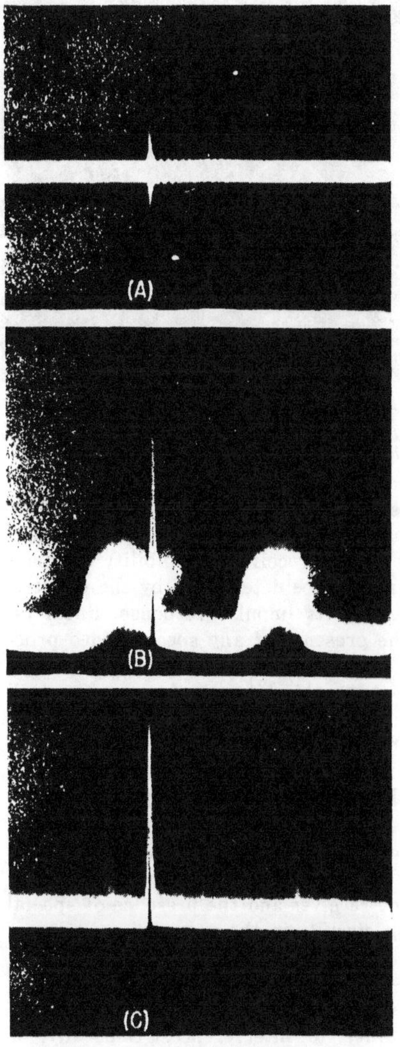

RADAR INTERFERENCE *as it appears on an A-scope presentation. Effect of a c-w signal (A), effect of swept-frequency jamming (B), and the effect of nonsynchronous pulse interference (C)—Fig. 1*

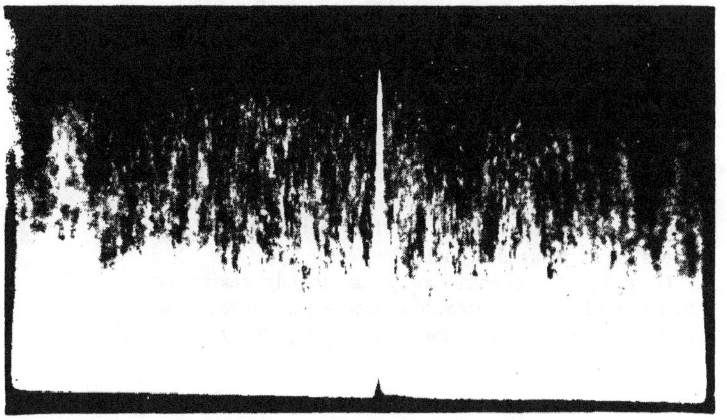

NOISE JAMMING *may saturate the video portion of the receiver making the desired signal barely discernible—Fig. 2*

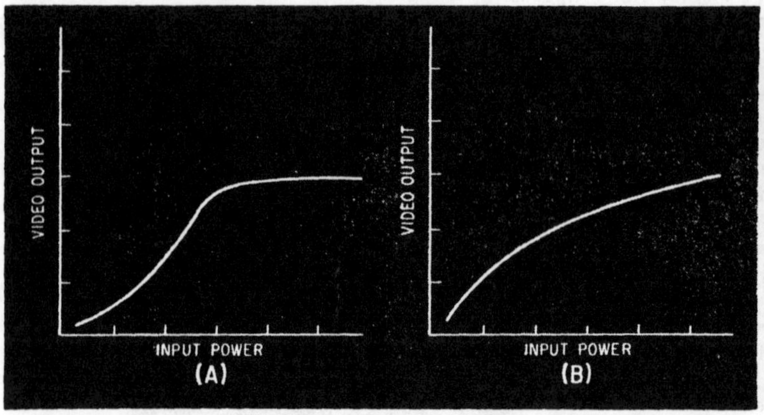

INPUT *versus output characteristic of normal receiver (A) and receiver using a log i-f amplifier (B) show how a large dynamic range may be obtained—Fig. 3*

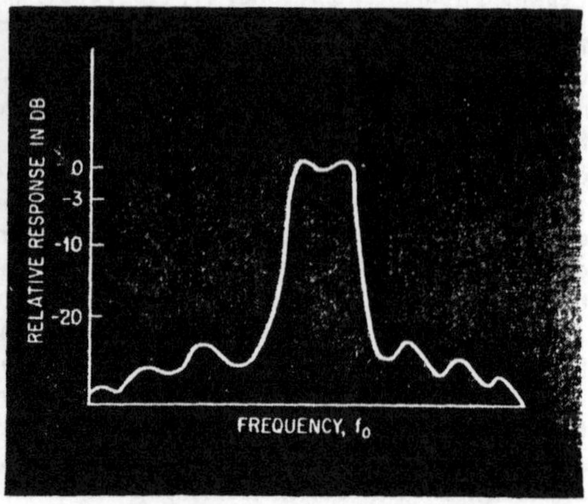

RESPONSE *curve of a typical receiver in which undesired responses are down 20 to 40 db—Fig. 4*

In general, the ability of the undesired signal to cause saturation will be determined by the pulse gain characteristics of the receiver. Its ability to mask the desired signal will be determined by the selectivity of the receiver, its impulse response, linearity, stability and the presence of any special video processing circuits or filters. The tendency of the undesired signal to cause blocking of the receiver will be determined by the recovery time characteristic of the receiver. The tendency of the receiver to lengthen the time duration of the interfering signal will be determined by the impulse response and recovery time of the receiver. The tendency of the interfering signals to cause spurious signals to appear in the output of the receiver will be determined by the selectivity of the receiver and the presence of special video-processing circuits in the receiver.

PULSE GAIN—The pulse gain, or input versus output, characteristics of a receiver have a major effect on its susceptibility to interference. Generally, a large dynamic range and good linearity of response tend to reduce this susceptibility.

The i-f and video dynamic ranges of a receiver may be different and will have different effects on susceptibility. For example, the c-w or carrier component of jamming signals may cause saturation or overload of the i-f amplifier. This will result in the suppression of desired target echoes and receiver noise. The c-w signal itself will cause a d-c voltage to appear at the output of the receiver's second detector. This d-c voltage will be blocked by the usual coupling capacitor between the second detector and the video amplifier. Thus, a large c-w signal can cause a decrease in the video output of the receiver and a suppression of desired signals.

Output video dynamic range is usually restricted to 10 or 15 db so as to accommodate the characteristics of the radar indicators. Noise, Fig. 2, or other

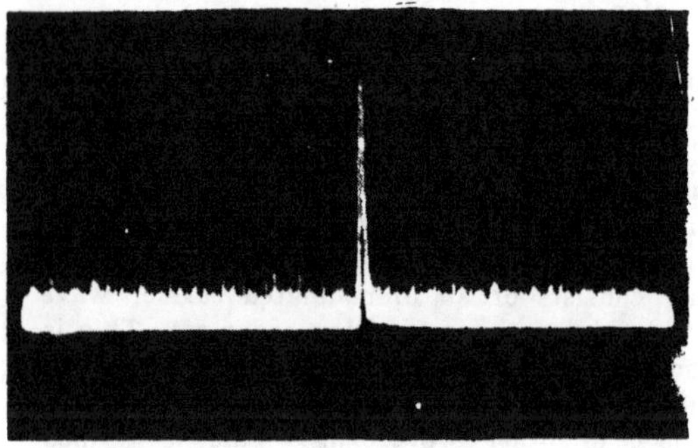

THE JAMMING PROBLEM AND WHAT TO DO ABOUT IT

This is a normal radar video presentation on an A-scope. The object of the following check list is to enable you to get a picture that looks like this as often as possible and under most conditions

Antenna

Side and back lobes should be 25-40 db down from main lobe

R-F Section

Image rejection should be 20 db or more

I-F Amplifier

Input dynamic range is 40 to 80 db

Bandwidth is 1.2 to 2.0 times the reciprocal of pulse width

Spurious responses should be more than 40 db down

Recovery time is less than the pulse width

Free from oscillation in the presence of both pulse and c-w signals

Capable of accommodating input signals with a 100 percent duty cycle without damage to components

Automatic Gain Control

Free from oscillation in the presence of signals, with frequencies of d-c up to ½ the i-f bandwidth, at the input to the agc loop

Agc voltage is proportional to both pulsed and c-w r-f signal amplitude

System Trigger Generator

Variation in inter-pulse interval (jitter) should not increase with high-level c-w and noise signals present at the receiver input terminals

---

jamming modulation, may saturate the video portion of the receiver if the received jamming power is as little as 10 or 15 db above the minimum discernible signal level. Desired signals may not be discernible even though they may be greater in amplitude than the jamming signal and may be present in the output of the receiver's i-f amplifier.

Cross-modulation products may be generated in the receiver by interaction of the jamming signal with the desired signal and the receiver local oscillator. These cross-modulation products will tend to mask the desired signal. The formation of the cross-modulation products will be minimized by a linear pulse gain characteristic.

A large dynamic range may be desirable from a performance viewpoint as well as from a susceptibility viewpoint. Several methods are commonly employed, and these tend to improve the susceptibility of the receiver as well as its normal operating characteristics, although there are several pitfalls.

Automatic gain control (agc) is sometimes employed to adjust receiver gain and avoid saturation by strong signals. The receiver dynamic range may be extended from a typical value of 20 db without agc to 60 to 100 db with agc. Unfortunately, most agc circuits are a-c coupled and will not pass d-c produced by a c-w signal at the output of the second detector. The amplitude of the c-w signal will, then, be ineffective in controlling the receiver gain, and the agc will not prevent saturation. In fact, the agc may make matters worse. The c-w will cause suppression, or a decrease in amplitude, of the desired signal. This decrease will be sensed by the agc circuit which will then increase the receiver gain in an attempt to maintain the signal at a constant level. The increase in receiver gain will hasten saturation, but a reduction in gain would have avoided saturation. Direct coupling should be used throughout the agc loop so that jamming signals as well as the desired signal will control receiver gain and saturation will be avoided.

Another difficulty with agc is oscillation of the agc loop. Jamming modulation or cross-modulation products formed in the receiver may occur at frequencies where oscillation will be supported by the agc loop. The agc voltage will fluctuate and cause receiver gain to change in a random manner. These loops should be designed to be stable in the presence of a wide range of signal frequencies.

A second method of obtaining a large dynamic range is by the use of a logarithmic or linear-logarithmic i-f amplifier, where output is proportional to the logarithm of the input as shown in Fig. 3. These amplifiers allow a large input dynamic range to be compressed to a small dynamic range compatible with the requirements of the radar indicators. The difficulties that may be encountered in the design of agc circuits are avoided, since gain is maintained constant.

From a susceptibility viewpoint, the large dynamic range of the log i-f amplifier is desirable, but the decrease in incremental gain at the higher input power levels may cause desired signals to be suppressed to below the usable amplitude by c-w jamming even though complete saturation is avoided. The nonlinear pulse gain characteristic may also increase generation of cross-modulation products in the presence of jamming.

A third method, and perhaps the most desirable from a susceptibility viewpoint, is the use of a second detector capable of operating at a low i-f voltage, and high-power output tubes or transistors as i-f amplifiers. This combination of a low minimum detectable signal voltage and a large output voltage capability results in a large input and output dynamic range. This is desirable because a linear response may be attained with the large dynamic range, and possible difficulties with agc circuits avoided.

SELECTIVITY—Noise, or other undesired signal modulation may cause masking or hiding of desired signals. As a general rule, random noise is more effective than other waveforms in masking the desired signal.

The most efficient means of avoiding the masking of desired signals by undesired ones is to keep the two signals separated in the receiving process. In other words, good receiver selectivity is desirable. Receiver selectivity may be divided into three categories, frequency selectivity, spatial selectivity and time-domain selectivity.

FREQUENCY SELECTIVITY—Frequency selectivity of a receiver determines the extent to which it will respond to signals at frequencies other than the radar system's operating frequency. Ideally, the radar receiver's response curve should exactly match the spectrum of the radar's transmitted pulse.

Care should be taken to insure that undesired responses at the lower portion of the receiver's frequency response curve (20 to 40 db down points) are not present as in Fig 4. These unwanted responses or spurious responses will allow off-frequency undesired signals to enter the receiver. These responses may be more troublesome than expected because one-way transmission of undesired signals give them an advantage over desired signals which must follow a two-way path. High levels of undesired signals may still be received. Spurious responses will also cause wider video pulses to result from the passage of a swept-frequency jammer through the receiver passband. The receiver video output will increase and decrease in step with the frequency of the jammer.

A replica of the receiver response curve (a pulse-like signal) will appear in the video output of the receiver. The width of the video pulse will be $T = B/S$, where T = width of video output pulse in μsec; S = jammer sweep speed in Mc per μsec; B = receiver bandwith in Mc. Thus, the width of the induced jamming pulse is proportional to receiver bandwidth and inversely proportional to jammer sweep speed. If receiver bandwidth is unnecessarily wide, then the effective duty cycle of the jammer will be increased and the jamming signal will be more effective in masking desired signals. The duty cycle of the jammer is the time the jammer is in the receiver pass-band divided by the jammer's repetition period, or the ratio of the receiver bandwidth to the jammer's sweep width.

The minimum width of the induced jamming pulse is approximately equal to the inverse of the receiver bandwidth. This minimum width is the impulse response time of the receiver.

The receiver may respond to undesired signals at the receiver's image frequency. The image frequency allows undesired signals from two parts of the frequency spectrum to enter the receiver, thereby increasing the amplitude of undesired signals. Image frequency response also causes twice as many induced pulses to be generated by swept-frequency jamming signals than would be the case if image response were not present.

A good image-rejection ratio is desirable to minimize the amount of interference that enters the receiver. Image-rejection ratios of 20 to 40 db may be obtained by the use of r-f filters or preselectors in the r-f portion of the radar receiver.

SPATIAL SELECTIVITY—Ideally, a radar system transmits and receives energy only along the axis

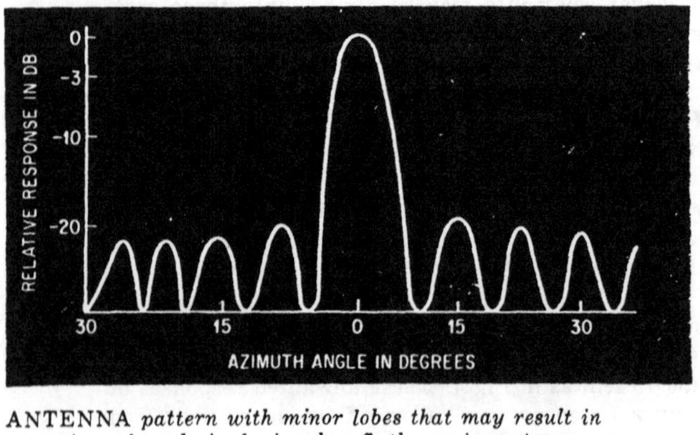

ANTENNA *pattern with minor lobes that may result in reception of undesired signals off the main antenna—Fig. 5*

of the radar antenna's main-beam response. Unfortunately, all antennas have a number of minor lobes in their beam patterns, Fig. 5.

The minor lobes allow undesired signals from sources at directions other than the direction of the main lobe to enter the receiver. Minor lobe responses are usually about 20 db down from main-lobe response. However, due to one-way transmission of the undesired signals, there may be sufficient amplitude to affect the reception of desired signals.

A minor-lobe response of 20 db down is adequate for normal radar operation. Energy from targets in the direction of the minor lobe will be 40 db down from the main-lobe response, since both the transmitted energy and the received energy will be attenuated by 20 db. Each reduction in side-lobe amplitude of 3 db will have the same effect as halving the jammer's transmitter power.

TIME-DOMAIN SELECTIVITY—The ability of the receiver to discriminate against signals having pulse lengths and prf's different from those of the radar transmitter will decrease its susceptibility to undesired signals.

Filters, including the matched-filter radar systems, will reject signals of longer duration than the desired signal. An example is the highpass filter or the fast time constant (ftc) circuit. A simple version of this circuit is an *R-C* coupling network with a time-constant slightly greater than that of the transmitter pulse. Signals of longer duration, such as ground clutter or low-frequency jamming modulation components, will be attenuated by the ftc circuit. Video processing circuits, such as video integrators or interference blankers will reject signals whose repetition frequency is not identical to that of the radar transmitter.

RECOVERY TIME—The time required for the receiver to regain normal sensitivity following the re-

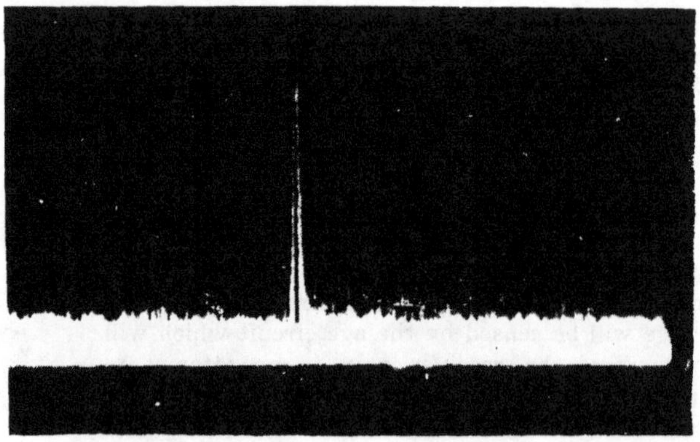

UNSTABLE *operation caused by unstable trigger—Fig. 6*

ception of a long-duration signal of high amplitude, recovery time, is proportional to the time constants in the receiver. For example, coupling capacitors in the i-f or video amplifiers may be charged, during the reception of a strong signal, by the grid (or base) current of the associated amplifier stage. When the signal has ceased, the capacitors will remain charged until they have had time to discharge through the grid (or base) network to ground. During the discharge period, a negative voltage will appear at the grid of a tube-type amplifier. This reduces stage gain until the capacitor has discharged.

A rapid discharge path should be provided for coupling capacitors. The inductive portion of the interstage coupling network should be placed on the grid (or base) side of the coupling capacitor to provide a low d-c resistance path for the discharge. High-resistance bias networks should be avoided. The tendency to use very large coupling capacitors in transistor i-f and video amplifiers should be minimized. Storage-time characteristics should be considered in the selection of transistor i-f amplifiers.

STABILITY—A receiver may tend to oscillate in the presence of undesired signals, even though the i-f is free from oscillation with no jamming present. Oscillation may occur because the c-w signal is of longer duration than the desired signals, and normal damping of oscillations is prevented. The c-w signal may also move the operating point of the i-f amplifier to a region of inherent instability where oscillation may occur. The receiver should be tested for the presence of undesired oscillation both with and without c-w jamming present.

Possible corrective measures include selecting dimensions of the i-f amplifier chassis so that waveguide propagation of signal energy from the output of the amplifier to its input cannot occur, interstage decoupling, power supply filtering, shielding and stable agc circuits.

Among other factors that may affect the susceptibility of the receiver, is the presence of long-duration signals such as c-w jamming. These may exceed the designed duty factor of the receiver. Larger-than-normal values of power will be dissipated in the amplifier stages.

Power ratings of some of the tubes, transistors or resistors in the i-f amplifier may be exceeded and component failure may occur. Components in i-f amplifiers should be rated for a 100 percent duty cycle.

Crystal diodes in the receiver mixer are also subject to burnout by strong signals. Crystals should have the highest burnout power rating obtainable and should be protected by limiters or other means. The usual TR tube provides no protection to the crystals except during the firing time of the transmitter.

The minimum time duration of the output signal of the receiver, with a signal of shorter duration than the transmitter pulse present at its input, is approximately equal to the reciprocal of the 3 db bandwidth of the receiver. This is also known as the impulse response time of the receiver. For example, the shortest pulse that can be reproduced by a receiver with a bandwidth of 1 Mc is a 1 μsec pulse. Pulses shorter than 1 μsec at the receiver's input will appear as 1 μsec pulses, reduced in amplitude, at the output. Thus, the effective time-duration of short-duration undesired signals will be increased as will their effectiveness in masking desired signals from being adequately received.

As a result, impulse response time, including transient responses, of the receiver should be as short as possible.

Undesired signals may be coupled, by the power supply, stray coupling paths or a number of other means, to the trigger pulse generator of the radar system. The trigger generator may then be triggered by the undesired signals and unstable operation of the complete system as shown in Fig. 6, or even non-operation, may result. This effect is most likely to occur in moving-target indication (MTI) systems, where a recirculating trigger is employed. The effect may not occur under normal operating conditions, but may occur in the presence of interference because of the higher-than-normal signal levels that are involved.

Desensitization in Transistorized PM/FM Receivers
F. Hansen
Proceedings of the 18th IEEE Vehicular Communications Conference, 1967
Pages 78-86

6.3

Summary

An analysis of the mechanism of desensitization - or blocking - in the RF- and mixer stage of a transistorized PM/FM receiver is given, and it is shown, that noise figure degradation because of noise conversion is the principal source of receiver desensitization. A simple expression for the noise factor increase as a function of the level of the interfering signal is derived.

Using this result an expression for the overall blocking level of a receiver, where blocking may take place in the RF-stage and/or in the mixer stage, is developed, and the influence of RF-stage and mixer stage parameters on the overall receiver blocking level is analyzed.

The theoretical results are supported by results of measurements on practical receiver circuits.

Introduction

Before the advent of the transistor - when communications receivers were equipped with vacumm tubes - receiver desensitization - or blocking - in the presence of a strong interfering signal was never really a problem, but since transistors have been introduced in the input and mixer stages of the receivers, this phenomenon has become a major problem in receiver design.

The problem is very often so serious, that the capabilities of the modern semiconductor devices with respect to sensitivity cannot be fully utilized, if at the same time the type approval requirements for blocking immunity are to be met.

As the mechanism of blocking in a multistage circuit is rather complex, it was felt that a closer investigation of the whole phenomenon was needed, and the present paper will describe some of the results of this investigation.

Definition of blocking level and method of measurement.

A slightly modified version of the method used by the british Post Office for type approval measurements was found to be convenient for definition and measurement of the receiver blocking level.

According to this modified method (Fig.1) two 5oΩ signal generators, A and B, are connected to the receiver input terminal through a 3-port matching network - which has an insertion loss of 6 dB. With signal generator B switched off, the level from signal generator A is adjusted to produce a signal to noise ratio of 1o dB at the receiver frequency. Signal generator B is then switched on, its frequency is adjusted to the receiver frequency plus or minus 15o kHz, and the output is increased, until the signal to noise ratio is reduced to 7 dB.

This level is designated the 3 dB blocking level, V'_o.

The minimum requirement is that the blocking level should not be less than 52 mV at any interfering frequency separated from the receiver frequency by more than $\pm$ 15o kHz.

As the slope of the signal to noise ratio curve is approximately 2:1 near the threshold level, a 3 dB decrease of signal to noise ratio corresponds to 1,5 dB noise figure increase.

During the following calculations it will sometimes be convenient to use the interference level which corresponds to 6 dB signal to noise ratio decrease, or 3 dB noise figure increase. This level is designated the 6 dB blocking level V_o.

Analysis of the mechanism of single stage blocking.

From the general expression

for the noise factor of cascaded stages:

$$F = F_1 + \frac{F_2 - 1}{G_1} + \frac{F_2 - 1}{G_1 G_2} \qquad (1)$$

it is evident, that an increase of the overall noise factor can be the result of:

either: a reduction of one or more of the individual stage gain factors, G_n

or: an increase of one or more of the single stage noise factors, F_n

or: a combination of gain-decrease and noise factor increase.

In a well designed receiver, the ratio of the noise factor of a given stage to the power gain of the preceeding stage will normally be so chosen, that the overall noise factor of the receiver will be practically determined by the input stage noise factor while the second stage contributes only little.

It is therefore to be expected, that a relatively large gain reduction - something like 6 dB or more - will be necessary in order to produce a 1,5 to 3 dB noise figure increase. Measurements on many receivers have indicated, however, that the actual gain reduction at the 3 dB blocking level is much smaller than this - typically a few tenth of a dB. It is thus reasonable to believe that single stage noise figure increase is the principal cause of blocking.

In a receiver where channel selectivity is placed immediately after the first mixer - as in modern receivers using quartz filters - the dominating scurce of blocking will thus be RF- or mixer stage noise figure degradation.

Mechanism of noise increase

The noise figure of transistors at high signal levels has been treated by several authors.

Webster[1], Vogel and Strutt[2], and others have dealt with mixer noise figures, and Suziki and Toya[3] have used the result of this work to calculate the noise figure of a bipolar transistor amplifier in the presence of a strong interfering signal.

These authors all assume that the noise figure is increased at high signal levels because of nonlineartities which cause noise power at frequencies outside the receiver frequency band to be converted into this band.

Suziki and Toya find that their theoretical results are in good agreement with practical measurements, and their results also indicate, that the amplifier noise factor, F, can be expected to increase linearly with the power level of the interfering signal, that is: follow the expression

$$F = F_o \left[1 + \left(\frac{V_i}{V_o} \right)^2 \right] \qquad (2)$$

where F_o is the small signal noise factor V_i is the interfering signal voltage, and V_o is the value of V_i which doubles the noise factor (6 dB blocking).

Suziki and Toya do not make any attempt to calculate the noise figure increase of the mixer stage, where the situation is quite different from what it is in the amplifier, mainly for 2 reasons.

The first reason is, that in the mixer stage 2 strong signals will be present at the same time: namely the local oscillator signal and the interfering signal.

The second reason is, that the noise spectrum of the mixer is considerably modified by noise power supplied from the local oscillator and from the multiplier chain.

Measurements on mixer stage blocking have indicated though, that the mixer noise figure varies with the interfering signal level in very much the same way, as the amplifier noise figure does. A very simple consideration will show that this allways will be the case, as long as the oscillator noise contribution is large enough to play an important role in the blocking mechanism. (Fig.2).

If we assume that noise conversion is the principal source of blocking in the mixer stage - like it is in the amplifier - and if we assume that the main part of converted noise power is converted directly from RF to IF, it will be obvious that IF-noise can be produced through conversion of noise power from all frequency bands, centered around the frequencies n times the interfering frequency plus or minus the intermediate frequency.

If we further assume, that the noise level supplied from the local oscillator dominates in this mechanism - an assumption which has been supported by results from experiments with different oscillator types and different multiplier bandwidth - the major part of the excess IF-noise will be oscillator noise converted through mixing with the interfering signal frequency F_i, and we can develope a qualitative expression for the mixer noise figure increase in the following way.

Not including the oscillator and signal frequencies in our considerations, we can assume the equivalent base emitter voltage to be the sum of the induced oscillator noise voltage, V_n, and the interfering signal voltage, V_i:

$$V_{be} = V_i + V_n$$

$$= V_i \cdot \sin\omega_i t + V_n \sin\omega_n t \qquad (3)$$

We can further assume the transistor characteristic to be given by its Taylor expansion:

$$I_c = I_{dc} + a_1 \cdot V_{be} + \underline{a_2 \cdot V_{be}^2} + \ldots . \qquad (4)$$

Substitution of (3) into the second order term of equation (4) yields

$$I_{c2} = a_2 \cdot (V_i \cdot \sin\omega_i + V_n \cdot \sin\omega_n t)^2$$

$$= a_2 \cdot (V_i^2 \cdot \sin^2\omega_i t + V_n^2 \cdot \sin^2\omega_n t +$$

$$\underline{2V_i \cdot V_n \cdot \sin\omega_i t \cdot \sin\omega_n t)} \qquad (5)$$

The third term of this equation represents the sum and difference frequencies produced by the mixing. It can be rewritten as follows:

$$2 \cdot a_2 \cdot V_i \cdot V_n \cdot \sin\omega_i t \cdot \sin\omega_n t$$

$$= a_2 \cdot V_i \cdot V_n \cdot \underline{(\cos\,(\omega_i - \omega_n)t}$$

$$- \cos\,(\omega_i + \omega_n)t) \qquad (6)$$

The term of interest in equation (6) is the first term, which represents the amount of collector noise current converted from ω_n to the intermediate frequency.

The excess IF-noise power can be calculated from

$$P_n = \tfrac{1}{2}\, a_2^2 \cdot V_i^2 \cdot V_v^2 \cdot \frac{1}{R_L}$$

$$= \text{constant x } V_i^2 \qquad (7)$$

Where R_L is equal to the IF-load resistance of the mixer stage. Equation (7) shows that the excess IF noise will be proportional to the square of the interfering voltage level, V_i, and this leads to the conclusion that the mixer noise factor will vary with the interference level according to

$$F = F_o \left(1 + \left(\frac{V_i}{V_o}\right)^2\right) \qquad (8)$$

which is identical to (2).

This means that we can expect the mixer noise figure to vary with the interference level in exactly the same way as the amplifier noise figure does.

As we have included only the second order term in these calculations, we can expect (8)(and (2)) to be valid for field effect transistors as well as for bipolar transistors. We may even expect the result to be more accurate for the field effect devices.

<u>Measurements on amplifier and mixer circuits.</u>

To check the validity of these results a number of measurements were made on RF-amplifiers and mixers, equipped with bipolar transistors as well as with field effect transistors, (Fig.3).

The circuit under investigation was made part of a 158 MHz PM-receiver, designed for 5o kHz channel separation and ± 15 kHz frequency deviation. Since channel selectivity in this receiver was obtained by means of a quartz crystal filter following the first mixer, blocking could take place in the mixer or RF-stage only.

During the measurements on the RF-stage a 5-circuit helical filter was inserted between RF-stage and mixer, and the blocking level was measured at 6 MHz frequency separation, ensuring that no blocking could occur in the mixer stage. Mixer measurements were made at 15o kHz frequency separation.

The relative noise figure increase was determined indirectly by measurement of receiver sensitivity corresponding to 1o dB signal to noise ratio.

Fig. 4 shows the result of some of the amplifier measurements. It is seen that bipolar transistors as well as field effect transistors follow equation (2) reasonable well. The two curves taken with the same bipolar transistor at different collector currents, show that the blocking level increases with collector current for these devices. This is of course due to the improved linearity at the higher current level.

The mixer measurements (Fig.5) show a similar agreement with the theoretical results for both types of devices.

The two curves showing results from measurements on the same bipolar mixer in connection with either a semiconductor or a vacuum tube local oscillator chain indicate that the blocking level of the mixer stage is determined to a large extent by the local oscillator source.

Blocking in cascaded stages

In a receiver where blocking may take place either in the RF-stage or in the mixer stage - or in both stages - it is of interest to investigate how the resulting blocking level can be determined, and how this level depends on the RF-stage and mixer stage parameters. We will assume the parameters indicated on Fig. 6 to be given.

The noise factor of the RF- and mixer stages are F_1 and F_2 respectively.

The power gain of the RF-stage is G_1, which is equal to the square of the voltage gain, A_1, when we refer to the same impedance level. (5o Ω).

The signal levels corresponding to 6 dB blocking are V_{o1} and V_{o2}, and the levels corresponding to 3 dB blocking are V'_{o1} and V'_{o2}. The relation between V_o and V'_o is that V'_o is equal to o,6 times V_o.

The interfering signal level at the RF-stage is V_{i1}, and the level at the mixer is V_{i2}, which is equal to A_1 times V_{i1}.

Using these parameters, we can calculate the receiver blocking level in the following way:

If we assume that the noise factors follow equations (2) and (8), they will be given by

$$F_1 = F_{o1} \cdot \left(1 + \left(\frac{V_{i1}}{V_{o1}}\right)^2\right) \qquad (9)$$

and
$$F_2 = F_{o2} \left(1 + \left(\frac{V_{i2}}{V_{o2}}\right)^2\right) \qquad (1o)$$

Substituting (9) and (1o) in (1) and setting

$$\alpha = \frac{F_{o2}}{F_{o1}} \qquad (11)$$

$$\beta = \frac{V_{o2}}{V_{o1}} \qquad (12)$$

and
$$F_o = F_{o1} + \frac{F_{o2} - 1}{G_1} \qquad (13)$$

and assuming $F_{o2} \gg 1$, we get

$$F = F_o \cdot \left[1 + \left(\frac{V_{i1}}{V_{o2}\sqrt{\dfrac{1 + \frac{\alpha}{G_1}}{\beta^2 + \alpha}}}\right)^2\right] \qquad (14)$$

or

$$V_o = V_{o2} \cdot \sqrt{\frac{1 + \frac{\alpha}{G_1}}{\beta^2 + \alpha}}$$

$$= V_{o1} \cdot \sqrt{\frac{1 + \frac{\alpha}{G_1}}{1 + \frac{\alpha}{\beta^2}}} \qquad (15)$$

Equation (14) shows that the overall noise factor varies with the interference level in exactly the same way as the single stage noise factors do, and equation (15) shows how the overall blocking level depends on the single stage parameters.

According to (15) the overall blocking level will be determined by the RF and mixer stage blocking levels, V_{o1} and V_{o2}, by the RF-stage power gain, G_1, and by the ratio, α, of mixer stage noise factor to RF stage noise factor. It will be noted that the overall blocking level does not depend on the absolute noise factor values, F_{o1} and F_{o2}.

In fig.7, the relative blocking level

$$\frac{V_o}{V_{o1}} = \sqrt{\frac{1 + \frac{\alpha}{G_1}}{1 + \frac{\alpha}{\beta^2}}}$$

has been plotted versus the ratio G_1/α with α/β^2 as parameter.

The curves indicate, that the relative blocking level may exceed unity for small values of G_1/α, and that it drops asymptotically to

$$\frac{V_o}{V_{o1}} = \frac{1}{\sqrt{1 + \frac{\alpha}{\beta^2}}}$$

for large values of G_1/α.

In a practical receiver, a typical value of G_1/α will be about 6 dB which corresponds to

$$F_o = F_{o1} \cdot 1{,}25$$

which means that the mixer contributes with about 1 dB to the overall noise-figure.

With this value of G_1/α the relative blocking level will usually be slightly less than unity.

As indicated on the curves, blocking will usually take place mainly in the RF-stage for small values of α/β^2 and in the mixer stage for large values of this ratio. In case the dynamic range of the RF-stage is not sufficiently wide, this stage may not follow the $(1 + X^2)$ - law at large input signal levels, and the curves will tend to flatten out around $V_o/V_{o1} = 1$, indicating that the RF-stage noise rises more rapidly than the mixer noise. This is indicated by the dotted lines.

In case β^2 is small compared to α - which is very often the case - RF-stage blocking can be neglected, and it seems obvious to express the overall blocking level relative to the mixer stage blocking level, V_{o2}.

In fig.8 V_o/V_{o2} has been plotted against G_1/α, with α as parameter, giving a set of curves very similar to those in fig.7.

A typical value of G_1/α will still be about 6 dB, and it can be seen from the curves, that if a high blocking level is to be obtained α must be as small as possible, meaning that we shall design for the lowest possible mixer noise figure.

Measurements of overall receiver blocking level.

In order to verify the theoretical results of this investigation a number of measurements were made on different receiver input configurations. A few of the most typical results of this part of the work, will be presented here.

The measurement were carried out in the following way:

The RF- and mixer stages were combined with the IF- and discriminator units used in connection with the single stage measurements, and a 50 Ω RF-attenuator was inserted between RF-stage and mixer stage in order that the power gain of the RF-stage could be varied without affecting other circuit parameters.

The interfering signal level corresponding to 3 dB blocking was measured for different values of RF-stage power gain. For comparison values were also calculated using the formulae derived above and the measured parameters of the individual stages.

The first example (fig.9) shows the performance of the combination of a bipolar RF-amplifier and a bipolar mixer.

The RF-stage operates at 6,6 mA collector current, the noise figure is 6 dB, and the 3 dB blocking level is 200 mV.

The mixer operates at an oscillator voltage of 250 mV and a collector current of 1 mA. The mixer noise figure is 10 dB, and the blocking level is 75 mV.

These parameters give $\alpha = 2{,}5$ and $\beta = 0{,}4$.

The calculated values of the overall blocking level agree reasonably well with the measured values. Blocking takes place in the mixer stage.

In the second example (fig.10) both stages are identical to those used in the first example, but the semiconductor oscillator-multiplier chain has been replaced be a vacum tube chain, resulting in an increase in the mixer blocking level from 75 mV to 140 mV. The effect of this is a corresponding increase of the overall blocking level. α is still equal to 2,5 and β equal to 0,7.

Again blocking takes place in the mixer stage, and there is relatively good agreement between measured and calculated values, except for a minor deviation at high values of RF-stage power gain, due to heavy mixer stage overload.

In the last example (fig.11) the combination of an MOS dual gate RF-stage and an FET triode mixer has been investigated.

The RF noise figure in this case is 3 dB, and the mixer noise figure is 6 dB. The blocking levels are 2oo mV and 11o mV respectively. These figures give α equal to 2 and β equal to o,5.

The calculated and measured blocking levels again agree well. Blocking again takes place in mixer.

Note that, because of the very low noise figures of the field effect devices, a much smaller value of RF-stage power gain may be chosen for a given sensitivity, and that this may lead to better blocking properties for the field effect input stage.

Conclusion

We can conclude from the examples which have been given, that in most practical circuits blocking will take place in the mixer stage. Exceptions from this rule may be receivers, designed for minimum power consumption, in which case RF-stage blocking may dominate.

To obtain maximum protection against desensitization, the ratio of mixer noise figure to RF-stage noise figure should be kept as low as possible, and the ratio of mixer stage blocking level to RF-stage blocking level should be made as high as possible. The local oscillator source should be carefully designed for optium signal to noise ratio, and the RF-stage should operate at a current level which ensures sufficient dynamic range.

The examples which we have chosen also indicate, that field effect transistors tend to yield better blocking properties than bipolar transistors do. This is partly due to their square law characteristic, and partly to the very low mixer noise figure that can be obtained with these devices - which by the way is also due to the square law characteristic.

A reasonable good agreement has been found between the calculated and measured results. This proves that the simplified theory developed above will be for most practical applications sufficiently accurate to permit calculation of the 3 dB blocking level of the receiver.

A more important result of the investigation is perhaps that it has provided a very valuable tool for a qualitative analysis of the mechanism of reveiver blocking.

References

1. Roger R. Webster, "The Noise Figure of Transistor Converters", IRE Transaction on Broadcast and Television Receivers, November 1961, pp. 5o.....65.

2. I.S. Vogel and M.I.O. Strutt, "Noise in Transistor Mixers", Proc. IEEE February 1963, pp.34o.....349.

3. T. Suziki and M. Toya, "Desensitization of Transistorized PM Receivers ", Electronics and Communications in Japan 1965, pp. 85.....9o.

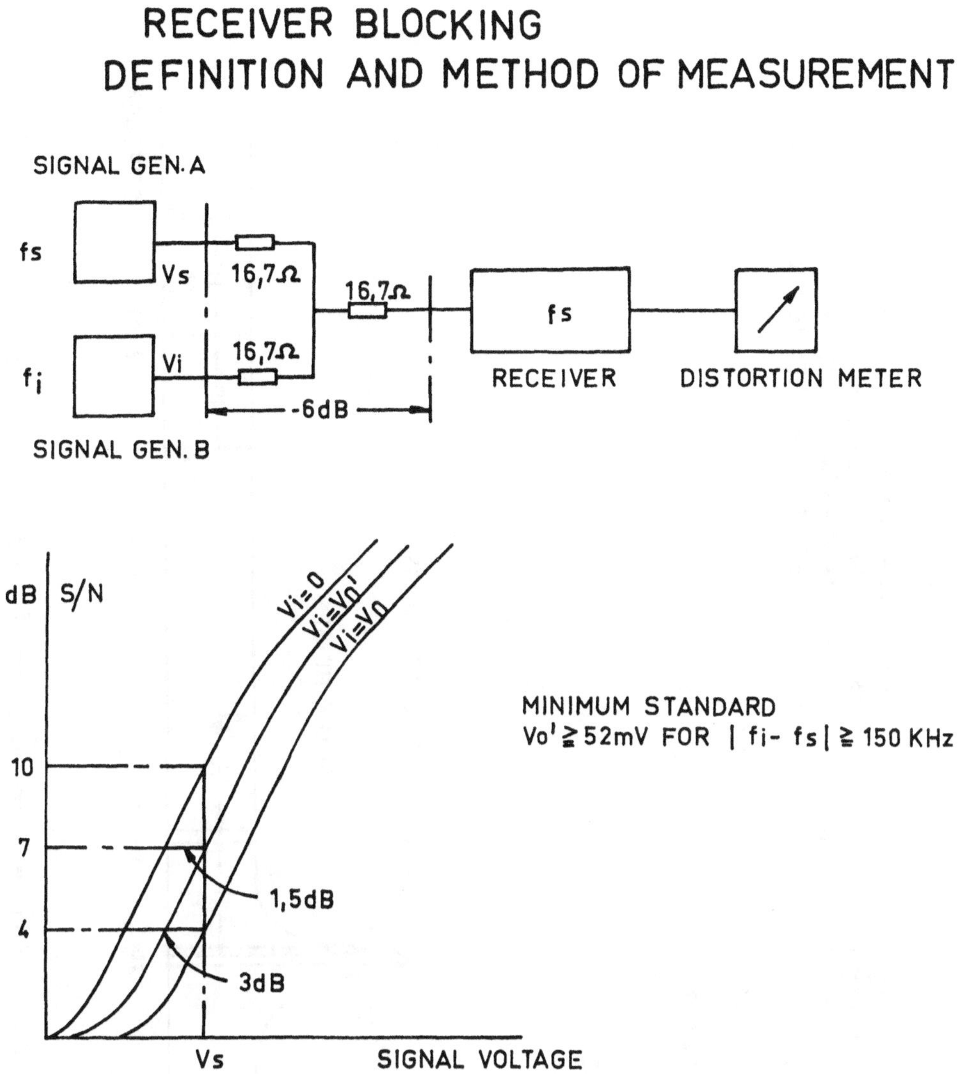

Fig. 1. Receiver blocking. Definition and method of measurement.

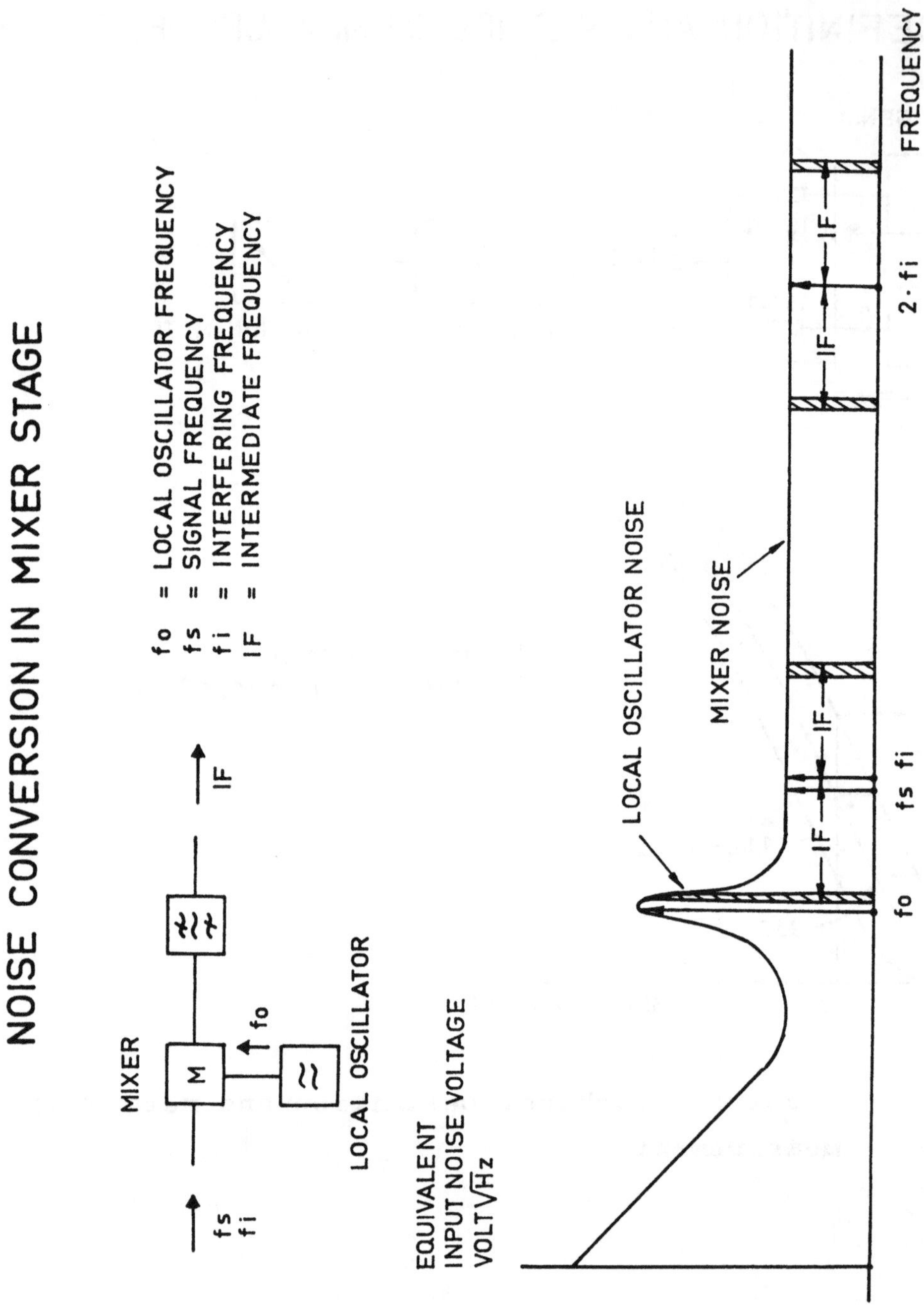

Fig. 2. Noise convertion in mixer stage.

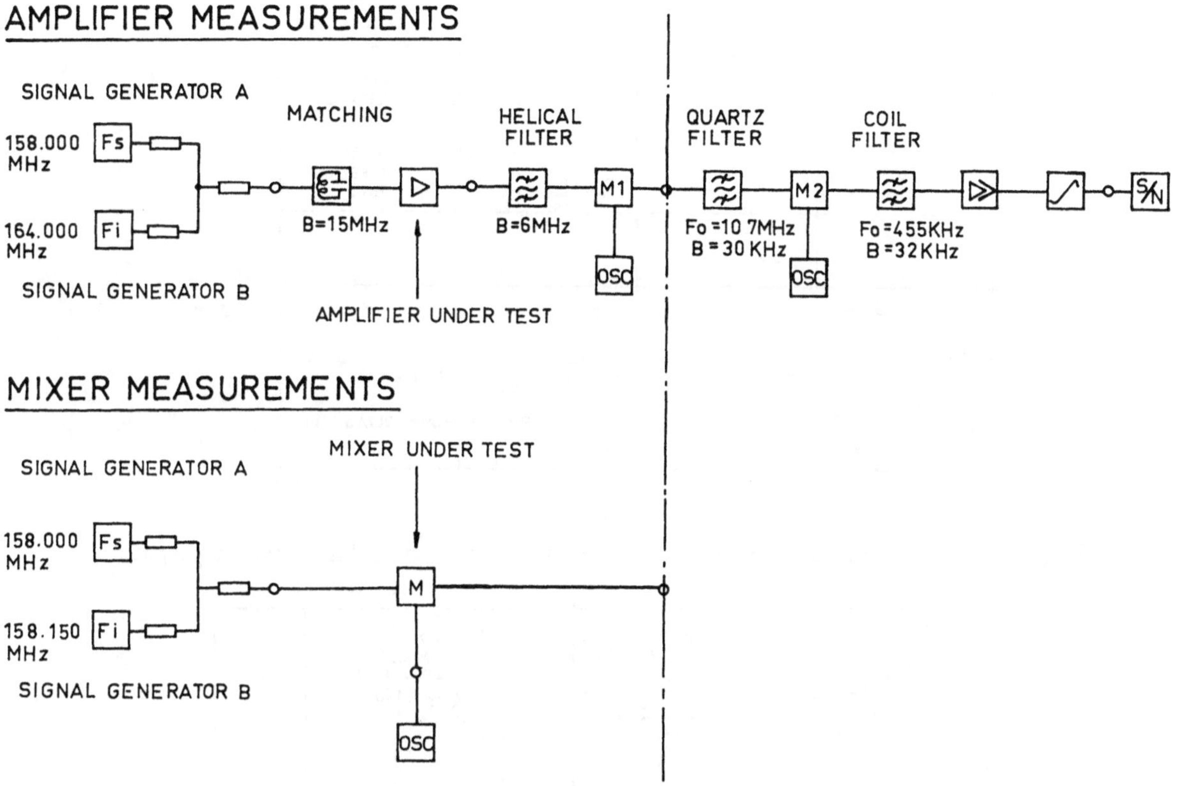

Fig. 3. Measurement of amplifier and mixer blocking levels.

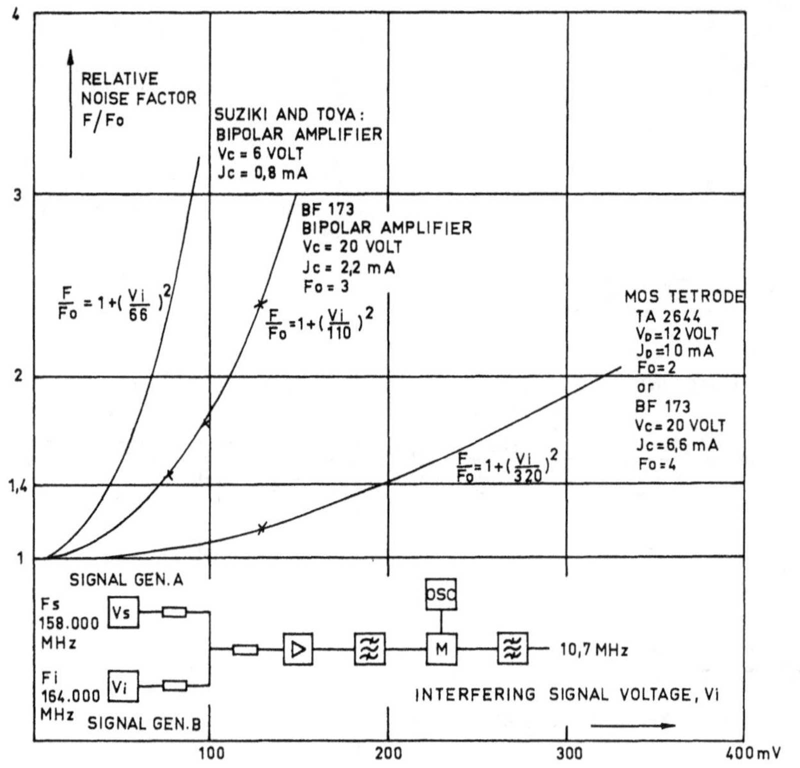

Fig. 4. Noise factor increase of amplifier stage.

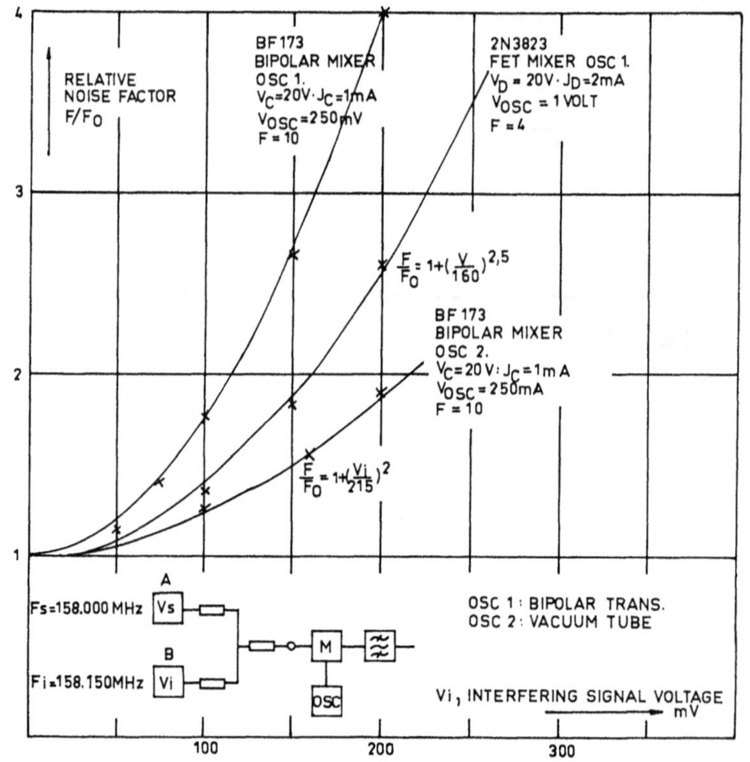

Fig. 5. Noise factor increase of mixer stage.

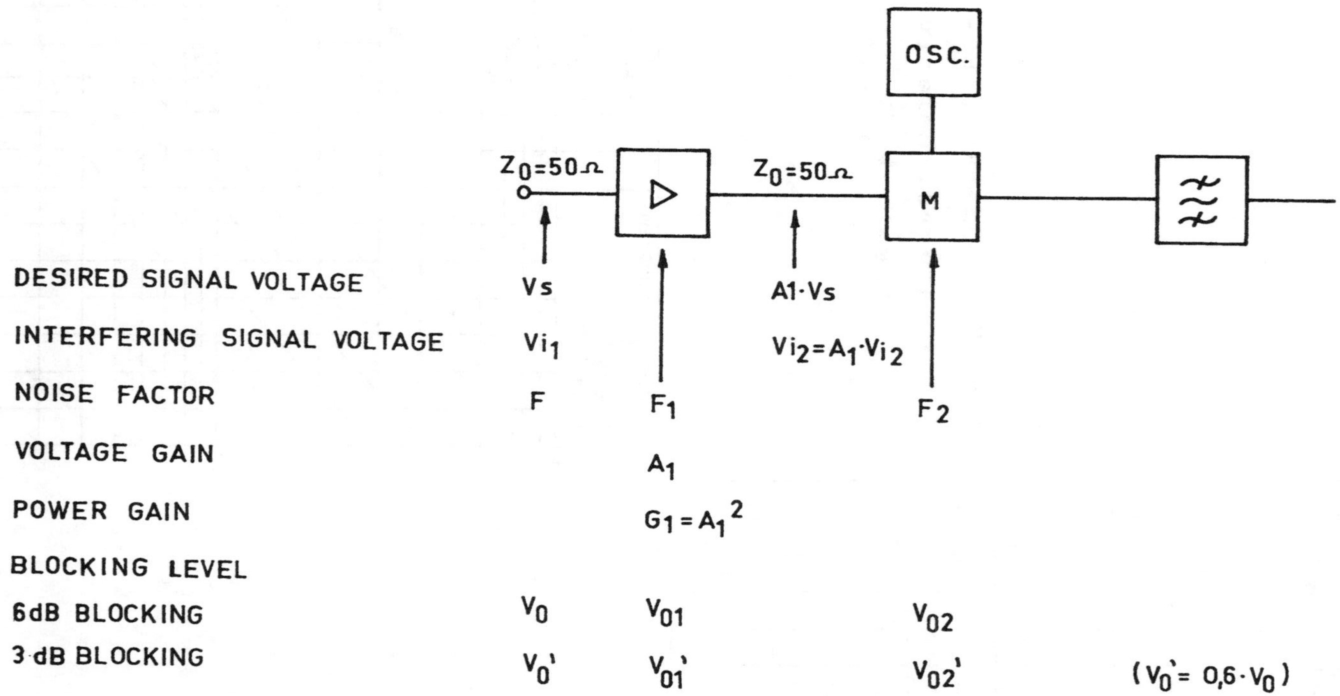

Fig. 6. Cascaded RF- and mixer stages. Parameters used for calculation of overall blocking level.

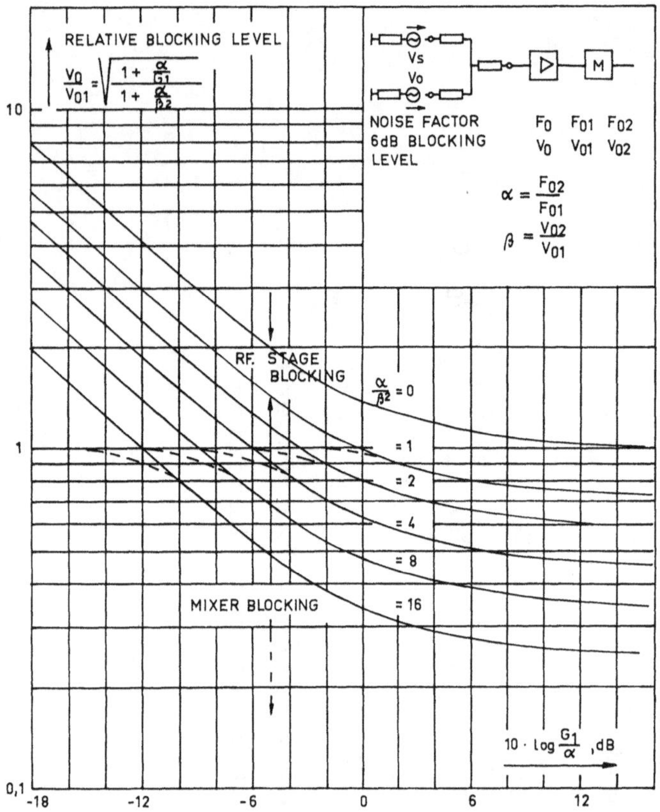

Fig. 7. Cascaded RF- and mixer stages. Relative blocking level, V_o/V_{o1}, versus G_1/α

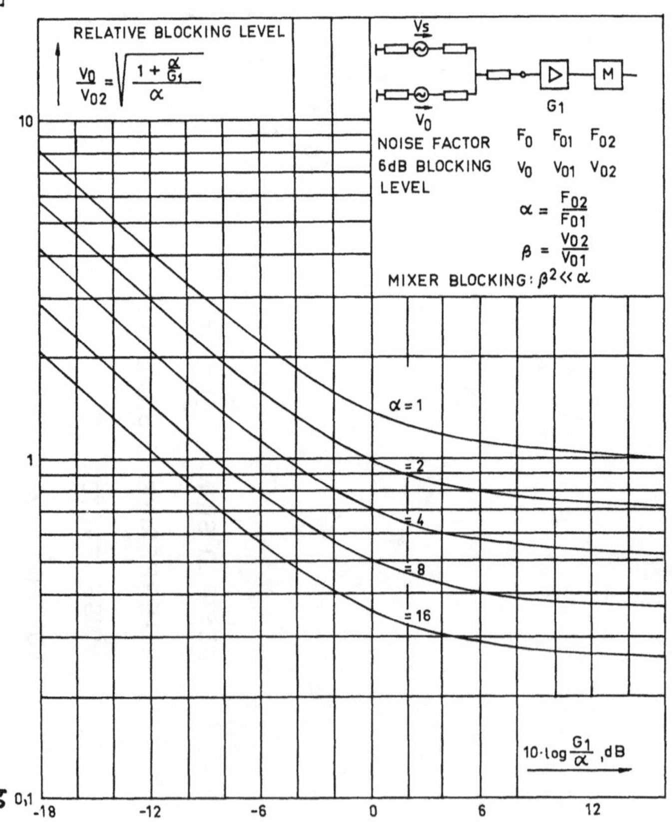

Fig. 8.

Cascaded RF- and mixer stages. Relative blocking level, V_o/V_{o2}, versus G_1/α. No RF-stage blocking

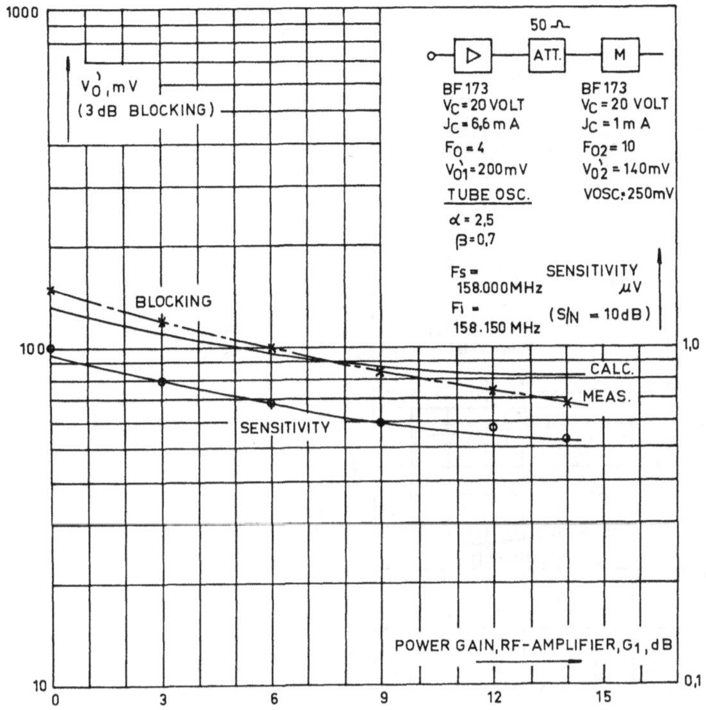

Fig. 9.

Bipolar RF-stage and bipolar mixer in connection with bipolar local oscillator/ multiplier chain.
3 dB blocking level, V_o', versus RF-stage power gain, G_1.

Fig. 1o.

Bipolar RF-stage and bipolar mixer in connection with vacuum tube oscillator/ multiplier chain.
3 dB blocking level, V_o', versus RF-stage power gain, G_1.

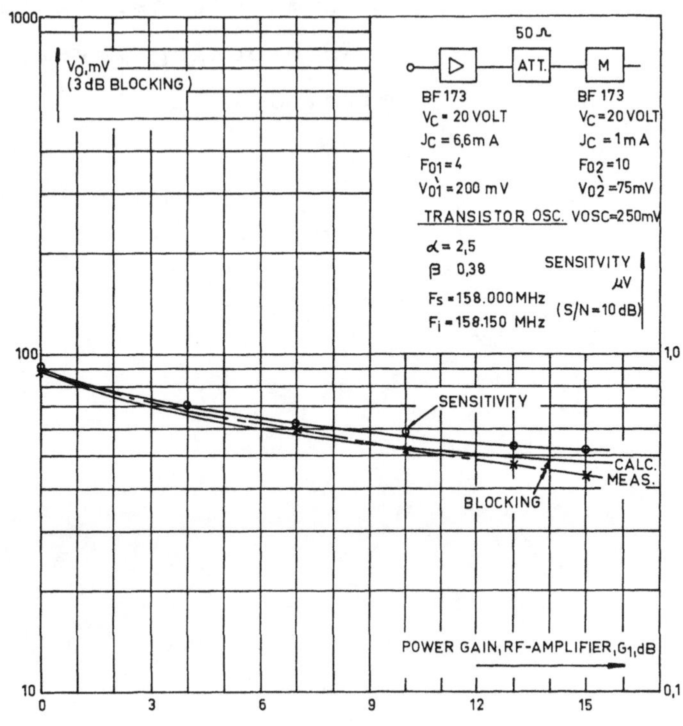

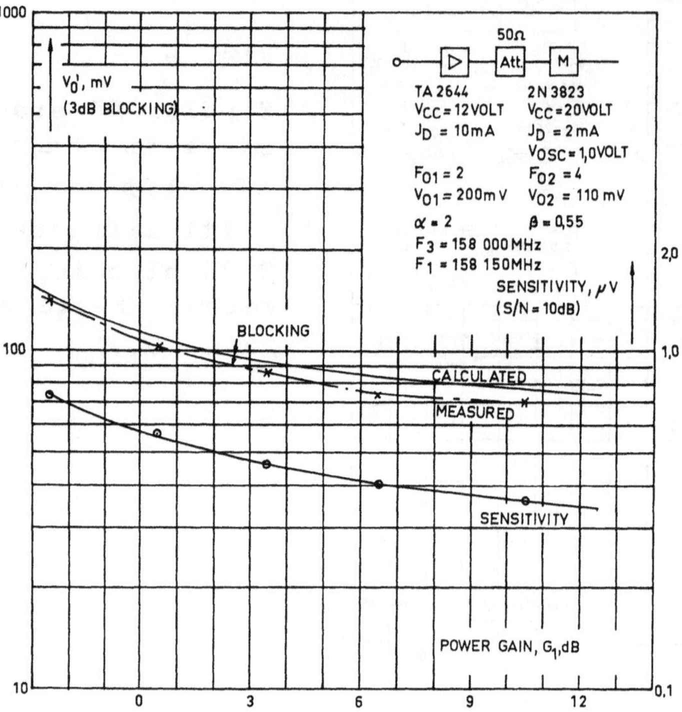

Fig.11. MOS RF-stage and FET mixer in connection with vacuum tube oscillator/multiplier chain. 3 dB blocking level, V'_o, versus RF-stage power gain, G_1.

Tackle Noise Variations at Video
R.W. Cope
Electronic Design Magazine, Vol. 20, September 26, 1968
Pages 54-56

6.4

Maintaining background noise at a constant level is of crucial importance in certain systems, such as radar and pulse-type navigation. In radar, circuitry that keeps the noise level fixed is called constant-false-alarm-rate (CFAR) circuitry.

Until very recently, true CFAR processing could only be performed at i-f (before signal detection), although requirements existed for post-detection processing. Now a new technique for performing CFAR processing at video is available; it should have wide applicability in radar, communications, and navigation systems, where jamming is a problem.

In the case of radar receivers, a constant noise background results in a constant amount of snow on the indicator—an important reference in determining targets. In the case of navigation and transponder receivers, detection is usually performed with a threshold level. Accordingly, with a *constant* threshold level, an increase in noise level would increase the detected false-alarm rate, and would reduce maximum attainable receiver sensitivity.

Until now, techniques that provide CFAR signal processing have been limited to:

- Wide bandwidth limiting followed by narrow bandwidth filtering.
- Conventional logarithmic receiver processing.
- Automatic gain control (agc).

Dicke fix works on wideband noise

The wide bandwidth limiter followed by a narrow bandwidth filter, sometimes called a Dicke fix, is shown in Fig. 1. It normalizes the power levels of all signals, after which the signals are processed through a narrowband filter whose bandwidth is optimized for the desired signal.

The noise or undesired signals are assumed to be of wide bandwidth. The narrowband filter rejects all of the noise that falls outside of its passband and thus suppresses the noise by an amount equal to the ratio of the narrow and wide bandwidths. In this way, the desired signals are effectively enhanced by the bandwidth ratio.

The Dicke fix displays CFAR characteristics because the wideband limiter is designed to hard limit on the lowest level of expected noise. Any additional noise is noticeable only insofar as it masks the signal. Therefore, its effect is to lower the signal without rising itself.

Log amplifier works on narrowband noise

The log receiver takes an entirely different approach to providing CFAR characteristics: It performs CFAR processing on narrowband noise. The noise bandwidth is considered narrowband when it coincides with the desired signal bandwidth. As shown in Fig. 2, it consists of a number of cascaded linear i-f amplifiers with a detector on the output of each. The outputs of the detectors are added together and their sum is accoupled to the output terminal. The transfer function of this device (i-f input volts to video output volts) is a piecewise linear logarithmic approximation.

In the log receiver, the dc portion of the output signal is produced by contributions from all the detecting stages, the limiting as well as the nonlimiting, whereas the ac portion is produced by contributions from the nonlimiting detecting stages only. This is because in those stages where the amplitude of the i-f is held constant due to limiting action, no detectable ac is produced. In

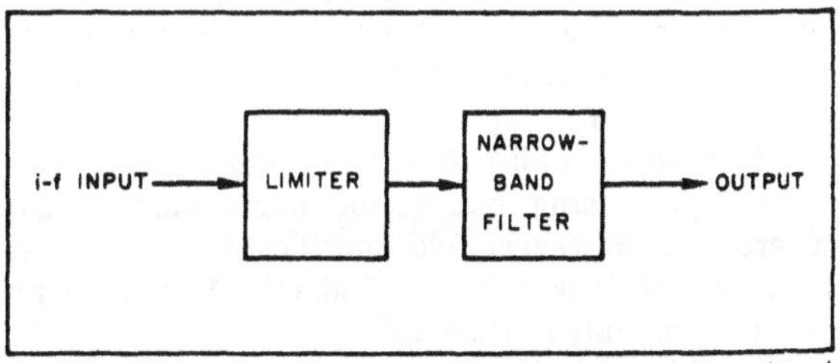

1. **The Dicke fix,** improves the signal-to-noise ratio of its output by the ratio of the input to output bandwidths. If the wideband limiter is designed to hard limit on the lowest expected noise level, the Dicke fix will have constant-false-alarm rate characteristics. This can't be done at video for the bandwidth will already be narrow.

Robert Cope, Senior Engineer, The Bendix Corp., Baltimore, Md.

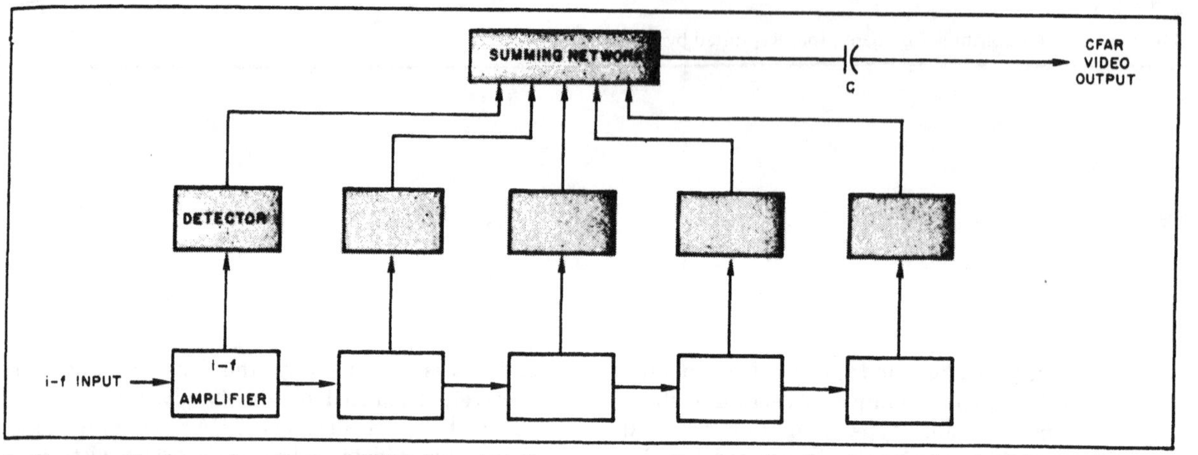

2. **The logarithmic receiver** displays CFAR characteristics because the stages that produce the ac portion of the output always contribute the same amount of total signal regardless of the input noise level. This condition is maintained so long as at least one, but not all, of the amplifier stages are limiting.

the nonlimiting stages both dc and ac components appear in the detected output. The early stages, containing signals too small for detection, contribute nothing to the video output.

The key to CFAR performance in this case is that the stages that contribute the ac portion of the output always contribute the same amount of total signal. Because the stage gains are all identical, the amount of total signal produced by the nonlimiting detecting stages is always the same, regardless of the input noise level to the log receiver. To obtain CFAR performance, the output signal is ac-coupled to eliminate the dc portion of the detected signal.

Agc suffers from time lag

The agc approach to CFAR processing is as follows: The system senses the average amplitude of the output noise of an amplifier and develops a voltage that is fed back to control the gain of the amplifier. In this way the average level of the output noise remains constant.

The major problem with the agc technique is that the integration time required to determine the average noise level produces a time delay in adjusting the gain. This time delay makes the approach an unsatisfactory way to obtain CFAR characteristics.

Up to now, in all CFAR techniques except agc, CFAR processing had to be performed at i-f. There are, however, two specific cases in radar alone where it is essential that CFAR processing be at video rather than i-f.

The first case is when CFAR capability is desired in a moving target indicator (MTI) system. The MTI processing is done first and includes detection, so that CFAR processing must be at video. The second case is when one wishes to combine CFAR processing with differentiated linearly detected video. The differentiation process discards any dc component in the a-m-detected video that may be prevalent in ground clutter, weather return or jamming. However, it is necessary that the detection process be linear so that large cw components do not suppress small signals occurring at the same time.

This obviously precludes the use of conventional logarithmic receivers because they are inherently nonlinear.

It is also not possible to use the Dicke-fix because the bandwidth has already been optimized to that of the signal. It is necessary that this be done prior to video detection to prevent degradation of the signal-to-noise ratio (S/N) in the case of low-level signals. The loss in S/N in a detector is 2:1 in dB for values of S/N less than unity.

In addition to the detector problem, there is additional loss of S/N degration in the limiting process. Limiters produce approximately 1 dB of S/N degradation for values of S/N less than 0 dB.

Log approach at video solves problem

The recently developed video CFAR, shown in Fig. 3, can handle both of these cases through its unusual ability to perform CFAR processing on narrowband bipolar video noise. Its concept is similar to that of the log receiver in that outputs from each stage are summed. The stages that are limiting produce only dc in the output, the stages containing large signals but are not limiting produce both ac and dc, and the earlier stages containing small signal levels contribute nothing to the output.

This similarity to the log receiver can be seen in a comparison of the block diagrams. The major differences are that the cascaded amplifiers handle video and not i-f, and that the detectors are full-wave and not simple peak or half-wave

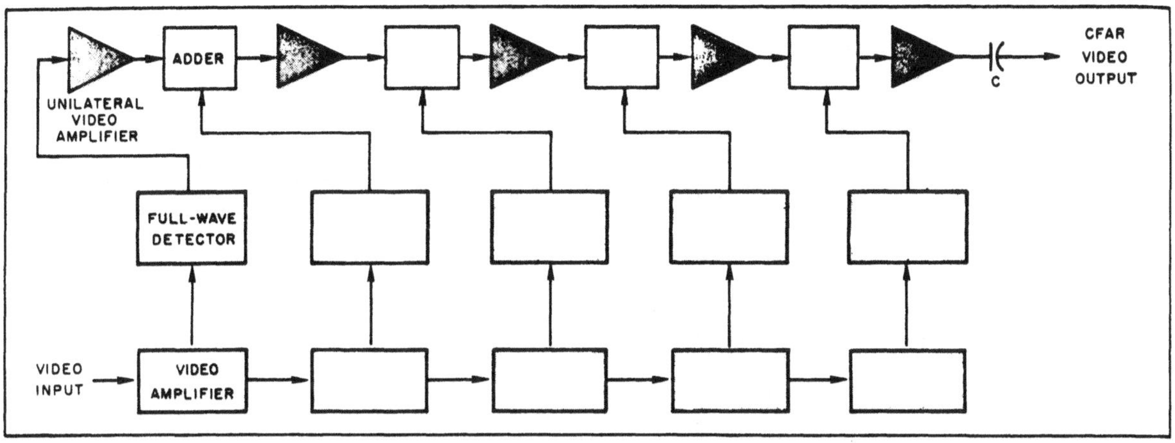

3. **True CFAR at video** is obtained with a log amplifier similar to that of Fig. 2. The detectors must be full-wave because the video signal is bipolar. The summing network employs unilateral video amplifiers to prevent undesirable feedback between the input and the output. The capacitor blocks the dc portion of the detected signal.

averaging type detectors.

The CFAR performance of this new device is due to the use of full-wave detectors. The full-wave detection of the noise can be looked upon as signal unipolarization such as is used in processing conventional MTI video. In the amplifier stages, where the noise is hard-limiting, the composition of the noise is square waves of varying frequency. When these square waves are full-wave detected, the result is a dc voltage. This particular phenomenon, the production of a pure dc voltage from the full-wave rectification of constant-amplitude square waves, is the main factor in the operation of the video CFAR.

How many stages do you need?

In designing a video CFAR, probably the most important factors you should consider are total dynamic range and the constancy of the noise level over the dynamic range. The CFAR dynamic range is equal to the gain of the individual stage times (n—1) when n is the number of stages, assuming that all the stages have equal gain. The n—1 factor permits the last stage to be in limit.

The second consideration, noise-level constancy, is dependent on the gain per stage. For any given dynamic range, fewer stages mean higher gain per stage, and thus fewer straight line segments to approximate the logarithm. Poor approximation of the true logarithm means variation in the noise level over the dynamic range. Thus, lower gain per stage results in a better approximation of the logarithmic function. A good nominal value for gain per stage is 10 dB.

The output of the CFAR circuit is fed through a capacitor to block the dc component. The size of this blocking capacitor is based on the signal bandwidth and the terminating resistor. For example, a 1.0 μF capacitor will permit 10 per cent droop of a 10 μs pulse when fed into a 100-ohm load. A capacitor larger than 1.0 μF will produce less droop, but will increase the decay time of output transients caused by sudden changes in input noise level.

The only design considerations of significance in the full-wave detectors and the adder network are these: The detectors on all stages should be identical and the gains measured from all detectors to the output terminal should be identical. It is important also that the adder network be unilateral to prevent undesirable feedback between output and input.

A circuit was designed and built to test the new video CFAR concept. It used six stages of amplification with 12 dB of gain per stage. This circuit preserved CFAR performance to within one decibel over a nearly 60-dB input dynamic range. ■■

Test your retention

Here are questions based on the main points of this article. Their purpose is to help you make sure you have not overlooked any important ideas. You'll find the answers in the article.

1. What is CFAR processing? When is it desirable?

2. How does a piecewise linear logarithmic amplifier obtain its CFAR characteristic?

3. Why is agc not always a satisfactory way to obtain CFAR characteristics?

Digital MTI, A New Tool for the Radar User
J.G. Ellis
Marconi Review, Vol. 36, No. 191, 4th Quarter, 1973
Pages 237-248

6.5

After a brief history of analogue MTI processors, the article outlines the areas of difficulties which are encountered using digital techniques. Consideration is given to the speed and accuracy of conversion as well as the type of canceller employed.

There follows a discussion of techniques for overcoming the effects of blind speeds and phases, in order to optimise the response of the MTI filter.

The problem of moving clutter is discussed and two methods are given for combatting this, these are log conversion, followed by a short time constant and Doppler compensation.

Correlation and integration techniques are considered as methods for eliminating impulsive interference and enhancing the signal to be displayed.

Throughout the article particular consideration is given to the problem of interfacing a processor with a plot extraction system.

Finally illustrations showing a modern processor, and the video display, before and after processing are included.

History

In the early days of radar a system consisted simply of a transmitter, a receiver and a display. Such simple systems as this suffered from several disadvantages, such as ground returns, interference and jamming; however they were virtually impervious to the effects of weather returns due to the comparatively low radiation frequency employed.

To overcome the above difficulties is was necessary to find some method of storing the information between radar pulses so that some form of comparison could be made between consecutive returns.

The first device to be used for this storage was the water delay line. This device was cumbersome and combined with thermionic valve circuitry it suffered greatly from instabilities and was very liable to failure.

The next device to be used for storage was the mercury delay line; this was more reliable but still very cumbersome. One technique that was also used with this type of storage was that of running two systems in cascade (Fig. 1) so that the response to different Doppler frequencies shown in Fig. 2(b) was obtained, in lieu of the response from a single cancellation system shown in Fig. 2(a). The other advantage of the double cancellation

*Marconi Radar Systems Limited.

system was that any drift in one of the cancellers was partially masked by the second.

However such a system was very large and complex and still suffered from unreliability. The next advance in technique was the use of quartz

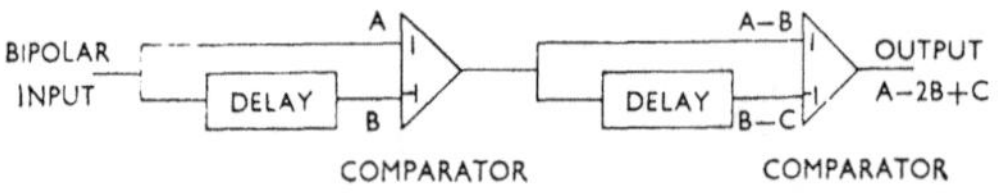

Fig. 1.

crystal delay lines. These had greater stability and a wider dynamic range. It was also possible to time multiplex the trigger generation with one of the signal channels and use the delay line in this way to determine the basic p.r.f. (see Fig. 3).

A system which cancelled stationary returns by the above techniques was termed a "moving target indicator" (MTI).

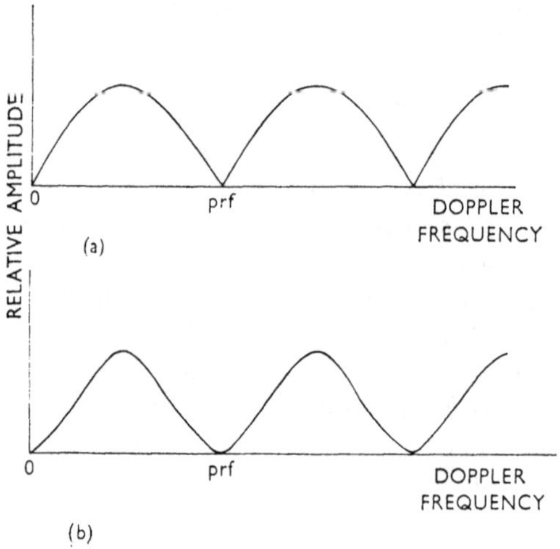

Fig. 2. (a) Response of a single cancellation system; (b) Response of a double cancellation system.

MTI processors using quartz lines and semiconductor techniques were much more reliable than previous systems, however they are expensive and extremely inflexible. This made it very difficult to operate with a staggered p.r.f. system or in a frequency diversity application. (That is a system where two transmitters, operating on different radiation frequencies in the same r.f. band, feed into one aerial).

Digital MTI

In order to improve on analogue systems digital techniques were considered, but before they could be used two problems had first to be overcome, namely, the conversion and the storage of the data.

Analogue to Digital Conversion

The conversion process posed problems because of the relatively wide bandwidth of radar signals and the amplitude accuracies required. If we consider a received pulse of 2 microseconds duration, it is necessary to sample the pulse at least twice in order to be able to reconsititute it

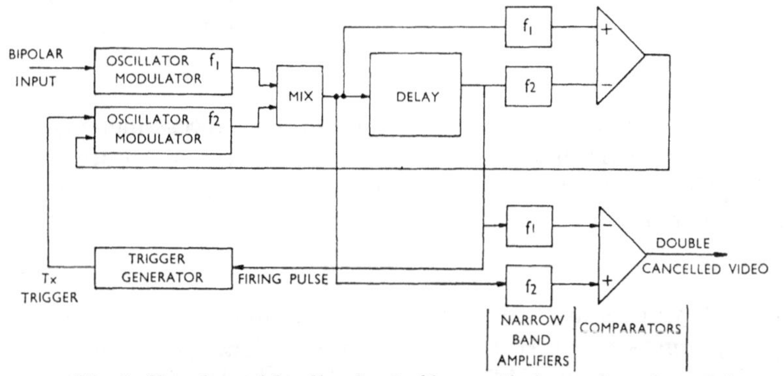

Fig. 3. Use of one delay line for double cancellation and p.r.f. control.

faithfully, at the same time making sure that there is negligible loss of detectability due to the sampling process. The "analogue to digital" converter (A/D) should therefore be able to operate at a 1 MHz sampling rate.

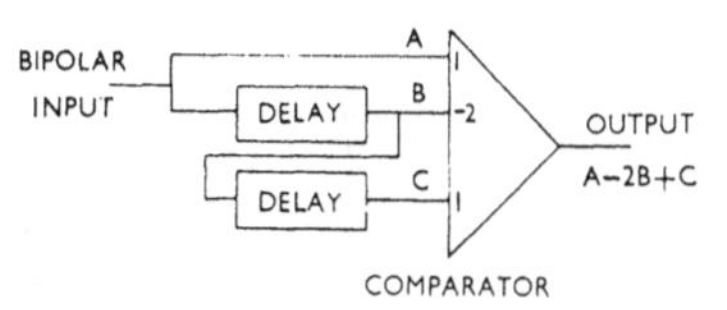

Fig. 4.

When cancellation capabilities in the order of 40 dB are required, it is necessary to digitise the bipolar radar information to at least 8 bit accuracy (256 discreet levels).

Before a digital MTI could become commercially feasible an A/D had to be developed which could be manufactured in quantity, with an 8 bit resolution in under 1 microsecond.

Storage

The problem of information storage was finally solved with the advent of MOS (Metal–Oxide–Silicon) storage techniques. (Before this, core stores were available which worked at the correct speed, but they were bulky, power consuming and very prone to radio interference).

The MOS shift register is cheap and getting cheaper, easily manufactured in quantity and, because it stores the information in a serial manner, ideally suited to the simple MTI application.

The shift register may, however, soon be superseded by random access memories, which can easily fulfil the MTI requirement and at the same time provide access to any specific data during the time when the store is

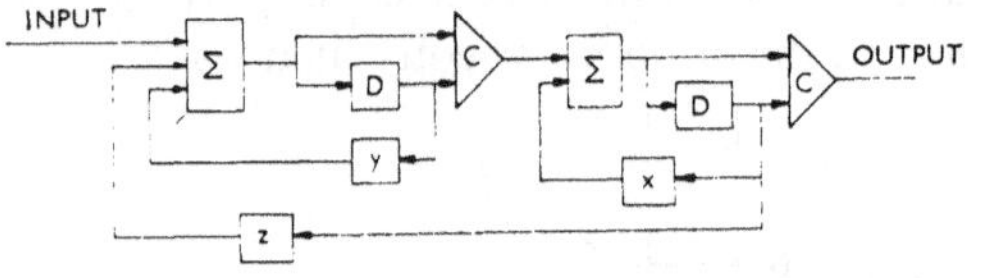

Fig. 5. A typical double canceller with feedback.

not being used for the MTI process. In fact the random access memory may be used to perform part of the non-real time plot extraction process, or to provide Doppler information for use with an extracted output, thus helping a computer in a tracking programme.

The Canceller

The early digital systems used two single cancellers in cascade to make a double canceller. Some systems still use this technique. There is, however, no need for this in a simple *digital* MTI system as the stability of the stored information is high enough to allow the use of the multiple comparator as shown in Fig. 4. In fact as explained below, it is inadvisable to use the system of cascaded single cancellers in a digital signal processor.

Fig. 6. Typical response of a system using feedback techniques.

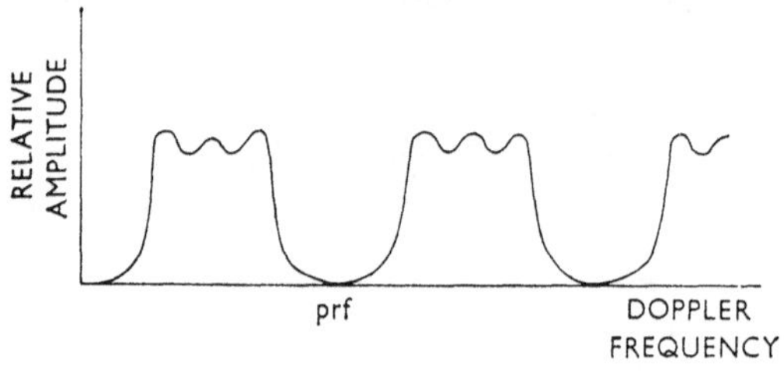

If the bipolar input to the canceller in Fig. 4 is quantised into 8 bits, then both the signals (A) and (B) are in 8 bit form. For an optimum velocity maximum amplitude target the output of the comparator (A–B) will be of 9 bit amplitude.

In order to pass this signal into the second cancellation channel, we must either use a different store, throw away the least significant bit (l.s.b.) or limit the signal amplitude to 8 bits.

The use of a different store is not advisable from the spares holding aspect. Throwing away the l.s.b. limits the cancellation performance of the system, and whilst the limiting of the signal is quite feasible in a simple MTI system, it does reduce the degree of super clutter visibility that may be obtained by any form of post MTI processing.

With the cascaded canceller system it is possible to use feedback techniques as shown in Fig. 5. By these means a velocity characteristic can be obtained such as that shown in Fig. 6. However this causes the impulse response of the system to be modified drastically so that for a characteristic with steep sides as in Fig. 6, thirty or more pulses can be bred in the system. In any situation where interference may be encountered the system using feedback becomes unusable and as most radar installations have to cope with interference, either from nearby radars or intentional jamming, there is a growing feeling among radar users that it is not worth installing such a system.

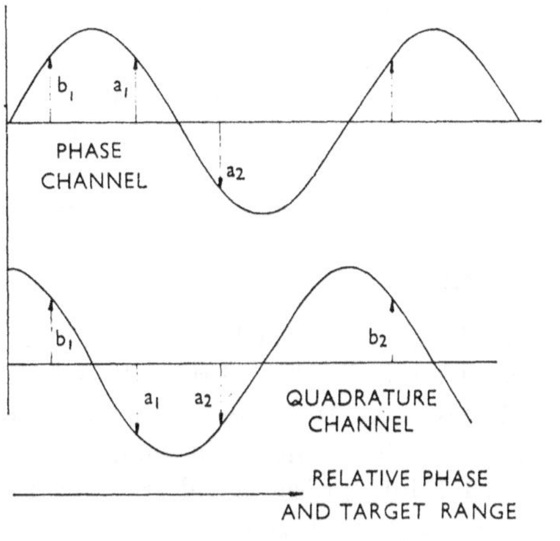

Fig. 7. Blind speeds and phase in a quadrature MTI system.

Blind Speeds and Phases

So far we have only considered simple MTI systems, the response of which leaves much to be desired because it is subject to blind speeds and blind phases. Blind speeds occur whenever the phase change pulse to pulse is equal to a multiple of 360 degrees. Blind phases occur when the output from the phase detector is the same pulse to pulse even though the angle between them is some value other than 360 degrees.

These effects can be seen in Fig. 7; if we consider the returns from target (a) it can be seen that in the *phase* channel the detected signal is alternately positive and negative, whereas in the *quadrature* channel the two signals are of equal amplitude and will cancel when passed to the comparator. This signal in the quadrature channel is an example of a blind phase. Target (b) is the same amplitude pulse to pulse in both channels and is an example of a blind speed.

If an MTI system consists of two such parallel cancellation circuits which are combined after cancellation and rectification, then the blind phase phenomena will be almost completely alleviated.

Although it would appear at first sight that by measuring the absolute phase of the returned signal, the blind phase problem can be resolved without the use of a separate quadrature MTI channel, this is not in fact so. In order to carry out the measurement of true phase with respect to the reference it is necessary to limit the amplitude of both the reference and the incoming signal, thus all amplitude information is destroyed so that a true phase output can be obtained. Therefore a target which is alternately in phase and anti-phase with a clutter return, will, if the clutter is larger than the target, always give the same phase signal and hence it will cancel as a blind phase.

Another way to attempt to reduce the hardware in a phase and quadrature system is to multiplex the two signal channels, switching alternately from one channel to the other, on the assumption that we have sufficient data on the two channels in the alternate samples.

Although when combined with a staggered p.r.f. this system may produce an acceptable output for a plan position indicator (PPI), the signal will not be acceptable to a modern plot extractor.

Staggered P.R.F.

In the past staggered p.r.f. systems intended to reduce the effects of blind speeds were limited to three different interpulse periods, due to the problem of realigning the video before cancellation.

In a digital system this realignment of video is not necessary as the returns for each interpulse period are stored with reference to the time of transmission, so that cancellation is achieved whilst the video is still in staggered form and realignment is only necessary when other equipment such as displays etc. require it.

As the stored information is always referred to the transmitter time, more elaborate stagger arrangements are possible than with analogue MTI. In practice 5, 6 or 7 different periods are employed in a modern system and using 6 intervals it is a simple matter to raise the first true blind speed to 100 times the natural blind speed of the system.

Another form of staggering that is now possible is to delay the transmitter pulse on the basic trigger in a random manner. Although the velocity response of such a system can not be easily defined, the system has obvious advantages in an impulsive jamming environment.

Plot Extractors

Before we consider any form of processing to follow the MTI system, it is worthwhile considering the requirements of a plot extractor.

The main purpose of a plot extraction and validation unit is to seek and locate all targets contained in the radar video, and to pass these targets in narrow band form either directly to a computer complex or over a narrow-band link to a remote site.

The three points to remember are that any processing must not reduce the overall probability of detection, nor bring about any positional inaccuracy, either in range or azimuth. At the same time the probability of false alarm must be kept as low as possible.

The signal must be sampled finely enough to allow the extractor to seek small (targets that would otherwise be lost in noise) by using some form of integration technique.

This need for accurate location and detection is sufficient to warrant the expense of both in-phase and quadrature channels because, if we only use one channel or switch in turn between one channel and the other, there will be a tendency for the return to become broken up in amplitude and for it to develop ragged leading and trailing edges.

Another cause of loss of positional accuracy is the use of any form of range or azimuthal integration. This can follow the MTI system, or when feedback techniques are employed, actually be in the MTI system itself.

Targets can also be lost or misplaced due to the use of circuits for "Pulse Repetition Frequency Discrimination" (PRFD), or "Pulse Length Discrimination" (PLD or FTC or STC); which can cause the amplitude of the target to become broken up. In the new Marconi Processor all these facts were considered when deciding on the types of processing which should be employed before the signals are sent from the processor to the extractor, and which types should be left until later, for use with an output to PPI displays etc.

The Problem of Moving Clutter

In any MTI system the clutter spectrum may contain components which can pass through a canceller. These can take two forms:

(a) Coherent variations due to a definite Doppler component of the clutter returns.

(b) Random or incoherent changes due to internal fluctuation of the return.

There are two techniques which can be used to overcome these effects. One is to compensate for the Doppler shift, the other is to use some form of short time constant (STC) circuit.

If we consider first the STC system, it will operate against both types of variation, but only if the clutter return is long enough for selective action to be taken. It also only provides super-clutter visibility, this means the signal to be detectable must be greater than the clutter residue by at least the minimum signal-to-noise ratio necessary for detection in system noise.

As many of the returns which pass through an MTI system, such as rain cloud or jamming, are of a noise like character, a simple STC circuit is not sufficient and a more sophisticated system must be used. It has long been known that the use of a logarithmic receiver can greatly alleviate the effects of noise like returns, when used in conjunction with some form of pulse length discrimination (PLD). Until this time this technique has been limited to intermediate frequency log amplifiers, but using a digital log conversion technique we can now apply it to the video signals from an MTI system.

Referring to Fig. 8(a). we see an approximate representation of a radar video return containing noise-like clutter. In Fig. 8(b) this has been converted to logarithmic form. the point to note being that the fluctuations

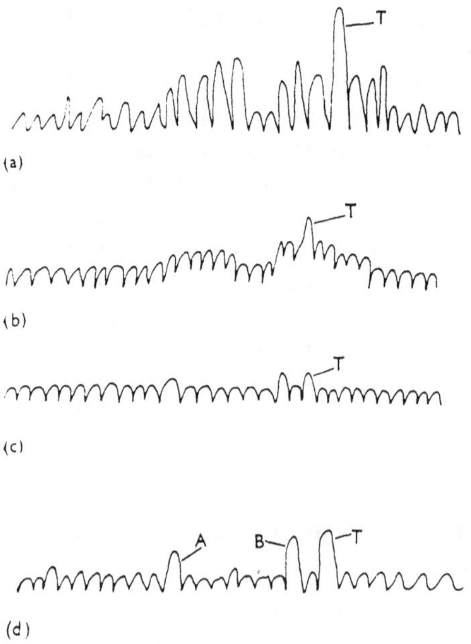

Fig. 8. (a) Linear noise containing noise like clutter and a target "T"; (b) The logarithmic equivalent of (a); (c) The logarithmic response after the application of a short time constant; (d) The signal restored to linear form showing the target "T" and the leading edges of the clutter, A and B.

due to the input noise are of a constant amplitude; only the d.c. or mean level varying.

In Fig. 8(c) this mean level has been eliminated by subtracting an integrated version of the signal from the signal itself, such that only the signal and the leading edges of the clutter blocks are left above the system noise level.

If the STC action is limited to only those returns which are longer than some pre-set value the majority of the noise will not be distorted, and hence the input to a plot extractor will still contain the structure needed to allow detection of small targets.

It should be stressed that this technique only provides super-clutter visibility, and does not seek targets which are smaller than the clutter in which they are contained, whereas the next system does seek such targets, however one advantage of the Log STC system is that it is independent of the Doppler content of the return.

Doppler Compensation

In order to achieve a measure of sub-clutter visibility we must find some way of eliminating the Doppler component of the clutter, at the same time allowing wanted signals to pass through the system with as little attenuation as possible. For instance the normal returns associated with clouds are such that if we can eliminate the radial velocity component, then the pulse to pulse changes are small enough to allow cancellation.

In the past techniques have been developed for cancelling this component by modifying the reference signal so that it contains a pulse to pulse phase shift equal to that contained in the return. This modification was controlled by an operator who had to set up sectors on the PPI display and then find by trial and error the speed and direction of motion of the centre of gravity of the clutter. This compensation was then modified by the rotational information.

In a complex modern radar system this continuous use of a highly skilled operator is not practical, nor is it economical, for it means that although the radar is in a fully operational state, someone must always be present to monitor and adjust the equipment as the weather changes its condition and velocity.

Digital techniques, and in particular the availability of both phase and quadrature information, mean that is is possible for this complex process to be performed automatically and a processing system produced, which in the absence of any form of moving clutter continues to work as a simple MTI system; cancelling only those returns with a Doppler frequency close to zero. However when moving, that is non-cancelling clutter, is present the system first calculates the necessary compensation, and applies it, after detection in combination with a subsequent Log/STC system to eliminate the unwanted returns. In this way we can first obtain the maximum sub-clutter visibility, and should there be any residual clutter, continue to achieve at least super-clutter visibility therein.

As the compensation is performed after phase detection using digital techniques, the system can monitor the output of a simple MTI system at the same time as the normal MTI process is performed. A choice can then be made as to whether to use the Doppler system in lieu of, or in series with the simple system. In this way the problem of moving clutter which is coincident with stationary clutter, can be eliminated.

This system shows to great advantage the use of the multi-comparator technique, for not only is the full dynamic range of the system maintained, but the current, delayed once and delayed twice etc., information is available in its original form; allowing compensation to be achieved at the same time as the simple cancellation process without the use of extra storage.

Clutter Switching

A major problem in any system which seeks to switch from one video to another automatically, is the decision process which controls the switching. Conventional processors monitor the duration of the return in range, and using this information switch from one video to another which is operationally more suitable. This system is dependent upon the duration of the return, and in regions of solid clutter will operate admirably. However natural clutter does not always lend itself to such a process because especially around the edges, the clutter block tends to have a broken nature, giving a number of closely spaced but separate returns. In such conditions it is impossible using conventional techniques to differentiate between the clutter and multiple targets.

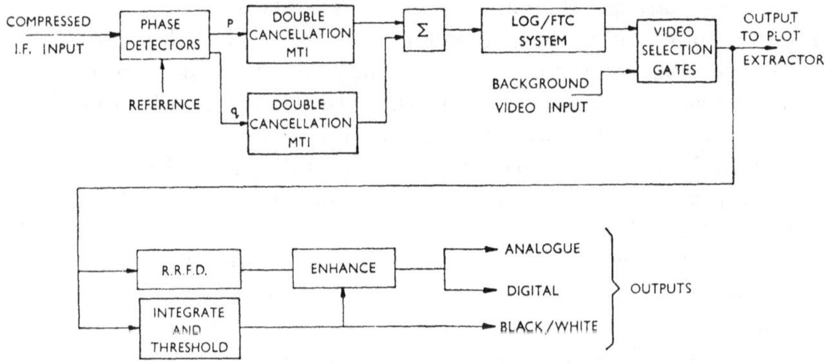

Fig. 9. Simplified block diagram of a modern processing system.

However modern systems can capitalise on cheaper storage techniques in order to provide stored information in azimuth as well as in range. In this way one can store the average clutter value over small segments of the complete PPI area, and using this information the processing system becomes adaptive to varying clutter patterns modifying its response, scan by scan, so that the processing employed is continually optimised to the current data.

Interference and Correlation

In any practical system, interference is a major problem, for on a civil site there is a high probability of co-sited radars, whilst a military system must be able to cope with impulsive jamming.

One technique which is employed to counter interference is "Pulse Repetition Frequency Discrimination" (PRFD). This system stores information from one period to the next, and then seeks correlation between the two signals, eliminating those returns which do not correlate.

Another way of using the correlating properties of signal returns is to integrate the returns azimuthally. In this way the signal-to-noise ratio of the return can be improved and interference reduced; however, the character of the noise after integration is not as suitable to a PPI display as normal noise. A technique for overcoming this problem is to set a threshold on the integrator output, such that a target is declared when the integrator output exceeds the set value.

Fig. 10. Marconi S–7100 signal processor.

The threshold crossings can then be passed out to a display which does not require noise or can be used to "enhance" the normal video, that is increase the amplitude of the output to maximum whenever targets are present. A drawback of any integration system is the tendency for the return to be stretched in azimuth; this can, however, be minimised by first limiting the amplitude of the input signal to the integrator.

Power Supplies and Heat Dissipation

A major problem in complex systems is that of heat dissipation, and in a processor currents of 20 amps are not uncommon, whilst a computer often demands 100 amps or more. However high efficiency power supplies and logic are likely to be readily available in the near future. This together with the use of large scale integration techniques in semiconductors mean that more and more sophisticated processing techniques should soon become economical.

Concluding Remarks

It has been found that the digital processing system can provide increased reliability and, due to the cheaper components such as basic logic and storage much more processing can be achieved at a cost comparable to older techniques. The system is also highly adaptable, not only to meet current requirements but also for future requirements and developments.

As techniques improve the interface between radar video and a computer system will become less difficult so that far more reliable systems can be developed which combine the advantage of primary radar, secondary radar and a computer complex into a much improved system for monitoring aeroplanes and other fast moving objects in even the most adverse conditions.

Figure 9 shows, in block diagram form, an MTI system; while Fig. 10 shows such a modern signal processor, the S–7100 developed by the

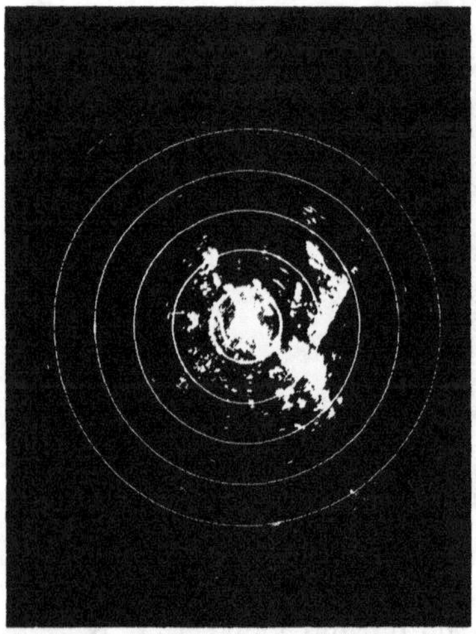

Fig. 11. PPI display before processing.

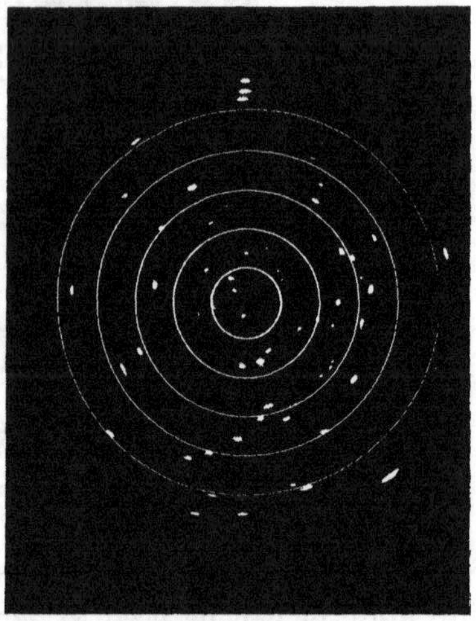

Fig. 12. PPI display after processing

Marconi Company, which can contain 2 MTI systems (phase and quadrature), circuits for moving clutter rejection, destaggering of video, clutter switching, PRFD, integration and enhancement. This equipment can give outputs in both analogue or digital form, with a special digital output available for use with a plot extractor. It also contains trigger distribution circuits and a trigger generation system which can operate in a basic mode, a staggered mode where there can be 2, 3, 4 or 6 different interpulse periods, or a staggered mode where the interpulse period is selected at random.

Figures 11 and 12 show a PPI display of video before and after processing.

Conical Scan in Ground Clutter with Non-coherent and Coherent MTI

C.S. Sorkin and B.D. Steinberg

Proceedings of the IRE East Coast Conference on Aeronautical Navigational Electronics, 1958 **6.6**

Pages 35-45

Abstract

This report is a first order investigation of the performance of a tracking loop system using a conical scanning antenna with noncoherent and coherent MTI signal processing. The major emphasis is with regard to the influence of the interference due to ground clutter echoes. The analysis is carried out in the frequency domain using the techniques suggested by S.O. Rice. The quantitative results describe the general superiority of the coherent system with respect to the noncoherent system except for the performance in the vicinity of the ground track. Furthermore, the results indicate a significant improvement in the performance provided by the double delay and subtraction MTI over and above single delay and subtraction.

A significant result of this study is that a noncoherent system with double delay and subtraction MTI provides adequate tracking performance to be useful in a clutter environment.

Introduction

This report has been extracted from Ref. 1 and includes what the authors believe to be the essential features of the much larger report of Ref. 1.

Fig. 1 includes a block diagram of the systems to be discussed. The analytical techniques used in the present study have been reviewed in the literature in great detail (Ref. 2). Therefore, the details of this analysis are not presented. Instead we shall be primarily concerned with the basic approach and assumptions reviewing the physical principles and discussing the results. Fig. 1a is a simplified block diagram of a standard pulsed radar system with a conical scanning antenna. Although the poor performance of this system in a clutter environment is all too familiar, this system is included for completeness. Furthermore, since this system is familiar to all of us it provides a convenient means of presenting the basic analytical approach used in the investigation of the more sophisticated systems of Fig. 1.

Noncoherent Radar

Fig. 1A shows a noncoherent receiver providing video to a narrowband filter tuned to the conical scan frequency. The filter output is compared in phase with the conical scan reference wave to yield the angular error signals of the angle tracking servos. No MTI or clutter reduction filtering is included.

Fig. 2 defines the geometry of the scan pattern. The instantaneous target strength is determined by ϕ and the beam pattern $G(\phi)$. From Fig.2,

$$\phi^2 = \lambda^2 + e^2 + 2\lambda e \cos \alpha t \tag{1}$$

To facilitate the analysis the two-way voltage pattern of the antenna is assumed parabolic with the half power point at the crossover, i.e., $G/G_0 = .5$ when $\phi = \lambda$:

$$\frac{G(\phi)}{G_0} = 1 - \phi^2/2\lambda^2 \tag{2}$$

Therefore, from Eq. (1) and (2),

$$\frac{G(\phi)}{G_0} \approx \frac{1}{2} - \frac{e}{\lambda} \cos \alpha t, \quad e \ll \lambda, \tag{3}$$

and the return target signal may be expressed as an amplitude modulated wave of the form,

$$s(t) = \sqrt{2T_0}\left(1 - \frac{2e}{\lambda} \cos \alpha t\right) \cos \omega t, \tag{4}$$

where T_0 = IF target signal power when $e = 0$.

The following artifice is used to simplify the analysis. The IF signal spectrum is composed of a fine structure of energy packets spaced by the reciprocal of the radar interpulse period throughout the IF passband. Each packet contains target and clutter signal components. Since these packets are replicas of one another it is only necessary to consider a single grouping to determine the relative magnitude of the signal and clutter after detection. Furthermore, it is assumed that there is phase coherence between successive echoes in order to readily determine the IF spectrum. This is a valid assumption since the coherent property is lost in the envelope detector and the video output is independent of the assumption. Finally, the spectrum of the clutter return about each harmonic in the IF is assumed Gaussian. This assumes the clutter source to be a large number of scatterers, each contributing to the total clutter signal independently. The standard deviation of the spectrum is set equal to the differential doppler between scatters subtending half the antenna beamwidth. The assumed spectrum, centered about some arbitrary IF frequency is,

$$C(\omega) = \frac{C_0}{\sqrt{2\pi}\,\sigma} \exp\left[-(\omega - \omega_0)^2/2\sigma^2\right], \tag{5}$$

where, $C_0 = \int_{-\infty}^{\infty} C(\omega)\, d\omega$, $\sigma = \frac{V}{\lambda} \sin\theta \cdot \Delta\theta$, and $\theta \gg H/R$

$C(\omega)$ = clutter power density spectrum
σ = standard deviation of
V = radar ground speed
θ = angle off the ground track
$\Delta\theta$ = antenna beamwidth
λ = RF wave length
ω_0 = spectrum center
H = aircraft altitude
R = target range

Fig. 3 is a picture of the IF spectrum including the first order sideband components introduced by the conical scan modulation of the clutter and target. Also shown in the figure is the video spectrum following the square law detector.

The antenna angle tracking servo is driven by the signal passed through the narrowband filter (bandwidth B) centered at the scan frequency.

Only those spectral components that contribute to the output of the audio filter at the scan frequency α need be studied. The audio filter output is described vectorially as follows:

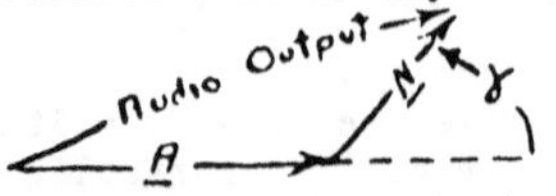

The target signal is $\underline{A}\cos(\alpha t)$ where $\underline{A} = T_0 e/r$ and the noise waveform at the output of the filter can be approximate by :

$$N(t)\cos(\alpha t+\gamma), \quad \frac{\overline{N^2}}{2} = \frac{C_0^{\,2}}{4\sqrt{\pi}\,\sigma}\exp\left[-(\alpha/2\sigma)^2\right], \qquad (6)$$

where N(t) and γ(t) are slowly varying with respect to $1/\alpha$ and $\overline{N^2}/2$ is the average noise power in B. To establish an equivalent rms angular error signal due to the output clutter alone when a target is on the bore site we set the output target error signal power in the absence of clutter equal to the output clutter power when the target is boresighted. Thus setting $A^2 = \overline{N^2}$, we get

$$\left(\frac{e}{r}\right)_{rms} = \frac{1}{\sqrt{2}\,\pi^{1/4}}\,\frac{C_0}{T_0}\left(\frac{B}{\sigma}\right)^{1/2}\exp\left[-\frac{1}{2}\left(\frac{\alpha}{2\sigma}\right)^2\right] \qquad (7)$$

This quantity is the rms error stimulus to the servo system in units of the offset nutation angle $\underline{r}$ when target is on boresight. Thus, for an X-band radar moving at Mach 1 having a beamwidth of 3°, a conical scan frequency of 60 cps and a narrowband filter width of 10 cps, e/r is on the order of 20 when the input target to clutter ratio is -10db and the antenna is oriented at 30° from the ground track. This means that for these conditions the rms error signal is about 20 times as large as the error signal obtained in the absence of clutter when the target is displaced from boresight by the nutation or offset angle.

When the antenna scans a nonhomogeneous ground patch the clutter signal is amplitude modulated at the conical scan frequency. For simplicity of analysis we can assume that this variation in the signal strength is sinusoidal, i.e.,

$$n(t) = f(t)\left[1 + b\cos(\alpha t + q)\right]$$

where f(t) is the unmodulated clutter signal, $\underline{b}$ is the modulation index and q the phase of the modulation.

Fig. 4A is a plot of Eq. (7) showing the open loop rms angular error signal in milliradians due to clutter interference and normalized by the received clutter to target ratio for zero modulation of the clutter by the conical scan process. Fig. 4B is similar with the clutter modulation index increased from zero to one. In this case the open loop rms angular error signal in milliradians is larger than the target to clutter ratio for all conditions cited and the additional interference introduced by the modulation of the clutter return essentially eliminates the improvement provided by the higher scan rates displayed in Fig. 4A.

It can be concluded that only under extremely restricted conditions is satisfactory angle tracking performance obtainable from a noncoherent receiver when targets are submerged in ground clutter.

Noncoherent Radar with MTI

In this system as shown in Fig. 1B, an MTI filter is added, following the envelope or second detector of the noncoherent radar receiver. The analysis of this system is an extension of the previous discussion and therefore the same assumptions apply. The MTI filter referred to is a delay and subtraction unit that makes a pulse-to-pulse comparison of the amplitude of the video signal returned from a particular radar range element. Ideally, a stationary target signal return would completely cancel in this filter comparitor; a moving target would give rise to a signal out of the delay and subtraction unit that would fluctuate at the doppler frequency rate established by the relative motions of the target with respect to the radar. This MTI filter has a sinusoidal frequency transfer characteristic and a $(\sin)^2$ power transfer function. Double delay and subtraction refers to a simple cascading of two such cancellation units.

Between the MTI filter and the narrowband filter centered at the scan frequency is a full wave rectifier. This is normally present in an MTI radar between the canceller and the indicator to convert the bipolar output of the filter into a unipolar signal suitable for display. In the present instance it serves also to produce the error signal at the scan frequency required to drive the tracking servos. This signal is developed by the intermodulation of the video target components which have passed through the canceller. Because the filter cancels a substantial part of the video clutter signal the useful intermodulation terms turn out to be sufficiently strong to control the angle tracker over a wide range of conditions.

For successful utilization of MTI it is essential that the error signal for the servos be derived from the final rectifier. Mistakingly using the output of the canceller directly leads identically to the results of the previous discussion.

Fig. 5A shows the video spectrum following the detector of the receiver (pt. A in Fig. 1B). This spectrum is characterized by two main groups. The first group appears about zero frequency and about all harmonics of the repetition frequency (not shown). It is generated by the mutual beating of the IF clutter components in the detection process. This group is referred to as the C-X-C group.

The second group is spaced from the first group by the difference in the doppler shifts between the target and clutter returns in the IF. The second group is derived from the beating between the target and clutter IF components. This group is referred to as the C-X-T group.

The C-X-T group finally gives rise to the desired tracking loop error signal as discussed below. When an account is made of the output target to clutter ratio it is important to note that two C-X-T members (symmetrically located with respect to the C-X-C members) are generated for every C-X-C member. The video signal at A in Fig. 1B is passed through the MTI filter, the second rectifier and then band limited by a narrow audio filter centered at the conical scan frequency. As in the case of the earlier detector, the rectification process gives rise to a number of continuous spectra. When the phase of the clutter modulation is random these continuous spectra prove to be noncoherently related and therefore add powerwise. In this case (when the phase of the clutter modulation is random) the only addition to the continuous spectra contributions is the target error signal spike. It is exactly at the conical scan frequency. Its formation is due to the coherent relationships that exist between components in the C-X-T group. This spike is the audio signal that provides honest error information to the tracking loop. In the event that the phase of the clutter modulation is fixed an additional signal spike is generated at the conical scan frequency following the second detector. This spike is an interference signal due to the coherent relationships between the components of the C-X-C group that exist when the phase of the clutter modulation is fixed.

As presented above, the interference signal components at the output of the narrowband filter following the rectifier is generated by the clutter echoes. This signal has two components. One component is described by the continuous spectrum of the interference and the second component is the power spike at the conical scan frequency (i.e., center of the narrowband filter). The relative magnitude of these two components depends on the flight geometry and the distribution of the clutter scatterers in the ground patch. These two components may be described most conveniently for two limiting cases; when the spike does not exist (b=0) and when the spike is at a maximum level (b=1) with respect to the continuous part.

As mentioned above, the spike component of the interference signal occurs when the phase of the clutter modulation is fixed. This can occur when the scatters are nonuniformly distributed throughout the ground patch or when the target is at an altitude which is a significant fraction of the radar altitude. In this latter case, the scanning phase imparted to each scatterer; i.e., that part of the scanning cycle at which each scatterer reflects the maximum energy, is approximately the same. It should be noted that the resultant spike is not random and introduces a bias or fixed angular error in the tracking loop. As a result the system tracks between the target and the ground patch, seeking a weighted mean angular position between the two.

In general, the tracking loop interference shall consist of both a continuous and spiked spectral component. The power in a continuous spectrum is proportional to the narrow bandwidth of the tracking loop and may be controlled within limits (these limits are dictated by the required dynamic response of the tracking loop system) by varying the bandwidth. The spike component, however, is located at the target signal frequency and cannot be reduced by filtering after the rectifier that generates both the target spike and the clutter spike. Filtering before rectification is significant and the system performance may be improved by more efficient MTI filtering such as double delay and subtraction.

Tracking Performance of the Noncoherent MTI System

The system performance may be discussed independently for small and large clutter ratios. In the case of small clutter to target ratios, the interference introduced in the final output due to the C-X-C term is negligible with respect to the contributions of the C-X-T terms. Both the desired tracking signal and the interference originate from these terms and therefore, their relative magnitudes are independent of the radar parameters and the filtering action before the rectifier including the MTI filter. Thus when the clutter to target ratio is relatively small the limitations of the tracking loop system are well defined and largely out of the hands of the designer.

On the other hand, when the clutter to target ratio is large, the resulting tracking loop interference is due to C-X-C terms while the desired target signal is still due to the C-X-T terms. In this case the magnitude of the interference fed to the servos is a function of the MTI filter characteristics and the spectral spread of the clutter. Furthermore, the relative magnitude of the tracking signal also depends upon the target doppler as well as the shape of the passband of the MTI filter. For the curves included in this report it was assumed that all target doppler frequencies were equally likely; the values calculated represent an average value weighted by the transfer characteristics of the MTI filter. For large clutter to target ratios the relative magnitude of tracking signal and interference signals are dependent upon the MTI filter characteristics.

Fig. 6 shows the normalized tracking loop error due to clutter plotted against the deviation of the clutter spectrum in the IF for large clutter to target ratios. These results are divorced from a particular radar except that they pertain to a PRF of 4,000 cps. More explicitly, Fig. 8 includes a plot of the maximum input clutter to target ratio allowed for satisfactory tracking performance (i.e., satisfactory tracking performance being defined as the tracking loop rms error less than 0.2 of the conical scan offset angle). Fig. 8 refers to a particular noncoherent radar system as defined by the parameters given in the figure. Furthermore, Fig. 8 is plotted for the case when b=1 and q is constant, which generally applies for target aircraft at altitudes that are a significant fraction of the radar altitude. This latter case is most unfavorable and imposes the greatest restriction on the tracking radar performance. Since this is not an unusual situation, the essential conclusions are discussed for this case.

The curves of Fig. 8 summarize the significant results of this analysis. The increase in the tracking capabilities afforded by the double delay subtraction MTI unit is greatest in the vicinity of the ground track. Furthermore, it appears from the figure that a double delay and subtraction MTI can provide useful tracking in clutter.

Coherent MTI

Fig. 10 is a block diagram of the coherent MTI system. Fig. 5 shows the signal spectra at various stages throughout the system in the same fashion as in the previous analysis. Note that the output of the detector, in this case a phase detector, is simply the IF signal spectrum shifted in frequency from the IF center frequency to the frequency origin. This may be viewed as a linear frequency translation in which the most important consequence is the preservation of the spectral shape. In contrast, noncoherent detection results in a considerable change in the signal spectrum due to the intermodulation or crossbeating of all the components in the envelope detector. This introduces additional spectral components, as well as causing a spread in the main elements.

The signals in the coherent system experience only one nonlinear demodulation process (envelope detection or square law rectification, i.e., following the MTI filter). Thus the resulting clutter interference is determined by a power summation of the terms generated by the self beating of the spectra in Fig. 5B. This is the most significant difference between the spectral histories of the coherent and noncoherent systems. It was shown in the earlier discussion that a spike due to the clutter interference exists after the final rectifier in the noncoherent case. This is not true for the coherent system.

The estimated angle tracking performance of a conical scan, coherent, airborne MTI radar against ground clutter is comparable to that of the previous estimates for the noncoherent MTI system for angles close to the ground track, but is considerably superior elsewhere. Fig. 7 is a universal plot of the tracking performance of the pulsed coherent MTI radar similar to Fig. 6 for the noncoherent system. The normalized tracking error due to the clutter interference is plotted against the deviation of the IF clutter spectrum. Thus the curves may be applied to any set of parameters defining a pulsed coherent MTI radar for any desired aircraft velocity with the single restriction on the PRF (PRF = 4000 cps).

Fig. 8 is a comparison of the coherent system and the noncoherent system. Except in a restricted sector about the ground track, coherent detection gives tracking performance superior to noncoherent detection. Furthermore, with coherent detection and good MTI filtering the conical scan angle tracking radar is relatively insensitive to target azimuth.

Conclusions

MTI filtering can reduce clutter interference so that conical scan tracking in clutter is satisfactory. It is shown in Ref. 1 that the performance of the MTI systems are fairly insensitive to the conical scan speeds and the second time around echoes. Fig. 8 plots the tolerable input clutter to target ratio for which satisfactory tracking performance is obtained as a function of target azimuth. The curves pertain to the radar system defined by the parameters in the figure. It gives a comparison of noncoherent and coherent detection, for both single delay MTI filtering and the more efficient double delay filtering.

The results of this study are encouraging. However, there has been no opportunity to experimentally check the conclusions. It is hoped that there is enough interest in this area that such a test program would be performed.

References

1. "Airborne Conical Scan Angle Tracking in a Ground Clutter Environment" by C.S. Sorkin and B. D. Steinberg, General Atronics Report 222-139-6, October, 1957.
2. "Mathematical Analysis of Random Noise" by S. O. Rice, Bell System Technical Journal, Vols. 23 and 24.

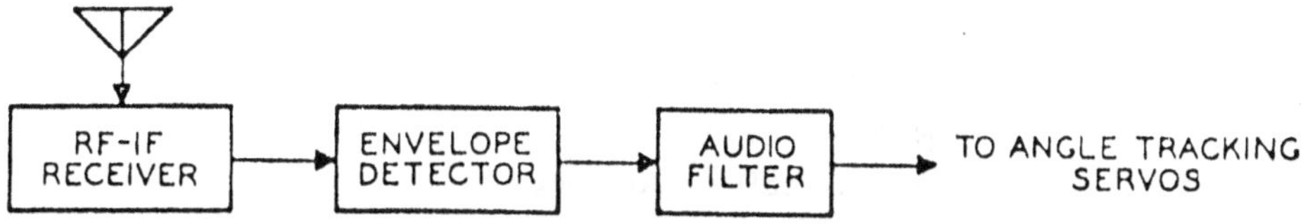

a. NONCOHERENT RADAR

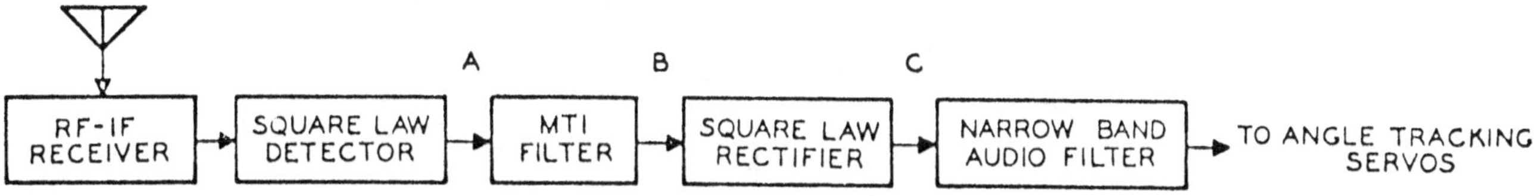

b. NONCOHERENT MTI

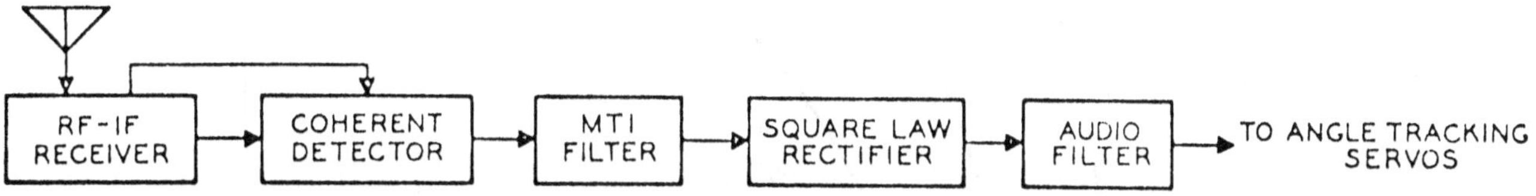

C. COHERENT MTI

FIGURE 1. SYSTEMS BLOCK DIAGRAMS

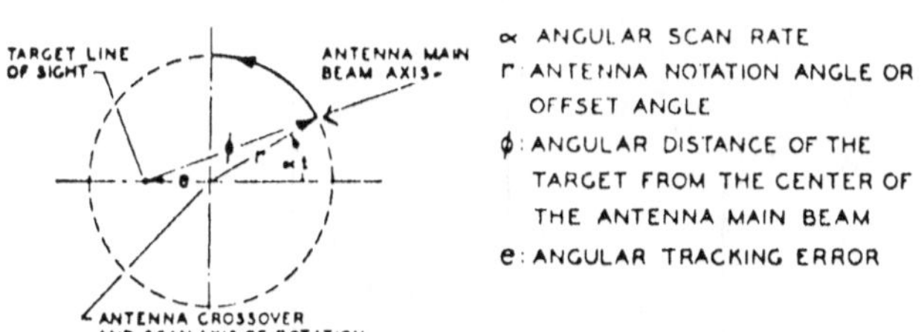

FIGURE 2. ANTENNA SCAN PATTERN

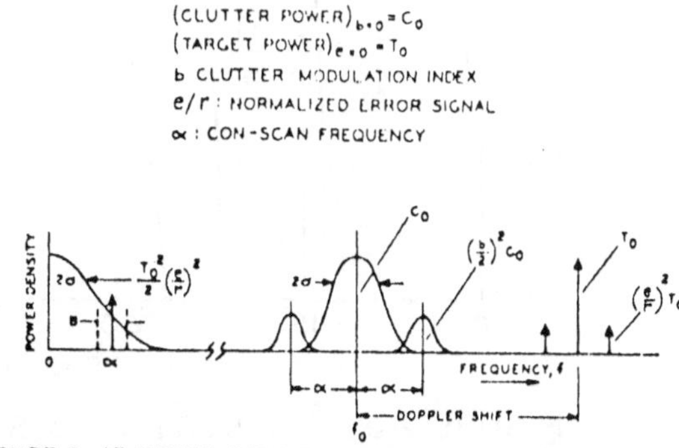

FIGURE 3. IF POWER DENSITY SPECTRUM AND VIDEO SPECTRUM AFTER A SQUARE LAW DETECTOR

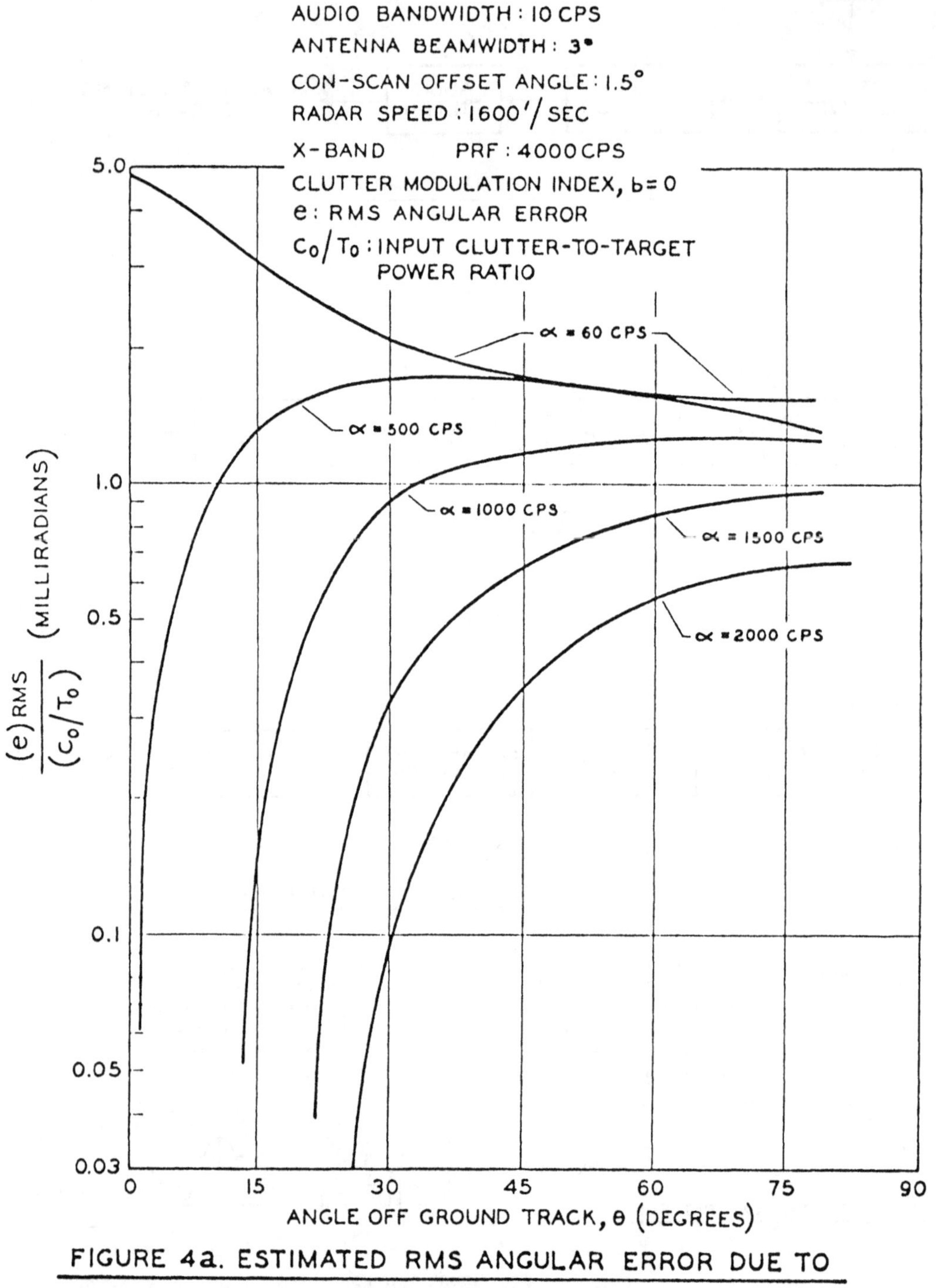

FIGURE 4a. ESTIMATED RMS ANGULAR ERROR DUE TO CLUTTER INTERFERENCE

(NONCOHERENT DETECTION)

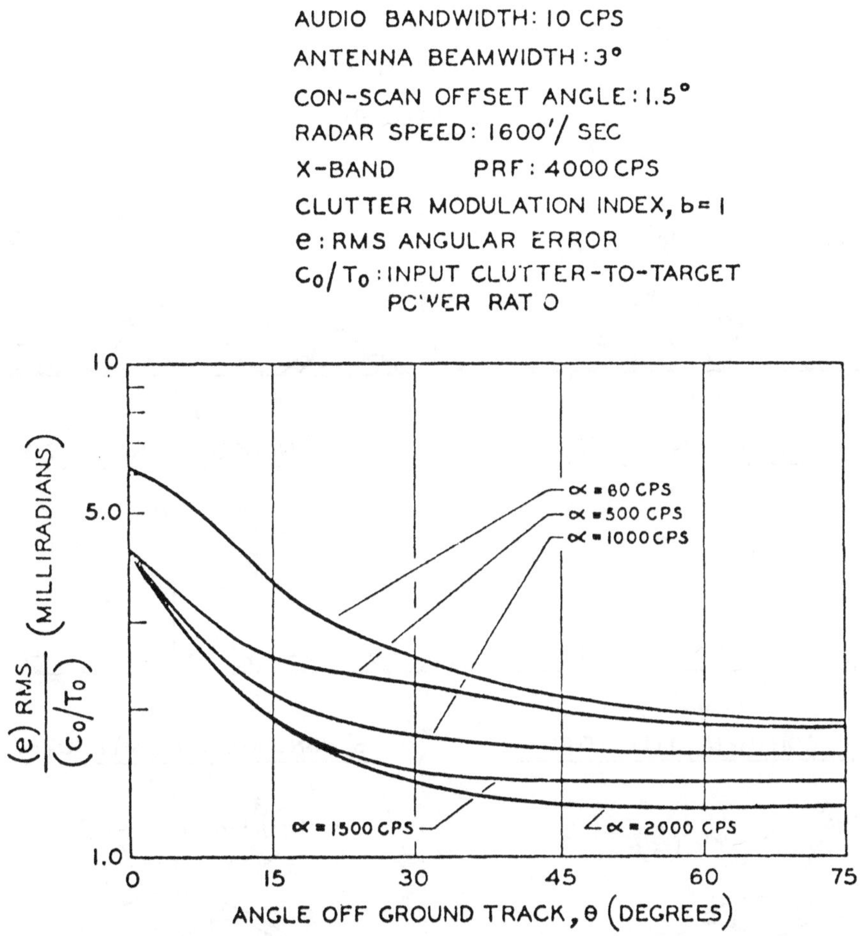

FIGURE 4b. ESTIMATED RMS ANGULAR ERROR DUE TO CLUTTER INTERFERENCE

(NONCOHERENT DETECTION)

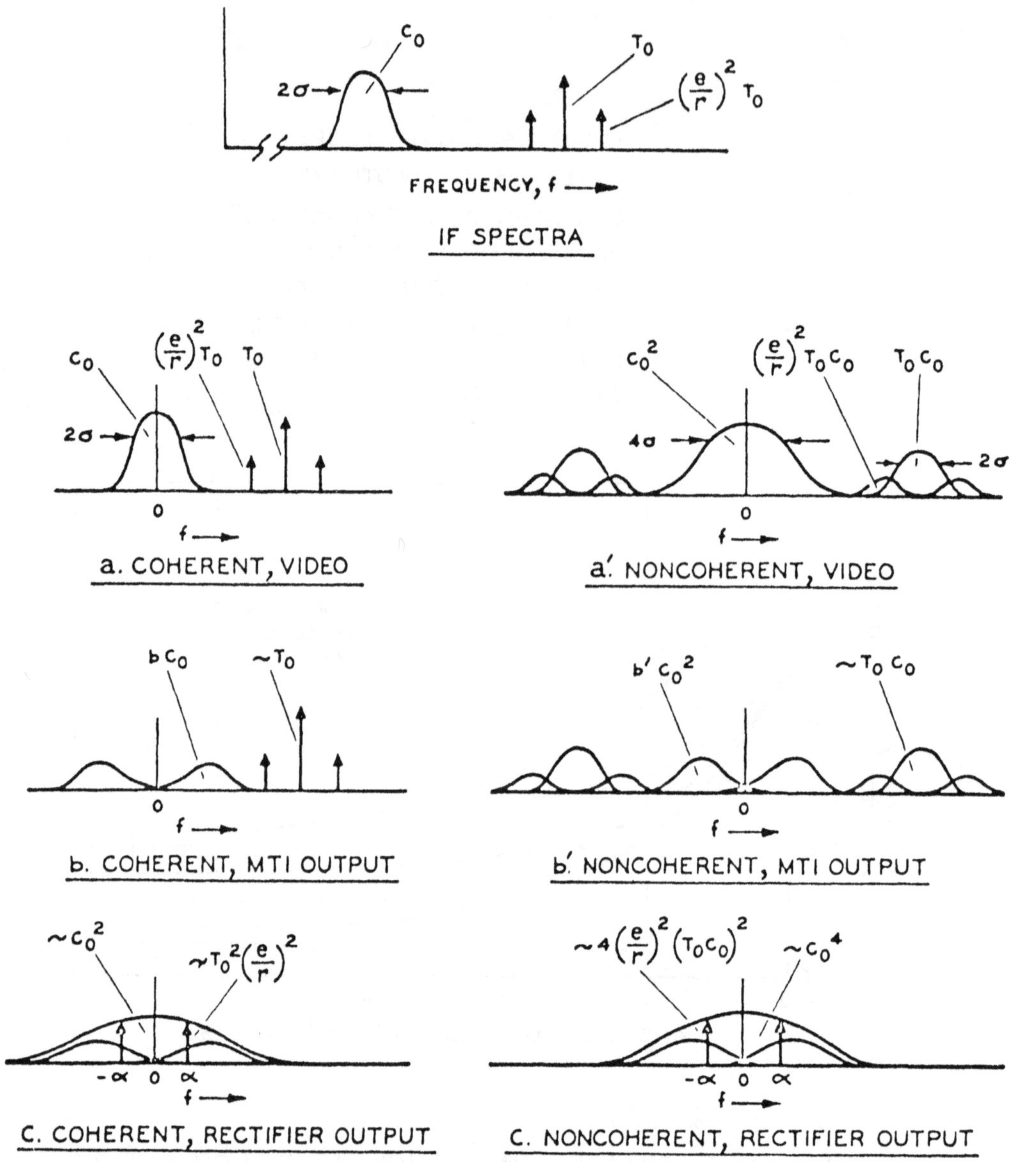

FIGURE 5. POWER DENSITY SPECTRA AFTER COHERENT AND NONCOHERENT DETECTION

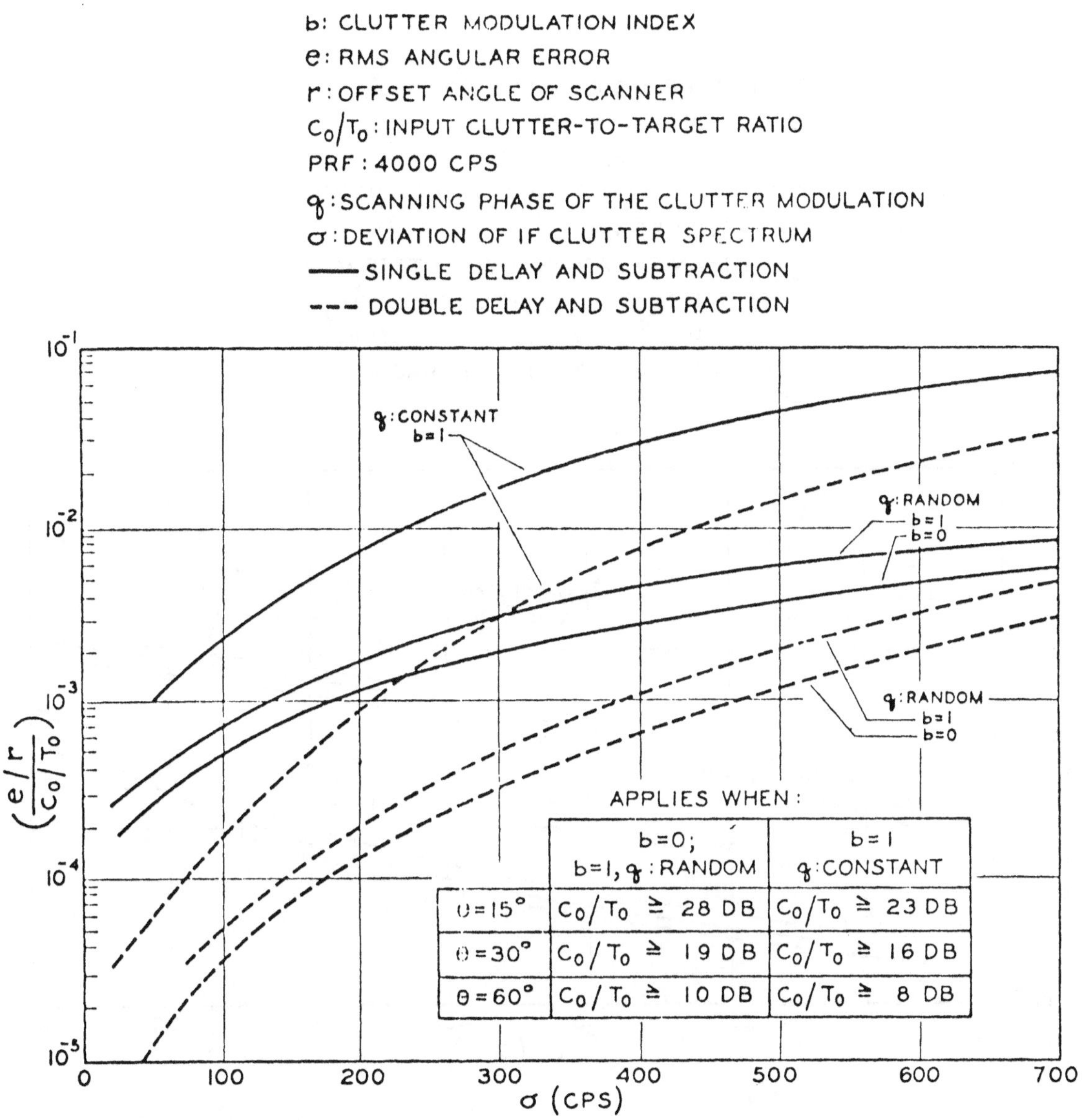

FIGURE 6. ESTIMATED RMS ANGULAR ERROR DUE TO CLUTTER INTERFERENCE

(NONCOHERENT DETECTION WITH MTI)

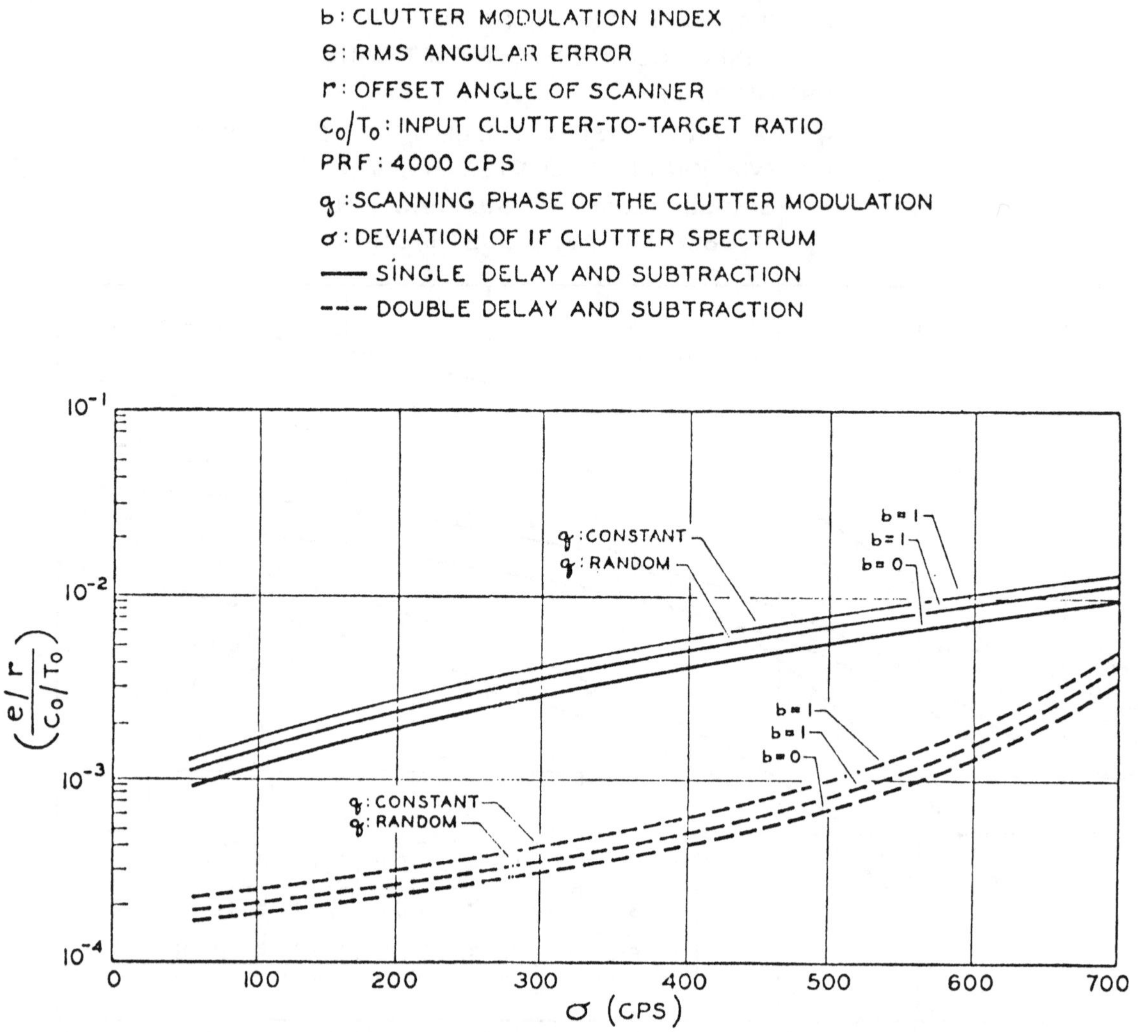

FIGURE 7. UNIVERSAL ESTIMATE OF RMS ANGULAR ERROR DUE TO CLUTTER INTERFERENCE

(COHERENT DETECTION WITH MTI)

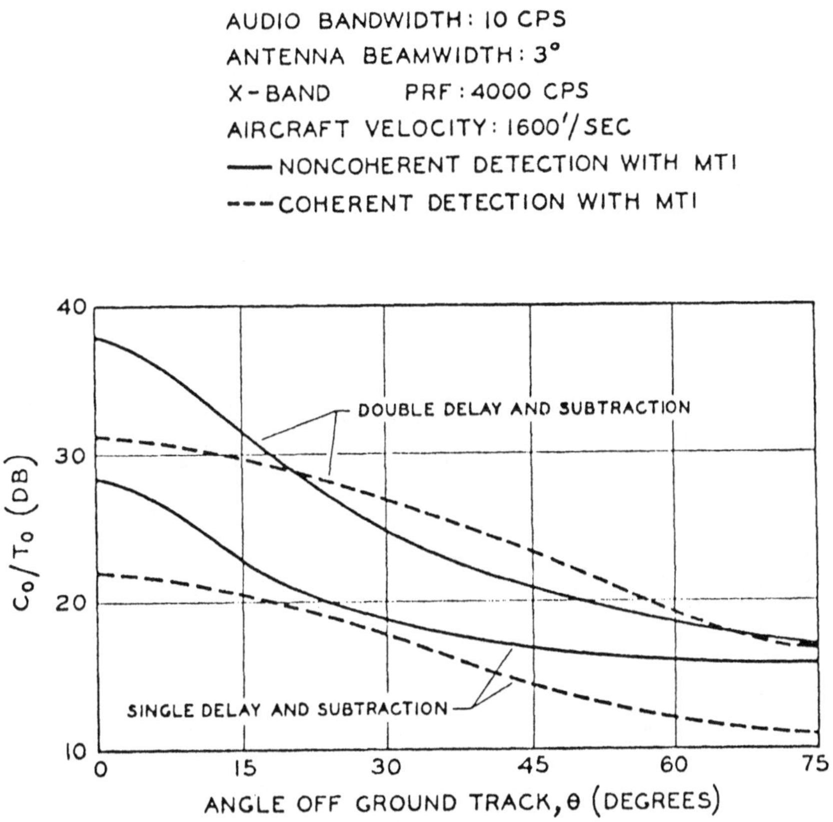

FIGURE 8. MAXIMUM INPUT C_0/T_0 FOR SATISFACTORY TRACKING PERFORMANCE

(i.e RMS ERROR <0.2 OF CON-SCAN OFFSET ANGLE)

Design Considerations for a Coherent (On Receive) MTI System for Conically Scanned Tracking Radar
A.K. Subramanian
Electro-Technology Magazine, Vol. 18, No. 4, December 1974
Pages 95-98

6.7

INTRODUCTION

Tracking Radars of the type which use conical scan for tracking, produce a scan modulation at a frequency corresponding to the conical scan frequency on all signals received by the radar during its operation. Such a scan modulation superimposed on the received signal will cause serious deterioration in the performance of any Moving Target Indicator (MTI) system conceived for this type of radars.

This paper deals with the salient features of design of a practical MTI system when it is conceived for operation with a conical scan tracking radar. Further, when such a radar is extended in its operational role (due to MTI capability), from tracking to that of search, the almost conflicting considerations faced by the designer, for optimum search and for optimum tracking are discussed in this paper.

SALIENT FACTORS FOR CONSIDERATION BEFORE THE MTI CAN BE FITTED FOR THE ABOVE RADAR SYSTEM

These can be dealt under the following heads :

(a) *The Antenna* : Tracking Radars employ narrow pencil beams to provide accurate auto-follow of the targets and as such their role in the search-mode, especially in the antenna coverage, is never adequate. Due to the very nature of the narrow beam, one resorts to either a 360° search or a sector search in bearing, with the antenna also covering the elevation plane by having stepped or continuous sector search. This is not a very efficient way of energy management, especially when we consider the effect of such a scanning procedure on MTI system.

Under the circumstances, to improve the detection capability in the search role, one can explore the possibility of restricting the scan widths in the two principal planes as well as increasing the scan rates. Both have practical limitations. However, to be effective against the modern aerial attacks which are delivered at very low heights, there is no need to scan over a very wide sector in the elevation. A sector search width of $\pm 1°$ around a mean elevation of 1° is adequate to cover the detection of targets flying at heights of 100′ to 200′ at ranges from 5 to 50 km. Since the antenna beam is essentially covering a low elevation sector, it becomes imperative to design an antenna having very low side lobe levels.

(b) *The Transmitting Source* : For practical radar systems with emphasis on mobile role, the choice of a transmitting source usually falls on a magnetron, which packs the maximum power for a given weight and size. A magnetron is not a coherent source and hence in such radar systems, the design of

* Electronics & Radar Development Establishment, Bangalore

a MTI is only feasible for coherent (on receive) mode of operation.

The magnetron has been improved in the last two decades, to have better stability and peak power figures but it still cannot compete with a coherent transmitter. It serves as a good technical compromise for the operational role, but it is to be emphasised here that with the advent of coaxial magnetron in the microwave field, the picture has improved very much in as much as these magnetrons have better stability figures, lower pulling factors, better spectrum characteristics; and it is further strengthened by its longevity of service. Hence coaxial magnetrons as a class of replacement device for the existing roles played by conventional magnetrons is a good technical decision to make under the MTI scheme. It has also been possible these days to get coaxial magnetrons which are a one to one retrofit for some of the popular conventional magnetrons and as such their replacement in existing radar sets have not posed any problem.

Injection locking of the magnetrons with a stable microwave source have been attempted but as yet no large systems have come out due to high injection ratio requirements.

(c) *Local Oscillator* : It is well known that for achieving coherent (on-receive) mode of MTI operation, one should go in for a Stable Local Oscillator (STALO) system instead of the conventional klystron and Gunn source devices, which do not have the required type of frequency stability. Quite a few schemes are available for the design of Stalos at microwave frequencies, which cater to the required stability figures of the order of 1 in 10^{-8} (short term) and 1 in 10^{-6} (long term). In fact conventional crystal oscillator-frequency multipliers schemes have been successfully used but these can only select the frequency of the STALO in steps rather than in a continuous manner. This is a limitation which is fundamental to the operation of such a scheme. It is however possible to achieve continuous tracking of magnetron frequency vis-a-vis the stalo frequency by going through a scheme using cavity stabilisation. Such a method would yield good ECCM capability which was available with the magnetron-klystron combination but was not feasible with schemes containing stalos of switchable frequencies.

The remarks contained in (b) and (c) above are not peculiar to radar systems having conical scanning, but they have been discussed in relation to the MTI signal processing which must take into consideration the stability figures that can be achieved through the RF signal.

(d) *Design considerations of the MTI system*: Coming in particular to the MTI signal processing system, the designer faces quite a few difficult considerations. This mostly relates to meeting the requirements of the search as well as tracking roles of the radar under clutter environment.

In the search mode, the radar does either a helical scan or sector scan depending on the requirement of a given situation. In normal Fire Control radar operation covering a vulnerable area or a point, sector search in azimuth is common as quite a few radars will be deployed together. In such a case, the radar signal has antenna scan modulation components due to sector search in azimuth as well as in elevation.

During tracking, though the antenna scan is eliminated in both the planes, the conical scan is switched on. Thus in this role, the signal gets an amplitude modulation at the scan frequency containing the track information. It is necessary to preserve this modulation on the signal without distortion right upto the angular error correcting units of the tracking circuits to ensure proper auto follow of the target.

The above scan modulations introduce many problems as far as the MTI processor is concerned, especially during tracking when the conical scan has to be necessarily preserved and reproduced.

It is in this context the design of the IF and phase detector stages in such a MTI scheme becomes rather critical and important. In a typical MTI system, the phase detector stage used for detecting the pulse to pulse phase difference of any input signal, is seriously impaired if the input signal has an amplitude variation apart from the phase shift introduced in it. It will be easily seen that such deterioration in the performance of the phase detector is inevitable especially when the radar is in its tracking process, producing amplitude modulation of the signal.

Obviously in a tracking radar, it is not possible to employ a limiter amplifier stage before the phase detector, as it will remove the amplitude modulation information necessary for tracking. Conversely, a linear amplifier would also be not suitable since it will faithfully transmit the amplitude modulation on a linear scale on to the phase detector.

Thus a true logarithmic amplifier working at IF frequencies before the phase detector is found to be a good technical compromise. In such a stage, the dynamic range of amplitude variation in the signal due to scan modulation can be reduced through appropriate logarithmic compression and at the same time preserving the amplitude modulation envelope information to provide the necessary auto-follow. Thus, a design of the true IF logarithmic amplifier is found to be a must for such a system.

In the design of Log IF stage, critical considerations have to be given in deciding the factors like dynamic range, logarithmic compression and gain of the amplifier. It must also be noted that the Log stages provide higher gains for low inputs and as such noise level in these stages can climb up if amplifiers are not designed with proper gain and dynamic range for a given sensitivity. These considerations further go through much variations of parameters, when considered for roles like short and long range radars. In tracking radars, they have to be adjusted along with any AGC scheme that may be available for tracking purposes. It may also be necessary to include a provision for STC in such Log IF amplifiers to reduce effects of clutter at very short ranges from the radar. Indeed, it is a prerequisite for radars working with antenna beams almost grazing the ground most of the times.

(e) *Phase Detector*: The design of the phase detector is important due to the scan modulated signal, although it is not rendered critical due to the Log IF input.

Phase detector design using conventional Sum and Difference technique is found quite suitable and faithful conversion of phase to amplitude information at the output is achieved. It must be emphasised here that though the scan rate of 20-40 Hz, corresponding to conventional scan modulation frequencies, is very low in the frequency spectrum compared to the doppler shift frequencies for targets flying at 1 mach and above, its effect on clutter is still considerable. Further, without the logarithmic compression, the effect of amplitude modulation occuring at these low scan rates can seriously distort phase detector response.

(f) *MTI canceller circuit*: No particular mention of the type of canceller will be made as the cancellers by themselves have pros and cons which are not peculiar to the radar systems they are developed for. However, a comment can be made on the cancellers required for processing pulse signals used in Fire Control Radars. It should be capable of handling pulse lengths less than a microsecond. Digital type cancellers for extremely low pulse lengths may present problems in this context.

(g) *Display circuits for search and track signals*: During the search mode of operation, it is necessary to have the targets showing up on the display as bright echoes for easy detection. Such a possibility exists in the MTI mode since the scope is not saturated with the clutter picture. Independent of the echo strength of the target at the input of the radar receiver, which can vary depending on the target being detected at the lowest to the maximum range possible on the radar, one can use a saturating type of amplifier for the search video. Provision of STC along with this facility will go a long way in enabling the operator to detect targets at the minimum to maximum possible range with greater efficiency and ease. However, this is only possible when the system cancellation is such that the residue of the clutter is not unduly highlighted on the display.

During the track phase, there is the inherent requirement of AGC to adjust to the target echo strength to provide a constant amplitude for smooth functioning of the auto-follow. Further, the AGC bias developed in relation to the strength of the target is applied to the other unwanted echoes to suppress them from being displayed or auto-followed. It is obvious that a saturating amplifier designed for the search video cannot be employed in this role. Thus, in a scheme employing MTI for such dual roles, one is constrained to make a dual channel video circuit, one catering to the exclusive need for search and the other for the optimum need for track. Such a combination can give the best performance figures in a radar combining search and track roles.

CONCLUSION

The outline design considerations governing the provision of a MTI system for a conically scanned track radar have been dealt in this paper. These, in practice, have rendered realisable solutions.

In conclusion it must be emphasised that the MTI provision to a radar system of this type can become complex in nature, unless each factor affecting the system has been duly weighed and suitable compensation in the processing of the MTI circuits has been given. Otherwise, it will not be possible to realise the improvement due to the MTI system on such radars which typically employ narrow beamwidths, fast scan with short dwell time on targets, low altitude coverage and lower powers compared to normal search radars.

Chapter 7
ECCM in the Receiver Signal Processor

Summary

These papers address various aspects of current emphasis on signal processing functions in radar receivers.

Synopsis of Reprint Papers in Chapter 7

7.1 *"Description and Performance Evaluation of the Moving Target Detector"*—
Cartledge and O'Donnell

The MTD was intended for application to Federal Aviation Administration (FAA) air traffic control radars to improve radar detection of aircraft in regions of ground and precipitation clutter, particularly when aircraft are moving tangentially to the radar. This is implemented by a special digital signal processor which uses adaptive thresholding and fully coherent processing. It also includes the functions previously performed by the analog Dicke Fix. A periodically updated disc memory stores a clutter map. The MTD processor utilizes about 2,900,000 range-azimuth-Doppler cells (48 nmi range and 360 degrees of azimuth over 55 knots Doppler with 1/16 mi. range resolution, 3/4° azimuth and eight Doppler cells). The MTD includes a hard wired circuit to eliminate non-synchronous pulsed interference.

Although intended for ground and precipitation clutter, the MTD could be used against chaff and jamming.

7.2 *"The Application of SAW Technology to Radar ECCM"*—Carr

Surface Acoustic Wave (SAW) devices have important applications to radar ECCM. This paper describes the SAW device and three SAW ECCM functions:

- delay lines for signal storage and integration,
- waveform coding and decoding for correlation with phase-shift keying (pseudo noise), chirp and frequency hop,
- real-time spectral analysis for adaptive ECCM (actually an ESM).

An increase in jam resistance (more properly an ECCM Improvement Factor) of 40 dB can be achieved. This paper was presented at the DDRE Radar ECCM Symposium in 1976.

7.3 *"Anticlutter and ECCM Design Criteria for a Low Coverage Radar"*—
Petrocchi, Rampazzo, and Rodriguez

A low altitude radar requires the use of a moving target indicator (MTI) to permit detection of aircraft in the presence of heavy clutter. If this aircraft employs active denial ECM (jamming), frequency agility would be useful as an ECCM. Usually MTI is incompatible with frequency agility. This paper describes an "agile MTI" and an "adaptive frequency selector" which result in a frequency agile MTI radar. The P.R.F. stagger sequence is optimized in order to obtain the best compromise between a good MTI Improvement Factor and a flat MTI filter velocity filter. The agile MTI utilizes a double canceller to provide sweep correlation. This paper was presented at the International Radar Conference in 1978.

7.4 *"Performance of CFAR Devices in ECM Environments"*—Tong and Steichen

Control of false alarm rate in automatic detection and tracking radars is essential to prevent processor saturation and to retain the ability to detect targets and maintain tracks. Some workers assume a wide-band noise jammer is statistically the same as thermal noise; i.e., a narrow-band Gaussian process with mean zero. This paper considers the effect of amplitude modulated noise and non-Rayleigh noise on the false alarm rate of three types of CFAR devices—cell averaging CFAR, log-CFAR, and the distribution-free CFAR. Use of too coarse quantization is shown to change the FAR by orders of magnitude. The cell averaging CFAR is shown to be the most target sensitive. The LOG CFAR is particularly sensitive to amplitude modulated blinking jamming. The distribution-free CFAR, although it must work with a binary integrator (with consequent lower efficiency in Gaussian noise), is indifferent to jammer statistics. This paper was presented at the DDRE Radar ECCM Symposium in 1976.

7.5 "Noise Quality Optimized Jammer Performance" – Turner, Ottoboni, Imada

Most ECCM analysts assume that jamming has the same spectral distribution, white Gaussian, as receiver noise; e.g., Johnson noise. This is not true however. A new parameter, "noise quality" which is a measure of uniformity of the spectral content of jammers, is defined in this paper. A method for its measurement is presented along with noise quality results for a typical jammer. The effect of noise quality on the performance of both amplitude and intensity modulated displays is shown. The Dicke-fix is postulated to be an effective ECCM against jammers which have a low noise quality. This paper is presented in this Chapter because of its close relation to the previous paper.

Bibliography Topics for Chapter 7

Signal Processing

BIBLIOGRAPHY FOR CHAPTER 7

SIGNAL PROCESSING

1. General Bibliographies

Barton, D.K., "International Cumulative Index on Radar Systems," *IEEE Pub. Cat. no. JH 4675-5,* 1978: extensive bibliography on signal processing, pp. 64-8; extensive bibliography on smoothing and filtering, pp. 69-70.

2. Journal Articles

Fowler, C.A., A.P. Uzzo, Jr., and A.E. Ruvin, "Signal Processing Techniques for Surveillance Radar Sets," *IRE Trans.,* vol. MIL-5, no. 2, April 1961, pp. 103-8 (reprinted in Schleher, D.C., *MTI Radar,* Artech House, Dedham, MA, 1978.)

Spafford, L.J., "Optimum Radar Signal Processing in Clutter," Phd thesis, Brooklyn Polytechnic Institute, 1967; also published in *IEEE Trans.,* vol. IT-14, no. 5, Sept 1968, pp. 734-743.

Fisher, J.R., "Digital Techniques in Radar and Sonar," *IEEE Electron. and Aerosp. Syst. Rec,* Sept 9-11, 1968, p. 615, (Abstract only).

Thompson, J.E., "Adaptive Signal Tracking System," *Proc. Nat. Electron. Conf.,* vol. 24, Dec 9-11, 1968, pp. 903-5.

Carpentier, M.H., "New Trends in Signal Processing," *Trans. S. Africa Inst. Elec. Engrs.,* vol. 60, pt. 5, May 1969, pp. 105-111.

Groginsky, H.L. and G.A. Works, "Pipeline Fast Fourier Transforms," *IEEE Computers and Communications Conf. Rec.,* Rome, New York, Sept 30 – Oct 2, 1969, Session IV, pp. 197-208.

Novosel'tsev, L. Ya., and N.A. Pungin, "Digital Devices for the Primary Processing of Radar Information Using Detectors with Slave Channels," *Izv VUZ Electron.,* no. 12, Dec 1969, pp. 1313-6 (in Russian, trans. abstract SA 70-32293).

Helds, H.D., "Applications of Digital Waveform Processing in Radars," *IEEE Int. Conv. Digest,* paper 4, B. 4, Mar 23-26, 1970, pp. 180-181.

Groginsky, H.L. and G.A. Works, "Pipeline Fast Fourier Transforms," *IEEE Trans.,* vol. C-19, no. 11, Nov. 1970, pp. 1015-1019.

Skenderoff, C., "Optical Correlation for Radar Signal Processing," *IE* (India) *J. E & T.E.,* vol. 51, May 1971, pp. 189-92.

Hartt, J.K. and L. Sheats, "Application of Pipeline FFT Technology in Radar Signal and Data Processing," *Proc. IEEE EASCON,* 1971, pp. 216-221.

Gray, G.A. and F.E. Nathanson, "Digital Processing of Radar Signals," *Proc. IEEE EASCON,* 1971, pp. 208-15.

Gottlieb, M., "Opto-acoustic Signal Processors Extend Radar and Communication System Capabilities," *Westinghouse Engr.,* vol. 33, no. 3, May 1973, pp. 76-82.

Edwards, T.R., and R.D. Knight, "Fast Digital Noise Filter," *Instrum. and Control Syst.,* vol. 47, no. 9, Sept 1974, pp. 73-6.

Carr, P.H., "The Application of SAW Technology to Radar ECCM," *DDRE ECCM Conf.,* 1976 (reprinted in this volume).

Muehe, C.W., "Moving Target Detector, an Improved Signal Processor," *AGARD Conf. Proc. CP-197, New Devices, Techniques and Systems in Radar,* 16 June 1976.

International Specialist Seminar on "The Impact of New Technologies in Signal Processing", *IEE Conf. Pub. No. 144,* Sept 1976, (Contains numerous papers on this theme including future trends).

Cartledge, L., and R.M. O'Donnell, "Description and Performance Evaluation of the Moving Target Detector," 8 Mar 1977, *Lincoln Lab Rpt ATC-69* (FAA-RD-76-190) (Major portion reprinted in this volume).

Grimm, J.D., and R.B. Steves, "Radar Signal Processing Technology," *IEEE Electro-78 Conv. Rec., pt. 28,* May 1978, Boston, pp. 28/2 - 1 thru 11.

Roberts, J.B.G., B.J. Darby, and R. Eames, "Signal Processing with CCD and SAW Devices," *Proc. Int. Conf. on Radar,* 4-8 Dec 1978, Paris, pp. 263-78.

Description and Performance Evaluation of the Moving Target Detector
L. Cartledge and R.M. O'Donnell
Project Report ATC-69, (Lincoln Laboratory), Report No. FAA-RD-76-190, March 8, 1977
Pages 16-44, 210
Reprinted by permission.

7.1

III. DESCRIPTION OF THE MTD AND THE PHILOSOPHY OF ITS DESIGN*

A. Introduction

Airport Surveillance Radars (ASR's) generally give good service in a manual control environment. However, until the present time there has not been a "solution" to the problem of providing high probability of detection with low false alarm rate in an automated aircraft surveillance system. The principal difficulties are an excessive number of false alarms, which will overload the automation computers of the ARTS-III system, and aircraft detection which is so spotty that the computer cannot make long continuous tracks of all aircraft.

To be truly useful any solution which is chosen must work not only when a single type of clutter or background is present, but also when several types of clutter are present simultaneously, for example, both in ground clutter and precipitation. In addition, the false alarms sent to the tracker must be spatially and temporally uncorrelated, or even a relatively small number of false alarms will give rise to an excessive false track rate out of the tracking computer.

We will discuss the basic radar problems along with the techniques to solve them and, finally, describe the MTD (Moving Target Detector) digital signal processor whose design incorporates features which attack all of the radar problems simultaneously.

B. Radar Problems and Solutions

The poor performance previously mentioned is due to competition of aircraft returns with the so called "clutter" returns and the response of the circuits used in the radar to overcome these clutter returns. It is convenient to classify the problems according to the types of the return. These problems are fixed ground clutter, second-time-around ground clutter, precipitation clutter, angels and surface vehicles, and interference from other radars.

1. Fixed Ground Clutter

By far the largest undesired radar reflections come from fixed objects on the ground. Ground clutter usually extends out to about 20 nmi. In very hilly or mountainous areas it may extend out to the maximum radar range (∿60 nmi). In ASR radars its natural or intrinsic spectrum is very narrow compared to the spectral spread caused by antenna scanning motion.

*Portions of this section are taken from References 1, 2, and 3.

Ground clutter varies appreciably from spot to spot in the area of coverage. Typical distributions of the mean backscatter coefficient σ_o are shown in Figure III-1. It tends to be highest from cities.

In the present ASR radars, ground clutter is reduced by three mechanisms: MTI filtering, antenna tilt and by mounting the antenna close to the ground to take advantage of the shielding effect of nearby objects. Figure III-2 shows the MTI filtering performance achievable using two cascaded delay line cancellers with and without limiting. Previous ASR radars have all employed limiting in the IF followed by a phase detector. The purpose of the limiting is to normalize the video output so that clutter residue from the MTI filter is reduced to the average noise level. This allows the video gain to be adjusted so the clutter will not show up on the controller's scope. Unfortunately, this limiting action spreads the clutter spectrum so that considerably poorer subclutter visibility (SCV) is achieved than if the normalization had been done by some other mechanism not involving nonlinearities.

If we consider the parameters of a typical terminal radar (ASR-7, see Table III-1) at 15 nmi and σ_o from Figure III-1 that is exceeded only 5 percent of the time, we find that for a 1 m^2 target (typical small aircraft) the input signal-to-clutter ratio is -31 dB. Since an output signal-to-clutter ratio of about 15 dB is needed for adequate target visibility, an improvement factor of about 46 dB is required. We see from Figure III-2 that this is not achievable with the present configuration. It is, thus, common practice to achieve greater signal-to-clutter advantage by tilting the antenna upward (see Figure III-3) by 2° to 5° depending on the local clutter situation. If tilted, as shown in Figure III-3, there is a 17-dB advantage (maximum range divided by zero elevation range to the fourth power) in input signal to clutter for an aircraft flying in the peak of the antenna pattern. This advantage is degraded as the aircraft gets out of the peak of the antenna patterns so that, typically, detection gets spotty due to competition with ground clutter for small aircraft below about 1.5^o or above 9°. These angles change depending on the antenna tilt and ground clutter intensity.

Another undesirable feature of the improvement curves in Figure III-2 is the very wide notch around the blind speeds. The effects of higher order blind speeds are usually reduced by using a staggered PRF. However the wide notch around zero

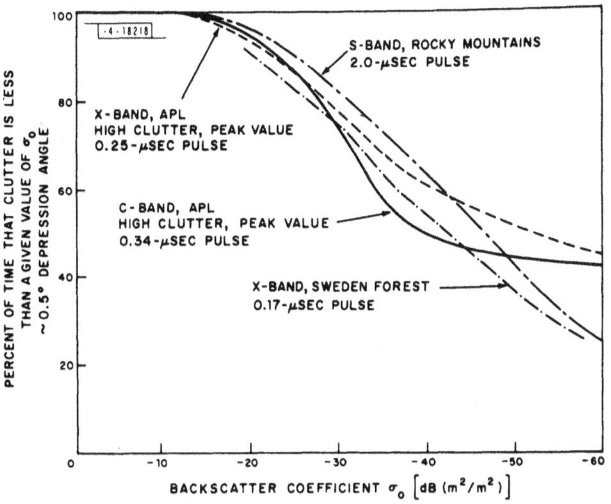

Fig. III-1. Land clutter backscatter distribution from surface radars (from Reference 4).

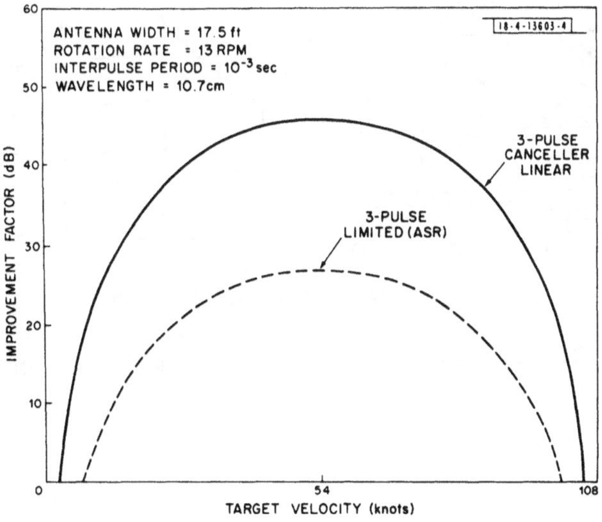

Fig. III-2. Performance of S-Band 3-pulse cancellers.
(Improvement factor is the ratio of signal-to-clutter out of the canceller to signal-to-clutter in. The latter ASR radars use 3-pulse cancellers following a limiter shown above by the curve "3-pulse limited (ASR)").

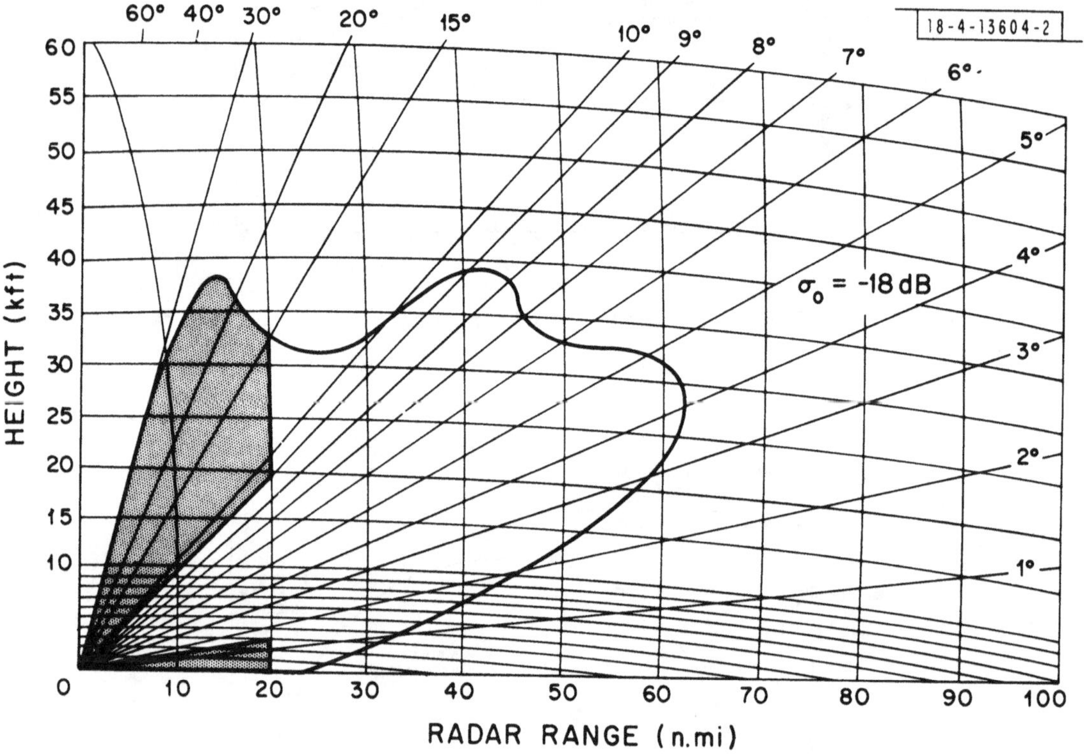

Fig. III-3. Coverage of ASR-7 radar against a 2-M^2 target.

means that targets will be lost for a considerable distance on the scope when the aircraft flies tangential to the radar. The 3-pulse canceller with limiting is worse in this respect than the 2-pulse canceller with limiting.

A further limitation in performance of existing ASR's is the presence at many sites of buildings or hills which limit the minimum elevation visible to the radar. Increasing the height of the antenna to overcome this limitation causes an undesirable increase in ground clutter level which could be overcome by improvement in SCV.

To give complete flexibility in siting and tilting the antenna while still rejecting ground clutter, an approximate 20-dB improvement in performance over that of existing ASR radars is needed. It is obvious that this requires linear processing of clutter and target signals. Hence, one must avoid all nonlinearities in the receiver. To narrow the blind speed region, one should process coherently groups of many pulses.

In order to assess quantitatively what could be considered a "good" MTI processor for improving the performance of ASR radars against fixed ground clutter, calculations were made of the performance of the so called "optimum processor". Given the initial conditions, the optimum processor has by definition the highest improvement in target-to-interference (interference is defined as clutter plus front-end noise) ratio of any processor. By knowing the performance of such a processor, one can judge how closely any other (i.e., suboptimum) processor approaches the theoretical limit.

The processor considered here can be defined as a device that takes M complex signal returns V_i, multiplies these returns by a complex filter weight $\bar{W}_i$, adds them and then takes the square of the amplitude

$$R = \left| \sum_{i=1}^{M} \bar{W}_i V_i \right|^2$$

V_i is composed of target, noise and clutter. The theory of optimization will not be shown here but follows that of DeLong and Hofstetter[11]. The clutter spectrum, which in this case is essentially all caused by the antenna scanning motion, is modeled by an antenna having a Gaussian beam shape as in Emerson[12].

Figure III-4 shows the target-to-interference improvement in decibels that is possible (optimum) for the mechanically rotating ASR antenna whose parameters are given in Table III-1. These results assume the use of a sufficiently stable coherent transmitter. Results almost this good have been obtained in tests using the magnetron-type transmitters in the ASR-5 and ASR-7. These results are reported in Section V-J.

The maximum clutter-to-noise ratio which can be handled will be set by the dynamic range of available analog-to-digital converters. Under the assumption that the rms thermal noise level is set at one lsb (least significant bit) the maximum improvement ratio that can be obtained is given by the relationship

$$IR_{max} = 6(n-1) + 10 \log_{10} (N)$$

where

IR_{max} = the best obtainable improvement ratio in dB

n = the number of bits in each of the two A/D converters used

N = the number of radar pulses integrated per decision

Clearly, the A/D limited level of performance will not be achieved if the analog hardware preceding the A/D converters limits the dynamic range of the system. The usable dynamic range can be reduced by nonlinearity or spurious modulations anywhere in the radar system. Usable dynamic range may also be limited by the effects of truncation or round-off in the digital computations.

The solid curve in Figure III-4 is the improvement obtained when the optimum filter is tuned to the doppler frequency of the target as the target doppler is varied. The dashed curve represents the frequency response of the particular optimum filter tuned to 300 Hz.

The following general characteristics of the optimum processor should be noted:

(a) The upper curve (Figure III-4) levels out at about $M \times C/N = 10^5$ where M is the number of pulses processed and C/N is the clutter-to-noise ratio, unless M is small. This points up the need for wide dynamic range A/D converters as explained above.

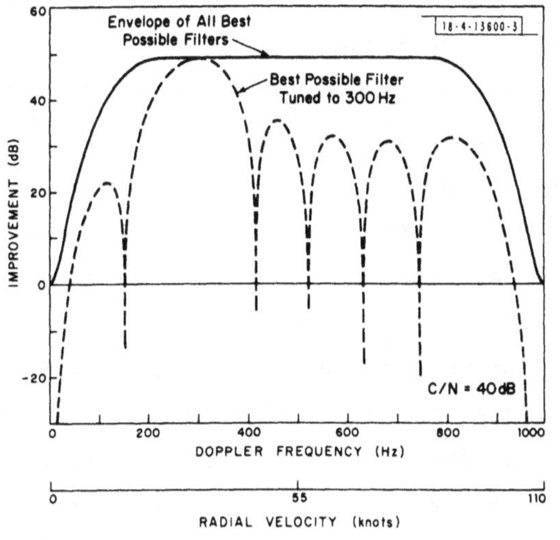

Fig. III-4. Improvement of target-to-interference ratio, scanning antenna.

TABLE III-1

ASR RADAR PARAMETERS

Antenna Width	5.25 m
Antenna Rotational Speed	1.36 m/sec
Wavelength	0.107 m
PRF	1000 pulses/sec
Number of Pulses Processed Per Look	10
Clutter-to-Noise Ratio	40 dB

(b) At the so called "blind speeds" 0 and 1000 Hz), there is no improvement but there is no deterioration either, thus, a target whose cross section is sufficiently above that of the clutter can be seen.

(c) For filters that are not tuned on or close to blind speeds, there are very deep nulls at the blind speeds.

(d) The width of the notch about the blind speeds increases with antenna rotational speed when all other parameters are held constant.

(e) The filter can be approximated by a discrete Fourier transform (DFT) only in certain special cases.

(f) Because the optimum weights, $\bar{W}_i$, are a function of clutter-to-noise ratio, the optimum processor requires some a priori knowledge. However, this ratio can be determined in principle by the application of a proper algorithm in the receiver together with ground clutter memory from scan to scan.

In Figure III-5 we compare the optimum processor results with those of conventional MTI processors (Figure III-2). We see that the amount of clutter rejection achieved in conventional MTI systems is far less than the best that can be done.

The implementation of the optimum processor for every range-azimuth cell calls for M complex multiplications for each target velocity examined. Usually, if M pulses are being processed, a filter bank with M filters will give adequate coverage for all target velocities. Thus, M^2 complex multiplications must be performed for every range cell. For a typical ASR, 800 range cells per sweep must be sampled on 10 sweeps and processed every 10 ms. If optimum filters were used, 8,000,000 complex multiplications per second would be required or 32 million simple multiplications.

A simpler processor can be built. The optimum processor can be broken into two parts; a clutter filter followed by a target filter. The filter used to reduce clutter multiplies the signal vector by the antenna weighting and by the inverse of the interference covariance matrix. The target filter used to enhance the target is a DFT. The near-optimum processor could consist of a digital filter which approximates as closely as possible the frequency response of the clutter filter followed by a DFT for the target filter. This combination gives an improvement factor within a few decibels of the optimum shown in Figure III-4 and

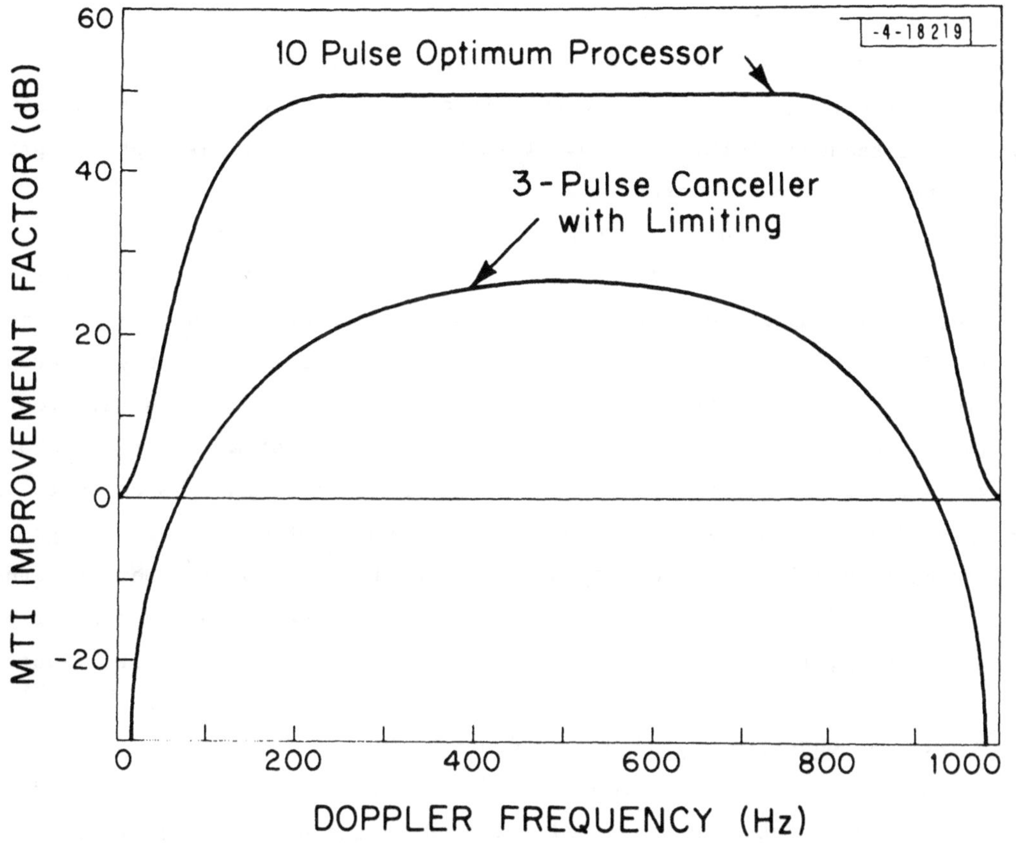

Fig. III-5. Comparison of optimum processor with conventional 3-pulse canceller.

is much less complex than implementation of the optimum processor. It should be noted that either processor provides target doppler information.

Through direct comparison it has been found that a simple three-pulse canceller without feedback forms the clutter filter portion of a near-optimum processor for the scanning antenna.

The near-optimum ground clutter processor would not be complete without adequate thresholding. For a typical ASR utilizing a near-optimum processor, ground clutter will appear only in the zero doppler filter and the filters immediately adjacent on each side. Geometrically ground clutter is very heterogeneous in character. It varies greatly in size from one resolution cell to the next. Thus, averaging nearby cells will not give a good estimate for thresholding purposes. A practical way to accurately set ground clutter thresholds is to use a digital ground clutter map which remembers the ground clutter in every range-azimuth resolution cell averaged over a sufficient time period (number of scans). Whenever an aircraft's cross section is sufficiently larger than the clutter over which it is flying, it will be seen even if it has zero radial velocity (tangential target).

2. Second-Time-Around Ground Clutter

The so called "second-time-around" clutter effect is caused by radar returns from ground clutter which is illuminated by the next-to-last radar pulse which was transmitted. This ground clutter is beyond the unambiguous range of the radar. These returns are prevalent where conditions of anomalous propagation exist such that the radar waves intercept the ground at great distances (greater than that corresponding to the interpulse period). This effect is also prevalent in mountainous regions where clutter normally extends beyond the unambiguous range.

The presently deployed ASR 3,4,5,6 and 7's use magnetron transmitters which transmit pulses with random phase from pulse to pulse. With this type of radar it is impossible to maintain the phase relation between the first- and second-time-around clutter returns and the two cannot be filtered out simultaneously.

Further, the present ASR's use pulse trains with staggered interpulse periods so as to avoid blind speeds. This causes the second-time-around clutter return to fall in different range cells on succeeding returns so there is no hope of filtering it out. To effectively filter out second-time-around clutter, a fully coherent transmitter (one coherent from pulse to pulse) and a constant PRF must be used. The PRF need not be constant forever, but only over an interval sufficient to collect a group of pulses for processing. If the PRF is changed from group to group of pulses, the radar is said to use "multiple PRF".

The ASR-8, which will be entering service soon, has a klystron transmitter that transmits pulses which are coherent from pulse to pulse. Thus, if a "multiple PRF were used with the ASR-8, second-time-around ground clutter could be filtered out adequately. This will be the case when MTD processors are used with the ASR-8's.

3. Precipitation Clutter

The backscatter from precipitation has been studied extensively. Figure III-6 shows the mean volume reflectivity from rain at 15 mm/hr. This is considered a heavy rain found only 0.04 percent of the time at New Orleans[(4)]. This heavy rainfall is usually found only in relatively small size cells in the center of storms. The radar should be designed to reject at least this level and as much higher a level as possible.

Also marked on Figure III-6 is the point where the volume reflectivity is such as to cause a 1 m^2 return at 30 miles in an ASR radar (rain return from a typical cell with precipitation extending from the surface to 10,000 ft). Rain at 15 mm/hr is about 13 dB above this value. Remembering that these are average reflectivities and that ∿15 dB signal-to-noise ratio is required for automatic detection, we need about 30-dB rain rejection for good performance.

The rain clutter spectrum is spread around some mean value determined by the wind velocity. The spectral spread observed by the radar is fixed by wind shear conditions[(4)]. The standard deviation of the rain velocity spectrum typically reaches values of 25 knots at 30 nmi and increases with range. This spread is due chiefly to wind shear. A typical heavy rain storm spectrum is presented in Figure III-7. The center velocity of the storm may be anywhere from -60 to +60 knots depending on wind conditions relative to the pointing direction of the antenna. Because the mean velocity of the storm is not centered at zero radial velocity for most antenna azimuths, the storm return will leak through the MTI filter for many azimuth angles of the antenna.

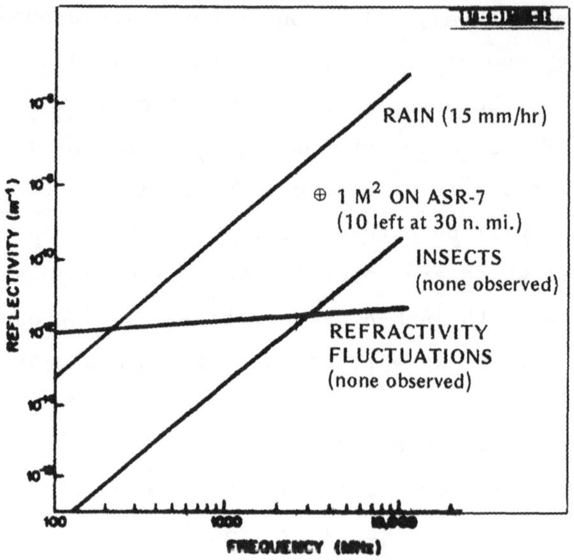

Fig. III-6. Reflectivity of various moving clutter sources.

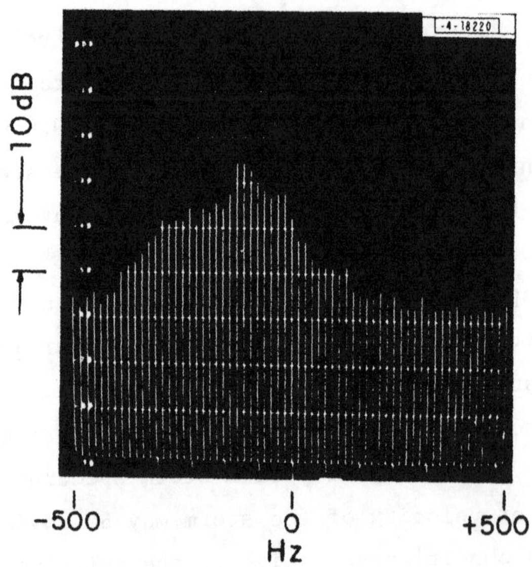

Fig. III-7. Typical spectrum from a heavy rainstorm taken using an S-Band radar (receiver noise level is -57 dB).

Circular polarization is normally used to reduce rain clutter by about 15 dB while reducing the aircraft signal level to a much lesser extent. The use of MTI helps reduce rain clutter except when the antenna is looking toward or away from the wind direction. In these directions the rain clutter spectrum is such that all of the rain clutter signals may pass through the MTI filters.

Log-FTC-antilog circuits[5,6] reduce the receiver gain in proportion to the average level of rain clutter for about a mile in range surrounding the cell of interest. This normalizes the rain clutter level, just as limiting is used to normalize ground clutter, at the output of the MTI circuit. Its purpose is to suppress the rain clutter on the scope. At the same time, of course, it suppresses the signal. For adequate detection, the signal amplitude must be appreciably above the clutter residue from the MTI filters.

Previous attempts to build a digitizer which works well in rain have failed due to a lack of recognition of the correlation properties of the clutter. Rain clutter signals are partially correlated from pulse to pulse so that noncoherent integration in the azimuth direction produces random signals whose variance is much greater than if the rain signals were noise-like[4] (uncorrelated). This greater variance requires a large increase in the detection threshold over that set for noise and a consequent loss in detectability of aircraft targets. If the threshold is not reset, a high false alarm rate in rain results. A typical correlation time for a rain storm is about 2 to 3 msec at S-band. This means that the <u>coherent</u> integration time should be <u>significantly greater</u> than 2 or 3 msec. This will insure that, after coherent integration, multiple threshold crossings from the same target may be added noncoherently without undue loss due to partial correlation of the samples.

It is easy to show that, since the detected rain clutter residue is partially correlated from azimuth to azimuth, the statistical spread of the noncoherently added returns is much greater than from receiver noise which is uncorrelated. This is shown diagramatically in Figure III-8. If a threshold is established as some multiple of the mean of the receiver noise (see Figure III-8) the probability of false alarm will be the area under the curve above the threshold. Correlated noise such as rain changes the curve to the one marked "correlated" without a

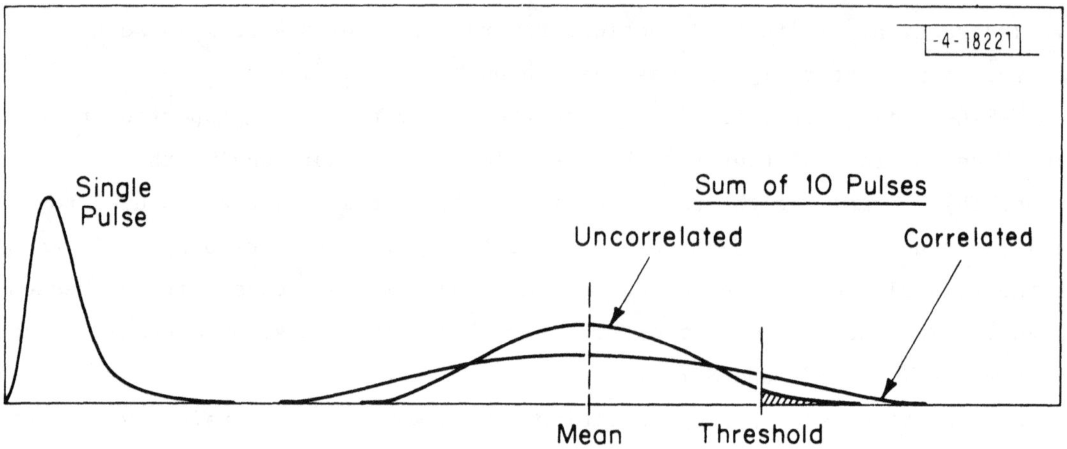

Fig. III-8. Noncoherent integration (probability distributions).

change in the mean so that the area above the threshold, and thus the probability of false alarm, goes up by orders of magnitudes. The correlation of the clutter can be measured adaptively and the threshold raised accordingly, but it has been shown by Nathanson (see reference 4, pages 88-92) that if this is done, most of the signal-to-noise improvement due to noncoherent integration is forfeited. Thus, it is clear that noncoherent integration in the presence of correlated clutter such as rain can only be done if the time between samples being non-coherently integrated is much greater than the correlation time of the correlated clutter.

It is fortunate that the optimum or near-optimum filtering against ground clutter described in the previous section utilizes a filter bank since this is a good approach to eliminating weather clutter. About 30 dB weather rejection is needed, 15 dB of which can be provided by circular polarization. Filtering of some sort is a viable solution to obtaining the remaining 15 dB. The filtering could be near-optimum as in the case of ground clutter except for the fact that the weather clutter spectrum (see Figure III-7) changes with time. This change could be measured and the filter adapted to the spectrum, but this would result in intolerable hardware complexity.

A good alternative is to use the filter bank produced by the near-optimum ground clutter filter. It is only necessary to set the threshold on each filter adaptively. A so called "mean-level" thresholding algorithm is employed. Since storms are rarely less than about one mile in extent, the moving clutter is averaged over a half mile on either side of the cell being examined for a target. Each velocity is averaged separately so that filters containing only noise will not be penalized.

Further, a multiple PRF system rather than a staggered PRF system(1) is used so that high speed aircraft typically fall in different filters in the filter bank on successive PRF's. There is a very high likelihood that the target return will be competing with noise only and not weather on one of two PRF's. Only for aircraft whose true (not aliased) radial velocity coincides with that of the rain will there be serious degradation in detection performance.

Precipitation clutter often appears in the zero radial velocity doppler filter. Sometimes relatively large zero velocity returns are received from a weather front which is moving quite rapidly through the radar's coverage. When this happens, changes in the zero velocity signal, for particular range azimuth cells, can occur faster than the ground clutter map can react. In this situation, large numbers of zero velocity threshold crossings can occur in areas which are spatially correlated from scan to scan. This, in turn, would give rise to false tracks if nothing were done about it. At NAFEC this problem was solved by adding instructions in the post-MTD software which prevents the tracker from initiating tracks on target reports whose radial velocity is zero.

4. Angel Clutter

The so called "angel clutter" refers to all returns which cannot be explained as being ground or precipitation clutter or targets. Much effort has been spent in studying angels. It is now believed that nearly all, if not all, angels are caused by bird flocks.

Returns from single birds(7,8) at S-band range in size between 10^{-4} and 10^{-2} m^2. A typical distribution of angel radar cross section is shown in Figure III-9. The radar return is principally from the body with very little from the wings. For large birds, the body is resonant near L-band (1300 MHz) and is in the Rayleigh region at UHF. Typically, there may be anywhere from one to several hundred birds in a resolution cell. Although the mean return from a typical flock of birds may be low ($\sim 10^{-2}$ m^2), the tail of the distribution has been observed to return up to 10 m^2. Birds have been seen as high as 12,000 ft altitude, but they usually fly less than 7,000 ft. The usual appearance on the scope is as so called "dot angels". "Ring angels" are also caused by birds as a large group leaves their nesting place at sunrise.

Of particular interest are the bird migrations in spring and fall. These have been described as "night effect", "falling leaves", "seasonal AP angel clutter", and have been reported by many terminals in the eastern section of the United States. The appearance on the scope when the radar is using MTI is that of two well defined lobes.

These migrations occur at night when there is a favorable wind. Migration will be very heavy on favorable nights so that most of the migration occurs on

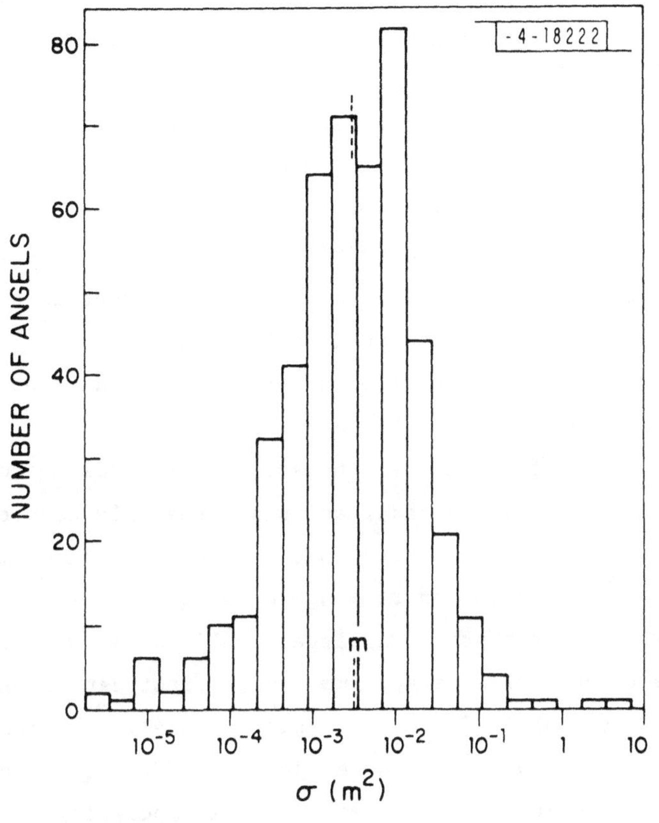

Fig. III-9. Distribution of "angel" cross sections (from Reference 7).

relatively few nights (5 to 15) each spring and fall. The number of birds associated with these migrations may be very large. One author estimated that a few million birds crossed a 100-mi front during one of the busy nights of the autumn migration over the Cape Cod region[9].

Birds fly between 15 and 45 knots true air speed. Taking into account winds, radial velocities over the range $\pm$ 80 knots or so may be observed.

A fairly effective radar improvement used against bird clutter is a carefully tailored sensitivity time control (STC)[9]. The STC varies the radar gain with range and is adjusted so that the minimum detectable target is a specific value, say 1 m^2. This calls for an R^{-4} attenuation law. Even with an R^4 STC applied, during the migratory season flocks of birds have caused many short false tracks in ASR radars (see reference 10).

The R^4 STC is most effective when the antenna provides uniform gain over the elevation coverage. However, ASR radar antennas generally have cosecant-squared elevation patterns. Under this conditon STC applied to reduce bird return has the undesired effect of reducing the short range coverage against high altitude airplanes. The MTD processor provides target amplitude as well as range, azimuth and doppler of the target. The target amplitude output can be used to set a secondary (post-MTD) threshold for each doppler channel. Schemes using such thresholds in post-MTD software were developed at NAFEC. These have proven to be quite effective at filtering out bird returns while operating with no observable degradation in detection sensitivity of high altitude aircraft.

It was noticed empirically that most of the radar returns from birds fall in the lower radial velocity doppler filters and give rise to a single threshold crossing as the radar antenna scanned by. By contrast, aircraft returns mainly consist of multiple threshold crossings, usually with all doppler filters being equally likely. The area CFAR (Constant False Alarm Rate) adaptive thresholding scheme which is used at NAFEC is described in Section III-C-3.

5. Surface Vehicles

The cross section of ground vehicles is in the same range as aircraft; namely, from 1 to 100 m^2. Radial velocities can range over $\pm$ 60 knots.

Some reduction in ground vehicle returns is achieved by tilting the antenna upward. One solution is to censor areas on the scope known to contain visible roads carrying cars with radial velocities outside the notch at zero velocity. The ARTS-III system starts these detections in track and by measuring the velocity of the target while in tentative track applies a speed threshold of about 50 knots before declaring it a firm track. This technique has proven reasonably effective and results, at worst, in a few false tracks from surface vehicles.

In summary, a radar which is used for air traffic control and uses a scanning beam antenna should have a fully coherent transmitter; a linear, large dynamic range receiver; a signal processor containing a near-optimum ground clutter filter bank; a fine-grained ground clutter map to set ground clutter thresholds; mean-level thresholding on weather. It would employ multiple PRF's for elimination of blind speeds and would output target range, azimuth, doppler velocity and amplitude for use in higher level thresholding and other software for filtering clutter and interference. These are the signal processing and adaptive thresholding techniques which are employed in the MTD processor.

C. Moving Target Detector (MTD) Description*

1. Implementation of Processing Algorithms

The MTD processor is implemented as a hard-wired, pipeline digital signal processor. It processes the full 360° coverage of an ASR radar out to a nominal range of about 48 nmi**. The azimuth coverage of the radar is broken into 480 Coherent Processing Intervals (CPI's) each about one half an antenna beam width ($3/4^{\circ}$) in extent. The range cell size is 1/16 nmi. During one CPI, ten pulses are transmitted at a constant PRF. These are processed by the MTD into eight doppler cells. Thus, there are about 2,900,000 range-azimuth-doppler cells in the MTD's coverage (see Figure III-10).

A block diagram of the processor was presented in Figure II-2. The I and Q (In-phase and Quadrature) signals are sampled at a 2.6-MHz rate by 10-bit A/D converters. The I and Q channels are then added coherently, two at a time, to

*A detailed description of the MTD hardware can be found in reference 13.

**The 48-nmi range was chosen for convenience. The MTD could have been implemented out to a range of 60 nmi.

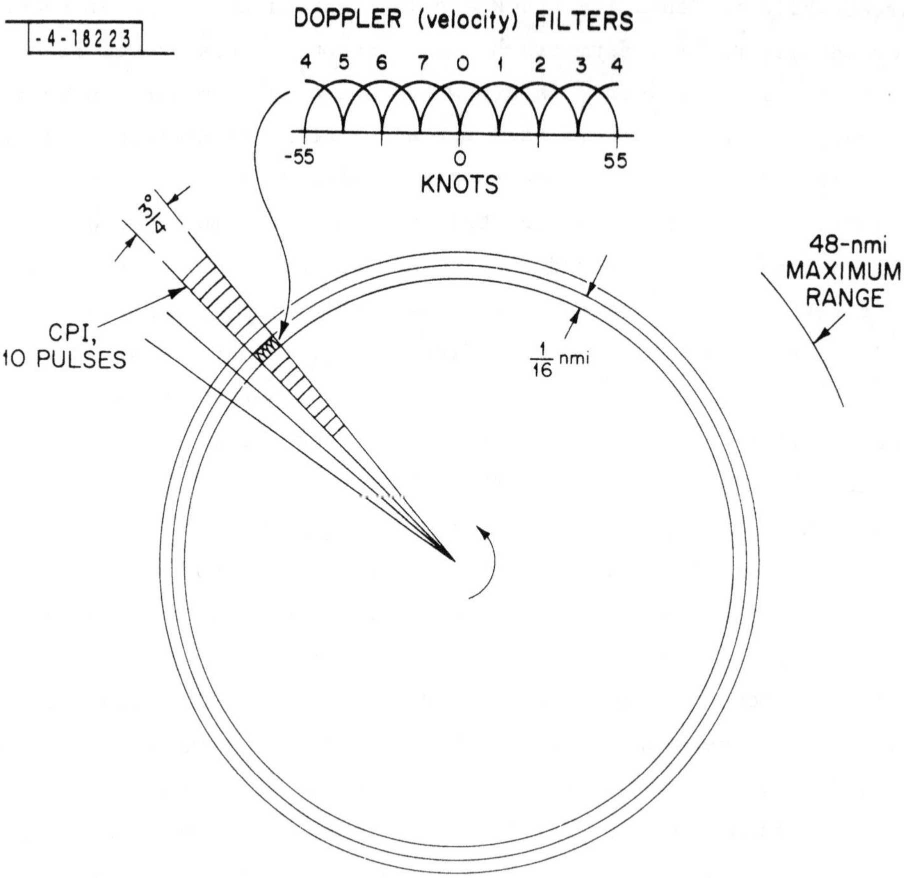

Fig. III-10. MTD resolution.

produce 11-bit I and Q channel words at a 1.3-MHz rate. Samples of both the I and Q channels for each of 760 range gates (47.5 miles of range) from each of 10 consecutive radar pulses are stored in a 8192-word memory. These 7600 words of memory are then processed sequentially (ten samples for each range cell) by a 3-pulse MTI canceller. The I and Q channels are each processed by separate hardware in the 3-pulse canceller section of the processor. Note that the ten samples of 11-bit I and Q channel samples are transformed by the 3-pulse canceller to eight 13-bit words. The output of the 3-pulse canceller for both the I and Q channels (real and imaginary parts of the signal) is fed into an 8-point Discrete Fourier Transform (DFT). The DFT produces a complex frequency coefficient for each of eight doppler cells.

The 8-point Fast Fourier Transform algorithm requires four real multiplications by $1/\sqrt{2}$. They are performed in a fixed-wired multiplier which approximates $1/\sqrt{2}$ as $1/2 + 1/8 + 1/16 + 1/64 + 1/256$ and requires only four adders. The remainder of the MTD is configured so that all multiplications are by integral powers of 2 and can be computed by simply shifting the binary data.

Weighting of the I and Q channel signals to reduce the side lobe level is done after the DFT. Subtracting 1/4 of the output of the two doppler filters adjacent to the one of interest is equivalent to a cosine on a pedestal weighting in the time domain.

Since the 3-pulse canceller has poor low doppler frequency response, a zero velocity filter (ZVF) is employed to see low radial velocity targets. This low-pass filter is implemented by coherently adding the first five samples of each of the I and Q channels, respectively, taking their magnitude and adding to this the magnitude of the sum of the last five samples. This gives a broader frequency response than simply adding coherently all ten samples and then taking the magnitude. The magnitudes of the signals which come out of the 3-pulse canceller-DFT-weighting chain are then taken.

After magnitudes are taken, adaptive thresholds are set and threshold crossings (detections) are noted and output. The adaptive thresholds are set depending on the clutter phenomena which are present. The doppler domain is divided into three domains: doppler cell 0, doppler cells 2 through 6 and doppler cells 1 and 7.

In doppler cell 0, the clutter is generally due to ground backscatter. The average ground backscatter cross section varies from range-azimuth cell to range-azimuth cell. The average backscatter signal level for each cell is measured and stored on a disc memory (see Figure II-2). To build up an accurate map, careful registration of the clutter map with the true pointing direction of the antenna is essential. This is achieved by breaking up each revolution of the antenna, marked by 4096 Azimuth Change Pulses (ACP's) into 240 units containing 17 or 18 ACP's each. Each unit contains two CPI's. The disc is accessed every two units (34 to 36 ACP's, approximately 44 msecs), during which time four CPI's worth of ground clutter data is read onto and off of the disc. The disc has a maximum access time of 18 msecs. Two, 3000-word MOS buffers are used to store the data for use in the processor. A recursive filter is used to update on a scan-to-scan basis the average signal level stored on the disc. This disc holds 480 x 768 = 368,640 clutter words, one for each range-azimuth (CPI) cell in the coverage. The words are stored in 10-bit floating point format to preserve the large dynamic range of the clutter signal.

On each scan, 1/8 of the stored clutter level is subtracted from the stored level. One-eighth of the signal level output from the ZVF is added to the value remaining after subtraction. This new level is then stored on the disc for thresholding in the next scan. The threshold for the zero doppler cell is a fixed value between 4 and 7 1/2 times the level stored on the disc. This fixed value may be altered by changing a plug.

In the doppler cells 2 through 6 the clutter is due chiefly to rain. For each doppler and range cell, the average signal level is measured by averaging the received signal over 14 range cells centered on the range cell of interest. This average is multiplied by a constant "a" to form the threshold for that range-azimuth-doppler cell for that scan. Since only coherent integration has been performed, the 14 outputs are statistically independent Rayleigh distributed numbers regardless of whether from noise or rain clutter so that the false alarm rate associated with multiplying constant "a" doesn't change as weather comes into the area. Figure III-11 shows the false alarm rate as a function of threshold setting when averaging over various numbers of statistically independent samples. Note that for 14 samples and a probability of false alarm of 10^{-6} the threshold is

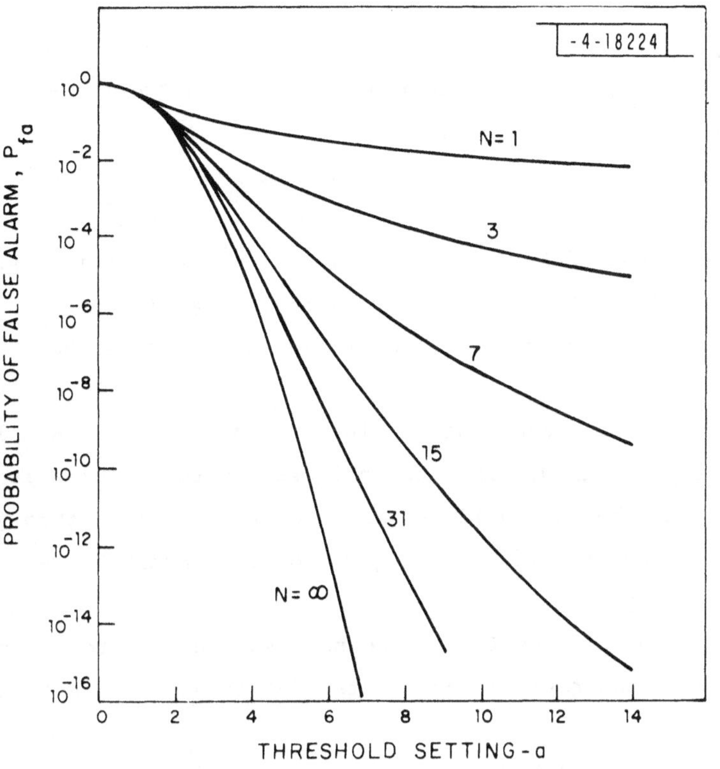

Fig. III-11. False alarm rate as a function of threshold setting. (Parameter N is the number of uncorrelated samples averaged to determine the threshold).

only about 2 dB above that derived assuming perfect knowledge ($N = \infty$) of the rain or noise conditions. The use of this type of mean-level threshold assumes that the rain level is constant over 14 cells (∿ 1 nmi) of interest. The experience with rain to date indicates an inconsequential increase in the false alarm rate when rain enters the region of interest.

In the actual implementation of the MTD the PRF is changed by about 20% between groups of 10 pulses so that a higher velocity target will be aliased into a different filter on each PRF (see Figure III-12). Thus, it is highly probable that all targets will be free of weather clutter on at least one of the two PRF's, except for those whose true radial velocity is equal within about 25 knots to that of the rain.

Doppler cells 1 and 7 can contain clutter due to rain or spillover from the ground backscatter in cell 0. The threshold set in these cells is the greater of two thresholds: (a) the threshold set as in doppler cells 2 through 6, or (b) a fixed binary fraction - $(1/2)^n$ n = integer - of the threshold set in cell 0, n is set by changing a plug.

Finally, note that if any I and Q channel sample is noted to have all the bits on (i.e., be in saturation), any target detections for that range cell are deleted.

An interference eliminator circuit has been hard-wired into the MTD to eliminate non-synchronous pulsed interference. The magnitude of each of the 10 pulses in each CPI is taken by adding the absolute values of I and Q. This algorithm though primitive is accurate to within 3 dB. The average is multiplied by five and stored. The 10 magnitudes are also stored until the average has been computed. They are then compared sequentially with five times the average. If any one exceeds this number any threshold crossings that might have occurred in that range gate during that CPI interval are inhibited.

The output message of circuitry of the MTD contains double buffering for up to 38 detections per CPI. Double buffering is required because the MTD and the Input/Output Processor (IOP) are asynchronous devices. At the start of each CPI, a PRF/azimuth (PAZ) word is entered into the first buffer. When a threshold crossing takes place a velocity/range/strength (VRS) word for that detection is entered in the first buffer. At the end of each CPI the PAZ word and any VRS words for that

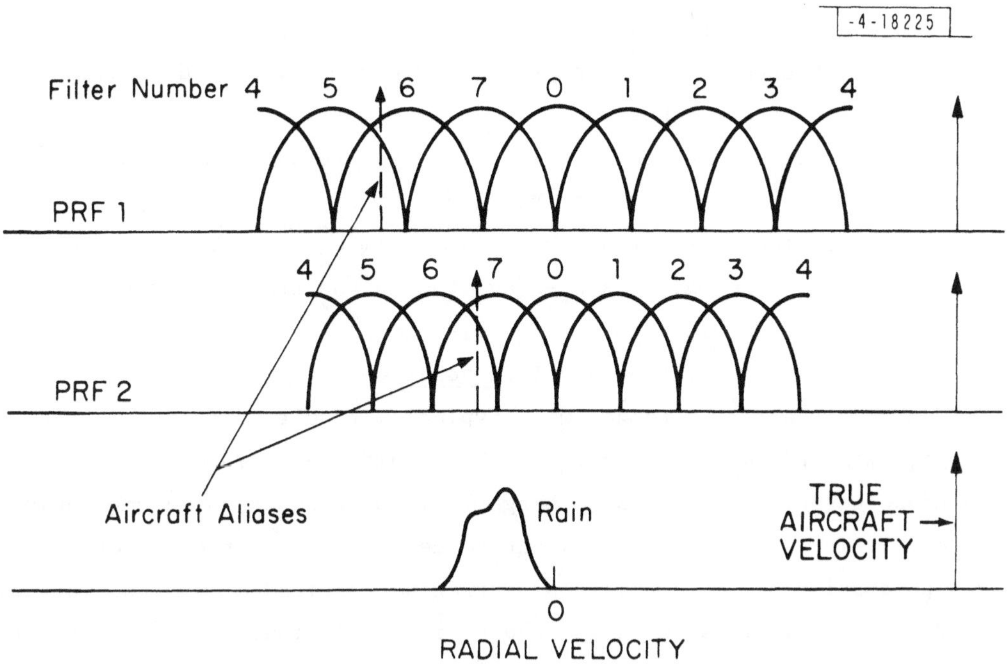

Fig. III-12. Detection in rain using two PRF's.
(Schematic diagram of multiple PRF response to rain and aircraft. The doppler foldover occurs whenever the sampling rate is too low).

CPI are transferred from the first to the second buffer. During the next CPI the contents of the second buffer are transferred to the IOP. A NOVA minicomputer is also connected to this output and receives the data in parallel with the IOP.

Figure II-1 presents a picture of the MTD hardware. Standard TTL circuits are used except for the MOS buffer to the disc memory. The construction utilizes wire-wrap boards to hold the integrated circuits. Besides the input core memory and the clutter map disc memory, the MTD consists of approximately 1000 integrated circuits. The MTD includes all of the digital timing for the radar and generates a 31-MHz, 1.0 μsec pulse which is up-converted in the transmitter to S-band to become the transmitted pulse.

2. Report Correlation and Interpolation Software

A single aircraft target will typically result in threshold crossings in many range, azimuth and doppler cells. These multiple threshold crossings must be associated as belonging to the same target. This process has become known as report correlation*. Also, a more accurate measurement of the target range, azimuth and doppler may be obtained by averaging amplitudes over the multiple threshold crossings. This is called report interpolation. These two functions are performed at NAFEC in the ARTS-III software. In addition, the target doppler returns received at the two different PRF's are used to calculate the aircraft's unambiguous radial velocity. If the threshold crossings are from only one PRF, then an ambiguous velocity flag is set.

Range and azimuth adjacency are the sole criteria used to correlate target reports. Doppler adjacency was not used as a criteria because it was found that amplitude modulation of the target cross section can give rise to split target reports.

For each range-azimuth cell (CPI) doppler report consolidation is performed. The doppler filter number with the largest strength response is noted. Since different doppler filters have different gains, the amplitude of each threshold crossing is divided by a specific normalization factor[15]. After range and

*A detailed description of the radar report correlation and interpolation algorithms used is given in references 15 and 16.

azimuth adjacency are used to consolidate in range and azimuth, target interpolation is performed. The range of the target is reported as the range cell of the largest response. The azimuth of the target report is calculated by taking the first moment of target intensity averaged over azimuth. The doppler interpolation is performed by taking the ratio of the strengths of the doppler filter with the largest strength and the adjacent doppler filter with the second largest strength. This ratio forms an estimate of the target's radial velocity. It is made with a precision of 1/64 of the PRF. This interpolation is done using a look-up table[15,16]. If there are no threshold crossings from a doppler filter adjacent to the doppler filter with the largest strength, then doppler interpolation is bypassed and the doppler is assumed to be the center frequency of the filter with the largest strength. Doppler interpolation is done separately for each PRF. The interpolated doppler numbers at each of the PRF's are then used to calculate the unambiguous radial velocity of the target.

3. Post-MTD Thresholding Software

a. Introduction

Early in the testing of the MTD at NAFEC it became evident that several environmental phenomena were causing more false alarms than initially predicted. Furthermore, these false alarms were partially correlated both spatially and temporally and, thus, were causing false tracks to be initiated by the ARTS-III tracker. Typically there were 50 to 100 false alarms per scan due to noise and as many as several hundred false alarms per scan from all environmental phenomena when they were present.

A series of thresholding algorithms, developed by W. Goodchild at NAFEC, have been particularly successful in eliminating almost all of the non-noiselike false alarms. These algorithms have been incorporated in the ARTS-III software before the radar reports are sent to the tracker.

b. Environmental Phenomena

As noted previously the zero velocity returns from fast moving weather systems can give rise to threshold crossings which are spatially and temporally correlated and thus cause false tracks. These zero velocity false alarms occur along the leading edge of a rain cloud which happens to be passing tangentially to the radar. The threshold crossings occur when the weather moves from one range azimuth cell to another faster than the clutter map can react.

Threshold crossings in the non-zero doppler channels are prevented by the mean-level threshold. The zero radial velocity cell has no mean-level threshold however. Implementation of a mean-level threshold in the zero-doppler channel would result in a 1-mile band of reduced sensitivity around each isolated ground clutter return. This, in turn, would drastically impair the detection performance of tangential aircraft in areas of ground clutter.

A second source of unwanted threshold crossings is the combination of non-synchronous pulsed interference with various forms of clutter. The interference detection circuits in the MTD operate by calculating the average amplitude of the returns from the 10 pulses in each range gate for each CPI. If any single return exceeds the average by some preset multiple, the large pulse is declared to be interference, and threshold crossings from that range gate are ignored. This system works well when the background is white noise, but it breaks down when the background is clutter (i.e., colored noise). In the latter situation, the total clutter energy is not equally divided among the doppler filters, whereas the single interfering pulse energy is non-synchronous interference which passes the amplitude threshold and may be detected in a doppler cell which is subject to low background power spectral density (receiver noise).

Finally, the bird problem is exacerbated by the requirement of maximum sensitivity in the high elevation portions of the CSC^2 antenna pattern. In effect this leads to excess sensitivity at the lower elevations. A consequence is large numbers of threshold crossings from bird echoes.

c. Threshold Algorithms

The post-MTD thresholding algorithms as implemented at NAFEC are as follows:

1. The coverage is divided into 22.5° x 4 nmi sectors.
2. From 0 to 16 nmi, and filters 0-7:
    a. Prior to correlation and interpolation, any replies are discarded which exceed the sector threshold for its corresponding doppler filter.

b. Following correlation and interpolation 1.99 is added to the sector threshold of each filter which contains a single CPI report. (Nominal threshold is 0).

c. If no single CPI reports occur in a sector on the following scan(s), the threshold is decremented for each filter by 1/4.

d. Single CPI reports passing the threshold test are sent to the tracker for possible track updating but <u>not</u> for track initialization.

3. From 16 to 48 nmi:

a. Correlation and interpolation is performed first.

b. Single CPI reports are discarded if they do not exceed the threshold.

c. Single CPI reports increment and lack of them decrement the threshold as above.

d. Single CPI reports passing the threshold test, are sent to the tracker for either track updating <u>or</u> track initialization.

4. Interference is removed from azimuthal sectors 5° wide and with range extent of 0 to 48 nmi as follows:

a. If there is an excess of 10 single CPI reports in a sector, they are prevented from initiating or updating tracks.

b. Once interference is detected, the single CPI reports involved are <u>not</u> used to increment the thresholds.

This set of four thresholding algorithms adequately controls the number of false alarms input to the tracker. Careful study has revealed no noticeable degradation in detection performance due to the use of these algorithms.

REFERENCES

1. C. E. Muehe, et al., "New Techniques Applied to Air Traffic Control Radars," Proc. IEEE, Vol. 62, No. 6, pp 716-723 (June 1974).

2. C. E. Muehe, "Digital Signal Processor for Air Traffic Control Radars," NEREM 1974, Boston, Mass. (29-31 October 1974).

3. R. M. O'Donnell, et al., "Advanced Signal Processing for Airport Surveillance Radars," EASCON 1974, Washington, D.C. (October 1974).

4. F. E. Nathanson, Radar Design Principles (McGraw-Hill, 1969).

5. "A Brief Description of the Solid-State ASR-7 Airport Surveillance Radar," Texas Instruments, Inc., (May 1970).

6. J. Croney, "Clutter on Radar Displays, Reduction by Use of Logarithmic Receivers," Wireless Eng., pp 83-96 (April 1956).

7. E. Eastwood, Radar Ornithology (Methuen & Co., London, England, 1967).

8. G. Pollon, "Distribution of Radar Angels," IEEE AES, Vol. AES-8, No. 6 (November 1972).

9. R. E. Richardson, et al., "Elimination of a Type of Natural Clutter in L-band Radars," Technical Report 178, Lincoln Laboratory, M.I.T. (15 April 1958), AD133851.

10. J. R. Barry, et al., "Angel Clutter and the ASR Air Traffic Control Radar," Applied Physics Laboratory, Johns Hopkins University (Feb. 1973), FAA-RD-73-158.

11. D. F. DeLong, Jr. and E. M. Hofstetter, "On the Design of Optimum Radar Waveforms for Clutter Rejection," IEEE Trans. Informat. Theory, Vol. IT-13, pp 454-463 (July 1967).

12. R. C. Emerson, "Some Pulse Doppler MTI and AMTI Techniques," Report 274, Rand Corp., p 104 (1 March 1954).

13. W. H. Drury, "Improved MTI Radar Signal Processor," Project Report ATC-39, Lincoln Laboratory, M.I.T. (3 April 1975).

14. P. B. McCorison, private communication.

15. K. D. Tschetter, "Coding Description of ARTS Implementation of Lincoln Laboratory Radar and Moving Target Detector," Report PX11136, Rev. A, Sperry

16. R. M. Johnson, Design Data for ARTS Implementation of Lincoln Laboratory Radar and Moving Target Detector, Sperry Univac, PX10512, July 1975.

The Application of SAW Technology to Radar ECCM
P.H. Carr
Proceedings of the DDRE Radar ECCM Symposium, 1976
Pages 28-1 — 28-13
Reprinted by permission.

7.2

SUMMARY

Surface acoustic wave (SAW) technology is particularly important in providing compact, low-cost, reliable components to implement signal processing techniques, which can increase the jam resistance of radars by as much as 40 dB. These include (1) delay lines for signal storage and integration, (2) waveform coding and decoding or correlation with phase-shift keying, chirp, and frequency hop, and (3) real-time spectral analysis for adaptive ECCM. For example, SAW devices can generate and decode such waveforms as linear and nonlinear FM chirps, phase-shift coded waveforms (such as a Barker code) on a chip with typical dimensions of one-half by three centimeters. This is orders of magnitude smaller than equivalent devices fabricated by conventional lumped constant circuits. Size reduction is particularly important for airborne radar. In addition, these devices are reliable and low-cost, as they are fabricated by the same photolithographic techniques used for integrated circuits.

1. INTRODUCTION

The operation of a Rayleigh surface acoustic wave delay line is shown in Figure 1. The electromagnetic input is converted to an elastic surface wave by the piezoelectric effect in the interdigital transducer. The elastic surface wave has the property that the displacement amplitudes decrease exponentially into the medium. This means that the propagating elastic energy is confined to within about one acoustic wavelength of the surface. (At 1000 MHz, one acoustic wavelength is about 4 μm.) When the surface wave arrives at the output transducer, it is reconverted

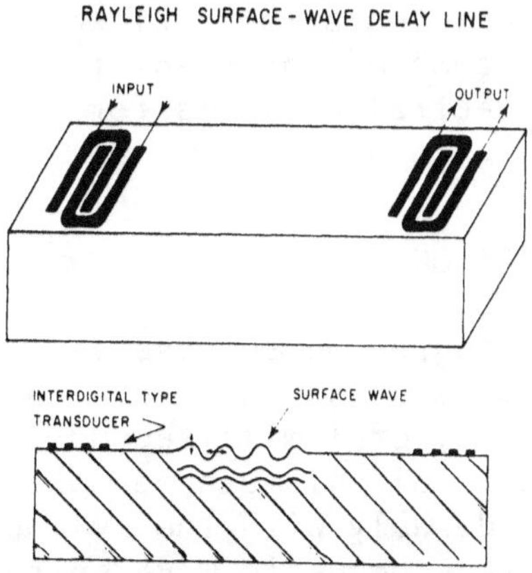

Figure 1. Operation of a Rayleigh surface wave delay line.

into an electromagnetic wave, which has been delayed by the time that it takes for the elastic or acoustic wave to travel from the input transducer to the output. The fact that the velocity of sound is five orders of magnitude slower than the velocity of electromagnetic waves gives rise to corresponding wavelength and size reductions. For example, an acoustic path-length of one centimeter gives a delay of a few microseconds; an equivalent electromagnetic path-length would require a kilometer to get the delay in coaxial cable or waveguide. These size-weight reductions of acoustic delay lines make new operational capabilities feasible for radar ECCM by means of storage and integration of successive radar waveforms.

An equal, if not more important application, is the use of surface acoustic wave devices as filters. Let us place the output transducer of Fig. 1 very close to the input transducer so that the time delay is small. As the surface acoustic wave velocity of the piezoelectric substrate is nondispersive, the frequency-dependent insertion loss is determined by the properties of the transducer. One can think of the transducer operating as an end-fire array: each pair of lines generates surface wave energy adding in phase at the frequency where the grating periodicity corresponds to the surface wave wavelength. The bandwidth of the array is inversely proportional to the number of lines, or sources in the transducer. By adding more lines, one adds more sources and increases the efficiency, but this is done at the expense of bandwidth.

The most important feature of the SAW device is that the surface acoustic wave energy is accessible along the entire surface of the substrate. The transducer lines may be placed at any point along the surface to implement a compact transversal filter, which can act as a matched-filter for correlating coded waveforms. These waveforms provide anti-jam ECCM protection as well as improving the range resolution of the radar. We shall now discuss in order the following ECCM applications: (1) delay lines for signal storage and integration, (2) waveform coding and decoding (or correlation) by means of phase-shift keying, chirp, and frequency hop, and (3) real-time spectral analysis for adaptive ECCM.

2. DELAY LINES FOR SIGNAL STORAGE AND INTEGRATION

Delay lines can be used for storing and integrating successive radar returns, a process which gives ECCM againse noise jammers. The length of a linear delay line having a time delay of 1 millisecond is 1.7 meters on a slow velocity (1700 m/sec) material, bismuth germanium oxide. It is possible to fold this path considerably with the cylindrical delay line shown in Fig. 2. Here the surface acoustic wave spirals in "barber pole" fashion from the input to the output transducer.[1] A photograph of an operational 300 MHz delay line on bismuth germanium

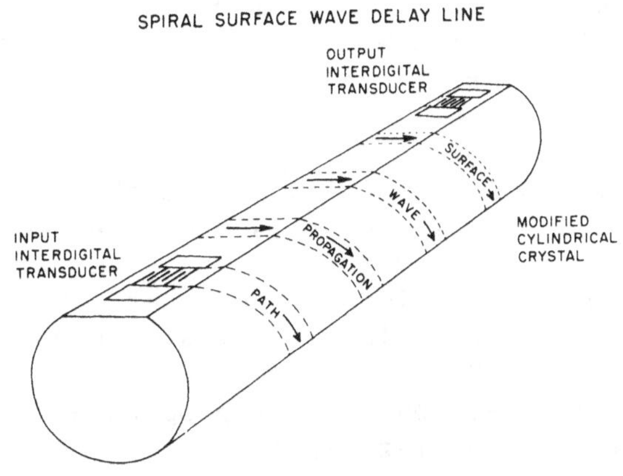

Figure 2. Operation of a compact spiral surface wave delay line.

Figure 3. Photograph of a spiral bismuth germanium oxide delay line. The time delay in propagating around one circumference of the cylinder is 29.2μsec. The center frequency is 300 MHz.

oxide is shown in Figure 3.[2] A delay time per revolution is 29.2 microseconds for the 16.76mm diameter bismuth germanium oxide cylinder. The observed loss at 300 MHz was 0.32 dB/μsec. This loss is too great to attain 1 millisecond of delay in a single delay line.

There are two approaches towards reaching millisecond time delays. The first involves the cascading of short delay lines with amplifiers. The second is to lower the frequency, as the attenuation is approximately equal to the inverse of the frequency squared.[1] The latter is only applicable where large bandwidths are not required. In the first approach, RADC/ET, at a center frequency of 840 MHz has obtained the following preliminary results: a time delay of 250 μsec over a bandwidth of 250 MHz for a time-bandwidth product of 6×10^4.[1] At a center frequency of 85 MHz where the propagation loss is lower, Fortunko and Shaw[3] have obtained a time delay of 0.91 msec and a bandwidth of 65 MHz for a time-bandwidth product of 6×10^4.

3. WAVEFORM CODING AND DECODING

Surface acoustic wave (SAW) technology is particularly important in implementing compact, low-cost, reliable components for the generation and correlation of coded waveforms which increase both the jam resistance of the radar and its range resolution. The gain against an uncoded jammer is the time-bandwidth product of the coded waveform. Time-bandwidth products of 10,000 can be achieved with compact SAW devices.

A great variety of coded waveforms can be implemented with SAW. We shall now discuss the following examples: phase-shift coded waveforms or pseudo-noise, chirp, and frequency-hop or discrete frequency sequences.

The coding and decoding or correlation of a binary, phase-shift coded waveform is illustrated in Figure 4.[4] The code is generated by a 5-nsec pulse. This pulse nearly instantaneously excites the seven wavelength long input transducer, which in this example generates a seven cycle, 70 MHz surface wave. As this 100 n-sec wave packet propagates along the piezoelectric substrate, it is detected either in-phase or out-of-phase by the 31 single-pair interdigital transducer taps spaced seven wavelengths apart. For positive phase detection, the second line of the transducer tap is grounded; and for negative phase detection, the first line is grounded. Six bits of the sequence can be seen in the top part of Fig. 4. This bi-phase modulated pseudo-random sequence can be used directly or mixed or multiplied to a higher frequency.

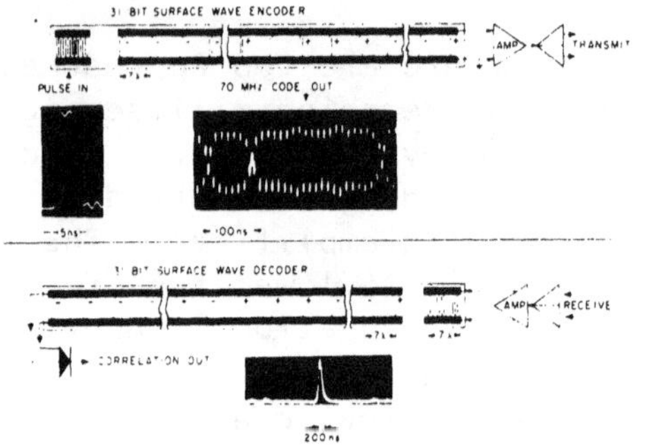

Figure 4. The generation of a pseudo-noise coded waveform is illustrated in the top and the decoding or correlation of this waveform in the bottom.

The received pulse is correlated or decoded on a device which is the mirror image of the encoder and fabricated with the same photolithographic master on an identical substrate. The correlation peak and sidelobes after rectification of the rf signal are shown in the bottom part of Fig. 4. We see that only one distinct output pulse corresponds to the initial pulse. Note that there is no problem synchronizing the correlator to the received pulse.

A signal-to-noise enhancement equal to the compression ratio is obtained through the encoding-decoding process. This is because the information amplitudes, A, add coherently, giving 31A or $(31)^2P$ (P is the power of one bit), while the noise or an uncoded jammer adds incoherently, yielding 31P. The number of taps is of course not limited to 31. Devices with as many as 1000 taps have been built with correspondingly larger anti-jam margins.[5] Fig. 5 shows a 31 tap surface wave device together with two lumped constant inductor-capacitor circuits having the same 31 bit capacity. These were used in the past to perform the same tapped delay line decoding. The size and weight reduction of the acoustic surface wave device (1 inch long, 0.35 in. wide and 0.16 in thick) is obvious. Since surface wave devices are made by the same photolithographic processes as integrated circuits, they have similar reliability

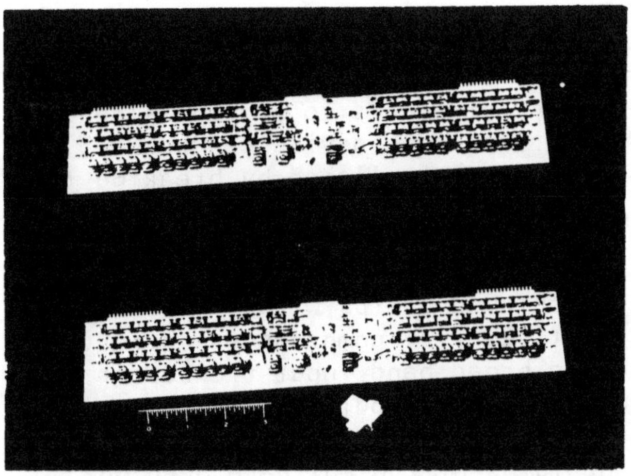

Figure 5. Size and weight advantage of a surface acoustic wave (SAW) device. At the top are shown two lumped constant inductor-capacitor circuits having the same 31-bit capability as the 0.35 by 1 inch SAW device shown in the bottom.

and cost advantages. The cost of the lumped constant circuits in Fig. 5 is about $3100, while the cost of the surface wave device is about $30. Here, we have assumed that the latter were produced in lots of about 1,000.

So far we have limited the discussion to fixed binary sequences. One can also make programmable or variable sequence or code correlators by changing the manner in which the taps are connected to the output. This can be accomplished by switching the polarity of the taps with diodes or other appropriate logic circuitry.[6,7] A very large number of discrete codes can be generated and correlated in this manner. Quadriphase and polyphase codes may also be implemented[8] together with complex analogue waveforms.

A method for generating and correlating both linear or nonlinear frequency modulation or "chirp" waveforms used to increase the range-resolution and ECCM of radars is illustrated in Fig. 6. The high frequency or short wavelength component of the narrow input pulse is generated at the location where the spacing between the lines of the input transducer is small, and the low frequency or long wavelength component is generated where the spacing is large. The propagation path to the wide-bandwidth output transducer is short for the high frequencies and long for the low frequencies. Thus, the expanded-frequency-modulated waveform shown in the top of Fig. 6 is generated. In the receiver channel, the "chirp" waveform is correlated as shown in the bottom of the figure, where both the low and high frequencies are detected at the same time to produce

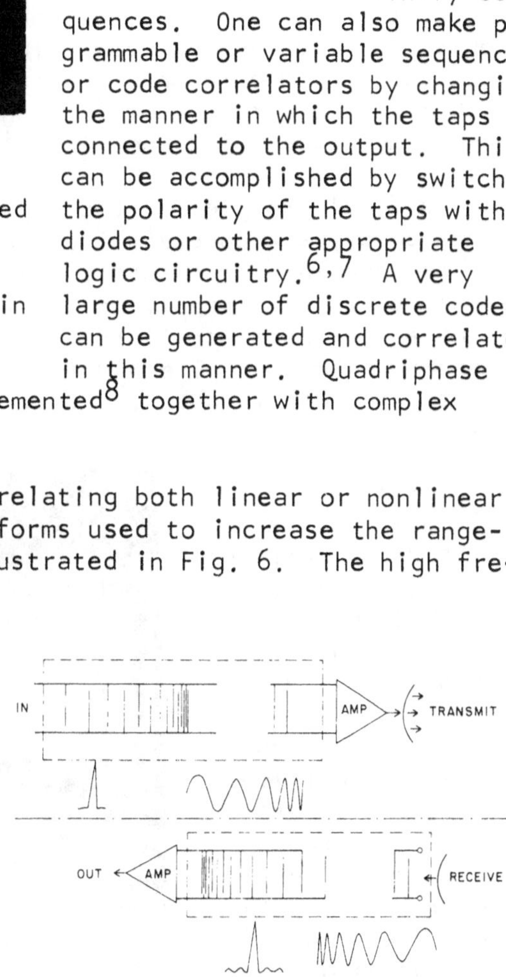

Figure 6. Generation and correlation of a chirp waveform with SAW devices.

the correlation peak and sidelobes. The sidelobes could be confused with the radar return from another nearby object. The solution is to amplitude-weight the different frequency components to produce sidelobes in the 30 to 40 dB range. The surface wave device can be designed to suppress the sidelobes. This is accomplished by breaking the transducer lines as shown in the bottom of Fig. 6 so that the transducer detects a smaller fraction of the energy.[9]

An alternate approach which has achieved chirps with time-bandwidth products of 10,000 is the reflective array compressor. Here the narrow generating pulse is applied to a broad band input transducer. The pulse is dispersed by reflection from nonperiodically spaced groves on the substrate and reconverted to chirped electromagnetic pulse by a broad-band output transducer.[10] Bandwidths of 250 MHz have been obtained with 40 μsec of dispersion.[11]

As a final example for the implementation of coded waveforms, we will now describe a coherent frequency synthesizer suitable for generating frequency-hop or discrete frequency sequences.[12] Frequency synthesis is achieved by applying a uniform comb-spectrum, which is a multiple of a reference oscillator or clock, to the contiguous SAW filter-bank shown in Figs.7&8. The SAW filter bank then selects, at each of its m outputs, the Nth through the (N + m) th harmonic of the reference oscillator.

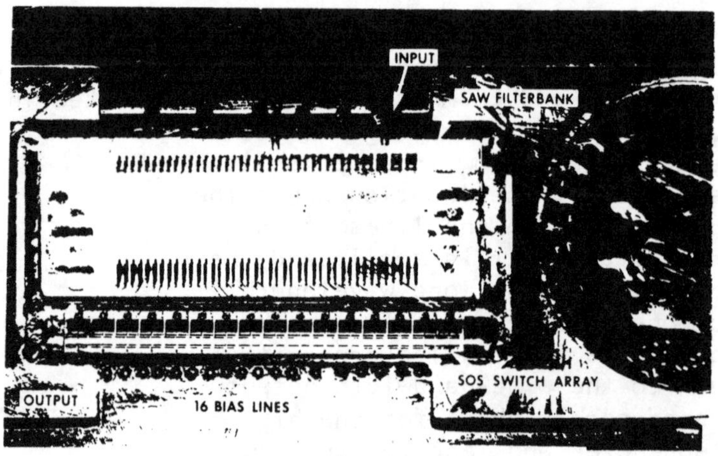

Figure 7. Photograph of the compact UHF frequency synthesizer.

The filter bank is comprised in this example of sixteen pairs of interdigital transducers (IDTs) which have been designed to constitute 16 orthogonal (sin x)/x filters with nulls spaced at identical frequency intervals. The filter pass frequencies are separated by integer multiples of the null spacing interval. Thus, the pass frequency of one channel corresponds to the nulls of all other channels. This response of a single channel is of the form

$$\frac{\sin x}{x} \cdot \frac{\sin 2x}{2x} \qquad (1)$$

when $x = \pi (f-f_o)/f_c$ and f is the applied frequency, f_o is the filter pass frequency, and f_c (the separation of the deep nulls) is chosen to be equal to the reference clock rate. The first factor in Eq. (1) is

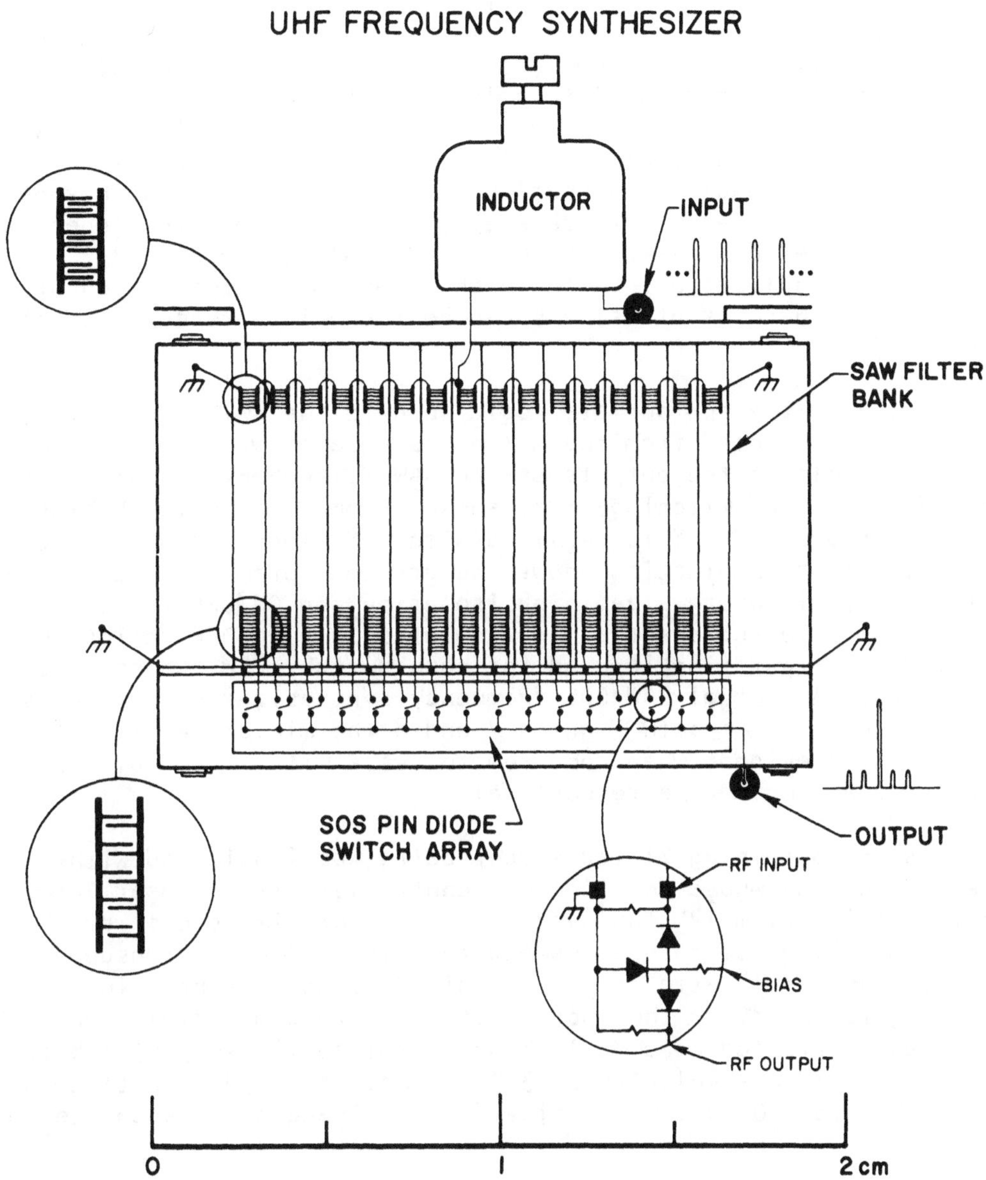

Figure 8. Schematic of a UHF frequency synthesizer for generating frequency-hop coded or discrete frequency coded waveforms.

determined by the input (short) transducer and the second factor is due to the output (long) transducer. When both input and output transducers have a null, the experimental insertion loss was observed to be almost 60 dB down from the pass frequency. It is noted that the frequency synthesizer application is ideally suited to implementation with SAW filters. That is, for such filters low insertion loss, deep nulls and/or low sidelobes are the primary requirements, with ripple and pass band distortion due to triple transit, mass loading (if applicable) and/or electrical loading not usually critical. Finally, with apodized transducers one can readily trade shape factor for low side lobes; with a Hamming weighted cosine-squared-on-a-pedestal apodization[13,14] being ideal for this application. A thorough treatment of the design of SAW filters for frequency synthesizers can be found in a recently published reference.[14]

By applying the appropriate logic word to the switching array, a single tone is selected from the m (in the case shown m = 16) tones constantly available at the outputs of the SAW filterbank; as shown in Figures 7 & 8. The switching array shown schematically (with biasing omitted) in Figures 7 & 8 is realized with PIN diodes using the SOS technology. This combination shows the greatest promise of achieving simultaneously miniature size, high isolation and ON/OFF ratios in excess of 60 dB. To achieve the high isolation and ON/OFF ratios the three diode configuration is used for each switch. Switching times are under 5 nsec. An actual SAW/SOS frequency synthesizer prototype is shown in Figure 7. In both Figures 7 and 8 the miniature size of the device is quite evident when compared to scale with a 2cm rule and the smallness of a single dime respectively.

A master containing 21 transducer pairs was fabricated with a step-and-repeat camera, whose position was controlled with a laser interferometer to within 0.1μm.[15] The overall transducer dimensions were held to approximately 0.6μm tolerances and the width of each transducer line is 1.5μm. The final patterns were obtained by contact printing, using the parting method[16], on the appropriate substrate material. When this pattern was fabricated on a YZ lithium tantalate (2.5x0.9cm) substrate having a surface wave velocity of 3230 m/sec, it yielded contiguous filters from 520 to 650 MHz, with relative pass frequencies accurate to 0.02%.

The requirement for higher frequencies ($f \sim 1$ GHz) may be met by producing sub-micrometer[17] width interdigital transducer electrodes using either electron-beam fabrication or direct optical projection[13,17] from 10X masters. To reduce the demands on the device fabrication, these high fundamental operating frequencies may also be achieved by operating filters at third overtone resonances or by using high velocity[18] substrate materials such as AℓN-on-sapphire and beryllium oxide (velocity ~ 6000 m/sec). Recently[17] frequency synthesizer SAW channels with 0.7μm linewidths and spaces (on YZ $LiTaO_3$), yielding a pass frequency of 1032.3 MHz, were fabricated using direct optical projection (Fig. 9). The primary advantage of this technique is the fact that 10X masters are less expensive than one-to-one masters. The final one-to-one image is obtained with a high resolution reducing lens directly in photoresist on the substrate. These examples illustrate the versatility of SAW technology to implement coded waveforms for radar ECCM.

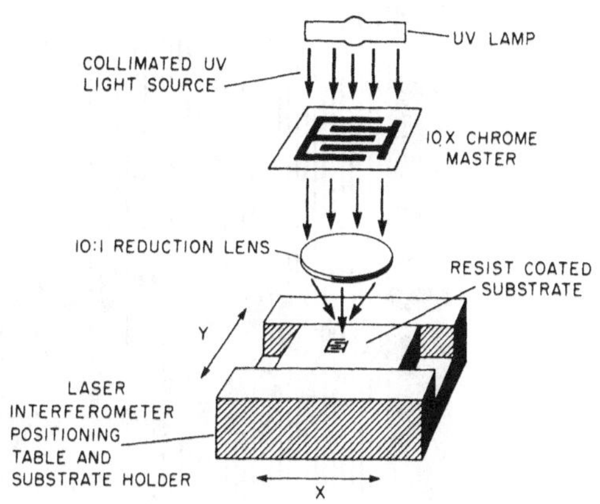

Figure 9. Illustration of the production of micrometer linewidth transducers by direct optical projection on the photoresist coated substrate.

4. REAL-TIME SPECTRAL ANALYSIS FOR ADAPTIVE ECCM

The first step in any adaptive ECCM is the real-time spectral analysis of the hostile electronic warfare environment. A channelized receiver consisting of contiguous filters maintains a high probability of intercept in a dense electromagnetic environment. However, such fast spectral analyzers have been difficult to implement due to the large size and cost of the hundreds of filters required. This is illustrated in Fig. 10, where the largest components in the rack of equipment on the upper left are the eight UHF lumped-constant electromagnetic filters. These can be replaced by the equivalent SAW filters on the one-inch-long-chip shown on the right. In addition to the advantages of smaller size, lower weight, and greater reproducibility, the cost savings of the SAW filters is projected to be $30 per unit when mass produced by the same photolithographic techniques used for integrated circuits.

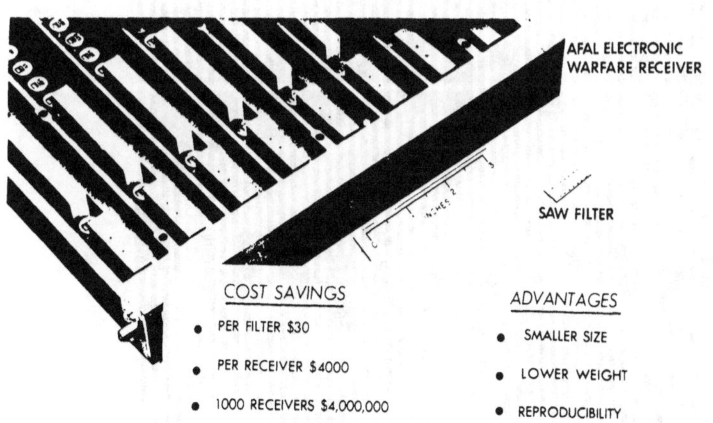

Figure 10. Implementation of a compact real-time spectral analyzer with SAW filters. The eight rectangular boxes at the upper left are UHF filters which can be replaced by eight equivalent SAW filters on one inch long chip at the right.

One is most often interested in measuring the frequency of

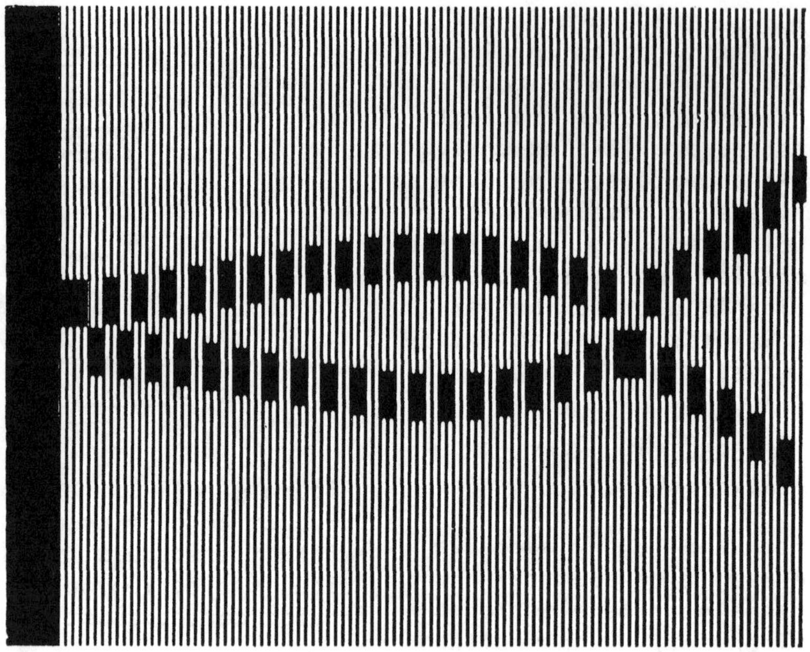

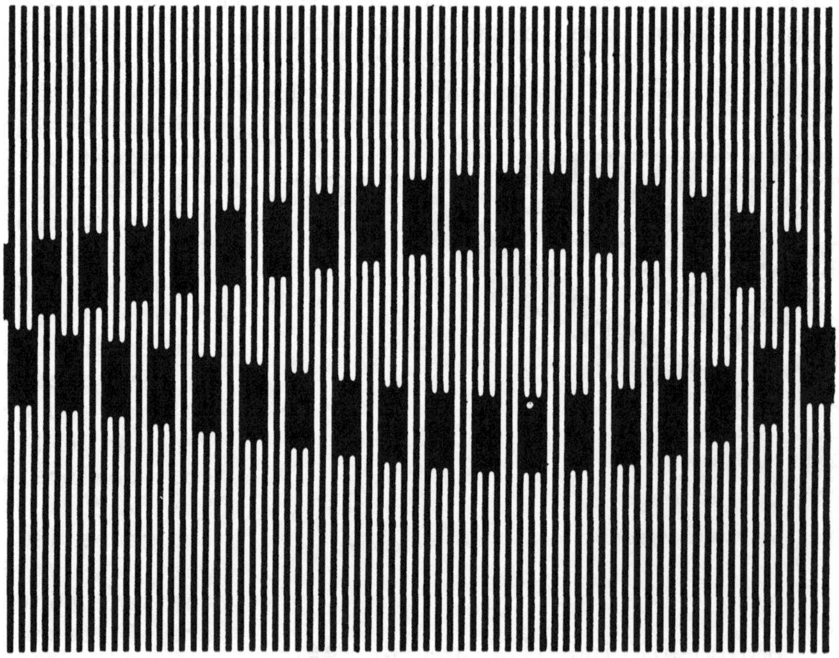

Figure 11. Microscope photographs of portions of interdigital SAW transducers fabricated using direct optical projection. Linewidths are 1.2μm.

radar pulses rather than continuous waves. In this application, filters are sought to be both highly selective in the frequency domain and relatively distortionless in the time domain. Obviously these conflicting requirements can never be fully satisfied. One approach is to use contiguous Butterworth filters in a double-detection signal processing scheme.[19] Photographs of a section of the transducers of the type which produce a Butterworth bandpass is shown in Fig. 11. The sharp, clear 1.2μm wide lines were fabricated with direct optical projection. The center frequency, determined by the spacing of the lines, is 350 MHz on the minimum diffraction cut of lithium tantalate.[19] The lines are spaced on quarter wavelength centers to minimize reflections due the slight impedance difference between the substrate and the metal transducer lines.[20] The break pattern is used to "apodize" the transducer, a technique for controlling the filter bandpass. The Fourier transform of this apodization function is the filter bandpass. In principle, any filter function can be synthesized with a SAW filter. Butterworth filters have conventionally been synthesized with lumped constant circuit elements. The SAW Butterworth filters have the unique property of being linear-phase, a property very difficult to achieve with lumped constant elements.

Recently, RADC/ET has discovered a new filter function, the flat exponential filter (FEF), which is straightforward to implement with surface acoustic waves.[21] This filter has a flat response in the center of the band and exponential fall-off out-of-band. The insertion loss vs. frequency response of an experimental FEF filter is shown in Fig. 12, for a center frequency of 325 MHz. Excellent agreement between theory

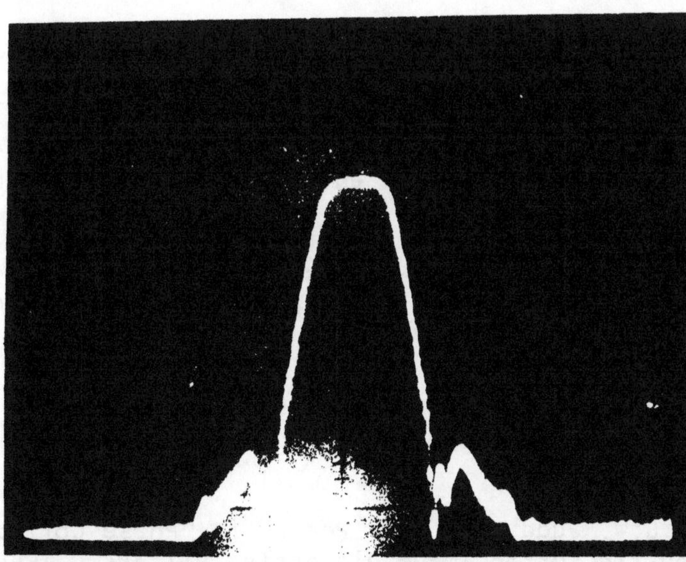

Figure 12. Experimental insertion loss versus frequency characteristic of a Flat Exponential Filter (FEF). Insertion loss = 26 dB. Vertical scale: 10 dB/div. Horizontal scale: 20 MHz/div with center cross-hatched line corresponding to 325 MHz.

and experiment is observed down to a level 35 dB below the pass band, below which there are sidelobes due to second-order effects. When compared to a two pole Butterworth filter, the FEF has better frequency selectivity which is surprisingly not sacrificed in the time domain. This may well be the first of a series of new filter functions which are easier to implement with SAW rather than conventional filters. RADC/ET is currently developing SAW contiguous filters suitable for significant size and cost reduction of real time spectral analyzers.

REFERENCES

1. A.J. Slobodnik, Jr., "A Review of Material Tradeoffs in the Design of Acoustic Surface Wave Devices at VHF and Microwave Frequencies," IEEE Trans. Sonics and Ultrasonics, vol SU-20, 315, October 1973.

2. A.J. Slobodnik, Jr., W.J. Kearns, J.H. Silva, and T.E.Fenstermacher, "A Comparison of SAW Wrap-Around Delay Line Geometries," 1974 Ultrasonics Symposium Proceedings, IEEE Cat. No. 74 CHO '96-1SU, p 185.

3. C.M. Fortunko and H.J. Shaw, "One Millisecond Surface Acoustic Wave Delay Line," 1975 Ultrasonics Symposium Proceedings, IEEE Cat. No. 75 CHO 994-4SU, p 537.

4. P.H. Carr, P.A. DeVito, and Thomas L. Szabo, "The Effect of Temperature and Doppler Shift on the Performance of Elastic Surface Wave Encoders and Decoders," IEEE Trans. Sonics and Ultrasonics, Vol. SU-19, p. 357, July 1972.

5. C.F. Vasile and R. LaRosa, "1000 Bit Surface Wave Matched Filter," Electronics Letters, vol 8, No. 19, 21 September 1972.

6. B.J. Hunsinger and A.R. Franck, "Programmable Surface-Wave Tapped Delay Line," IEEE Trans. Sonics and Ultrasonics, vol. SU-18, pp 152-154, July 1971.

7. W.C. Fifer, R. LaRosa, and J.F. Crush, "Switchable Acoustic Matched Filter," Hazeltine Corp., Final Technical Report, Rome Air Development Ctr, TR-72-72, April 1972.

8. H. van de Vaart, "Generation and Correlation of Quadriphase-Coded Signals Using Acoustic Surface Waves," Proc. IEEE, vol. 60, pp 454-455, April 1972.

9. H.M. Gerard. G.W. Judd, and M.E. Pedinoff, "Phase Corrections for Weighted Acoustic Surface-Wave Dispersive Filters," IEEE Trans. Microwave Theory and Techniques, vol. MTT-20, pp 188-192, February 1972.

10. R.C. Williamson and H.I.Smith, "The Use of Surface-Elastic-Wave Reflection Gratings in Large Time-Bandwidth Pulse-Compression Filters," IEEE Trans. Microwave Theory and Technique, vol MTT-21, p 195, April 1973.

11. H.M. Gerard, O.W. Otto, R.D. Weglein, "Wideband Dispersive Surface Wave Filters, First Interim Report," ECOM-0110-1 Report, March 1974.

12. A.J. Budreau, K.R. Laker, and P.H. Carr, "Compact Microwave Acoustic Surface Wave Filter Bank for Frequency Synthesis," Proceedings of the Symposium on Optical and Acoustical Micro-Electronics, Polytechnic Institute of New York, April 16-18, 1974, p 471.

13. W.J. Kearns and A.J. Slobodnik, Jr., "Fabrication of Micrometer Line Width SAW Filters Using Direct Optical Projection," Tech. Report; AFCRL-TR-75-0055, Air Force Cambridge Research Labs., AFSC, Hanscom AFB, MA., January 1975.

14. A.J. Slobodnik, Jr., "Surface Acoustic Wave Filters at UHF: Design and Analysis," Tech.Report, AFCRL-TR-75-0311, June 1975.

15. Electromask, Inc., Van Nuys, California

16. H.I. Smith, F.J. Bachner, and N. Efremov, "A High-Yield Photolithographic Technique for Surface Wave Devices," J.Electrochemical Society, vol 118, (1971), pp 821-825.

17. A.J. Budreau, W.J. Kearns and P.H. Carr, "State-of-the-Art in Microfabrication of SAW Devices," 1975 Ultrasonics Symposium, Los Angeles, California, September 1975, Paper C-1.

18. A.J. Budreau, J.H. Silva and K.R. Laker, "Microwave Frequency SAW Filters on Aluminum Nitride-on-Sapphire and Beryllium Oxide," 1974 Ultrasonics Symposium, November 1974, pp 299-301.

19. A.J. Slobodnik, Jr., T.E. Fenstermacher, and W.J. Kearns, G.A. Roberts, and J.H. Silva, "A Minimum Diffraction Lithium Tantalate Substrate for Contiguous SAW Butterworth Filters," 1975 Ultrasonics Symposium Proceedings, IEEE Cat No 75 CHO 994-4SU, September 1975.

20. P.H. Carr, "Reduction of Reflections in Surface-Wave Delay Lines with Quarter Wave Taps," Proc. IEEE, vol 60, p 1103, September 1972.

21. A.J. Slobodnik, Jr., K.R. Laker, and T.E. Fenstermacher, "A SAW Filter with Improved Frequency and Time Domain Characteristics for the Frequency Measurement of Narrow RF Pulses," 1975 Ultrasonics Symposium Proceedings, IEEE Cat. No. 75, CHO 994-4SU, p327, September 1975.

Anti-clutter and ECCM Design Criteria for a Low Coverage Radar

G. Petrocchi, S. Rampazzo, and G. Rodriguez

International Conference on Radar, Session VI, Paris, December 1978
Pages 194-200

7.3

Introduction

The environment in which a low coverage radar system operates is characterized by the simultaneous presence of heavy clutter and intentional jamming (active and passive ECM). Strong clutter signals are present in a very large sector of the sweep since the sweep duration is generally short. The possibility of rapidly changing the radiated frequency and the pulse-repetition frequency leads to a completely agile structure which is able to confuse jammers.

The transmission of white noise therefore constitutes the jammer's last resource against this radar; such transmission, however, represents a considerable dilution of the jamming power to the advantage of the radar.

Good performance against clutter and jamming may thus be obtained through the capability of changing the transmission frequency from sweep to sweep in a very large bandwidth simultaneously performing MTI clutter cancellation. However the frequency agility mode of operation, in the conventional MTI configurations, involves the loss of the doppler filtering capability and thus the visibility in the clutter area is possible only in terms of super clutter visibility. For a low coverage radar it is therefore necessary to guarantee in every operative condition a very effective cancellation without losing the ECCM performance inherent to pulse-to-pulse frequency agility. Moreover the frequency agility should be performed in an adaptive manner to enhance the ECCM performance.

This paper describes how these anticlutter and ECCM performances can be achieved by the joint action of a special device called "AGILE MTI" (patent pending) which provides the necessary clutter cancellation even in the presence of pulse to pulse frequency agility, together with an "ADAPTIVE FREQUENCY SELECTOR" (A. F. S.) which detects the presence and type of jammer and selects the least disturbed frequency among those available in the radar prior to each sweep.

This is possible, starting from the principle that even if the jammer is radiating white noise, a coloured noise whose spectrum has minimum and maximum levels in the radar bandwidth is received by the antenna system, due to the frequency response of the antenna sidelobe structure and to the multipath phenomena.

The AGILE MTI with the AFS thus ensures the necessary sub-clutter visibility even in a jamming environment, forcing the jammer to dilute its power over the whole radar bandwidth.

AGILE MTI description

In fig. 1 is represented the schematic functional block diagram of the proposed solution.

The AGILE MTI consists of a double canceller MTI filter without feedback, a memory stack, a multiplexer and a microprogrammable control logic which is part of the A. F. S.

In principle, having stored the sweeps at different frequencies in the memory stacks, it is possible to select through the multiplexer the sweeps at the same frequency and perform the MTI processing.

The control logic has the following functions:

- selection of the frequency to be transmitted, as a consequence of the measurement and evaluation of the interference received by the environment in the radar bandwidth;
- storage of the sweep in reception in memory stacks together with the information related to the frequency transmitted;
- selection of the last two sweeps having the same frequency of the sweep actually in reception and their feeding through the multiplexer, to the double canceller together with the sweep in reception.

An example of correlation of sweeps at the same frequency is shown in fig. 2.

Obviously, to achieve acceptable figures of Improvement Factor, a limitation on the maximum number of frequencies to be used in the time-on-target is necessary. As already mentioned, the transmission frequency selection is adaptive to the ECM environment hence the pattern of transmitted frequencies cannot be known a priori. Thus in this operative mode, only the statistical distribution of the available Improvement Factor can be considered.

In fact the Improvement Factor is not deterministic, depending upon the separation in time between the correlated returns (i. e. the transmission frequency sequences).

Moreover, to avoid blind speeds, the cancellation is performed on successive radar returns staggered in time according to a particular p. r. f. staggering logic.

The main problems therefore to be examined in the design of such a device are:

a) Optimization of the action of the "AGILE MTI" with the "ADAPTIVE FREQUENCY SELECTOR" in order to obtain the best MTI Improvement Factor, in presence of pulse to pulse frequency agility;
b) Optimization of the best p. r. f. stagger sequence in order to have a flat as possible velocity response of the MTI filter.

Optimization of the MTI Improvement Factor

The Improvement Factor of a double canceller is defined as:

$$IF = CA + G_N$$

where G_N is the noise gain and CA is the clutter attenuation represented by the following formula:

$$CA = \frac{\int_{-\infty}^{+\infty} S(f)\, df}{\int_{-\infty}^{+\infty} |H(f)|^2 S(f)\, df}$$

S(f) is the clutter spectrum and H(f) is the transfer function of the double canceller

$$H(f) = 1 - 2\exp[-j2\pi f K_1 T] + \exp[-j2\pi f (K_1 + K_2) T]$$

K_1 and K_2 are the number of sweeps respectively between the transmission of the second and third pulse and between the transmission of the first and second pulse at a certain frequency.

T is the pulse repetion period.

Assuming a gaussian land clutter spectrum centered around f = o

$$S(f) = So \exp\left(\frac{-f^2}{2\sigma^2}\right)$$

and

$$K_A = \sqrt{2}\,\pi\, K_1 T \sigma$$

$$K_B = \sqrt{2}\,\pi\, K_2 T \sigma$$

$$CA = \frac{1}{6 - 4\exp(-K_A^2) - 4\exp(-K_B^2) + 2\exp(-(K_A + K_B)^2)}$$

since $G_N = 6$ for the double canceller.

$$IF = \frac{3}{3-2\exp(-K_A^2)-2\exp(-K_B^2)+\exp-(K_A+K_B)^2}$$

Typical values of p.r.f. for a low coverage radar are comprised in the range from 1000 Hz to 5000 Hz.
The standard deviation of the clutter spectrum considered, is referred to an environment characterized by wooded hills with a wind speed of about 40 km/h and with a clutter spectrum broadening due to the antenna scanning. The overall standard deviation of the clutter spectrum in m/sec $\sigma = \sqrt{\sigma_v^2 + \sigma_{scan}^2}$ is assumed to be less than 1.5 m/sec.
The Improvement Factor can be evaluated statistically by means of the function P {IF} which represents the probability to achieve a certain I. F.
In the case of random equiprobable frequency transmission, P {IF} can be easily evaluated because of the unequivocal relationship of the I. F. with K_1 and K_2

$$I.F. = I.F. (K_1, K_2)$$

$$P\{K_1, K_2\} = P\{K_1\}\, P\{K_2\}$$

The probability of achieving a certain Improvement Factor P {IF} = P $\{IF\}_o$ is obtained as the sum of the functions P $\{K_1, K_2\}$ relative to all the configurations of K_1, K_2 which give IF = IFo.
P $\{K_1\}$ and P $\{K_2\}$ can be easily determined considering the case of equiprobable frequency transmission.

$$P\{K_i\} = \frac{1}{n}\left(\frac{n-1}{n}\right)^{i-1}$$

where n is the number of available frequencies.
In fig. 3 is represented the I. F. distribution function p {IF ≥ IFo} calculated from P IF in the case of 3, 4 and 5 equiprobable frequencies.
p {IF ≥ IFo} is the probability of having an I. F. greater than IFo.
From fig. 3 it can be seen that in corrispondence to certain transmission patterns, the I. F. is undesirably low. It is therefore necessary to find a solution to this problem especially when the frequency selection is adaptive to the ECM environment.
Actually in the case of transmission controlled by the A. F. S., P {K} cannot be defined in a univocal way.
In fact, in this case P {K} depends as above mentioned upon all the parameters connected with the frequency selection, such as the jammer noise spectrum, multipath phenomena, antenna radiation pattern etc.
It is therefore difficult to resolve the problem analytically and hence the analysis has been performed by means of a simulation.
The results have pointed out that the best compromise between the ECCM and MTI performances is achieved using a number of indipendent frequencies such that the ratio between this number and the number of the pulses in the time on target is approssimately 1/8.
One of these frequencies can be changed with another frequency chosen among those available in the radar during the operation, at intervals multiple of the time-on-target, as a result of the A. F. S. evaluation on the spectrum structure and level of the interference.
A first solution to the problem connected with low I. F. in the case of frequency selection controlled by A. F. S. may be obtained by fixing a minimum value of I. F. to be guaranteed and forcing the transmission of a certain frequency, by means of the control logic, before the delay becomes too long.
To guarantee for instance an I. F. ≥ 25dB, in the case of 24 pulses in the t.o.t., it is necessary to force the transmission of the 3 frequencies in order to obtain the following configurations of K_1, K_2:

K_1	K_2
1	2
1	3
2	1
2	2
2	3
3	1
3	2
3	3
3	4
4	3
4	4

It is interesting to observe that in this case the mean value of I. F. is 30 dB.
In this case we have a too severe limitation due to the fact that the configuration K_1, K_2 = 1.1 to which an MTI I. F. ≅ 50 dB corresponds, is not allowed (the configuration 1.1 corresponds to the transmission of the same frequency for three successive sweeps).
On the other hand to allow the 1.1 configuration a 4 sweep delay forcing should be necessary. This solution can be possible but the MTI I. F. would be too low because of the increased delay with respect to the previous solution.
Another solution consists in a free frequency selection in transmission. The A. F. S. action on a noise-like jammer is such as to provoke patterns leading to a high I. F. with the exception of some cases in which the delay is too long between the transmission of the same frequency. Due to the fact that the receiver dynamic range has to be as large as possible, but fixed, while the MTI I. F. can vary, distributed clutter residues can appear in corrispondence of long delays between the transmissions of the same frequency.
The resulting non CFAR condition can be eliminated either by pulse compression techniques using phase coded pulses or by using a correlation technique in the time on target.
In this case, using a correlation technique in the time-on-target an MTI I. F. of 40 dB has been achieved as mean value.
Eliminating the A. F. S. action on the transmission and supposing a random selection of the three frequencies in the time-on-target, the MTI I. F. results from the simulation to be 28 dB.
Consequently we can see that the action of the A. F. S. with respect to the random frequency transmission is effective both from the ECCM and anticlutter points of view.

Optimization of the p. r. f. stagger

Also for the solution of this problem a simulation has been necessary.
In correspondence of Pd = 80%, a certain number of diagrams have been drawn representing the losses introduced by the MTI filter on the S/N ratio as a function of the doppler frequency.
In fig. 4(a), (b) are shown the diagrams of $L_{MTI} = f\left(\frac{fd}{p.r.f.}\right)$ for two different sextuple p. r. f. stagger sequences in the fixed frequency transmission case.
The criteria used to determine these sequences

are those commonly used for stagger a certain p. r. f. with a determined percentage.
The same p. r. f. stagger sequences used in the fixed frequency case have been used to calculate the diagrams of $L_{MTI} = f\left(\frac{fd}{p.r.f.}\right)$ in the AGILE MTI case with 3 frequencies in the time-on-target selected by the A. F. S. in the frequency forcing and non-forcing modes, using a moving window azimuth correlation for the CFAR on clutter residues.
In fig. 5 (I), (II) and fig. 6 (I), (II) are shown the diagram of $L_{MTI} = f\left(\frac{fd}{p.r.f.}\right)$ relative to the two above mentioned possible modes of operation of the AGILE MTI, using the p. r. f. stagger sequences used for diagrams (a) and (b) in fig. 4.
From the above mentioned figures we can see that in the case of frequency agility due to the AFS with frequency forcing, the staggering created by the configurations of K_1, K_2 has the effect of reducing the losses in correspondence of the first blind speed which is lower than in the fixed frequency case, infact the equivalent average p. r. f. is lower than the basic one.
In the case of frequency non-forcing (free A. F. S. mode), the same frequency can be transmitted for several successive sweeps, so that the effect of basic stagger is dominant. The frequency response of MTI filter in this case, is quite similar to that of fixed frequency case with the same stagger.

Conclusions

We can conclude from the above considerations on the I. F. and M. T. I. filter frequency response for the two main operation modes that the most effective one is the free frequency A. F. S. selection mode (non forcing). In fact this operation mode provides the best ECCM performances allowing a frequency selection always adaptive to the environment yet maintaining a good M. T. I. I. F. with an M. T. I. frequency response similar to the fixed frequency one.

Acknowledgements

We thank Dr. G. Bettini for his suggestions and Dr. S. Dicuonzo for his cooperation in carrying out the simulations.

Bibliography

1) G. Bettini, L. Penso, G. Rodriguez "Cancellatore MTI in presenza di agilità pura" Technical Report RT-76142 In. Selenia.

2) Nathanson, "Radar Design Principles".

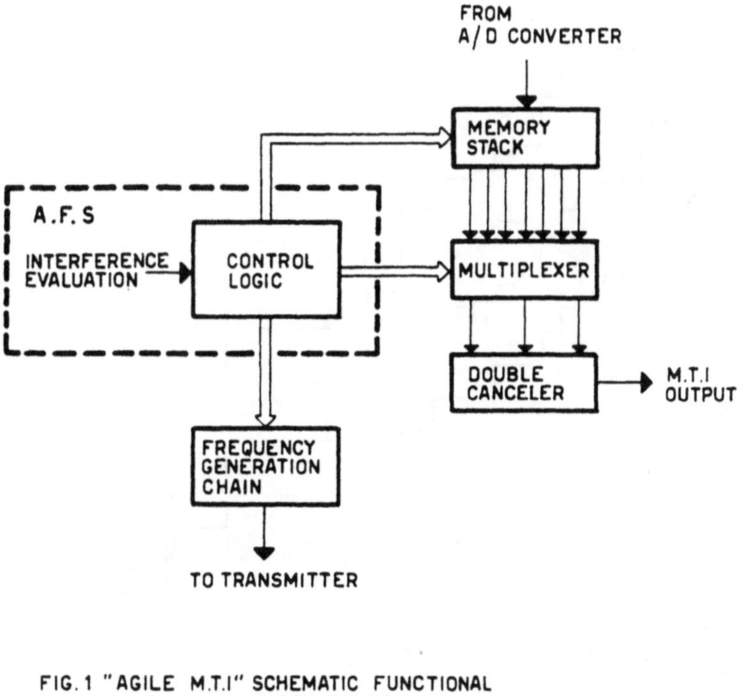

FIG. 1 "AGILE M.T.I" SCHEMATIC FUNCTIONAL BLOCK DIAGRAM

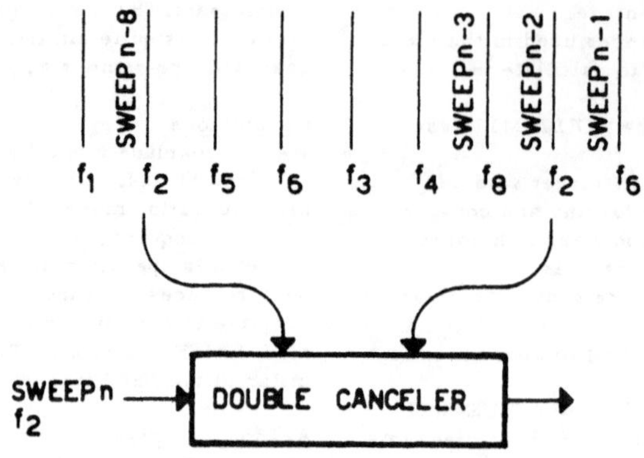

FIG. 2 EXAMPLE OF "AGILE M.T.I" SWEEP CORRELATION

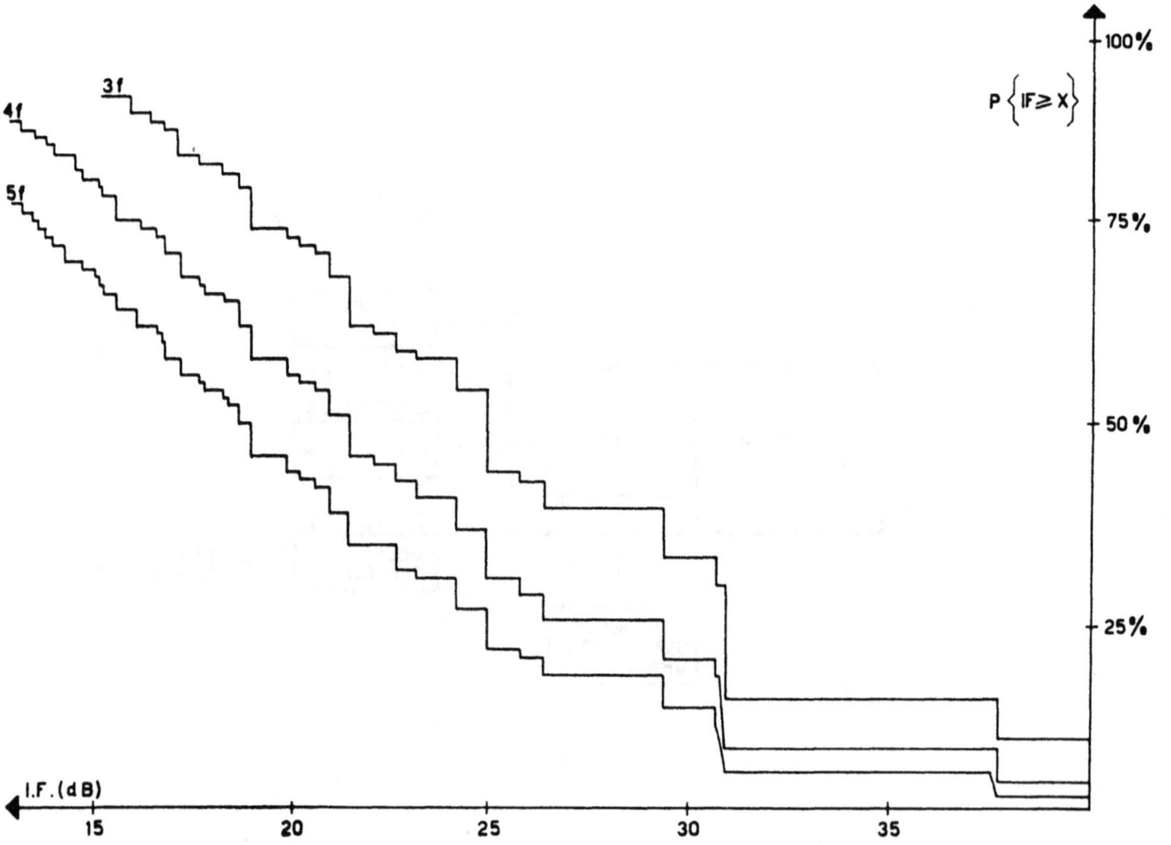

FIG. 3 IF DISTRIBUTION FUNCTION

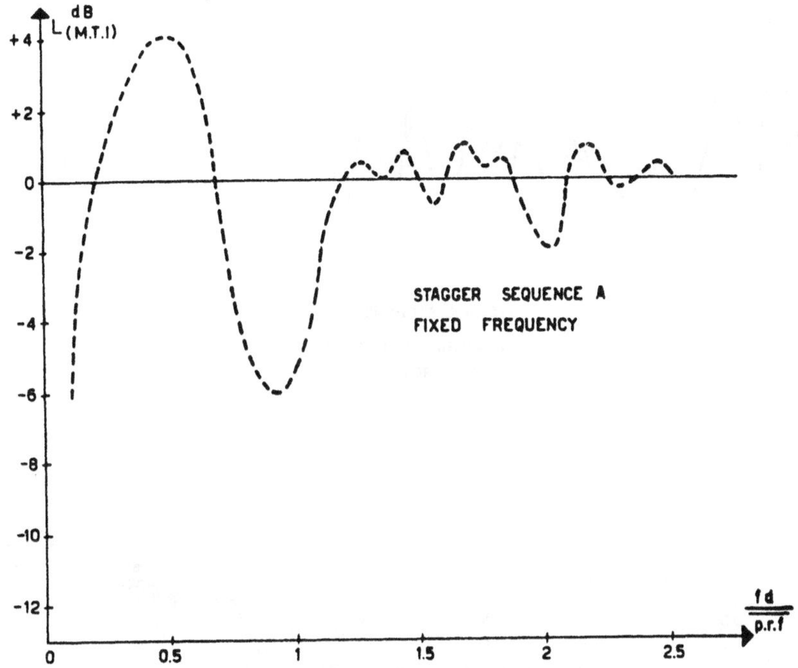

FIG. 4a DOUBLE CANCELER LOSSES

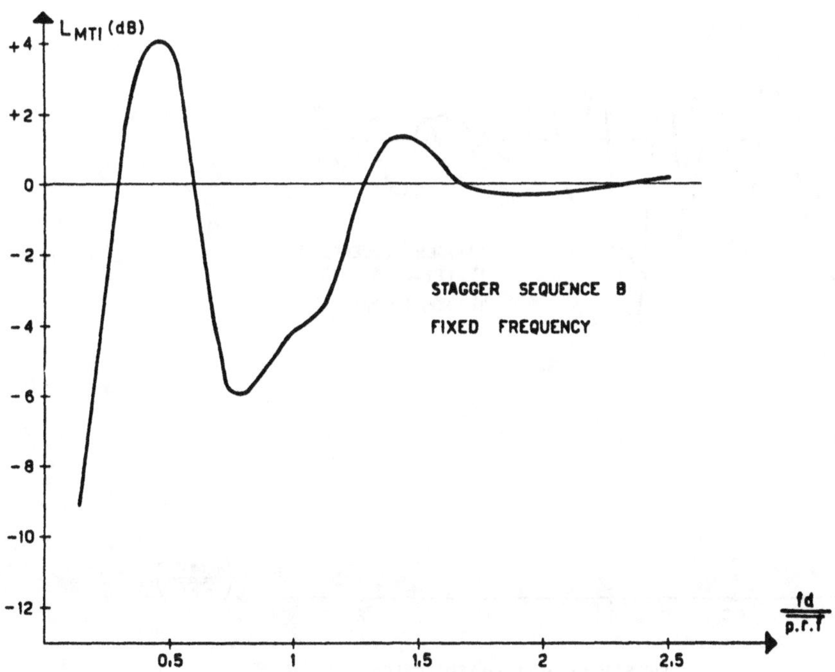

FIG. 4b DOUBLE CANCELER LOSSES

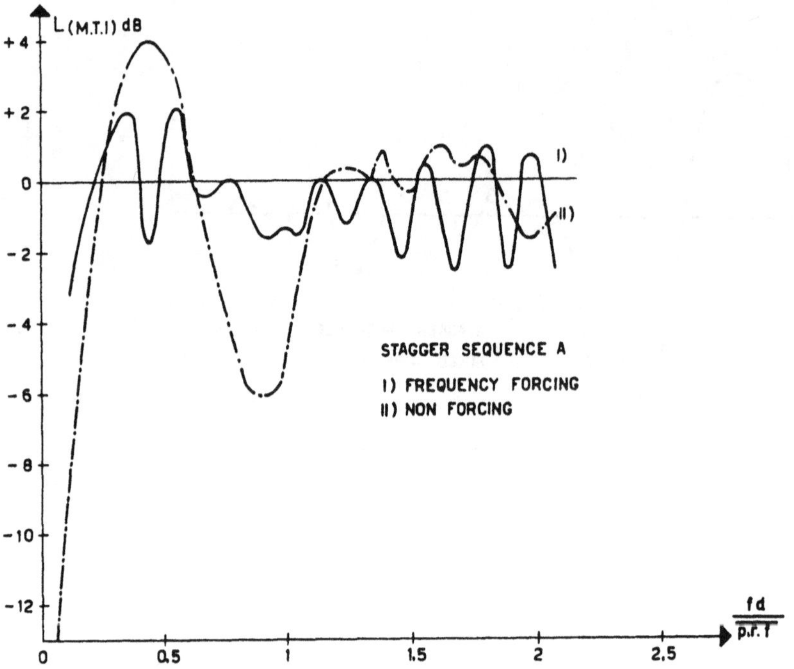

FIG. 5 AGILE M.T.I FILTER LOSSES

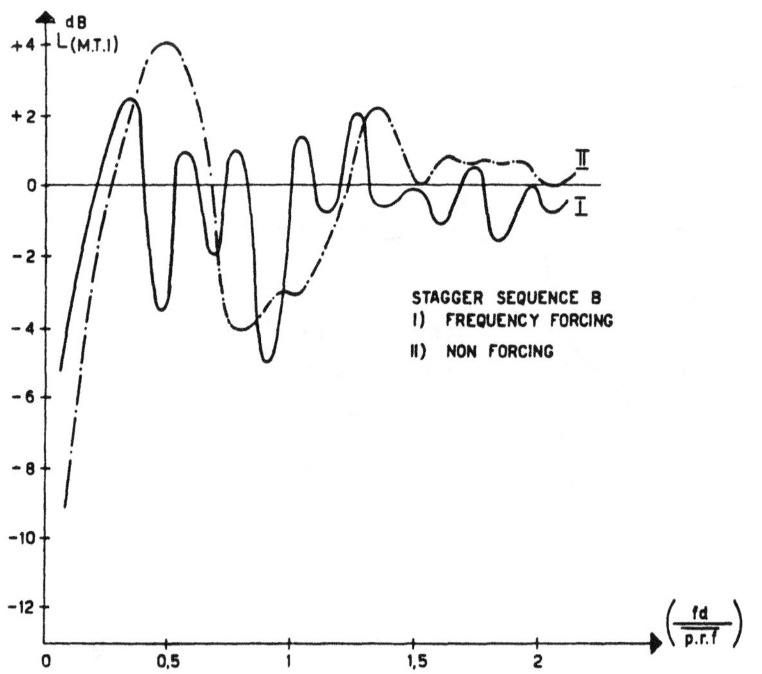

FIG. 6 AGILE-M.T.I FILTER LOSSES

Performance of CFAR Devices in ECM Environments
P.S. Tong and P.E. Steichen
Proceedings of the DDRE Radar ECCM Symposium, 1976
Pages 23-1 – 23-15
Reprinted by permission.

7.4

SUMMARY

Most CFAR devices cannot maintain constant false alarm probability when certain generally held assumptions about the environment are violated. Such weaknesses can be manipulated by the enemy to his advantage and defeat the radar system. This paper explores some of these possibilities for three types of CFAR devices, namely the cell averaging CFAR, the Log-CFAR and the distribution-free CFAR.

1. INTRODUCTION

Failure to control false alarm rate in an automatic detection-and-tracking radar system will saturate the processor and impair its ability to detect targets and maintain tracks. Therefore, a well designed CFAR circuit is a critical element of such a system. And since it is in jamming or heavy clutter environments that a CFAR is most needed, it is pertinent to examine their performance in these stressed conditions. This paper summarizes some of the results dealing with the more common CFAR circuits and the more damaging types of jamming.

Three types of CFAR circuits will be considered. These are the cell averaging CFAR, the log cell averaging CFAR and the distribution-free CFAR. Figure 1a, 1b, and 1c show schematics of the three CFAR circuits. The basic properties of these circuits are discussed in Mitchell and Walker [1], Hansen [2], and Dillard and Antoniak [3] respectively. The types of jamming to be considered are wide-band noise, amplitude-modulated noise and non-Rayleigh noise. The performance against wide-band noise is well understood for all the CFAR circuits, but the effect of quantization on these circuits is not as well understood. It will be shown below that the false alarm probability can change by orders of magnitude if the quantization is too coarse.

Throughout our discussion it will be assumed that the radar return is demodulated by a square-law demodulator followed by one of the CFAR circuits whose output is detected and then processed with a binary-integrator. A sketch of this arrangement is shown in Figure 2. In one instance we shall consider a linear detector also. The choice of the less-efficient binary integrator is motivated by the form of the distribution-free CFAR and the fact that a comparative analysis is greatly simplified by this choice. It is easy to show that the false alarm probability P_{FA} of the system is given by

$$P_{FA} = \sum_{j=K}^{M} \binom{M}{j} (p_f)^j (1-p_f)^{M-j} \tag{1}$$

where M is the number of hits-on-target and K is the final decision threshold at the output of the binary integrator. p_f is the false alarm probability of a single target return at the CFAR output, i.e., it is the probability of a (+1) at the CFAR output. For given M and K, the probability P_{FA} is uniquely determined by p_f. Therefore, under this assumed system configuration, performance comparisons can be made on the basis of p_f alone.

As a baseline example for illustration purposes, it will be assumed that M is 25 and the number of range cells in the CFAR circuit is N = 16. From [3], p_f of the distribution-free CFAR is 1/N+1= 0.0588. To obtain $P_{FA} = 10^{-6}$, K must be 10 as given by (1). The same K value will be assumed for the cell averaging CFAR and log-CFAR. The CFAR constants, c, in the latter circuits will be chosen so that p_f is 0.0588 in all cases. Figure 3 shows the change in P_{FA} as a function of p_f. As seen, a factor of two increase in p_f will result in a change of 3 orders in P_{FA}.

In the sections below, we shall first consider the masking effect of a noise jammer and the change of p_f due to quantization. These will be followed by the performance of the CFAR circuits in amplitude-modulated noise jamming and non-Rayleigh noise jamming.

2. THE MASKING EFFECT OF A NOISE JAMMER

As assumed commonly, a wide-band noise jammer is statistically the same as thermal noise; it is a narrow-band Gaussian process with mean zero. Hence, the sum of the jamming signal and thermal noise will again be narrow-band Gaussian with mean zero and the results obtained in References [1], [2], and [3] will apply. As shown in these references, all circuits will maintain constant false alarm probability p_f under this type of input. However, the signal detection capabilities of the three circuits do differ. Single pulse detection probabilities for these circuits are shown in Figure 4 as a function of input signal-to-(jamming + noise) ratio. In obtaining these curves, we have used the results of Hansen for the log CFAR. In the later case, we have to interpolate to obtain the desired performance curve at $p_f = 0.0588$. In all cases, the target is assumed to be non-fluctuating.

For the distribution-free CFAR, the detection probability, p_d, is given by

$$p_d = \sum_{j=0}^{N} \binom{N}{j} (-1)^j \frac{1}{j+1} \exp\left(-\frac{S}{N}\left(\frac{j}{j+1}\right)\right) \qquad (2)$$

The derivation of this equation may be found in Tong [4]. It is quite obvious that the cell averaging CFAR is the best of the lot in this type of environment. The other circuits are, nevertheless, not too far behind. It should be pointed out that the comparison is made on the basis of a single pulse. When there are more hits-on-target, the distribution free CFAR must use a binary integrator whereas the others may use a more efficient linear integration scheme. Dillard and Antoniak [3] shows that the distribution-free CFAR is about 1.5 dB less sensitive than the cell averaging CFAR.

3. EFFECT OF QUANTIZATION IN NOISE JAMMING

Many CFARs are implemented digitally. The demodulator output is quantized before it is stored. The quantization effect will cause the false alarm probability p_f to change as the input noise level varies. In practice, there is also the problem of dynamic range which will not be considered here. It will be assumed that the quantizer is uniform with an infinite number of output levels at 0, 1, 2... In the log-CFAR case, the quantizer has negative levels as well. The slicing threshold spacing of the quantizer is denoted by Δ. Figure 5 shows the input-output relationship of the uniform quantizer.

Cell Averaging CFAR

The exact p_f is difficult to compute except in very special cases. As an approximation, it will be assumed that the adaptive threshold Z defined by

$$Z = \frac{1}{N} \sum_{i=1}^{N} y_i \qquad (3)$$

is Gaussian with the proper mean and variance as given below. y_i is the quantized output of the demodulator. Therefore,

$$p(Z) = \frac{1}{\sqrt{2\pi}\,\sigma_2} e^{\frac{(Z-m_Z)^2}{2\sigma_z^2}} \qquad (4)$$

Our comparison with known cases shows that this approximation is reasonable for large N(16) and high false alarm probability (.01). The mean and variance of Z are obviously related to the mean and variance of $\{y_i\}$ by the relationship

$$m_z = m_y$$

$$\sigma_z^2 = \frac{\sigma_y^2}{N} \tag{5}$$

Since y assumes discrete values, its mean and variance are given by

$$m_y = \sum_{i=0}^{\infty} i(f_i - f_{i+1})$$

$$\sigma_y^2 = \sum_{i=0}^{\infty} i^2(f_i - f_{i+1}) - m_y^2 \tag{6}$$

where f_i is the probability that the input to the quantizer is above the ith slicing threshold, i.e.,

$$f_i = \Pr(x \geq i\Delta)$$

For square law demodulators,

$$f_i = \int_{i\Delta}^{\infty} \frac{1}{2\sigma^2} \exp\left(-\frac{x+S^2}{2\sigma^2}\right) I_o\left(\frac{Sx}{\sigma^2}\right) dx$$

$$= Q\left(\frac{S}{\sigma}, \frac{i\Delta}{\sigma}\right) \tag{7}$$

where $S^2/2$ is the input signal power if there is a signal and σ^2 is the (noise + jamming) power. $Q(\alpha,\beta)$ is the Marcum-Q function and $I_o(x)$ is the Modified Bessel function. In the absence of a signal, f_i is given by

$$f_i = \exp\left(-\frac{i\Delta}{2\sigma^2}\right) \tag{8}$$

Had a linear detector been used, f_i would be given by

$$f_i = \int_{i\Delta}^{\infty} \frac{x}{2\sigma^2} \exp\left(-\frac{x^2+S^2}{2\sigma^2}\right) I_o\left(\frac{Sx}{\sigma^2}\right) dx \tag{9}$$

and in the absence of any signal,

$$f_i = \exp\left(-\frac{(i\Delta)^2}{2\sigma^2}\right) \tag{10}$$

The probability of declaring a signal when there isn't any for a digitized cell averaging CFAR is given by

$$p_f = \Pr(y_T \geq c\,Z)$$

$$= \sum_{i=0}^{\infty} \Pr(y_T = i)\,\Pr(cZ \leq i) \tag{11}$$

Here, y_T denotes the quantized value of the target cell and c is the multiplicative CFAR constant. In deriving (11), it has been arbitrarily decided that a "tie" is an indication of a target. Based on the Gaussian assumption of Z,

$$\Pr\ (cZ \leq i) = \int_{-\infty}^{(i+1)\Delta} \frac{1}{\sqrt{2\pi}\,\sigma_z} \exp\left(-\frac{(cZ-m_z)^2}{2\sigma_z^2}\right) dZ \tag{12}$$

and $\Pr\ (y_T=i)$ is simply

$$\Pr\ (y_T = i) = f_i - f_{i+1} \tag{13}$$

p_f for the square-law detector is shown in Figure 6 for a number of different Δ/σ^2 and c. c has been chosen such that p_f in the limiting situation of extremely fine quantization is close to 0.0588. Observe that as Δ/σ^2 becomes much greater than unity, the probability of a single hit false alarm p_f approaches unity. This is a consequence of our rule for resolving "tie". Had we chosen to ignore "ties", then the false alarm probability would rise to a certain level and then decrease to zero. But regardless of the 'tie' rule, the change of p_f cannot be ignored in the design of digital CFAR.

Log CFAR

The analysis of the digital log-CFAR is treated in the same manner as the digital linear CFAR. Again, the density function of Z will be assumed to be Gaussian with certain mean and variance. Equations (3), (4), and (5) will hold but (6) will be replaced by

$$m_y = \sum_{i=-\infty}^{\infty} i(f_i - f_{i+1})$$

$$\sigma_y^2 = \sum_{i=-\infty}^{\infty} i^2(f_i - f_{i+1}) - m_y^2 \quad (14)$$

because ln x is both positive and negative. f_i will be given by

$$f_i = \int_{e^{i\Delta}}^{\infty} \frac{1}{2\sigma^2} \exp\left(-\frac{x+S^2}{2\sigma^2}\right) I_o\left(\frac{S\ x}{\sigma^2}\right) dx \quad (15)$$

and if S = 0,

$$f_i = \exp\left(-\frac{e^{i\Delta}}{2\sigma^2}\right) \quad (16)$$

where the use of a square-law demodulator has been assumed. p_f is given by

$$p_f = Pr(y_T \geq Z + c)$$

$$= Pr(y_T = i)\ Pr(Z + c \leq i) \quad (17)$$

$Pr(y_T = i)$ is simply $f_i - f_{i+1}$ and $Pr(Z + c \leq i)$ is given by the integral of p(Z) with the proper m_z and σ_z^2 and the proper limits.

After the logarithmic transformation, the variance in the absence of target signal is always $\pi^2/6$ regardless of the input power. This property will render p_f almost totally insensitive to input power variations. Figure 7 shows p_f as a function of σ^2 with Δ (normalized to $\pi^2/6$, the variance of the log amplifier output) as a parameter. As seen, up to $\Delta = 0.5$, p_f is almost a constant but when $\Delta = 1.0$, p_f exhibits periodic fluctuations.

Distribution-Free CFAR

The false-alarm probability, p_f, for the digital distribution-free CFAR can be evaluated without any approximation. Assuming a square law demodulator, the probability of y equal to i or greater is given by (10). p_f is

$$p_f = \sum_{i=0}^{\infty} (f_i - f_{i+1})\ Pr^N(y_j \leq i) \quad (18)$$

for j = 1, 2 ... N. Hence,

$$p_f = \sum_{i=0}^{\infty} \left[\exp(-\frac{i\Delta}{2}) - \exp(-\frac{(i+1)\Delta}{2}) \right] \left[1-\exp(-\frac{(i+1)\Delta}{2}) \right]^N \tag{19}$$

This equation has been evaluated for N = 16 in Figure 8. As compared to the cell averaging CFAR, the fluctuation in p_f is about the same. Among the three, the log-CFAR seems to be the most desirable CFAR device in terms of tolerance to digitization. It should be pointed out, however, that we have neglected the dynamic range effect which must be considered in the implementation of a practical CFAR.

4. FALSE ALARM INCREASE DUE TO AMPLITUDE MODULATED NOISE JAMMING

Although we have not treated the more basic aspects of the CFAR circuits here, it is shown in the references cited above that successful CFAR behavior depends on a number of basic assumptions of the input noise, jamming signal or clutter. If these assumptions are violated, then the false alarm probabilities may vary in rather unpredictable manners. Three of these assumptions are

- A1 stationarity of range samples
- A2 Gaussian properties of input
- A3 azimuth input independence

The third assumption concerns basically the video integration process. It will not be treated in this paper. This section will analyze the behavior of the three selected CFAR in non-stationary amplitude modulated noise and the next will treat their performance in non-Gaussian noise.

In our analysis, it will be assumed that the CFAR circuits are strictly analog without any dynamic range limitations. It will be further assumed that the variable thresholds Z of the cell averaging CFAR and log CFAR are exactly equal to the mean of the CFAR input x.

Cell Averaging CFAR

Assuming a square demodulator is being used with this CFAR, the mean of x is

$$m = \int_0^{\infty} \frac{x}{2\sigma^2} \exp\left(-\frac{x}{2\sigma^2}\right) dx = 2\sigma^2 \tag{20}$$

Since it is known that the single pulse decision threshold T is given by the solution of

$$p_f = e^{-T/2\sigma^2} \tag{21}$$

We have

$$T = (-\ln p_f)m \tag{22}$$

The quantity $(-\ln p_f)$ is the CFAR constant for infinitely many cells. This constant will be larger for fewer cells.

As a model for the amplitude modulated noise jammer, it will be assumed that the jammer is a blinker whose power level is J_1 50% of the time and J_2 50% of the time. System noise will be treated as a part of the jammer noise. Thus, when the jammer is on, the mean of x is given by

$$m = 1/2\ (2J_1 + 2J_2) \tag{23}$$

From (20), $2J_i$ is the mean of the demodulator ouput when the jammer power is J_i. The false alarm probability p_f' under jamming is

$$p_f' = \frac{1}{2}\exp\left[-\frac{m(-\ln p_f)}{2J_1}\right] + \frac{1}{2}\exp\left[-\frac{m(-\ln p_f)}{2J_2}\right] \tag{24}$$

where m is given by (23). If $J_1 >> J_2$, then

$$p_f' \simeq \frac{1}{2}\exp\left[-\frac{m(-\ln p_f)}{2J_1}\right] \simeq \frac{1}{2}(p_f)^{1/2} \tag{25}$$

For $p_f = 0.0588$, $p_f' = 0.1212$ and P_{FA} of the baseline example will be 5×10^{-4} instead of 10^{-6}. The increase is more than two orders. Figure 9 shows p_f' versus the ratio of J_1/J_2.

More generally, p_f' is given by

$$p_f' \simeq \alpha p_f^{\ \alpha} \tag{26}$$

where α is the blinker duty cycle. This equation may be maximized with respect to α. For $p_f = 0.0588$, α_{max} is about 35% and the maximum p_f' is .13.

Log - CFAR

Carrying out the analysis as done in the preceding section, it is found that the mean of x is $(\ln(2\sigma^2) - \gamma)$ where γ is the Euler constant equal to .577. From (16), p_f is given by

$$p_f = \exp\left(-\frac{e^T}{\exp(-m+\gamma)}\right) \tag{27}$$

Hence,

$$T = m + \gamma + \ln(-\ln P_f) = m + c \tag{28}$$

where c is the additive constant. Suppose the jammer is a 50% duty cycle blinker; the measured mean of x is

$$m = \frac{m_1 + m_2}{2} \tag{29}$$

where $m_1 = \ln 2J_1 - \gamma$ and $m_2 = \ln 2J_2 - \gamma$. Combining (27), (28) and (29), p_f' is given by

$$p_f' = \frac{1}{2} p_f^{\left(\frac{J_2}{J_1}\right)} + \frac{1}{2} p_f^{\left(\frac{J_1}{J_2}\right)} \tag{30}$$

If $J_1 << J_2$, it is seen that p_f' can be very nearly 0.5. p_f' versus J_1/J_2 for this CFAR is also shown in Figure 9.

Distribution-free CFAR

The false alarm probability of this CFAR is given by [3]

$$p_f = \frac{1}{N+1} \tag{31}$$

Suppose the jammer power level is high and its duty cycle is α, the false alarm probability is given by $p_f' = \Pr$ (false alarm$|$E) Pr(E) where E is the event that the target cell power level is high. Obviously Pr(E) is equal to α. The conditional probability is determined by (31) and the number of cells with high power level. This number is equal to $\alpha(N+1)$. Therefore,

$$p_f' = \frac{\alpha}{\alpha(N+1)} = \frac{1}{N+1} \tag{32}$$

The false alarm probability of this CFAR is unaffected by this type of non-stationary jamming.

5. PERFORMANCE IN NON-GAUSSIAN NOISE

A jammer capable of transmitting non-Gaussian noise is unusual even though it can be done in principle. There is, however, much experimental evidence of non-Gaussian clutter. Thus, in this final section, we shall consider the false alarm probability changes as the input violates the Gaussian assumption. It will be assumed that the thermal noise is neglible in the analysis. As a jamming model, it will be assumed that the probability density function at the demodulator output is Weibull rather than the usualy exponential which is a consequence of the Gaussian assumption. The cumulative distribution of this family has the form

$$\Pr(x \le X) = 1 - e^{-(X/2\sigma^2)^\eta} \tag{33}$$

The inverse of the parameter η is called the Weibull parameter. When $\eta^{-1}=1$, this is reduced to the Gaussian case.

The mean of the variable x is the power of the jammer. In this case, it is

$$E[x] = 2\sigma^2\Gamma\,(1 + \eta^{-1}) \tag{34}$$

In the following, we shall consider p_f as a function of η^{-1} assuming that E[x], the power of the jammer, is held fixed. Thus, the cumulative distribution of the jammer signal at the demodulator output is

$$Pr(x \leq X) = 1 - \exp[-\frac{X\,\Gamma(1+\eta^{-1})}{2\sigma^2}]^{\eta} \tag{35}$$

Cell Averaging CFAR

As shown in the preceding section, the decision threshold is $T = cm = (-\ln p_f)m$. If η^{-1} varies from unity, the false alarm probability p_f' is

$$p_f' = e^{-(-\ln p_f)^{\eta}} \tag{36}$$

which is obtained from (35) using the proper mean value for x. Figure 10 shows p_f' as a function of η^{-1}

Log CFAR

From (35), it is obvious that

$$p_f = 1-Pr(x \leq e^T) = \exp\,[-\frac{e^T\Gamma(1+\eta^{-1})}{2\sigma^2}]^{\eta} \tag{37}$$

The mean value of x is given by

$$m = \ln[2\sigma^2/\Gamma(1+\eta^{-1})] - \gamma/\eta \tag{38}$$

The decision threshold under normal conditions is given by (28), i.e.,

$$T = m + \gamma + \ln\,(-\ln p_f)$$

Combining (28), (37) and (38), p_f' is

$$p_f' = \exp - [e^{-\gamma(1-\eta)}\,(-\ln p_f)^{\eta}] \tag{39}$$

This result is plotted in Figure 10. As compared to the cell averaging CFAR, the log CFAR is more severely affected by the change of the parameter η^{-1}.

Distribution-free CFAR

As its name implies, this CFAR is totally insensitive to the change of input statistics. As shown by Dillard and Antoniak [3], the probability of one random variable being the largest of a group of identically distributed random variables is always 1/(N+1) where N+1 is the size of the group. It does not matter whether the random variables are Gaussian, Weibull, Log-normal or any other form.

6. CONCLUSIONS

We have examined the CFAR behaviors under various types of jamming for three different CFAR devices. Being the most target-sensitive, the cell averaging CFAR is best under the usually assumed conditions. However, even in these conditions, it is affected by the digitization process.

The Log-CFAR performance is very nearly equal to the cell averaging CFAR under perfect conditions. It is also the least affected by the digitization processes. But, as the input jamming statistics change, the false alarm probability becomes exceedingly large. It is particularly sensitive to amplitude-modulated blinking type of jamming.

The Distribution-free CFAR must work with a binary integrator. Therefore, in normal conditions, it is not as efficient as the other two CFARs discussed here. The difference is small for single-hit at high false alarm probability but it is about 1.5 dB in multiple-hit situations (see [3]). The merits of the distribution-free CFAR is its indifference to jamming input statistics.

REFERENCES

[1] Mitchell R. and J. Walker, "Recursive Methods for Computing Detection Probabilities," IEEE AES Transactions, July 1971, pp. 671-676.

[2] Hansen, V. G. and H. R. Ward, "Detection Performance of the Cell-Averaging LOG/CFAR Receiver," IEEE Transactions, Vol. AES-8, September 1972, pp. 648-652.

[3] Dillard G. M. and C. E. Antoniak, "A Practical Distribution-free Detection Procedure for Multiple-Range-Bin Radars," IEEE AES Transactions, September 1970, pp. 629-635.

[4] Tong, P, "Greatest-of-N Distribution-Free CFAR," Technology Service Corporation internal memo, TSC-W4-30, June 10, 1974.

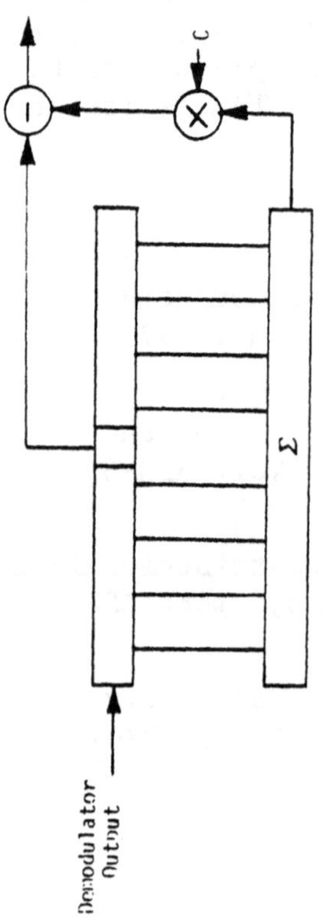

Figure 1a. CELL AVERAGING CFAR

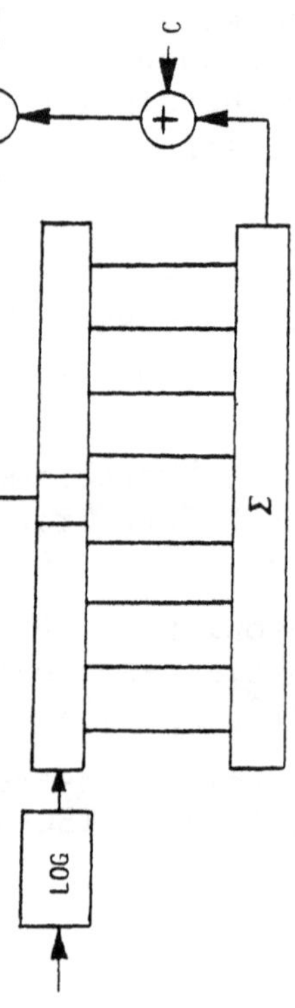

Figure 1b. LOG CFAR

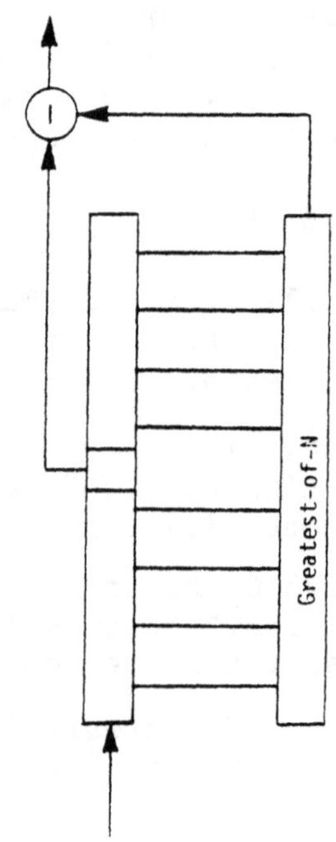

Figure 1c. DISTRIBUTION-FREE CFAR

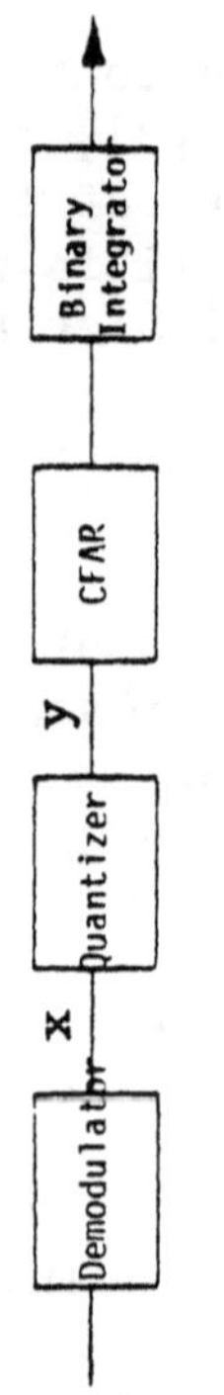

Figure 2. TYPICAL DETECTION SYSTEM

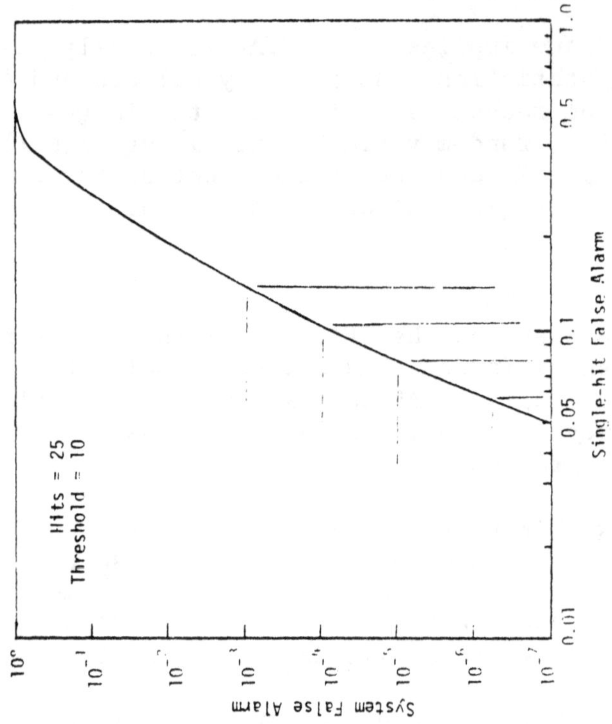

Figure 3. SYSTEM FALSE ALARM PROBABILITY VERSUS SINGLE HIT FALSE ALARM PROBABILITY

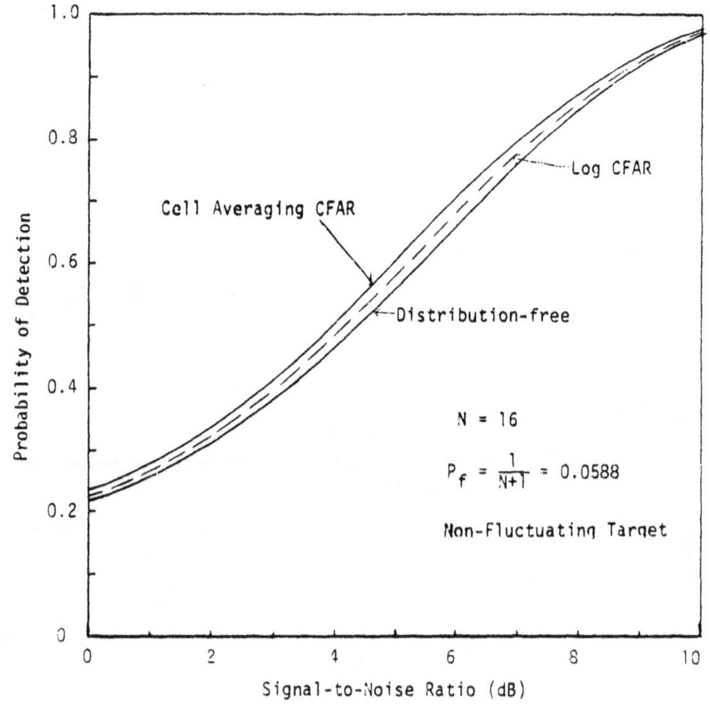

Figure 4. SINGLE-HIT p_d OF CFAR DEVICES

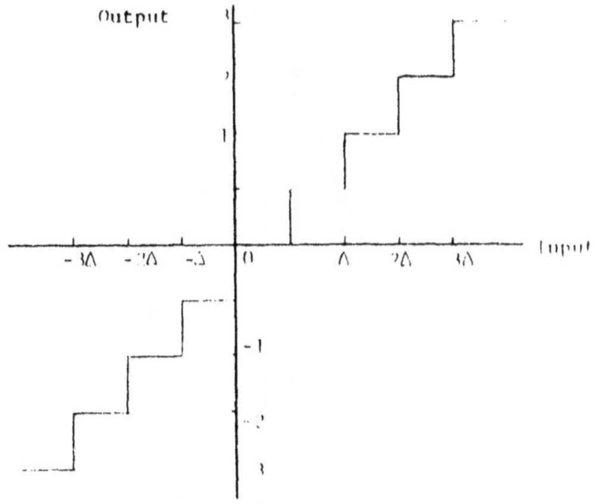

Figure 5. QUANTIZER

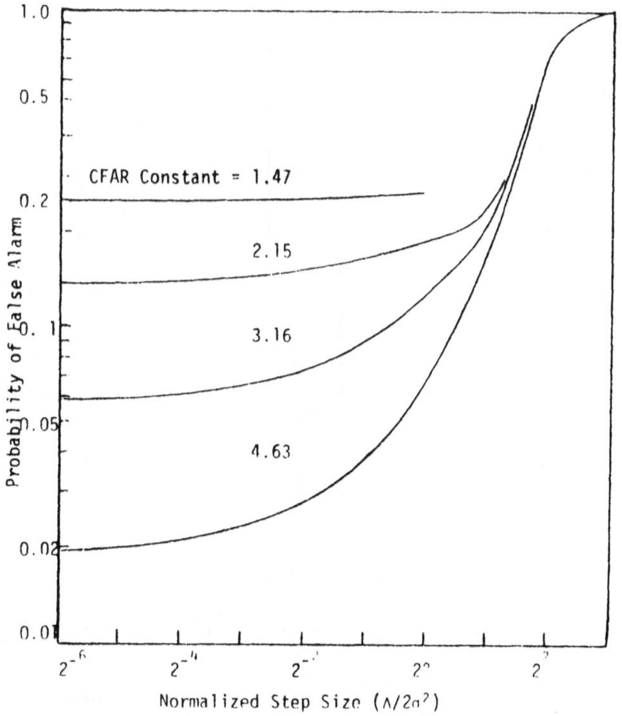

Figure 6. p_f OF A DIGITAL CELL AVERAGING CFAR

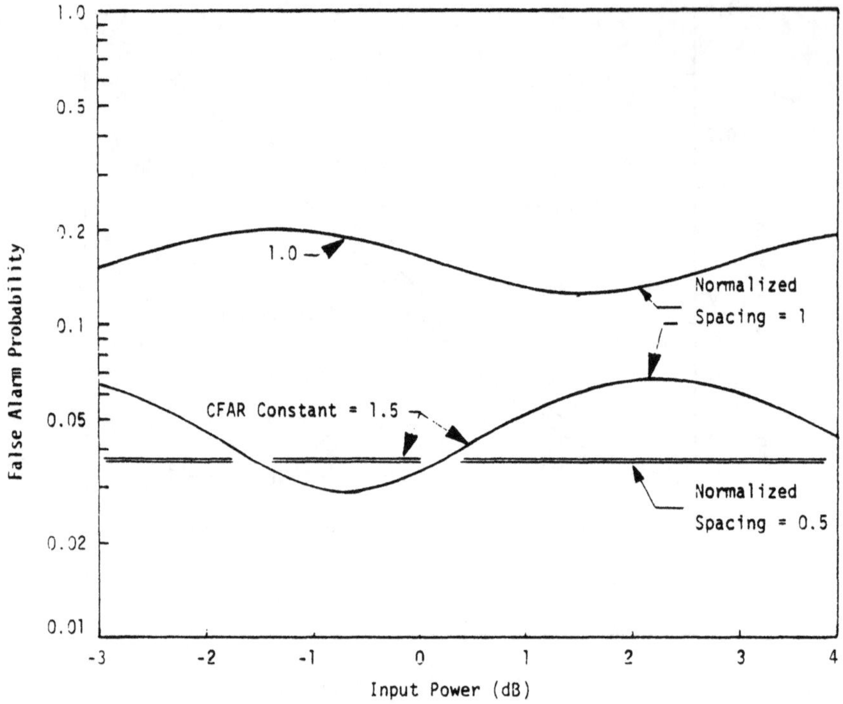

Figure 7. SINGLE-HIT FALSE ALARM PROBABILITY OF A LOG CFAR

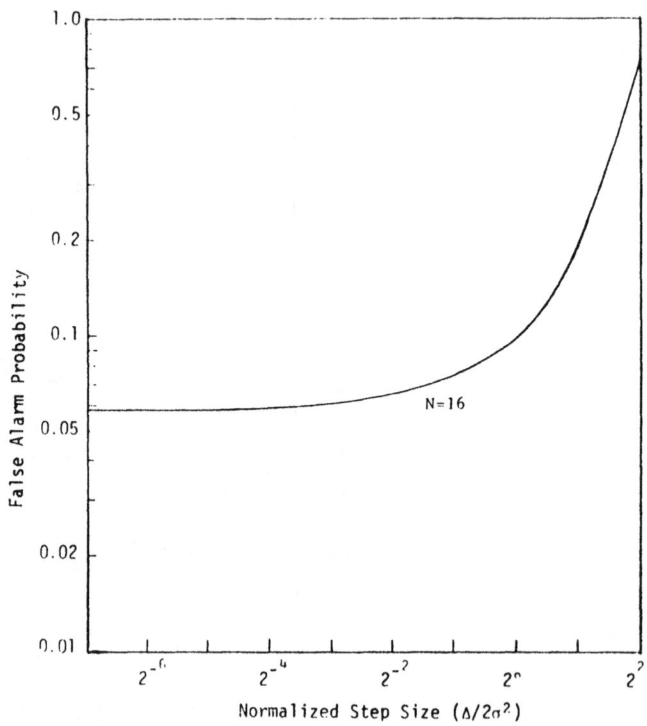

Figure 8. SINGLE-HIT FALSE ALARM PROBABILITY OF A DIGITAL DISTRIBUTION-FREE CFAR

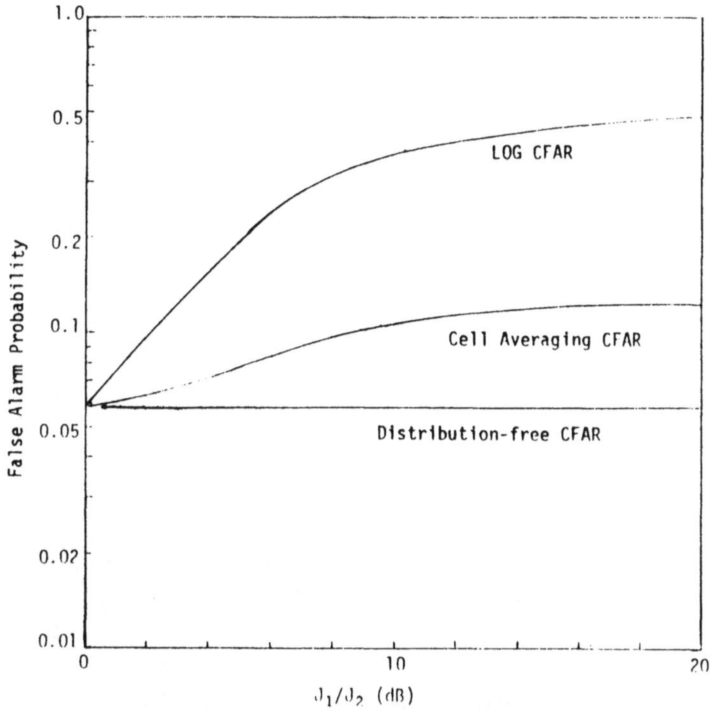

Figure 9. FALSE ALARM PROBABILITY IN AMPLITUDE MODULATED NOISE 50% DUTY CYCLE

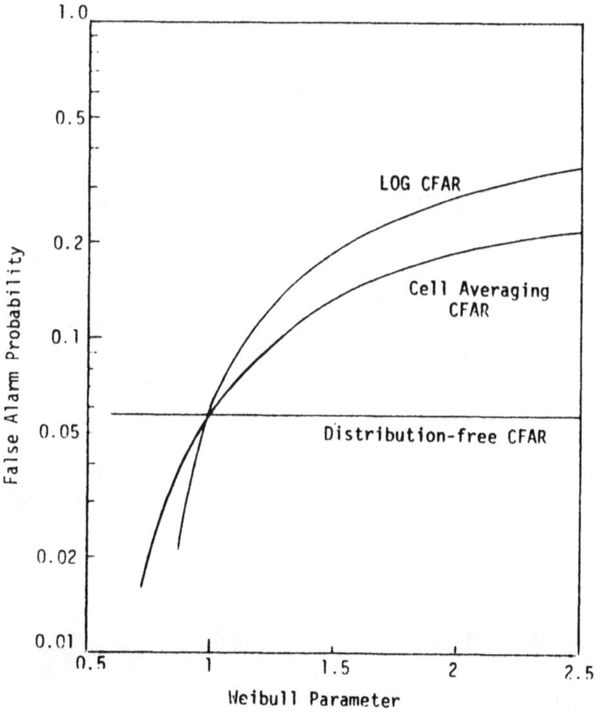

Figure 10. FALSE ALARM PROBABILITIES IN WEIBULL NOISE

Noise Quality Optimizes Jammer Performance
F.M. Turner, E.G. Ottoboni, and A.M. Imada
Electronic Warfare Magazine, Vol. 9, No. 6, November/December 1977
Pages 117-119

7.5

F.M. Turner
Senior Research Engineer

E.G. Ottoboni
Senior Research Engineer

A.M. Imada
Research Engineer

SRI International, Menlo Park, CA 94025

The noise quality of noise jammers can now be quantified with a minimum of costly simulator and/or flight testing during the various phases of a jamming systems' life cycle. Noise quality is as fundamental to a noise jammer as rf center frequency, rf spectral width, and power density. Further, noise quality can be measured with about the same ease and in about the same time as these more familiar noise jammer factors.

The concept of noise quality is as old as noise jamming itself; however, it has only been recently that technology has made it possible to rapidly measure and quantify it. This is expected to have significant impact whenever noise is employed for jamming, for training against noise ECM, and in selecting ECCM techniques against noise.

The number of ways one can produce noise for jamming radars is only limited by human ingenuity. Perhaps the most popular method employed today is modulation of an rf carrier—usually with frequency modulation. A wide variety of modulations exists among FM noise jammers, evidence of a somewhat chaotic situation with regard to specifications. Precise specifications govern power output and uniformity of the jamming rf power density spectrum. However, the content and effectiveness of the transmitted noise are less satisfactorily prescribed. In an advance in this area, the first order probability density function of the noise produced in victim receivers by jamming signals is electrically measured and characterized. This characterization is called noise quality. It has been measured for various operational and developmental radar jammers and correlated with ECM screening effectiveness via simulator, ground, and flight tests. Jammers should be specified with different values of noise quality to achieve optimum effectiveness against the victim radar ECCM equipment. The choice of the proper noise quality value will permit effective jamming to be accomplished with a fraction of the power otherwise required.

Noise Quality *Vs.* A Scopes. The effect of noise quality on the required jamming-to-signal power ratio (J/S) for a conventional-pulse radar with an A scope CRT display is shown in Fig. 1. The 17 dB variation in required J/S is obtained for a range of noise quality values which approximate those that have been measured for production and developmental jammers. There are 16 data points derived from 2,192 operator target detection decisions (non-scintillating targets), made by three experienced operators on a radar simulator (the Air Force Avionics Laboratory's Electronic Defense Evaluator). The highest noise quality values are on the right hand side, where the jamming signal is both white and Gaussian. Seventeen dB less jamming power is required to screen an aircraft target at high noise quality values, than at the lowest noise quality values where the jamming signal is white but non-Gaussian. The operators were able to readily "read through" the jamming with lower noise quality values.

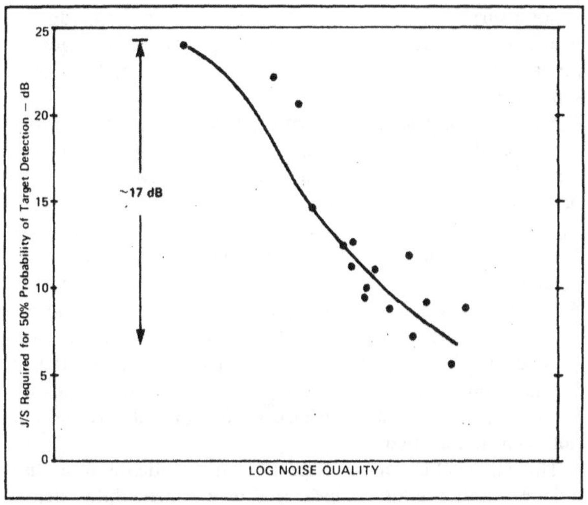

1. J/S *vs.* noise quality for linear receiver/A scope display. White Gaussian noise provides the highest quality, allowing a 17 dB reduction in jamming power.

It is apparent from Fig. 1 that noise quality is strongly correlated with target screening effectiveness and that a measure of "noise quality" has been achieved.

Maximum Entropy Yields Best Noise Quality. The theoretical basis for quantifying noise quality is rooted in the concept of entropy in the statistical (informational) sense, as opposed to the older thermodynamic entropy concept. Applied to electrical signals, entropy is a measure of the uncertainty or the "unpredictableness" of the signal. From an ECM designer's viewpoint, the signal in the radar receiver (consisting of radar echoes and jamming) should have maximum unpredictableness (maximum entropy) for highest-confidence target screening applications. When the signal is represented as a multidimensional random variable X with a continuous density function p (X), its entropy is:

$$H(X) = -\int_{-\infty}^{\infty} \cdots \int_{-\infty}^{\infty} p(x_1, x_2, \ldots x_n) \log p(x_1, x_2, \ldots x_n)dx_1\, dx_2 \ldots dx_n + \text{constant} \quad (1)$$

or if one considers just one dimension, the amplitude, its entropy is:

$$H(x) = -\int_{-\infty}^{\infty} p(x) \log p(x)dx + \text{constant} \qquad (2)$$

where p(x) is the first order (amplitude) probability density function of the random waveform amplitude.

Given that the mean square value of the signal amplitude is a specified constant (corresponding to a specified power for the waveform) it was shown by Claude Shannon (in 1948) that Gaussian noise has the maximum entropy of any random waveform for a specified average power. (Similarly, for the multidimensional random variable of Eq. (1), the maximum entropy is the multidimensional Gaussian distribution.)

Besides amplitude, the other description of noise important to ECM designers is the spectral content of the noise waveform. It has been shown that Gaussian noise with energy uniformly distributed over a given frequency band has the maximum possible entropy for any real waveform, *i.e.*, white Gaussian noise has the maximum possible entropy.[1] Thus, for highest-confidence radar target screening, the ECM goal is to produce white Gaussian noise in the victim radar receiver. This is not a surprising conclusion; white Gaussian noise is identical to the noise produced by the thermal agitation of charges in a resistor, and such noise determines the ultimate radar receiver sensitivity.

Noise Quality Measurements. The whiteness uniformity of the spectral content of the jamming noise produced in an i-f filter passband (victim radar receiver bandwidth), can be determined by a conventional high-quality spectrum analyzer. The spectrum analyzer resolution is set to one-tenth or less of the victim receiver bandwidth to achieve a sufficiently accurate determination of the whiteness. With very few exceptions, the noise produced in realistic receiver bandwidths by operational or developmental noise jammers has been observed to be almost white during the five-year period of sponsored investigations. In practice, non-white noise occurs so rarely (under realistic conditions) that its effect on target screening effectiveness has not been investigated.

The degree of closeness of the noise amplitude distribution to a Gaussian distribution is measured by a rf probability density analyzer. The probability density analyzer and its dedicated on-line, programmable calculator for computing a noise quality value in near real-time is shown in Fig. 2. For comparative purposes, the capital investment is approximately the cost of two good microwave spectrum analyzers.

The formula used to calculate noise quality is:

$$\frac{1}{\text{Noise Quality}} \triangleq \frac{1}{3}\left[\frac{\text{summed error} + \text{average error} + \text{rms error}}{3} + |\text{relative entropy in bits}| + \frac{|\text{kurtosis} - 3| + |\text{skewness}|}{2}\right] \qquad (3)$$

where $0 \leq$ noise quality $\leq \infty$.

The terms in this formula are calculated by comparing the measured amplitude probability density function to a theoretical Gaussian probability density function having the same mean and variance. Generally it has not been found necessary to compare the two density functions over a range exceeding plus or minus three standard deviations of the density functions. The formula assumes that the noise is white because it has not been necessary to determine the effects of non-white noise on "noise quality."

When Eq. (3) is utilized to calculate noise quality from measurements on high-quality video noise sources, such as those found in most laboratories, the noise quality values range from 10 to 70 (dimensionless). Near the other end of the scale, the noise quality of a sinusoid (cw signal) is found to have a value of approximately unity.

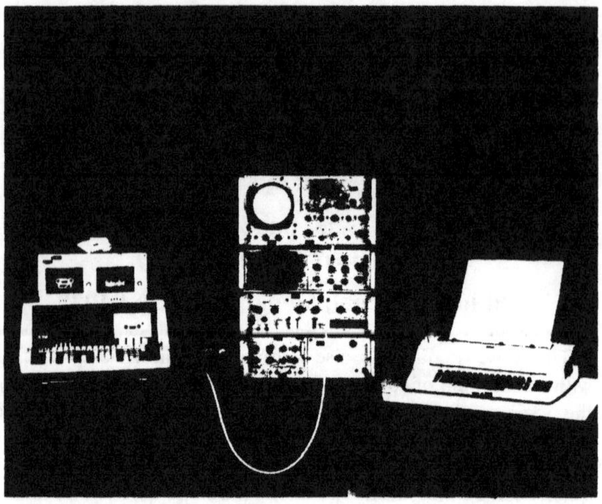

2. Probability density analyzer used in noise quality measurements, shown with dedicated on-line calculator.

Noise Quality For A Typical Jammer. When the same noise jamming signal is employed against victim radar receivers of differing bandwidths, the noise quality produced in the victim receiver varies with bandwidth, (Fig. 3). The noise quality value increases as the victim receiver bandwidth decreases, which is typical for FM-by-noise jammers. This behavior occurs because the responses produced in the receiver passband become longer in duration, and overlap to a greater degree with the responses produced by previous transitions of the sweeping rf energy of

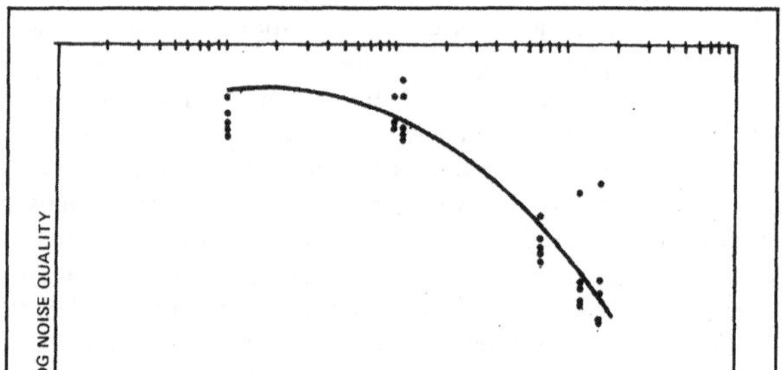

3. Noise quality *vs.* victim receiver bandwidth for one FM noise jammer. This effect occurs because the receiver passband responses increase and overlap previous signals.

the FM-by-noise jammer. As the number of overlapping independent responses is increased, the composite waveform has amplitude statistics with a more nearly Gaussian distribution (as predicted by the central limit theorem). Thus, the waveform has a higher noise quality value. This behavior permits one to state generally that if a specified noise quality is produced in a given receiver bandwidth, the noise quality values will be greater for smaller receiver bandwidths, and less for victim receiver bandwidths that are larger than the given bandwidth.

The effect of noise quality on the required J/S power ratio is shown in Fig. 4 for a conventional pulse radar with PPI and B

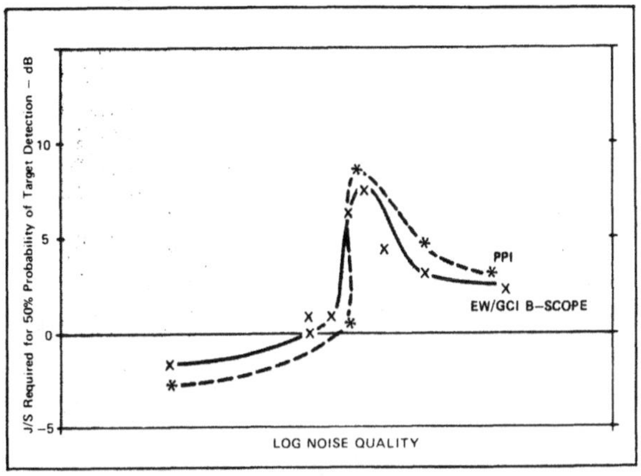

4. J/S *vs.* noise quality for the linear receiver used with PPI and B scope. The peak and subsequent decrease are not fully understood; with a possible explanation related to operator recognition patterns.

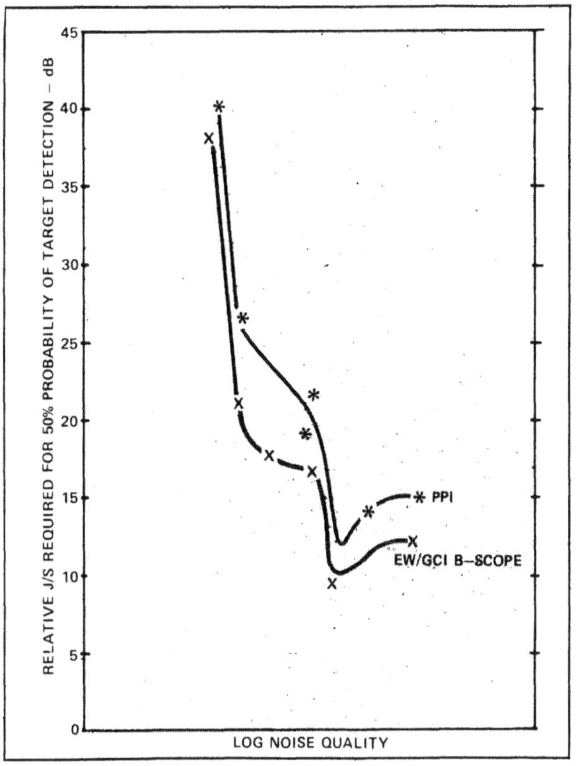

5. J/S *vs.* noise quality for a non-CFAR Dicke-Fix receiver used with PPI and B scope. The curve fluctuations may be caused by operator or display parameters.

scope displays. The approximately 10 dB variation in required J/S is obtained for the same range of noise quality values used in obtaining the A scope data, (Fig. 1). The 15 data points in Fig. 4 are derived from 2,280 operator detection decisions (non-scintillating targets) made by up to four experienced operators (two operators for the PPI, and four operators for the B scope). Both the PPI and B scopes presented an approximate 20 degree sector display of a simulated EW/GCI radar continuously scanning 360 degrees at 3 rpm. In contrast to the A scope results, the J/S required for screening the target was less at low noise quality values than at high noise quality values. The low noise quality jamming appeared as "random pulses" on these intensity modulated displays (PPI and B scopes); the random pulses are probably responsible for the reduced operator target detection performances with the intensity modulated displays. However, these random pulses did not reduce operator performance on intensity modulated scopes when a Dicke-Fix receiver was used, (Fig. 5), or on the A scopes (one type of deflection modulated display). This is one example of the ECCM capability of the A scope display, as compared to the intensity modulated displays. The A scope by itself is an ECCM. A detailed understanding of these effects awaits the development of a comprehensive theory on display/human operator interaction.

From the results presented in Figs. 1 and 4, noise jamming signals should be designed to produce high noise quality values against radars using deflection modulated displays (A scopes, J scopes, *etc.*); while the opposite is required for PPIs, B scopes, or other intensity modulated displays. If the above guidelines are followed, the required J/S is at least 10 dB less than if noise quality values at the opposite end of the noise quality scale are utilized. However, the effects of ECCM signal processing devices on low noise quality jamming waveforms will alter these guidelines.

Since the low noise quality waveforms produced random pulses with duty cycles less than 100 percent, it seems probable that low noise quality jamming is susceptible to various ECCM techniques. As already shown, if a radar operator is equipped with an A scope display, the low noise quality jamming is relatively ineffective. The effectiveness of low quality noise against the PPI and B scope displays can surely be negated by ECCM techniques such as the Dicke-Fix receiver.

Although only radar simulator results have been presented here, ground tests and flight tests have been also conducted which show that a useful method of quantifying "noise quality" for radar target screening purposes has been achieved. ■

This article is based on work supported by a succession of contacts issued by the Air Force Avionics Laboratory, Wright-Patterson AFB, OH. Although the author is responsible for the contents of this article, the results are obviously the product of many people. Key contributions were made by R. A. Bacca of AFAL/WRW, and R. C. Cumming and C. J. Shoens of SRI.

Reference

1. P.M. Woodward, *Probability and Information Theory with Applications to Radar,* McGraw Hill Book Co., New York, NY, 1953, p. 40.

Chapter 8

ECCM IN THE SYSTEM, OPERATIONAL AND OTHER ECCM'S

Summary

This chapter addresses ECCM in the System as well as Operational and other ECCMs. Tables IV and V from paper 1.1 (on page 16 of this volume) list ECCMs in the system, and operational and other ECCMs, respectively.
Due to space limitations only three representative papers pertaining to the listings cited in the tables will be included here.

Synopsis of Reprint Papers in Chapter 8

8.1 "Improved Radar Designs Outwit Complex Threats"— Ivanoff

Most ECM/ECCM papers address denial ECM (noise jamming) of search radars. This interesting paper actually concerns three air defense missile guidance systems, rather than conventional radars as might be assumed from the title of the paper. Some significant improvements to semiactive CW missile seekers (actually bistatic radars!) are described. These include the inverse receiver and elimination of the rear receiver.

8.2 "OADARS (Optical Aids to Detection and Ranging Systems)"— Jelalian, Kawachi, and Miller

Originally presented at ELECTRO-76, this paper deals primarily with laser ranging radars and makes extensive comparisons with microwave radars. Laser radars have several inherent ECCMs as compared to microwave radars; e.g., much narrower beamwidth (microradians vs milliradians) and a different (much higher) frequency. Laser radar here is intended as an adjunct to microwave radar such as a range-only-radar (ROR). Some topics related to this paper are laser range finders, laser designators (target illuminators) and the use of laser radars for angle tracking. Some European laser radars include a goniometer (typically a conical scan rotating reticle scheme) to provide angle measurement when required.

8.3 "Electronic Correlator for Passive Ranging"— Norsworthy

As pointed out in Paper 2.C.2, target range is necessary for defense systems to know whether the target is within open fire range. Denial jammers may prevent a radar from measuring target range. Under certain conditions range to the target may be determined from the jammer signal through correlation techniques. This paper describes two applications: 1) single airborne platform ranging to an airborne jammer involving measurement of multipath delay and 2) multiplatform ranging to airborne and ground-based jammers in a dense jammer environment. Experimental demonstration of the single system is described. The first application is a rare case of the beneficial use of multipath. When combined with elevation angle strobe measurement, the altitude of the jamming target may be determined. The second application (also referred to in this paper as cooperative air-to-air passive ranging) is actually passive angle triangulation. Correlation is used to ensure that each measuring receiver is using the same signal (jammer) source. This is very useful in multiple jammer environments. This paper was presented at the International Countermeasures Conference in 1976.

Bibliography Topics for Chapter 8

Laser Radar
Other ECCMs
Tracking Radar

BIBLIOGRAPHY FOR CHAPTER 8

LASER RADAR

1. Books

Petrovskiy, B.I., and O.A. Pozhidayev, *Laser Radars,* Pub. House of Soviet Ministry of Defense, 1969 (in Russian; trans. JPRS 50123).

Bachman, C., *Lasar Radar Systems and Techniques,* Artech House, Dedham, MA, 1979.

2. Journal Articles

Buddenhagen, D.A., *et. al;* "An Experimental Laser Ranging System," *IRE Int. Conv. Rec., Pt. 5,* 1961, pp. 285-90.

Jelalian, A.V., "Optical Aids to Detection and Ranging Systems," *Electron. Prog.,* vol. 16, no. 2, Summer 1974, pp. 7-12.

Khorol, D.M., and B.I. Shkurski, "Optimal Extraction of Optical Signals from an Interference Background in the Presence of Internal Noise at the Radiation Receiver," *Autom. and Remote Control,* vol. 35, no. 9, *pt. 1,* Sept 1974, pp. 1413-1423.

Hoge, F.E., "Integrated Laser/Radar Satellite Ranging and Tracking System," *App. Opt,* vol. 13, no. 10, Oct 1974, pp. 2352-58.

Werner, C.H., "The DFVLR LIDAR System IV," *Laser and Elektro-Opt* (Germany), vol. 7, no. 1, Mar 1975, pp. 44-5 (in German, trans. abstract SA 75-30139).

Taukchi, V.M., "Analysis of the Operation of a Pulse-phase Light Rangefinder," *Izv VUZ Priborostr,* vol. 18, no. 3, 1975, pp. 98-102 (in Russian, trans. abstract SA 75-30137).

"Recognizing Distant Objects Such as Aircraft," British Aircraft Corp., Ltd., *UK Patent no. 1406707,* 17 Sept 1975, abstract SA 76-21837.

Roth, J.A., "The WSMR Laser Ranging and Tracking System (LRTS) on Two POGO Tracking Tests," *Opt. Engr.,* vol. 14, no. 5, Sept-Oct 1975, pp. SR 155-7.

Courtenay, T.H., J.F. Boulter, and H. Henshall, "Active Imaging with a TEA CO_2 Laser," *Infrared Phys.,* (Great Brit.), vol. 16, no. 1-2; Jan-Mar 1976, pp. 95-102.

Jelalian, A.V., D.A. Kawachi, and C.R. Miller, "OADARS (Optical Aids to Detection and Ranging Systems)," *IEEE ELECTRO Rec., 1976, pt. 30,* pp. 30-33 thru 1-5 (reprinted in this volume).

Cooke, C., "High Accuracy Position Location Using a Laser Tracking and Ranging System," *Proc. IEEE PLANS (Position, Location and Navigation Symposium),* Feb 1977, pp. 1-6.

Cooke, C.R., "Laser Radar Systems, Some Examples," in *Proc. Effective Utilization of Optics in Radar Syst.,* Soc. Photo-Opt. Instr. Engrs. (SPIE), vol. 128, 1977, pp. 103-7, (discusses various laser radar applications to rangefinders, target designators, fire control missile beamrider, command guidance, etc.).

OTHER ECCMs

1. Books

Peterson, W., and F.J. Weldon, Jr., *Error Correcting Codes,* MIT Press, Cambridge, MA, 1961.

Tipugin, V.N., and V.A. Veytsel, *Radio Control,* Moscow, 1962 (in Russian; Translation—JPRS 15,381).

Maksimov, M.V., and G.I. Gorgonov, *Radio Control of Rockets,* Moscow, Soviet Radio, 1964 (in Russian; trans. NASA TT-F367).

Gutkin, L.S., Yu. P. Borisov, *Radio Guidance of Missiles and Space Vehicles,* Moscow, Soviet Radio, 1968 (in Russian).

Maksimov, M.V., *Noise Immunity of Multichannel Command Guidance Radio Links,* Moscow, Soviet Radio, 1970 (in Russian).

Medvedev, G.A., (ed.), *Adaptive Automatic Systems,* (collection of articles), Moscow, Soviet Radio, 1972 (in Russian).

2. Journal Articles

Nichols, L.W., *et. al.,* "Military Applications of IR Techniques," *Proc. IRE,* vol. 47, no. 9, Sept 1959, pp. 1611-1624.

Blokh, E.L., and A.A. Kharevich, "Coding Stable with Respect to Fading (Anti-fading Coding)," *Telecomm.,* no. 4, 1960, pp. 361-6.

Johnston, S.L., "A Survey of Navigational Measurement Methods for Missile Guidance Systems," *IRE Trans.,* vol. PGI-5, June 1960, pp. 60-65.

Eckler, A.R., "The Construction of Missile Guidance Codes Resistant to Random Interference," *BSTJ,* vol. 39, no. 4, July 1960, pp. 973-994.

Bird, W.E., "Testing Electromagnetic Warfare Devices," *Air Proving Ground Command, Elgin AFB Rpt. APGC TN 61-35,* Sept 1961, DDC AD 272787.

Glenn, A.B., "Jamming Vulnerability of Communication Systems," *IRE Trans.,* vol. CS-9, no. 3, Sept 1961, pp. 226-232.

Livingston, G.D., "Anti-jamming," *Infantry Magazine,* vol. 51, Sept-Oct 1961, pp. 51ff.

Leskinen, J.I., "Electronic Warfare Training in Anti-jamming Techniques," *7th Annual Human Factors Engrg. Conf.,* Ann Arbor, Mich., 5 Oct 1961, DDC AD 427065.

Glenn, A.B., and G. Lieberman, "Effect of Propagation and Antenna Fluctuations on Communication Systems in Jamming Environment," *IEEE Trans.,* vol. CS-10, no. 1, Mar 1962, pp. 43-60.

Abe, M., and T. Honma, "Radar Homing Device," *Toshiba Rev.,* vol. 17, no. 4, Apr 1962, pp. 403-9 (in Japanese, trans. abstract SA 62-11393).

Glenn, A.B. and G. Lieberman, "Performance of Digital Communications Systems in an Arbitrary Fading Rate and Jamming Environments," *IRE Trans.,* vol. CS-11, no. 1, Mar 1963, pp. 57-68.

Khvorostenko, N.P., "The Noise Immunity of Optimum Higher Binary Codes," *Telecomm. Radio Engrg., pt. 2,* vol. 18, no. 12, 1963, pp. 1ff.

Pew, R.W., J.C. Duffendack, and L.K. Fensch, "Sine-wave Tracking Revisited," *IEEE Trans.,* vol. HFT-8, no. 2, June 1967, pp. 130-5 (concerns manual tracking).

Brickey, W.A., "Electronic Warfare," *Infantry Magazine,* vol. 61, Sept-Oct 1971, pp. 28-30.

Johannessen, R., and J. Mun, "Range Rate Without Differentiation in FM/CW Radar," *IEE Conf. Pub. no. 105,* London, Oct 1973, pp. 62-7.

Gilbert, E.N., F.J. MacWilliams, and N.J.A. Sloane, "Codes Which Detect Deception," *BSTJ,* vol. 53, no. 3, Mar 1974, pp. 405-24.

Mullen, J.A., "Electronic Countermeasures Problems for Extremely Low Frequency (ELF) Communications," *IEEE Trans.,* vol. CO-22(4), 1974, pp. 496-503 (Adjunct to radar).

Lampos, N.J., "Electronic Warfare Planning and Tactics," *Electronic Warfare Magazine,* vol. 6, no. 6, Nov/Dec 1974, p. 44ff.

"Electro-Optical Systems," *AGARD Lecture Series,* no. 76, May 1975.

Garrett, P.H., "Advances in Low-frequency Radio Navigation Methods," *IEEE Trans.,* vol. AES-11, no. 4, July 1975, pp. 562-74 (contains ECM-ECCM discussion).

Kain, J.E., and D.J. Yost, "Target State Estimation in an ECM Environment," *J. Spacecraft and Rockets,* vol. 12, no. 8, Aug 1975, pp. 492-8 (proposes alternate means, i.e., other techniques, for range measurement in ECM environment).

Breitzmann, J.F., "Operation of a CCD Moving Target Indicator," *IEEE Int. Solid-State Circuit Conf. Proc.,* Philadelphia 18-20 Feb 1976, pp. 36-7 (optical moving target indicator).

Thomas, H.W., "Tracking in a Multiradar Environment," *Proc. IEE.,* vol. 123, no. 3, Mar 1976, pp. 191-4.

Norsworthy, K.H., "Electronic Correlation for Passive Ranging," *Proc. Int. Countermeasures Conf.,* London; EW Institute Ltd, Los Altos, CA, Sept 1976, pp. 290-319 (reprinted in this volume).

Marple, M.J., W.H. Land, and R.D. Lavin, "Extended Area Range Instrumentation Utilizing DME Links," *IEEE Trans.,* vol. AES-12, no. 5, Sept 1976, pp. 590-9 (DME could be used for missile navigation/guidance instead of radar).

Ohlsson, E., and H. Svensson, "Electro-optical Fire Control of Anti-aircraft Guns," *Proc. Mil. Electron Def. EXPO-76,* Oct 1976, Wiesbaden, Germany, pp. 574-90.

Knoppik, N, K. Brammer, F. Herzmann, and A. Kainzinger, "Simultaneous Automatic Tracking in Multiple Radar Networks," *Proc. Int. Conf. on Radar,* Paris, 4-8 Dec 1978, pp. 100-08 (describes passive tracking).

Backus, P.H., "Electro Optics," *J of Electron. Def.,* vol. 2, no. 2, Mar/Apr, 1979, pp. 24ff.

TRACKING RADAR

1. General Bibliographies and Books

Obrezkov, G.V., and V.D. Razevig, *Methods of Analyzing Tracking Interruption,* Moscow, Soviet Radio, 1972 (in Russian).

Barton, D.K., "International Cumulative Index on Radar Systems," *IEEE Pub. Cat. no. JH 4675-5,* 1978: extensive bibliography on tracking radar, pp. 71-2 (some items have ECM/ECCM implications).

2. Journal Articles

Brown, W.M., "Analysis of Range Tracking Radar in the Presence of Interference," *IRE 2nd Mil-E-Con Rec.,* 1958, pp. 230-4.

Aasen, M.D., "Methods for Predicting Interference Effects in Tracking Radars," *IEEE Trans.,* vol. EMC-6, Oct 1964, pp. 49-57.

Yurchenko, Yu. S., "Noise Immunity of Radar Servo Trackers in Conditions of Extraneous Interference," *Radio Engrg. and Electron. Phys.,* vol. 14, no. 10, Oct 1969, pp. 1641-3.

Chidambara, M.R., and A.P. Shivaprasad, "Digital Computer Study of the Performance of a Digital Filter for Radar Tracking," *Electro-Technol.,* vol. 18, no. 3, Dec 1974, pp. 82-85.

Fitzgerald, R.J., "Target Tracking Filters," *Electron. Prog.,* vol. 17, no. 1, Spring 1975, pp. 31-8.

Yurchencko, Yu. S. and M.V. Katikov, "Tracking Setting Optimization of a Pulse Signal Tracker," *Izv VUZ Radioelektron.,* vol. 18, no. 7, Jul 1975, pp. 59-63 (in Russian, trans. abstract SA 76-5905).

Katikov, V.M., "Optimization of Pulse Tracking Radars with Regular Breaks in the Received Signal," *Izv VUZ Radioelektron,* vol. 19, no. 7, Jul 1976, pp. 26-32 (in Russian, trans. abstract SA 76-48430).

Silvka, L.P., "Missile Post-multiple-target Resolution Guidance," US Patent No. 4,010,467, Mar 1, 1977 (Uses a feedback circuit to modify a transient signal present in tracking radars when resolution of multiple targets occurs.)

Improved Radar Designs Outwit Complex Threats

A. Ivanov

Microwaves Magazine, Vol. 28, April 1976
Pages 54-71

8.1

THE objectives of any air-defense missile system are to locate, track and destroy any hostile aircraft before it reaches its target. Over the years, as threat aircraft and other targets have become more sophisticated, the demands upon the design of air-defense systems have challenged the ingenuity of microwave design engineers.

Radar guided missiles today form one of the largest markets for microwave components and subsystems. In the "real world" of radar, where a quantity of more than two has been defined as production,[1] to think of over 50,000 systems may seem staggering. Yet in the past 20-25 years, this has been the magnitude of the radar guided air defense missile business. Except for phase shifters for phased arrays, no other microwave hardware is truly mass produced on such a scale.

There are many necessary functions which must be successfully performed in order to permit a lethal interception of the target by a guided missile. These include initial target detection, launching the missile, successful operation of the propulsion, guidance and control systems during flight and fuzing and detonation of the warhead at intercept. Only the guidance aspects will be discussed here.

There are three types of guidance techniques commonly used in air defense guided missiles: (1) beam riding, (2) command guidance and (3) homing. All can employ radar sensors. The most widely used is the semi-active radar homing, in which the target-tracking radar provides target illumination and the target reflected energy received by the missile is used for guidance (Fig 1).

Alex Ivanov, Staff Engineer, Division Headquarters, Raytheon Company, Missile Systems Division, Hartwell Road, Bedford, MA 01730.

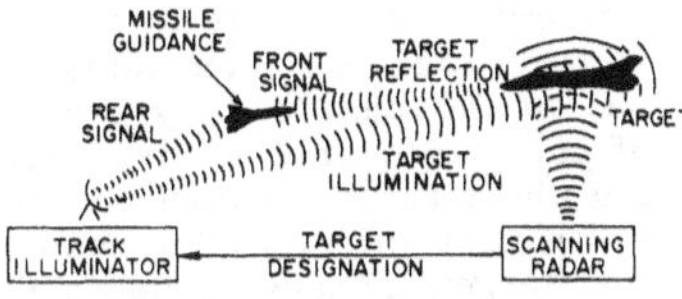

1. **In semi-active homing,** a tracking radar tracks and illuminates the target. The missile seeker then compares the target reflected illumination with its rear (reference) signal to extract guidance information.

Kamikazes spur missile development

Air-defense guided missile development began in the closing months of World War II. Radar fire control of guns was employed and could extract a relatively high attrition rate from an attacking force. In the face of determined penetration attempts at low altitudes, such as Kamikaze attacks on the U.S. Pacific Fleet, a much higher degree of success was necessary. Even a handful of successful penetrators could inflict great damage.

The objectives of air defense missile systems are to achieve virtually 100% effectiveness, to prevent even a single aircraft from reaching its target. In order to approach this level of performance with a reasonable number of missiles, the accuracy of any single missile must be very high, and a high single shot kill probability is essential.

To achieve this accuracy, a guidance system substitutes automatic closed-loop control for the open-loop calculation and prediction employed in conventional artillery control. The function of the guidance system is to continuously sense errors in the missile-to-target intercept geometry and translate them into corrective maneuvers until interception is accomplished.

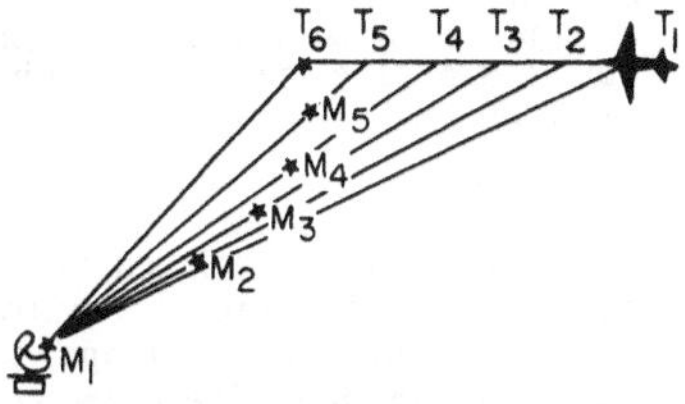

2. **The beam-rider missile** centers itself in the beam of the tracking radar and must maneuver even when the target course remains constant.

Three guidance techniques [2,3,4,5]

In the beam-rider system, Fig. 2, the target is tracked by an external radar. The missile itself does not perceive the target, but detects its own position relative to the radar-beam tracking the target. By keeping itself centered in the beam, it attempts, like the radar beam it rides, to pass through the target. Since the missile is always on a line between the radar and target, it must continuously maneuver to remain in this line even when the target is flying a straight-line course. This course requires severe corrective maneuvers near intercept. Another drawback is that any angular tracking error at the radar becomes a larger linear error as the range is increased. The accuracy of a beam-rider system is, therefore, inversely proportional to target range.

In a command-guidance system, (Fig. 3), the target is again tracked by an external radar. The missile itself does not perceive the target. A second radar tracks the missile. Measurements of both target and missile position are fed

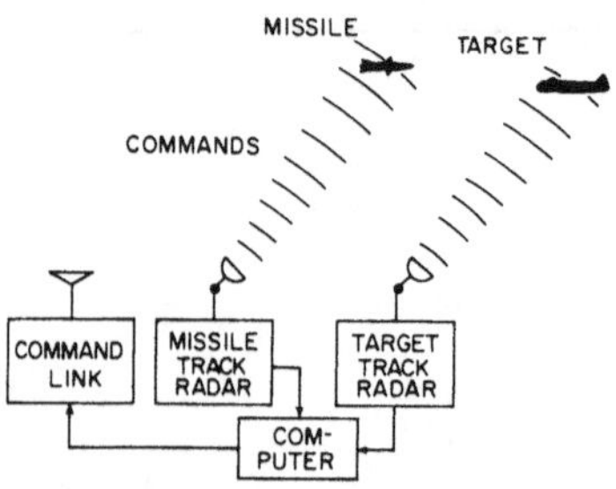

3. **Command-guidance systems** use data from separate target tracking and missile-tracking radars to compute the missile maneuver required for intercept.

to a computer which calculates the missile trajectory required for intercept and develops the needed commands. These are continuously transmitted to the missile. A more efficient trajectory can be used, since there is no need for the missile to fly along the radar target line-of-sight. A lead collision course can be plotted and prediction based on rate of change of position, rather than only-position information, can minimize the terminal maneuver. However, once again, accuracy is inversely proportional to range, since a fixed angular error at the radar is magnified at increasing ranges.

The homing missile is the most intelligent of the three, but also the most complex. It perceives the target with its own radar and computes its own control signals. Various types of homing systems exist, depending on the source of the energy which is sensed. Passive homing uses energy originating at the target. An active homing missile transmits its own radar energy at the target and receives target-reflected echoes. In a semi-active system, an external radar illuminates the target and the missile receives the reflected-target echoes.

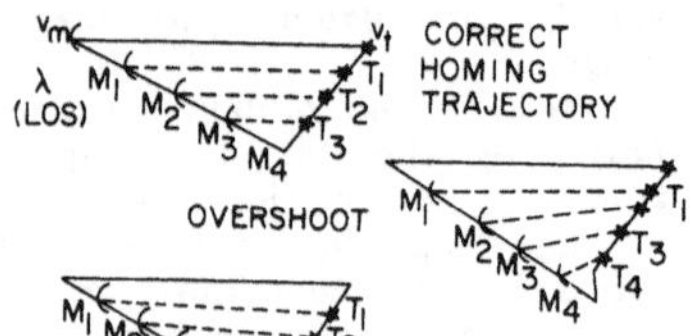

4. **Line-of-sight (LOS) rate** remains constant when missile and target are on a collision course.

Homing guidance[3,6]

In a homing missile, a collision course is characterized by a line-of-sight (LOS) from missile-to-target which does not rotate, (Fig. 4). Any rotation of the LOS is indicative of a deviation from a collision course which must be corrected by a missile maneuver.

In proportional navigation, the missile rate of turn is made proportional to the rate of turn of the LOS. The choice of the constant of proportionality (navigation ratio or N) affects the trajectory, such that increasing values of N cause early correction of collision course errors, reserving the missile's maneuver capability near intercept for countering target maneuvers and noise, the relationship being:

$$n_L = N' V_C \dot{\lambda}$$

where: n_L = lateral acceleration
N' = effective navigation ratio
V_C = closing velocity
$\dot{\lambda}$ = rate of change of the line-of-sight

As the missile gets increasingly closer to the target, the quality of the target information continually improves, since, as range decreases, an angle error results in a lesser linear error.

The fundamental limit on accuracy is the target's own angle noise (glint or scintillation) and is independent of intercept range.

The most widely-used systems have, over the years, employed semi-active homing. Since the active system differs only by virtue of the presence of the illuminator transmitter on board the missile, the following discussion of the semi-active system can be easily extended to cover the active system as well. Similarly, passive homing can be considered a subset of the semi-active.

Illumination radar waveforms

Semi-active systems employing various waveforms such as pulse-cw, pulse doppler (pd) and more sophisticated coded pulse techniques have been designed and built over the years.[5,7] The cw is the simplest and at the same time, provides low-altitude capability by discriminating against clutter on the basis of velocity or doppler frequency. Conventional pulse systems have been used but are not suitable for high clutter environments.

The illumination for a semi-active missile system can be provided by a cw tracking radar, a cw transmitter slaved to another tracking radar, or a pulse or pulse doppler tracking radar at another frequency with the cw illumination injected into the antenna system from a separate cw transmitter. The tracking cw illuminator is generally a two-dish radar, because sufficient receiver/transmitter isolation cannot usually be achieved in a single-dish system. Where space constraints preclude use of two separate antennas, such as in a fighter aircraft, the cw injection technique is used.

An alternate way to use a single dish tracking illuminator is to employ high-duty-cycle pd (30-50%) and achieve the isolation by conventional TR switching. In the receiver, however, only the central line of the pd spectrum is employed. The prf is chosen high enough to yield unambiguous doppler, thus when the receiver selects the central line, the spectrum is identical to the cw case. The receiver can use a range gate, or merely accept the loss resulting from using only the central line power of the pd spectrum. In either case, the rest of the receiver and signal processing is the same as for the cw system. The semi-active seeker implementation for a pd illuminator follows this same pattern. This is also the most common approach for an active seeker.

Low-duty-cycle pd (less than 10%) can also be used, but for this case, the loss would be prohibitive if only the central line power were used. Low-duty-cycle systems, therefore, use range gating to optimize performance. The range-gated system also provides range resolution, in addition to retaining the doppler resolution capability of the cw systems.

Phased array increases fire power

In a semi-active system, the target must be continuously illuminated, which means an illuminator must be constantly aimed at the target during an engagement.[5,7] This constrains the number of targets that a given system can engage. However, mechanically-gimballed radars are forced to live with the limitation of one radar tracking only a single target (or formation of closely-spaced targets).

The advent of phased-array radars allowed a single transmitter to illuminate many targets by sequentially jumping its agile beam from one target to the next one. Of course, the illumination would no longer be continuous, and the missile would have to operate in a sampled data mode, extracting information during the time its target was illuminated, and then holding the information until the next sample. The illumination could be cw during the dwell time of the radar beam on the target (interrupted or keyed cw) or use a pd or more complex waveform. By removing the one radar-one target constraint, an increase in firepower can be realized.

Whereas the simple cw systems generally home all the way from launch to intercept, in the more sophisticated systems, homing is generally used for only the last few seconds of flight. In these systems, a midcourse phase (inertial, beam rider or command) is employed to get the missile to an appropriate point on its trajectory, where it acquires the target and enters the terminal homing phase of its flight. This is more efficient from the standpoint of both missile trajectory and radar power. The missile can fly out to longer ranges by a commanded or inertial up-and-over trajectory, spending less time in the denser air at low altitude. The radar power needed for illumination is sized by the terminal phase of flight, a fraction of the total intercept range. Midcourse commands impose much less severe demands on the radar power since this is a one-way transmission path.

Pulse-compression radar was introduced to achieve very fine range resolution at long ranges without extending the peak power capability of transmitters. To retain compatability with these complex waveforms, a variation of semi-active homing called TVM (target-via-missile) evolved. The target-reflected illumination is received in the missile, but instead of being processed on board is retransmitted to the illuminating radar where the complex waveform is processed and guidance information extracted. The actual steering commands are then transmitted to the missile as in a command guidance system.

Cw semi-active guidance[2,5,7]

In the semi-active system, the missile receives the target reflected illumination in its front antenna and a sample of the directly received illumination (often through sidelobes of the illuminator antenna) in its rearward-looking reference antenna. The front and rear signals are coherently detected against each other, resulting in a spectrum which contains the doppler-shifted target signal at a frequency roughly pro-

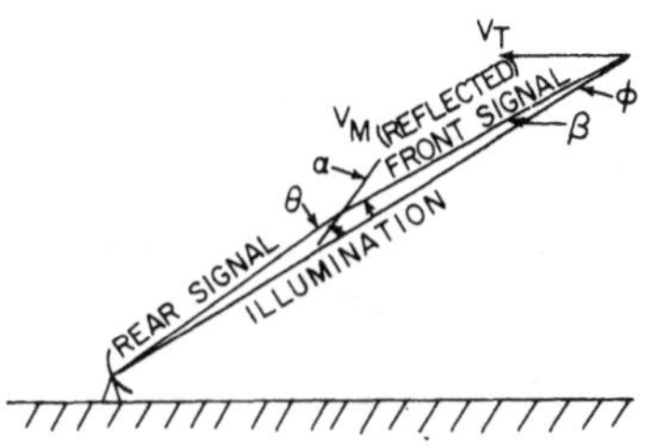

5. Semi-active system geometry results in a doppler-shifted target signal with a frequency roughly proportional to closing velocity.

portional to closing velocity. A narrowband frequency tracker searches the spectrum and is used to lock onto this target doppler and extract guidance information from it. The cw tracking illuminator functions the same way, except the rear signal is replaced by a sample of the transmitter output.

The doppler relationships for a ground-based semi-active system geometry (Fig. 5) are:

$$f_{d\,rear} = -\frac{f_o}{c} V_M \cos \Theta$$

$$f_{d\,front} = \frac{f_o}{c} (V_T \cos \phi + V_T \cos \beta + V_M \cos \alpha)$$

$$f_d = f_{d\,front} - f_{d\,rear} = \frac{f_o}{c} (V_M \cos \Theta + V_M \cos \alpha + V_T \cos \phi) + V_T \cos \beta)$$

$$\text{Closing Velocity} = V_c = V_M \cos \alpha + V_T \cos \beta$$

for the head on case, $\alpha, \beta, \Theta, \phi = 0°$

$$V_c = V_M + V_T$$

$$f_d = \frac{f_o}{c} 2\, V_c$$

$$\approx 20 \text{ Hz/ft/sec at X-band}$$

For a ship-based or airborne illuminator, an additional component—illuminator velocity—must be considered. It does not fundamentally affect the spectrum except to broaden the clutter return (adding reflections from clutter patches illuminated by backlobes or sidelobes of the illuminator which now have a velocity component rather than all being at a fixed frequency). The doppler shift at X-band corresponds roughly to 20 Hz per 1 ft/sec of closing velocity. Scaling is a convenient way to handle other velocities or microwave frequencies.

A representation of the signal spectra seen by a semi-active seeker, for both fixed and airborne cases, is shown in Fig. 6. It should be noted that the target signal must compete with clutter and with feedthrough or spillover of the illumination into backlobes of the seeker antenna, which can be many orders of magnitude greater than the target signal.

There are three main problems which must be addressed in connection with these large interfering signals. One is the need to prevent lock on the clutter signal, which in some geometries may be a spectrally narrow signal, resembling a target signal. This is generally accomplished by limiting the portion of the doppler spectrum which is searched during the acquisition process, to exclude the clutter frequency.

The second problem is often termed the subclutter visibility (SCV) or subfeedthrough visibility (SFV) problem. In essence, this refers to the maximum ratio of clutter (or feedthrough)-to-signal with which the system can operate. This, in its simplest form, can be related to the dynamic range of the seeker receiver, including not only possible suppression of the target signal by the clutter or feedthrough, but also potential cross modulation or intermodulation effects.

The third problem is also related to SCV (and SFV), and is concerned with the spectral purity of the transmitter and the local oscillator. The spectrum of Fig. 6 will be broadened by fm noise, so that noise sidebands will appear at the target doppler frequency and mask the target signal. Considering the magnitude of feedthrough or clutter, it can be seen that very low fm noise is required to prevent performance degradation.

The spectra for a cw tracking illuminator are similar to Fig. 6 except that clutter occurs at zero doppler, or the same frequency

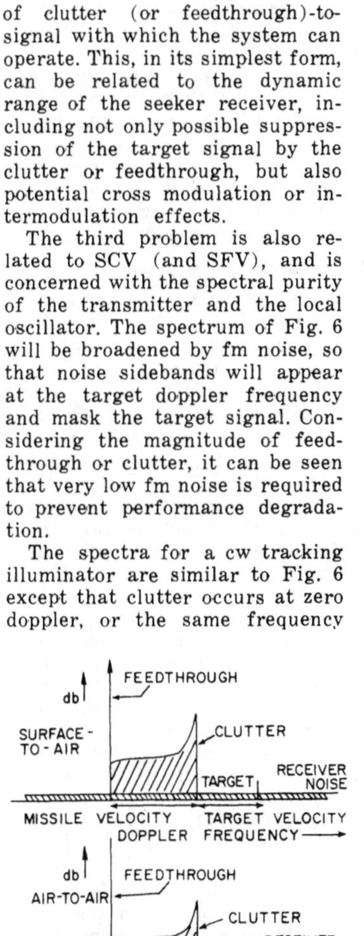

6. The doppler spectrum contains clutter at a maximum frequency corresponding to missile velocity. The feedthrough (or spillover) signal occurs at zero doppler.

RADAR IMPROVEMENTS

as the feedthrough. Thus, the first problem—locking on clutter—does not exist. The second and third problems, however, are worse since the transmitter leakage into the receiver will usually be much higher than feedthrough in the missile seeker. Consider, for instance, a feedthrough-to-signal ratio of 100 dB. Assuming a 10 dB S/N requirement for detection, a 1 kHz detection bandwidth and a doppler of 10 kHz, the fm noise in a 1 kHz band at 10 kHz from the carrier must be 110 dB down from the carrier. A sideband-to-carrier ratio (sb/c) of −110 dB corresponds to a modulation index of 4.48×10^{-6}, or an equivalent rms noise deviation of 0.0448 Hz/kHz at 10 kHz. As transmitter power is increased, the fm noise must be correspondingly reduced to prevent it, rather than receiver noise, from limiting detection performance. For the higher-power radars, therefore, some form of feedthrough cancellation as well as transmitter noise reduction techniques are employed.

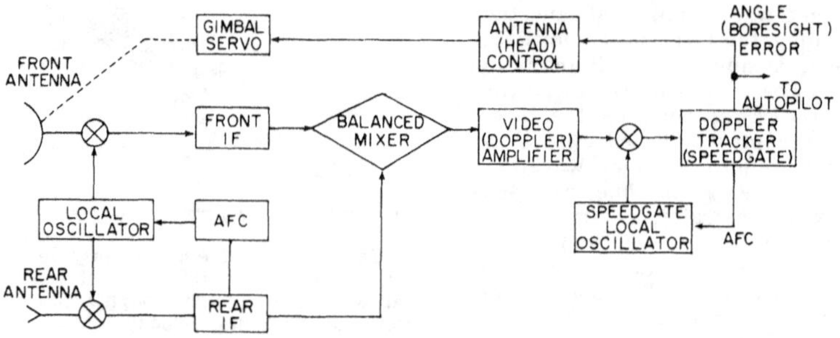

7. **The simplest cw semi-active seeker** extracts the target doppler by coherently detecting the front signal against the rear signal.

Missile seeker development

The simple missile seeker for a cw semi-active system, shown in Fig. 7, is typical of the earliest systems developed in the late 1940's and early 1950's.[5,7] It consists of rear receiver, a front receiver, a signal processor and a tracking loop to control the gimballed seeker (front) antenna. The missile also contains an autopilot to guide it and stabilize the airframe, a fuze to detonate the warhead at the optimum time, and a source of electrical and hydraulic (in most missiles) power.

The purpose of the rear receiver is to provide a coherent reference for detection of the front (target) signal. The rear signal, after conversion to i-f, closes the afc loop around the microwave local oscillator (LO) and acts as the reference for the i-f coherent detector. The target signal, received in the front antenna, is heterodyned to i-f and amplified in a relatively wideband amplifier. It is then converted to doppler in the coherent detector. The doppler signal is amplified in the video amplifier which has a bandwidth equal to the total range of possible dopplers, and is then heterodyned with the speedgate LO, which tracks the desired signal in the narrow speedgate (sometimes called the velocity gate or doppler tracker).

Target acquisition is accomplished by sweeping the frequency of the speedgate LO over the doppler bandwidth (or portion thereof). In essence, this examines the spectrum by moving it past the narrow frequency window of the speedgate. When a signal exceeds the detection threshold, the search is stopped and the signal is examined to verify that it is a coherent target rather than a false alarm due to noise. A valid target is then tracked in frequency and guidance commands are extracted from it.

The boresight angles are used to close the guidance loop. The front antenna employs a rotating subreflector to conically scan the received beam. The amplitude modulation of the received signal is recovered in the speedgate and resolved into the two orthogonal pitch and yaw gimbal axes. These signals then drive the antenna servos to keep the antenna beam on boresight. The antenna is space stabilized by feedback from rate gyros mounted on the antenna, so that only target-to-missile, line-of-sight rates will result in boresight errors.

The boresight error (proportional to $\dot{\lambda}$ in Fig. 4) is also used by the autopilot to steer the missile along a proportional navigation trajectory to intercept the target. As has been shown, proportional navigation is based on the fact that if two objects are closing on each other, they will collide if the LOS between them does not rotate in inertial space. The important fact to note is the entire guidance function is accomplished with only a measurement of angle. The doppler tracking loop (or range tracking loops used in pulse systems) is present merely to restrict the window-in frequency (or range)—from which angle error information is extracted. The angle information is what is needed for guiding the missile.

To normalize the guidance error gain (a constant scale factor of volts/degrees off boresight is required) over the full dynamic range of target signal amplitudes, AGC is required in the receiver. This may be done separately in the i-f, doppler amplifier and speedgate, or the agc may be derived down-stream in the receiver and used to agc an earlier section (for instance, agc derived at the doppler amplifier output may be used to control the gain of the i-f). The speedgate normally agc's on receiver noise, and can, therefore, expand its gain to some extent as the receiver noise and target are compressed earlier in the receiver. The considerations that dictate the specific implementation are to what degree the feedthrough and clutter shall control the gain for the target signal, while preventing saturation on the large signals. The dynamic range of these amplifiers and agc loops form one limitation on the achievable SCV and SFV. The other concern is that of possible cross modulation between the large interfering signal and small target which could cause false guidance commands. Prevention of such nonlinearities is a major design objective, both in the circuit design and physical layout, since very high gain is concentrated in a very limited space.

Rear receiver eliminated

These early systems were the first step in an evolution that has gone through several iterations to the systems in production today or being designed for future deployment. The technology and the threat have both advanced.

An example of the former is the coherent reference, provided in the early systems by a rear receiver afc'ing a klystron. Since the early illuminators were not precisely frequency controlled (initial frequency setability and long-term drift), the missile rear receiver would have to search for and lock onto its illuminator frequency and then track it to keep the LO properly positioned. Time delay matched front and rear i-f channels reduced the klystron fm noise to tolerable values. With the development of crystal-controlled transmitters, the illumination frequency could be known precisely prior to missile launch. Thus, the microwave LO could be preset before launch, and the requirement for tunability could be replaced by a crystal-controlled LO. A fixed-frequency LO could be made sufficinetly noise free to permit use of an "on-board reference," eliminating the rear receiver entirely. Coherent detection of the front signal still results because both transmitter and LO are precisely crystal controlled, and the target-doppler shift is properly recovered. A major design simplification and advancement thus results.

Threats complicate design, too

Most of the advances in the threat, however, have brought about increased complexity of the seeker as well as the tracking illuminator. From the simple threat of a subsonic aircraft the requirements grew to much faster targets, of smaller radar cross-sections, flying at lower and higher altitudes, possessing greater maneuverability and employing increasingly powerful and more sophisticated ECM.

Thus, the need grew for a wider range of doppler-frequency coverage, greater clutter and feedthrough rejection and more ECM immunity. At some points, hardware design requirements—such as circuit Q's or percentage tuning bandwidth of oscillators—dictated use of second i-fs to avoid going to baseband (video) frequencies, double conversion in the speedgate to achieve the final narrow banding at lower frequencies and moderate circuit Qs, clutter rejection schemes. Eventually, the basic system containing three frequency conversion in the signal path grew to a system containing seven frequency conversions. Separate circuits were added to provide ECCM fixes.

This level of complexity was inconsistent with a seeker, which fundamentally is a round of ammunition. It has to perform reliably under extremes of environment-shock and vibration, low and high temperature, long-term storage and field handling—and yet be sufficiently inexpensive to be a cost-effective weapon. Since it also has to fit within a limited volume, increasing complexity makes for even denser packaging. These reliability and cost requirements demanded less complexity —not more.

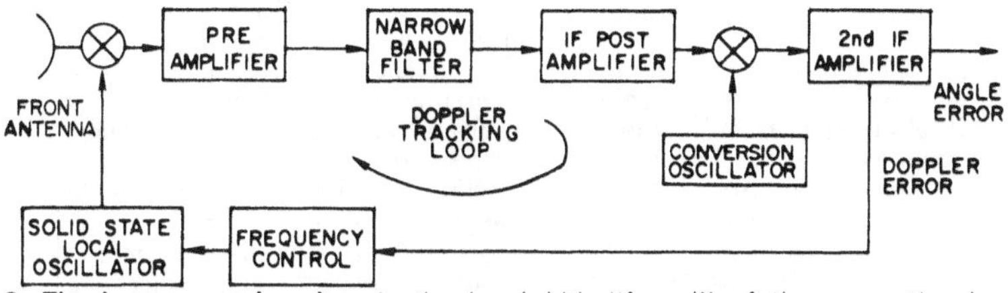

8. **The inverse receiver inverts** the bandwidth "funnel" of the conventional receiver to exclude interference very early in the seeker.

The introduction of solid-state circuitry to replace the vacuum tubes of the early systems provided a step towards more reliability, but the actual parts count increased, since, in general, to fulfill the function of a single tube required two or more transistors. What was needed was an inherently simpler approach which could match or better the performance of the conventional receiver.

Crystals key to inverse receiver

This inherently simpler approach came in the form of the "inverse receiver".[5,7,8] It gets its name from the fact that, unlike the conventional receiver consisting of a wide i-f, a somewhat narrower doppler amplifier, and finally a narrow speedgate—an ever-narrowing funnel—it performs the narrow banding or speedgating at i-f, thus inverting the "funnel" of the conventional receiver.

The critical components necessary for the inverse receiver are high selective crystal filters at i-f frequencies and low-noise tunable solid-state microwave sources. The simplicity of the inverse receiver coupled with a fully solid-state seeker produces the high reliability needed while giving improved performance.

The simplified block diagram of an inverse receiver is shown in Fig. 8. Whereas in the conventional receiver, the target signal must compete with feedthrough, clutter and jamming signals until the final stages with the dynamic range requirements of the receiver and its agc loops dictated by these large undesired signals, the inverse receiver excludes them virtually at the input. The narrow crystal filter, constituting the speedgate bandwidth, is placed in the i-f, after only a nominal amount of fixed preamplifier gain, sufficient to maintain noise figure, has been added. One additional conversion is used in the receiver to avoid the problem of too much gain at one frequency. In the resulting two-conversion system, complexity is significantly reduced and unwanted signals are eliminated very early in the signal path, thus reducing dynamic range requirements and avoiding any possible source of distortion.

The doppler tracking loop is closed through the microwave LO, which must, therefore, be tunable over the doppler frequency range of interest. The microwave LO, thus essentially fulfills the role of the speedgate LO in the conventional receiver of Fig. 7. The inverse receiver, can be thought of as a double-conversion speedgate with the speedgate afc loop closed around the microwave LO and the input to the speed gate being the microwave output of the seeker antenna.

The i-f spectrum at the mixer output will have the same form as Fig. 6. Sweeping the LO moves the spectrum past the narrow crystal filter to accomplish acquisition as in the conventional speedgate. Doppler tracking is similarly done by controlling the LO frequency to keep the target in the narrow filter. The angle error signals required for guidance are extracted after the second i-f amplifier. A single agc loop, required to cope with only the target-signal variations, is used to normalize the angle error signals.

The inverse-receiver concept can be adapted to either a conical scan or monopulse angle-tracking system. For a conical-scan system, the scan sidebands must fall within the crystal-filter bandwidth, thus limiting the maximum scan frequency. Use of a high-frequency scan would require wider bandwidth or separate processing channels for the directional information.

Use of a conventional monopulse angle tracking scheme normally requires three complete channels which must track in gain and phase to maintain the proper relative amplitude of the sum and difference channels. The well-known advantage of monopulse over conical scan,[9,10] however, is significant, since it eliminates externally-generated amplitude fluctuations (such as propeller modulation, fading noise or amplitude-modulated jammers) from the guidance signal. The monopulse system extracts all the angular information simultaneously rather than requiring a period of time to determine the position of a source of signals, as a conical-scan system does. The agc can, therefore, be made instantaneous and the amplitude variations since they affect sum and difference channels by the same relative amount, are never detected as erroneous guidance signals.

The inverse receiver permits the inherent capability of the monopulse and the basic simplicity of the conical-scan systems to be combined into one. Three identical mixers, preamps and crystal filters, first process the three monopulse signals. Immediately after the narrow-band filters, however, the difference channels are modulated to form sidebands at a multiplexing frequency higher than the filter bandwidth. Interference at the multiplexing frequency is, therefore, prevented from passing through the filters. The modulated-difference channels are combined with the sum signal and the composite signal is processed in a single channel, just as a conical-scan signal. The normalized monopulse error signals are then demultiplexed and used for closing the guidance loops.

Improved performance is inherent in such a seeker, while hardware is significantly less complex than for a conventional receiver.

PD sampled data and TVM

Either the conventional or inverse receiver can be configured to work with other than the simple cw waveforms.[5,7] The inverse receiver would retain its performance advantages over the conventional for these approaches.

For high-duty cycle and high prf pd, no fundamental receiver hardware changes are needed. The receiver only sees the central line of the spectrum and operates as in cw. Care must be taken to insure that the rear afc loop (if one is used) locks onto the central line. If range gating is used (a necessity for low-duty cycle pd), the range gates must, of course, precede the narrow speed-gate filter. Similarly, for pulse-compression waveforms, the pulse-compression lines would also have to precede the range gates and narrow-band filters. For all of these systems, the combination of i-f frequency and transmitter pulse spectrum must be chosen to avoid problems at the image frequency. A spectral line of the feedthrough or clutter occurring at the image may limit the SFV or SCV, sometimes necessitating use of image-rejection mixers or rf preselection.

For sampled data operation, the primary difference is in the doppler acquisition scheme. Since target illumination occurs only in short bursts, the use of a sweeping gate for acquisition would result in excessively long acquisition times. The doppler uncertainty region must, therefore, be examined in parallel by a bank of doppler filters. The illumination burst must be shaped or the received signal time gated to prevent the spreading of clutter through the target doppler spectrum due to the pulsed nature of the transmission. Finally, sample holds must be added in afc, agc and angle- track loops.

In the TVM mechanization, the block diagram is functionally the same as for the missile-borne seeker, except that the signal amplification and processing blocks are on the ground and a missile-to-ground data link and a ground to-missile command link replace the signal lines of the on-board seeker. ••

References

1. D. K. Barton, "Real World Radar Technology," *IEEE 1975 International Radar Conferences*, pp. 1-22, (April, 1975).
2. A. S. Locke, et al, *Guidance*, Van Nostrand Co., Princeton, NJ, (1955).
3. T. L. Phillips, "Anti-Aircraft Missile Guidance," *Raytheon Electronic Progress*, Vol. II, No. 5, pp. 1-5, (March-April, 1958).
4. F. A. Gross, W. M. Hall, D. K. Barton, "Detection and Tracking Systems: An Historical Overview," *Raytheon Electronic Progress*, Vol. XVI, No. 3, pp. 2-11, (Fall, 1974).
5. J. J Long, A. Ivanov, "Radar Guidance of Missiles," *Raytheon Electronic Progress*, Vol. XVI, No. 3, pp. 26-28, (Fall, 1974)
6. M. W. Fossier, B. A. Hall, "Fundamentals of Homing Guidance," (unpublished Raytheon seminar presentation), (1962).
7. A. Ivanov, "Radar Guidance of Missiles," *IEEE 1975 International Radar Conference*, pp. 331-335, (April, 1975).
8. R. M. Jaffe, "Monopulse Radar Receiver," U. S. Patent 3,713,155, (Jan. 23, 1973).
9. M. I. Skolnik, *Introduction To Radar Systems*, McGraw-Hill, New York, (1970).
10. M. I. Skolnik, et al, *Radar Handbook*, McGraw-Hill, New York, (1970).

OADARS (Optical Aids to Detection and Ranging Systems)
A.V. Jelalian, D.A. Kawachi, and C.R. Miller
IEEE ELECTRO-76 Record, Part 30, May 1976
Pages 30-3-1—30-3-5

8.2

This paper is concerned with the utilization of laser systems as adjuncts to microwave fire control radar systems, with the intent of increasing the fire control system's ECCM effectiveness. (Figure 1)

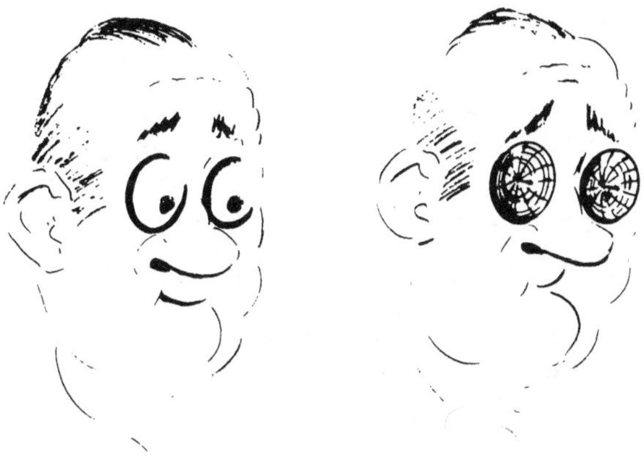

Figure 1.

In order to provide a reference point, Figure 2 illustrates the wavebands associated with the electromagnetic spectrum. The microwave system engineer is usually interested in the waveband to 300 GHz, associated with millimeter waves. The optical system engineer utilizes the electromagnetic spectrum from the visible (0.5μ) to the far IR (10μ). The wavebands in this portion of the electro-magnetic spectrum are determined by absorption and scattering effects, resulting in specific atmospheric windows.

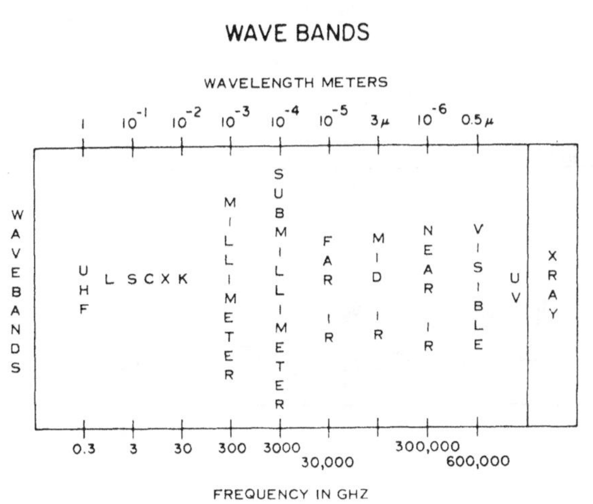

Figure 2.

Atmospheric propagation effects from the visible band through the entire microwave band are shown in Figure 3. Listed vertically in this figure is the attenuation, in dB per n.mi., associated with each of the atmospheric conditions noted relative to the visible wavelength of 0.7 microns, the near IR of 1.06 microns, the mid-IR of 3.8 microns, the far IR of 10.6 microns and that of the millimeter waves of .3 cm and the microwaves of 1 and 3 cm. One may see from this chart that operation from the visible to the near IR is significantly affected by haze conditions and that as the atmospheric haze increases, significant attenuation occurs. Operation in the mid- and far-IR results in significantly less atmospheric attenuation. Operation at the 10.6 micron wavelength indicates that operation in a tropical environment would result in a 5 dB per n.mi. attenuation coefficient. Correspondingly, one may see that operation in rainfall rates of as high as 12½ mm per hour results in the same atmospheric attenuation coefficient as exist with millimeter wave radar systems. Millimeter wave transmission would appear to be better under fog conditions than that of a CO_2 system. Moving further toward longer wavelengths, one observes that the attenuation diminishes with regard to rainfall and fog as the longer wavelengths are approached.

ATTENUATION VS. WAVELENGTH

DB/NMI	λ 0.7μ	λ 1.06μ	λ 3.8μ	λ = 10.6μ			0.3 CM	λ 1 CM	3 CM
0.4	EX. CLR.	EX. CLR.	CLEAR V=23KM		SUB ARCTIC WINTER				12.5 MM/HR
1.0	STD. CLR. V=25KM	S.C.	HAZE V=5KM	CLEAR				2.5 MM/HR	FOG V=100 FT
1.4		CLEAR					RAIN 2.5 MM/HR	FOG V=100 FT	
1.8	CLEAR V=15KM			HAZE	SUB ARCTIC SUMMER	LT. FOG (1) V=2KM	FOG V=400 FT		50 MM/HR
2.8		LT. HAZE							
3.2	LT. HZ. V=8KM				MID LAT. SUMMER	LT. (2) RAIN 2.5 MM/HR			
4.3		M. HAZE							
5.2	M. HZ. V=5KM				TROPICAL	FOG (1) V=0.5KM		12.5 MM/HR	
7.1		HAZE							
9.5	HAZE V=3KM							25 MM/HR	
11.0						MED. (2,3) RAIN 12.5 MM/HR	RAIN 12.5 MM/HR		
16.5						25 MM/HR	FOG V=100 FT	30 MM/HR	
					M. CLATCHEY	(1) GOODWIN (2) RENSCH (3) CHU		BARTON SKOLNIK	

Figure 3.

Figure 4 illustrates the advantages associated with operation with laser beams. For a diffraction limited beamwidth operation at a 1 cm wavelength with a 12" (30 cm) dish, yields a beamwidth of approximately 33 mrad. Correspondingly, operation at the 10 micron wavelength, associated with a CO_2 laser, reduces this beamwidth 3 orders of magnitude to 33 microradians. Because of this smaller beamwidth, significant angle tracking accuracy improvements may be realized with laser systems.

BEAMWIDTH OF A LASER TRANSMITTER

THE DIFFRACTION LIMITED BEAMWIDTH OF A TRANSMITTING SOURCE MAY BE EXPRESSED AS:

$$\theta = C \frac{\lambda}{D}$$

WHERE:

λ = WAVELENGTH IN (METERS)
D = TRANSMITTING APERTURE (METERS)
θ = ANGULAR BEAMWIDTH IN RADIANS

FOR

λ = 1 CM

AND

D = 30 CM (1 FT)

$\theta \approx$ 33 C MILLIRADIANS — RADAR

FOR

$\lambda = 10^{-3}$ CM (10.6 MICRONS)

AND

D = 30 CM

$\theta \approx$ 33 C MICRORADIANS — CO_2 LASER

C DEPENDENT UPON $\frac{1}{e}$; $\frac{1}{e^2}$; OR HALF POWER ANTENNA BEAMWIDTH.

Figure 4.

Figure 5 illustrates the accentuated Doppler shift associated with the operation at smaller wavelengths. Here the Doppler shift associated with the backscatter signal from a target is 20 cycles of Doppler shift per foot per second of target motion at a 3 cm wavelength and 600 kHz per foot per second of target velocity at the 1 micron wavelength. Because most 1 micron sources do not have good coherence length, one is typically concerned with using the carbon dioxide (CO_2) laser which has excellent coherence lengths and typical Doppler shifts of 57.5 kHz per foot per second of target motion. Because of this accentuated Doppler shift, instantaneous target velocity measurements may be made by CO_2 laser Doppler systems.

THE EQUATION FOR THE DOPPLER SHIFT MAY BE EXPRESSED AS.

$$F_D = \frac{2V}{\lambda} \cos \gamma$$

V = AIRCRAFT VELOCITY (RELATIVE)
λ = TRANSMISSION WAVELENGTH
γ = ANGLE BETWEEN RELATIVE VELOCITY VECTOR AND LINE OF SIGHT

BECAUSE OF THE SHORTER OPERATING WAVELENGTH ASSOCIATED WITH LASERS THERE ARE MORE CYCLES OF DOPPLER SHIFT PER FOOT/SECOND OF VELOCITY.

RADAR	= 20 CPS/FT/SEC AT λ_R	=	3 CM
LASER	= 600 KC/FT/SEC AT λ_L	=	1 μ
	= 60 KC/FT/SEC AT λ_L	=	10 μ

Figure 5.

Figure 6 illustrates the multipath problem which can occur when a threat aircraft attempts to penetrate defenses by flying at low altitude. The interaction of the radar tracking beam between the target and the terrain result in both an indirect and a direct path t the aircraft, as shown in the top of th figure. Here the microwave antenna bea is seen to illuminate the incoming target as well as the terrain. Energy bouncing off the terrain may be subsequently reflected up to the incoming threat and back to the microwave receiver. Because this path is reciproca significant elevation angle tracking errors can occur for microwave tracking radars. The use of the laser with its attendant very narrow beamwidth helps reduce this multipath effect by minimizing the amount of energy that may illuminate the terrain.

MULTIPATH ERRORS

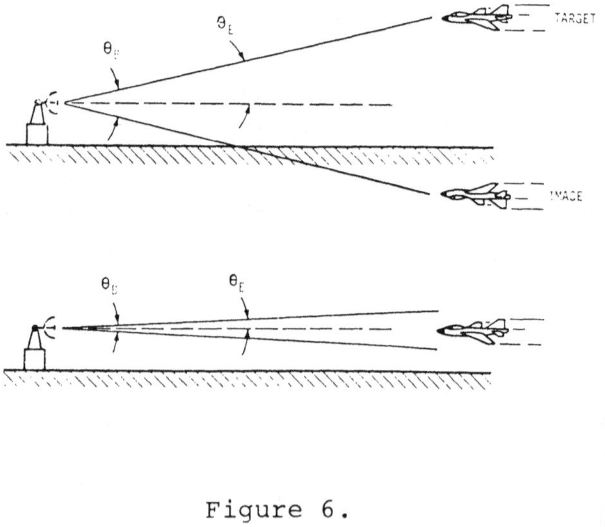

Figure 6.

Figure 7 indicates the two different types of detection techniques that may be utilized with laser systems. On the top, the return signal power, P_S, enters the receiver through an optical filter and is directed to an optical detector where video-detection occurs. Competing with this signal is the receiver noise and optical background power (P_B). The background noise may be optically filtered to reduce its intensity. The incoherent detection receiver may be contrasted to that of a coherent detection receiver noted on the bottom of Figure 7. The coherent detection receiver has a portion of the transmitter beam, noted in this figure as F_O, coupled over to the detector and referred to as the power of the local oscillator P_{LO}. This reference signal is now utilized to homodyne the received backscattered target signal. As a result of the received signal coherence, significant opto-electronic filtering results. Because of this, the coherent detection receiver has significant sensitivity improvement over the incoherent detection receiver.

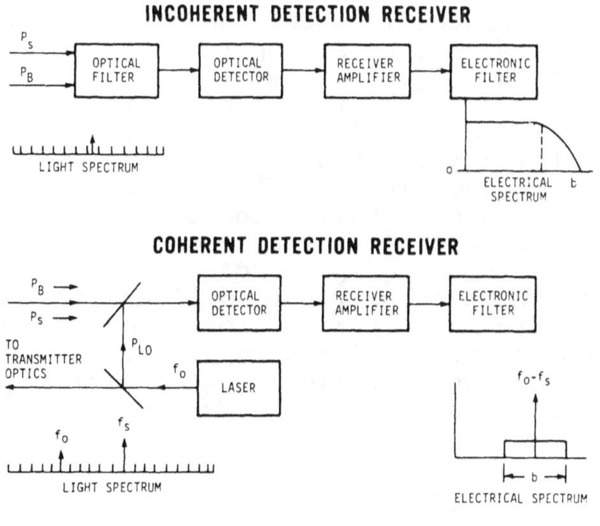

Figure 7.

S/N expressions associated with these systems are shown in Figure 8. The S/N ratio of the incoherent system is equal to: η, the quantum efficiency, (the conversion efficiency of photons to photoelectrons in the receiver); x, the received signal powers squared, divided by 2 x h, (Planck's Constant); f, the transmitter frequency; B, the electronic bandwidth and a summation of noise terms, composed of the received signal power, the background power, the effective dark current power and the thermal noise power associated with the receiver. This may be contrasted to that of the coherent detection system, noted in the bottom of this figure, where the S/N ratio is dependent upon: η times the product of the received signal power and the local oscillator power, divided by the same noise terms as above, included with the local oscillator power as a noise source. The interesting point from this S/N equation is that one can and does increase the local oscillator power associated with the receiver, such that it becomes the predominant noise source. When this happens, the S/N expression becomes: η times the received signal power divided by 2hνB, and may be seen to be totally background immune. As a result, therefore, of the increased sensitivity and the background immunity of the coherent detection system, significant counter-countermeasure capabilities are achieved.

COMPARISON OF INCOHERENT & COHERENT DETECTION RECEIVERS

THE S/N EQUATION FOR INCOHERENT DETECTION INCOHERENT DETECTION MAY BE EXPRESSED AS:

$$S/N \quad \frac{\eta P_s^{\,2}}{2h\ f\ B\left[P_s + P_B + KP_{DK} + KP_{TH}\right]}$$

WHERE:

- η = QUANTUM EFFICIENCY
- f = TRANSMISSION FREQUENCY
- B = ELECTRONIC BANDWIDTH
- P_s = RECEIVED SIGNAL RADIATION POWER
- P_B = BACKGROUND RADIATION POWER
- P_{DK} = EQUIVALENT DARK CURRENT RADIATION POWER
- P_{TH} = EQUIVALENT RECEIVER THERMAL NOISE RADIATION POWER

THE S/N EQUATION FOR COHERENT DETECTION MAY BE EXPRESSED AS:

$$S/N \quad \frac{\eta P_s\ P_{LO}}{2h\ f\ B\left[P_{LO} + P_s + P_B + KP_{DK} + KP_{TH}\right]}$$

WHERE:

P_{LO} = REFERENCE LOCAL OSCILLATOR POWER

Figure 8.

The parameters of a 1.06 micron 100 mJ YAG system may be seen in Figure 9. Ranges in excess of 5 n.mi. may be realized from this system in clear weather. This may be observed on Figure 10 which shows the probability of detection vs. S/N ratio as a function of range with a neodymium laser. Under the conditions of clear visibility a 50% probability of detection may be observed in excess of 7 n.mi. for the neodymium laser with silicon avalanche detector, and approximately 4 n.mi. for a silicon detector. In the advent of haze, the YAG system range performance is reduced.

YAG[A] LASER SYSTEM PARAMETERS

LASER:		
WAVELENGTH	1.06	MICRONS
PULSE ENERGY	100	MJ
PULSE LENGTH	0.02	μSEC
BEAM DIVERGENCE	0.4	MRAD
CROSS SECTION	1	M^2
S/N	22	dB
RANGE[B], CLEAR (VISIB. ~15 N.MI.)	5.7	N.MI.
RANGE[B], HAZE (VISIB. ~ 2 N.MI.)	2.5	N.MI.

NOTES: (A) WITH QUADRANT SILICON AVALANCHE PHOTODIODE DETECTOR

(B) AGAINST A 1 SQ. METER (σ) TARGET FOR INITIAL DETECTION, DETECTION PROBABILITY = 95% PER SCAN

Figure 9.

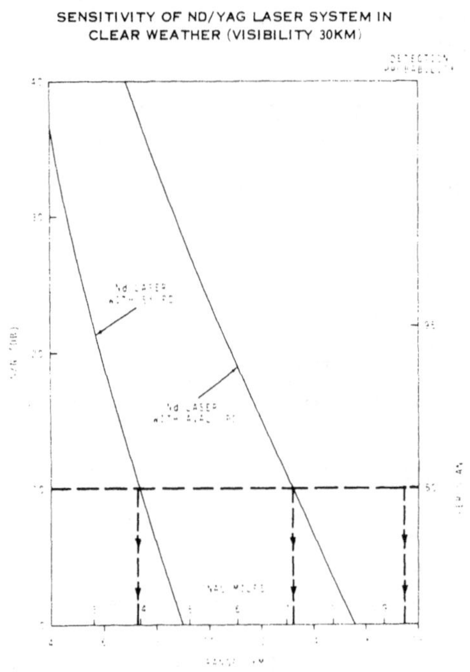

Figure 10.

Figure 11 illustrates the effect that atmospheric haze would have on the YAG/avalanche detector system performance.

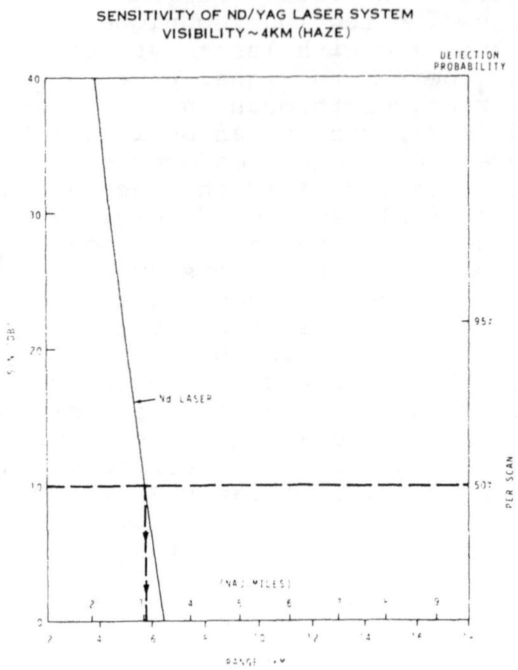

Figure 11.

Figure 12 shows the system parameters of a pulse Doppler CO_2 laser, having a 5 cm optical system. Note that in this slide cryogenic cooling is required of the detector. Utilizing a S/N ratio of 15 dB for the detection criteria, yields a 70% single pulse detection probability and system range performance illustrated in Figure 13. Here the range performance as a function of the atmospheric conditions for a one square meter target cross-section indicates the CO_2 laser radar system could have ranges in light fog in excess of 4 n.mi. and ranges in excess of 2 n.mi. under conditions of rain (10mm/hr).

CO_2 LASER SYSTEM PARAMETERS

TRANSMITTER PEAK POWER	10 KW
OPTICS DIAMETER	5.0 CM
TRANSMITTER/RECEIVER BEAMWIDTH	0.3 MR
PULSEWIDTH	1.0 TO 10 μS NOMINAL 4 μS
DETECTOR QUANTUM EFFICIENCY	0.5
DETECTOR TEMPERATURE	77°K
SYSTEM LOSSES	5 DB
S/N IN CLEAR WEATHER (1 DB/NMI)	15 DB
TARGET CROSS-SECTION	1.0 SQ. METERS
DETECTION TECHNIQUE	COHERENT
OPERATING WAVELENGTH	10.6μ
REPETITION RATE	HANDOVER VOLUME DEPENDENT

Figure 12.

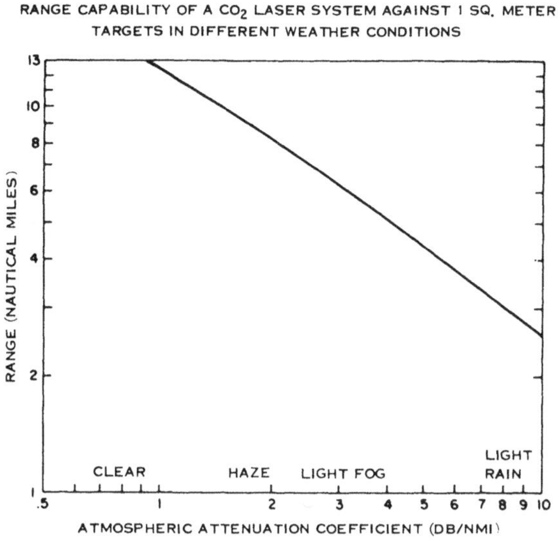

Figure 13.

Figure 14 illustrates an A-scope display of a Raytheon airborne pulse Doppler CO_2 laser radar, having 10KW of peak power. Here one may see, in the lower part of the figure, an A-scope display, where the X-axis is seen to have a range capability of 0 - 16 n.mi. At approximately 13 n.mi. an 8 dB S/N ratio may be seen on this A-scope, thereby illustrating that coherent CO_2 systems can be configured to operate in realistic environments and have sufficient coherence to perform long range measurement.

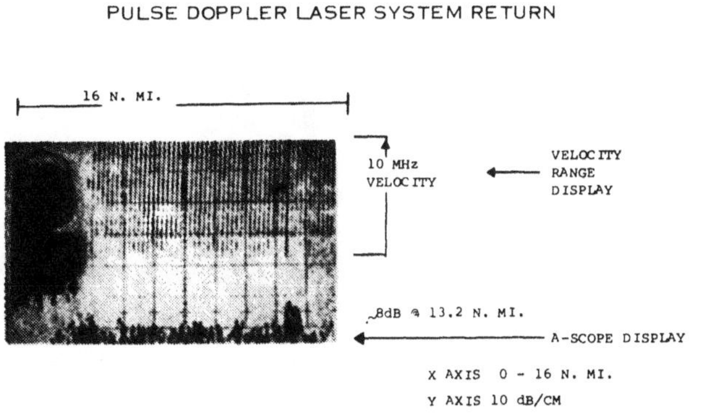

Figure 14.

In summary, both YAG (1.06 micron) and CO_2 (10.6 micron) laser systems can add significant capability to existing fire control systems by providing range, velocity and angle tracking of low altitude targets. Figure 15 suggests that the fire control weapon delivery system designer may have a significantly more effective countermeasure resistant system by considering the use of optical aids to detection and ranging systems.

Figure 15.

Electronic Correlator for Passive Ranging
K.H. Norsworthy
Proceedings of the International Countermeasures Conference, London, September 1976
Pages 290-320

1.0 INTRODUCTION

My paper describes the development and test of a wideband electronic correlator. One potential application of the correlator is to the problem of passive ranging to jamming targets.

For this application two different operations have been studied; 1) Single-platform ranging to an airborne jammer involving the measurement of ground bounce delay (chart 1), and 2 Multi-platform ranging to airborne and ground based jammers in a dense jammer environment (chart 2). An experimental correlator has been developed, and field tested, that includes design features and performance capabilities suitable for both applications.

The single-platform operation uses the correlator in its autocorrelation mode, since there is usually insufficient angular separation between the direct and multipath signals to allow separation of those signals by angle of arrival techniques. Delay accuracy of 100 nanosec. is sufficient (greater delay accuracy not warranted due to the uncertainty in bounce point altitude) but high process gain is required to overcome the low signal-to-interference ratio encountered by the ground bounce component. The delay range of interest is not large, and only small amounts of differential doppler are encountered in this mode.

A capability to measure the altitude of the jammer is provided by combining the conventional Kalman filtering of angle strobes with the correlator multipath delay measurements. This also provides faster convergence of the ranging algorithm.

The multi-platform application (chart 2) employs the correlator in its cross correlation mode, and seeks to make delay measurements with high accuracy (say 10 nanosec.) so that precision jammer location can be accomplished. High process gain is again necessary, if a high density of jammers is to be handled, and in this case the time delay and differential doppler parameters are likely to vary over a substantial range.

When supported with accurate platform position information the multi-platform concept can provide precision jammer location despite a sophisticated, dense jamming environment.

2.0 THE FUNCTION OF THE CORRELATOR

As described in this paper, the function of a correlator is to measure the differential time delay experienced by a radiated signal as it travels via two separate paths from a transmitter (T) to a common receiver location (L), see chart 3.

Because the receiving antennas may also receive a great deal of interference radiation, the process of delay measurement is not a trivial one. It is accomplished by forming the average cross product of the received signals at each of a number of delay displacements,

$$R_{12}(\tau) = \frac{1}{T} \int_0^T S_1(t)\, S_2(t+\tau)\, dt$$

At a delay displacement equal to the propagation path difference the average cross product exhibits a peak magnitude, see chart 3, distinguishable from the magnitudes surrounding it.

3.0 DESIRABLE FEATURES OF A CORRELATOR

Analyses performed at Boeing over the past three years have defined desired performance levels for a correlator applicable to the passive ranging applications. The correlator should accept input signal bandwidths up to 20 MHz (see chart 4) and should compute a substantial number of simultaneous (closely spaced) correlation coefficients; perhaps 256 coefficients separated by 25 nanosecond increments. Direct interface to a system computer is desirable so that interpolation between the measured correlation points can define the correlation peaks with a delay resolution as high as 5 nanoseconds.

To allow the computation region to be moved to high delay values, a large amount of precomputation delay should be selectable. This precomputation feature is usually easily implemented so the equivalent of up to 10,000 $\Delta\tau$ is not unreasonable to plan.

The most basic performance characteristic of the correlator is its "Process Gain," that is, its capability to measure the delay properties of small amplitude signals that may be buried well below the interference noise levels. A rigorous definition of the term "Process Gain" is difficult for the correlator because one is dealing with a processor that has two input channels (not one) and a single output channel. An approximate interpretation suffices here.

The "process gain" of a correlator is said to be +40 db if signal-to-noise conditions of -X db at each input give rise to an output correlation peak with signal-to-noise (40 - 2X) db. Alternatively, the process gain may be said to be 40 db if a very clean signal on one input and a signal-to-noise of -X db on the other input gives rise to an output signal-to-noise ratio of (40-X) db.

It has been known for many years that the theoretical process gain of a correlator is limited by the time-bandwidth product of the signals processed. Imperfections in the correlator usually cause this theoretical upper limited to be missed, especially when seeking high values of process gain.

To be of value in passive ranging a correlator should give close to theoretical performance, up to process gains of 40 db, and it should be capable of achieving this despite the existence of unknown doppler shifts in the two input signal channels.

To achieve the desired process gain capability it is necessary that the correlator perform MULTIPLICATION operations at very high speed. Fortunately, the input signals can be hard clipped without significantly affecting the time delay of the correlation peaks and this hard clipping allows the multiplier circuitry to be greatly simplified and replicated many times.

When using such simplified circuitry it is not infeasible for a correlator to perform up to one billion multiply operations per second (chart 5).

If the precise delay value of a correlation peak is to be derived by computer interpolation of the measure coefficients, it is important that the shape of the correlation curve be symmetrical about its peak despite doppler effects. This characteristic is obtained by appropriate processing of quadrative information channels.

4.0 BOEING EXPERIMENTAL CORRELATORS

In the past two or three years Boeing Aerospace Company, Research and Engineering Division, has fabricated and evaluated a number of experimental correlators that approach the performance characteristics listed.

Chart 6 shows a photograph of a first correlator that operates with $\Delta\tau$ = 100 nanosecond delay increments and 6×10^7 multiply operations per second. Field test results using this correlator are given in Section 5.

Chart 7 shows a photograph of a more recent correlator that operates with $\Delta\tau$ as low as 40 nanosecond and performs 3×10^8 multiply operations per second. To increase its general experimental usefulness, this latter correlator was designed to also perform other statistical measurements such as; Ensemble Averaging of signal transients, and Amplitude Probability Distributions, and also to include the capability for performing the Fourier Transform of the measured statistical functions.

5.0 FIELD TEST RESULTS

To experimentally verify the performance of the correlator shown in chart 6, and identify and solve any unexpected (practical) signal handling problems a ground based field test was performed as shown in chart 8.

Two classes of tests were performed; 1) against a ground-based radiator at precisely known locations and, 2) against an airborne radiator, thereby inducing an unknown amount of doppler into the problem.

For both series of tests the correlator was installed in a test trailer on a hill overlooking the Boeing facility at Kent, Washington, U.S.A. Separate antennas, see chart 9, were used for receiving the radiated signal via the direct path (A) (chart 8) and the relay path (C and B).

A photograph of the inside of the trailer (chart 10) shows the correlator with its associated receiver equipment, and computerized data acquisition equipment.

Chart 11 shows the antennas that were used at the relay site to receive and retransmit the S Band radiation.

5.1 Zero Doppler Tests

In the first series of tests the radiator used for simulating a jammer was mounted in a station wagon (chart 12) and positioned on a road approximately 2 miles from the receiving site, see chart 8.

Correlation results were first recorded, with high signal-to-interference conditions, for two positions of the station wagon separated by only 10 feet along the road. The change in the resulting correlation shape, see chart 13, was repeatable and indicated that interpolation between the measured correlation coefficients could be used to track movements of the radiation source with resolutions approaching 10 feet.

The next series of tests sought to evaluate the "process gain" of the correlator under various conditions of signal-to-interference noise. To accomplish this, various levels of uncorrelated interference noise were added into the relay signal channel and the repeatability of the correlator measurements determined for successive integration intervals of 0.030 seconds each. For the tests reported, the signal amplitudes over the direct and relay path were both set 30 db above receiver noise itself. Chart 14 shows that for relay path signal-to-interface ratios below 0 db the measurement process gain was approximately 40 db, as was expected from design considerations.

The above signal-to-noise and "process gain" measurements relate to the repeatability of the amplitude of the correlation peak. Of somewhat more practical concern to the passive ranging problem is the repeatability of the delay measurement at the (computer interpolated) peak. Chart 15 shows standard deviations of the delay parameter, and it is seen that delay values are measured with 1σ values of approximately 10 nanosecond for signal-to-interference ratios (on the relay channel) as low as -20 db. These test results were obtained using a signal bandwidth of 3.5 MHz and a 0.030 second integration period.

Comparison of the measured delay with laser instrument survey of the site showed agreement to within a very few nanoseconds.

5.2 Non-Zero Doppler Tests

To evaluate the correlator under conditions where different doppler effects were present on the two signal channels. The simulated jammer was flown in a light airplane in an approximate straight line over the ground reference location shown in chart 8.

A sequence of (amplitude) signal-to-noise measurements out of the correlator are shown in chart 16, and the sequence of delay measurements for two crossing tracks are shown in chart 17.. The results verify that the experimental correlator continued to operate satisfactorily despite the differential doppler conditions.

6.0 CONCLUSION (CHART 18)

Electronic correlators have been developed and field tested that exhibit performance characteristics suitable for use in passive ranging applications. Automatic data reduction (interpolation) procedures have been effective in extracting delay accuracies substantially better than the delay increment ($\Delta\tau$) between the measured correlation coefficients.

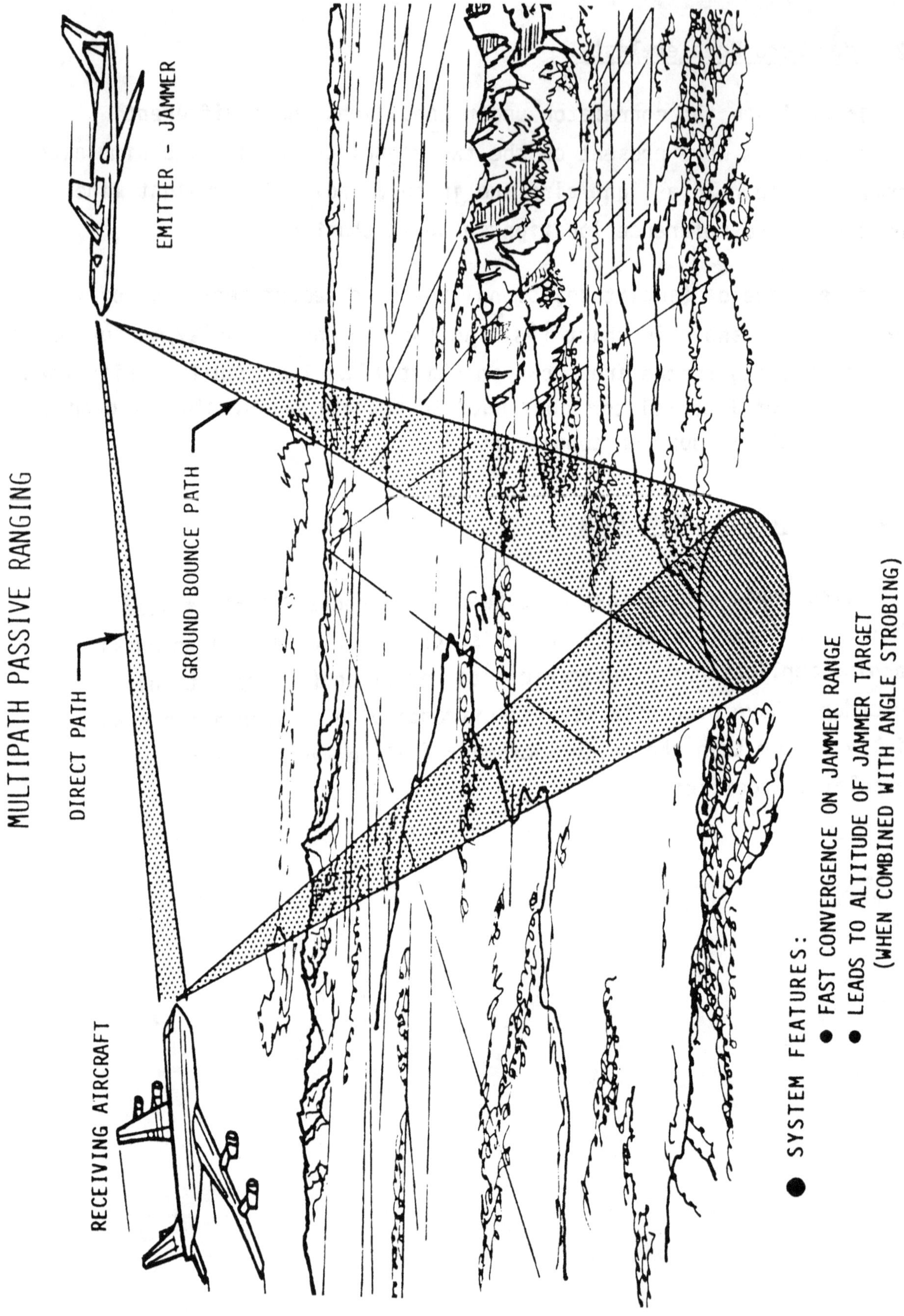
MULTIPATH PASSIVE RANGING
DIRECT PATH
GROUND BOUNCE PATH
EMITTER - JAMMER
RECEIVING AIRCRAFT
● SYSTEM FEATURES:
● FAST CONVERGENCE ON JAMMER RANGE
● LEADS TO ALTITUDE OF JAMMER TARGET
(WHEN COMBINED WITH ANGLE STROBING)

CHART 1

COOPERATIVE AIR TO AIR PASSIVE RANGING

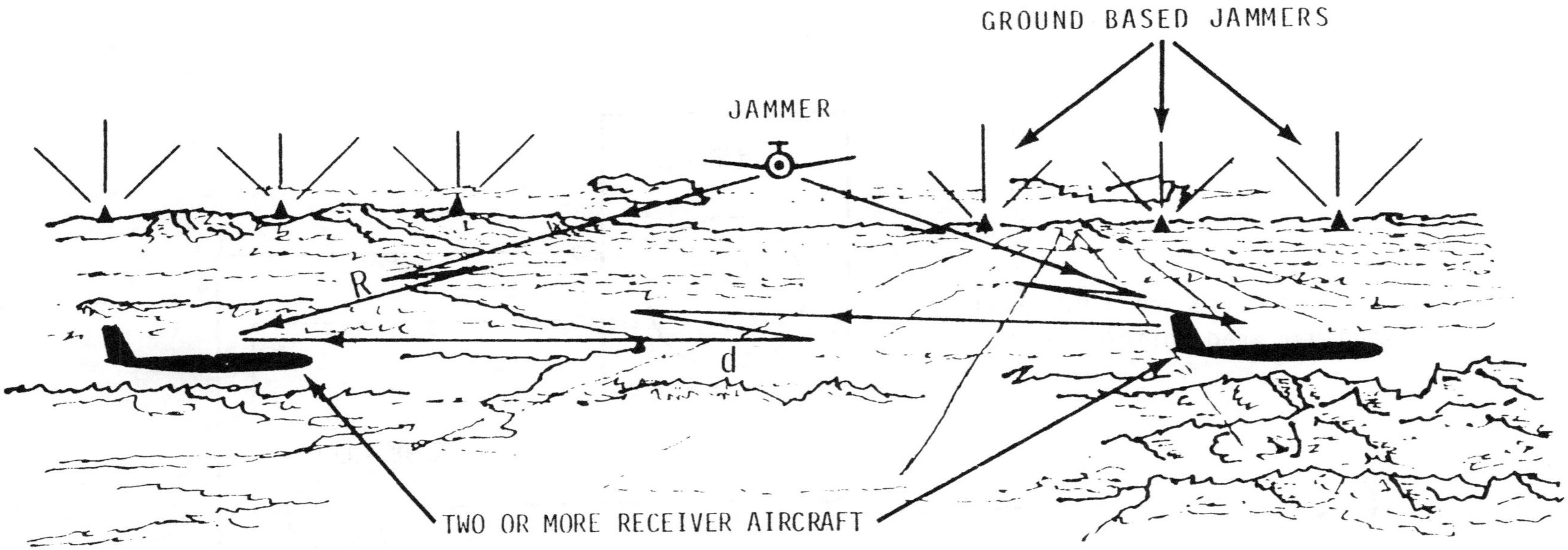

- Employs target azimuth, observer separation and signal propagation time difference to determine target range
- Handles increased density threat
- Permits rapid deghosting (reduced computer load)

CHART 2

FUNCTION OF THE CORRELATOR

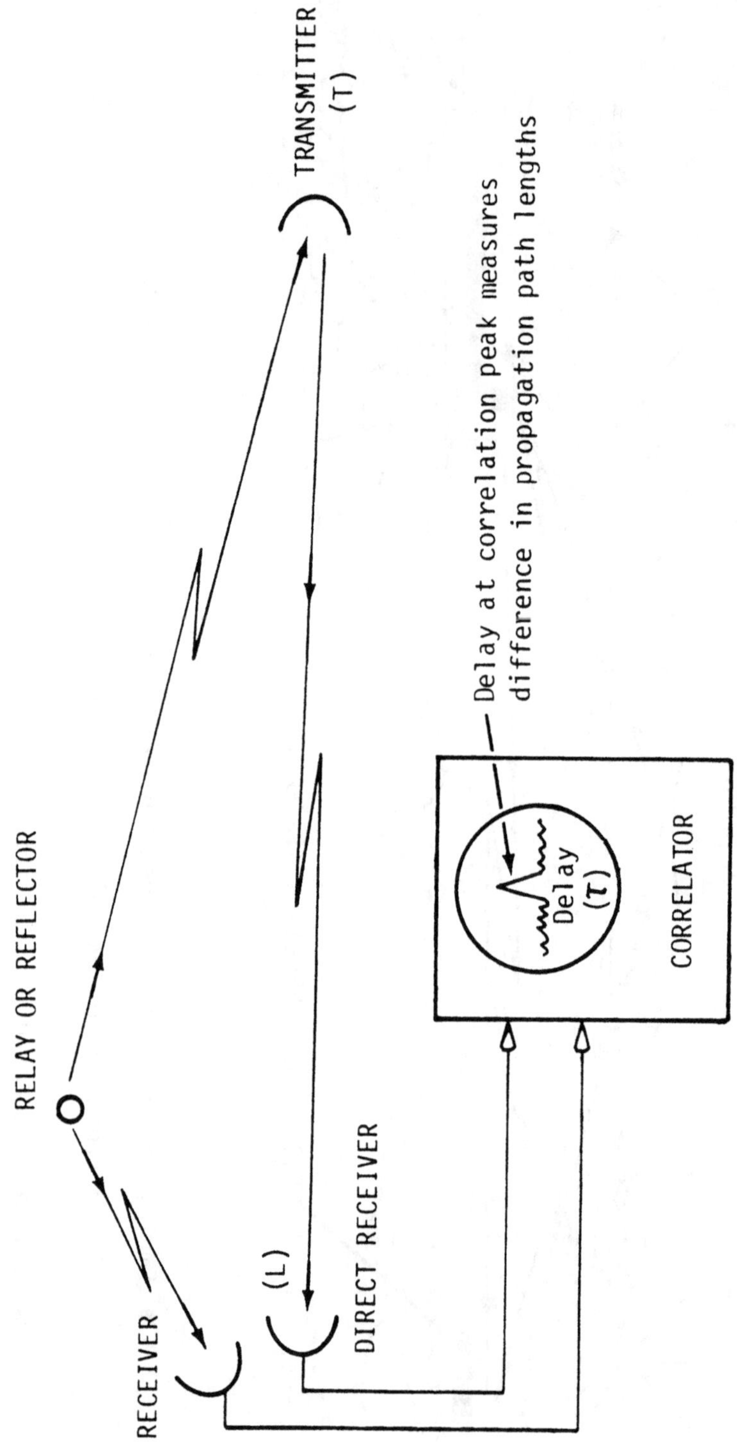

CHART 3

DESIRED FEATURES OF CORRELATION

- WIDEBAND INPUT SIGNALS (20 MHz)
- NUMEROUS DELAY POINTS DISPLAYED (e.g. 256)
- DELAY RESOLUTION - 25 NANOSEC. (COMPUTER INTERPOLATION TO 5 NANOSEC.)
- SUBSTANTIAL PRECOMP DELAY (10,000 $\Delta\tau$)
- PROCESS GAIN - 40 db (IN .03 SEC. INTEGRATION TIME)
- DOPPLER TOLERANCE ~ TO 1 KHz

CHART 4

DESIRED FEATURES OF CORRELATOR

- HIGH PRODUCT RATE ~ (1 BILLION MULTIPLIES PER SEC.)
    - PROVIDES IMPROVED PROCESS GAIN
- SYMMETRICAL CORRELATION PEAK
    - PERMITS DELAY INTERPOLATION
- PRECISION SAMPLING - ECL LOGIC
    - PROVIDES FINE DELAY RESOLUTION
    - PERMITS SAMPLING AND PROCESSING OF WIDE BAND SIGNALS
- DIGITAL INTERFACE
    - PROVIDES PRECISE INTERPRETATION AND FURTHER PROCESSING OF CORRELATOR OUTPUTS

CHART 5

MULTIPATH FIELD TEST - PRELIMINARY MEASUREMENTS
(ZERO DOPPLER)

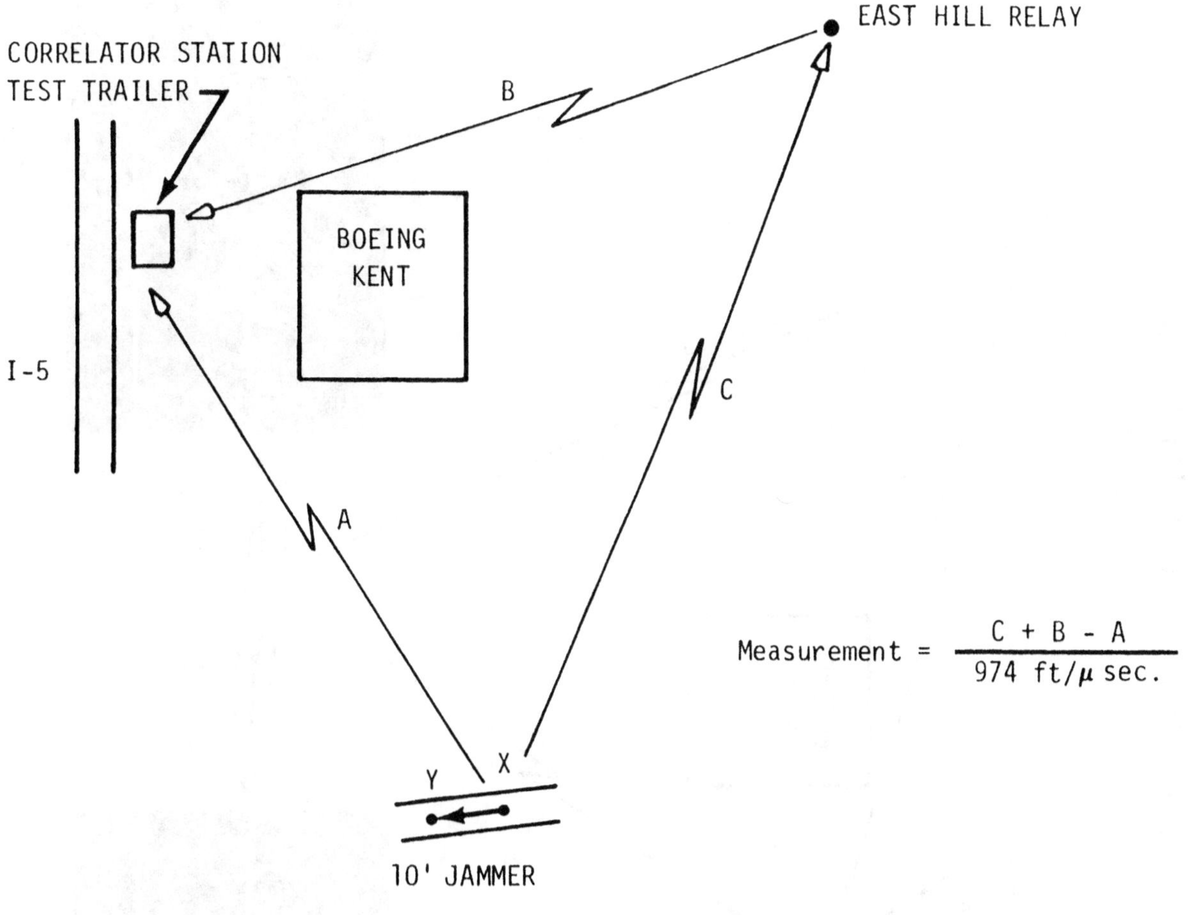

CHART 8

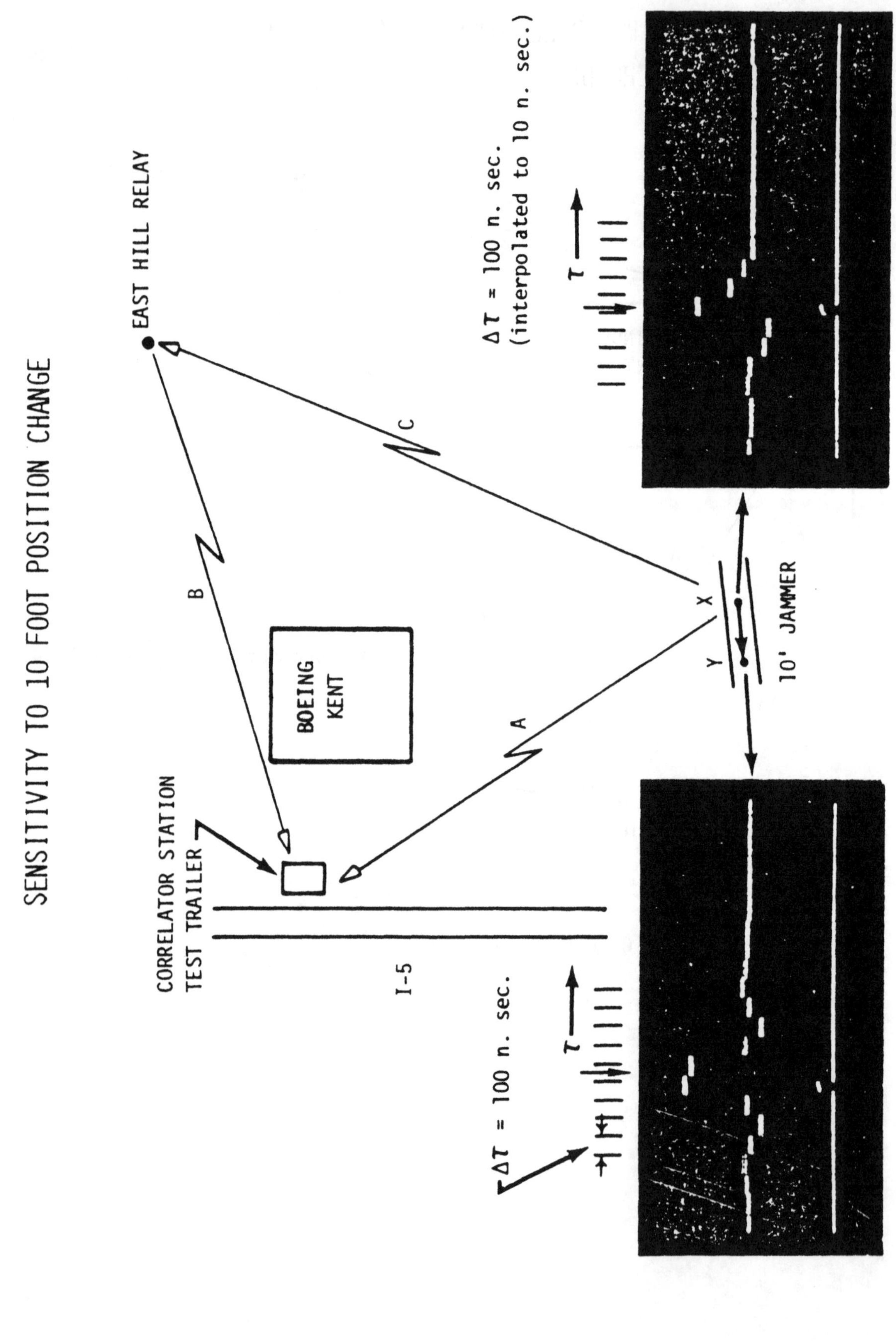
SENSITIVITY TO 10 FOOT POSITION CHANGE
EAST HILL RELAY
CORRELATOR STATION
TEST TRAILER
BOEING
KENT
I-5
A
B
C
X
Y
10' JAMMER
ΔT = 100 n. sec.
(interpolated to 10 n. sec.)
ΔT = 100 n. sec.
T
X - LESS DELAY
Y - LARGER DELAY

CHART 13

CORRELATOR PROCESS GAIN

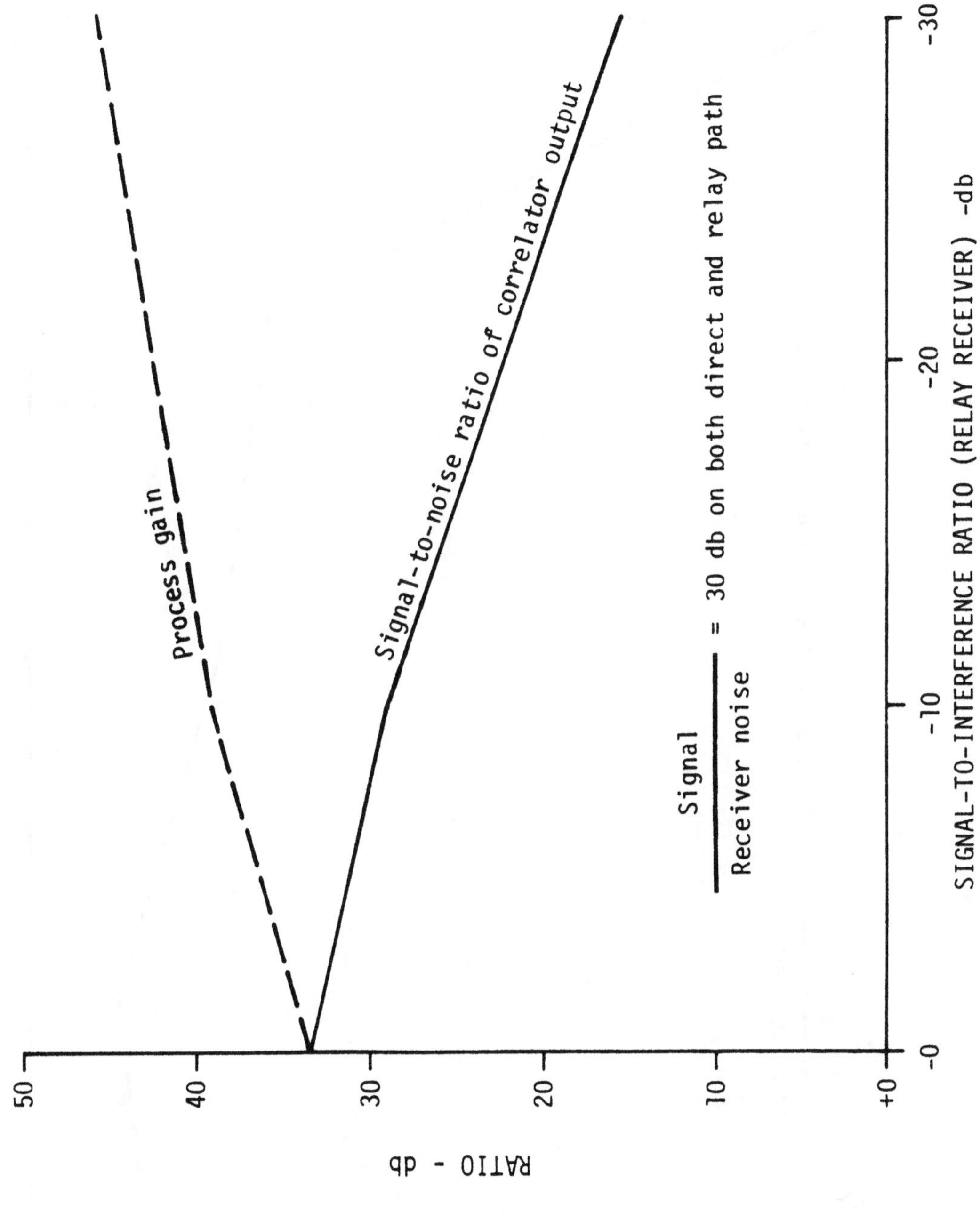

CHART 14

STANDARD DEVIATION - TDOA MEASUREMENT

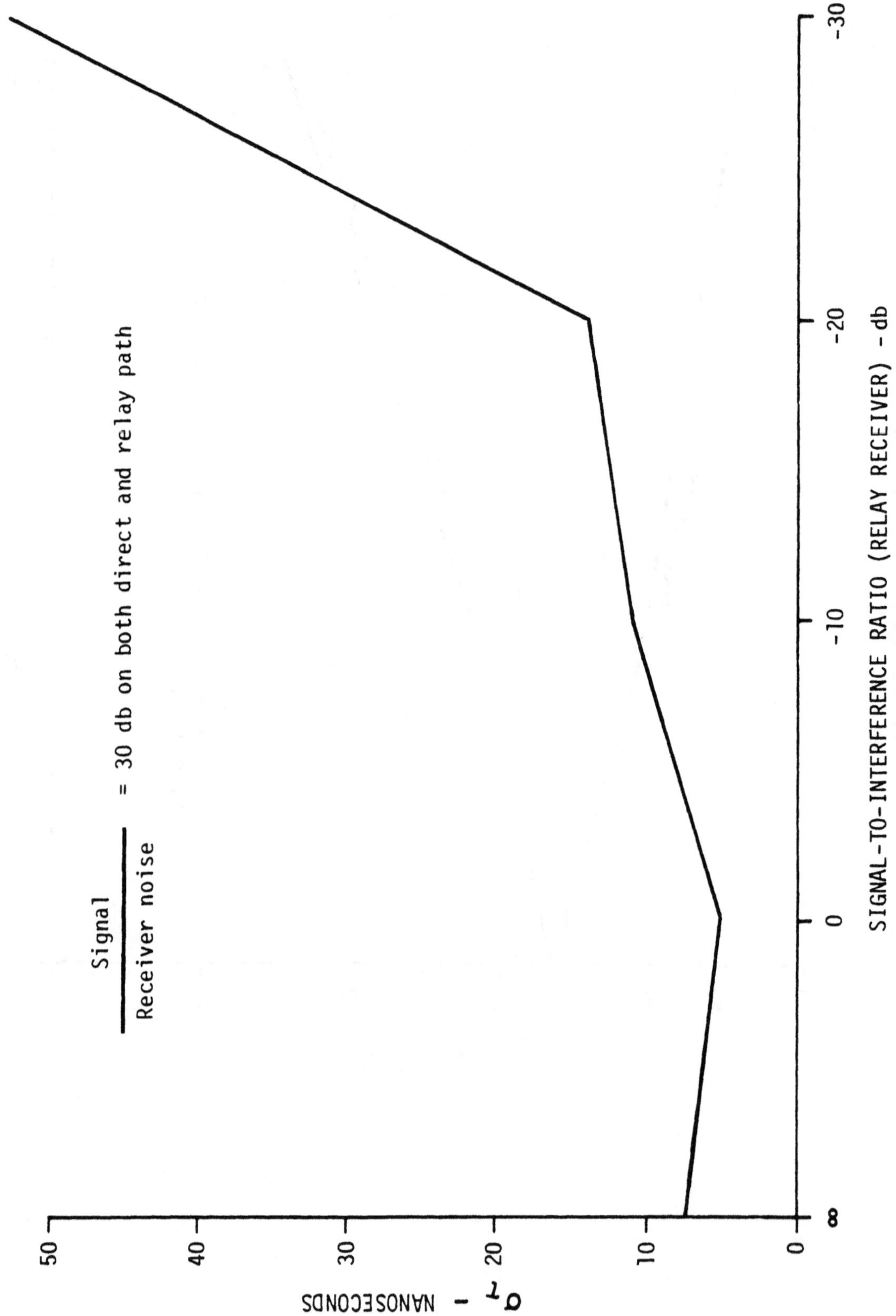

CHART 15

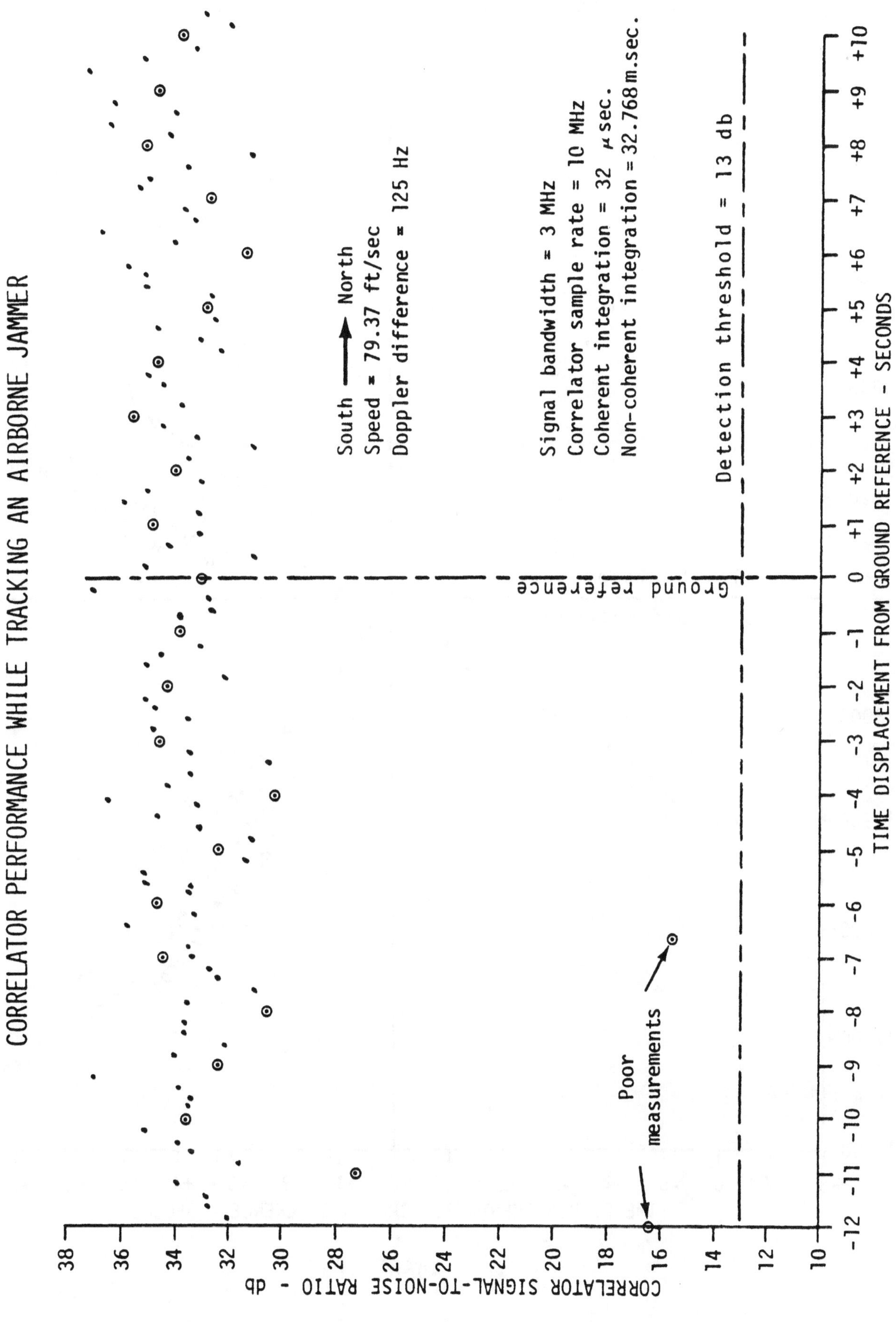
CORRELATOR PERFORMANCE WHILE TRACKING AN AIRBORNE JAMMER
South → North
Speed = 79.37 ft/sec
Doppler difference = 125 Hz
Signal bandwidth = 3 MHz
Correlator sample rate = 10 MHz
Coherent integration = 32 μsec.
Non-coherent integration = 32.768 m.sec.
Detection threshold = 13 db
Ground reference
Poor measurements
CORRELATOR SIGNAL-TO-NOISE RATIO - db
TIME DISPLACEMENT FROM GROUND REFERENCE - SECONDS

CHART 16

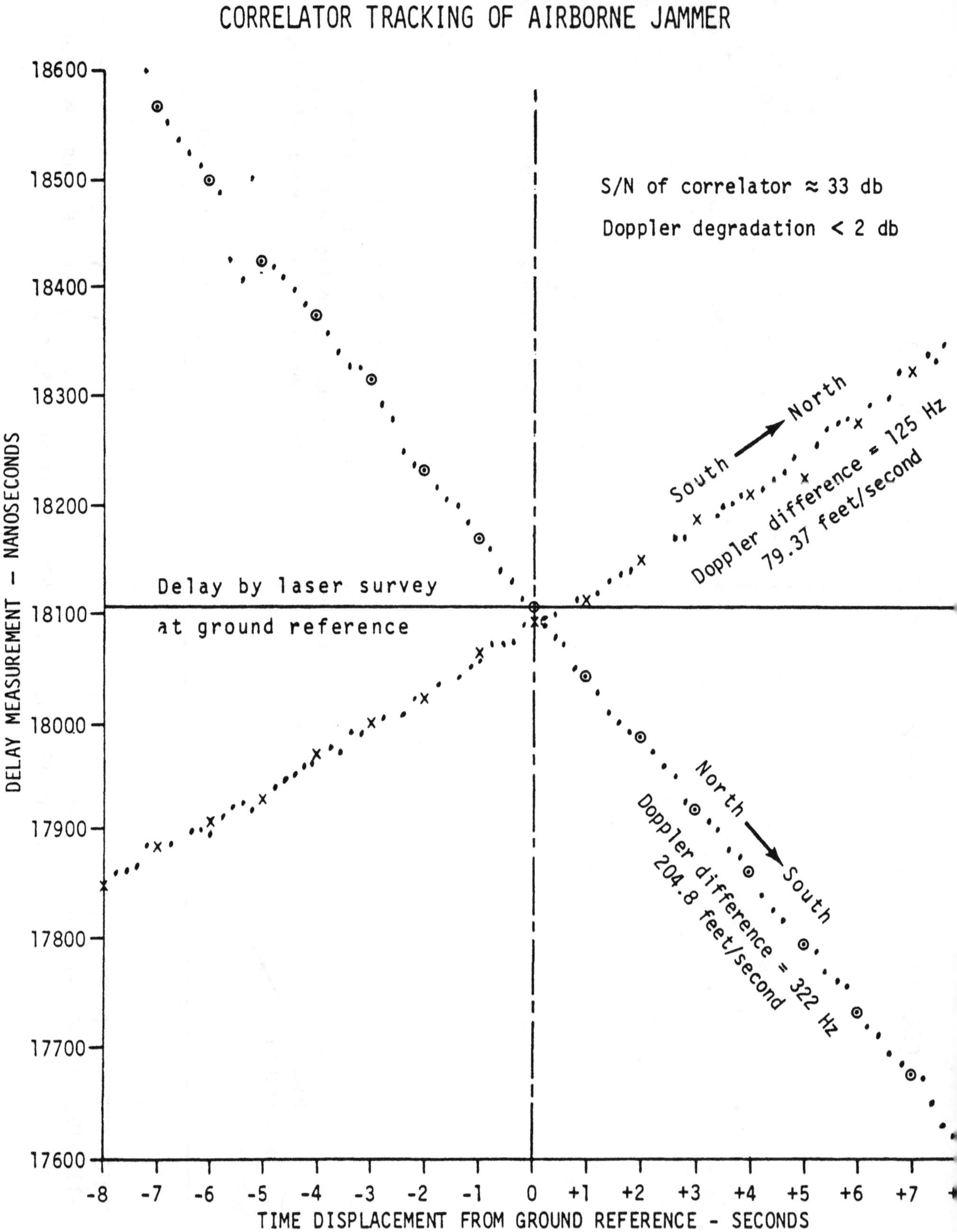
CORRELATOR TRACKING OF AIRBORNE JAMMER
S/N of correlator ≈ 33 db
Doppler degradation < 2 db
Delay by laser survey
at ground reference
South → North
Doppler difference = 125 Hz
79.37 feet/second
North → South
Doppler difference = 322 Hz
204.8 feet/second
DELAY MEASUREMENT - NANOSECONDS
18600
18500
18400
18300
18200
18100
18000
17900
17800
17700
17600
-8 -7 -6 -5 -4 -3 -2 -1 0 +1 +2 +3 +4 +5 +6 +7
TIME DISPLACEMENT FROM GROUND REFERENCE - SECONDS

CHART 17

CHAPTER 9 ECM/ECCM, ECCM EFFICACY, SIMULATION AND ANALYSIS

Summary

Three closely related topics are addressed in this chapter:

- The interaction between ECM and ECCM,
- The need for overall ECCM measures of effectiveness and some possible measures,
- Radar ECM/ECCM simulation and analysis techniques.

A. ECM/ECCM Interaction

The previous chapters have dealt primarily with various ECCM techniques without specific regard for their applicability either against specific ECMs or usefulness in specific radar types; e.g., search, track, etc. Table VI in Paper 1.1 (see page 16 of this volume) presents a matrix of various ECMs with corresponding appropriate ECCMs without regard to radar type.

Table I from Paper 2.C.1 (see page 122 of this volume) is an ECM/ECCM matrix for search/acquisition radars.

Similarly, Table II from Paper 2.C.1 (see page 123 of this volume) pertains to ECM/ECCM for tracking radars.

Synopsis of Reprint Papers in Chapter 9

B. ECCM Efficacy Assessment

9.1 "Burnthrough Range, an Overrated Parameter"— Johnston

Burnthrough range is a widely used but improper measure of effectiveness of ECM. This paper indicates reasons for its inappropriateness and suggests more appropriate measures such as change in kill probability, number of targets destroyed, etc. The jammer noise quality of Turner, *et. al.* (Paper 7.4) is one contributor to errors in burnthrough range calculation.

9.2 "ECCM Improvement Factors (EIF)"— Johnston

The term ECCM Improvement Factor was conceived as a measure of the performance (but *not* effectiveness) of an ECCM. This paper was first presented at the Joint Electronic Warfare Planning Conference, Air Force Academy, Colorado Springs, Co., February 1973. The term ECCM Improvement Factor has now been adopted by the IEEE in their Standard Radar Terms and Definitions (see [1] in Chapter 1). This term should be used instead of other terms such as processing gain, cancellation ratio, etc.

C. RADAR ECM/ECCM Simulation and Analysis Techniques

9.3 "Simulation to Obtain a Systems Measure of an Air Duel Environment"— Pritsker, VanBuskirk, Wetherbee

This combined analog-digital simulation of a ground controlled intercept (GCI) radar and interceptor aircraft includes the radar operator in the loop. The authors correctly indicate that the probability of survival of the bomber is the systems measure of the effectiveness of the countermeasures (noise denial jammer). This supports Paper 9.1 (above).

There are many factors which should be included in a proper simulation of this type: aircraft radar cross section patterns as a function of azimuth and elevation aspect angles, jammer noise quality (see Paper 7.4), ECCMs in the radar, jammer antenna pattern and radar operator performance. Some of these are discussed in this paper. Their effects on radar detection are shown in Paper 9.7 in this Chapter.

9.4 "Simulator Checks out Radio-radar AJ Operation"— Lambert

The behavior of a radio/radar ECCM can be very difficult to determine analytically. This paper describes a very powerful but simple method for simulation of a radar on a conventional general purpose analog computer using a frequency scaling technique. Various combinations of conventional analog computer apparatus (integrators, function generators, differentiators, operational amplifiers, etc.) are used to represent radar elements such as bandpass filters, AFC loops, discriminators, and so on. The principal drawback is that when frequency is scaled down, time is scaled up. This technique does not lend itself to study of the complete weapon system.

9.5 "Radar/ECM Computer Modeling"— Cikalo and Greenbaum

Digital simulation of an inverse gain deceptive jammer against a conical scan tracking radar is described. The methodology utilizes a readily available general purpose computer program in which the digital computer is made to emulate various functions of an analog computer. The illustrative example clearly demonstrates the effectiveness of inverse gain jammers against simple conical scan radars and the need for ECCMs.

9.6 "Recent Innovations in Radar Simulation"— Hancock, Krolak, Wilson, Cleary

This paper has the same objectives as the previous paper; however, the methodology is different. Here individual elements of the radar are directly programmed on a general purpose digital computer. The necessary blocks are then assembled as modular software to represent the desired radar. A wide variety of radar features can be represented: pulse compression with range sidelobes, matched filters, Fourier Transforms, envelope detectors, targets, clutter, antenna patterns, ECCMs, etc. Reference to simulation work of others is contained in the bibliography at the end of this chapter.

9.7 "Application of Computer Simulation Techniques to the Evaluation of Radar System Performance in ECM/ECCM Environments"— Gutsche, Bogusch, Guigliano

Computer drawn plots of simulated radar PPI displays are presented. These plots were generated by a radar simulation computer program. The ECM example was for a UHF chirp pulse air search shipboard radar employing CFAR and monopulse. The threat consisted of two flights of aircraft, five aircraft in each flight. The threats approached the ship from different directions. A standoff jammer was also included with self-screening jamming. Various effects are shown in the PPI plots: aircraft radar cross section variation with aspect angle, different jammer power levels, range resolution due to chirp, standoff versus self screening jamming, influence of signal-to-noise ratio on monopulse and standard PPI displays.

This paper illustrates the utility of a computer simulation to evaluate the effects of ECM, ECCM fixes, signal processing techniques and various scenarios on radar performance. This particular example apparently did not include aircraft engagement by a weapon as was done in Paper 9.3 This paper was presented at the DDRE Radar ECCM Symposium in 1976.

Bibliography Topics for Chapter 9

ECM Detection Range –J/S Computation
ECM/ECCM Efficacy Analysis
ECM/ECCM Simulation

BIBLIOGRAPHY FOR CHAPTER 9

ECM DETECTION RANGE – J/S COMPUTATION

1. Books and General Bibliography

Barton, D.K., *The Radar Equation: Radars, Vol. 2,* Artech House, Dedham, MA, 1974.

Barton, D.K., "International Cumulative Index on Radar Systems," *IEEE Pub. Cat. no. JH 4675-5,* 1978 (extensive bibliography on the radar equation pp. 54-5.)

Blake, L.B., *Radar Range Performance,* Lexington Books, Lexington, MA, 1978.

2. Journal Articles

Norton, K.A., and A.C. Omberg, "The Maximum Range of a Radar Set," *Proc. IRE Waves and Electrons.,* vol. 35, no. 1, Jan 1947, pp. 4-24 (reprinted in Barton, *The Radar Equation: Radars, Vol. 2,* 1974, Artech House, Dedham, MA).

Tetlebaum, S.I., "Basic Equations in Radiolocation," *Radiotekhnika,* vol. 2, no. 2, 1947, pp. 24-34 (in Russian, trans. abstract SA 47, p. 342).

Tiberio, U., "The Reduced Range in a Radar Subjected to an External Noise Generator," *Proc. IRE,* vol. 42, no. 12, Dec 1954, pp. 1791-1798.

Siebert, W.M., "Some Applications of Detection Theory to Radar," *IRE Nat. Conv. Rec.,* vol. 6, pt. 4, 1958, pp. 5-14.

Wall, R.A., "Radar Jamming Chart," *Electronics Magazine,* vol. 32, no. 49, Dec 4, 1959, pp. 116-118.

"ECM Data File," *Space/Aeronautics,* vol. 33, Apr 1960, pp. 143-5 (contains nomographs for J/S computation).

Dax, P.R., "The Exploration of a Given Volume with a High Accuracy Radar: Considerations of Power, Data Rate, and Accuracy," *IRE Nat. Conv.Rec.,* vol. pt. 5, 1961, pp. 250-257.

Lipnick, R., "ECM Nomograph," *Electronics Industries Magazine,* vol. 21, no. 9, Sept 1962, pp. 115-116.

Kholl, V., "Generalized Equation of a Radar for Comparing Characteristics of Various Radars," *Foreign Radio Electron.,* no. 4, 1963 (in Russian).

Boothe, R.R., "A Digital Computer Program for Determining the Performance of an Acquisition Radar through Application of Radar Detection Probability Theory," *US Army Msle. Cmd. Rpt. RD-TR-64-2,* Dec 1964, DDC AD 462473.

Blake, L.V., "A Guide to Basic Pulse Radar Maximum Range Equations, Pt 2," Naval Res Lab Rpt 7010, AD 703211, Dec 31, 1969.

Barton, D.K., "Radar Equations for Jamming and Clutter," *IEEE Trans.,* vol. AES-3, no. 6 (supp), Nov 1967, pp. 340-355 (reprinted in Barton, *The Radar Equation: Radars, Vol. 2,* Artech House, 1974.

Morchin, W.C., "Radar Range in a Jamming Signal Environment," *Microwave J.,* vol. 11, no. 6, June 1968, pp. 45-36 (reprinted in Barton, *op. cit).*

Friedman, H.J., "Jamming Susceptibility," *IEEE Trans.,* vol. AES-4, no. 4, July 1968, pp. 515-528 (1969, p. 561) (1969, p. 946).

Bijovet, J.A., "Standard Methods for Predicting and Specifying Performance of Air Surveillance Radar Systems," *SHAPE Tech. Center, Tech Rpt. TR-50-U,* Apr 1969.

Miamidian, L.R., "Comments on 'Jamming Susceptibility'," *IEEE Trans.,* vol. AES-5, no. 3, May 1969, pp. 561-562c.

Barton, D.K., "Simple Procedures for Radar Detection Calculations," *IEEE Trans.,* vol. AES-5, no. 5, Sept 1969, pp. 837-846.

Greenbaum, M., "Further Comments on 'Jamming Susceptibility'," *IEEE Trans.,* vol. AES-5, no. 6, Nov 1969, p. 996c.

Matsuo, M., and T. Kuroiwa, "Digital-detection Probability of Fluctuating Pulsed Radar Signals," *IEEE Trans.,* vol. IT-15, no. 6, Nov 1969, pp. 721-722c.

Young, C.W., "Jammer Equipment Performance Figure of Merit – Part 1," *Frequency Technology Magazine,* vol. 8, no. 5, May 1970, pp. 13-19 (pt. 2, June 1970, pp. 18-19).

Chrzanowski, E.J., "Tech Note: Detection Capability of an ECM Receiver versus Direction (Detection) Capability of a Radar," *Microwave J.,* vol. 14, no. 5, May 1971, pp. 58-60.

Brookner, R., "Cumulative Probability of Target Detection Relationships for Pulse Surveillance Radars," *IE* (India), *JE, and TE,* vol. 51, May 1971, pp. 125-134.

Davies, L.E., "Radar Burnthrough Performance Using Software Techniques," *Stanford Res. Inst. Tech. Note SED-TN-504,* Sept 1973.

Mehlis, J., "Search Radar Cumulative Probability of Detection," *IEEE Trans.,* vol. AES-9, no. 6, Nov 1973, pp. 948-950c (reprinted in Barton, *op. cit.*).

Mehlis, J.G., "Radar Range Performance Predictions," *Conf. Rec. – 7th Annual Asilomar Conf. on Circuits, Systems, and Components,* Nov 27-29, 1973, pp. 332-338.

Kanter, I., "A Generalization of the Detection Theory of Swerling," *IEEE EASCON Rec.,* 1974, pp. 198-205.

ECCM EFFECTIVENESS ANALYSIS

Bloch, F., and M. Hammermesh, "Anaysis of the Effectiveness of Countermeasures," *Radio Res. Lab. Harvard Univ. Rpt. 411-140,* Dec 20, 1944, DDC ATI 15369.

Nilsson, N.J., "An Application of the Theory of Games to Radar Reception Problems," *IRE Nat. Conv. Rec.,* vol. 7, pt. 4, 1959, pp. 130-40.

Oblinger, J.T., and R.P. Fisher, "An ECM System Analysis," *IRE Mil. Electron. Conf. Rec.,* 1961, pp. 303-10.

Zetterberg, L.H., "Signal Detection under Noise Interference in a Game Situation," *IEEE Trans.,* vol. IT-8, no. 5, Sept 1962, pp. 547-52.

Davies, L.E., "An Approach to Analysis of Vulnerability of a Friendly Weapon System to Enemy ECM," *Systems Evaluation Dept., Stanford Res. Inst. Tech. Note SED 19,* 24 April 1964.

Losec, G.L., "General Considerations and Check List for Use in an Analysis of a Defensive Weapon System's Capability," *Systems Evaluation Dept., Stanford Res. Inst. Tech. Note SED TN 52,* Jan 1965.

Strother, C., Jr., "Considerations of Comparative ECM Relationships in Weapons Systems Synthesis," *IEEE Trans.,* vol. AES-3, Sept 1967, pp. 829-30.

Lund, F.H., "Operational Effectiveness of Electronic Countermeasures," *Southwest IEEE Rec. of Tech. Pgm,* Dallas, 1967, pp. 1-6-1 thru 8.

Mayers, R.S., "A Quantitive Approach to ECM Effectiveness," *Navigator Magazine,* vol. XVI, no. 2, Winter 1968, pp. 27-9.

Johnston, S.L., "Military Radar Weapon System Analysis," presented Sept 18, 1970, as lecture in the short course, Principles of Radar, Georgia Institute of Technology, Atlanta, Georgia. Published as lecture notes.

Ablett, C.B., "On Effectiveness Methodologies for Electronic Warfare," *29th Military Operations Res. Symp.,* 29 June, 1972.

Lund, F.H., "Simulation's Role in OCM Vulnerability Evaluation," *IEEE NAECON–73 Rec.,* pp. 60-6.

Johnston, S.L., "ECCM Improvement Factors (EIF)," *Electronic Warfare Magazine,* vol. 6, no. 3, May/June 1974, pp. 41ff (reprinted in this volume).

Johnston, S.L., "Burnthrough Range, An Overrated Parameter," *Electronic Warfare Magazine,* vol. 6, no. 6, Nov/Dec 1974, pp. 49ff (reprinted in this volume). Comments and author's reply: *Electronic Warfare Magazine,* vol. 7, no. 1, Jan/Feb 1975, p. 48.

Mitchell, O.B., "ECCM: Effective Deployment and Employment," *Electronic Warfare Magazine,* vol. 6, no. 6, Nov/Dec 1974, pp. 36-8.

Zophy, F.G., "Methodology for the Evaluation of Air Defense System Effectiveness," Master's thesis, US Naval Post Grad School, Mar 1975, AD B004484.

Frank, J., and P.S. Tong, "ECCM Measures of Effectiveness," *Electronic, Electro-optical and Infrared Countermeasures Magazine,* vol. 3, no. 1, Jan 1977, pp. 37ff. (Editor's comment: This paper should be compared with the papers reprinted in this chapter.)

Steeg, G.F., "Principles of Airborne Self Defense ECM Systems; Pt. 1, External System Design Considerations; Pt. 2, Internal Systems Design Considerations," *J. of Electron. Def.,* vol. 1, no. 2, Sept/Oct, 1978, pp. 53ff.

ECM/ECCM SIMULATION

1. Books

Mitchel, R., *Radar Signal Simulation,* Artech House, Dedham, MA, 1976.

2. Journal Articles

Sternlicht, L., "A Radar Electronic Countermeasures Simulator," *IRE Nat. Conv. Rec.,* 1958, vol. 6, pt. 8, pp. 94-105.

Pritsker, A.A.B., R.C. Van Buskirk, and J.K. Wetherbee, "Simulation to Obtain a Systems Measure of an Air Duel Environment," *IRE Trans.,* vol. EC-8, no. 1, Mar 1959, pp. 55-9 (reprinted in this volume).

Lambert, J., and A.J. Heidrich, "Radar Systems Simulation Techniques," *IRENat. Conv. Rec.,* vol. 7, pt. 4, 1959, pp. 190-203.

Hanford, R.W., "Radar Testing for a War Environment," *IRE Nat. Conv. Rec.,* 1959, vol. 7, pt. 5, pp. 208-17 (also contains list of ECCMs for indicated ECMs).

Hanford, R.W., "Radar Testing for a Countermeasures Environment," *IRE NAECON Rec.,* May 1959 (also contains list of ECCMs for indicated ECMs).

Lambert, J., "Simulator Checks Out Radio-radar AJ Operation," *Space/Aeronautics,* vol. 33, no. 4, Apr 1960, pp. 157-66 (reprinted in this volume).

Leskinin, J.I., "Simulation of Radar Countermeasures," *Electronics Magazine,* vol. 33, no. 31, Jul 1960, pp. 98-9.

Hitt, W.D., and H.W. Ray, "A Laboratory Evaluation of the Effects of Electronic Countermeasures on System Performance," *Human Factors Magazine,* Aug 1960, pp. 128-35.

Norton, R.L., "Development of a Dynamic Target and Countermeasures Simulator," *IRE Trans.,* vol. MIL-4, no. 4, Oct 1960, pp. 587-90.

Fischer, L.G., and G. Frenkel, "A Simulator for the Evaluation of Electromagnetic Systems," *IRE Int. Conv. Rec.,* vol. 9, pt. 2, 1961, pp. 135-42.

Fischer, L.G., and G. Frenkel, "Search Radar Analog Simulator Reproduces Jamming Scintillation," *Electronics Magazine,* Aug 1961, pp. 58-60.

Gerhart, K.G., and L. Moses, "Radio Frequency Interference (Jamming) Prediction by the Use of Analog Computers and GEESE Techniques," *Proc. 8th Tri-Service Conf. on Electromag. Compat.,* Oct 1962, pp. 388-404.

Sumner, G.C., and G.H. Robertson, "Simulation Applied to Electronic Warfare Problem," *IEEE NAECON-63 Rec.,* pp. 114-9.

Cook, D.H., "ECM/ECCM Simulation Program," *IEEE Aerospace and Electronics Systems Conf. Rec.,* Sept 9-11, 1968, pp. 181-6.

Mayers, R.H., "The EDE Simulator," *Navigator Magazine,* vol. XVI, no. 2, Winter 1968, pp. 22-5.

Mukutmoni, M., and A. Rama Rao, "Simulation of a Radar on a Digital Computer," *Electro-Technol* (India), vol. 18, no. 1, Mar 1974, pp. 17-25.

Swisher, G.M., and C.W. Sigley, "Design of a Synchronization System to Simulate Electromagnetic Environments in Early Warning/Ground Control Intercept Radars," *Proc. 17th Midwest Symp. on Circuits and Systems,* Lawrence, Kansas, 16-17 Sept 1974, pp. 174-82.

Digiovanni, J.J., and G.D. Jordan, "EW Simulation — Advantages and Recent Advances," *Microwave J.,* vol. 17, no. 9, Sept 1974, pp. 45-8.

Hancock, R.J., P.D. Krolak, L.K. Wilson, and J.C. Cleary, "Recent Innovations in Radar Simulation," *IEEE Reg. III SOUTHEASTCON-75 Rec.,* pp. 6D-4-1 thru 9 (reprinted in this volume).

Cikalo, J., and M. Greenbaum, "Radar/ECM Computer Modeling," *IEEE NAECON-75 Rec.,* pp. 431-437 (reprinted in this volume).

Gutsche, S.L., R.L. Bogusch, and F.W. Guigliano, "Application of Computer Simulation Techniques to the Evaluation of Radar Systems Performance in ECM/ECCM Environments," *DDRE ECCM Conf. 1976* (reprinted in this volume).

Gasparini, O., "A Simulation Procedure for the Optimization of Jammer Efficiency Against Search Radars," *Proc. Mil. Electron. Def. EXPO-76,* Oct 1976, Wiesbaden, Germany, pp. 191-203.

Digiovanni, J.J., "EW Testing — The Gordian Knot; Pt. I," *Electronic Warfare Magazine,* vol. 9, no. 2, Mar/Apr, 1977, pp. 91ff.

Digiovanni, J.J., "EW Testing – The Gordian Knot; Pt. II," *Electronic Warfare Magazine,* vol. 9, no. 4, Jul/Aug, 1977, pp. 79ff.

Adamy, D.L., "EW Simulation: Preparation of a Dynamic Scenario," *Electronic Warfare and Defense Electronics Magazine,* vol. 9, no. 5, Sept/Oct, 1977, pp. 117ff.

Tickle, G., "Radar System Simulation and Performance Prediction," *IEE Conf. Pub. 155,* Radar 77, London, Oct 1977, pp. 451-553.

Cram, L.A., M.R. Gidday, and K.O. Rossiter, "Basic Radar Evaluation by Computer Simulation," *IEE Conf. Pub. 155,* Radar 77, London, Oct 1977, pp. 455-457.

Baird, P.D.A., "RF Simulation for Coordination of ECCM," *IEE Conf. Pub. no. 155,* Radar '77, Oct 1977, London, pp. 460-2.

Utsi, B., "Real Time Radar Signal Simulation," *IEE Conf. Pub. 155,* Radar 77, London, Oct 1977, pp. 478-81.

Anon., "Modern EW Simulation Techniques," pp. 390ff in *The International Countermeasures Handbook,* H.F. Eustace, ed., 4th ed. 1978-9, 1978.

Burnthrough Range, An Overrated Parameter
S.L. Johnston
Electronic Warfare Magazine, Vol. 6, No. 6, November/December 1974
Pages 49-54

9.1

Frequently used as measure of performance in ECM environments, this article shows why Burnthrough Range is a poor figure of merit.

Stephen L. Johnston
Army Missile Command
Redstone Arsenal, Alabama

Burnthrough range is frequently used as the measure of performance in an ECM scenario, but its value as a figure of merit or performance criterion is, in fact, very poor for five reasons:

- Burnthrough range equation is incomplete
- Parameter values used in the equation are often not properly chosen
- Burnthrough range is not a single number
- Burnthrough range applies to only one possible type of ECM
- In modern weapons systems, burnthrough range may mean absolutely nothing.

Figure 1. Increase in detection range with integration from Steinberg[7].

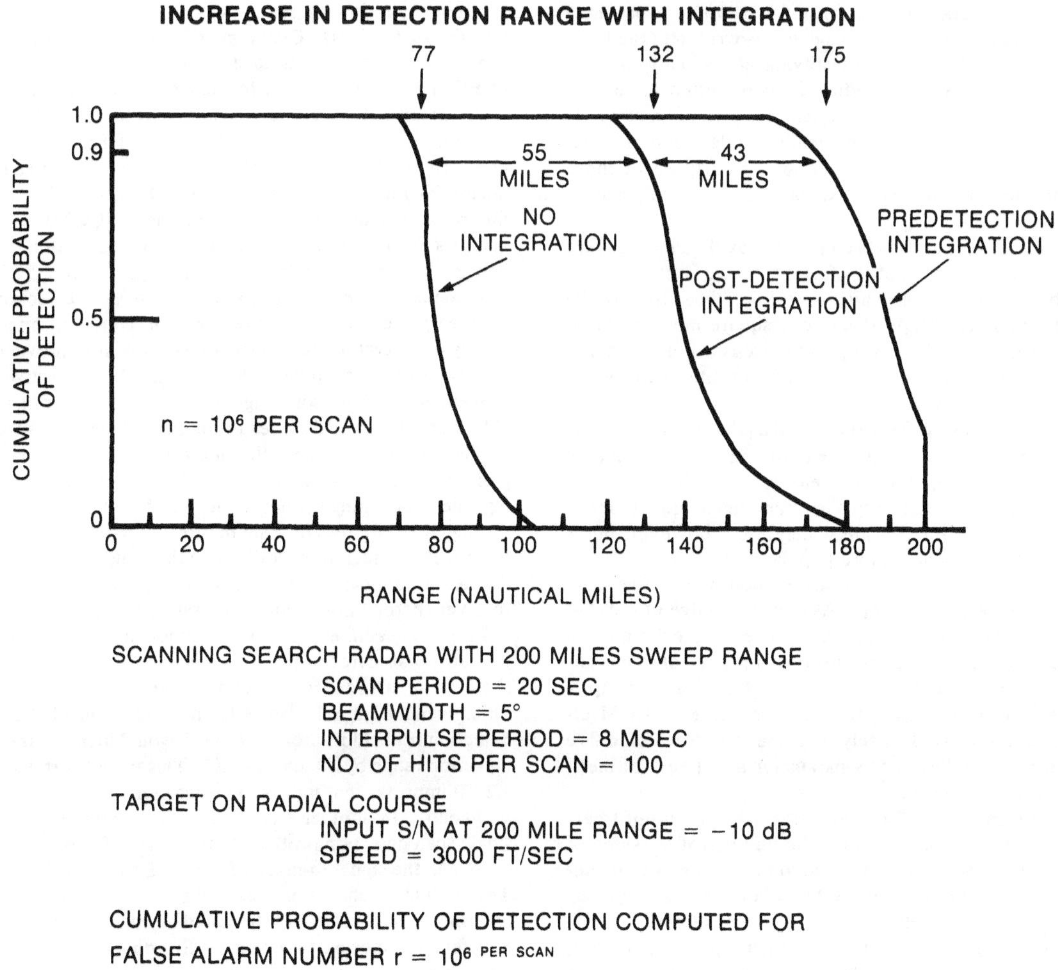

ECCM: SYSTEMS

By dissecting the classical burnthrough range equation and by closely inspecting its various factors, there is a clear indication that a single value for Burnthrough Range is a fallacy. At best, two values should be given, one when everything is most favorable to the defense, and one when everything is most favorable to the offensive force.

Consideration of several radar concepts shows detection in the clear, i.e., burnthrough, is not necessary in certain concepts. In fact use of noise jamming by the offense may be of advantage to the defense.

The role of radar in modern weapon systems is then treated to show that a radar is just one link in the chain and it may be paralleled by other radars and/or sensors.

Finally, a possible criterion is suggested as a substitute for burnthrough range as a radar/ECM performance measure.

Truth or Fiction? The occurrence of incorrect and misleading concepts or false beliefs is widely prevalent in practically all areas of human endeavor and interest including religion, politics, history and even science and engineering. You can recall many of these yourself: "The earth is flat." "Everything that goes up must come down." "Jet propulsion will not work in outer space, i.e., a vacuum." (Dr. Herman Oberth, the noted space pioneer, had his doctoral dissertation rejected because of his belief that jet propulsion would work in a vacuum.) And nearly 2000 years ago it was written in the Holy Bible: "There shall be prophets testifying falsely in my name"[1].

These incorrect and misleading concepts often arise as a result of a genuine but misguided attempt to simplify a principle or belief so that it may be more readily understood and employed by as many people as possible. Nearly all of these follow the rule that there is an exception to every rule except this one.

For many years radar has been plagued by the false concept of a one-square-meter target. Unfortunately we no longer have cannon balls in our military inventory. Even those that we did have were considerably less than one square meter in radar cross section. This false concept was discussed in depth at a previous NORAD EW Conference[2]. Electronic warfare has had a false concept also.

Burnthrough range is frequently used as the measure of performance of a radar in an ECM environment. Self-screening range is frequently used as the measure of performance of a jammer against a radar. The two are numerically equal for onboard jamming. Both terms are widely used; however, burnthrough range is probably more popular.

Burnthrough range is readily understood and in the early days was a fairly good concept. As our knowledge of radar—especially signal detection and signal processing techniques—has increased, this concept has become a prime candidate for discard. Two recent papers in the AOC Magazine[3,4] prompt a reappraisal of burnthrough range as a measure of ECM effectiveness. First, we'll closely examine the aforestated five reasons for the decline of this parameter as a figure of merit in an ECM environment.

Burnthrough Range Equation. Every ECM author treats the burnthrough range equation. The number of versions of this equation is approximately equal to the product of the factorials of the weight of all the ECM authors. Not all versions include the same variables. Most of these treatments do not properly define all of the variables in their equation. None of the versions reviewed included the effects of ECCM, although one version obliquely implied that this should be done.

The self-screening (burnthrough) range equation in the General Electric Company electronic countermeasures calculator GEN-188[5] is

$$R = \sqrt{\frac{1}{4\pi} \frac{P_T}{P_j} \frac{G_R}{G_j} \frac{B_j}{B_R} \frac{J}{S} \sigma}$$

where

P_T = radar transmitter peak power (watts)
P_j = jammer average power output (watts)
G_R = gain of radar antenna
G_j = gain of jammer antenna
B_j/B_R = ratio of jammer transmitted bandwidth to receiver bandwidth
J/S = ratio of jammer power to signal power at the radar antenna required to cause jamming
σ = effective target radar cross section area (square meters).

This equation is incomplete as it stands, although a careful definition of the above terms could incorporate the deficiencies given in the following sections.

Effects of ECCM. Every radar incorporates some type or types of ECCM. The purpose of ECCM is to offset the effects of ECM, i.e., improve performance of the radar in an ECM environment. A recent paper at the 18th Annual Joint EW Planning Conference[6] listed 150 types of ECCM's and proposed the parameter ECCM Improvement Factor (EIF) as a means for representing the effects of ECCM's. The effect of failure to include integration (a possible ECCM)[6] in a radar analysis and the effects of detection probability for a radar in a benign environment are illustrated in Figure 1 from Steinberg[7].

Three cases are shown: no integration, post detection (noncoherent) integration, and predetection (coherent) integration. The two integration cases provide equivalent improvements in S/N ratio of approximately 9 and 14.3 dB, respectively. Improvements in detection range of 132/77 or a factor of 1.72 and 175/77 or a factor of 2.28 result for cases 2 and 3, respectively. If this were for jamming, the improvements would have been greater: $(1.72)^2 = 2.93$ and $(2.28)^2 = 5.22$, respectively, since benign detection range follows the fourth root law while burnthrough range follows the square root law. This assumes that the two types of integration would have the same effectiveness with the denial jamming under consideration as with conventional receiver thermal noise. Such an assumption may not be valid.

Target Detection Process. Marcum[8] in his classical paper treated radar detection as a statistical process. Accordingly there is an associated detection probability and false alarm time. From Figure 1 also note that reduction of the desired detection probability from 0.9 to 0.5 would increase the benign detection range by a factor of 1.07. This would correspond to a 1.2 dB improvement in S/N. This was for cumulative detection probability. In another study[11], it was shown that a change of single scan detection probability from 0.1 to 0.9 resulted from a change in the signal-to-noise of 1 to 4 dB for various situations. The fluctuation characteristics of the target were also shown by Swerling to be involved[9,10]. It should be noted that it is incorrect to refer to a specific target as a "Swerling Type I" since the fluctuation characteristics are a function of viewing aspect angles. Radar detection theory further involves the statistical characteristics of the competing signal (internal noise in the benign environment and the jammer waveform in the ECM case). Even noise modulation may not be Gaussian. Special modulations such as frequency modulated noise require special treatments. This will be discussed later under the J/S parameter in the burnthrough equation.

Losses. Account must be made for system losses, e.g., waveguide loss, antenna efficiency, antenna scanning losses, propagation attenuation (atmospheric attenuation), and maintenance losses.

Accuracy of Parameters. Accuracy of the parameters in the burnthrough equation above must be considered. The accuracy of the computed value of burnthrough range (even if the above omissions are supplied) is no better than the accuracy of the values of the parameters used. A 3-dB error in radar or jammer power output (a factor of 2) will result in a 41% error in burnthrough range. Despite excellent maintenance it is impossible to completely eliminate variations in radar power output. It should also be recognized that values of parameters for a proposed radar should be discounted. History has noted that many radars have not performed as well as predicted by the design paper.

If the values of the eight parameters of the burnthrough range equation were each known to be $\pm$ 1 dB, the overall accuracy of burnthrough range would be 0 to 8 dB corresponding to a maximum error of a factor of 2.5 in burnthrough range.

Variability Of Values Of Parameters In The Burnthrough Range Equation. Even when accuracy of the values of the equation parameters is excluded, there is no single value of any of these parameters for a given target-radar case, hence there is no single value of burnthrough range. Consider the following parameters:

1) Transmitter Power—Some radars are programmed to vary transmitted power as a function of scanning elevation angle. It is also possible to vary transmitted power for different PRF's.

2) Jammer Power—It is possible to vary jammer power output with time, radar PRF, etc.

3) Bandwidth Ratio—Receiver saturation/overloading due to strong jammer signals can affect this parameter.

4) Radar Antenna Gain—Most search (acquisition) radars use a radar antenna whose gain varies with target elevation. Frequently, this is the well-known cosecant-squared shape or some variation thereof. In this pattern the antenna gain varies as the square of cosecant of the target elevation angle. Generally it is desired to provide maximum detection range (and hence antenna gain) on the most distant targets. Such targets are usually at zero elevation angle. As the target approaches (at a constant altitude), the elevation angle increases and hence antenna gain decreases. A variation of antenna gain with elevation angle of 30 dB may be experienced. The effect of ground reflections on the free space antenna pattern may also result in multiple lobes in the actual antenna elevation pattern.

5) Jammer Antenna Gain—It is very possible that the jammer does not use an isotropic antenna. If a jamming aircraft maneuvers, it is very likely that the fuselage will obscure the jammer antenna, thus lowering the antenna gain in the direction of the radar.

6) Jammer-to-Received Signal Ratio—The J/S ratio required to cause jamming may vary due to a number of causes. For example, a radar may have a number of ECCM's that can be selected by the radar operator. The J/S ratio required to cause jamming should be different from different ECCM's. Various workers assume various values of J/S required. In one example, General Electric[5] showed that a value of -5.5 dB was required. Jordan[4] however stated that 3 dB is generally adequate. A study several years ago presented some very interesting data on probability of detection as a function of J/S ratio for a number of different types of jamming waveform. These were experimentally derived from a scaled frequency (GEESE)[12] simulation of an acquisition radar. This is the only such data known to the author.

7) Target Radar Cross Section—Target radar cross section is a three-dimensional function of aspect viewing angle in pitch, yaw, and roll. Radar cross section is generally smallest at or near nose-on and is usually very large when the aspect angle is perpendicular to the wing (leading or trailing edge) or to the fuselage. Radar cross section is also a function of frequency. The use of a single value of radar cross section is a very serious error as was discussed in Reference 2; a variation of 30 dB in RCS over an engagement history may be expected.

In a benign environment radar simulation previously cited[11], data plots were given for two types of engagements: head-on approach and an offset flyby. Graphs of RCS time history, aspect, and elevation angle time history and S/N versus slant range were given. An RCS variation of 30 dB occurred. In one case over a change of range of a factor of 4 (corresponding to 24 dB since $40 \log_{10} 4 = 24$ dB for the benign environment case), the overall S/N ratio changed by 50 dB due to the composite effect of the changes in antenna gain, slant range, and RCS. Those data very strongly show the constant antenna gain and RCS treatments of Flaherty[3] and Jordan[4] are not proper as they now stand.

It can be shown that burnthrough range doubles or halves for a change of 6 dB (cf. 12 dB for the benign, i.e., nonjamming detection range case) in the value of any parameter or any combination of changes of parameters. This change of value may be due to either accuracy or to "natural" variation as discussed above. Unfortunately it is not meaningful to attempt to identify the two extremes, i.e., worst and best (maximum and minimum) values of detection or burnthrough range. It is extremely unlikely that all maximum or all minimum values of parameters will occur together. Similarly one should not attempt to establish an average, mean, or even root mean square (RMS) value of detection range.

Proper Method For Burnthrough Range Determination. The only meaningful method for determining burnthrough range requires:

- A complete burnthrough range equation
- Establishment of the accuracy of values of the parameters
- Establishment of functions for the equation parameters, e.g., use of RCS patterns as functions of aspect angles and variations of aspect angles with target-radar geometry.
- Establishment of an engagement mission profile
- Use of all the above in a proper computer simulation such as one of the better ones described in Reference 12, which meet the requirements discussed in Reference 13.

ECCM: SYSTEMS

This simulation should compute detection probability for each and every radar scan of the target by a scanning (search) radar or continuously for a tracking radar. For each scan a binary situation, i.e., burnthrough, yes or no, will result.

If the values of target-radar range are tabulated for those scans on which burnthrough occurs, it will be readily obvious that there is no single value of burnthrough range. It may result that the target will burn through on every scan after some initial burnthrough point on the mission profile (this initial point could be the usually meant burnthrough range). The target might burn through on one or a few scans then not again during that mission or possibly not for many scans.

It is possible to properly determine burnthrough ranges (or more properly burnthrough points); however, these will probably qualify the person who determines them for membership in that elite (but not limited) Greek Fraternal Order Σ φ N—literally translated—sigma phi's nothing.

Burnthrough range even if properly determined means nothing by itself. This will be illustrated later.

Burnthrough Range ECM Limitations. There are a host of ECM's[14,15]. These may be generally classified as denial, deception, and confusion, depending on the type of effect on the radar. Burnthrough applies only to one specific group of denial ECM's—the use of an active jammer whose purpose is to prevent target detection (i.e., screen the target.) It does not apply to chaff, active range detection jammers, confusion jammers, special effect jammers, anti-radiation missiles, target radar cross-section reduction, etc.

In modern electronic warfare, active target detection denial jamming is only a small (but nevertheless important) part of electronic warfare. It should be noted that a noise jammer may result in a number of effects on the radar—it may deny/delay detection, degrade radar range and/or angular accuracy, deceive the operator, and confuse the radar operator.

Burnthrough range is frequently used in or implied as a simple static one-aircraft-on-one-radar situation. In our real would we have a dynamic many aircraft against many radars case.

Clearly, burnthrough range, even if properly defined and determined, has severe limitations.

The Role Of Radar In A Modern Weapon System. The role of radar in a modern weapon system must be carefully considered. A radar is not an end item unto itself. Radars may be used for a host of functions: surveillance, target acquisition, target tracking, missile guidance, aircraft navigation, etc.

A modern weapon system may use several radars, each performing a different function. It may have several different radars for the same function to provide redundancy, reliability, and diversity. It may use a multifunction radar. Combinations of radar, optical detection/tracking of targets, infrared, or laser detection/tracking are possible.

In many of the ECCM's listed in Reference 6, denial or target detection range is negated. Several of these have been implemented in various US Army Air Defense Missile Systems and will be discussed below.

In a homing missile system, the seeker missile requires only precision measurement of target angle (both azimuth and elevation)[17]. Target range is required only to determine when or whether to launch a missile, i.e., is the target within range. Precision range measurement is not required. Use of jamming to deny/delay burnthrough would have doubtful utility against such a system; in fact, such jamming would announce presence of aircraft (or possible enemy intentions) at much greater than normal detection ranges (except at very low altitudes where terrain masking might prevent greater detection ranges) since jamming is a one-way propagation while radar is two-way propagation.

In some Army missile systems, there are several different radars operating in several frequency bands. The attacking aircraft must deny/delay detection to all four radars to prevent detection. Burnthrough for even one scan on any radar is generally sufficient.

Track-on-jam/home-on-jam is another very effective ECCM. Here the jammer itself becomes the source of radiation that the missile seeker homes on. Use of HOJ can actually reduce seeker miss distance (and hence increase missile lethality) since the largest source of homing miss distance—target glint —is eliminated in the HOJ mode.

A missile system may use passive IR homing in conjunction with active radar for detection and coarse location of an attacking aircraft. Attempt to prevent target burnthrough by the radar through use of jamming may be counter-productive. The jammer will serve to announce presence of hostile aircraft or hostile intentions. The radar may obtain coarse target position by strobe-on-jam and/or triangulation from several different netted radars. In one such system there may be number of netted defense missile systems, each of which can report target position.

A defense missile system may use an extremely large warhead such that target position inaccuracies due to ECM against the radar may be countered.

Usually jamming will only delay but not completely deny burnthrough by the radar. This delay may or may not reduce missile first engagement range. Very likely the missile system may retain a self-defense capability. This may be extremely important, especially for a multiple deployment case.

Usually defensive systems are not deployed singularly but are multiply deployed, often with overlapping coverage. ECM against one defensive site may be less or more effective against an adjacent site.

The aircraft may burn through on an adjacent radar before it burns through on the radar it is approaching. It may thus become vulnerable to destruction by the adjacent air defense system.

Frequently defense systems will be deployed in a mixture of types, e.g., AAA and SAM and/or different SAM types. Denial of burnthrough against any one site will probably be meaningless.

From the foregoing it will be appreciated that in modern weapon systems burnthrough may signify nothing. As indicated in Reference 6, at present we have no standard means for comparison of or a measure of the effectiveness of either ECM's or ECCM's. While general standards of performance of ECM and ECCM are not feasible, certainly we should have standard means for comparison and standard measures of ECM/ECCM effectiveness.

A Substitute For Burnthrough Range As An ECM Criterion. By now it should have been made perfectly clear that burnthrough range is a very poor ECM criterion whose use should be prohibited. Condemnation alone would not be sufficient; a substitute must be provided.

ECCM: SYSTEMS

At first glance it might seem that the ECCM improvement factor (EIF)[6] would be a good substitute. Indeed it is attractive but it has several drawbacks:

1) It does not apply to all ECM's and ECCM's

2) A radar that used an ECCM having an EIF of 25 dB would be of no real benfit, if at the required detection point there existed a J/S ratio of 40 dB. A value for an EIF alone could thus be misleading.

Clearly the substitute criterion must be all encompassing. It must include overall behavior of the entire weapon systems-target-engagement interaction. It might seem that such a criterion to be proposed can be used for either ECM alone, ECM and ECCM, offense or defense. It may be either qualitative or quantitative.

Simply stated, the criterion is given as a question: Is the outcome of the engagement affected? E.G., does the use of some specified ECM now permit the attacking aircraft to accomplish his mission and survive (i.e., prevent the defensive system from defending the intended target) or will the use of a specified ECCM permit the defense system to defend the intended target despite the presence of a specified ECM?

Note that this criterion involves a change of performance due to use of ECM or ECM and ECCM. It is based on the premise that without ECM the defensive system could destroy the attacking aircraft. ECM might prevent this. ECCM might negate ECM. This criterion requires answers to these conditional statements.

These questions can be answered properly only by use of a total weapon system-target aircraft-engagement simulation. It is essential that the effects of the radar operator be included. Such a simulation is relatively difficult because it requires an extensive amount of detailed technical characteristics of the radar, some of which may not be readily available. Some of the attendant problems are discussed in Reference 13, 18, and 19. One possible simulation is REDCAP, which was developed under US Air Force Program 691X, Electronic Warfare Technology Advanced Development Program[12].

Conclusions. Burnthrough range has been shown to be a meaningless criterion as a measure of ECM effectiveness because:

1) The burnthrough range equation usually used is incomplete

2) Values of parameters used in the burnthrough range equation are often not properly chosen

3) Burnthrough range is not a single value

4) Burnthrough range applies to only one possible type of ECM

5) In modern weapon systems, burnthrough may mean nothing.

Recommendations. It is recommended that an overall weapon system-attacking aircraft-engagement simulation analysis be developed for all weapon systems, including the radars. This will require establishment of many radar characteristics that are normally not available now, especially EIF. This effort should be expanded to provide standard means for comparison of and measures of ECM/ECCM effectiveness.

Acknowledgments. The author wishes to acknowledge a multitude of discussions of electronic warfare with numerous individuals both within and outside of the Government. These individuals are too numerous to mention. Their views have helped shape the ideas presented in this paper. ■

REFERENCES

1. St. Matthew 24:24, *The Holy Bible*.
2. Johnston, Stephen L., "Target Model Pitfalls (Illness, Diagnosis, and Prescription) (U), Presented at the 11th Annual NORAD Electro-magnetic Warfare Conference at US Naval Post Graduate School, Monterey, California, 27-29 October 1969. Published in Proceedings of the 11th Annual NORAD Electromagnetic Warfare Conference, pp. 133–147 (Confidential).
3. Flaherty, James W., "The ECM Chessboard," *Electronic Warfare Magazine*, Vol. 5, No. 2, April 1973, pp. 37–39.
4. Jordan, Gary Blake, "Electronic Countermeasures and Computer Modeling: Part I," *Electronic Warfare Magazine*, August 1973, Vol. 5, No. 4, pp. 67–75.
5. Electronic Countermeasures Calculator, GEN 188, General Electric Company Light Military Electronics Department, Utica, New York.
6. Johnston, Stephen L., "ECCM Improvement Factor (EIF)," Presented at the 18th Annual Joint Electronic Warfare Planning Conference, held at the US Air Force Academy, Colorado Springs, Colorado, 27–29 March 1973 (Uncl). Published in conference proceedings. Also EW Magazine, May/June 1974, Vol 6, No. 3, pp. 41–46.
7. Steinberg, B. D., "Coherent Integration of Doppler Echoes in Pulse Radar," General Atronics Report 118-112-1, April 1957, Contract No. NR-2132(00), AD 135520, also published in: Raymond S. Berkowetz (ed.), *Modern Radar Analysis and System Design*, John Wiley and Sons, Inc., New York, 1965.
8. Marcum, J. I., "A Statistical Theory of Target Detection by Pulsed Radar," RAND Research Memo RM-754, 1 December 1947, AD 101287 (also in IRE Transactions on Information Theory Special Monograph Issue IT-6, No. 2, April 1960, pp. 59–144) (Unclassified).
9. Swerling, P., "Probability of Detection for Fluctuating Targets," RAND Research Memo RM-1217, 17 March 1954, AD 080638 (also in IRE Transactions on Information Theory Special Monograph Issue IT-6, No. 2, April 1960, pp. 269–308) (Unclassified).
10. Swerling, p., "Probability of Detection for Some Additional Fluctuating Target Cases," Aerospace Corporation Report TDR-669 (9990)-14, March 1966, AD-488613 (Unclassified).
11. Johnston, Stephen L., "Forward Area Alerting Radar Simulation and Analysis(U)," US Army Missile Command, Redstone Arsenal, Alabama, Report RE-TR-70-8, 18 June 1970, AD 511079 (SECRET).
12. Johnston, Stephen L., "A Survey of Radar/ECM Simulation (U)", US Army Missile Command, Redstone Arsenal, Alabama, Report RE-TR-69-18, August 1969, AD 505445 (Confidential).
13. Johnston, Stephen L., "Military Radar-Weapon System Analysis," presented 18 September 1970 as lectures in the short course, *Principles of Radar*, Georgia Institute of Technology, Atlanta, Georgia, published as lecture notes.
14. Fitts, MAJ Richard E., "Fundamentals of Electronic Warfare," 1 March 1972, United States Air Force Academy (Unclassified).
15. *Electronic Warfare Design Handbook*, Vol. II, Electronic Countermeasures(U), Air Force Avionics Laboratory, February 1972, AD 519883L (SECRET).
16. *Electronic Warfare Design Handbook*, Vol. IV, Electronic Countermeasures(U), Air Force Avionics Laboratory, August 1972, AD 522328L (SECRET).
17. Locke, Arthur S., *Principles of Guided Missile Design*, D. Van Nostrand Company, Inc., Princeton, New Jersey, 1955.
18. Johnston, Stephen L., "Considerations on Electronic Warfare Simulation, System Characteristics and Target Models(U)," presented at the Electronic Warfare Simulation Symposium of Working Panel Q-8, TTCP, held at Rome Air Development Center, Griffiss Air Force Base, 19–20 November 1969. Published in EW Simulation Symposium Conference Proceedings, Rome Air Development Center Report 69S 1669, pp. 1–22 (Confidential).
20. Johnston, Stephen L., "Characteristics Required for Radar Simulation/Analysis(U)," presented at the 15th Annual Joint EW Planning Conference, held at Eglin Air Force Base, Florida, —4 December 1969. Published in 15th Annual Joint EW Planning Conference Proceedings, USAF AFSC EW Division Report ASD TR 70-3, August 1970, pp. 549–560 (Confidential).

ECCM Improvement Factors (EIF)
S.L. Johnston
Electronic Warfare Magazine, Vol. 6, No. 3, May/June 1974
Pages 41-45

9.2

Stephen L. Johnston
US Army Missile Command
Redstone Arsenal
Alabama 35809

"In the beginning God created heaven and the earth. Man was placed in the garden of Eden to enjoy it. . . . and it came to pass that evil entered man. The Lord observed this and decided to take corrective action."[1] There is a parallel in the history of radar: in the middle man created radar. It was placed in warfare to provide a remote, long distance, all-weather eye. It came to pass that ECM entered the radar picture. ECM was observed and man decided to take corrective action. ECCM was thus born.

During the early days of radar, observation of radar scopes was almost a pleasure as pictured in Figure 1. The increase of radar sensitivity to provide an increase of detection range had unwanted side effects. "Natural" ECM-clouds, ground returns (trees, mountains, etc.) and unintentional man made ECM-returns from buildings, water tanks, ground vehicles, etc., began to make radar scope watching unpleasant. The advent of intentional ECM, jammers and chaff, really tore things up. Pity the radar operator of Figure 2.

The white hat guys came to the rescue with a cornucopia of ECCM's of Figure 3. During the early days ECCM was known as antijam or A-J. A-J techniques were referred to as "alphabet soup" after the abbreviations for the early A-J's as DBB, CFAR, FTC, STC, etc.

Certainly radar required medicine to cure ECM. The radar operator has now largely recovered as shown in Figure 4. Unfortunately the radar operator has incurred some problems during his treatment: the ECCM's cost money to implement increasing the price of the radar. Radar complexity increased, affecting radar reliability and maintainability. Some ECCM's can actually degrade radar performance in the benign environment. Some ECCM's when used in combination can be worse than no ECCM. Some ECCM's are more effective than others. Some ECCM's are more effective against some ECM's than against others. Worst of all there is no yardstick for assessing the value of an ECCM. There is a question as to whether the cure is worse than the ill. The radar user of Figure 5 certainly has a dilemma. While the discussion of this paper centers around ECCM, it should be recognized that it also applies to IRCCM.

Need for measures of ECCM effectiveness. This introduction was certainly very elementary. The need for a measure of ECCM effectiveness should also be obvious. It is common practice at present to require that ECCM's be incorporated in a radar. Frequently there is no attempt to assess the effectiveness of the ECCM's. In fact the writer knows of no presently used measures of ECCM effectiveness. The absence of such measures would prevent assessment of the effectiveness of an ECCM. It is essential that effectiveness of ECCM be established to determine whether their utilization is cost effective.

ECCM performance is implicitly included whenever an actual radar with ECCM is used in tests or when the radar is simulated in a hardware type simulation as described in Johnston[2]. Often however, the tests or simulations are not performed with and without ECCM. Comparative data are thus not possible. The digital "model" type simulations, by Johnston[2], do not incorporate ECCM effects. This is due to the absence of ECCM performance data and to a measure of the effectiveness of ECCM.

Measures of ECCM effectiveness. The purpose of ECCM is to negate the effectiveness of ECM used by the enemy. ECM effectiveness is frequently given in the resulting J/S ratio at the radar. Sometimes the burnthrough range* of the radar or the self screening range of the jammer, i.e., its performance in an ECM environment, is given.

As discussed by Johnston[3], the only true measure of ECCM effectiveness is its effect on the performance of the

*In a forthcoming paper, the writer plans to discuss the fallacies of burnthrough range.

ECCM/EIF

associated *weapon system*, e.g., change in the number of targets killed, change in first engagement range, or the like in an air defense system. This requires that the performance of the associated weapon system (not the radar alone) be compared for two cases: without ECCM and with ECCM. The importance of this change in performance must be carefully assessed.

Weapon system performance assessment in an ECM-ECCM environment requires that the effects of ECM and ECCM be properly represented.

ECCM properties. ECCM are those actions taken to counter the effect of the use of ECM. Such things as chaff, decoys, anti-ARM's, maneuvers, and tactics are included. The radar operator can be a very effective ECCM. ECCM's are legion. Table I from Johnston,[3] lists approximately 150 types of ECCM. It will be noted that many of these may be placed into the general category of signal processing techniques. As such their function is to reduce the J/S ratio in the radar receiver. This is similar to the effect of radar MTI. In a clutter environment, MTI discriminates against large unwanted returns from clutter so as to permit detection of the smaller returns from the desired target.

ECCM representation. Representation of ECCM in simulations can be a very difficult task as discussed by Johnston[3]. Signal processing type ECCM's could be represented easily with few problems.

MTI effectiveness was first presented in terms of the cancellation ratio and SCV by Emslie and McConnell[4]. Barton[5] discussed these two terms and states his preference for the term "Improvement Factor" defined by Steinberg[6]. Reilly treated an (overall) MTI System Improvement Factor[7]. The relationship between the Improvement Factor (I) and SCV was given by Shrader[8].

MTI effectiveness measures can thus be used as a basis for ECCM representation and as a measure of ECCM effectiveness. The term EIF was proposed by Johnston[3] as a measure of ECCM effectiveness.

ECCM improvement factor (EIF). The proposed definition of EIF is the ratio (in dB) of the ECM signal required to produce the same output signal in the radar using ECCM to the ECM signal level producing the same output without the ECCM.

For example, assume that when a Dicke-Fix ECCM in a radar is switched on, the jammer power output must be increased 10 dB to retain the same receiver output J/S ratio that existed before the ECCM was turned on. The EIF would thus be 10 dB.

It should be recognized that an ECCM device could actually have an insertion loss, i.e., its use in a benign environment could cause a reduction in the benign detection range. The EIF must therefore be defined with and without ECM present.

Applications of EIF. The principal application of EIF should be in the assessment of ECCM effectiveness. EIF data should be used as input data in describing radar performance in digital radar simulations. As previously discussed, EIF can be properly used only in total weapon systems simulations to determine ECCM effectiveness. An ECCM with an EIF of 16 dB is not necessarily twice as good as that of another ECCM with an EIF of 10 dB. (A change of 6 dB will double or half the burnthrough range of a radar.) The only meaningful criterion is the change in overall weapon system effectiveness. An ECCM with an EIF of 10 dB which reduced the J/S ratio from 23 to 13 dB could be inadequate if this J/S ratio were obtained at the range at which detection was required to permit target destruction prior to the bomb release line.

A very important application of EIF will be in the comparison of the relative effectiveness of various types of ECCM. At present this is not done.

Another application will be in the comparison of the relative effectiveness of the same ECCM type in diferent radars. This should be very interesting when the two radars are designed by different design groups and/or are designed at different times. Such comparisons cannot be made now.

Methods for obtaining EIF data. The first method for obtaining EIF data should be theoretical analysis. Shrader[8] presented theoretical data for MTI improvement factors for various types of MTI. Theoretical analysis has been performed on some ECCM types. Results however have not been given in terms of EIF.

A large amount of radar ECCM test data, bench tests and flight tests, now exist. These test data could be analyzed to produce EIF data. These data should be compared with theoretical EIF data. This source is only applicable to existing radars and ECCM.

A number of ECCM types are not readily amenable to theoretical analysis. Empirical relations must be developed for these. This must be based on either radar test data or simulation data.

Simulation should be a very powerful tool to provide EIF data. The GEESE type and the EWES as described by Johnston[2] should be considered. Simulation is very important when either the ECCM cannot be analyzed theoretically or when an actual radar and ECCM do not exist.

EIF problems and limitations. Problems and limitations with EIF are as follows:

- The EIF is not an overall ECCM effectiveness measure as previously indicated.
- The EIF concept is not applicable to all types of ECCM.
- No EIF data are available and the EIF concept has not been used to date.
- Some types of ECCM can be expected to be sensitive to the type of ECM. Jamming waveforms such as barrage noise, CW, pulsed noise, etc., could also affect the EIF. The EIF would thus have to be defined appropriately.
- Some ECCM types may be sensitive to the level of ECM, in fact there may be a nonlinear relationship. This relationship could be represented by presenting the EIF as a curve, table, or equation as a function of receiver input J/S rather than as a single value.
- ECCM effectiveness should be expected to be sensitive to the presence of more than one ECM type or even to simultaneous use of two identical jammers. It should be noted that as Bragdon[9] pointed out by the effect of two uncorrelated noise sources such as two noise jammers, one of 20 dBm and the other of 10 dBm, is not 30 dBm. For some ECCM types careful analysis of the spectral content of two ECM types may be necessary.
- Some ECCM's when used in combination may be expected to be interactive. The resultant EIF of the combination of ECCM's may be less than the least EIF of any separate ECCM.
- ECM may result in the desensitization of some receivers, especially communication types. A strong signal slightly off the desired frequency may cause a poorly designed receiver to lose sensitivity and thereby prevent detection of the weak desired signal[10]. Proper representation of receiver desensitization and the appropriate ECCM could be difficult.

Conclusions. At present we have no measure of ECCM effectiveness and no means for comparison of EC-

CM's. The EIF concept, while limited, would permit representation of many ECCM's, thus enabling ECCM comparison. EIF is a useful concept which should be further developed and utilized.

Recommendations. The following work is recommended:

- Theoretical analysis should be perormed on all ECCM types to develop analytical expressions for EIF.
- All available radar test data should be analyzed to extract experimental EIF data.
- Experimental EIF data should be obtained on all radar ECCM's.
- Simulations should be performed in lieu of radar testing to obtain EIF data.
- Theoretical EIF expressions should be compared with the measured EIF data obtained and any differences should be resolved.
- Empirical expressions should be developed for EIF for those ECCM's which are not readily amenable to theoretical analysis; test and simulation data will be required for this.
- Methods should be developed for employing EIF data.
- Standard procedures should be established for future use of EIF.
- The preceding actions should also include infrared ECCM. ■

References

[1]Genesis I, The Holy Bible.

[2a]Johnston, Stephen L., *Simulation of the Effects of ECM on Radar Performance* (U), 14th Annual Joint ECM Planning Conference Proceedings, Fort Monmouth, New Jersey, 15-17 October 1968, pp. 2-29 through 2-39 (Confidential).

[2b]Johnston, Stephen L., *A Survey of Radar/ECM Simulation* (U), 10th Annual NORAD Electromagnetic Warfare Conference Proceedings at USAF School of Aerospace Medicine, Brooks Air Force Base, Texas, 13-15 November 1968, pp. 227-236 (Confidential).

[2c]Johnston, Stephen L., *A Survey of Radar/ECM Simulation* (U), US Army Missile Command, Redstone Arsenal, Alabama, Report No. RE-TR-69-18, August 1969, AD 505 445 (Confidential).

[3]Johnston, Stephen L., *Military Radar-Weapon System Analysis*, Presented 18 September 1970 as lectures in the short course "Principles of Radar," Georgia Institute of Technology, Atlanta, Georgia, Published as lecture notes.

[4]Emslie, A. G., and McConnell, R. A., "Moving Target Indication," *Radar System Engineering*, L. N. Ridenour, Ed. (New York: McGraw-Hill Book Company, 1947) Chapter 16, pp. 652-653.

[5]Barton, David K., *Radar System Analysis,* Englewood Cliffs, New Jersey, Prentice-Hall, Inc., 1965, pp. 198-199.

[6]Steinberg, Bernard D., "MTI Radar Filters," *Modern Radar: Analysis, Evaluation, and System Design,* R. S. Berkowitz, Ed. (New York: John Wiley and Sons, 1965) Chapter 2, pp. 494-495.

[7]Reilly, J. P., "Moving Target Indicators (MTI)," *Radar Design Principles: Signal Processing and the Environment*, Fred E. Nathanson, Ed. (New York: McGraw-Hill Book Company, 1969) Chapter 9, pp. 318-358.

[8]Shrader, William W., "MTI Radar," *Radar Handbook*, Merril I. Skolnik, Editor-Chief (New York: McGraw-Hill Book Company, 1970) Chapter 17, pp. 17-1 through 17-59.

[9]Bragdon, Clifford R., *Noise Pollution: The Unquiet Crisis*, Philadelphia: University of Pennsylvania Press, 1971, pp. 52-55.

[10]Moore, Roy, "Designing Communications Receivers for Good Strong-Signal Performance," *HAM RADIO Magazine,* February 1973, pp. 6-16.

ECCM/EIF

Table I
An ECCM Type Lexicon

- Acceleration Limitation
- Angle Sector Blanking
- Angular Resolution
- Audio Limiter
- Aural Detection
- Autocorrelation Signal Processing
- Automatic Cancellation of Extended Targets (ACET)
- Automatic Threshold Variation (ATV)
- Automatic Tuner (SNIFFER)
- Automatic Video Noise Leveling (AVNL)
- Back-Bias Receiver
- Baseline-Break (on A-Scope)
- Bistatic Radar
- Broad-Band Receiver
- Coded Waveform Modulation
- Coherent Long-Pulse Discrimination
- Compressive IF Amplifier
- Constant False Alarm Rate (CFAR)
    - Cross-Gated CFAR
    - Dispersion Fix (CFAR)
    - IF Dicke-Fix CFAR (Dicke-Fix)
    - MTI CFAR
    - Unipolar Video CFAR
    - Video Dicke-Fix CFAR (Dicke-Fix)
    - Zero Crossings CFAR
- Contiguous Filter-Limiter
- Cross Correlation Signal Processing
- CW Jamming Canceller
- Detector Back Bias (DBB) (Same as Detector Balanced Bias)
- Dicke-Fix
    - Clark Dicke-Fix (Cascaded Dicke-Fix)
    - Coherent MTI Dicke-Fix
    - Craft Receiver
    - Dicke Log Fix
    - IF Canceller MTI Dicke-Fix
    - IF Dicke-Fix CFAR (Zero Crossings Dicke-Fix CFAR)
    - Instantaneous Frequency Dicke-Fix
    - Noncoherent MTI Dicke-Fix
    - Video Dicke-Fix CFAR
- Diplexing
- Doppler-Range Rate Comparison
- Double Threshold Detection
- Electronic Implementation of Baseline-Break Technique
- Fast Manual Frequency Shift
- Fast Time Constant (FTC)
- Fine Frequency
- Frequency Agility
- Frequency Diversity
- Frequency Preselection (Narrow Bandwidth)
- Frequency Shift
- Gain Control
    - Automatic Gain Control (AGC)
    - Dual Gated AGC
    - Fast AGC
    - Gated FAGC
    - Instantaneous AGC
    - Manual Gain Control
    - Pulse Gain Control
    - Sensitivity-Time Control
- Guard-Band Blanker
- High PRF Tracking
- High Resolution Radar
- IF Diversity
- IF Limiter
- Image Suppressor
- Instantaneous Frequency Correlator (IFC–CRAFT)
- Integration
    - AM Video Delay Line Integration
    - Coherent IF Integration
    - Coherent (IF) Integration (Moving Target)
    - Coherent (IF) Integration (Stationary Target)
    - Display Integration
    - FM Delay Line Integration
    - Noncoherent (Video) Integration
    - Pulse Integration
    - Video Delay-Line Integration
- Inter-Pulse Coding (PPM)
- Jamming Cancellation Receiver
- Jittered PRF
- Kirbar Fix
- Least Voltage Coincidence Detector
- Linear Intra-Pulse FM (CHIRP)
- Lin-Log IF
- Lin-Log Receiver
- Lobe-on-Receiver Only LORO, Also (SORO)
- Log Fix (Also, Log FTC)
- Logarithmic Receiver
- Logical ECCM Processing
- Main Lobe Cancellation (MLC)
    - Monopulse MLC
    - Polarization MLC
- Manually Aided Tracking
- Manual Rate-Aided Tracking
- Matched Filtering
- Monopinch
- Monopulse Tracker
- MTI
    - Area MTI (Velocity Filter)
    - Cascaded Feedback Canceller (MTI)
    - Clutter Gating (MTI)
    - Coherent MTI
    - Noncoherent MTI
    - Pulse Doppler
    - Pseudocoherent MTI
    - Single-Delay Line (MTI Canceller)
    - Re-Entrant Data Processor
    - Three-Pulse Canceller
    - Two-Pulse Canceller (Single-Delay Line MTI Cancellation)
- Multifrequency Radar
- Multisimul Antenna
- Phased Array Radar
- Polarization Diversity
- Polarization Selector
- Post Canceller Log FTC
- PRF Discrimination
- Pulse Burst Mode
- Pulse Coding and Correlation
- Pulse Compression, Stretching (CHIRP)
- Pulse Edge Tracking
- Pulse Interference Elimination (PIE)
- Pulse Shape Discrimination
- Pulse-to-Pulse Frequency Shift (RAINBOW)
- Pulse Width Discrimination (PWD), Pulse Length Discrimination (PLD)
- Random-Pulse Blanker
- Random-Pulse Discrimination (RPD)
- Range/Angle Rate Memory
- Range Gating
- Range Rate Memory
- Scan-Rate Amplitude Modulation
- Short Pulse Radar
- Side-Lobe Blanker
- Side-Lobe Canceller
- Side-Lobe Reduction
- Side-Lobe Suppression (SLS)
- Side-Lobe Suppression by Absorbing Material
- Staggered PRF
- Transmitter Power
- Two Pulse Autocorrelation
- Variable Bandwidth Receiver
- Variable PRF
- Variable Scan Rate
- Velocity Tracker
- Video Correlator
- Wide-Bandwidth Radar
- Zero-Crossing Counter

Simulation to Obtain a Systems Measure of an Air Duel Environment
A.A.B. Pritsker, R.C. VanBuskirk, and J.W. Wetherbee
IRE Transactions, Vol. EC-8, No. 1, March 1959
Pages 55-59

9.3

Summary—**A combined analog-digital simulation of an air battle between an attacking bomber aircraft and a ground controlled interceptor, including the intermediate human radar operator, has been designed for the purpose of evaluating the effects of airborne electronic countermeasures upon a ground-based radar operator.**

Both real and nonreal time simulation are used in the experimental setup. The simulation encompasses as much as possible of the system as affected by the operators' performance in the hope that a systems measure could be obtained. It is hypothesized that probable success of the bomber is the systems measure of the effectiveness of the countermeasures.

Introduction

THIS paper describes the work done on a simulation study, the purpose of which was to evaluate the effects of electronic countermeasures (ECM). In this study, a combined analog-digital simulation of an air battle between an attacking bomber aircraft and a ground-controlled interceptor, including the intermediate human radar operator, has been devised. This simulation is regarded primarily as a tool for investigating the methodology and techniques for evaluating the effects of ECM. The paper deals mainly with the simulation aspects of the study.

Modern air warfare has generated many sophisticated electronic systems, such as radar systems for gathering information about an enemy. Opposing the electronic information-gathering systems are the ECM systems which attempt to defeat the former by denying, delaying, distorting, or falsifying the information. In the case of airborne ECM operating against ground-based radars, there has been a notable lack of system measures to assess and predict the effectiveness of ECM. On the other hand, there is no lack of immediate figures of merit to evaluate the effects of ECM on radar equipments and operator performance. However, these immediate figures of merit such as signal-to-noise ratio, blip-to-scan ratio, cumulative probability of detection, and detection time do not allow for the assessment of the effects of ECM on the system beyond the radarscope operator. For this study, it has been hypothesized that a better systems measure of effectiveness of electronic countermeasures is the probability of survival of the bomber.

Simulated System

The system simulated is shown pictorially in Fig. 1. A radar antenna picks up the radar return of a bomber which is generating ECM. The radar information is processed and displayed on a radar indicator to a human operator. The human operator tracks the bomber by determining the successive positions and headings of it on the radar indicator. Based on this track, the operator determines a command heading for an interceptor to fly. The interceptor follows the directions of the operator until its airborne intercept radar has acquired the bomber at which time the terminal air duel is conducted.

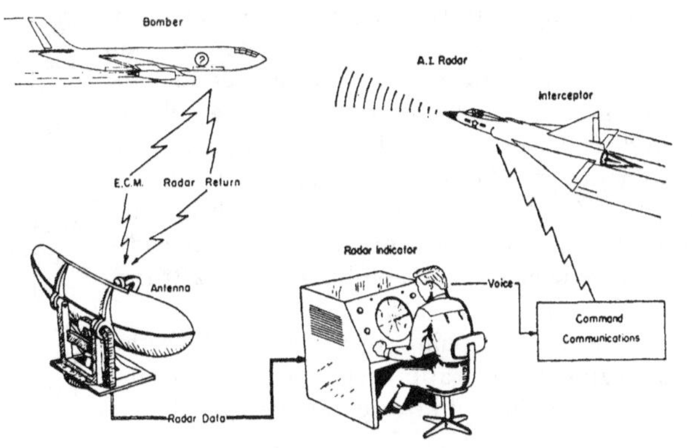

Fig. 1—Pictorial sketch of system simulated.

The radar return of the attacking bomber, the ECM signals, and the various portions of the radar detection process are simulated using the QRC-10(T) ECM Simulator. These simulated radar signals are used as inputs to a radar indicator to provide a PPI display. While tracking the target, the operator calls the position and heading of the bomber to a keypunch operator who records the information on an IBM punched card. The true bomber-position information and time are recorded on the same IBM punched card. Thus, a set of these cards describes completely the target's true and observed tracks.

Up to the keypunching step, the simulation is carried out in real time and is entirely analog except for the conversion to digital form for storage on the punched cards. These punched cards are then used as the input to a digital computer. A digital computer program has been written to simulate an aircraft controller directing an interceptor against the bomber and the final phase of the air battle. The probability of bomber survival is then computed and used as the system's measure of ECM effectiveness.

* Received by the PGEC, November 12, 1958. Work on which this paper is based done by Battelle Mem. Inst., Columbus, Ohio, under subcontract to General Mills, Inc., Minneapolis, Minn. Principal contract: AF 33(616)-3739.

† Battelle Memorial Inst., Columbus, Ohio.

Experiment

To ensure that the results obtained from this simulated system were reliable, an experimental design was constructed. This design, involving 500 trials, consisted of 50 different ECM displays presented to 10 different subjects (operators). Two displays consisted of the target signal only, with no ECM signal present. The remaining 48 displays were created by systematically varying the power level and mode of operation of the ECM. In this design, the effects due to ECM displays, subjects, bomber track, and sequence of presentation were experimentally controlled. Bomber courses were approximately straight lines with different azimuth orientations. The experiment was designed to enable analysis of the functional relationships between performance measures and the errors in the system. Throughout the digital programs, immediate and intermediate measures of effectiveness are computed. The immediate measures of effectiveness are defined as those measures which can be related directly to the performance of the human operator. These measures include target-detection time, number of reads made by the operator, and the mean and standard deviation of range azimuth, and target heading errors. The intermediate measures are defined as those measures which are derived from the operator's performance after the observed data have been operated on by a complex transformation. These measures include the position and heading errors at roll-out point (the point at which the interceptor begins its final approach), bomber-velocity error, and the probability of interceptor success.

Experimental Setup

The principal piece of equipment used in the analog portion of the system was the QRC-10(T) ECM Simulator,[1] used to generate the ECM signals and the radar return of the attacking bomber. In order to conduct the experiment, it was necessary to modify and to instrument this equipment. It should be noted that the QRC-10(T) was not designed as a vehicle for this type of experiment, but was selected as the most expedient choice. Units of the QRC-10(T) used in the experiment were the basic unit, multi-target generator, and the antenna pattern simulator. A block diagram outlining equipment interconnections is shown in Fig. 2. The basic unit was used to simulate the ECM signals which were fed into the antenna pattern simulator. The antenna pattern simulator produced amplitude variations of the applied ECM signal which simulated the azimuth directivity pattern of the radar antenna. The output of the antenna pattern simulator was at IF frequencies and was servo-driven so that it could follow the target aircraft that was transmitting the ECM signal. The program target generator produced video signals which simulated the bomber radar return. Programming of the bomber's course was accomplished by the use of cams which automatically actuated the range and azimuth control of the bomber. The orientation and shape of the cam determine the course of the simulated bomber. These signals were mixed with the ECM signals after the latter had been fed through an IF strip. This mixed video signal was displayed on the radarscope indicator. The program target generator was instrumented so that target azimuth and range could be automatically digitized. Range data were obtained from a time interval meter connected to a phantastron circuit which initiated the range gate for the target generator of the QRC-10. A converter was used to convert the binary output of the time interval meter to the decimal form suitable for input for an IBM 526 key punch. Azimuth data were obtained by attaching a Coleman digitizer shaft encoder to a shaft common to the rotor of the control transformer which initiated the azimuth gate for the target generator. Time information was generated from a synchronous motor driving a Veeder root mechanically actuated read-out counter. The radar operator views the display on the radarscope indicator and determines the position and heading of the simulated bomber. The operator relates this information to a keypunch operator who upon receipt of the signal initiates a switch which causes the QRC-10 to be interrogated and the information to be automatically punched on a punched card. On the same card is punched the information called out by the radar operator. Fig. 3 illustrates a sample ECM display obtained from the simulated system. A picture of the equipment used in the experiment is shown in Fig. 4.

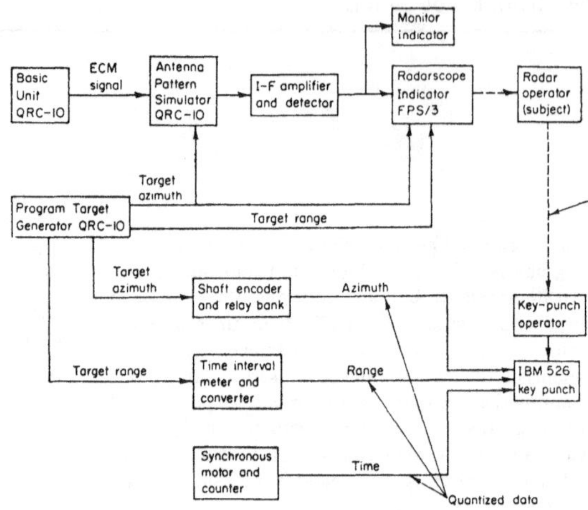

Fig. 2—Block diagram equipment setup.

A booth was designed to house the radar indicator and operator in order to simulate AC and W site conditions. The booth was soundproofed and incorporated a broadband blue lighting system.

Operator Performance and Task

During an experiment, it was the operator's task first to detect the target and then determine its range, azi-

[1] L. Sternlicht, "A radar electronic countermeasures simulator," 1958 IRE National Convention Record, part VIII, Communications; Military; Radio-frequency Interference, pp. 94–98.

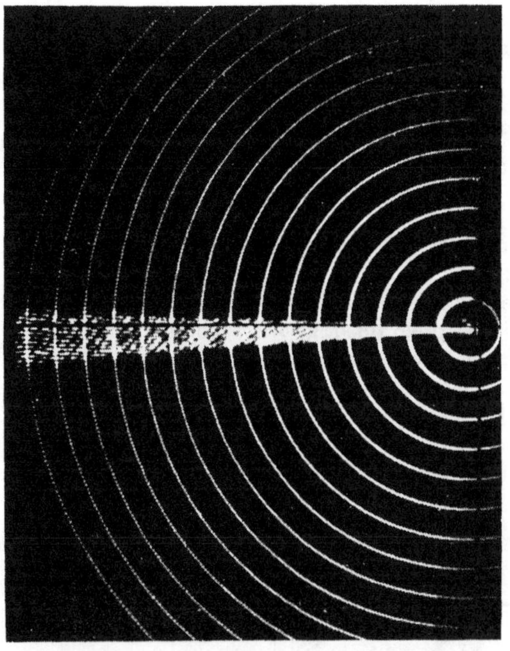

Fig. 3—Representative radarscope display.

Fig. 4—Equipment utilized for the simulation.

muth, and heading. Detection time was defined as the time required for the operator to establish a track, *i.e.*, to obtain two target points. As an aid in tracking, the operator used a grease pencil to mark each target point. In an effort to standardize the operator's task, call-outs were limited to an arbitrary maximum. The operator was instructed to call out as soon as possible following the sounding of a buzzer which was sounded 30 seconds after the recording of the last observed data. Communication between radarscope operator and keypunch operator was made by means of a two-way communication system which was monitored in order to reduce transmission errors and to verify the keypunching. At each reading, the keypunch operator would have the time of reading and true target data automatically read on the punched card before transcribing the radar operator's call-out observed data. The experiment was conducted for 22 minutes, which was defined as the maximum allowable bomber-penetration time. All indicator controls were preset and adjustment of controls by the operator was prohibited. The data collected during the experiment was then used as the input to the digital simulation.

Digital Simulation

A digital computer program was written to solve the equations of a mathematical model of an air duel between one bomber and one interceptor. The model covered the time period from interceptor takeoff to interceptor armament release. Because of the drum storage limitations of the IBM 650 (the digital computer used in this program), it was necessary to divide the program into two parts: 1) the vectoring of the interceptor against the observed bomber track, and 2) the terminal air battle between the interceptor and bomber. These parts were overlapped by storing the outputs of the vectoring phase on magnetic tape and using them as inputs to the terminal air-battle phase. The vectoring model was used to obtain the flight path of an interceptor that would have been directed against a bomber's path as seen by the experimental operator.

It was assumed that the errors involved in vectoring an interceptor were independent of the ECM used against the scope operator. This assumption implied that the intercept controller had fixed decision-making processes when vectoring the interceptor, irrespective of the countermeasures used. Thus, the vectoring model was viewed as a complex transfer function to provide an interceptor path, given a bomber path.

The interceptor path which the computer solved was based on the real-time, observed bomber data stored on punched cards. At a given point in time, the computer prescribed a path for the interceptor based on the data available up to that time. The interceptor's path was revised and recomputed, either as new data were presented or when a specified time period had elapsed. The interceptor path as determined by the computer had to conform to certain preset criteria. The criteria used in this experiment were satisfied when the interceptor had flown an initial approach leg, made a turn consistent with its aerodynamic limitations, and turned into a final closing straight leg at target altitude and at interceptor speed. The tactic of the interceptor was completely described by specifying the final leg distance and the angle between bomber and interceptor tracks during the final leg. The geometry of the situation and paths of the aircraft are shown in Fig. 5. The computations necessary to derive the interceptor's path included: 1) the interceptor's climb to bomber altitude, 2) the length of the approach leg, 3) the length of the arc during its turning phase, 4) the length of the final leg, and

5) the armament travel distance. A command heading for the interceptor was derived by equating the time for the interceptor to fly these five portions of its path to the time required for the bomber to fly an extrapolated straight-line course. It was simple enough to set up equations relating these times and distances which determined the geometry of the problem, but it was difficult to solve these equations directly for a specific solution. For this reason, solutions were obtained by an iterative procedure.

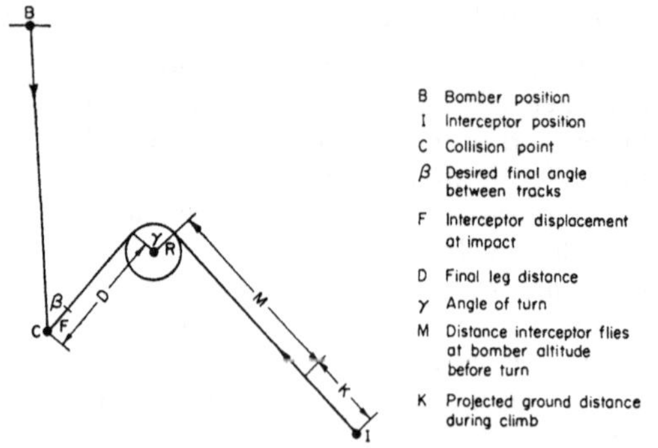

Fig. 5—Vectoring geometry.

This iterative procedure involved varying the time to collision which specified the collision point, *C*, since the bomber position and heading were known. An equation of distances along the interceptor path was set up for the estimated collision point. A second equation was then computed to determine the distance the interceptor could travel in the estimated time to collision with allowances for time to climb and time for armament travel. These two distances were then equated and the algebraic difference determined. Time was then varied by a combination of linear approximations and averaging techniques until the difference of the two distances was less than 1/10 of a nautical mile.

In the above computations, previous interceptor heading was not taken into account, but the interceptor's aerodynamic limitations in achieving the new command heading were considered. For this reason, it was necessary, in some instances, to recompute vectoring solutions at shorter time intervals in order to ensure that correct vectoring was being accomplished. New command headings of less than 5° from previous headings were not compensated, so that minor oscillations in the interceptor's path were nullified, *i.e.*, a 10° dead band in heading existed. If a vectoring solution could not be obtained to meet the preset criteria, the interceptor was directed on a collision-type course.

A collision-type course was also used when the interceptor completed its approach leg. In a collision solution, the interceptor was vectored to a point which led the target by a distance such that the armament impact occurred at that point. By using a collision solution at the beginning of the turn phase, the interceptor command heading changed radically. However, when the interceptor attempted to achieve the new command heading, it was actually flying the specified turn, providing it turned in accordance with a prescribed load factor. As the turn progressed, new command headings were computed. When the interceptor heading was the same as the command heading, the turn phase had been completed and the interceptor began its final leg. During the final leg, the interceptor was continually vectored and revectored on the basis of the observed data. A complete description of the final leg was stored on magnetic tape and was used as the input to the second part of the digital program. Fig. 6 is a point-by-point plot of the time history of a vectoring solution.

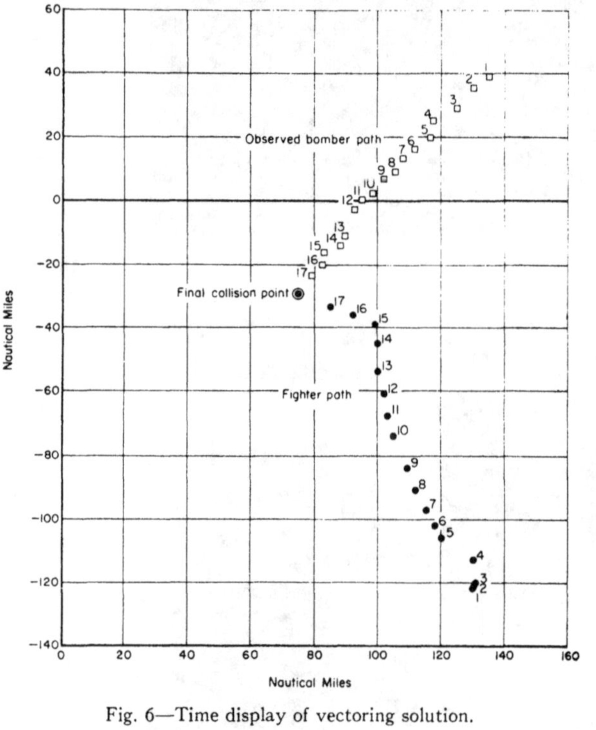

Fig. 6—Time display of vectoring solution.

The interceptor base was always located in the same position with respect to the true bomber track to eliminate the effects of initial interceptor position on probability of survival scores between experimental trials. However, as detection time increased, the bomber approached the interceptor base, causing probability of survival to be a function of detection time. After detection, the interceptor flew a course in which the heading vector was parallel and opposite to the bomber's heading vector. The interceptor was also prohibited from turning until it had reached a specified altitude. During the interceptor's climb period, its ground speed and rate of climb were obtained from a table stored in the computer, defining the interceptor's mission profile. The interceptor had to achieve bomber altitude before completing its initial approach phase. Bomber altitude was not varied in this experiment.

The second part of the program simulated the terminal air battle and was initiated at the beginning of the final leg. The first operation of the interceptor during this phase was to initiate airborne intercept (AI) radar search. This was accomplished in the computer by determining whether the bomber was within detection range and azimuth scan limitations of the interceptor's radar. If these two conditions were satisfied, a probability of AI detection was computed as a function of the range between interceptor and bomber, the aspect angle between the interceptor and bomber, the AI radar system parameters, the radar echo area of the bomber, and the visibility factor. Since probability of AI detection was a function of time, it was computed for each radar scan period until detection occurred. Detection occurred if the computed probability of detection was greater than a number drawn from a uniform distribution. The uniform distribution was composed of pseudo-random numbers generated by a multiplicative congruence scheme suggested by Lehmer.[2] In this scheme, the $i+1$ random number was generated by

$$U_{i+1} = 23U_i \bmod (10^8 + 1). \tag{1}$$

If the range, scan angle, or detection criteria were not satisfied, the interceptor and bomber were flown for the next scan period. The interceptor was flown in accordance with the simulated operator's commands and the bomber was flown with respect to the recorded data from the QRC-10(T). If detection did not occur within the 22-minute time limit, an unsuccessful intercept was considered to have occurred.

After detection had occurred as described above, a lock-on phase was initiated. The interceptor and bomber were flown until the range between the bomber and interceptor was less than the maximum lock-on range. The time to achieve lock-on was then computed by drawing a number from a log-normal distribution characterized by the 40 and 90 per cent fractiles. The bomber and interceptor were then flown for this lock-on time.

The third phase of the terminal air duel was that of conversion, *i.e.*, redirection of the interceptor by its fire-control system, after AI radar acquisition of the bomber, to a position from which the interceptor's weapon could be fired. Conversion was accomplished in the program by solving the fire-control equations and determining if the interceptor's heading was within the heading prescribed by the fire-control system. If it was not within the prescribed limits, the interceptor was flown with a change of heading to reduce this error. This process was continued until either an unsuccessful intercept occurred or a collision-type course was obtained.

When the entire process outlined above was completed without occurrence of an unsuccessful intercept, the interceptor was flown to its point of armament launch. At this time, the aspect angle relative to the bomber's path was calculated, and an interceptor success was recorded as a function of this angle. The simulated terminal air duel as described above was repeated N times for each of the 500 trials, since it was a stochastic process producing different results which were dependent on the random numbers generated. A sequential analysis of the probability of detection-conversion was corporated into the program so that the number of times the trial was repeated was a function of the probability of detection-conversion. After the iterations were completed, a probability of bomber survival was computed by (2).

$$P_s = 1 - \frac{\sum_{j=1}^{18} N_j P_{jk}}{N} \tag{2}$$

where

P_s = probability of survival of the bomber
N_j = number of interceptor successes in the jth bin, $j = 1, 2, \cdots, 18$.
P_{jk} = kill probability of the interceptor armament as a function of aspect angle
N = total number of iterations.

These probabilities of bomber survival scores were then used to evaluate the effectiveness of the simulated ECM patterns.

Conclusions

This study was primarily concerned with the methodology involved in assessing ECM. Simulation was regarded as a tool by which hypothesized measures could be tested and evaluated and upon which future research could be based. The results of the study will be in the form of the relationships and interactions existing among the parameters of the system simulated.

Acknowledgment

The basis for the vectoring model was taken from one developed by Dr. Robert A. Porter (now Chief Scientist, NORAD) while he was at Eglin Air Force Base. The basis for the terminal air-battle model was obtained from Hughes Development Company, Culver City, Calif.

[2] D. H. Lehmer, "Mathematical methods in large-scale computing units," *Proc. Second Symp. on Large-Scale Digital Calculating Machinery*, Harvard University, Cambridge, Mass., pp. 141–146; 1951.

Simulator Checks Out Radio-Radar A J Operation
J. Lambert
Space/Aeronautics Magazine, Vol. 33, No. 4, April 1960
Pages 157-166

9.4

- Simulation before breadboarding
- Ultimate accuracy of 0.1 db foreseen

by John Lambert, Project Engineer, Electronic Systems Evaluation, Defense Systems Dept., General Electric Co.*

ANTI-JAMMING (AJ) circuit design in aerospace electronic systems is complicated by the fact that jamming modulation and receiver non-linearities are extremely difficult to analyze. Many existing systems are marginal in an electronic countermeasures environment. The system designer can only rely on experimental data on AJ circuit performance, yet often the design of a new system must be frozen before the AJ circuits can be optimized experimentally.

To get out of this dilemma, you can build a low frequency analog model of your system and use it to evaluate AJ circuits and jammer effectiveness. This is the approach that led to the General Electric Electronic System Evaluator (GEESE). This unit can simulate new systems long before any electronic equipment is breadboarded. When it is used, all subsystems, including those covering AJ performance, can be optimized early in the design cycle.

In one application, a model of an FM receiver was simulated on GEESE (*Figs. 1 & 4*). The IF frequency was assumed to be 300 kc with a 15-kc bandpass and was scaled down 10,000:1 (to 30 cps with a 1.5-cps bandpass). The audio rate was assumed to be three

more on next page

* Defense Systems Dept., General Electric Co., Atlantic Bldg., Syracuse, N. Y.

FIGURE 1: Computer diagram for the simulation of the FM receiver of Figure 4.

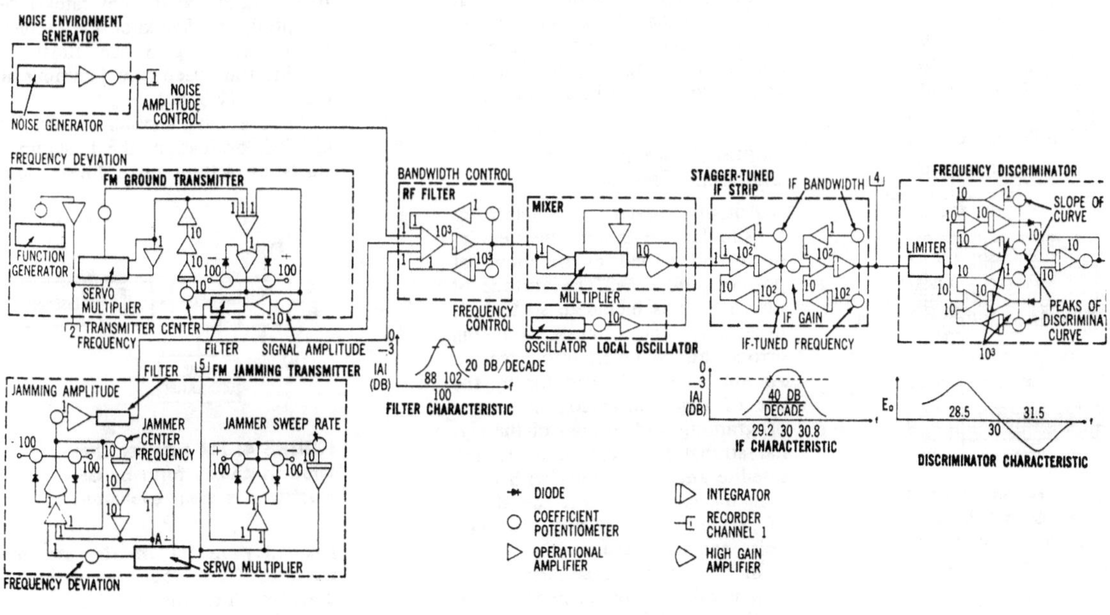

kilocycles and scaled down to 0.3 cps.

Since the carrier frequency is arbitrary, a 100-cps carrier was chosen. It was frequency-modulated at 0.3 cps, with a maximum deviation of 0.7 cps. The local oscillator operated at 130 cps, and the 10^{-4} system scaling was carried from the mixer through the entire system.

Simple model represents satellite receiver

Frequency scaling is important in the use of GEESE since the system simulation at low audio frequencies makes it possible to record all signals and observe all transient- and steady-state phenomena while interference signals are fed in or system parameters are changed.

With our simplified model, we can represent a satellite receiver. Because of the orbital motion of the satellite, the signal from the ground transmitter (and from any ground-based enemy jammers) in such a case, will vary as a function of position, antenna pattern, antenna tracking errors, and time.

The orbit can be programed directly into GEESE, since the dynamic orbit of the satellite can be solved by analog computer. The satellite is programed with respect to the ground transmitter and the jammer so that the signal strengths at the satellite of both transmitter and jammers will vary continuously as functions of position and time.

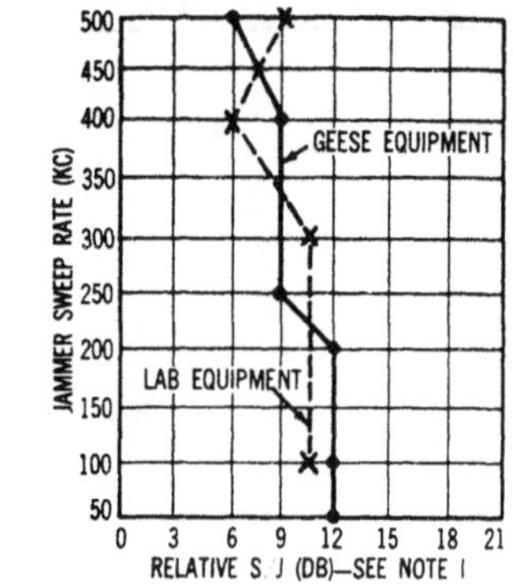

FIGURE 3: Jamming effects on the lab model of a sophisticated phase-locked AFC receiver and on the same receiver simulated on the GE electronic system evaluator (GEESE). The circles indicate breakdowns of the phase-lock loop. The frequency deviation of the CW jammer equals that of a 100-kc jammer centered on the signal frequency. For security reasons, the absolute S/J ratio is not shown.

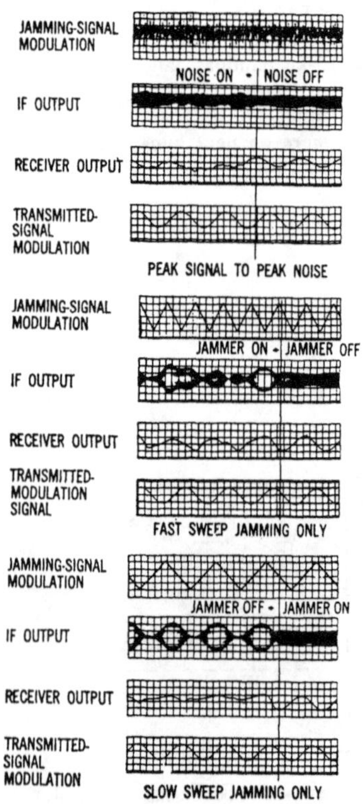

FIGURE 2: Receiver waveform recordings. The IF output is amplitude calibrated for signal strength; the ratio of peak signal to peak noise is —3 db.

Signal-to-jam ratios were changed discretely

Antenna pattern characteristics are simulated with non-linear equipment and varied as functions of position and time. The attitude of the satellite is included, so that, if necessary, the minute tracking errors can be taken into account. Anomalies are allowed for by the use of digital-to-analog automatic programing and so control the signal strengths at the input to the satellite antenna. Along with these dynamic characteristics, the effects of atmospheric attenuation and noise are simulated.

In our sample case, the signal-to-jam ratios were changed in discrete steps. When this method is used a variety of anti-jam devices and techniques can be inserted, and their interference can be evaluated in each case.

The recordings of *Figure 2* illustrate GEESE's ability to record any signal within the system as well as the effects of natural friendly, and enemy-produced interference environments.

Channel 1 shows the audio signals carrying the information. Channel 2 is the output of the simulated receiver; ideally, it should be a reproduction of channel 1. Channel 3 is an amplitude plot of the IF strip. Channel 4 is the modulating signal of the jammer. The jammer in the second and third cases is a frequency-modulated transmitter comparable to a carcinotron. It has a maximum deviation of 10 cps and is swept at 0.28 and 0.5 cps. These values correspond to a 100-kc deviation at 2.8 and five kilocycles.

Jam and no-jam periods are compared

The recordings show the interference signals and the effects of the interference in the IF stages and at the output. There is a portion of time on each recording when the jamming is not turned on and another showing the effects of successful jamming. Recording 1 shows the effect of noise jamming; recording 2, the effect of carcinotron jamming at 0.5 cps rate (representing a five-kilocycle sweep rate); recording 3 the effect of carcinotron-sweep-type jamming at 0.28 cps (2.8 kc).

In developing a system model for GEESE evaluation, the programer starts by setting up a linear, low frequency model of the system, working from the system block diagram. This model contains the exact linear transfer and is usually

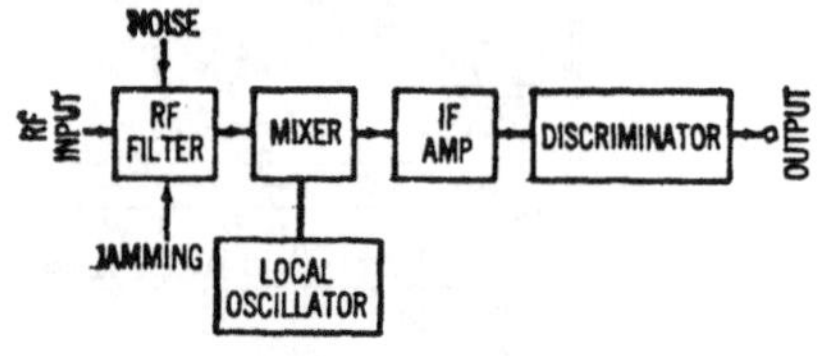

FIGURE 4: FM receiver simulated in analog form by GE's electronic system evaluator.

within six decibels of experimental results.

From the linear model, the most important system parameters are determined. Next the linear model is refined by programing into it the two or three most important non-linear and second-order functions (e. g., the actual first-mixer-characteristic curve, local-oscillator harmonics, non-linearities in the preamplifier, and AGC response). These non-linear characteristics are estimated on the basis of circuit diagrams and/or of discussions with the circuit designer.

After a few weeks, accuracy should be three db

The programing of non-linear functions is straightforward. After several weeks of intensive effort, the system model should be accurate to within three decibels.

So far, GEESE has not been used in any case requiring an accuracy better than three decibels. Test repeatability on actual system hardware is normally within three decibels.

What would be the ultimate accuracy if all second-order and non-linear equipment characteristics were measured and programed into the model? No direct experimental work has been done to answer this question. From the accuracy limits of the basic GEESE components (mixer, phase detectors, function-generating equipment, etc.) one can infer that a model of a unit like a receiver could be made accurate to within 0.1 db.

Accuracy of one decibel is hard to get

A complete system containing 10 major units like receivers, transmitters, data links, etc., should be reproducible to within one decibel. In practice, though, an accuracy of one decibel or less is very hard to get.

Analytic checks are made in every evaluation to show that the results are within the required accuracy. By combining communications and control theories, a system analysis of the steady-state performance as a function of signal-noise ratio can be achieved.

Figure 3 shows the close correlation between results predicted by GEESE and actual prototype test results. In the case shown, the criterion for successful operation of the jammer is whether or not it breaks the phase-lock loop (frequency tracking loop).—End

Radar/ECM Computer Modeling
J. Cikalo and M. Greenbaum
IEEE NAECON-75 Record
Pages 431-437

9.5

ITT Avionics Division, Nutley, N. J.

Abstract

This paper describes a digital simulation technique for analyzing a radar's angle tracking performance in the presence of an inverse gain ECM signal. A brief discussion on the computer modeling methodology is introduced since a number of applicable programming languages are currently available. A computer model of a generic conical scanning radar was synthesized using the continuous system modeling program (CSMP). The radar model was subjected to a time sequence of pulses, constituting an inverse gain ECM signal, and its angle tracking performance examined for a range of jamming-to-signal (J/S) ratio values. The methodology described can easily be used for general evaluation of the performance of a variety of electronic countermeasures techniques and radars.

INTRODUCTION

In the development of ECM techniques to degrade threat radar operation, the evaluation of ECM effectiveness is of paramount importance. Methods of evaluation run the gamut from theoretical analyses to laboratory simulation and finally to field test. Each of these methods progressively approach the "real world", but such advantage however, is offset by additional cost and time. One way to reduce the expense and effort required at a particular level of evaluation, is to ensure that sufficiently detailed and accurate data is obtained at the preceding level. Since simulation and field test are physical methods and are generally determined by available facilities, the approaches used during the theoretical phase are most amenable to the use of digital computation. This type of computation can range from the solution of simple equations to full scale system modeling. This latter type of computation is a direct digital representation of the basically analog (or physical) evaluation methods to be used later. To support this type of analysis, a number of modeling languages are available and sufficiently flexible to support a wide range of uses; these include Continuous System Modeling Program (CSMP) and Continuous System Simulation Language (CSSL). Such languages allow the working engineer to model radar and ECM systems with one-to-one transformation simplicity, and obtain outputs fully consistent with physical methods. Moreover modeled system parameters can easily be changed and output effects quickly noted.

As an example of this type of computer modeling, this paper presents an analysis performed in the evaluation of inverse gain (IG) ECM versus a generic conical scanning radar. As a measure of ECM effectiveness the radar boresight deviation from the target position was used, with the J/S ratio as the parameter of interest. In addition, so that the approach can be applied to other problems, the computer methodology is also described to indicate the specific programming language compatibility to a general closed loop ECM/radar interaction evaluation analysis. Thus, a brief discussion is provided on the general computer modeling methodologies, their advantages and disadvantages.

COMPUTER MODELING AND SIMULATION LANGUAGES

Computer methods of studying the performance of systems by means of their transfer functions have been in use for some time. In this method of analysis a significant portion of effort has often been expended for computer program development. In addition, the ability to easily transform a given system into a computer model is often confused by mathematical detail, e. g. differential/integral equations. This difficulty is compounded when complex systems are modeled. The transformation of differential equations that describe these systems into operational programs, usually requires skilled computer programmers. Thus, the computer program development, and often its utility, has been in the domain of the computer programmers. The programmer interface between system analyst and the program operation on the computer often turned out to be an inefficient arrangement. Out of this, the need for problem-oriented programs structured to help prepare problems for solution on large and small scale digital machines emerged.

The pressing need to involve the engineers directly in the process of setting up and running the problem with an on-line interaction culminated in simulation approaches. Now, simulation is a firmly established tool for investigating phenomena ranging from information flow in business organizations to the dynamic behavior of complex closed loop mechanical and electronic systems. There are many simulation languages, most of them are application oriented. For example, the business problems have been treated as discrete processes on digital computers through the use of such discrete system simulation program as the General Purpose Simulation System (GPSS) [1]. However, the problems encountered by control system engineers which are inherently continuous processes have been simulated on analog computers. As the systems became more and more complex, requiring greater accuracy and flexibility, interest has grown in the applications of digital computers to modeling continuous systems. To ease the transition from analog to digital computation, allowing the engineer a direct interaction, a family of simulation languages were evolved, such as MIMIC, MIDAS, CSMP, CSSL, etc. Each of these languages affords certain advantages depending on the specific applications as well as the size of an available digital computer. To cite a few, the S/360 CSMP (IBM) [2] and CSSL (CDC) [3] are strictly equation oriented while 1130 CSMP (IBM) [4] and Tymeshare CSMP [5] are primarily block language oriented.

The CSMP is a complete program for modeling the system dynamics (response/behavior) by an analog computer method on a digital computer. As in any analog computer simulation, the program language is set of functional blocks, representing the elements and organization of an analog computer. This allows problems to be prepared directly and simply from either a block-diagram representation or a set of ordinary differential equations. The program language consists of a complement of standard functional blocks and a number of special purpose blocks, whose characteristics are specified for particular simulation needs. These functional blocks are the components by which a continuous system may be represented. Included are provisions for accepting application oriented statements for defining the connections between the interacting blocks. Each element block type is provided with a diagramatic symbol and a language symbol. This provides a simple and straight forward programming via a block oriented approach. The block types include such functions as integrators, limiters, summers, pulse generators, harmonic generators, jitter, relays, sample-hold, etc. There exists an almost one-to-one correspondence between most elements of a physical system and selectable language blocks. The special block functions are written in Fortran. They are used to represent the blocks not part of the standard complement, or to perform the composite function of several standard blocks.

In the last five years, a number of simulation languages have been developed. It has not been generally and sufficiently emphasized that these analog computing techniques are applicable to both small and large scale system modeling and analysis. The use of these simulation techniques can simplify the programming task, as many of them employ a block oriented language that can be applied directly to a system under study. The powerful features of the computer simulation program is conveyed in this paper by considering the problem of modeling a conical scanning radar response to an inverse gain ECM signal waveform.

SIMULATION PROGRAM DESCRIPTION

The simulation language employed in the modeling task is the Continuous System Modeling Program (CSMP) which is operational on the IBM 1130 digital scientific computer or any other computer that has a Fortran IV compiler. The selection of this particular program was based on its software package availability at the ITT Avionics computer facility.

A simulation program is initiated by the preparation of a system block diagram representing all the elements (or functions) and showing their inter-connections. The system may be visualized as a complex of sub-systems. Individual blocks of the subsystem are then translated into corresponding sets of CSMP language statements. Each of the CSMP element types specifies a functional relationship involving a maximum of three (3) input variables and as many parameters (or initial conditions). The output of each element is a single valued quantity defined by the particular functional relationship. The analyst need not be concerned with sequences of computations since an automatic sorting operation and proper order of calculations is provided by the CSMP executive routine. Within the program a centralized integration is provided with all integrations performed in parallel. A second order Rung Kutta technique is employed in the integration, with time the independent variable, advanced by a fixed amount with each evaluation of the derivative. The key modeled system responses can be plotted on a line plotter (in small digital systems) or printed out and/or stored in a disk file and later plotted or used as inputs to other simulation programs.

This is a very practical means for studying or evaluating ECM effectiveness. When a small digital computer is available, the essential operational flexibility of on-line monitoring and direct interacting are obtained by the use of keyboard console. The salient system responses can be graphically represented on a plotter, the printer, or the keyboard console. During the run, the modeling program may be interrupted for the purpose of modifying the inter-connection in the configuration,adjustments of parameter values, etc. From the interrupt point, the simulation may be instructed to continue the computations or begin a new run. This is especially convenient when an analyst uses a terminal hooked to a time-shared computer.

RADAR MODELING

To determine the effectiveness of a given ECM technique, a closed loop process is involved. An ECM signal is generated and applied to the input of the subject radar. The radar tracking response perturbation is then used as a measure of the ECM system effectivity for a given set of the signal parameters. To realize the required closed loop operation, a comprehensive model of a conical scan target tracking radar was developed. Using the CSMP approach, the angle tracking functions of the radar easily modeled. A schematic representation of a generic conical scan tracking radar modeled in the simulation study is shown in Fig. 1 (Skolnik, pp 167 [6]).

The development of a mathematical model of the conscan radar is an extensive task in itself. However, the model was primarily intended to be used as a scientific tool in the evaluation of IG ECM (which is primarily an angle degradation technique) , many of its features were simplified. Only the radar devices and functions which directly participate in the target echo and ECM signal angle processing were modeled.

Other portions of the radar which are needed for a closed loop type operation were idealized in their representations. Of specific importance are such system elements as the angle tracking servo, IF amplifier, AGC, antenna beam pattern, squint angle, angle demodulator, and antenna mount inertia. In this case, the target was assumed to be perfectly tracked in range.

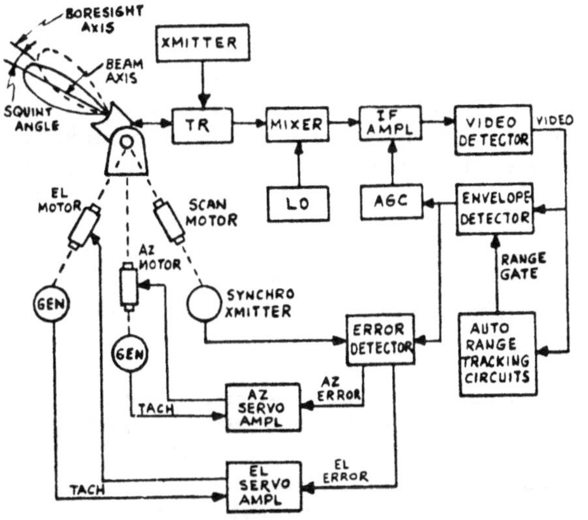

Figure 1
Typical Con Scan Radar

The angle tracking operation was always started by a set of initial antenna beam and target angular position values which allowed the radar to lock-on in angle.

Characteristics and parameters of some of the pertinent functions chosen for the radar model are:

Antenna Pattern: $A(\theta) = \dfrac{\cos \theta\pi}{1 - 4\theta^2}$

Beamwidth: $BW = 1.2^\circ$

Squint Angle: $\theta_S = 0.6^\circ$

Angle Track Network: Improved Type I Servo

Servo Transfer Function: $F(s) =$

$$\frac{125\,(1 + s/2.1)}{(1 + s/0.1)(1 + s/20)}$$

Servo Response Time = 0.50 sec.

Scan Frequency f_s = 50 Hz

AGC Bandwidth = 5 Hz

AGC Dynamic Range = 80 db

Pulse Repetition Rate PRF = 1 kpps

ECM MODELING

There are many ECM techniques for deceiving or degrading the angle tracking function of a conical scan target tracking radar. One particular technique, described in the open literature (Vakin Ref. [7] pp 144), is capable of producing angle errors and breaklocks when the ECM signal waveform is properly shifted in phase with respect to the radar transmitted conscan modulated signal. The ECM technique is called inverse gain (IG).

The required IG characteristics and the way it affects the angle tracking function can best be explaned in terms of the radar operation. A basic conscan radar generally uses a single dish antenna with a mechanically rotated feed. The feed is off-set, and as it is rotated the illuminated dish generates a pencil beam which scans the space about the boresight. When a target lies off-boresight, the echo signal at the radar receiver output contains amplitude modulations due to the beam scanning motion. The spatial beam pattern and the resultant signal modulation are shown in Figure 2 (a) and (b) respectively. To track the target in angle, signal is first stripped out of the nutation frequency, passed through a low pass filter, and applied to two angle tracking servo networks, which drive the boresight axis to coincide with the line-of-sight. As the boresight axis approaches the line-of-sight, the amplitude in the received signal decreases. When a constant signal level is reached, it is indicative of the antenna boresight being coincident with the target line-of-sight.

To counter the tracker, an ECM equipment aboard the target is required for intercepting the transmitted signal, detecting the scan modulation, and retransmitting back to the radar the signal with the appropriate modulation characteristics. The ECM signal is on-off modulated at the scan rate, and must be 180 degrees out of phase with respect to that of the radar as illustrated in Figure 2 (c). In the presence of this ECM signal, at sufficiently high Jam-to-Signal (J/S) values, the closed loop angle tracking servo is driven into an unstable operating region, resulting in a loss of angle lock.

SIMULATION MODEL DEVELOPMENT

To proceed to the actual simulation model synthesis, it is helpful to compose an overall system block diagram, including all the subsystems. Such a composite system block diagram is shown in Figure 3. Included in the diagram are provisions for generating target positions, target echo signal, and the ECM signal. Some of the

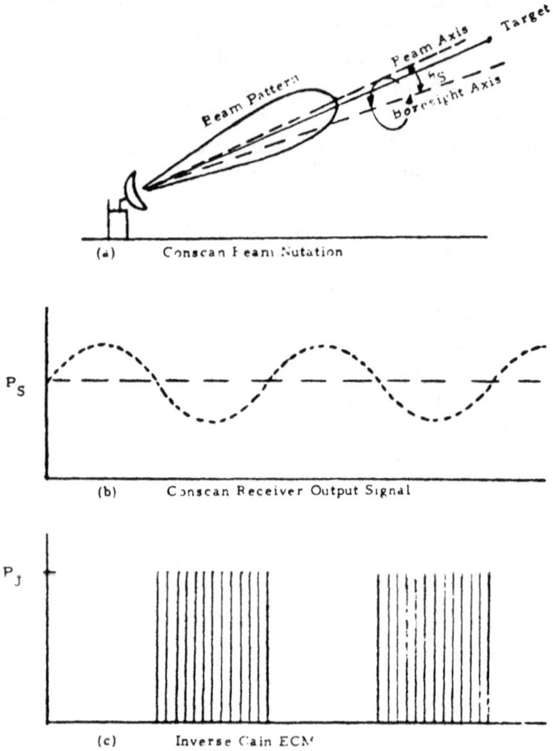

Figure 2

radar networks and other subsystems, in practice, are further expanded into individual elements before proceeding to the translation into CSMP blocks. When all the elements are defined, in either functional blocks (IF amplifier, AGC, boxcar detector, etc..) or in mathematical expressions (antenna beam pattern, servo network) the system is then converted to the CSMP form. Some networks are directly synthesized by replacing the analog elements with their counterparts found in the standard complement of the CSMP blocks (one for one replacement) while other networks take several CSMP blocks to synthesize their transfer functions. The AGC network, though nonlinear, is easily handled by a block for block replacement approach. The servo is readily synthesized by blocks on the basis of its root-locus transfer function expression. Such network functions as the antenna pattern spatial response, $\cos \theta \pi / (1 - 4\theta^2)$ expression, takes twelve (12) CSMP blocks for implementation. However, to conserve the source of blocks, this expression is best handled by one special block, which is written in the Fortran language. For the same reasons, many blocks whose characteristics are known and are not subject to modifications, have their transfer functions computed in a composite form in the special blocks. It is important to stress that all the system elements

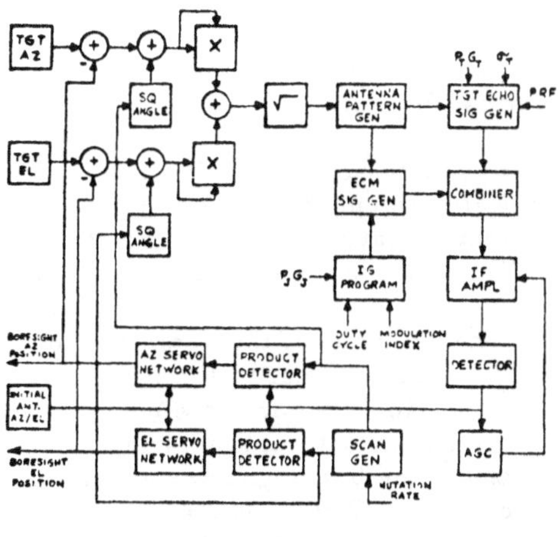

BLOCK DIAGRAM
ANALOG SIMULATION OF
IG ECM PROGRAM VS CON SCAN RADAR

Figure 3

subject to modifications and/or whose behavior is to be monitored should be explicit modeled by the individual CSMP blocks. To conserve space, the system CSMP block diagram for modeling the radar angle tracking response to the ECM signal input is shown in a condensed version in Figure 4. This diagram can now be easily converted to useful software. For programming convenience, the simulation system is viewed as a complex of subsystems. Each subsystem may be individually converted to the CSMP form, debugged, tested, and made operational as an entity in itself. For example, the servo frequency and time responses and stability for a given set of parameter values can be separately analyzed. Having determined the desired characteristic/performance, all the individual subsystems are then combined into one complete CSMP program to be used in the simulation study.

The entire simulation program is prepared in key punched cards and assembled into a source deck. The configuration cards contain block type symbols, block number, and the number of the input blocks. The initial condition cards contain the parameter values (constants, etc.). At the end of the deck are cards carrying such information as the integration interval, total run time, block numbers to be plotted, block numbers to be printed, etc. The source program is read by the computer which converts this higher level simulation language into Fortran statements and passes them to the Fortran compiler. The resulting object program is then loaded along the CSMP run time and executed. On a terminal, the simulation program can be prepared on a paper tape or directly entered on a teletype.

To conduct the simulation by means of the prepared CSMP program, one needs to specify the

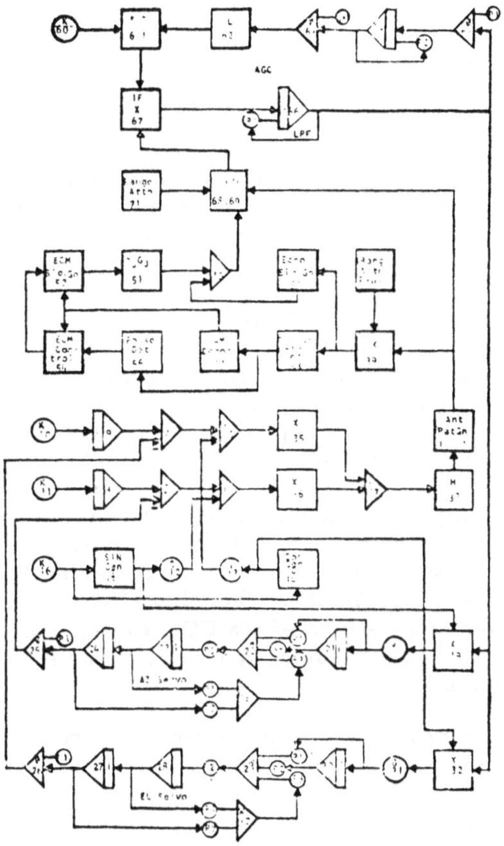

Figure 4- CSMP SIMULATION BLOCK DIAGRAM

the integration interval. The specification of the integration time interval value is very important as it impacts the system response/behavior and the execution time. In a pulsed radar system analysis, the pulse repetition interval (PRI) is usually taken for the integration step interval. For most conscan radar models the integration step is set equal to perhaps $1/10\ f_s$ to $1/20\ f_s$, where f_s is the nutation frequency. For pulsed doppler systems, where the PRF may be on the order of 10 to 200 kpps, the interval value selection is based on the network with the highest frequency response; otherwise the computation time is prohibitively long if the PRI interval is used.

Conducting a simulation study using the CSMP program approach is elegantly simple. The analyst is provided with an unlimited opportunity for observing the responses of any network in the system by means of on-line monitoring. Normally, (on a small computer) one system output is plotted on a plotter, outputs from five blocks are printed, and one to two outputs can be plotted on the keyboard console. In the course of the simulation run, the analyst can interrupt the computation and request a different set of blocks to be interrogated in printed or plotted form on the keyboard console. It is almost analogous to observing an actual system in operation on expanded time scale as one second of real time operation often, depending on the system complexity, results in several minutes of time on a small computer.

This modeling approach can be applied to study power management problems. It would, of course require the use of a simulation language with a larger complement of blocks (on the order of several hundred) and expansion from one-on-one to one-on many capability. All that would be required is an expansion from one to several target tracking radars and modification of the ECM receiver to handle the threat signal analysis, selecting one or more radars for countering. As more threat signals are selected and countered, the power management of the ECM system power can thus be macroscopically observed and its performance assessed.

ECM PERFORMANCE EVALUATION AND RESULTS

When a computer modeling study is conducted by means of the indicated approach, numerous radar networks are readily accessible to monitoring and their responses become available for establishing the ECM system/signal effectiveness. In the particular simulation study being used as an example the radar angle tracking performance was of a primary importance. Indicative of the angle tracking radar susceptibilities to the inverse gain ECM signal are the responses of the AGC network and both angle tracking servo networks which result in moving the antenna beam in space. All these responses may be observed as a function of time and compared to those produced by the radar when it tracked the target in a non-ECM environment.

Subjecting the conscan radar to an inverse gain ECM signal waveform causes the AGC network to reduce gain in the IF amplifier. The reduction in gain is directly related to the ECM signal power. Due to the narrow bandwidth frequency characteristics, the AGC network cannot respond to the echo signal during the half cycle (nutation frequency) when the ECM signal is absent (or at a reduced level). This AGC supression effect allows the modulated signal to be passed to the angle demodulators and farther to the servo network. In a closed loop type operation, the servo networks attempt to reduce the amplitude modulation which results in moving the antenna beam away from the target position. The antenna beam azimuth and elevation positions are indicative of the target location in space. Thus, an erroneous measure of the target position is attained.

A convenient way to represent the inverse gain ECM signal effectiveness is in terms of the antenna boresight deviation from the true target position. The antenna boresight loci at various J/S input levels (from 0 to 20 db) are shown in Figure 5. They represent the beam motion mapped in the plane normal to the beam boresight axis, involving six complete computer runs. Each run started out with identical initial conditions. Initially, the boresight was placed at an angle of about 0. 5 beamwidths away from the target position. Angle lock-on began on the target echo alone. After 0. 25 seconds of lock-on tracker mode of operation, the ECM signal generator was turned on A programmed device was provided within the ECM generator for determining the nutation frequency and the phase of the radar transmitted signal modulation and turning on the ECM signal transmission.

As illustrated by the computer plotted loci, the radar was able to maintain angle track, however, with an error, up to about 15 dB J/S value. The error progressively increased with the J/S values. For larger values, the radar could not sustain angle lock on the target. No attempts were made to reacquire the target after track break. All runs were of equal time durations. At the higher J/S values, the track break occurred in shorter times.

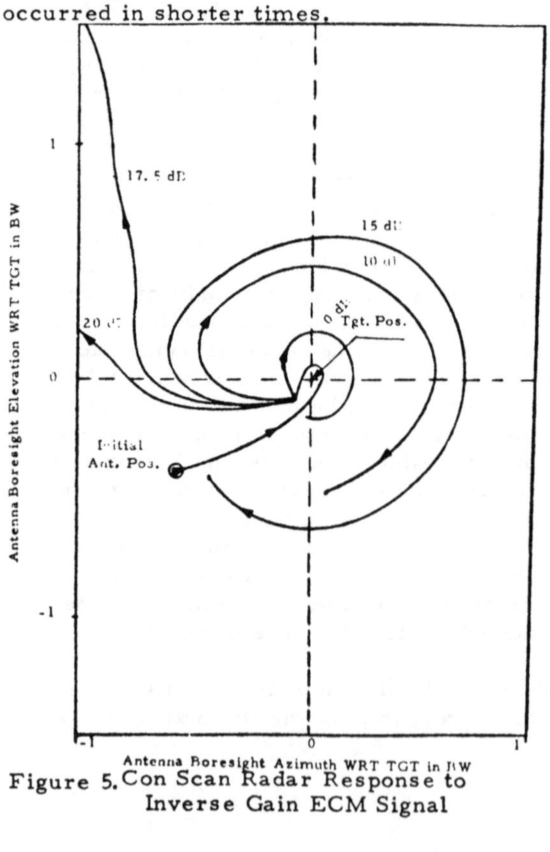

Figure 5. Con Scan Radar Response to Inverse Gain ECM Signal

CONCLUSIONS

This paper has considered the use of simulation languages as a method of analyzing and evaluating ECM techniques under development. The ease and flexibility of modeling radars and ECM has been illustrated by an example based on the use of the CSMP package. That this method applies to a wide range of digital computer and program libraries indicates it is available to a large number of engineers. Further, because of direct relationship to system block diagrams, the presented approach is simple to learn and use and is ammenable to parameter variation and analysis of its effects.

References

1. IBM General Purpose Simulation Systems Users Manual, IBM SH 20-0851, 1970.

2. IBM System/360 Continuous System Modeling Program, Users Manual, Application Program H20-0367-3, 1969.

3. Control Data Continuous System Simulation Language III (CSSL-3) Users Guide, Version 1, 1973.

4. IBM 1130 Continuous System Modeling, Program Description and Operations Manual, GH20-0282-2-, 1969.

5. Tymeshare, CSMP, TYMCOM-IX MANUALS, May 1972.

6. M. Skolnik, Introduction to Radar Systems, McGraw Hill Book Co., 1962.

7. S. A. Vakin, et al -Principles of Jamming and Electronic Reconnaissance. FTD, Wright-Patterson AFB, O, May 1969.

Recent Innovations in Radar Simulation
R.J. Hancock, P.D. Krolak, L.K. Wilson, and J.C. Cleary
IEEE Region III SOUTHEASTCON-75 Record, Volume 2, April 1975
Pages 6E-4-1 – 6E-4-9

9.6

Summary

Digital simulation of radar systems offers an economical approach to system performance prediction by combining measured or theoretical radar subsystem characteristics in order to determine total system performance.

This paper describes an advanced radar simulation model which permits the inclusion of "real world" data in the simulation for validation purposes.

Introduction

Radar design and fabrication techniques have reached a sophistication and complexity which make prediction of radar system performance difficult, if not impossible, by conventional analysis techniques. During the past decade much of the increase in sophistication of operational radar systems has been directed toward improving the ability of a radar to discriminate between targets of interest and interference produced by noise, ground clutter, chaff, and ECM. A few of the factors contributing to increased radar system sophistication are the following:

1. Improved techniques for manufacturing matched filters give the radar designer a wider choice of waveforms and allow better processor matching to received waveforms.
2. Improved digital storage techniques allow larger time spans of data to be processed with improved frequency resolution and increased integration times.
3. Solid state devices exhibit improved stability which in turn contributes to better system resolution properties.
4. Antenna apertures greater than 1000 wavelengths result in improved angular resolution properties.

The ability of radars to discriminate between targets and clutter, both natural and manmade, is sensitive to fine scale variations in subsystem parameters such as antenna sidelobes, Doppler filter characteristics, and sidelobe performance of pulse compression systems. The relationship between such subsystem parameters and system performance is not easily discernible through conventional analysis techniques. Therefore, to determine their effect on radar system performance, one of the following two approaches must be used:

1. Model the system by using a special purpose and highly flexible hardware system, or,
2. Model the system by using mathematical expressions and use an analog or digital computer to simulate the system according to the mathematical model.

In the remainder of this paper the RADC radar simulation computer model which resulted from implementation of the second approach is described.

Digital simulation offers an economical approach to system performance prediction by combining measured or theoretical radar subsystem characteristics in order to determine total system performance. The ability to include "real world" data in the simulation for validation purposes significantly enhances the acceptability of simulation as a radar system design tool. In addition, controlled simulation has one distinct advantage over field testing of a radar system; that is, all parameters can be rigorously controlled. Conditions can be repeated exactly so that parametric studies can easily be done.

The fact that simulation models have, at one time or another, been constructed to solve problems does not mean that a generalized radar simulation model is passè. Rather, a comprehensive model structured to minimize user set-up time, yet flexible enough to simulate a number of different radar systems, is a new capability.

A functional block diagram of the RADC radar simulation computer model is shown in Figure 1. The simulation model is capable of simulating a fixed ground based monostatic or bistatic radar operating in environments which are free of interference or which contain clutter, jamming, and/or chaff. It would be a simple matter of including a coordinate system transformation to allow simulation of a radar in a satellite, aircraft, or balloon.

Both ground and weather clutter may be simulated with the clutter radar cross section (RCS) determined from measurements data processed to derive the RCS cumulative probability distribution. The simulation model has the capability of generating phased array antenna patterns. The transmitter waveforms which can be generated are single frequency, interpulse frequency diversity, step-frequency diversity, linear FM swept frequency and discrete intrapulse phase codes. The model has provisions for the introduction of nonlinearities, weighting functions, and filters at any point in the simulation. The transfer functions of these devices are defined by user supplied input data. The input data may be either measurements data or theoretical approximations. Targets may be simulated by use of (1) a discrete scatter model, (2) RCS probability distributions based on measurement data, or (3) theoretical solutions for simple targets such as geometrical theory of diffraction, physical optics approximation, and Ufimtsev's fringe wave theory. A double delay line MTI processor is implemented for use in predicting radar system clutter rejection performance. The output data available from the simulation model will be in the form of simulated A-scope displays, probability density functions, cumulative distribution functions, power spectra, and autocorrelation functions.

Simulation Model General Description

In developing the radar simulation model, the intent was to develop a software package of individual simulation modules which represents specific elements of a radar and to develop the necessary framework to connect them together. The basic concept was that large complex radar systems can be accurately modeled by combining models of the elements which form the system. This approach allows one to develop the mathematical models of individual elements and then implement each in a separate software module. Each module may then in turn be tested against data of proven validity to determine if the module is correct.

In order to ensure that a useful design tool would result, certain requirements were imposed during the development effort: First, the radar simulation model must be efficient in terms of the time expended by the user in performing system evaluations. Second, the simulation model must be flexible so that a number of different radar systems may be evaluated without computer program modifications. Third, the simulation model must be accurate. The use of idealistic approximations should be minimized and the use of empirical data emphasized. In addition the accuracy of the model should be verifiable in a simple straightforward manner. Fourth, the simulation model should be structured such that new capabilities may be easily incorporated as the state of the art improves.

There are instances in which the generalized approach used in this simulation model is not the most efficient in terms of computer processor time. But, the use of a standardized approach for connecting various transfer functions and stimuli does minimize the time expended by the user in setting up a problem. The intent here is to expedite the translation of a radar system description into a meaningful simulation model computer program.

The development of a radar simulation model should logically be performed in a manner similar to that used in building an actual radar system. A radar system consists of a group of subsystems which are in turn composed of various functional elements. These functional elements are such items as amplifiers, analog-to-digital converters, and filters. If one is to construct a generalized multimode radar system simulation model then emphasis should be placed on accurately simulating the functional elements. The reasoning behind this approach is that the same functional elements are used in all radars; only the parameters change. For example, all radars have amplifiers, filters, and detectors, but the transfer functions of these devices vary from one system to another

Therefore, the first task in developing a generalized and multimode simulation model was to develop a library of mathematical models of the functional elements. These models were structured in such a manner that the parameters are easily varied. An example is given to illustrate this point. All analog-to-digital converters serve the function of converting sampled analog data to a digital representation. The primary differences between any two A-to-D converters are the value of the least significant bit (LSB), the number of output bits, and the sampling rate, f_s. Therefore, the parameters of an A-to-D converter model would be the LSB, number of bits, and sampling rate. This data would be provided by the user and entered into the computer through data cards.

In order to be useful the mathematical models of the functional elements must be implemented in the form of a computer program. This constituted the second task in the development of a comprehensive simulation model. The portion of the computer program which corresponds to a specific functional element was assigned the generic name, "simulation module." Typically a simulation module is a subroutine or a function subprogram. This task received careful attention because limitations imposed at this point could not easily be remedied at a later time after a number of subroutines were written. Therefore, a set of guidelines was established for specifying module parameters, the input data form and the output data form.

Once the library of simulation modules was constructed the next task was to provide a means of connecting the simulation modules together in order to form subsystem models and in turn to implement a radar system model. The analogous function of connecting hardware elements together is performed by coaxial cable, waveguide, or propagation through free space. Therefore, the third task was to develop a superstructure for connecting the modules together in a manner prescribed by the user. Simulation control cards serve as the vehicle by which this linkage is communicated to the simulation model. The superstructure is composed of a simulation controller and assorted support modules which do not have counterparts in radar systems.

The fourth task was to implement a data initializer which reads the user supplied data in the form of a punch card deck. This deck completely defines the simulation modules to be used, the manner in which they are connected, and the parameters which characterize the devices represented by the simulation modules. The punch cards which make up this deck are of two types, simulation control cards and module parameter input cards. The format of these cards is structured to minimize the user's task in setting up a simulation. In many cases the format of the user specified data is not suitable for direct use by the simulation modules and therefore must be converted to a different form. A simple example is the conversion of angles specified by the user in degrees to radians, a form suitable for use by the sine and cosine functions of the computer. In addition to processing module parameter input cards, the loader reads the simulation control cards and from these determines what modules are to be used and in what order they will be executed by the computer.

The following example is given in order to illustrate the relationship between a radar system block diagram and the corresponding simulation model. Figure 2(a) is the block diagram of a radar system. Figure 2(b) is the block diagram of the simulation used to represent that radar system. The names used in the blocks are those of the simulation modules. Those portions of the radar system which are linear and time invariant are simulated in the frequency domain; e.g., the IF amplifier bandpass and the matched filter. The clutter and target models are treated as linear time invariant processes and are represented by their respective impulse responses, c(t) and e(t). Multiplication in the frequency domain is used to introduce the effect of the linear time invariant processes on the radar waveform. The simulation module CONV performs the multiplication of the signal by the transfer functions. The alternative to modeling a linear time invariant process in the frequency domain is to perform a discrete convolution of the sampled time domain signal with the sampled impulse response of the linear system through which the signal passes. The nonlinear devices such as the saturating IF amplifier A( ) and envelope detector B( ) are simulated in the time domain. The Fourier transform ZFFT and its inverse ZIFFT are used to perform the conversions between the time and frequency domain representations.

In defining the simulation of the radar system, some simulation modules correspond on a one-for-one basis to radar subsystems. For example, the module FGENXY represents the transmitter and generates the transmitted waveform. In other instances two or more simulation modules are combined to synthesize one radar subsystem. For example, the IF amplifier is represented by a bandpass filter followed by a nonlinearity representing the saturation characteristics of the amplifier. In Figure 2(b) alternative representations of radar subsystems are shown with dashed outlines. For example, the matched filter can be modeled using the impulse response of the device, the measured transfer function in the frequency domain, or an s-plane specification.

System Elements

The functions performed by each functional element of a radar may be categorized into one of the following classes: (1) signal source, (2) passive device transfer function, or (3) modulator. The properties of these devices which are of primary concern to the radar system designer are those which can be measured at the terminals.

In order to simulate a signal source the envelope and phase modulation functions and the center frequency must be specified. In the radar simulation model provisions are made for generating the commonly used radar transmitter waveforms. Another signal source is noise which is simulated by a random number generator. In addition, if desired, the stimulus used in the model can be generated by some external source and introduced into the model through magnetic tape.

Transfer functions of passive devices may be of two types, theoretical or measured. The theoretical solutions are used for such devices as full-wave and square-law detectors, and filters specified in the complex frequency plane as poles and zeros. The approach used with theoretical transfer functions is to develop the equations which relate input to output with parameters used to specify those terms that vary from one device to another. When measurement data are to be used, a table of input vs. output is constructed and read into the simulation model. The subroutine within the model which simulates the passive device simply performs a table look up for each input sample of the signal. The data can be in the form of volts-in vs. volts-out or gain/phase shift vs. frequency.

The third type of device is the modulator. The operation of this device consists of multiplying the signal time function by the modulating function. This device is presently mechanized to use empirical data only. If desired, subroutines can be written to generate modulating functions according to some theoretical relationship.

Modular Software

Modular software was a central concept in the development of the computer program for the radar simulation model. Through the use of this concept, the program is structured in functional blocks which correspond to physical devices in a hardware system. This one-for-one correspondence facilitates the setup of simulations and allows independent checkout and verification of the accuracy of each module.

The executable statements which make up each module are based on a mathematical model of the device to be simulated. If the device is one which is easily characterized by an equation and parameters then the amount of input data specified by the user will be quite small. An example would be the specifying of pole and zero locations in order to define a filter. There are other devices which cannot be easily represented by an equation. An example is the transfer function of an amplifier which has a nonlinear response for weak and strong signals and an intermediate range which is linear. This device is most easily represented by a table of measurements data taken on the device. The subroutine for simulating such a device would, for each input sample, look up the corresponding output value from the transfer function table.

Each software module is allocated a unique Module Reference Number which is used by the data initializer in reading the proper data cards and in scheduling execution of the module. The Module Reference Number is used by the simulation controller to transfer control to the proper module.

Simulation Control

If the radar system simulation model is to have broad application in the analysis of varied radar systems, a means must be provided by which the user can arrange the logical structure of the simulation to correspond to the block diagram of the radar to be analyzed. This is accomplished by simulation control cards which schedule the execution of simulation modules. This determination of simulation model configuration is accomplished with no modification to the FORTRAN program.

The following example is given to illustrate the procedure followed in setting up a simulation for a radar system. The steps involved in evaluating a radar system are shown in Figure 3. The first phase of the process is the determination of the general characteristics of the radar based on the mission requirements. This step must be performed in any radar system design task. Typically, the results of Phase 1 are a tabulation of system level performance parameters. In Phase 2 a block diagram is generated which represents a radar system configuration that should satisfy the requirements specified in Phase 1. In order to evaluate the tentative radar system design, the simulation model must be configured to simulate the system represented by the block diagram. Simulation control cards are prepared to define the configuration of the simulation with each control card representing one step in the simulation. Once the control cards are prepared the radar system simulation model structure is defined. The next step in the design process is to define the characteristics of each block in the system block diagram. This is shown as Phase 3 in Figure 3. The specification of the parameters of each block is accomplished by module parameter data cards. There are some modules such as the envelope detector and square-law device which do not require input data. At this point in the process a set of punch cards which represents the tentative design of the radar system has been generated. The next step is to perform the simulation and compare the results to the requirements specified in Phase 1.

The execution of the simulation model is divided into two activities. The first activity, defined by the initializer load module, converts the data contained in the simulation control and data cards to a new format suitable for use in the simulation activity that follows. The reason for this preliminary step is that the format of the input punch card data was selected to minimize the effort required of the user in preparing a simulation. The reformatted input data from the input cards are stored on a temporary file for subsequent use in the simulation activity. A listing is printed which tabulates the modules to be executed and the input parameters of each. Upon completion of the data initialization activity, the simulation activity is performed.

In many cases it is desirable to investigate the effect of one parameter on the radar system performance.

In Figure 3 this is noted as Phase 4. Provisions have been incorporated into the model to facilitate the variation of parameters. When a parametric study is to be performed, the following sequence of events occurs: First, the basic radar system is defined with control and data cards as described in the previous paragraphs of this subsection. This defines the configuration and the first execution of the simulation. A MODIFY control card is then used to designate those blocks which are to be changed. A data card placed immediately behind the MODIFY control card changes the parameter(s) of the designated block.

Example Simulation

In this section a simple simulation will be described to illustrate usage of the computer model. The block diagram of the system to be simulated is shown in Figure 4(a). The corresponding block diagram of the simulation model is shown in Figure 4(b). Each block represents one step in the simulation and contains the Module Reference Number for the operation to be performed and the name of the simulation module to be used in the step. The module FGENXY generates the phase encoded waveform represented in the time domain. The Fourier transform module ZFFT converts from time to frequency domain. The module XYTODB converts the complex frequency domain representation to Energy Spectral Density and the results are plotted using the routine PLOTTR. The transmitted waveform is processed by the simulation module FILT which represents the receiver bandpass filter. The transmitted waveform is now ready for processing by the phase decoder.

In this simulation job the transfer function of the phase decoder is obtained by using the module FGENXY. This is done by simply running the transmitter backwards to get the desired matched filter impulse response. There are other modules available in the simulation computer model for representing surface acoustic wave devices and digital phase decoders.

The module CONV multiplies the filtered transmitted waveform by the phase decoder transfer function. The waveform is then converted from frequency domain to time domain by the inverse Fourier transform module ZIFFT. The resulting complex envelope is converted to its corresponding real envelope by the module XYTOM and the output plotted by the module PLOTTR.

The object in performing this simulation was to determine the effect of receiver bandwidth on the peak response and sidelobes of the phase decoder output waveform. The parameters of the transmitted phase code are the following:

Subpulsewidth = 100.0 nanoseconds
Switching time = 5.0 nanoseconds
Number of Subpulses = 15
Binary Phase Code = -+-++--+---++++

The receiver bandpass filters used in the simulation were:

Butterworth Design (4 Pole)
100 MHZ bandwidth
10 MHZ bandwidth

Chebyschev Design (4 Pole, 3dB passband ripple)
100 MHZ bandwidth
10 MHZ bandwidth

Both filter designs were normalized to give unity gain for the center frequency of the passbands. The parameters of the phase decoder are the following:

Bandwidth = 200.0 MHZ
Tap Spacing = 100.0 ns
Tap Phase Code = ++++---+--++-+-

The transmitter output spectrum generated by the simulation is shown in Figure 5. The envelopes of the decoder output waveform for the 100 and 10 MHZ bandwidth Butterworth receiver filter designs are shown in Figures 6 and 7, respectively. Similarly, the output for the 100 and 10 MHZ bandwidth Chebyschev design is shown in Figures 8 and 9, respectively. By comparing Figures 7 and 9, the effect of filter transient responses on time sidelobes can be observed. The Q of the Chebyschev poles is appreciably higher than that of the Butterworth poles and therefore the Chebyschev filter has a transient response which damps out more slowly than that of the Butterworth filter. The peak response for a Chebyschev filter is less than that of the corresponding Butterworth filter of the same bandwidth because less signal energy is passed by the Chebyschev filter.

Conclusion

The RADC radar simulation computer model is an application oriented software package which will allow radar system design engineers to configure simulation models of various radar systems by combining models of functional elements which are commonly found in radars. The computer model is composed of the following: (1) A set of modules which simulates the functional elements used to construct radar systems; (2) a set of input/output modules for reading and interpreting user supplied data and for converting simulation results into a form suitable for interpretation by the user; and (3) a framework for connecting the modules together to form the simulation model of a radar.

In summary, we have not a model of just one radar, but rather, the models of the basic ingredients required to construct the model of virtually any radar system. While no claim is made that this computer model can simulate every radar system, it is claimed that the superstructure and the elements provided will satisfy most of the requirements for simulating any system. As necessary, the computer model can be expanded by adding special modules developed to represent those aspects not currently available.

Acknowledgment

The radar simulation computer model described in this paper reflects the results of work funded by the Rome Air Development Center Laboratory Directors' (Discretionary) Fund, Customer Project Order 72-RADC-65 (Job Order No. 01707201) under Contracts F30602-72-C-0393 and F30602-73-C-0380. The Input/Output characteristics of the computer model are being enhanced by development of a Dedicated User Interface under Contract F30602-75-C-0063.

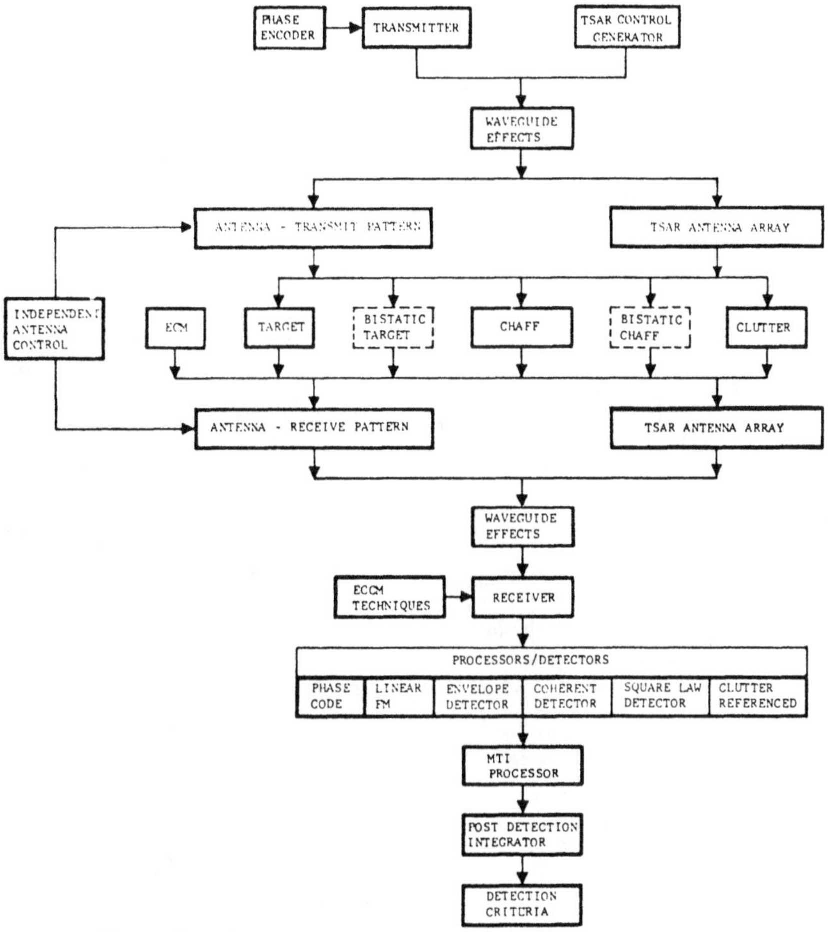

Figure 1: RADAR SYSTEM SIMULATION MODEL BLOCK DIAGRAM

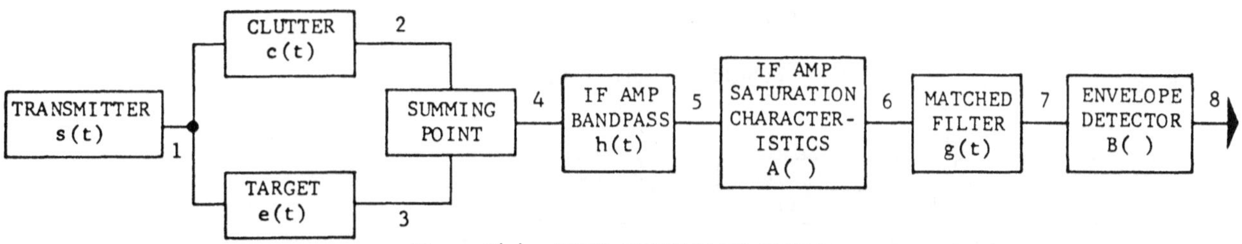

Figure 2(a): RADAR SYSTEM BLOCK DIAGRAM

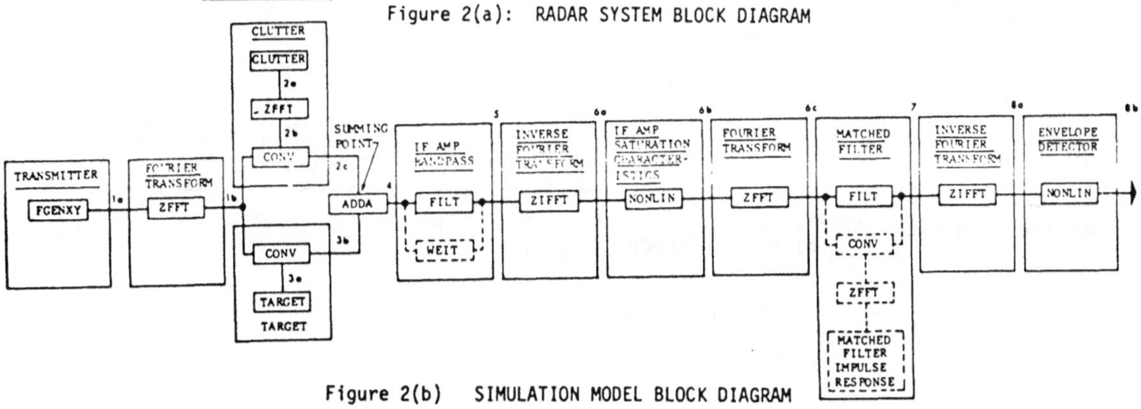

Figure 2(b) SIMULATION MODEL BLOCK DIAGRAM

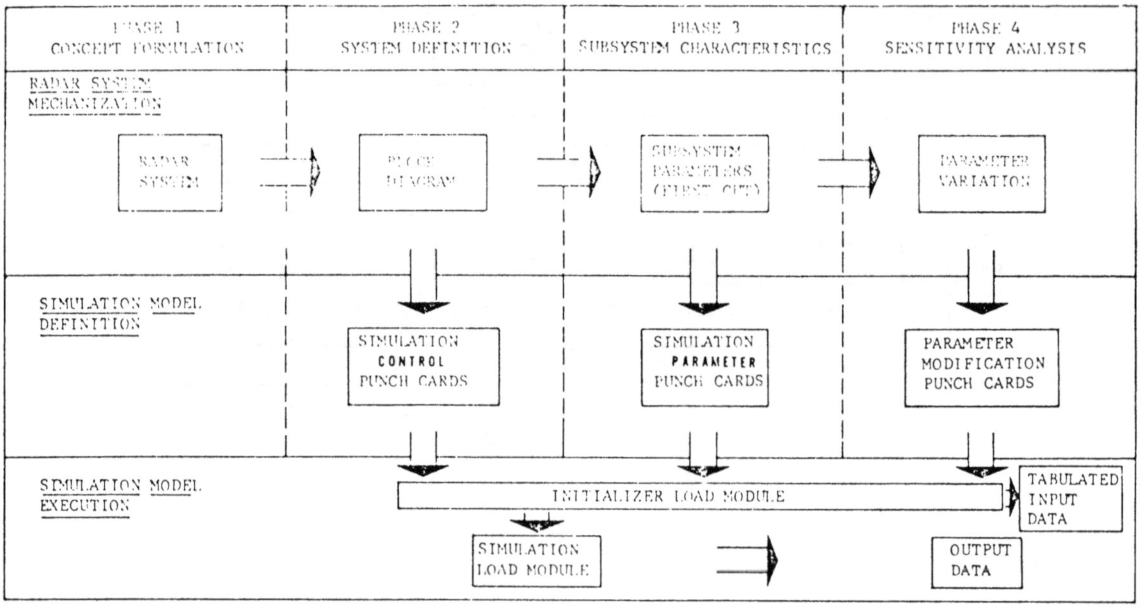

Figure 3: RADAR SYSTEM EVALUATION

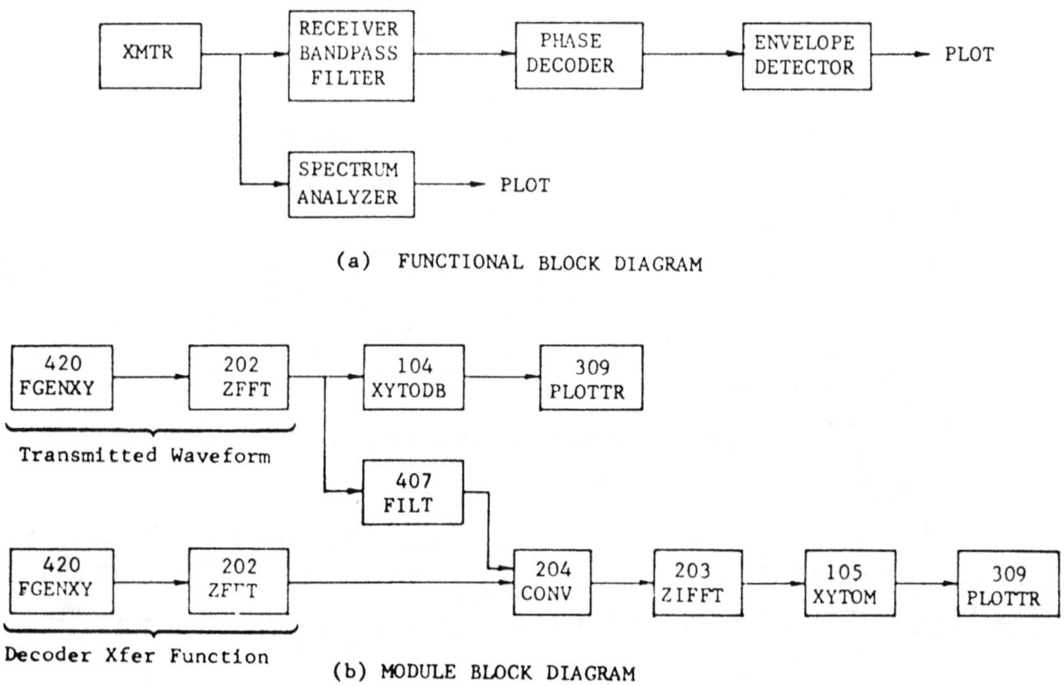

(a) FUNCTIONAL BLOCK DIAGRAM

(b) MODULE BLOCK DIAGRAM

Figure 4: EXAMPLE SIMULATION

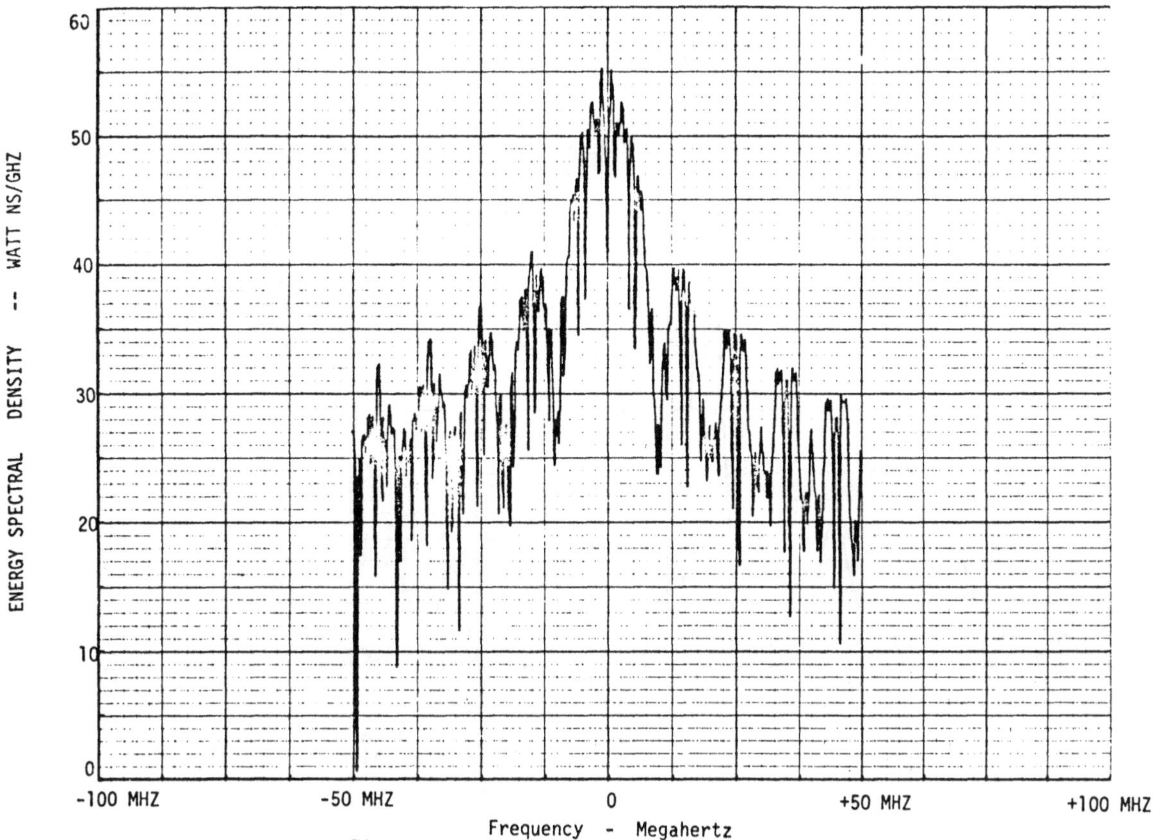

Figure 5: SPECTRUM OF 15 BIT BINARY PHASE CODE

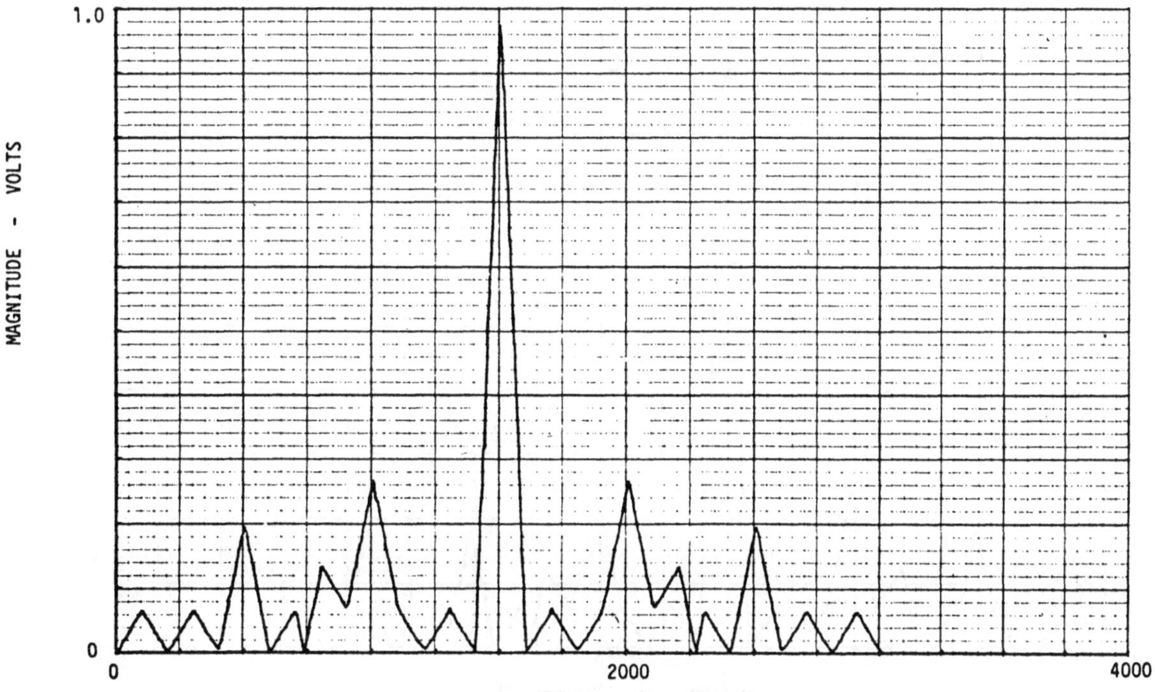

Figure 6: DECODER OUTPUT WAVEFORM ENVELOPE
(100 MHZ Butterworth Filter)

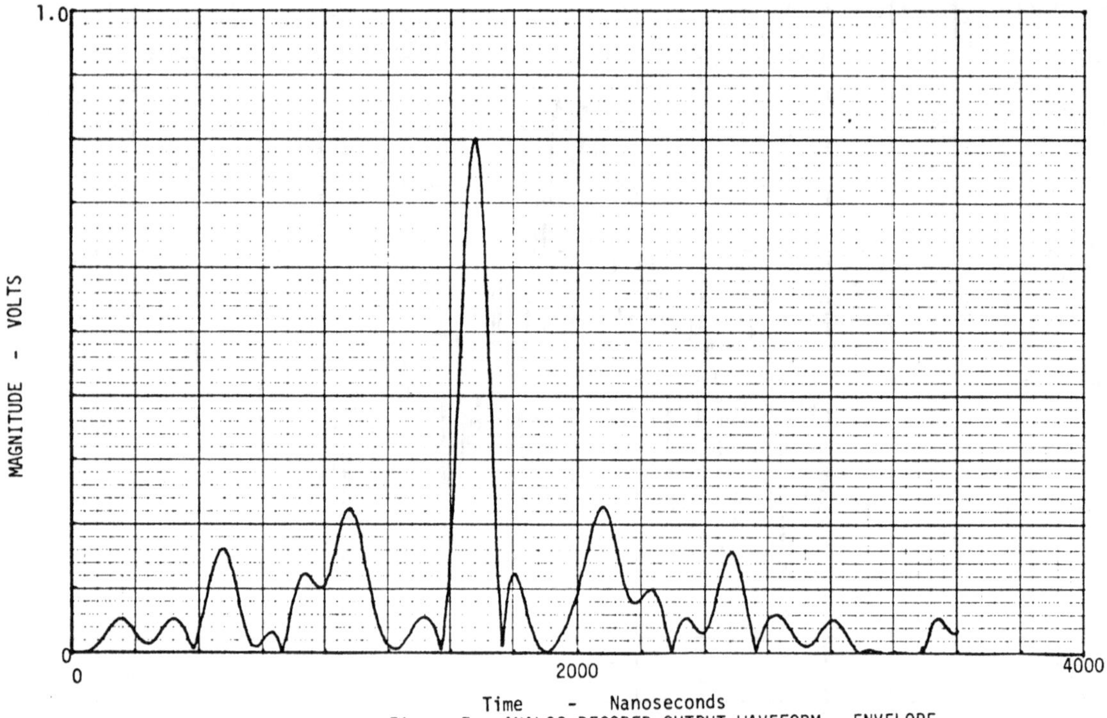

Figure 7: ANALOG DECODER OUTPUT WAVEFORM - ENVELOPE (10 MGZ BW Butterworth Filter)

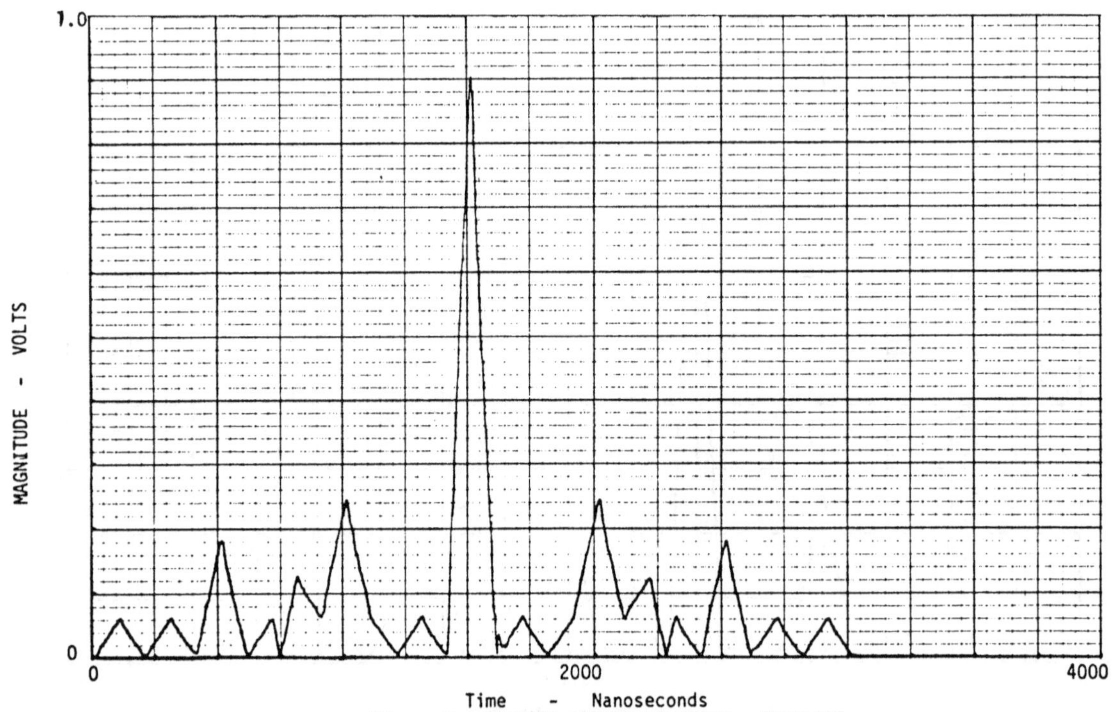

Figure 8: DECODER OUTPUT WAVEFORM - ENVELOPE (100 MHZ Chebyschev Filter)

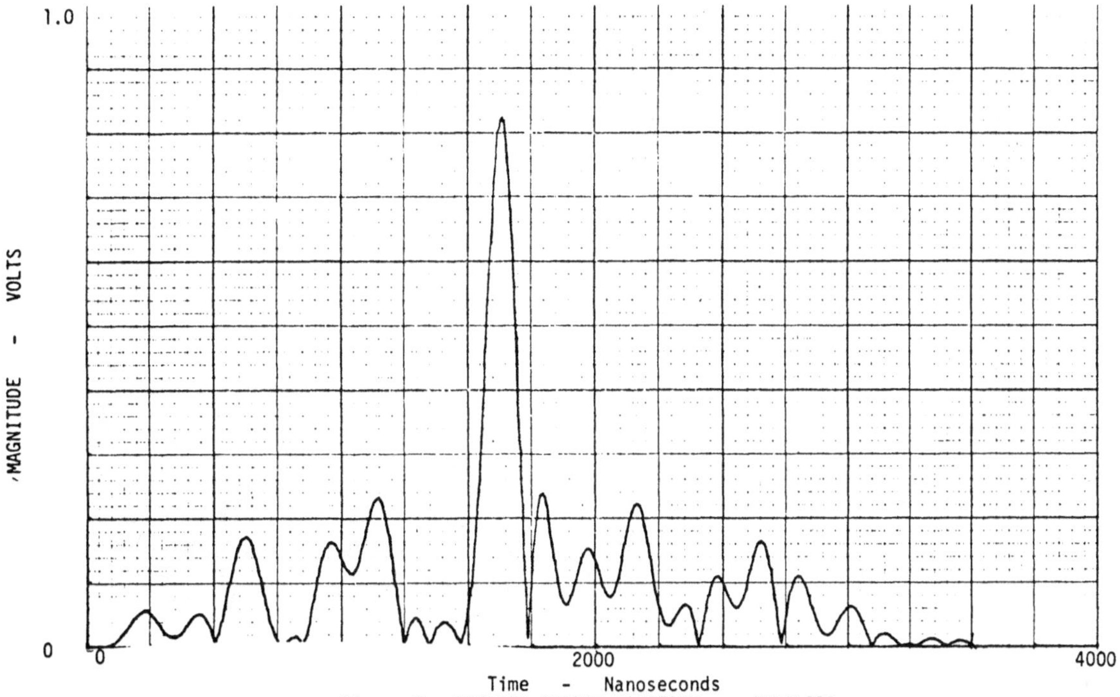

Figure 9: DECODER OUTPUT WAVEFORM - ENVELOPE
(10 MGZ Chebyschev Filter)

Applications of Computer Simulation Techniques to the Evaluation of Radar System Performance in ECM/ECCM Environments
S.L. Gutsche, R.L. Bogusch, and F.W. Guigliano
Proceedings of the DDRE Radar ECCM Symposium, 1976
Pages 39-1 – 39-16
Reprinted by permission.

9.7

SUMMARY

In this paper, the utility of detailed digital computer simulations to analyze radar performance in realistic combat electromagnetic environments is discussed. The paper first briefly describes the methodology of simulation augmented analysis in addressing the problem of evaluating the effectiveness of radar system and subsystem performance. A general description of the simulation capabilities is presented, followed by an example of the application of the simulation to a specific ECM problem.

Results from the simulation to date have shown that the interaction between a modern radar and a hostile electromagnetic environment is quite complex and that a realistic assessment of radar performance in these environments requires detailed modeling of the radar, environment, and all important interactions. Furthermore, analysis and examination of simulation results has been shown to improve the performance of friendly radars (and ECCM techniques) as well as to detect and exploit the vulnerabilities of hostile radars.

I. COMPUTER SIMULATION AS AN ECM/ECCM ANALYSIS TOOL

There exists a need to evaluate vital weapon, radar, optical, and communications systems performance in realistic combat electromagnetic environments. We are aware of the difficulties involved in performing such evaluations, and believe that the great complexity of the problem suggests that the usual straightforward, analytical "paper study" approach, by itself, cannot adequately represent or treat all of the important details. For example, in the case of a radar system the signal processing, monopulse techniques, treatment of angular and range errors, response to multiple targets, use of doppler information, ECCM fixes, actual system parameters (antenna gains and rotation rates, frequency, polarization, modulation techniques, power levels, emission bandwidths, pulse characteristics, receiver sensitivities, etc.), all combine in an interactive way to determine the vulnerability of that particular radar to certain electronic countermeasures likely to be employed. In consideration of these complexities, we discuss a method to augment the analytical approach by the use of powerful, detailed, existing computer simulations in order to best evaluate the response of each key system in each scenario and electromagnetic environment of interest. Thus, the simulation can allow rapid identification and investigation of radar sensitivity to key problem parameters. This can be done in conjunction with experimental programs, greatly reducing their cost and increasing their utility.

The simulation-augmented approach inherently contains several advantages not found in other approaches. The generalized nature of the simulation implies that its application is not limited to specific radar types, but can be readily extended to include nearly any system of interest with only minor modifications. Secondly, the simulation approach provides for the generation of real-world operational results (as opposed to those produced by a textbook treatment) which can be viewed as "data" in the absence of, or in conjunction with, actual experimental verification. A third major advantage involves the relative ease of performing variational parametric analyses to determine optimum system performance in a hostile, countermeasure environment. Indeed, the variation of important problem parameters to determine the sensitivity of radar systems operation and response to changes in those parameters has been extensively employed to evaluate the effectiveness of many various radars under realistic conditions.

The general question to be addressed is usually the following: Can the radar system complete its intended mission in the hostile countermeasure environment anticipated in a realistic combat scenario? Specific questions to be answered include: Is the reduction of system or subsystem effectiveness permanent? If not, what is the duration of the ineffectiveness? Are cost-effective mitigants available? Can alternate systems or subsystems be employed to complete the mission? Are existing or proposed ECCM fixes effective against the threats of interest?

Figure 1 presents a general statement of the problem as well as a general outline of how it can be approached using the simulation as an aid. This methodology has been successfully employed on several diverse radar systems and subsystems.

II. THE SIMULATION

Limitations of space do not allow an adequate description of the capabilities of the simulation we have employed in this problem, so only a brief discussion is presented.

PROBLEM: EVALUATE THE EFFECTIVENESS OF VITAL RADAR SYSTEMS AND SUBSYSTEMS IN A REALISTIC COMBAT EM ENVIRONMENT

GENERAL METHOD OF SOLUTION:

- DEFINE SYSTEM'S MISSION
- DEFINE THE EM ENVIRONMENT IN WHICH IT MUST OPERATE
- DEFINE THE THREAT SCENARIO
- OBTAIN IMPORTANT SYSTEM PARAMETERS
- CONSTRUCT NECESSARY MATHEMATICAL MODELS (IF THEY DON'T ALREADY EXIST)
- EXERCISE SIMULATION FOR CASES OF INTEREST
- PERFORM PARAMETRIC ANALYSES
- ANALYZE RESULTS, IF UNACCEPTABLE SUGGEST AND EVALUATE ALTERNATIVES
- REACT TO INFORMATION OBTAINED

FEEDBACK LOOP

FIGURE 1. SIMULATION-AUGMENTED ANALYSIS

The development of the simulation was originally initiated to allow the evaluation of sophisticated Ballistic Missile Defense (BMD) radars in complex electromagnetic (EM) environments. This development has continued over a four year period under sponsorship by various DoD activities. To date it has been applied to a wide spectrum of problems and radar types.

The simulation includes detailed models of complex EM environments, including provisions for ground, air, and space platforms; multiple radar types (surveillance, track, etc.); and multiple sources such as other radars, jammers, chaff clouds, clutter regions, nuclear weapons effects, etc. Important radar signal propagation effects, signal processing, and radar software functions are also modeled in great detail. As an example, the mean propagation effects modeled include absorption, refraction, multipath, focusing/defocusing, phase shifts and dispersion. The statistical propagation effects of phase, amplitude, and angular scintillations are also included.

Figure 2 presents a few major features of the signal processing simulation and certain radar software functions.

Figure 3 shows typical simulation output for a particular radar in a particular EM environment. Pulse-to-pulse signal-to-noise ratios are generated, as are corresponding errors in elevation angle. Similar results are available for azimuthal angular errors and range errors. In this particular example, up to a problem time of about 219 seconds there

ELEMENTS OF SIGNAL PROCESSING SIMULATION

- MATCHED FILTER SIMULATION
  - COMPLEX AMBIGUITY FUNCTIONS (CW, CHIRP PULSE, STEPPED-FM, ETC)
  - SINGLE PULSES, PULSE TRAINS, MULTIPLE SIGNALS
  - RANDOM NOISE
  - JAMMER NOISE
  - COMPUTES COMPLEX SIGNALS + NOISE SAMPLES
  - SUBSEQUENT PROCESSING CAN BE COHERENT OR NONCOHERENT
- DOPPLER FILTER BANK
- MONOPULSE ANGLE MEASUREMENT
- VIDEO SIGNAL SAMPLING
- SIGNAL PROCESSING LOGIC
  - VIDEO SAMPLING TIMES
  - DOPPLER FILTER SELECTION
- FALSE ALARM SIMULATION

RADAR SOFTWARE FUNCTIONS

- DATA PROCESSING MANAGEMENT (RADSOF)
  - DATA INTERFACES
  - FUNCTION SELECTION
  - FUNCTION TERMINATION
- SURVEILLANCE FUNCTION (RSURV)
  - SURVEILLANCE INITIATION
  - DETECTION
  - KNOWN OBJECT RECOGNITION
  - SURVEILLANCE CONTINUATION
- VERIFICATION FUNCTION (RVERFY)
  - FALSE ALARM RECOGNITION
  - DETECTION/VERIFICATION
  - REQUEST TRACK INITIATION
- TRACK FUNCTION (RTRACK)
  - TRACK INITIATION
  - TRACK DATA FILTERING
  - COAST TRACK DECISION
  - LOST TRACK DECISION
  - REQUEST SURVEILLANCE

FIGURE 2. MAJOR ELEMENTS OF RADAR SIMULATION SIGNAL PROCESSING AND SOFTWARE FUNCTIONS.

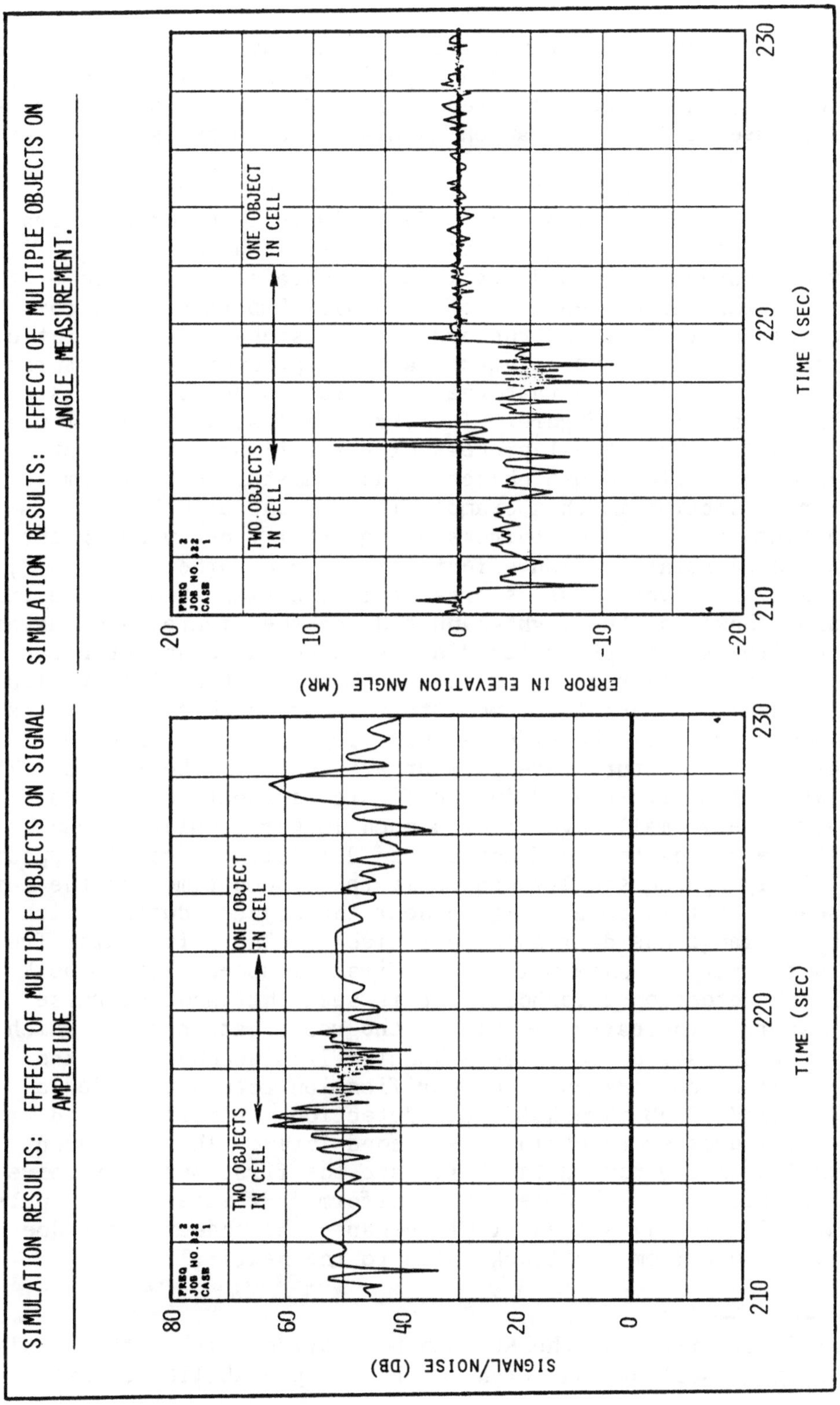

FIGURE 3. SAMPLE SIMULATION RESULTS.

were two targets in the radar range cell, causing amplitude beating by the two RCS contributors as well as a negative bias in elevation angle. After 219 seconds, the targets are separable and the rate of S/N scintillation decreases and the elevation angular error bias is eliminated.

Figure 4 is another example of simulation output for a different radar and a different EM environment. This figure is a computer-drawn plot of the simulated video signal out of the sum channel matched filter for a few radar surveillance and track pulses obtained from some of the simulation runs that have been made of a complex BMD radar. In this plot, the abscissa is range and the ordinate is signal-to-noise ratio. In these simulation runs, the objects (targets) producing the returns are approaching the radar, so that the first return pulse shown is at the right hand side of the figure. The figure illustrates the simulated video waveforms for the overall process of surveillance, verification, and tracking of a single smooth object in an undisturbed environment. The object cross section in this example is constant at 0.5 square meters (constant cross section targets are generally not used except to focus attention on scintillation effects produced by other sources)*, and normal radar receiver noise is present in the simulation. The first four waveforms shown, at the right-hand side of the figure, are surveillance returns from the target which did not exceed a 15-db detection threshold. The surveillance PRF is 40 Hz. Many additional surveillance pulses are transmitted over the time interval during which these first four returns occur — only the video signals for the scan raster positions in which target returns are present are shown in the figure. The total simulation time ilustrated in the figure is about 10 seconds (this is only a very small segment of a much longer simulation run). The surveillance pulses that did not exceed detection threshold appear relatively widely spaced simply because it takes some time for the surveillance scan to return to a position near the object, during which time the object range has decreased appreciably. The fifth pulse from the right is the surveillance pulse that first produced a video peak above the 15-db detection threshold. It is seen that another pulse occurs very shortly thereafter — this is the verification pulse which the simulated radar system software scheduled 1/40 second after the surveillance detection. Because the verification return exceeded a software-specified 12-db threshold, the detection was verified and track initiation was requested to begin 0.25 second later. The subsequent pulses shown (advancing toward the left) are the first set of received track pulses, which were scheduled at a uniform 4 Hz rate. These pulses appear relatively closely spaced simply because the object range does not change very much from one track pulse to the next.

---

*Statistical fluctuations of the RCS can be provided within the simulation by sampling Swerling, Rayleigh, or Rician probability density functions.

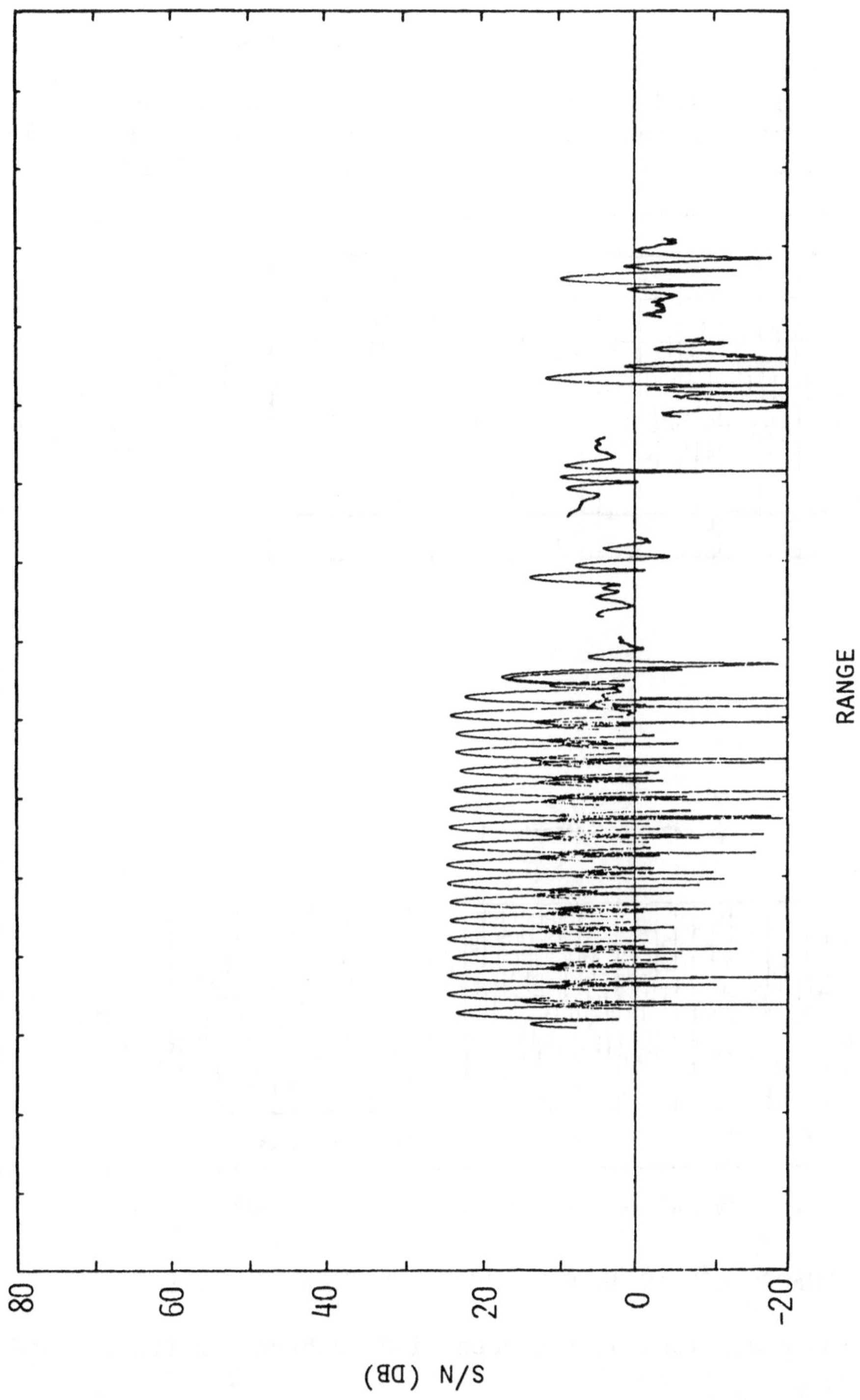

FIGURE 4. SIMULATION RESULTS: SURVEILLANCE, VERIFICATION, AND TRACK VIDEO WAVEFORMS.

Figure 5 shows an example of the comparison of simulation-produced S/N (or RCS) histories to data which was later taken. In this particular example, the tracked target was a satellite and the parameter of interest was the radar signal amplitude scintillation. Good statistical agreement was observed between the predicted and observed data.

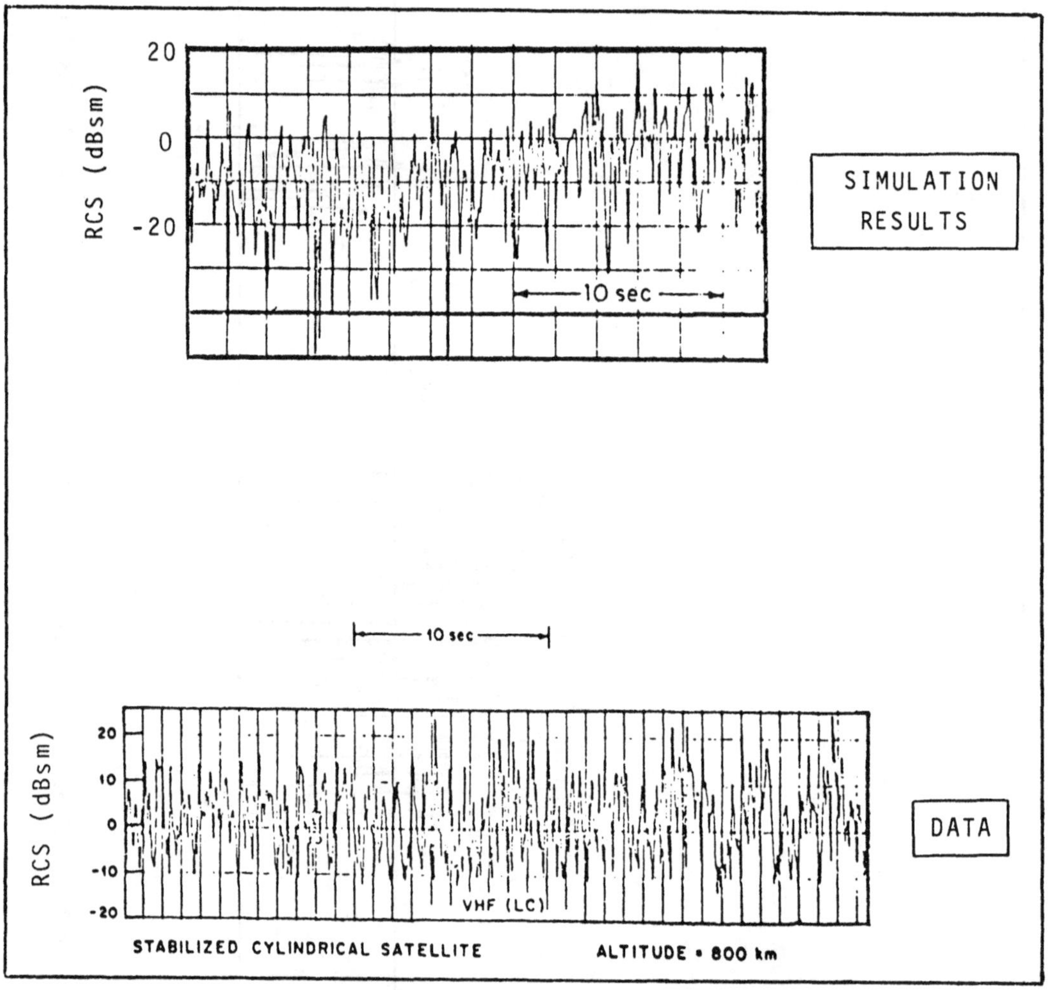

FIGURE 5. COMPARISON OF PREDICTED AND OBSERVED SIGNALS.

III. RADAR PERFORMANCE IN AN ECM ENVIRONMENT: AN EXAMPLE

The initial conditions of the scenario* we have constructed and

---

* This scenario is chosen for demonstration purposes only. Generation of results from a more detailed threat scenario or task force scenarios is also readily accomplished.

examined specifically for this paper are presented in Figure 6. A shipboard radar is indicated in the center of the plan view figure. The radar is assumed to be a UHF chirp pulse (compression ratio of 100) air search type. Specific radar parameters chosen include a 300 pps PRF; 50 μs pulse width; 10° azimuthal beamwidth, 20° elevation beamwidth, 6 rpm revolving 21 dB gain antenna; and a CFAR, 10 dB threshold-for-display PPI output.

The threat is posed by two flights of aircraft, five aircraft in each flight. Flight 1 is initially located at a distance of 100 km (54 nm) from the ship at an angle of 30° north of west. This formation is at an altitude of 10 km (33,000 ft) flying due east at a speed of 600 m/sec (1350 mph). The aircraft are in a linear formation as seen in Figure 6, with an average separation distance between aircraft of 100 meters.

The second threat is Flight 2, initially positioned at a distance of 80 km (43 nm) from the ship at an angle of 20° south of west. These aircraft are at an altitude of 5 km (16,500 ft) flying directly at the ship at a speed of 300 m/sec (670 mph). They are in a V formation as seen in Figure 1, with an average separation distance of 100 meters.

The third threat feature is the stand-off jammer aircraft located at an initial distance of 120 km due west (65 nm) from the ship. The altitude is 12 km (40,000 ft) and the aircraft is flying at a speed of 200 m/sec (450 mph) at an angle 20° east of south.

We shall examine the development of this engagement, using the simulation, for an elapsed time of 180 seconds, at which time the threatening aircraft formations will be at locations noted by the dark triangles of Figure 6. We shall present some representative simulation results during this three minute time frame.

Figures 7 through 10 are computer drawn plots of simulated PPI displays for the radar. The radius of the circle shown is 120 km (65 nm) chosen because that is the furthest range of any of the target complexes. Except where otherwise stated, all plots are standard PPI displays where the angular position corresponds to the antenna beam center azimuth. The radial distance from the center of each display is linear with range from the ship's radar. The intensity of the "blips" is a function of the received signal-to-noise ratio from that target complex on that scan.

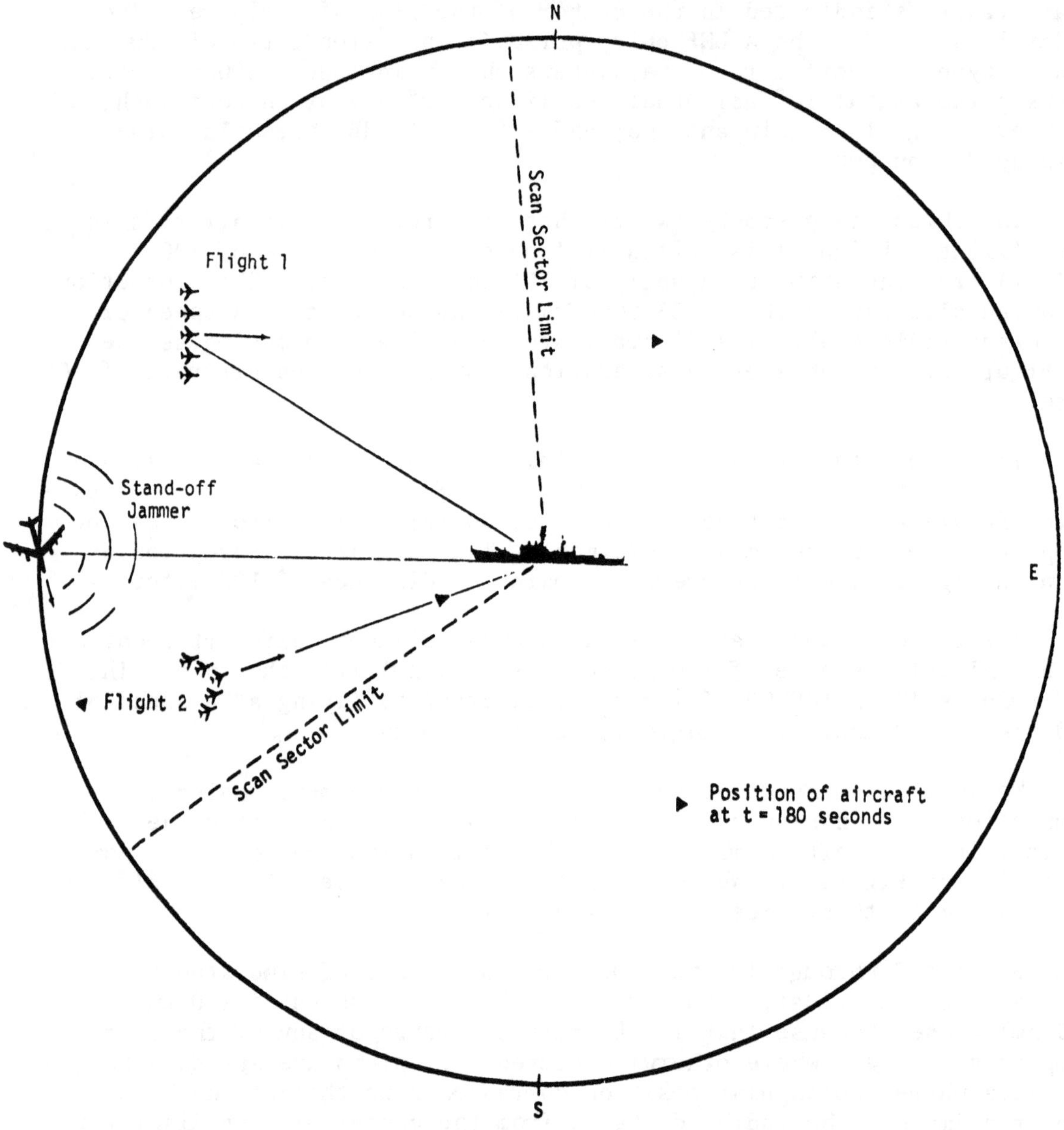

FIGURE 6. SCENARIO — INITIAL CONDITIONS.

Figure 7a shows the computer simulated PPI display for the engagement scenario previously described, without the stand-off jammer. Data is shown for times of t=0, 20, 40, 60, 80, 100, 120, and 140 seconds. Each antenna revolution requires 10 seconds at the specified 6 rpm rate and results are presented for every other scan (every 20 seconds). From the figure it is seen that Flight 1 is detected on the first scan, while Flight 2 is not detected until t=20 seconds. This is a result of Flight 2 being at an initially lower elevation angle with respect to the radar (about 3.2° at t=0), which places it in the side lobes of the antenna pattern, the center of which is taken to be placed at an elevation angle of 15° for the examples shown here. As the problem progresses and both flights approach the ship, the elevation angles, of course, increase. The returns from the two flights are spread in angle on the simulated PPI displays due to the 10° azimuthal beamwidth of the radar antenna. The returns are also spread in range, although very narrowly due to the good range resolution of the chirp pulse, providing about 100 meter range resolution. The individual aircraft within the two flights are not resolvable by the untrained eye, due to the relatively wide azimuthal beamwidth.

Figure 7b shows how the detection of the flights is unaffected by the stand-off jammer operating at the low power level of .1 w/MHz over the radar's band. Figure 8 shows how the stand-off jammer becomes more effective as the jammer power is increased. In Figure 8b we observe that Flight 1 is not detected until t=100 seconds, and Flight 2 not at all during the 140 seconds of problem time.

There are a number of factors that affect the target returns on a pulse-to-pulse and a scan-to-scan basis in addition to the fact that the range to the aircraft is decreasing with time over the period shown. First, random noise in the simulated radar receiver causes pulse-to-pulse and scan-to-scan variations in the target signal-to-noise ratio. Second, the aircraft separation in range is comparable to or less than the chirp pulse range resolution, and hence, the simulated radar receiver signal processor has multiple signals present which interfere either constructively or destructively, depending on the relative phases of the signals. This also produces pulse-to-pulse and scan-to-scan variations in signal video amplitude at the output of the receiver matched filter. Third, the RCS of each aircraft continually changes with time as the aspect angle changes, which can cause scan-to-scan variations in the PPI displays. This is clearly evident in the weak return from the 60 second scan of Flight 1. At that particular moment, the alignment of the aircraft with respect to the radar is such that the RCS of the target complex (and probably the individual aircraft) is in a relative null. This is a very real effect, and can only be generated by a simulation using realistic RCS data. (For this problem, anechoic chamber data on the threat aircraft was digitized and entered into the computer code).

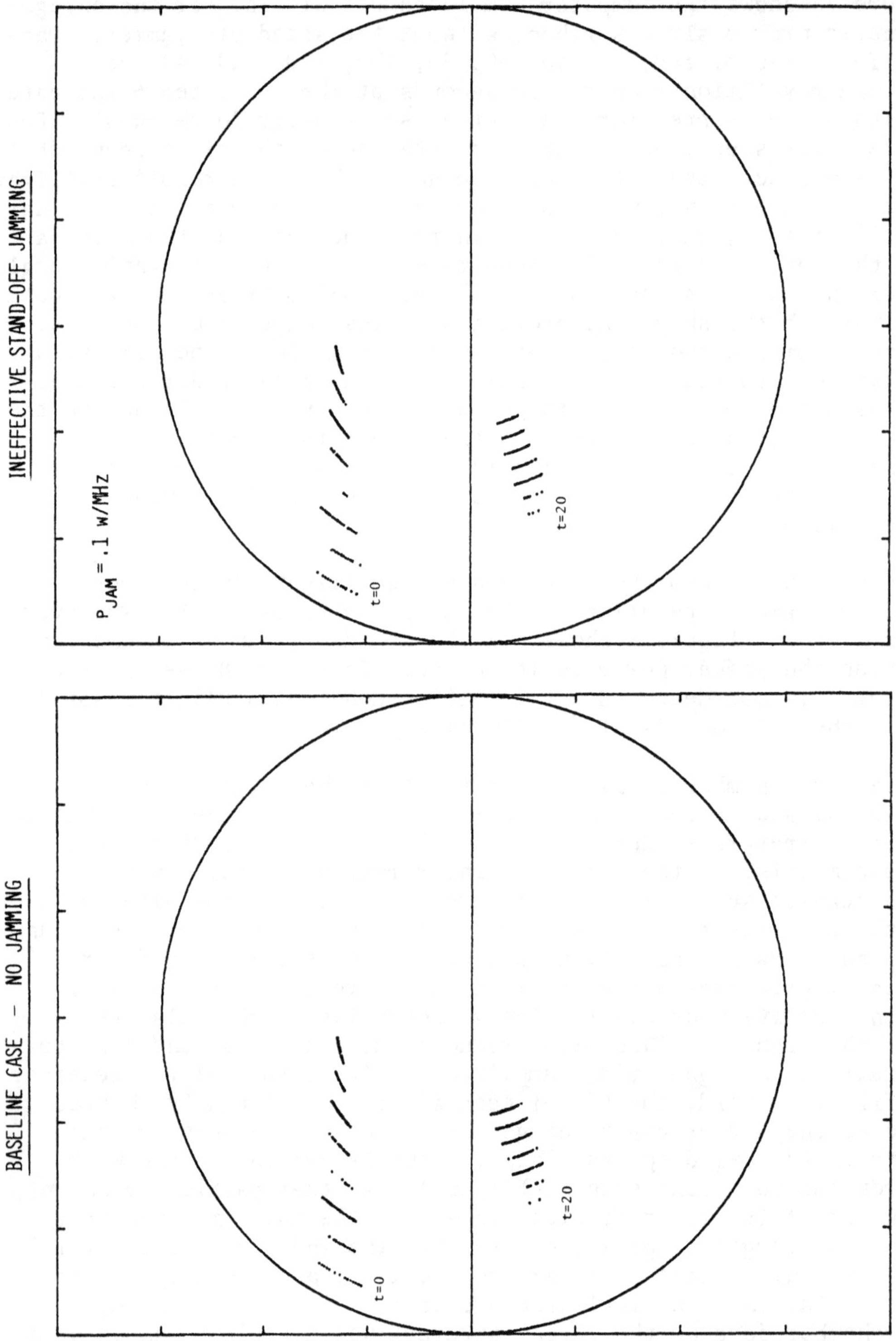

FIGURE 7. BASELINE CASE AND INEFFECTIVE STAND-OFF JAMMING RESULTS.

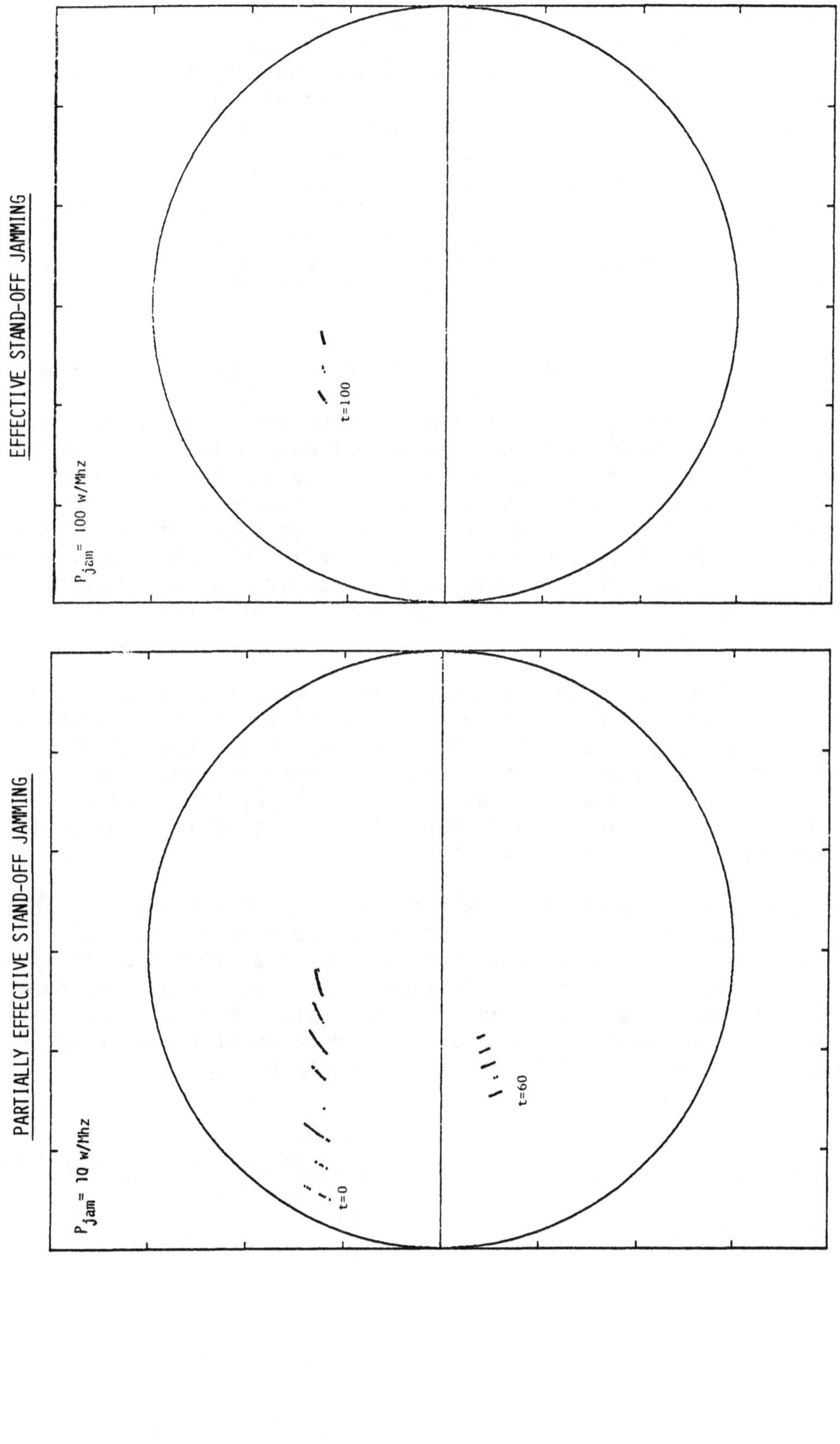

FIGURE 8. PARTIALLY EFFECTIVE AND EFFECTIVE STAND-OFF JAMMING RESULTS.

Figure 9 presents results from a case where penetration jamming is employed. Whereas the noise-jamming in Figures 7 and 8 was due to a slow flying stand-off aircraft, the results in Figure 9 are for the case where a jammer is placed aboard the lead aircraft of Flight 2. (The stand-off jammer is not deployed in this case). We see the results of increasing the self-screening or penetration jammer power density from .005 to .010 w/MHz. Detection of Flight 2 is delayed two more scans (or 40 seconds) due to the increased power level. As expected, detection of Flight 1 is unaffected. This figure is an example of the usefulness of performing parametric analyses to determine susceptibility of individual radars to particular parameters affecting the ECM, ECCM environment.

Figure 10a shows the results for the same situation as presented in Figure 7a (no jamming), but with the azimuth of the displayed target returns derived from the azimuth monopulse measurement instead of the beam center azimuth. This is simply one way of displaying the results obtained from the detailed simulation of monopulse angle measurements. At early times when the received signal-to-noise ratios are low, the monopulse PPI displays at these early times are not much different from the preceding standard PPI displays. At later times when the signal-to-noise ratios are higher, the monopulse angle measurements are more accurate, and the angular spread of the displayed target returns is seen to decrease, as expected.

Figure 10b shows what the PPI display would look like if the pulse of the radar was not chirped. The range spread of the target returns is seen to be much greater, reflecting the poor range resolution of an uncompressed pulse. The nominal range resolution here is about 8 km, and at later times when the signal-to-noise ratio is high, target video signals out of the receiver matched filter can be spread in range up to nearly twice the nominal range resolution.

We present these results to show the utility of the simulation in examining the isolated effects of different features of individual radars. This is extremely useful when evaluating the effectiveness of ECCM fixes, new signal processing techniques, etc., in a countermeasure environment. We again emphasize that the simulation readily treats far more complex problems than the one examined. More complex scenarios, radars, countermeasures, and ECCM fixes are easily examined.

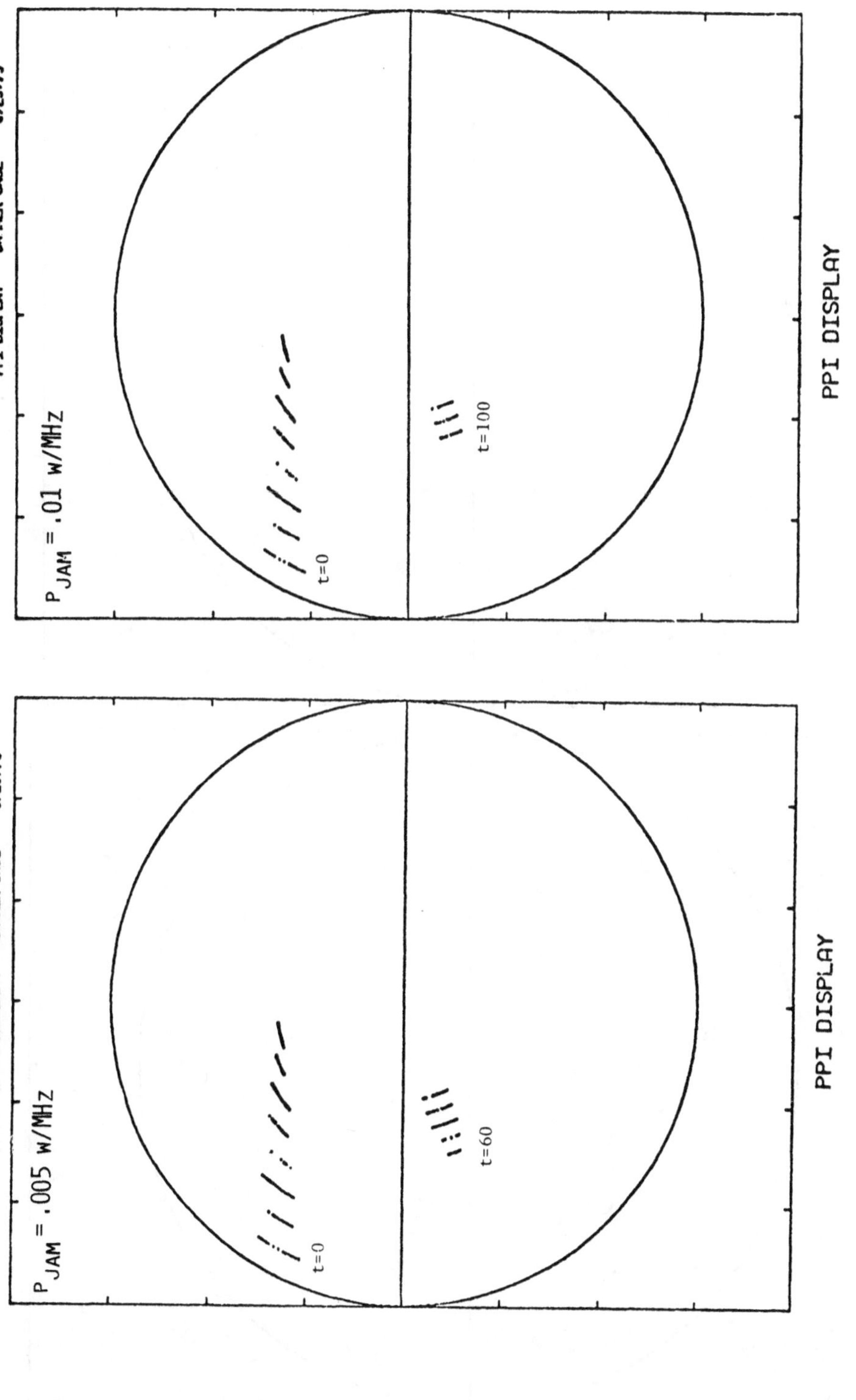

FIGURE 9. PENETRATION JAMMING EFFECTIVENESS VS JAMMING POWER.

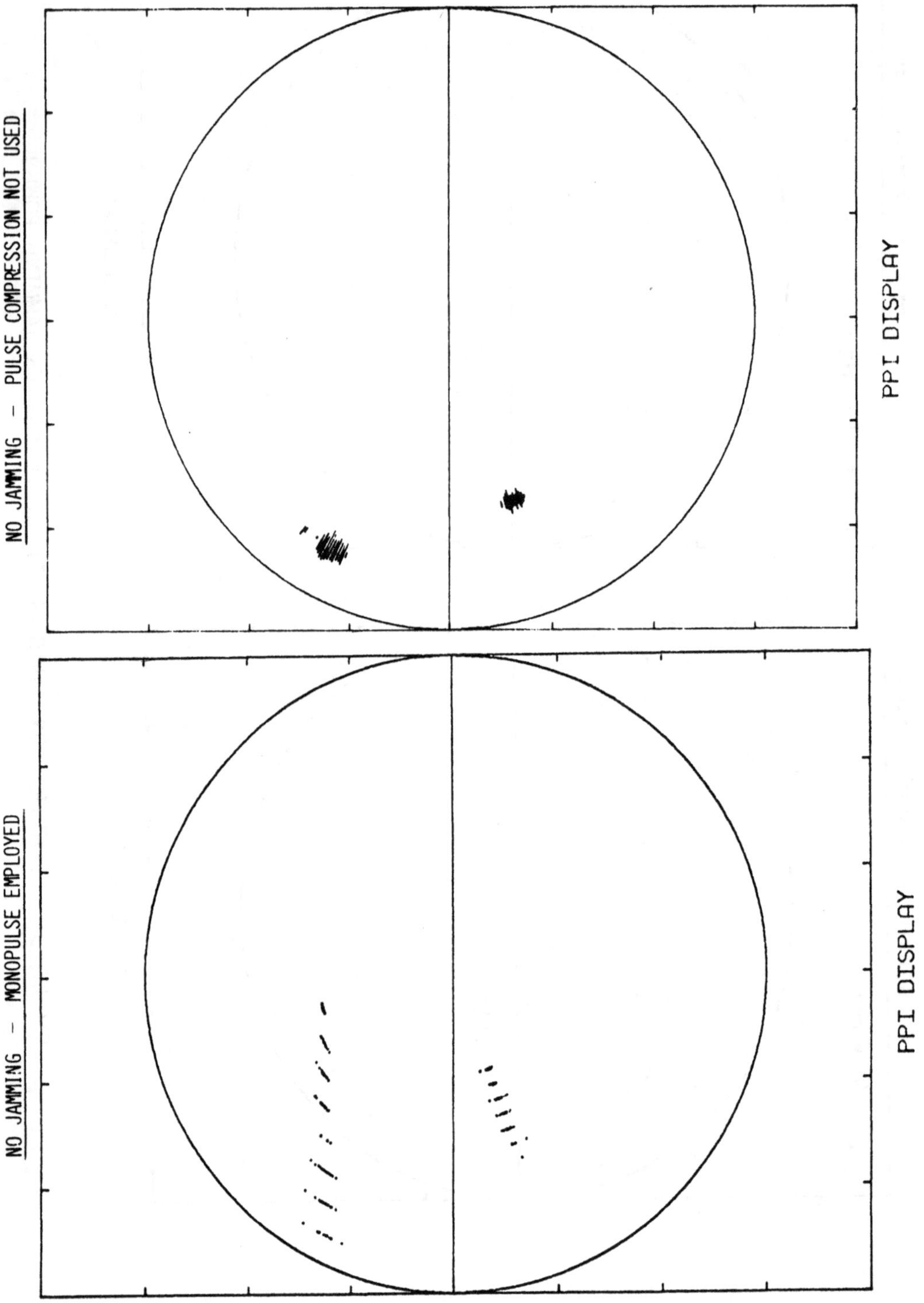

FIGURE 10. EFFECTS OF MONOPULSE AND PULSE COMPRESSION ON RESULTS.

www.ingramcontent.com/pod-product-compliance
Lightning Source LLC
LaVergne TN
LVHW082017200926
848481LV00010B/2876
* 9 7 8 0 8 9 0 0 6 0 7 9 7 *